BALANCE EQUATION APPROACH TO ELECTRON TRANSPORT IN SEMICONDUCTORS

Frontiers of Research with the Chinese Academy of Sciences

Vol. 1 Landau–Lifshitz Equations
by Boling Guo and Shijin Ding

Vol. 2 Balance Equation Approach to Electron Transport in Semiconductors
by X. L. Lei

Frontiers of Research with the Chinese Academy of Sciences — Vol. 2

BALANCE EQUATION APPROACH TO ELECTRON TRANSPORT IN SEMICONDUCTORS

X L Lei
Shanghai Jiaotong University, China

World Scientific

NEW JERSEY · LONDON · SINGAPORE · BEIJING · SHANGHAI · HONG KONG · TAIPEI · CHENNAI

Published by

World Scientific Publishing Co. Pte. Ltd.
5 Toh Tuck Link, Singapore 596224
USA office: 27 Warren Street, Suite 401-402, Hackensack, NJ 07601
UK office: 57 Shelton Street, Covent Garden, London WC2H 9HE

British Library Cataloguing-in-Publication Data
A catalogue record for this book is available from the British Library.

Frontiers of Research with the Chinese Academy of Sciences
BALANCE EQUATION APPROACH TO ELECTRON TRANSPORT IN SEMICONDUCTORS

Copyright © 2008 by World Scientific Publishing Co. Pte. Ltd.

All rights reserved. This book, or parts thereof, may not be reproduced in any form or by any means, electronic or mechanical, including photocopying, recording or any information storage and retrieval system now known or to be invented, without written permission from the Publisher.

For photocopying of material in this volume, please pay a copying fee through the Copyright Clearance Center, Inc., 222 Rosewood Drive, Danvers, MA 01923, USA. In this case permission to photocopy is not required from the publisher.

ISBN-13 978-981-281-902-4
ISBN-10 981-281-902-9

Printed in Singapore.

Preface

With the size shrinking down to the submicrometer scale and working frequency rising up to the subterahetz range modern semiconductor devices are performing under strong and/or rapidly time-varying electromagnetic fields. In these circumstances not only fails the conventional linear relationship between the current density and the electric field, but the drift-diffusion model, which has so far been the main basis for most device simulations, is no longer valid for describing the carrier transport in devices. The development of modern microfabrication techniques has been forcing people to bring the quantal condensed-matter physics directly into the arena of electronics yet intending the transport process to be expressed in a way as concise as possible.

Since 1950s physicists have been proposed many transport theories based on the many-particle quantum and statistical mechanics and using innovative techniques such as superoperators, path integrals, correlation functions, Feynman diagram expansions, nonequilibrium Green's functions, Wigner distribution functions, etc. These theories have been great success, but are often formidable to deal with accurately and few of them seems to offer a simple transport dynamics embodying scattering mechanisms in a realistic semiconductor. Over the years, with quantal ingredients putting in from time to time, the Boltzmann-type equations for the distribution function of microscopic particles remain almost the only theoretical tool capable of generating useful performance-oriented information in a device. As a statistical but complete description of carrier transport the distribution function can, in principle, be solved from the Boltzmann equation under the condition of given electric and magnetic fields and given impurity, phonon and intercarrier scatterings, and then all the transport quantities are obtained. In practice, solving the Boltzmann integro-differential equation is a very hard job for realistic scattering mechanisms especially in the presence of electron–electron interactions. Drastic approximations, such as relaxation-time treatment, are usually taken in order to obtain quasi-analytical results.

With the development of modern computational capacity, Monte Carlo and other numerical techniques have been employed in transport calculations. They allow numerical solution to the Boltzmann equation with less approximations and have become powerful tools in investigating semiconductor transport and device modeling. However, pure numerical methods having intrinsic shortcomings in offering physical insights and systematic analyses, can never replace an analytical or quasi-analytical theory.

In many cases, for the calculation of concerned transport quantities it is sufficient to known a partial relevant information of transport rather than the distribution function containing the complete information. With this in mind people derive moment equations of finite orders from the Boltzmann-type equation and, under some further approximations, establish hydrodynamic transport theories. These theories have been playing very important roles in simulating carrier dynamics in various semiconductor devices.

In 1984, C.S. Ting and the author of the present book proposed a balance-equation scheme based on the separation of the integrative mechanic motion of the whole carrier system (the center of mass) from the statistical motions of individual carriers and developed a transport theory using the average carrier drift velocity as the control parameter. Contrary to most existing dynamic response theories (such as Boltzmann-type theory) in which the current density and other transport quantities (e.g. the conductivity) are calculated from the distribution function obtained as the response to a given electric field, in the balance-equation theory the electric field is derived to balance the frictional force for a given carrier drift velocity thus the transport property (e.g. the resistivity) is obtained as a function of the drift velocity or the current density, a physical quantity most affecting the conduction behavior and most direct controlled in the experiment.

Because of technical difficulties, electron–electron interactions, which can be of vital importance, are often ignored in most of transport treatments. The admission to including electron–electron interaction introduces considerable complication. However, there is a substantial simplification in the situation when electron–electron interactions are large enough to induce rapid thermalization of the carriers about their drifted transport state. This opens the possibility of a quasi-analytic description of transport in the strong electron–electron scattering limit in the balance-equation theory.

Due to its simplicity and effectiveness, the balance-equation approach has become a useful tool to deal with many realistic transport phenomena in semiconductors, and provided a reliable basis for developing theories, modeling devices and explaining experiments.

This book is devoted to a systematic, comprehensive and up-to-date description of the physical basis of the balance-equation theory and its applications to various transport problems in various types of semiconductors. The different aspects of the balance-equation method were reviewed in the volume entitled "Physics of Hot Electron Transport in Semiconductors" (edited by C.S. Ting, World Scientific, 1992). Since then, the balance-equation theory has been further developed and extensively applied to investigate various new transport phenomena, such as transport in nonparabolic systems, spatially inhomogeneous transport and device modeling, miniband transport of semiconductor superlattices, radiation-induced transport and magnetotransport, effects of impact ionization in high-field dc and high-frequency transport, microwave-induced magnetoresistance oscillation, radiation-induced electron cooling, etc. The present book complements the volume of "Physics of Hot Electron Transport in Semiconductors" and part of the material is a direct extension of the first review article written by me and N.J.M. Horing in that book.

I am thankful to C.S. Ting, N.J.M. Horing, H.L. Cui, M. Lax, W. Cai, D.Y. Xing, M. Liu, Z.B. Su, and Y.L. Chen for their kind support and allowing me to use relevant parts of their review articles collected in the 1992 volume in the present book. I am grateful to my colleagues J.Q. Zhang, B. Dong, S.Y. Liu, J.C. Cao, M.W. Wu, X.F. Wang, Z.Q. Zou, X.M. Weng, J. Cai, W.M. Shu, R.Q. Yang, X.J. Lu, W. Xu, etc., for their contributions to the balance-equation approach and their helps in preparing this book. Finally I would like to thank my respectful senior colleagues for their constant encouragement and all my friends with whom we have shared the development of this field.

X. L. Lei
November 2007
Shanghai Jiaotong University, China

Contents

Preface		v
1.	Main Physical Considerations and Transport Balance Equations	1
	1.1 Center-of-Mass and Relative Variables of Electrons	1
	1.2 Center-of-Mass Treated as a Classical Particle	4
	1.3 Initial Density Matrix	7
	1.4 Density Matrix and Statistical Averages to Linear Order in H_I	9
	1.5 Force Balance Equation	10
	1.6 Energy Balance Equation	12
2.	DC Steady-State Transport	15
	2.1 Balance Equations for DC Steady-State Transport in Three-Dimensional Systems	15
	2.2 Degenerate Case, Impurity and Phonon Scatterings	19
	2.2.1 Ohmic resistivity, Bloch–Grüneissen formula	19
	2.2.2 Warm-electron conduction	21
	2.2.3 Electron cooling	22
	2.2.4 Nonlinear zero-temperature limit	23
	2.3 Acoustic and Optic Phonon Scatterings	25
	2.3.1 Acoustic phonon scatterings	25
	2.3.2 Optical deformation-potential scatterings	28
	2.4 Effect of Dynamic Screening	29
	2.5 Balance Equations for Quasi-Two-Dimensional Systems	31

2.6	Hot Electron Transport in GaAs/AlGaAs Heterojunctions		37
2.7	Transverse Transport in Quantum-Well Superlattices		45
2.8	Balance Equations for Transverse Transport in Tunneling Superlattices		52
2.9	Transport in Quantum Wires		59
	2.9.1	Cylindrical quantum wires	59
	2.9.2	Harmonically confined quantum wires	68

3.	Time-Dependent and High-Frequency Transport		71
3.1	Balance Equations for Slowly Time-Dependent Transport		71
3.2	Memory Effects in Transient Response for the Weak Current Case		76
3.3	High-Frequency Small-Signal Conductivity		80
	3.3.1	The case of zero dc bias	82
	3.3.2	The case of finite dc bias	84
3.4	Plasmon Contribution to High-Frequency Conductivity in Quasi-2D Systems and Superlattices		90
3.5	High-Frequency Conductivity of Quantum Wires		94
3.6	Nonlinear High-Frequency Conductivity		101
	3.6.1	Harmonic expansion method	101
	3.6.2	Nonlinear calculation without memory effect	102

4.	Center-of-Mass Velocity Fluctuations, Noise and Diffusion			109
4.1	Langevin-Type Equation			109
4.2	Thermal Noise and Diffusion: General Relations			112
4.3	Thermal Noise in Three-Dimensional Semiconductors			115
	4.3.1	3D degenerate systems, impurity and acoustic phonon deformation potential scatterings		115
		4.3.1.1	Finite (nonzero) lattice temperature	116
		4.3.1.2	Zero lattice temperature	117
	4.3.2	Nondegenerate 3D system, optic deformation potential scattering		118
4.4	Quasi-2D Systems and Superlattices			119
4.5	Effects of Velocity Fluctuation on the Drift Motion			124

5. **Effects of Nonequilibrium and Confined Phonons**   127

   5.1   The Energy-Loss Rate of Electron Systems, Nonequilibrium Phonons 127
   5.2   Nonequilibrium Phonons in 3D Bulk Systems 129
   5.3   Semiconductor Superlattices, 3D Phonon Model 135
   5.4   Nonequilibrium Phonons in Quasi-Two-Dimensional Systems 136
   5.5   Semiconductor Superlattices, Quasi-2D Phonon Model . 142
   5.6   Effect of Nonequilibrium Phonons on Hot-Electron Transport 147
   5.7   Confined Phonons in Quasi-2D Semiconductors 155

6. **Systems with Several Species of Carriers**   161

   6.1   Balance Equations for Type-II Superlattices 161
   6.2   DC Steady State Transport 165
   6.3   Negative Minority-Electron Mobility in Electron–Hole Plasma 169
   6.4   Coulomb Drag between Two Semiconductor Layers, Transfer Resistivity 171
   6.5   High-Frequency Small-Signal Transport in Type-II Superlattices 175
   6.6   Multivalley Semiconductors, Systems of Several Carrier Species with Particle Interchange 179
   6.7   Balance Equations and Langevin Equations for Systems of Several Carrier Species 183
   6.8   DC Steady-State Transport in GaAs 189
   6.9   Quasi-Two-Dimensional Multiple Subband Systems .. 195
     6.9.1   Multisubband system treated as composed of multiple species of carriers 195
     6.9.2   Chemical potential difference considered for different subbands 199
     6.9.3   Effect of intersubband Coulomb interaction on transport 203
     6.9.4 δ-doped structures, electric-field induced real-space transfer 206
     6.9.5   Quasi-one-dimensional multiple subband systems 213

	6.10	Fluctuations and Diffusion Coefficients in Systems of Several Carrier Species with Particle Exchange	216
		6.10.1 Velocity and particle-number fluctuations	216
		6.10.2 Diffusion coefficients	218
7.	Balance Equation Transport Theory and Electron Correlation		221
	7.1	Full Quantum-Mechanical Treatment in the Laboratory Reference Frame	221
	7.2	Perturbation Expansion and Steady-State Balance Equations	224
	7.3	Isothermal and Adiabatic Resistivities	230
	7.4	Resistivity of Two-Subband System versus Interband Coulomb Interaction	233
	7.5	Boltzmann Equation with Electron–Electron Scattering	236
	7.6	Distribution Function in the Balance Equation Theory	239
	7.7	Coupled Force-Balance and Particle-Occupation Rate Equations	245
8.	Balance Equation Approach to Magnetotransport		251
	8.1	Balance Equations of Hot-Electron Transport in the Presence of a Magnetic field	251
	8.2	General Results of Steady-State Magnetotransport in Isotropic Bulk Systems	258
	8.3	Longitudinal Configuration	261
		8.3.1 Linear resistivity due to impurity scattering	261
		8.3.2 Resistivity due to optic phonon scattering, magneto-phonon resonance	262
	8.4	Transverse Configuration	265
		8.4.1 The logarithmic divergence of linear impurity resistivity and its nonlinear elimination	266
		8.4.2 Optic phonon scattering, magnetophonon resonance	267
	8.5	Small-Signal Transport in a Magnetic Field, Cyclotron Resonance	270
		8.5.1 Balance equations for high-frequency small-signal transport in a magnetic field	270

8.5.2 High-frequency complex resistivity and
complex conductivity in a magnetic field ... 272
8.5.3 Cyclotron resonance 274
8.6 Magnetotransport in Quasi-Two-Dimensional Systems . 277
8.6.1 Nonlinear resistivity tensor of 2D systems in
Faraday configuration 277
8.6.2 Density correlation function of 2D electron
systems in a magnetic field, Landau level
broadening 279
8.6.3 Linear magnetoresistivity at low temperatures,
Shubnikov de-Haas oscillation 283
8.6.4 Magnetoresistance oscillations induced by
direct current 285
8.6.5 Linear magnetoresistance of GaAs
heterojunctions in wide ranges of temperatures
and magnetic fields 289
8.7 Balance Equations of Two Species of Carriers in a
Magnetic Field 292
8.7.1 Type-II superlattices, electron–hole plasma .. 292
8.7.2 Weak magnetic-field expansion of
two-dimensional density correlation function . 295
8.7.3 Transport behavior of two-subband systems
and inter-subband Coulomb interaction 296

9. Higher Order Scatterings and Alternative Formulations
of Balance Equation Theory   301

9.1 General Forms of Balance Equations, Intracollisional
Field Effect 301
9.1.1 The frictional force and energy-loss rate in an
electron–atom system, intracollisonal field
effect 302
9.1.2 General forms of balance equations without
intracollisional field effect 305
9.1.3 Evaluation of the intracollisional field effect .. 307
9.2 Balance Equations in the CTPG Representation, Effects
of Higher Order Scatterings 309
9.2.1 Balance equations in the CTPG function
representation 310
9.2.2 Effect of an electric field on weak localization . 314

	9.2.3	The phonon–plasma coupled mode and electron energy-loss rate 318
9.3		Derivation of Force Balance Equation from Dielectric Response 324
9.4		Nonequilibrium Statistical Operator Method 326
9.5		Generalized Quantum Langevin Equation Approach . . 328
9.6		Drifted Electron-Temperature Model 330

10. Weakly Nonuniform Systems, Hydrodynamic Balance Equations   333

    10.1   Hamiltonian of Small Fluid Elements 334
    10.2   Rates of Change of Particle Number, Momentum and Energy 337
    10.3   Hydrodynamic Balance Equations 338
    10.4   Thermoelectric Power 340
    10.5   Effect of Phonon Drag 344
    10.6   Onsager Relations and Hydrodynamic Balance Equations 349
    10.7   Carrier Transport in Semiconductor Devices 353
          10.7.1   Hydrodynamic equation and Poisson equation 353
          10.7.2   Simulation of transport process in 1D semiconductor devices 357

11. Balance Equations for Hot Electron Transport in an Arbitrary Energy Band   361

    11.1   Heisenberg Equation of Motion for Electrons in a Brillouin Zone 362
          11.1.1   Electrons in a periodical potential 362
          11.1.2   Single band subspace, rates of changes of momentum and energy 365
          11.1.3   The rate of change of the lattice momentum . 369
          11.1.4   Derivation using effective Hamiltonian 372
    11.2   Initial Density Matrix and Distribution Function 374
    11.3   Momentum Balance Equation and Lattice Momentum Balance Equation 379
          11.3.1   Momentum and energy balance equations . . . 379
          11.3.2   Lattice momentum and energy balance equations 381

	11.4	Boltzmann Equation and Balance Equation, Approximate Distribution Functions	383
		11.4.1 Derivation of momentum, lattice momentum and energy balance equations from Boltzmann equation	383
		11.4.2 Approximate distribution functions	387
12.	Miniband Transport in Semiconductor Superlattices		391
	12.1	Superlattice Miniband	391
	12.2	Esaki–Tsu Model of Miniband Transport	393
		12.2.1 The original analysis of Esaki and Tsu	393
		12.2.2 Calculation with carrier statistical distribution	394
	12.3	Boltzmann Equation with Relaxation Time Approximation	397
		12.3.1 One-dimensional theory	397
		12.3.2 Three-dimensional theory	399
	12.4	Superlattice Miniband Transport Treated with Momentum and Energy Balance Equations	403
	12.5	Superlattice Miniband Transport Treated with Lattice Momentum and Energy Balance Equations	409
	12.6	Comparison of Using Two Different Initial Distribution Functions	412
	12.7	Laterally Confined Superlattices	415
		12.7.1 The momentum and energy balance equations of a laterally confined superlattice	416
		12.7.2 DC steady state transport	419
	12.8	Transient and High-Frequency Transport in Superlattices	423
		12.8.1 The Transient response of a superlattice to a step electric field	423
		12.8.2 High-frequency small-signal transport	428
		12.8.3 High-frequency large-signal response	431
	12.9	Miniband Transport of Superlattices in a Quantized Magnetic Field, Magneto-Phonon Resonance	435
	12.10	Balance Equations for One-Dimensional Superlattices, Relaxation Time Approximation	441
		12.10.1 Balance equations for one-dimensional superlattices	441

12.10.2　DC and small-signal solutions of
　　　　　　　one-dimensional balance equations for
　　　　　　　superlattices 444

13. **Nonparabolic Systems with Magnetic field, Impact Ionization or under Nonuniform Condition** — 449

　13.1　Balance Equations for Electron Transport in an
　　　　Arbitrary Energy Band in Crossed Magnetic and
　　　　Electric Fields 449
　13.2　Examples of Quasi-Classical Magnetotransport in
　　　　Arbitrary Energy Bands 454
　　　　13.2.1　3D tight-binding systems 454
　　　　13.2.2　Quantum-wire arrays 458
　　　　13.2.3　Superlattice vertical transport in the presence
　　　　　　　　of a transverse magnetic field 461
　13.3　Electronic Transport in Nonparabolic Kane Bands . . . 462
　　　　13.3.1　The Kane band model 462
　　　　13.3.2　Hot-electron transport in Kane band systems . 464
　　　　13.3.3　Hall resistivity of hot-electron transport in
　　　　　　　　nonparabolic Kane bands 467
　　　　13.3.4　Transport of a Kane band system under a
　　　　　　　　strong longitudinal magnetic field 470
　13.4　Transport Balance Equations for Nonparabolic
　　　　Multiband Systems 473
　13.5　Applications of Nonparabolic Multiband Balance
　　　　Equations 478
　　　　13.5.1　High electric field transport in Si 478
　　　　13.5.2　High electric field transport in GaAs 482
　　　　13.5.3　Effect of high-lying minibands on superlattice
　　　　　　　　longitudinal transport 484
　13.6　Balance Equations with Impact Ionization and Auger
　　　　Recombination Processes 489
　　　　13.6.1　Impact ionization and Auger recombination
　　　　　　　　processes in semiconductors 489
　　　　13.6.2　Balance equations with ionization and
　　　　　　　　recombination processes 493
　　　　13.6.3　Quasi-steady and steady transport under a
　　　　　　　　constant electric field 496

	13.6.4	Multivalley balance equations with impact ionization	499
13.7		Hydrodynamic Balance Equations for Systems with Arbitrary Energy Dispersion	504
13.8		Modeling of Transport in Semiconductor Devices from Nonparabolic Band Material	507
13.9		Instabilities in Superlattice Miniband Transport and Spatiotemporal Domains	510
	13.9.1	Hydrodynamic balance equations for superlattice miniband transport	511
	13.9.2	Analyses with small wave-like perturbations, convective and absolute instabilities	513
	13.9.3	Spatiotemporal domains in voltage-biased superlattices	520

14. Carrier Transport in Semiconductors Driven by THz Radiation Fields 527

14.1		Balance Equations for Electron Transport in Semiconductors Driven by a THz Electric Field	528
	14.1.1	Center-of-mass and relative-electron variables in a uniform high-frequency electric field	528
	14.1.2	Momentum- and energy-balance equations under a high-frequency electric field	532
14.2		Transport in Quasi-Two-Dimensional Semiconductors Irradiated by a THz Field	536
	14.2.1	Balance equations for THz-driven quasi-2D systems	536
	14.2.2	Transport of a GaAs-based quantum well irradiated by THz fields	538
14.3		Miniband Transport in Semiconductor Superlattices Irradiated by THz Fields	543
	14.3.1	Balance equations for THz-driven electron transport in an arbitrary energy band	543
	14.3.2	Superlattice miniband transport subject to a strong THz field	548
14.4		Nonlinear Free-Carrier Absorption of Intense THz Radiation in Semiconductors	550
	14.4.1	Nonlinear absorption coefficient of a THz radiation in a bulk semiconductor	550

14.4.2 Absorption rate of a THz radiation passing through a 2D sheet 554
14.5 Transport in Quasi-2D Semiconductors with Impact Ionization Irradiated by a Strong THz Field 556
14.6 THz-Field-Driven Miniband Transport of a Superlattice in a Longitudinal Magnetic Field 561
14.7 Extension to Radiation Fields of Lower Frequency . . . 565

15. Radiation Driven Magnetotransport in Two-Dimensional Systems in Faraday Geometry   573

15.1 Balance Equations for Radiation-Induced Magnetotransport in Two-Dimensional Electron Systems in Faraday Configuration 574
    15.1.1 Hamiltonian in terms of center-of-mass and relative electron variables 574
    15.1.2 Force- and energy-balance equations 577
    15.1.3 Density correlation function of 2D electrons in a magnetic field, Landau level broadening . . . 582
15.2 Nonlinear Cyclotron Resonance in Quasi-2D Systems . 584
    15.2.1 Cyclotron resonance of drift velocity and dynamic conductivity 584
    15.2.2 Incident electromagnetic field, selfconsistent field and radiative decay 586
    15.2.3 Transmittance and Faraday effect 587
    15.2.4 Cyclotron resonance of electron heating and cooling 592
    15.2.5 Cyclotron resonance of THz photoconductivity at high temperatures 593
15.3 Radiation-Induced Magnetoresistance Oscillations in High-Mobility Two-Dimensional Electron Systems at Low Temperatures 597
    15.3.1 Magnetoresistance oscillations under monochromatic irradiation 598
    15.3.2 Magnetoresistivity, electron temperature and energy absorption rate 603
    15.3.3 Vanishing linear photoresponse at cyclotron resonance and its harmonics at low temperatures 606
    15.3.4 Multiple and virtual photon processes 609

　　　　15.3.5　Nonlinear magnetoresistivity and differential
　　　　　　　　magnetoresistivity 611
　15.4　Magnetoresistance Oscillations under Bichromatic
　　　　Irradiation in Two-Dimensional Electron Systems . . . 617

Bibliography　　　　　　　　　　　　　　　　　　　　　　　　　623

Index　　　　　　　　　　　　　　　　　　　　　　　　　　　　633

Chapter 1

Main Physical Considerations and Transport Balance Equations

1.1 Center-of-Mass and Relative Variables of Electrons

We consider a semiconductor system under the influence of a uniform electric field $\boldsymbol{E}$. The system consists of ions vibrating around their individual equilibrium positions in the lattice and numerous mobile carriers having total number N, which, in most cases, will be simply called "electrons" without distinguishing their charge type. These electrons, interacting with each other through the Coulomb potential, are coupled with lattice vibrations and scattered by randomly distributed impurities. Assuming a single-band effective-mass description for electrons, we can write the Hamiltonian of this electron–ion system as

$$H = \sum_i \frac{p_i^2}{2m} + \frac{e^2}{4\pi\epsilon_0\kappa} \sum_{i<j} \frac{1}{|\boldsymbol{r}_i - \boldsymbol{r}_j|} - e \sum_i \boldsymbol{r}_i \cdot \boldsymbol{E}$$
$$+ \sum_{i,a} u\left(\boldsymbol{r}_i - \boldsymbol{r}_a\right) - \sum_{i,l} \boldsymbol{u}_l \cdot \boldsymbol{\nabla} v_l\left(\boldsymbol{r}_i - \boldsymbol{R}_l\right) + H_{\text{ph}}. \qquad (1.1)$$

Here, $\boldsymbol{r}_i$ and $\boldsymbol{p}_i = -\mathrm{i}\boldsymbol{\nabla}_i$ $(i = 1, \cdots, N)$ are the coordinate and momentum of the ith electron with effective mass m and charge e. The first term on the right-hand-side of Eq. (1.1) is the kinetic energy of electrons; the second term is the Coulomb interaction between electrons in the host semiconductor having a dielectric constant κ; the third term is the potential energy of electrons in the uniform electrical field $\boldsymbol{E}$; $v_l(\boldsymbol{r} - \boldsymbol{R}_l)$ denotes the potential produced by the lth ion on the lattice site $\boldsymbol{R}_l$, and $\boldsymbol{u}_l$ is its displacement from the regular (equilibrium) position; $u(\boldsymbol{r} - \boldsymbol{r}_a)$ denotes the additional scattering potential due to an impurity located at $\boldsymbol{r}_a$, which is assumed randomly distributed; and the last term H_{ph} represents the Hamiltonian of lattice vibration itself (phonons).

To deal with the electronic transport in semiconductors we first focus our attention on electrons. The initial starting point of the balance equation approach is the separation of the motion of the center of mass, i.e. the motion of all electrons as an integrity, from the relative motion of the electrons in this many-particle system. We define (Ting, Ying and Quinn, 1976a; Lei and Ting, 1984) the center-of-mass momentum and coordinate of the electron system, $\boldsymbol{P}$ and $\boldsymbol{R}$, by

$$\boldsymbol{P} = \sum_i^N \boldsymbol{p}_i, \quad \boldsymbol{R} = \frac{1}{N}\sum_i^N \boldsymbol{r}_i, \quad (1.2)$$

and the momentum and coordinate of the ith electron relative to the center of mass, $\boldsymbol{p}'_i$ and $\boldsymbol{r}'_i$, by

$$\boldsymbol{p}'_i = \boldsymbol{p}_i - \frac{1}{N}\boldsymbol{P}, \quad \boldsymbol{r}'_i = \boldsymbol{r}_i - \boldsymbol{R}. \quad (1.3)$$

For brevity, we will simply call $\boldsymbol{p}'_i$ and $\boldsymbol{r}'_i$ the relative momentum and coordinate of the ith electron, or the momentum and coordinate of the ith relative electron. It is easily seen that $\boldsymbol{P}, \boldsymbol{R}$ are canonical conjugate variables:

$$[R_\alpha, P_\beta] = \mathrm{i}\,\delta_{\alpha\beta} \quad (1.4)$$

($\alpha,\beta = x,y,z$), and they commute with $\boldsymbol{r}'_i, \boldsymbol{p}'_i$:

$$[\boldsymbol{r}'_i, \boldsymbol{P}] = [\boldsymbol{R}, \boldsymbol{p}'_i] = 0. \quad (1.5)$$

The N relative-electron momenta and coordinates so defined are not completely independent, but subject to constraints:

$$\sum_i^N \boldsymbol{p}'_i = 0, \quad \sum_i^N \boldsymbol{r}'_i = 0, \quad (1.6)$$

and their commutation relations involve an noncanonical $1/N$ term:

$$[r'_{i\alpha}, p'_{j\beta}] = \mathrm{i}\,\delta_{\alpha\beta}(\delta_{ij} - \frac{1}{N}). \quad (1.7)$$

The original derivation of the balance equation theory (Lei and Ting, 1984, 1985a,b) proceeds with a reasonable assumption that the $1/N$ term in Eq. (1.7) is negligible for a macroscopic large N system, such that one can essentially treat the relative-electron momenta and coordinates as canonical conjugate variables:

$$[r'_{i\alpha}, p'_{j\beta}] \simeq \mathrm{i}\,\delta_{\alpha\beta}\delta_{ij}. \quad (1.8)$$

While the approximate equality (1.8) is very useful and justifiable in dealing with most macroscopic transport problems, it is an approximation. Therefore, care must be taken to avoid using it in a way that can result in a macroscopic error. For example, the summation of the right-hand-side of (1.7) over index j for all N particles is zero, while that of (1.8) yields an nonzero value $i\delta_{\alpha\beta}$. This indicates that it is impossible to construct a translation generator of the system by summing the individual momenta of relative electrons $\boldsymbol{p}'_j$ over all particles, since this sum is zero rather than the total momentum of the system. Nevertheless, it has been shown (Lei, 1990a), that the neglect of the noncanonical $1/N$ term in Eq. (1.7) induces no error within the framework of balance equation theory if the center-of-mass coordinate $\boldsymbol{R}(t)$ is treated classically.

The system, consisting of N independent particles (relative electrons) which are characterized by the set of canonical conjugate variables $\boldsymbol{p}'_i$ and $\boldsymbol{r}'_i$ ($i = 1, \cdots, N$), is called the relative electron system. In the second quantization representation of plane wave basis, the relative electron system can be described by means of the creation and annihilation operators $c^\dagger_{\boldsymbol{k}\sigma}$ and $c_{\boldsymbol{k}\sigma}$ for a relative electron having wave vector $\boldsymbol{k}$, spin σ, and energy $\varepsilon_{\boldsymbol{k}} = k^2/2m$. In terms of the center-of-mass variables and the relative-electron variables we can separate the Hamiltonian (1.1) of the electron-ion system into a center-of-mass part H_cm, a relative-electron part H_er, a phonon part H_ph, together with electron–impurity and electron–phonon interactions H_ei and H_ep:

$$H = H_\text{cm} + H_\text{er} + H_\text{ph} + H_\text{ei} + H_\text{ep}, \tag{1.9}$$

$$H_\text{cm} = \frac{P^2}{2Nm} - Ne\boldsymbol{E}\cdot\boldsymbol{R}, \tag{1.10}$$

$$H_\text{er} = \sum_i \frac{\boldsymbol{p}'^2_i}{2m} + \frac{e^2}{4\pi\epsilon_0\kappa} \sum_{i<j} \frac{1}{|\boldsymbol{r}'_i - \boldsymbol{r}'_j|} \tag{1.11}$$

$$= \sum_{\boldsymbol{k},\sigma} \varepsilon_{\boldsymbol{k}} c^\dagger_{\boldsymbol{k}\sigma} c_{\boldsymbol{k}\sigma} + \sum_{\boldsymbol{q}} \tfrac{1}{2} v_c(q) \left(\rho_{\boldsymbol{q}}\rho_{-\boldsymbol{q}} - N\right), \tag{1.12}$$

$$H_\text{ph} = \sum_{\boldsymbol{q},\lambda} \Omega_{\boldsymbol{q}\lambda} b^\dagger_{\boldsymbol{q}\lambda} b_{\boldsymbol{q}\lambda}, \tag{1.13}$$

$$H_\text{ei} = \sum_{\boldsymbol{q},a} u(\boldsymbol{q})\, e^{i\boldsymbol{q}\cdot(\boldsymbol{R}-\boldsymbol{r}_a)} \rho_{\boldsymbol{q}}, \tag{1.14}$$

$$H_\text{ep} = \sum_{\boldsymbol{q},\lambda} M(\boldsymbol{q},\lambda)\, \phi_{\boldsymbol{q}\lambda} e^{i\boldsymbol{q}\cdot\boldsymbol{R}} \rho_{\boldsymbol{q}}. \tag{1.15}$$

Here, ρ_q is the density operator of the relative electrons,

$$\rho_q = \sum_i e^{i\boldsymbol{q}\cdot\boldsymbol{r}'_i} = \sum_k \rho_{kq}, \quad \rho_{kq} = \sum_\sigma c^\dagger_{k+q\sigma} c_{k\sigma}, \qquad (1.16)$$

$\phi_{q\lambda} \equiv b_{q\lambda} + b^\dagger_{-q\lambda}$ is the phonon field operator with $b^\dagger_{q\lambda}$ and $b_{q\lambda}$ being creation the and annihilation operators for a phonon of wavevector $\boldsymbol{q}$ in branch λ having frequency $\Omega_{q\lambda}$, $\nu_c(q) = e^2/\epsilon_0\kappa q^2$ is the Coulomb potential (κ is the background dielectric constant), and $u(\boldsymbol{q})$ and $M(\boldsymbol{q},\lambda)$ are respectively the electron–impurity potential and the electron–phonon matrix element in the plane wave representation, satisfying $u(\boldsymbol{q}) = u^*(-\boldsymbol{q})$ and $M(\boldsymbol{q},\lambda) = M^*(-\boldsymbol{q},\lambda)$.

1.2 Center-of-Mass Treated as a Classical Particle

As is seen from Hamiltonian (1.9), our system comprises three parts: the center of mass, relative electrons and phonons. The center of mass (CM), described by the canonical conjugate variables $\boldsymbol{R}$ and $\boldsymbol{P}$ ($\boldsymbol{R}$ is the average position of all electrons and $\boldsymbol{P}$ is the total momentum of all electrons) and considered as a single particle having mass Nm and charge Ne, is moving under the influence of the acceleration due to the applied electric field $\boldsymbol{E}$ and the damping due to electron–impurity and electron–phonon interactions H_{ei} and H_{ep}. Because of its enormous mass, the motion of the center of mass is essentially classical, and we can treat the CM-related operators classically, regarding the coordinate variable $\boldsymbol{R}$ in H_{ei} and H_{ep} as the real, time-dependent expectation position of the center of mass, $\boldsymbol{R}(t)$. The relative electron system, which is composed of a large number of interacting particles and described by the set of canonical conjugate variables $\boldsymbol{r}'_i$ and $\boldsymbol{p}'_i$, or by creation and annihilation operators $c^\dagger_{k\sigma}$ and $c_{k\sigma}$, is treated fully quantum-mechanically. The significance is that the relative electron system does not directly sense the electric field but it is coupled to the center-of-mass motion through the time-varying CM position $\boldsymbol{R}(t)$ appearing in both H_{ei} and H_{ep}. In the following we will denote

$$H_{\text{It}} = H_{\text{ei}} + H_{\text{ep}} \qquad (1.17)$$

as a time-dependent interaction containing $\boldsymbol{R}(t)$ as a time-dependent classical parameter. Note that the error arising from treating $\boldsymbol{R}(t)$ as a classical variable, i.e. from neglecting the noncommutability of the CM coordinate operators at different times, $\boldsymbol{R}(t)$ and $\boldsymbol{R}(t')$, is of order of $1/N$, and has been shown (Lei, 1990a) to be exactly canceled by that arising from the neglect of the noncanonical $1/N$ terms in the commutation relations of the relative electron variables, Eq.(1.7). Therefore, the two major approximations made above in formulating the balance equation theory, i.e. treating relative electrons as an unconstrained canonical system and treating the center of mass as a classical single particle, induce no error save in regard to the intracollisional field effect (see Sec. 10.1). In accordance with this Hamiltonian, it is apparent that the CM velocity $\boldsymbol{V}$, or the time derivative of the CM position, is given by

$$\boldsymbol{V} \equiv \dot{\boldsymbol{R}} = -\mathrm{i}\,[\boldsymbol{R}, H] = \frac{\partial H}{\partial \boldsymbol{P}} = \frac{\boldsymbol{P}}{Nm}. \tag{1.18}$$

Note that the center of mass, which is accelerated by the electric field and damped by the relative-electron–phonon bath medium, is a Brownian particle due to the random force associated with the electron–impurity and electron–phonon interactions. Therefore the CM velocity $\boldsymbol{V}$ is composed of a drift or average part $\boldsymbol{v}$ and a fluctuation or random part $\delta\boldsymbol{V}$:

$$\boldsymbol{V} = \boldsymbol{v} + \delta\boldsymbol{V}. \tag{1.19}$$

Here we denote $\boldsymbol{v} = \langle \boldsymbol{V} \rangle$ or $\langle \delta\boldsymbol{V} \rangle = 0$, with the $\langle \cdots \rangle$ symbol understood as an average either over an appropriate time period or over a statistical ensemble which eliminates the fluctuations. The forces experienced by the center of mass are obtained by calculating the time derivative of the CM momentum $\dot{\boldsymbol{P}} = -\mathrm{i}\,[\boldsymbol{P}, H] = -\partial H/\partial \boldsymbol{R}$, yielding

$$\dot{\boldsymbol{P}} = Ne\boldsymbol{E} + \hat{\boldsymbol{F}}, \tag{1.20}$$

with $\hat{\boldsymbol{F}} = \hat{\boldsymbol{F}}_\mathrm{i} + \hat{\boldsymbol{F}}_\mathrm{p}$, and

$$\hat{\boldsymbol{F}}_\mathrm{i} = -\mathrm{i}\sum_{\boldsymbol{q},a} u(\boldsymbol{q})\,\boldsymbol{q}\,\mathrm{e}^{\mathrm{i}\,\boldsymbol{q}\cdot(\boldsymbol{R}-\boldsymbol{R}_a)}\rho_{\boldsymbol{q}}, \tag{1.21}$$

$$\hat{\boldsymbol{F}}_\mathrm{p} = -\mathrm{i}\sum_{\boldsymbol{q},\lambda} M(\boldsymbol{q},\lambda)\,\boldsymbol{q}\,\phi_{\boldsymbol{q}\lambda}\mathrm{e}^{\mathrm{i}\,\boldsymbol{q}\cdot\boldsymbol{R}}\rho_{\boldsymbol{q}}, \tag{1.22}$$

are identified as forces respectively due to impurity and phonon scatterings. Being quantum-mechanical operators in the relative-electron and phonon space, $\hat{\boldsymbol{F}}_\mathrm{i}$ and $\hat{\boldsymbol{F}}_\mathrm{p}$ also comprise an average part and a fluctuating part. The latter contributes a random force exerting on the center of mass. Eq.(1.20) is an operator force balance equation of the system.

Similarly, we can obtain the rate of change of the energy of the phonon system and that of relative electron systems from the Heisenberg equation of motion:

$$\dot{H}_{\text{ph}} = -\mathrm{i}\,[H_{\text{ph}}, H] = \hat{W} \equiv -\sum_{\bm{q},\lambda} M(\bm{q}, \lambda)\,\phi_{\bm{q}\lambda}\,\mathrm{e}^{\mathrm{i}\bm{q}\cdot\bm{R}}\rho_{\bm{q}}, \qquad (1.23)$$

$$\dot{H}_{\text{er}} = -\mathrm{i}\,[H_{\text{er}}, H] = \hat{S} \equiv -\mathrm{i}\sum_{\bm{q},\lambda} M(\bm{q}, \lambda)\,\phi_{\bm{q}\lambda}\,\mathrm{e}^{\mathrm{i}\bm{q}\cdot\bm{R}}\sum_{\bm{k}}(\varepsilon_{\bm{k}+\bm{q}} - \varepsilon_{\bm{k}})\rho_{\bm{k}\bm{q}}$$
$$-\mathrm{i}\sum_{\bm{q},a} u(\bm{q})\,\mathrm{e}^{\mathrm{i}\bm{q}\cdot\bm{r}_a}\sum_{\bm{k}}(\varepsilon_{\bm{k}+\bm{q}} - \varepsilon_{\bm{k}})\rho_{\bm{k}\bm{q}}, \qquad (1.24)$$

with

$$\dot{\phi}_{\bm{q}\lambda} \equiv -\mathrm{i}\,[\phi_{\bm{q}\lambda}, H_{\text{ph}}] = -\mathrm{i}\,\Omega_{\bm{q}\lambda}(b_{\bm{q}\lambda} - b^{\dagger}_{-\bm{q}\lambda}). \qquad (1.25)$$

Energy balance for the electron system requires that the energy supplied per unit time by the electric field, $N e \bm{E} \cdot \bm{V}$, matches the sum of the energy increase rate of the center of mass, $d(\tfrac{1}{2} N m V^2)/dt$, the increase rate of the relative electron internal energy, $\dot{H}_{\text{er}}$, and the energy loss rate of the electron system to the phonon system $\hat{W}$:

$$N e \bm{E} \cdot \bm{V} = N m \bm{V} \cdot \frac{d\bm{V}}{dt} + \dot{H}_{\text{er}} + \hat{W}. \qquad (1.26)$$

Eq. (1.26) is an energy balance equation of the electron system in operator form. Of course, Eq. (1.24) which states the energy balance for the relative electron system, is also an operator energy balance equation. All these equations are in operator form in the relative electron and phonon spaces.

The balance equations can be derived by taking the ensemble averages of these operator equations over the density matrix of the relative electron and phonon systems. The statistical average of a dynamical variable at time t can be evaluated in the Schrödinger picture by

$$\langle \mathcal{O} \rangle = \text{tr}\{\hat{\rho}\mathcal{O}\}. \qquad (1.27)$$

Here $\text{tr}\{\cdots\}$ indicates the trace of the operator in the bracket, $\mathcal{O}$ is the operator (possibly time dependent) representing the dynamical variable in the Schrödinger picture and $\hat{\rho}$ is the density matrix satisfying the Liouville equation

$$\mathrm{i}\,\frac{d\hat{\rho}}{dt} = [H_{\text{er}} + H_{\text{ph}} + H_{\text{It}}, \hat{\rho}] \qquad (1.28)$$

subject to an initial condition at the initial time t_0

$$\hat{\rho}\big|_{t_0} = \hat{\rho}_0. \qquad (1.29)$$

We will use $t_0 = -\infty$ or $t_0 = 0$ as the initial time.

1.3 Initial Density Matrix

Physically, we expect that external conditions (the applied electric field, magnetic field, bath temperature, etc) uniquely determine the final steady state of a system in transport. This means that the final steady state to which the system evolves under the influence of a given electric field, for instance, does not depend on the particulars of the initial condition. After a sufficiently long time the system will arrive at a unique steady state no matter what an initial state it starts from. However, the time required for the system to go from an initial state to the final state does depend strongly on the initial state. As a parametrized theory, the balance equation approach allows us to choose an initial state, which should be simple enough to handle easily and has its major features resemble, as closely as possible, the final state, such that the system, starting from this virtual initial state, can reach the real final state through a short evolution process. When dealing with a time-independent (steady) problem, the final state is of course the steady transport state we are seeking for. When dealing with a time-dependent problem, the "final" state can be the transport state at a time t during the transport process. To figure out the key features of the final state we image to turn off the electron–impurity and electron–phonon interactions H_I together with the electric field $\boldsymbol{E}$ at a certain time t suddenly during the temporal development of the transport process (but keep the electron–electron interaction intact). Then the center of mass, the relative electrons and phonons are decoupled from each other. Being a free particle, the center of mass moves inertially with the constant velocity $\boldsymbol{v}$ equal to the drift velocity $\boldsymbol{v}(t)$ of the transporting electron system at time t. The relative electron system, as an isolated object, will approach a thermal equilibrium state after a certain relaxation time τ_th, having a well-defined thermodynamic temperature $T_\mathrm{e}(t)$. This temperature apparently depends on the state of relative electrons at time t, and will be called the electron temperature of this transport state, even though it may not be possible to define a thermodynamic temperature directly with a transport state (it is not a thermal equilibrium state). In the same vein, the phonon system, which has been decoupled from the electron system, can also be considered to reach a thermal equilibrium state jointly with the bath at the lattice temperature T. This is justifiable if phonons relax faster than electrons. If this is not the case one may need to consider that during the time period after turning off the interactions when relative electrons have arrived at an equilibrium state, phonons only reach a quasi-equilibrium state with a mode $q\lambda$- and time-dependent temperature $T_{q\lambda}(t)$. Further discussion of this will be given in Chapter 5 (Nonequilibrium and Confined Phonons). For the most part in this book, we assume the former case for phonons and correspondingly choose the initial state as the one in which the center of mass moves with constant velocity $\boldsymbol{v} = \boldsymbol{v}(t)$, the relative electrons are in a thermal equilibrium state with temperature $T_e = T_e(t)$ and phonons are in its thermal equilibrium state with lattice temperature T. The initial density matrix for the relative-electron–phonon system is then (Lei and Ting, 1985a; Xing and Ting, 1987)

$$\hat{\rho}_0 = \frac{1}{Z} e^{-H_{er}/T_e} e^{-H_{ph}/T}, \qquad (1.30)$$

where Z is a normalization coefficient and H_{er} is implicitly measured from the Fermi level corresponding to a grand canonical ensemble for the relative electron system. This initial density matrix depends on time t parametrically, $\hat{\rho}_0 \equiv \hat{\rho}_{0t}$.

It should be noted that there are at least two distinct time scales involved. The motion of the center of mass, the time variation of the applied electric field $\boldsymbol{E}(t)$ and the temporal development of statistically averaged quantities such as $\boldsymbol{v}(t)$ and $T_e(t)$, occur on a macroscopic time scale, which is assumed to be much larger than the microscopic relaxation time, or the evolution time scale of the Liouville equation. In other words, we assume that the evolution from the initial state to the final state in accordance with the Liouville equation is a fast microscopic process, whereas the temporal development of the transport quantities is a slow macroscopic process. The latter is included in the parametric time-variation of the initial density matrix $\hat{\rho}_{0t}$ and the center-of-mass velocity $\boldsymbol{v}(t)$.

The initial state described by Eq. (1.30) is one in which the electrons are characterized as being in thermal equilibrium with respect to their center of mass which is moving at a velocity equal to their average drift velocity in the presence of the electric field. Such a selection of the initial state is significant (Lei, 1990b). Since the final state bears the same center-of-mass velocity as the initial state, the time evolution from the initial state to the final state is solely a thermalization process with no need for momentum relaxation. This process involves not only impurity and phonon scatterings, but in most cases, it is determined mainly by intercarrier scatterings.

For systems in which there exists a strong carrier–carrier scattering the microscopic relaxation is very fast, lending validity to the time-dependent description of transport. Even for steady state transport in a constant electric field, a short thermalization time τ_{th} also plays a key role in making it possible to develop a simple perturbative expansion (Lei, 1990b). The underlying phenomenology is that the strong intercarrier scattering promotes a tendency towards rapid thermalization of the electrons in their drifting frame, such that the final steady state is only "a small perturbation away" from the thermal equilibrium state of relative electrons as given by (1.30). We will return to this matter in Chapter 7.

1.4 Density Matrix and Statistical Averages to Linear Order in H_{I}

The statistical average of a dynamical variable $\mathcal{O}$ at time t, Eq. (1.27), can alternatively be written in the interaction picture as

$$\langle \mathcal{O} \rangle = \text{tr}\{\hat{\rho}(t)\mathcal{O}(t)\}, \tag{1.31}$$

where the operators in the interaction picture are defined as

$$\mathcal{O}(t) \equiv e^{i(H_{\text{er}}+H_{\text{ph}})t}\mathcal{O}e^{-i(H_{\text{er}}+H_{\text{ph}})t}, \tag{1.32}$$

$$\hat{\rho}(t) \equiv e^{i(H_{\text{er}}+H_{\text{ph}})t}\hat{\rho}\,e^{-i(H_{\text{er}}+H_{\text{ph}})t}. \tag{1.33}$$

In view of the Liouville equation (1.28) and the initial condition (1.30), $\hat{\rho}(t)$ satisfies the equation

$$i\frac{d}{dt'}\hat{\rho}(t') = \big[H_{\text{I}t'}(t'), \hat{\rho}(t')\big] \tag{1.34}$$

and is subject to the initial condition

$$\hat{\rho}(t')\big|_{t'=-\infty} = \hat{\rho}_0. \tag{1.35}$$

Or, equivalently, it obeys the integral equation

$$\hat{\rho}(t') = \hat{\rho}_0 - i \int_{-\infty}^{t'} dt''\big[H_{\text{I}t''}(t''), \hat{\rho}(t'')\big]. \tag{1.36}$$

This integral equation facilitates a perturbative expansion in orders of the interaction H_I. The first order iteration of Eq. (1.36) yields

$$\hat{\rho}(t') = \hat{\rho}_0 - \mathrm{i}\int_{-\infty}^{t'} dt''\left[H_{\mathrm{I}t''}(t''), \hat{\rho}_0\right]. \tag{1.37}$$

The statistical average of a dynamical variable $\mathcal{O}$, Eq. (1.31), can thus be written, to the linear order in H_I, as

$$\langle \mathcal{O} \rangle = \langle \mathcal{O} \rangle_0 - \mathrm{i}\int_{-\infty}^{t} dt'\langle[\mathcal{O}(t), H_{\mathrm{I}t'}(t')]\rangle_0, \tag{1.38}$$

where

$$\langle\cdots\rangle_0 \equiv \mathrm{tr}\{\hat{\rho}_0(\cdots)\}. \tag{1.39}$$

Note that $\hat{\rho}_0$, and the average $\langle\cdots\rangle_0$, may be dependent on time t.

Eq. (1.38), containing only to the linear order of impurity and phonon scatterings, is the formula mostly often used for calculating the statistical averages of physical quantities in the balance equation approach. Of course, one can continue to iterate Eq. (1.36) to obtain an expansion series with arbitrarily higher order terms of H_I. In principle, a perturbation treatment is valid when the resultant expansion series converges. In this case it is reasonable to take the partial sum of the first finite-number terms or to sum over a partial array of infinite terms as an approximation for the full series, even we are not able to sum over the entire series. On the contrary, if the resultant series of the iterative expansion does not converge or contains divergent high order terms, the perturbative treatment will generally no longer work. Even if a renormalization through summing up the divergent terms seems possible, the physical nature of the issue may be changed completely.

1.5 Force Balance Equation

Taking the statistical average of operator equation (1.20) in the relative electron and phonon space to the lowest order in H_I in accordance with (1.38), and using expressions (1.14) and (1.15) for H_{ei} and H_{ep}, we obtain the equation of motion of the center of mass, or the time-dependent momentum or force balance equation of the system (Lei and Ting, 1984,

1985a; Xing and Ting, 1987)

$$Nm\frac{d}{dt}\boldsymbol{v}(t) = Ne\boldsymbol{E}(t) + \tilde{\boldsymbol{f}}_\mathrm{i} + \tilde{\boldsymbol{f}}_\mathrm{p}, \qquad (1.40)$$

by identifying the average rate of the CM momentum change as

$$\langle \dot{\boldsymbol{P}} \rangle = Nm\frac{d}{dt}\boldsymbol{v}(t), \qquad (1.41)$$

$\boldsymbol{v}(t)$ being the drift part of the center-of-mass velocity. In Eq. (1.40), $\tilde{\boldsymbol{f}}_\mathrm{i}$ and $\tilde{\boldsymbol{f}}_\mathrm{p}$, the frictional forces experienced by the center of mass respectively due to electron–impurity and electron–phonon interactions, are given by

$$\tilde{\boldsymbol{f}}_\mathrm{i} = -\mathrm{i}\, n_\mathrm{i} \sum_{\boldsymbol{q}} \boldsymbol{q}\, |u(\boldsymbol{q})|^2 \int_{-\infty}^{\infty} dt'\, A(\boldsymbol{q}, t, t')\, \Pi(\boldsymbol{q}, t - t'), \qquad (1.42)$$

$$\tilde{\boldsymbol{f}}_\mathrm{p} = -\mathrm{i} \sum_{\boldsymbol{q},\lambda} \boldsymbol{q}\, |M(\boldsymbol{q}, \lambda)|^2 \int_{-\infty}^{\infty} dt'\, A(\boldsymbol{q}, t, t')\, \Lambda(\boldsymbol{q}, \lambda, t - t'). \qquad (1.43)$$

In this, n_i is the impurity density, and the retarded correlation functions are defined as

$$\Pi(\boldsymbol{q}, t - t') = -\mathrm{i}\, \theta(t - t')\langle [\rho_{\boldsymbol{q}}(t),\, \rho_{-\boldsymbol{q}}(t')]\rangle_0, \qquad (1.44)$$

$$\Lambda(\boldsymbol{q}, \lambda, t - t') = -\mathrm{i}\, \theta(t - t')\langle [\phi_{\boldsymbol{q}\lambda}(t)\rho_{\boldsymbol{q}}(t),\, \phi_{-\boldsymbol{q}\lambda}(t')\rho_{-\boldsymbol{q}}(t')]\rangle_0, \qquad (1.45)$$

with

$$\rho_{\boldsymbol{q}}(t) = \mathrm{e}^{\mathrm{i} H_{\mathrm{er}} t}\rho_{\boldsymbol{q}}\, \mathrm{e}^{-\mathrm{i} H_{\mathrm{er}} t} = \sum_{\boldsymbol{k}} \mathrm{e}^{\mathrm{i}(\varepsilon_{\boldsymbol{k}+\boldsymbol{q}} - \varepsilon_{\boldsymbol{k}})t}\rho_{\boldsymbol{k}\boldsymbol{q}}, \qquad (1.46)$$

$$\phi_{\boldsymbol{q}\lambda}(t) = \mathrm{e}^{\mathrm{i} H_{\mathrm{ph}} t}\phi_{\boldsymbol{q}\lambda}\, \mathrm{e}^{-\mathrm{i} H_{\mathrm{ph}} t} = b_{\boldsymbol{q}\lambda}\mathrm{e}^{-\mathrm{i}\Omega_{\boldsymbol{q}\lambda} t} + b_{-\boldsymbol{q}\lambda}^{\dagger}\mathrm{e}^{\mathrm{i}\Omega_{\boldsymbol{q}\lambda} t}, \qquad (1.47)$$

and the step function $\theta(t) = 1$ for $t \geqslant 0$ and $\theta(t) = 0$ for $t < 0$.

The expressions (1.42) and (1.43) of frictional forces $\tilde{\boldsymbol{f}}_\mathrm{i}$ and $\tilde{\boldsymbol{f}}_\mathrm{p}$ have important physical connotation. Firstly, the center-of-mass motion enters (1.42) and (1.43) through the function

$$A(\boldsymbol{q}, t, t') \equiv \exp\bigl[\mathrm{i}\boldsymbol{q} \cdot (\boldsymbol{R}(t) - \boldsymbol{R}(t'))\bigr], \qquad (1.48)$$

which plays a crucial role in the balance equation theory. Since the CM velocity relates to its position by $\boldsymbol{V} = \dot{\boldsymbol{R}}$, when neglecting effects associated with the velocity fluctuation $\delta\boldsymbol{V}$ of the center of mass in (1.19), the CM velocity V can be approximated by its average part v and $A(q,t,t')$ function is determined only by the drift velocity $v(t)$ of the electron system:

$$A(q,t,t') = \exp\left[iq \cdot \int_{t'}^{t} v(s)\, ds\right]. \tag{1.49}$$

In this way the frictional forces $\tilde{f}_i$ and $\tilde{f}_p$ depend on the time-dependent average velocity of the center of mass, i.e. the time-dependent drift velocity of the electron system, $v(t)$. This indicates that the behavior of the frictional forces (thus the major transport properties) of the system is direct affected by the drift motion of carriers rather than by the electric field, which shows up only as a force acting on the center-of-mass in the force balance equation (1.40). The functionally dependence involving in Eqs. (1.42) and (1.43), however, is rather complicated and generally hard to handle. Approximations can be made in different cases to simplify the treatment.

The second point connoted in the expressions (1.42) and (1.43) is related to the correlation functions $\Pi(q, t-t')$ and $\Lambda(q, \lambda; t-t')$ given by (1.44) and (1.45), which involve the statistical average $\langle \cdots \rangle_0$ with respect to the initial density matrix $\hat{\rho}_0$ given by (1.30) containing the electron temperature $T_e(t)$ as a parameter. Thus, the frictional forces $\tilde{f}_i$ and $\tilde{f}_p$ expressed in Eqs. (1.42) and (1.43) also depend on the time-dependent electron temperature $T_e(t)$ of the system.

1.6 Energy Balance Equation

Taking the statistical average of the right hand side of the operator expression (1.23) in the relative electron and phonon space to the lowest order in H_I, we obtain the energy increase rate of the phonon system due to electron–phonon interaction, i.e. the energy transfer rate from the electron system to the phonon system:

$$\tilde{w} = \langle \hat{W} \rangle = -\sum_{q\lambda} |M(q,\lambda)|^2 \int_{-\infty}^{\infty} dt'\, A(q,t,t') \Gamma(q,\lambda, t-t'), \tag{1.50}$$

with

$$\Gamma(q, \lambda, t-t') = -i\theta(t-t') \langle [\dot{\phi}_{q\lambda}(t)\rho_q(t),\, \phi_{-q\lambda}(t')\rho_{-q}(t')] \rangle_0 \tag{1.51}$$

and

$$\dot{\phi}_{q\lambda}(t) = e^{iH_{ph}t}\dot{\phi}_{q\lambda}e^{-iH_{ph}t} = -i\Omega_{q\lambda}\left(b_{q\lambda}e^{-i\Omega_{q\lambda}t} - b^{\dagger}_{-q\lambda}e^{i\Omega_{q\lambda}t}\right). \tag{1.52}$$

This energy-transfer rate $\tilde{w}$ also functionally depends on the CM drift velocity $\boldsymbol{v}(t)$ through the $A(\boldsymbol{q},t,t')$ function and on the electron temperature $T_{\mathrm{e}}(t)$ through the statistical average of the correlation function $\Gamma(\boldsymbol{q},\lambda,t-t')$ with respect to the initial density matrix $\hat{\rho}_0$.

An energy balance equation is derived from the statistical average of the operator equation (1.26). Retaining only the drift part of the center-of-mass velocity (i.e. neglecting the contribution from the CM velocity fluctuation) and identifying $\langle \dot{H}_{\mathrm{er}} \rangle$ as the rate of increase of the internal energy of relative electrons,

$$\langle \dot{H}_{\mathrm{er}} \rangle = \frac{dU}{dt} \tag{1.53}$$

with $U \equiv \langle H_{\mathrm{er}} \rangle$, we have

$$Ne\boldsymbol{E} \cdot \boldsymbol{v} = Nm\boldsymbol{v} \cdot \frac{d\boldsymbol{v}}{dt} + \frac{dU}{dt} + \tilde{w}. \tag{1.54}$$

Eliminating the term $Nm\boldsymbol{v} \cdot d\boldsymbol{v}/dt$ by means of the force balance equation (1.40), we can write the energy balance equation (1.54) in the form

$$-\frac{dU}{dt} = \boldsymbol{v} \cdot \tilde{\boldsymbol{f}} + \tilde{w}, \tag{1.55}$$

where $\tilde{\boldsymbol{f}} = \tilde{\boldsymbol{f}}_{\mathrm{i}} + \tilde{\boldsymbol{f}}_{\mathrm{p}}$. Eq. (1.54) or Eq. (1.55) is the time-dependent energy balance equation of the system (Lei and Ting, 1984, 1985a; Xing and Ting, 1987).

It should be noted that energy balance equation can also be derived by taking the statistical average of the operator expression (1.24) to the lowest order in H_{I} (Lei and Ting, 1984, 1985a),

$$\langle \hat{S} \rangle = -\mathrm{i}\, n_{\mathrm{i}} \sum_{\boldsymbol{q}} |u(\boldsymbol{q})|^2 \int_{-\infty}^{\infty} dt'\, A(\boldsymbol{q},t,t') \sum_{\boldsymbol{k}} (\varepsilon_{\boldsymbol{k}+\boldsymbol{q}} - \varepsilon_{\boldsymbol{k}}) \Pi(\boldsymbol{k},\boldsymbol{q},t-t')$$

$$-\mathrm{i} \sum_{\boldsymbol{q},\lambda} |M(\boldsymbol{q},\lambda)|^2 \int_{-\infty}^{\infty} dt'\, A(\boldsymbol{q},t,t') \sum_{\boldsymbol{k}} (\varepsilon_{\boldsymbol{k}+\boldsymbol{q}} - \varepsilon_{\boldsymbol{k}}) \Lambda(\boldsymbol{k},\boldsymbol{q},\lambda,t-t'), \tag{1.56}$$

with

$$\Pi(\boldsymbol{k},\boldsymbol{q},t-t') = -\mathrm{i}\,\theta(t-t')\langle [\rho_{\boldsymbol{k}\boldsymbol{q}}(t), \rho_{-\boldsymbol{q}}(t')] \rangle_0, \tag{1.57}$$

$$\Lambda(\boldsymbol{k},\boldsymbol{q},\lambda;t-t') = -\mathrm{i}\,\theta(t-t')\langle [\phi_{\boldsymbol{q}\lambda}(t)\rho_{\boldsymbol{k}\boldsymbol{q}}(t), \phi_{-\boldsymbol{q}\lambda}(t')\rho_{-\boldsymbol{q}}(t')] \rangle_0, \tag{1.58}$$

$$\rho_{\bm{k}\bm{q}}(t) = \mathrm{e}^{\mathrm{i}\,H_{\mathrm{er}}t}\rho_{\bm{k}\bm{q}}\,\mathrm{e}^{-\mathrm{i}\,H_{\mathrm{er}}t}, \tag{1.59}$$

and identifying $\langle \dot{H}_{\mathrm{er}}\rangle = dU/dt$. This results in another energy balance equation:

$$\frac{dU}{dt} = \langle \hat{S}\rangle. \tag{1.60}$$

This equation is equivalent to Eq. (1.54) for steady state transport (Lei and Ting, 1985a) and for time-dependent drift motion involving negligible memory effects (Chapter 3). However, there may be some differences when discussing nonlinear processes in which memory effects are significant, or discussing processes involving effects of velocity fluctuations in higher orders (see Chapter 4).

Using the perturbation expansion (1.38) to calculate the relative electron internal energy $\langle H_{\mathrm{er}}\rangle$, we find that only the $\hat{\rho}_0$ term contributes, i.e.

$$U = \langle H_{\mathrm{er}}\rangle_0, \tag{1.61}$$

which depends on the electron temperature T_{e}. Expressing the rate of change of the relative electron internal energy in terms of its specific heat $C_{\mathrm{e}} = \partial U/\partial T_{\mathrm{e}}$ and the rate of change of the electron temperature, we can write

$$\frac{dU}{dt} = C_{\mathrm{e}}\frac{dT_{\mathrm{e}}}{dt}. \tag{1.62}$$

Chapter 2

DC Steady-State Transport

2.1 Balance Equations for DC Steady-State Transport in Three-Dimensional Systems

When the applied electric field $\boldsymbol{E}$ is constant in time, the system approaches a steady state with a constant drift velocity $\boldsymbol{v}_\text{d}$ and a constant electron temperature T_e after transients die out. For steady-state transport we can take the velocity function $\boldsymbol{v}(s)$ inside the integral of the expression (1.49) of $A(\boldsymbol{q}, t, t')$ function as the constant drift velocity $\boldsymbol{v}_\text{d}$ such that

$$A(\boldsymbol{q}, t, t') \approx \exp\left[\mathrm{i}\boldsymbol{q} \cdot \int_{t'}^{t} \boldsymbol{v}(s)\,ds\right] = \exp\left[\mathrm{i}\boldsymbol{q} \cdot \boldsymbol{v}_\text{d}(t - t')\right]. \tag{2.1}$$

Substituting (2.1) into (1.42), (1.43) and (1.50), we obtain the force and energy balance equations for the steady state, written for unit volume system with n denoting the carrier number density (Lei and Ting, 1984, 1985a):

$$n e \boldsymbol{E} + \boldsymbol{f} = 0, \tag{2.2}$$

$$\boldsymbol{v}_\text{d} \cdot \boldsymbol{f} + w = 0. \tag{2.3}$$

Here the frictional force experienced by the center of mass due to impurity and phonon scatterings, $\boldsymbol{f} = \boldsymbol{f}_\text{i} + \boldsymbol{f}_\text{p}$, and the electron energy-loss rate to the lattice due to electron–phonon interaction, w, can be expressed as

$$\boldsymbol{f}_\text{i} = n_\text{i} \sum_{\boldsymbol{q}} \boldsymbol{q}\, |u(\boldsymbol{q})|^2\, \Pi_2(\boldsymbol{q}, \omega_0), \tag{2.4}$$

$$\boldsymbol{f}_\text{p} = \sum_{\boldsymbol{q},\lambda} \boldsymbol{q}\, |M(\boldsymbol{q},\lambda)|^2\, \Lambda_2(\boldsymbol{q}, \lambda, \omega_0)$$

$$= 2 \sum_{\boldsymbol{q},\lambda} \boldsymbol{q}\, |M(\boldsymbol{q},\lambda)|^2\, \Pi_2(\boldsymbol{q}, \Omega_{\boldsymbol{q}\lambda} + \omega_0) \left[n\!\left(\frac{\Omega_{\boldsymbol{q}\lambda}}{T}\right) - n\!\left(\frac{\Omega_{\boldsymbol{q}\lambda} + \omega_0}{T_\text{e}}\right)\right], \tag{2.5}$$

$$w = \sum_{q,\lambda} \Omega_{q\lambda} |M(q,\lambda)|^2 \Gamma_2(q,\lambda,\omega_0)$$

$$= 2 \sum_{q,\lambda} \Omega_{q\lambda} |M(q,\lambda)|^2 \Pi_2(q, \Omega_{q\lambda}+\omega_0) \left[n\left(\frac{\Omega_{q\lambda}}{T}\right) - n\left(\frac{\Omega_{q\lambda}+\omega_0}{T_e}\right) \right]. \tag{2.6}$$

In these equations $\omega_0 \equiv q \cdot v_d$, $n(x) \equiv 1/(e^x - 1)$ stands for the Bose function, $\Pi_2(q,\omega)$, $\Lambda_2(q,\lambda,\omega)$ and $\Gamma_2(q,\lambda,\omega)$ are the imaginary parts of the Fourier spectra of the correlation functions $\Pi(q,t)$, $\Lambda(q,\lambda,t)$ and $\Gamma(q,\lambda,t)$ defined respectively by (1.44), (1.45) and (1.51). For instance,

$$\Pi(q,\omega) = \int_{-\infty}^{\infty} e^{i\omega t} \Pi(q,t)\, dt. \tag{2.7}$$

The function $\Pi(q,t)$, defined by Eq. (1.44), is the density–density correlation function or simply the density correlation function of the relative electron system described by the Hamiltonian (1.12),

$$H_{\text{er}} = \sum_{k,\sigma} \varepsilon_k c_{k\sigma}^\dagger c_{k\sigma} + H_{\text{ee}}, \tag{2.8}$$

in a thermal equilibrium state with temperature T_e. That is the thermodynamic correlation function of an electron system without impurity nor phonon scattering and without electric field. This function has been widely investigated for different types of systems (Keldysh, Kirzhnitz and Maradudin,1989). For a single band three-dimensional (3D) system in the absence of Coulomb interaction between carriers, this density correlation function has a well known expression (δ is a small positive real number):

$$\Pi_0(q,\omega) = 2 \sum_k \frac{f(\varepsilon_{k+q}, T_e) - f(\varepsilon_k, T_e)}{\varepsilon_{k+q} - \varepsilon_k + \omega + i\delta}, \tag{2.9}$$

with its real part $\Pi_{01}(q,\omega)$ and imaginary part $\Pi_{02}(q,\omega)$ given by

$$\Pi_{01}(q,\omega) = -2 \sum_k f(\varepsilon_k, T_e) \left(\frac{1}{\varepsilon_{k+q} - \varepsilon_k + \omega} + \frac{1}{\varepsilon_{k+q} - \varepsilon_k - \omega} \right), \tag{2.10}$$

$$\Pi_{02}(q,\omega) = 2\pi \sum_k [f(\varepsilon_k, T_e) - f(\varepsilon_{k+q}, T_e)] \delta(\varepsilon_{k+q} - \varepsilon_k + \omega). \tag{2.11}$$

Here

$$f(\varepsilon_k, T_e) = 1/\left[\exp(\varepsilon_k - \mu)/T_e + 1\right] \tag{2.12}$$

is the Fermi distribution function at electron temperature T_e, and μ is the chemical potential. For systems of parabolic energy band, $\varepsilon_k = k^2/2m$, the summation over $\boldsymbol{k}$ can easily be carried out, yielding the following analytical result for $\Pi_{02}(\boldsymbol{q},\omega)$:

$$\Pi_{02}(\boldsymbol{q},\omega) = -\frac{m^2 T_e}{2\pi q} P_2(\boldsymbol{q},\omega), \qquad (2.13)$$

$$P_2(\boldsymbol{q},\omega) = \ln\left\{\frac{1+\exp\left[-\frac{1}{2mT_e}\left(\frac{q}{2}-\frac{m\omega}{q}\right)^2 + \frac{\mu}{T_e}\right]}{1+\exp\left[-\frac{1}{2mT_e}\left(\frac{q}{2}+\frac{m\omega}{q}\right)^2 + \frac{\mu}{T_e}\right]}\right\}.$$

The expression of $\Pi_{01}(\boldsymbol{q},\omega)$, (2.10), which contains a principal value of an integral, generally can not be carried out analytically except in the degenerate limit. For a degenerate system at sufficient low electron temperature T_e, the real part and imaginary part of $\Pi_0(\boldsymbol{q},\omega)$ can be expressed as

$$\Pi_{01}(\boldsymbol{q},\omega) = -\frac{mk_F^2}{4\pi^2 q}\left\{4a + \left[1-(a+b/a)^2\right]\ln\left|\frac{1+(a+b/a)}{1-(a+b/a)}\right|\right.$$
$$\left. + \left[1-(a-b/a)^2\right]\ln\left|\frac{1+(a-b/a)}{1-(a-b/a)}\right|\right\}, \qquad (2.14)$$

$$\Pi_{02}(\boldsymbol{q},\omega) = -\frac{mk_F^2}{4\pi q}\left\{\left[1-(a-b/a)^2\right]\theta\left(1-(a-/ba)^2\right)\right.$$
$$\left. - \left[1-(a+b/a)^2\right]\theta\left(1-(a+b/a)^2\right)\right\}, \qquad (2.15)$$

in which $\theta(x)$ is the Heaviside unit step function: $\theta(x) = 0$ for $x < 0$ and $\theta(x) = 1$ for $x \geqslant 0$, the symbols $a \equiv q/2k_F$ and $b \equiv \omega/4\varepsilon_F$ with k_F the Fermi wavevector and $\varepsilon_F = k_F^2/2m$ the Fermi energy at zero temperature.

The introduction of carrier–carrier Coulomb interactions H_{ee} may substantially modify the correlation function. In the random phase approximation (RPA), the electron density correlation function with Coulomb interactions, $\Pi(\boldsymbol{q},\omega) = \Pi_1(\boldsymbol{q},\omega) + i\Pi_2(\boldsymbol{q},\omega)$, is obtained as

$$\Pi(\boldsymbol{q},\omega) = \frac{\Pi_0(\boldsymbol{q},\omega)}{1-v_c(q)\Pi_0(\boldsymbol{q},\omega)} = \frac{\Pi_0(\boldsymbol{q},\omega)}{\epsilon(\boldsymbol{q},\omega)}, \qquad (2.16)$$

where

$$\epsilon(\boldsymbol{q},\omega) \equiv 1 - v_c(q)\Pi_0(\boldsymbol{q},\omega) \qquad (2.17)$$

is the RPA dielectric function (complex) of the electron system with $\epsilon_1(\boldsymbol{q},\omega)$ and $\epsilon_2(\boldsymbol{q},\omega)$ representing its real and imaginary parts. The imaginary part of the electron RPA density correlation function, $\Pi_2(\mathbf{q},\omega)$, is given by

$$\Pi_2(\mathbf{q},\omega) = \frac{\Pi_{02}(\mathbf{q},\omega)}{|\epsilon(\mathbf{q},\omega)|^2}. \qquad (2.18)$$

We may consider the factor $|\epsilon(\mathbf{q},\omega)|^2$ as a kind of dynamic screening on the electron–impurity potential $|u(\mathbf{q})|^2$ and on the matrix element of electron–phonon interaction $|M(\mathbf{q},\lambda)|^2$. The frictional forces $\mathbf{f}_\text{i}$ and $\mathbf{f}_\text{p}$, as well as the electron energy-transfer rate w, all can be written in terms of the dynamically screened potentials:

$$u(\mathbf{q})/\epsilon(\mathbf{q},\omega_0), \qquad (2.19)$$

$$M(\mathbf{q},\lambda)/\epsilon(\mathbf{q},\Omega_{\mathbf{q}\lambda}+\omega_0), \qquad (2.20)$$

together with the density correlation function Π_{02} in the absence of intercarrier Coulomb interaction. This screening is dynamic in that the dielectric functions appearing in (2.19) and (2.20) take values at the frequency equal to $\omega_0 \equiv \mathbf{q}\cdot\mathbf{v}_\text{d}$ (for impurity scattering) or equal to $\Omega_{\mathbf{q}\lambda}+\omega_0$ (for phonon scattering), which are dependent on the drift velocity $\mathbf{v}_\text{d}$.

For an isotropic system the current density $\mathbf{J} \equiv ne\mathbf{v}_\text{d}$ is in the same direction as the electric field, $\mathbf{E} = \rho\mathbf{J}$, and the total electric resistivity ρ can be defined by

$$\rho \equiv \frac{\mathbf{E}\cdot\mathbf{v}_\text{d}}{nev_\text{d}^2}, \qquad (2.21)$$

The force-balance equation (2.2) is rewritten as

$$\rho = \rho_\text{i} + \rho_\text{p}, \qquad (2.22)$$

where

$$\rho_\text{i} \equiv -\frac{\mathbf{f}_\text{i}\cdot\mathbf{v}_\text{d}}{n^2e^2v_\text{d}^2} \qquad (2.23)$$

is the resistivity due to impurity scattering, and

$$\rho_\text{p} \equiv -\frac{\mathbf{f}_\text{p}\cdot\mathbf{v}_\text{d}}{n^2e^2v_\text{d}^2} \qquad (2.24)$$

is the resistivity due to phonon scattering. In terms of an equivalent resistivity associated with the electron energy-transfer (energy-loss) rate,

$$\rho_\text{E} \equiv \frac{w}{n^2e^2v_\text{d}^2}, \qquad (2.25)$$

the energy-balance equation can be written in the form

$$\rho - \rho_{\rm E} = 0. \qquad (2.26)$$

We first consider the solution of balance equations (2.2) and (2.3) in the limit of weak current density at finite lattice temperature T. In the case of small $v_{\rm d}$ the energy-balance equation has the solution

$$T_{\rm e}/T = 1 + o(v_{\rm d}^2). \qquad (2.27)$$

Therefore the impurity- and phonon-induced resistivities in the limit $v_{\rm d} \to 0$ are obtained from expressions (2.23) and (2.24) with $T_{\rm e} = T$, whence

$$\rho_{i0} = \rho_{\rm i}(v_{\rm d} \to 0) = -\frac{n_{\rm i}}{n^2 e^2} \sum_{\bm{q}} q_x^2 |\bar{u}(\bm{q})|^2 \frac{\partial}{\partial w} \Pi_{02}(\bm{q}, \omega)\big|_{\omega=0}, \qquad (2.28)$$

$$\rho_{p0} = \rho_{\rm p}(v_{\rm d} \to 0) = -\frac{2}{n^2 e^2} \sum_{\bm{q},\lambda} q_x^2 |\bar{M}(\bm{q},\lambda)|^2 \Pi_{02}(\bm{q}, \Omega_{\bm{q}\lambda}) \left[-\frac{1}{T} n'\left(\frac{\Omega_{\bm{q}\lambda}}{T}\right)\right], \qquad (2.29)$$

where we have used the symbols

$$|\bar{u}(\bm{q})|^2 \equiv |u(\bm{q})/\epsilon(\bm{q},0)|^2, \qquad (2.30)$$

$$|\bar{M}(\bm{q},\lambda)|^2 \equiv |M(\bm{q},\lambda)/\epsilon(\bm{q}, \Omega_{\bm{q}\lambda})|^2. \qquad (2.31)$$

Therefore, as far as the linear resistivity is concerned, the RPA of intercarrier interaction is equivalent to a screening of the scattering potentials.

2.2 Degenerate Case, Impurity and Phonon Scatterings

In this section we consider nonpolar semiconductors in the degenerate case at low temperatures (i.e. the electron temperature $T_{\rm e} \ll \varepsilon_{\rm F}$, $\varepsilon_{\rm F}$ is the Fermi level at zero temperature) and assume that impurities and longitudinal acoustic phonons are the main scatterers.

2.2.1 *Ohmic resistivity, Bloch–Grüneissen formula*

In the low-temperature degenerate case we can use expressions (2.14) and (2.15) for Π_{01} and Π_{02} to obtain the well-known result for the impurity-induced Ohmic (linear) resistivity:

$$\rho_{i0} = \frac{m^2 n_{\rm i}}{12\pi^3 n^2 e^2} \int_0^{2k_{\rm F}} dq\, |\bar{u}(\bm{q})|^2 q^3. \qquad (2.32)$$

Here
$$\bar{u}(\boldsymbol{q}) = u(\boldsymbol{q})/[1 - \nu_c(q)\Pi_{01}(q,0)] \qquad (2.33)$$

is the statistically screened impurity potential.

For the evaluation of phonon-induced resistivity at low temperatures, we consider long-wavelength longitudinal acoustic phonons with a Debye-type spectrum

$$\Omega_{\boldsymbol{q}} = v_s q, \qquad (2.34)$$

v_s being the longitudinal sound velocity. The longitudinal acoustic phonons are coupled to electrons through the deformation potential. If $v_s/v_F \ll 1$ ($v_F \equiv k_F/m$ is the Fermi velocity), it is appropriate to use the following approximate expression for $\Pi_{02}(\boldsymbol{q}, \Omega_{\boldsymbol{q}\lambda})$ in (2.29),

$$\Pi_{02}(\boldsymbol{q},\omega) \approx -\frac{m^2}{2\pi q}\theta(1-(q/2k_F)^2)\omega, \qquad (2.35)$$

yielding a succinct result for phonon-induced linear resistivity:

$$\rho_{p0} = \frac{m^2 v_s}{6\pi^3 n^2 e^2}\int_0^{2k_F} dq\, q^4 |\bar{M}(\boldsymbol{q},\lambda)|^2 \left[-\frac{1}{T}n'\left(\frac{\Omega_{\boldsymbol{q}\lambda}}{T}\right)\right]. \qquad (2.36)$$

This is the famous Bloch-Grüneissen formula (Ziman 1960, 1972). For acoustic deformation-potential scattering the electron–phonon matrix element (neglect screening) is approximately proportional to the wavevector,

$$|\bar{M}(\boldsymbol{q},\lambda)|^2 = C_d q, \qquad (2.37)$$

with $C_d = \Xi^2/(2v_s d_c)$ (d_c is the mass density of the lattice and Ξ is the deformation potential constant, i.e. the energy band shift induced by the unit lattice dilation), and ρ_{p0} can be written as

$$\rho_{p0} = \rho^* g(t_F), \qquad (2.38)$$

where $t_F \equiv T/\Theta_F$, $\Theta_F \equiv 2k_F v_s$,

$$g(t) \equiv \frac{1}{t}\int_0^1 dy \frac{y^5}{(e^{y/t}-1)(1-e^{-y/t})}, \qquad (2.39)$$

and ρ^* is a coefficient of resistivity dimension (depending on C_d and other properties of the system), which will be used as a reference scale of the resistivity in the next few subsections. The function $g(t)$ has the following limit behavior:

$$g(t) = \begin{cases} 124.43\,t^5, & \text{for } t \ll 1 \\ t/4, & \text{for } t \gg 1, \end{cases} \tag{2.40}$$

indicating that the resistivity induced by acoustic-phonon deformation-potential scattering is proportional to T^5 at low temperatures, and proportional to T at high temperatures.

The above formulas for impurity- and phonon-induced linear dc state-steady resistivities are known as isothermal resistivities, i.e. the results of the force-balance equation when the electron density is sufficient high that there is strong coupling between carriers. In the case of dilute carrier system, the resistivity expressions may be different (so called adiabatic resistivity, see Sec. 7.3 for further discussion).

2.2.2 Warm-electron conduction

In the case of $v_s/v_F \ll 1$ it is easy to obtain the solution of the balance equation (2.26) to the order of $(v_d/v_s)^2$. The result can be written as

$$T_e = T\big[1 + A(v_d/v_s)^2\big], \tag{2.41}$$
$$\rho_p = \rho_{p0}\big[1 + B(v_d/v_s)^2\big], \tag{2.42}$$

with

$$A = \frac{1}{3}\left[-1 - \frac{n'(1/t_F)}{2g(t_F)t_F} + \frac{\rho_{i0}}{\rho^* g(t_F)}\right], \tag{2.43}$$

$$B = A\left[4 + \frac{n'(1/t_F)}{g(t_F)t_F}\right] + \frac{1}{10}\left[24 + \frac{4n'(1/t_F)}{g(t_F)t_F} - \frac{n''(1/t_F)}{g(t_F)t_F^2}\right]. \tag{2.44}$$

Here ρ_{i0} is the impurity-induced Ohmic resistivity. In the case of $v_s/v_F \ll 1$ the total resistivity $\rho = \rho_i + \rho_p \approx \rho_{i0} + \rho_p$. For a degenerate system $(T \ll \varepsilon_F)$, when $t_F \gg 1$

$$A \approx \frac{1}{3}, \quad B \approx 0; \tag{2.45}$$

when $t_F \ll 1$

$$A \approx \frac{1}{3}\left[-1 + \frac{\rho_{i0}}{\rho^*}\frac{1}{g(t_F)}\right], \tag{2.46}$$

$$B \approx \frac{4}{3}\left[\frac{4}{5} + \frac{\rho_{i0}}{\rho^* g(t_F)}\right]. \tag{2.47}$$

The interesting feature is that when impurity resistivity ρ_{i0} is small enough, the coefficient A is negative, i.e. the electron temperature is lower than the lattice temperature in a certain temperature range. Figure 2.1 shows the calculated coefficients A and B as functions of reduced temperature $t_F \equiv T/\Theta_F$ in the cases of different values of ρ_{i0}/ρ^* (Lei and Ting, 1985c).

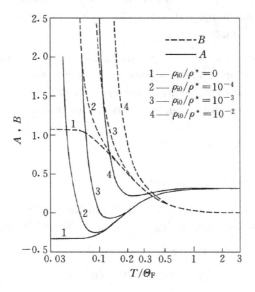

Fig. 2.1 Temperature variation of coefficients A and B of $(v_d/v_s)^2$ terms in the expansions (2.41) and (2.42) for T_e/T and ρ_p/ρ_{p0}. From Lei and Ting (1985c).

2.2.3 Electron cooling

An interesting consequence of the force and energy balance equations, which has been seen in the warm-electron conduction analysis in the preceding subsection, is the possible cooling of relative electrons, i.e. the electron temperature T_e may be lower than the lattice temperature T in the presence of a finite drift velocity v_d (Lei and Ting, 1985a,c). Such a lowering of the electron temperature below the lattice temperature occurs in low-impurity samples. At low lattice temperatures, when acoustic phonons dominate the scattering, the maximum cooling occurs at a current density for which the electron drift velocity is near the sound velocity. Fig. 2.2 provides an example of electron cooling in the case of nonlinear conduction with dominant acoustic phonon deformation potential scattering.

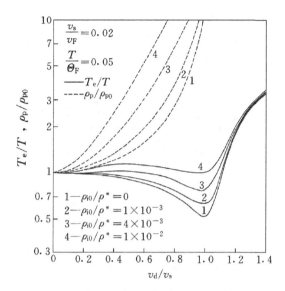

Fig. 2.2 The temperature ratio T_e/T and dimensionless phonon-induced resistivity ρ_p/ρ_{p0} (dashed lines) are shown as functions of v_d/v_s at several lattice temperatures for a system with $v_s/v_F = 0.02$ at $T/\Theta_F = 0.05$. Zero-field phonon resistivity $\rho_{p0} = 1.94 \times 10^{-3} \rho^*$. From Lei and Ting (1985a).

In the case of dominant optic phonon scatterings, electron cooling is most likely to occur at higher drift velocity and at higher lattice temperature (Sec. 4.4).

2.2.4 Nonlinear zero-temperature limit

It is of particular interest to examine nonlinear conduction in the zero lattice temperature limit. When $T \to 0$, $n(\Omega_{q\lambda}/T)$ vanishes and only the electron temperature T_e is involved in the phonon-induced frictional force $\boldsymbol{f}_p$ and the electron energy transfer rate w in Eqs. (2.5) and (2.6). The force and energy balance equations require T_e to approach a finite value which depends on the applied field and the impurity-related resistivity. As a simple illustration, we assume $M(\boldsymbol{q}, \lambda)$ to be of the deformation potential type, (2.37). For $v_s \ll v_F$, the energy-loss related resistivity ρ_E (2.25) and the phonon-induced resistivity ρ_p (2.24) can be written, respectively, at

$T \to 0$ limit, as $(t_F^* \equiv T_e/\Theta_F)$

$$\frac{\rho_E}{\rho^*} = \frac{3}{2}\left(\frac{v_s}{v_d}\right)^2 \int_0^1 dy\, y^4 \int_{-1}^1 dx \left(1 + \frac{v_d}{v_s}x\right) n\left[\frac{y}{t_F^*}\left(1 + \frac{v_d}{v_s}x\right)\right], \quad (2.48)$$

and

$$\frac{\rho_p}{\rho^*} = -\frac{3}{2}\frac{v_s}{v_d} \int_0^1 dy\, y^4 \int_{-1}^1 dx\, x\left(1 + \frac{v_d}{v_s}x\right) n\left[\frac{y}{t_F^*}\left(1 + \frac{v_d}{v_s}x\right)\right]. \quad (2.49)$$

If $v_d/v_s < 1$, the integrations in (2.48) and (2.49) are easily carried out, yielding

$$\left(\frac{T_e}{\Theta_F}\right)^5 = 0.0134\frac{\rho_{i0}}{\rho^*}\frac{(v_d/v_s)^2}{[1 - (v_d/v_s)^2]^2}, \quad (2.50)$$

$$\frac{\rho_p}{\rho^*} = \frac{4}{3}\frac{\rho_{i0}}{\rho^*}\frac{(v_d/v_s)^2}{1 - (v_d/v_s)^2}, \quad (2.51)$$

and the impurity resistivity ρ_i is almost equal to its weak-field value ρ_{i0}. The electric field $E = (\rho_p + \rho_i)J$ can be expressed as

$$\frac{E}{E^*} = \frac{\rho_{i0}}{\rho^*}\frac{v_d}{v_s}\left[1 + \frac{4}{3}\frac{(v_d/v_s)^2}{1 - (v_d/v_s)^2}\right], \quad (2.52)$$

with $E^* = \rho^* nev_s$. It is easily seen from Eqs. (2.50) and (2.51) that for $v_d/v_s \ll 1$ or $E \ll E^*$,

$$\frac{T_e}{\Theta_F} = 0.422\left(\frac{\rho^*}{\rho_{i0}}\right)^{1/5}\left(\frac{E}{E^*}\right)^{2/5}, \quad (2.53)$$

$$\frac{\rho_p}{\rho^*} = \frac{4}{3}\frac{\rho^*}{\rho_{i0}}\left(\frac{E}{E^*}\right)^2. \quad (2.54)$$

Equation (2.53) is in agreement with the result obtained by Arai (1983). It should be noted that Eqs. (2.53) and (2.54) are valid only in a rather narrow electric field range, depending on the impurity resistivity ρ_{i0}/ρ^*. They are completely invalid when $\rho_{i0} = 0$. However, Eqs. (2.50) and (2.51) hold as long as $v_d/v_s < 1$ irrespective of impurities. Therefore, in the absence of impurity scattering ($\rho_{i0} = 0$), we have $T_e = 0$, $\rho_p = 0$ and $E = 0$ when $v_d < v_s$. This means that in the absence of impurity scattering at zero lattice temperature $T = 0$, the current can flow without resistance up to a drift velocity $v_d = v_s$. Resistance appears once $v_d > v_s$. The calculated results of T_e/Θ_F and the phonon-induced resistivity ρ_p/ρ^* are shown in

Fig. 2.2 as functions of the electric field for different values of impurity scattering strength (Lei and Ting, 1984). It is worth noting that in the presence of a finite electric field the phonon contribution to resistivity does not vanish at $T = 0$.

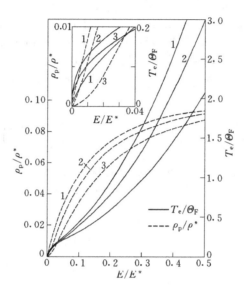

Fig. 2.3 The electron temperature T_e and phonon resistivity ρ_p in the zero lattice-temperature $T = 0$ are shown as functions of the dimensionless electric field E/E^* for several values of impurity resistivity in a degenerate electron system with acoustic phonon and impurity scatterings. $\rho_{i0}/\rho^* = 0, 0.03$, and 0.1 for curves 1, 2, and 3, respectively. From Lei and Ting (1984).

2.3 Acoustic and Optic Phonon Scatterings

In this section we shall discuss acoustic-phonon scattering and optic-phonon scattering in the wide ranges of temperature. For a single-band semiconductor with fixed electron density n the chemical potential μ is related to the electron temperature T_e through the equation $n = 2\sum_{\bm{k}} f(\varepsilon_{\bm{k}}, T_e)$.

2.3.1 *Acoustic phonon scatterings*

We first assume that acoustic phonons are the only scatterers and $M(\bm{q})$ is of the deformation potential type (2.37). For convenience, we introduce a temperature-dimension parameter

$$\Theta_s = mv_s^2, \qquad (2.55)$$

which is about $1\,\text{K}$ for $v_s = 5\times 10^5\,\text{cm/s}$ and $m = 0.6 m_e$ (m_e is the free electron mass), and denote $t_s \equiv T/\Theta_s$ and $\alpha \equiv T_e/T$.

We first consider nondegenerate semiconductors, in which carriers in thermal equilibrium obey a Maxwell–Boltzmann distribution even at low temperatures. In this case the chemical potential μ is negative and $|\mu|/T_e \gg 1$, such that the function $P_2(\boldsymbol{q},\omega)$ in (2.13) becomes

$$P_2(\boldsymbol{q},\omega) = \mathrm{e}^{-|\mu|/T_e} \times$$

$$\left\{ \exp\left[-\frac{1}{2mT_e}\left(\frac{q}{2}-\frac{m\omega}{q}\right)^2\right] - \exp\left[-\frac{1}{2mT_e}\left(\frac{q}{2}+\frac{m\omega}{q}\right)^2\right] \right\}. \qquad (2.56)$$

Under the condition $\alpha t_s \gg 1$, the energy-balance equation (2.26), expanded to the leading order of $(\alpha t_s)^{-1}$ in the nondegenerate case, can be written as

$$\left(\frac{v_d}{v_s}\right)^2 = \frac{6\alpha^3 t_s^3}{S(\alpha, t_s)} - 3, \qquad (2.57)$$

where

$$S(\alpha, t_s) = \int_0^\infty dy\, \mathrm{e}^{-y^2/2\alpha t_s} y^4 \coth\left(\frac{y}{t_s}\right). \qquad (2.58)$$

If the condition $\alpha/t_s \gg 1$ also holds, we have

$$S(\alpha, t_s) \approx 3\sqrt{\frac{\pi}{2}} (\alpha t_s)^{5/2}, \qquad (2.59)$$

and then Eq. (2.57) yields

$$\alpha = \frac{\pi}{8 t_s}\left(\frac{v_d}{v_s}\right)^4. \qquad (2.60)$$

On the other hand, in the case of $\alpha t_s \gg 1$ the resistivity ρ_p given by (2.24) can be written as (ρ_{p0} stands for the resistivity when $v_d \to 0$)

$$\frac{\rho_p}{\rho_{p0}} = \frac{S(\alpha, t_s)}{2\alpha^{3/2} g_s(t_s)}, \qquad (2.61)$$

with

$$g_s(t_s) = -\int_0^\infty dy\, \exp\left[-\frac{(y-1)^2}{2t_s}\right] n'(2y/t_s), \qquad (2.62)$$

which, for $t_s \gg 1$, reduces to

$$g_s(t_s) \approx t_s^3. \qquad (2.63)$$

Therefore, when $\alpha/t_s \gg 1$ and $t_s \gg 1$, we have

$$\frac{\rho_p}{\rho_{p0}} \approx 1.88 \frac{\alpha}{t_s^{1/2}}. \qquad (2.64)$$

Combining Eqs. (2.60) and (2.64), we can easily obtain a relation between the temperature ratio $\alpha = T_e/T$ and the electric field $E = \rho_p n e v_d$:

$$\frac{T_e}{T} = 0.5 \left(\frac{T}{\Theta_s}\right)^{1/5} \left(\frac{E}{E_0}\right)^{4/5}, \qquad (2.65)$$

which is valid for $T/\Theta_s \gg 1$ and $E/E_0 \gg 1$. Here $E_0 = \rho_{p0} n e v_s = v_s/\mu_0$, μ_0 being the carrier mobility in the zero-field limit. Equation (2.65) differs from the result of the carrier temperature model analysis (Seeger, 1982). The latter predicts a linear dependence of T_e/T on the electric field with a temperature-independent coefficient.

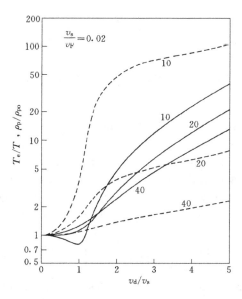

Fig. 2.4 The temperature ratio T_e/T (solid lines) and dimensionless phonon-induced resistivity ρ_p/ρ_{p0} (dashed lines) are shown as functions of v_d/v_s at several lattice temperatures for a system with $v_s/v_F = 0.02$. The scattering is entirely due to acoustic phonons. The numbers near the curves are the values of T/Θ_s. From Lei and Ting (1985a).

Numerical calculations for pure acoustic phonon scattering are carried out in the general case directly using the expression (2.13) for $P_2(\boldsymbol{q},\omega)$ by Lei and Ting (1985a). Fig. 2.4 is an example of their results.

2.3.2 Optical deformation-potential scatterings

The case in which the optic phonon scattering dominates is also interesting. In a nonpolar semiconductor the deformation-potential scattering induced by longitudinal optic phonons plays a major role. As is usually done in discussing optic-phonon scattering, we employ the Einstein model for the optic phonon spectrum, $\Omega_{\boldsymbol{q}\lambda} = \Omega_\mathrm{o}$, and assume its effective electron–phonon matrix element to be a constant:

$$|\bar{M}(\boldsymbol{q},\lambda)|^2 = \frac{D^2}{2d_\mathrm{c}\Omega_\mathrm{o}}, \qquad (2.66)$$

where d_c is the mass density of the lattice and D is the nonpolar optic deformation potential. It is convenient to introduce a velocity scale v_o defined by

$$mv_\mathrm{o}^2 \equiv \Omega_\mathrm{o}, \qquad (2.67)$$

and denote

$$t_\mathrm{o} \equiv T/\Omega_\mathrm{o}. \qquad (2.68)$$

The salient feature of optic-phonon scattering is the saturation of the current density at high electric fields. That is, the solution of the energy-balance equation exists only when $v_\mathrm{d} < v_\mathrm{m}$ with a saturation maximum velocity v_m. When v_d approaches v_m, the electron temperature $T_\mathrm{e} \to \infty$. To determine the saturation value of the drift velocity, we examine the asymptotic behavior of the energy-balance equation at large αt_o. To the leading term of its expansion in $(\alpha t_\mathrm{o})^{-1}$, the energy-balance equation yields

$$\left(\frac{v_\mathrm{d}}{v_\mathrm{o}}\right)^2 = 3\left[2 - \frac{\mathrm{K}_1'(1/2\alpha t_\mathrm{o})}{\alpha t_\mathrm{o}\mathrm{K}_1(1/2\alpha t_\mathrm{o})}\right]^{-1} \tanh\left(\frac{1}{2t_\mathrm{o}}\right), \qquad (2.69)$$

where $\mathrm{K}_1(x)$ and $\mathrm{K}_1'(x)$ are the modified Bessel function and its derivative (Wang and Guo, 1989). The saturation value v_m can be obtained by letting $\alpha t_\mathrm{o} \to \infty$ in Eq. (2.69), whence

$$\left(\frac{v_\mathrm{m}}{v_\mathrm{o}}\right)^2 = \frac{3}{4}\tanh\left(\frac{1}{2t_\mathrm{o}}\right), \qquad (2.70)$$

which is temperature dependent, having a maximum $v_{\rm m} = 0.866v_{\rm o}$ at $T = 0$. In the limit $E \to \infty$ the above result is the same as that obtained from the Boltzmann equation (Seeger, 1982). It is also consistent with the result of Thornber and Feynman (1970) in regard to the saturation at high field. Fig. 2.5 shows the dimensionless resistivity and electron temperature calculated by Lei and Ting (1985a) as functions of drift velocity for the case of pure nonpolar optic phonon scattering.

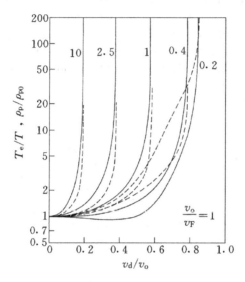

Fig. 2.5 The temperature ratio $T_{\rm e}/T$ (solid lines) and dimensionless phonon-induced resistivity $\rho_{\rm p}/\rho_{\rm p0}$ (dashed lines) are shown as functions of $v_{\rm d}/v_{\rm o}$ at several lattice temperatures for a system with $v_{\rm o}/v_F = 1$. The scattering is entirely due to nonpolar-optic phonons. The number near the curves is the value of $T/\Theta_{\rm o}$. From Lei and Ting (1985a).

2.4 Effect of Dynamic Screening

One of the most important effects of the Coulomb interaction between electrons is to induce screening. In most earlier investigations of transport, carrier–carrier Coulomb interactions were taken into account only in the form of static screening. On the contrary, the balance equation transport theory naturally incorporates the dynamic effects of carrier–carrier interactions in the electron density correlation function $\Pi_2(\boldsymbol{q},\omega)$, which is expressed by Eq. (2.18) in the random phase approximation. The role of this dynamic screening in impurity-induced dc resistivity has been examined by

Lei and Ting (1985b) in some detail. The full effect of the RPA density correlation function on the behavior of resistivity and electron temperature as functions of drift velocity is also investigated numerically by the same authors (Lei and Ting, 1985a). Although the inclusion of the dynamic Coulomb interaction generally does not qualitatively change the transport behavior at relatively small drift velocity (current density), as is obvious from the analytical expression, it has significant quantitative influence at high current density in the following two aspects.

Firstly, since $\epsilon(\boldsymbol{q},\omega) \to 1$ (or $\Pi_{01}(\boldsymbol{q},\omega) \sim 0$ and $\Pi_{02}(\boldsymbol{q},\omega) \sim 0$) for very large ω, in the case of large drift velocity $v_{\rm d}$, $\epsilon(\boldsymbol{q},\omega_0)$ and $\epsilon(\boldsymbol{q},\omega_0 + \Omega_{\boldsymbol{q}\lambda})$ can be replaced by 1 for the dominant $\boldsymbol{q}$-integration region. Therefore the contributions from very large ω should be equivalent to that without screening. This is a high-field descreening effect.

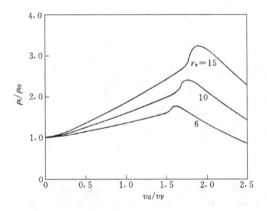

Fig. 2.6 Impurity-induced resistivity $\rho_{\rm i}/\rho_{\rm i0}$ is plotted as a function of $v_{\rm d}/v_{\rm F}$ in the degenerate case at several values of $r_{\rm s}$ (the radius of the electron sphere in atomic units) for charged-impurity scattering. For large value of $v_{\rm d}/v_{\rm F}$ the plasmon pole of the electron gas contributes to the resistivity. In the figure ρ stands for $\rho_{\rm i}$, and ρ_0 stands for $\rho_{\rm i0}$. From Lei and Ting (1985b).

The other dynamic screening effect comes from plasmon contributions. It is seen from the RPA expression for $\Pi_{02}(\boldsymbol{q},\omega)$, Eq. (2.18), that significant contributions to the frictional forces and energy transfer rate may occur in the region where both $[1 - \nu_c(q)\Pi_{01}(\boldsymbol{q},\omega)]$ and $\Pi_{02}(\boldsymbol{q},\omega)$ nearly vanish, the region of plasmon excitations. This effect on dc steady state transport is significant in the impurity-induced resistivity at low temperatures, as shown in Fig. 2.6. The plasmon contribution to impurity-induced resistivity shows up as a broad bump in the resistivity in the range of $v_{\rm d}/v_{\rm F} \sim 1.5$–$2.0$.

2.5 Balance Equations for Quasi-Two-Dimensional Systems

The balance equation approach expounded in the preceding sections for isotropic three-dimensional systems can be easily extended to different low-dimensional structures, such as inversion layers, heterojunctions, quantum wells, superlattices and quantum wires. In this section we will derive balance equations for quasi-two-dimensional systems following the discussion of Lei, Birman and Ting (1985).

We consider mobile carriers in, for example, a GaAs/AlGaAs heterojunction. Here the mobile carriers are electrons released from donors located somewhere in the heavily doped AlGaAs region ($z < 0$) and are essentially confined in the GaAs region ($z > 0$), formed by the band discontinuity of GaAs/AlGaAs interface at $z = 0$. These carriers form a quasi-two-dimensional electron system. In the effective-mass approximation this electron system can be described generally by the following Hamiltonian (Ando, Fowler and Stern, 1982)

$$H = \sum_i \left[\frac{\boldsymbol{p}_i^2}{2m} + \frac{p_{zi}^2}{2m_z} + U(z_i) \right] + \sum_{i<j} V(\boldsymbol{r}_i - \boldsymbol{r}_j, z_i, z_j), \qquad (2.71)$$

where $\boldsymbol{p}_i = (p_{xi}, p_{yi})$ and $\boldsymbol{r}_i = (x_i, y_i)$ are the two-dimensional momentum and coordinate of the ith electron parallel to the heterojunction interface (x–y plane) and p_{zi} and z_i are those perpendicular to the plane; m and m_z are, respectively, the electron band effective masses parallel and perpendicular to the interface; $U(z)$ is the confining potential, and the last term is the electron–electron interaction. Under the influence of a uniform electric field applied in the x–y plane the electron transport takes place only within the plane, and it is convenient to introduce two-dimensional (2D) center-of-mass variables $\boldsymbol{P} = (P_x, P_y)$ and $\boldsymbol{R} = (X, Y)$:

$$\boldsymbol{P} = \sum_i \boldsymbol{p}_i, \quad \boldsymbol{R} = \frac{1}{N} \sum_i \boldsymbol{r}_i, \qquad (2.72)$$

and two-dimensional relative electron variables

$$\boldsymbol{p}'_i = \boldsymbol{p}_i - \frac{1}{N}\boldsymbol{P}, \quad \boldsymbol{r}'_i = \boldsymbol{r}_i - \boldsymbol{R}, \qquad (2.73)$$

where N is the total number of electrons. In terms of these variables the Hamiltonian of this electron system in the presence of a uniform electric field $\boldsymbol{E}$ parallel to the layer plane can be written as the sum of a 2D center-of-mass part $H_{\rm cm}$ and a relative electron part $H_{\rm er}$, with

$$H_{\rm cm} = \frac{\boldsymbol{P}^2}{2Nm} - Ne\boldsymbol{E}\cdot\boldsymbol{R} \qquad (2.74)$$

and

$$H_{\rm er} = \sum_i \left[\frac{\boldsymbol{p}_i'^2}{2m} + \frac{p_{zi}^2}{2m_z} + U(z_i)\right] + \sum_{i<j} V(\boldsymbol{r}_i' - \boldsymbol{r}_j', z_i, z_j). \qquad (2.75)$$

Note that $H_{\rm er}$ is identical to Hamiltonian (2.71), indicating that under the influence of a uniform electric field the relative electron system satisfy the same Hamiltonian as the original electron system in the laboratory reference frame without the electric field. Hereafter we shall omit the primes of $\boldsymbol{p}_i'$ and $\boldsymbol{r}_i'$ in (2.75) and other formula, and always understand $\boldsymbol{p}_i$ and $\boldsymbol{r}_i$ with reference to the relative motion of the electrons.

In the absence of carrier–carrier interaction the single electron state of the system is characterized by a subband index n and a two dimensional wavevector $\boldsymbol{k}_\| = (k_x, k_y)$ with the eigen energy

$$\varepsilon_{n\boldsymbol{k}_\|} = \varepsilon_n + k_\|^2/2m \qquad (2.76)$$

and the wave function

$$\psi_{n\boldsymbol{k}_\|}(\boldsymbol{r}, z) = \frac{1}{\sqrt{S}} e^{i\boldsymbol{k}_\|\cdot\boldsymbol{r}} \zeta_n(z), \qquad (2.77)$$

where S is the area of the quasi-2D system and $\zeta_n(z)$ is the envelope function (Stern and Howard, 1964).

Using this set of single electron states as basis and denoting the corresponding creation (annihilation) operators as $c_{n\boldsymbol{k}_\|\sigma}^\dagger$ ($c_{n\boldsymbol{k}_\|\sigma}$), we write the Hamiltonian of the relative-electron system in the form

$$\begin{aligned} H_{\rm er} &= \sum_{n,\boldsymbol{k}_\|,\sigma} \varepsilon_{n\boldsymbol{k}_\|} c_{n\boldsymbol{k}_\|\sigma}^\dagger c_{n\boldsymbol{k}_\|\sigma} \\ &+ \frac{1}{2} \sum_{\substack{m',m,n',n \\ \boldsymbol{k}_\|,\boldsymbol{k}_\|',\boldsymbol{q}_\| \\ \sigma,\sigma'}} V_{m'mn'n}(q_\|)\, c_{m'\boldsymbol{k}_\|+\boldsymbol{q}_\|\sigma}^\dagger c_{n'\boldsymbol{k}_\|'-\boldsymbol{q}_\|\sigma'}^\dagger c_{n\boldsymbol{k}_\|'\sigma'} c_{m\boldsymbol{k}_\|\sigma}, \end{aligned} \qquad (2.78)$$

where $V_{m'mn'n}(q_\|)$ is the matrix element of the Coulomb potential between electrons:

$$V_{m'mn'n}(q_\|) = \frac{e^2}{2\epsilon_0 \kappa q_\|} H_{m'mn'n}(q_\|), \qquad (2.79)$$

$$H_{m'mn'n}(q_\parallel) = \iint dz_1 dz_2 \zeta_{m'}^*(z_1)\zeta_m(z_1)\zeta_{n'}^*(z_2)\zeta_n(z_2)(e^{-q_\parallel|z_1-z_2|}+\text{images}), \tag{2.80}$$

with κ the dielectric constant of the background material containing electrons. The "images" term in (2.80) depends on the difference of the dielectric constants on the two sides of the interface in the usual way. In the case of GaAs/AlGaAs heterojunction the dielectric constants on both sides are almost equal, such that the "images" term can be neglected, and we will use this approximation throughout the following discussion for simplicity.

We assume that the elastic scatterings in the system are mainly due to ionized impurities, and the impurity potential takes the form (neglecting images)

$$U(\mathbf{r}-\mathbf{r}_a, z-z_a) = \frac{Ze^2}{4\pi\epsilon_0\kappa}\left[(\mathbf{r}-\mathbf{r}_a)^2 + (z-z_a)^2\right]^{-1/2}, \tag{2.81}$$

where $(\mathbf{r}_a, z_a)$ stands for the location of an impurity and Z is its charge number. The contribution from all impurities gives rise to the electron–impurity interaction of the form

$$H_{\text{ei}} = \sum_{\substack{n',n,\sigma \\ \mathbf{k}_\parallel, \mathbf{q}_\parallel, a}} U_{n'n}(q_\parallel, z_a)\, e^{i\mathbf{q}_\parallel\cdot(\mathbf{R}-\mathbf{r}_a)} c^\dagger_{n'\mathbf{k}_\parallel+\mathbf{q}_\parallel\sigma} c_{n\mathbf{k}_\parallel\sigma}, \tag{2.82}$$

in which

$$U_{n'n}(q_\parallel, z_a) = \frac{Ze^2}{2\epsilon_0\kappa q_\parallel} F_{n'n}(q_\parallel, z_a), \tag{2.83}$$

with

$$F_{n'n}(q_\parallel, z_a) = \int_0^\infty e^{-q_\parallel|z-z_a|}\zeta_{n'}^*(z)\zeta_n(z)dz. \tag{2.84}$$

For the impurities locating on the AlGaAs side, $z_a < 0$, $F_{n'n}(q_\parallel, z_a) = \exp(q_\parallel z_a)I_{n'n}(q_\parallel)$, with the form factor

$$I_{n'n}(q_\parallel) = \int_0^\infty e^{-q_\parallel z}\zeta_{n'}^*(z)\zeta_n(z)dz. \tag{2.85}$$

We consider that the phonon modes in the GaAs/AlGaAs heterojunction are essentially the same as in a bulk GaAs, characterized by a three-dimensional wavevector $\mathbf{q}$ and a branch index λ having frequency $\Omega_{\mathbf{q}\lambda}$ that the phonon Hamiltonian in the heterojunction is the same as (1.13) in terms of phonon creation and annihilation operators $b_{q\lambda}^\dagger$ and $b_{q\lambda}$:

$$H_{\text{ph}} = \sum_{q\lambda} \Omega_{q\lambda} b_{q\lambda}^\dagger b_{q\lambda}. \tag{2.86}$$

In the quasi-two-dimensional representation, the electron–phonon interaction takes the form:

$$H_{\text{ep}} = \sum_{\substack{n',n \\ q_\parallel, q_z, \lambda}} M_{n'n}(q, \lambda) \, e^{i q_\parallel \cdot R} (b_{q\lambda} + b_{-q\lambda}^\dagger) \sum_{k_\parallel, \sigma} c_{n'k_\parallel + q_\parallel \sigma}^\dagger c_{n k_\parallel \sigma}, \tag{2.87}$$

with

$$M_{n'n}(q, \lambda) = M(q, \lambda) I_{n'n}^*(i q_z), \tag{2.88}$$

$I_{n'n}(x)$ is the same form function as defined by Eq. (2.85) and $M(q, \lambda)$ is the electron–phonon matrix element in the 3D plane-wave representation.

Following the procedure outlined in Chapter 1 we can derive force- and energy-balance equations similar to Eqs. (1.40) and (1.55). For steady-state dc transport these equations, written for the quasi-2D system of unit area, are of the form

$$N_s e \boldsymbol{E} + \boldsymbol{f} = 0, \tag{2.89}$$

$$\boldsymbol{v}_d \cdot \boldsymbol{f} + w = 0. \tag{2.90}$$

Here N_s is the electron sheet density, and $\boldsymbol{f}$ is the frictional force experienced by the center of mass and w is the energy-loss rate, of the electron system with unit area:

$$\boldsymbol{f} = \boldsymbol{f}_i + \boldsymbol{f}_p, \tag{2.91}$$

$$\boldsymbol{f}_i = \sum_{n',n,\boldsymbol{q}_\parallel} \boldsymbol{q}_\parallel |U_{n'n}(q_\parallel)|^2 \Pi_2(n', n, \boldsymbol{q}_\parallel, \omega_0), \tag{2.92}$$

$$\boldsymbol{f}_p = 2 \sum_{n',n,\boldsymbol{q},\lambda} \boldsymbol{q}_\parallel |M_{n'n}(\boldsymbol{q}, \lambda)|^2 \Pi_2(n', n, \boldsymbol{q}_\parallel, \Omega_{q\lambda} + \omega_0)$$

$$\times \left[n\left(\frac{\Omega_{q\lambda}}{T}\right) - n\left(\frac{\Omega_{q\lambda} + \omega_0}{T_e}\right) \right], \tag{2.93}$$

$$w = 2 \sum_{n',n,\boldsymbol{q},\lambda} \Omega_{q\lambda} |M_{n'n}(\boldsymbol{q}, \lambda)|^2 \Pi_2(n', n, \boldsymbol{q}_\parallel, \Omega_{q\lambda} + \omega_0)$$

$$\times \left[n\left(\frac{\Omega_{q\lambda}}{T}\right) - n\left(\frac{\Omega_{q\lambda} + \omega_0}{T_e}\right) \right], \tag{2.94}$$

in which

$$|U_{n'n}(q_\parallel)|^2 = \int dz\, n_\mathrm{i}(z)|U_{n'n}(q_\parallel,z)|^2 \tag{2.95}$$

is the effective impurity potential and $n_\mathrm{i}(z)$ is the impurity volume density at z (the impurity density is assumed uniformly distributed in the x–y plane). In Eqs. (2.92)–(2.94) $\Pi_2(n',n,\boldsymbol{q}_\parallel,\omega)$ is the imaginary part of the Fourier spectrum of the diagonal electron density correlation function

$$\Pi(n',n,\boldsymbol{q}_\parallel,t) = -\mathrm{i}\theta(t)\sum_{\boldsymbol{k}_\parallel \sigma}\left\langle \left[c^\dagger_{n'\boldsymbol{k}_\parallel+\boldsymbol{q}_\parallel\sigma}(t)c_{n\boldsymbol{k}_\parallel\sigma}(t),\, c^\dagger_{n\boldsymbol{k}_\parallel\sigma}c_{n'\boldsymbol{k}_\parallel+\boldsymbol{q}_\parallel\sigma}\right]\right\rangle, \tag{2.96}$$

where the interaction-picture operator is defined as $c(t) \equiv \mathrm{e}^{\mathrm{i}H_\mathrm{er}t}c\,\mathrm{e}^{-\mathrm{i}H_\mathrm{er}t}$ and the statistical average is taken over the relative electron equilibrium density matrix $\hat{\rho}_0 \sim \mathrm{e}^{-H_\mathrm{er}/T_\mathrm{e}}$ at the electron temperature T_e. The other nondiagonal parts of the subband density correlation average are assumed small and neglected.

The total (nonlinear) resistivity of a quasi-two-dimensional system is given by

$$R \equiv \frac{1}{N_\mathrm{s}e\mu} \equiv \frac{\boldsymbol{E}\cdot\boldsymbol{v}_\mathrm{d}}{N_\mathrm{s}ev_\mathrm{d}^2}, \tag{2.97}$$

where $\mu = v_\mathrm{d}^2/(\boldsymbol{E}\cdot\boldsymbol{v}_\mathrm{d})$ is the (nonlinear) mobility of the system. According to force-balance equation (2.89), R is the sum of the impurity-induced resistivity $R_\mathrm{i} = 1/(N_\mathrm{s}e\mu_\mathrm{i})$ and the phonon-induced resistivity $R_\mathrm{p} = 1/(N_\mathrm{s}e\mu_\mathrm{p})$,

$$R = R_\mathrm{i} + R_\mathrm{p}, \tag{2.98}$$

or, written in terms of inverse mobilities,

$$\frac{1}{\mu} = \frac{1}{\mu_\mathrm{i}} + \frac{1}{\mu_\mathrm{p}}, \tag{2.99}$$

where μ_i is the impurity-limited mobility,

$$\frac{1}{\mu_\mathrm{i}} \equiv -\frac{\boldsymbol{v}_\mathrm{d}\cdot\boldsymbol{f}_\mathrm{i}}{N_\mathrm{s}ev_\mathrm{d}^2}, \tag{2.100}$$

and μ_p is the phonon-limited mobility,

$$\frac{1}{\mu_\mathrm{p}} \equiv -\frac{\boldsymbol{v}_\mathrm{d}\cdot\boldsymbol{f}_\mathrm{p}}{N_\mathrm{s}ev_\mathrm{d}^2}. \tag{2.101}$$

In limit of $v_{\rm d} \to 0$, the energy-balance equation (2.90) requires $T_{\rm e} = T$ (to the linear order of $v_{\rm d}$), the impurity- and phonon-limited linear mobilities are given by

$$\frac{1}{\mu_{\rm i}} = -\frac{1}{N_{\rm s}e} \sum_{n',n,\bm{q}_\|} q_x^2 \, |U_{n'n}(q_\|)|^2 \frac{\partial}{\partial \omega} \Pi_2(n',n,\bm{q}_\|,\omega)\Big|_{\omega=0}, \qquad (2.102)$$

$$\frac{1}{\mu_{\rm p}} = -\frac{2}{N_{\rm s}e} \sum_{n',n,\bm{q},\lambda} q_x^2 \, |M_{n'n}(\bm{q},\lambda)|^2 \Pi_2(n',n,\bm{q}_\|,\Omega_{q\lambda}) \left[-\frac{1}{T} n'\!\left(\frac{\Omega_{q\lambda}}{T}\right)\right]. \qquad (2.103)$$

If electrons occupy only the lowest subband (with envelope function $\zeta_0(z)$ and eigen energy $\varepsilon_{0\bm{k}_\|} = \varepsilon_{\bm{k}_\|}$) and the influence of all higher-lying subbands is negligible, we can take only $n = n' = 0$ and $m = m' = 0$ for the subband-index summations in all related equations. This corresponds to a single 2D isotropic system, for which all the formulas and equations are similar to those of an isotropic 3D system except there exists an extra form factor reflecting the subband wave function in the quasi-2D case. For instance, the electron density correlation function in the presence of inter-carrier Coulomb interaction has an expression in the RPA similar to (2.18) of 3D case:

$$\Pi_2(\bm{q}_\|,\omega) = \frac{\Pi_{02}(\bm{q}_\|,\omega)}{|\epsilon(\bm{q}_\|,\omega)|^2}. \qquad (2.104)$$

Here the RPA dielectric function is given by

$$\epsilon(\bm{q}_\|,\omega) = 1 - V(q_\|) \Pi_0(\bm{q}_\|,\omega) \qquad (2.105)$$

with

$$V(q_\|) = \frac{e^2}{2\epsilon_0 \kappa q_\|} H(q_\|), \qquad (2.106)$$

$$H(q_\|) = \int_0^\infty dz_1 \int_0^\infty dz_2 \, {\rm e}^{-q_\| |z_1 - z_2|} \, |\zeta_0(z_1)|^2 \, |\zeta_0(z_2)|^2. \qquad (2.107)$$

The density correlation function of a pure 2D electron system in the absence of intercarrier Coulomb interaction,

$$\Pi_0(\bm{q}_\|,\omega) = 2 \sum_{\bm{k}_\|} \frac{f(\varepsilon_{\bm{k}_\|+\bm{q}_\|},T_{\rm e}) - f(\varepsilon_{\bm{k}_\|},T_{\rm e})}{\varepsilon_{\bm{k}_\|+\bm{q}_\|} - \varepsilon_{\bm{k}_\|} + \omega + {\rm i}\delta} \qquad (2.108)$$

has its real and imaginary parts in the form

$$\Pi_{01}(\boldsymbol{q}_\|,\omega) = -2\sum_{\boldsymbol{k}_\|} f(\varepsilon_{\boldsymbol{k}_\|}, T_e)\left(\frac{1}{\varepsilon_{\boldsymbol{k}_\|+\boldsymbol{q}_\|} - \varepsilon_{\boldsymbol{k}_\|} + \omega} + \frac{1}{\varepsilon_{\boldsymbol{k}_\|+\boldsymbol{q}_\|} - \varepsilon_{\boldsymbol{k}_\|} - \omega}\right), \tag{2.109}$$

$$\Pi_{02}(\boldsymbol{q}_\|,\omega) = 2\pi \sum_{\boldsymbol{k}_\|} \left[f(\varepsilon_{\boldsymbol{k}_\|}, T_e) - f(\varepsilon_{\boldsymbol{k}_\|+\boldsymbol{q}_\|}, T_e)\right]\delta(\varepsilon_{\boldsymbol{k}_\|+\boldsymbol{q}_\|} - \varepsilon_{\boldsymbol{k}_\|} + \omega). \tag{2.110}$$

The impurity- and phonon-limited linear mobilities can be written as

$$\frac{1}{\mu_i} = -\frac{1}{N_s e}\sum_{\boldsymbol{q}_\|} q_x^2 |\bar{U}_{00}(q_\|)|^2 \frac{\partial}{\partial \omega}\Pi_{02}(\boldsymbol{q}_\|,\omega)\Big|_{\omega=0}, \tag{2.111}$$

$$\frac{1}{\mu_p} = -\frac{2}{N_s e}\sum_{\boldsymbol{q},\lambda} q_x^2 |\bar{M}_{00}(\boldsymbol{q},\lambda)|^2 \Pi_{02}(\boldsymbol{q}_\|,\Omega_{\boldsymbol{q}\lambda})\left[-\frac{1}{T}n'\left(\frac{\Omega_{\boldsymbol{q}\lambda}}{T}\right)\right], \tag{2.112}$$

where

$$|\bar{U}_{00}(\boldsymbol{q}_\|)|^2 \equiv \frac{|U_{00}(\boldsymbol{q}_\|)|^2}{|\epsilon(\boldsymbol{q}_\|,0)|^2}, \tag{2.113}$$

$$|\bar{M}_{00}(\boldsymbol{q},\lambda)|^2 \equiv \frac{|M_{00}(\boldsymbol{q},\lambda)|^2}{|\epsilon(\boldsymbol{q}_\|,\Omega_{\boldsymbol{q}\lambda})|^2}. \tag{2.114}$$

Therefore, for linear mobility the effect of electron–electron interaction in RPA is equivalent to a screening of the scattering potentials.

Things become more complicated when effects of higher subbands have to be taken into account together with electron–electron interaction. Up to now there has been no universally accepted recipe for calculating the electron density correlation function with Coulomb interaction in RPA in a multi-subband systems. The next section will introduce a possible scheme.

2.6 Hot Electron Transport in GaAs/AlGaAs Heterojunctions

The balance equations outlined in the preceding section have been applied to the investigation of hot-electron transport in GaAs/AlGaAs heterojunctions by Lei (1985) and by Lei et al (1986). They determine the multi-subband electron density correlation functions in the presence of intercarrier Coulomb interaction within RPA, $\Pi(n',n,\boldsymbol{q}_\|,\omega)$, using the following set of equations:

$$\Pi(n', n, \boldsymbol{q}_{\parallel}, \omega) = \Pi_0(n', n, \boldsymbol{q}_{\parallel}, \omega) \left[1 + \widetilde{V}_{n'nn'n}(q_{\parallel}) \Pi(n', n, \boldsymbol{q}_{\parallel}, \omega) \right], \quad (2.115)$$

with the renormalized Coulomb potential $\widetilde{V}_{n'nn'n}(q_{\parallel})$ satisfying the set of RPA equations,

$$\widetilde{V}_{n'nm'm}(q_{\parallel}, \omega) = V_{n'nm'm}(q_{\parallel}) + \sum_{l',l} V_{n'nl'l}(q_{\parallel}) \Pi_0(l', l, \boldsymbol{q}_{\parallel}, \omega) \widetilde{V}_{l'lm'm}(q_{\parallel}, \omega). \quad (2.116)$$

In the above equations, $V_{n'nm'm}(q_{\parallel})$ is the "bare" Coulomb potential given by (2.79), and

$$\Pi_0(n', n, \boldsymbol{q}_{\parallel}, \omega) = 2 \sum_{\boldsymbol{k}_{\parallel}} \frac{f(\varepsilon_{n'\boldsymbol{k}_{\parallel}+\boldsymbol{q}_{\parallel}}, T_{\mathrm{e}}) - f(\varepsilon_{n\boldsymbol{k}_{\parallel}}, T_{\mathrm{e}})}{\varepsilon_{n'\boldsymbol{k}_{\parallel}+\boldsymbol{q}_{\parallel}} - \varepsilon_{n\boldsymbol{k}_{\parallel}} + \omega + \mathrm{i}\delta} \quad (2.117)$$

is the density correlation function for electrons in subbands n and n' in the absence of Coulomb interaction.

The RPA equation set (2.116) consists of infinite chains of higher-order unknown potentials. To proceed one has to truncate them by considering only a finite number of lowest subbands. In addition, we need the envelope functions $\zeta_n(z)$ for carrying out the relevant form factors. A convenient and useful choice is to employ the Fang–Howard–Stern variational functions

$$\zeta_0(z) = \left(\frac{b^3}{2} \right)^{\frac{1}{2}} z \, \mathrm{e}^{-bz/2}, \quad (2.118)$$

$$\zeta_1(z) = \frac{3}{2} \left(\frac{b_1^5}{b^2 - bb_1 + b_1^2} \right)^{\frac{1}{2}} z \left[1 - \left(\frac{b + b_1}{6} \right) z \right] \mathrm{e}^{-b_1 z/2}, \quad (2.119)$$

$\cdots$, $\zeta_s(z)$ as the envelope functions for total $s+1$ subbands, respectively. The parameters $b, b_1, \cdots, b_s$ are determined by minimizing the energies $\varepsilon_0, \varepsilon_1, \cdots, \varepsilon_s$ for given carrier and depletion-layer charge densities N_{s} and N_{dep}. For the ground subband the variational parameter is given by $b^3 = (33me^2/8\epsilon_0\kappa)\left(N_{\mathrm{s}} + \frac{32}{11} N_{\mathrm{dep}}\right)$ (Ando, Fowler and Stern, 1982).

With these envelope functions the integrations in Eq. (2.80) are performed to give a closed expression for $H_{n'n,m'm}(q_{\parallel})$. For instance,

$$H_{0000}(q_{\parallel}) \equiv H(q_{\parallel}) = \frac{1}{8}(8b^3 + 9b^2 q_{\parallel} + 3b q_{\parallel}^2)(b + q_{\parallel})^{-3}. \quad (2.120)$$

The density correlation functions are then obtained from Eq. (2.115) and the truncated RPA equation (2.116).

The form of the effective elastic scattering potential $|U_{n'n}(q_\parallel)|^2$ in the frictional force $\boldsymbol{f}_i$ of (2.92) depends on the detailed distribution $n_i(z)$ of the scatterers. We assume that there are two different kinds of scatterers in the semiconductor heterojunctions. One kind is from the remote ionized dopants which are located within a narrow layer at a distance s from the interface in the barrier of the AlGaAs side with a sheet density N_r. Another kind is from the background charged impurities which are existing in the conduction channel with nearly a uniform volume density n_b throughout the GaAs region. The effective impurity scattering potential has the form (assume that all the impurities have same charge number Z for brevity)

$$|U_{n'n}(q_\parallel)|^2 = \left(\frac{Ze^2}{2\epsilon_0 \kappa}\right)^2 \left[N_r q_\parallel^{-2} e^{-2q_\parallel s} I_{n'n}^2(q_\parallel) + n_b q_\parallel^{-3} J_{n'n}(q_\parallel)\right], \quad (2.121)$$

in which $I_{n'n}(q_\parallel)$ is defined by (2.85), and

$$J_{n'n}(q_\parallel) = q_\parallel \int_0^\infty dz_1 \left|\int_0^\infty dz\, e^{-q_\parallel |z-z_1|} \zeta_{n'}^*(z)\zeta_n(z)\right|^2. \quad (2.122)$$

Their explicit expressions for the lowest subband are

$$I_{00}(q_\parallel) \equiv I(q_\parallel) = b^3(b+q_\parallel)^{-3}, \quad (2.123)$$

$$J_{00}(q_\parallel) \equiv J(q_\parallel) = \frac{1}{4}\left(2b^6 + 24b^5 q_\parallel + 48b^4 q_\parallel^2 + 43b^3 q_\parallel^3 \right.$$
$$\left. + 18b^2 q_\parallel^4 + 3b q_\parallel^5\right)(b+q)^{-6}. \quad (2.124)$$

Both acoustic phonons (via piezoelectric coupling and deformation potential coupling) and polar optic phonons (via Fröhlich coupling) can interact strongly with electrons in GaAs-based systems, and their roles in electron transport have to be taken into account carefully. At low temperatures only acoustic phonons contribute to the scattering. For a given acoustic phonon branch, the matrix elements of the piezoelectric interaction and those of the deformation potential are $\pi/2$ out of phase so that their contributions to $|M(\boldsymbol{q}, \lambda)|^2$ are additive (Mahan, 1981). Only the longitudinal mode gives rise to the deformation potential coupling with matrix element (Vogl, 1980)

$$|M(\boldsymbol{q}, l)_{\text{def}}|^2 = \frac{\Xi^2 q}{2d_c v_{\text{sl}}}, \quad (2.125)$$

where d_c is the mass density of the crystal, v_{sl} is the velocity of the longitudinal sound wave and Ξ is the deformation potential, i.e. the shift of the band edge per unit dilation. Both longitudinal and transverse phonons contribute to the piezoelectric interaction. In GaAs (zinc-blende structure) there is only one nonzero independent piezoelectric constant $e_{14} = e_{25} = e_{36}$ and the dielectric constant and sound speed are isotropic. The piezoelectric matrix element is (Mahan, 1972)

$$M(\boldsymbol{q}, \lambda) = \frac{8\pi e e_{14}}{\kappa q^2} \sqrt{\frac{1}{2 d_c v_{s\lambda} q}} \left[q_x q_y e_z(\boldsymbol{q}, \lambda) + q_z q_x e_y(\boldsymbol{q}, \lambda) + q_y q_z e_x(\boldsymbol{q}, \lambda) \right], \tag{2.126}$$

in which $v_{s\lambda}(\lambda = l, t)$ is the velocity of the λth branch phonon and $\boldsymbol{e}(\boldsymbol{q}, \lambda)$ stands for the unit polarization vector of the phonon with wavevector $\boldsymbol{q} = (q_x, q_y, q_z)$ in branch λ. For the longitudinal acoustic branch the piezoelectric contribution

$$|M(\boldsymbol{q}, l)_{\text{piez}}|^2 = \frac{32\pi^2 e^2 e_{14}^2}{\kappa^2 d_c v_{\text{sl}}} \frac{(3 q_x q_y q_z)^2}{q^7} \tag{2.127}$$

is additive to the deformation potential one of (2.125). There are two independent transverse acoustic branches. Their total contribution is (Lei, Birman and Ting, 1985)

$$\sum_{j=1,2} |M(\boldsymbol{q}, t_j)|^2 = \frac{32\pi^2 e^2 e_{14}^2}{\kappa^2 d_c v_{\text{st}} q^5} \left[q_x^2 q_y^2 + q_y^2 q_z^2 + q_z^2 q_x^2 - \frac{(3 q_x q_y q_z)^2}{q^2} \right]. \tag{2.128}$$

In GaAs, the transverse sound speed, $v_{\text{st}} = 2.48 \times 10^5$ cm/s, is much smaller than the longitudinal sound speed $v_{\text{sl}} = 5.29 \times 10^5$ cm/s. Therefore, it is expected that the transverse acoustic phonons, thus the piezoelectric interactions, play a major role in determining the phonon-limited mobility and energy transfer rate at very low temperatures. The role of longitudinal acoustic phonons and the deformation potential couplings will increase with elevated temperature. On the other hand, when the temperature rises above 40 K, polar optical scattering begins to contribute significantly. Polar interaction between longitudinal optical phonons and electrons has coupling matrix element of the Fröhlich type (Mahan, 1972):

$$|M(\boldsymbol{q}, \text{LO})|^2 = \frac{e^2}{2\epsilon_0 q^2} \left(\frac{1}{\kappa_\infty} - \frac{1}{\kappa} \right) \Omega_{\text{LO}}, \tag{2.129}$$

in which κ_∞ is the optical dielectric constant and Ω_{LO} is the longitudinal optical phonon frequency.

Numerical calculations are performed for carrier linear mobilities μ_o, μ_a, and μ_i respectively due to polar-optical-phonon scattering, acoustic-phonon scattering (including longitudinal and transverse acoustic phonons),

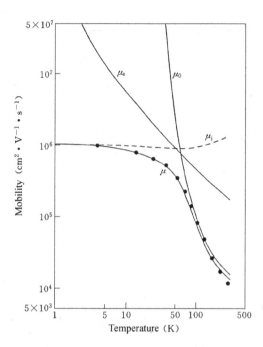

Fig. 2.7 Linear mobilities μ_i due to impurity scattering, μ_a due to acoustic phonon scattering, μ_o due to polar optic phonon scattering, and the total linear mobility μ are shown as functions of lattice temperature. The system is a GaAs-AlGaAs heterojunction with carrier sheet density $N_s = 2.2 \times 10^{11}$ cm^{-2}. From Lei (1985).

and impurity scattering (including the remote impurities at a distance $s = 37$ nm from the interface and the randomly distributed background impurities, with equal contributions to the resistivity at zero temperature), according to Eqs. (2.102) and (2.103) for a GaAs/AlGaAs heterojunction having carrier sheet density $N_s = 2.2 \times 10^{11}$ cm^{-2}, by taking into account the lowest and next lowest subbands and deriving the frequency-dependent RPA electron density correlation function $\Pi(n',n,\boldsymbol{q}_\parallel,\omega)$ with the help of Eqs.(2.115) and (2.116). The calculated results are plotted in Fig. 2.7, together with the total mobility obtained by $\mu^{-1} = \mu_o^{-1} + \mu_a^{-1} + \mu_i^{-1}$ (Lei, 1985), showing good agreement with the experimental data by Mendez, Price, and Heilblum (1984). Note that the only adjustable parameter is the impurity density, which is so chosen that the total mobility at low temperature is in agreement with the measurement.

This has been the first mobility calculation in the literature for quasi-2D electron system using a full temperature-, wavevector- and frequency-

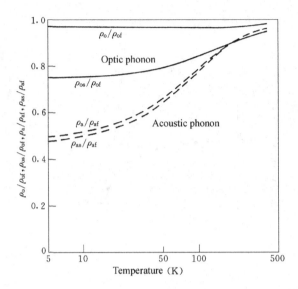

Fig. 2.8 Ratios of the optic-phonon induced resistivity and acoustic-phonon induced resistivity, ρ_o and ρ_a, obtained with dynamical screening, and ρ_{os} and ρ_{as} obtained with static screening, to ρ_{of} and ρ_{af} obtained without screening. The system is a GaAs-AlGaAs heterojunction with carrier sheet density $N_s = 2.2 \times 10^{11} \, \text{cm}^{-2}$. From Lei (1985).

dependent screening in the random phase approximation. Before that, this kind of theoretical investigations was carried out either without including screening, or using static screening or zero-temperature screening.

To see the difference between the dynamic and static screenings we show in Fig. 2.8 the ratios of the polar-optic-phonon induced resistivity and acoustic-phonon induced resistivity, ρ_o and ρ_a, obtained by using dynamical screening, and ρ_{os} and ρ_{as} obtained by using static screening, to ρ_{of} and ρ_{af} obtained without screening. It is seen that compared with those without screening, the static screening reduces both acoustic-phonon and optic-phonon induced resistivities. In comparison with those of static screening, the dynamic screening significantly increases the optic-phonon induced resistivity but only slightly increases the acoustic-phonon induced resistivity. It is particularly interesting to note that, in the case of polar-optic-phonon scattering, the resistivity enhancement due to plasma resonance in the case of dynamic screening almost completely compensates the resistivity decrease resulting from the static screening, such that the result obtained by including dynamic screening is almost equivalent to that without screening.

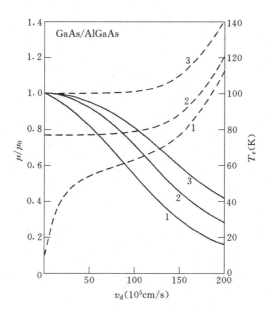

Fig. 2.9 Calculated normalized mobility μ/μ_0 (solid curves) and electron temperature T_e (dashed curves) are shown as functions of drift velocity v_d for a GaAS/AlGaAs system with $N_s = 4.0 \times 10^{11}\,\text{cm}^{-2}$ and $\mu_0(0) = 2.2 \times 10^5\,\text{cm}^2/\text{Vs}$. The elastic scatterings are assumed entirely due to remote impurities located at a distance $s = 12.5\,\text{nm}$ from the interface. The lattice temperatures are respectively: 1 — 10 K; 2 — 77 K; 3 — 100 K. From Lei et al (1986).

Starting from balance equations (2.89) and (2.90) Lei et al (1986) calculate the nonlinear mobility and the electron temperature in GaAs/AlGaAs heterojunctions. In Fig. 2.9 the calculated mobility normalized by its linear value, μ/μ_0 (solid curves), and the electron temperature T_e (dashed curves), are shown as functions of the drift velocity v_d for a GaAs/AlGaAs system with carrier sheet density $N_s = 4.0 \times 10^{11}\,\text{cm}^{-2}$ and zero temperature linear mobility $\mu_0(0) = 2.2 \times 10^5\,\text{cm}^2/\text{Vs}$ at lattice temperatures $T = 10, 77$ and 100 K, respectively. Here μ_0 is the Ohmic mobility at temperature T.

In Fig. 2.10 the calculated normalized mobility μ/μ_0 (solid curves) and electron temperature T_e (dashed curves) are shown as functions of the applied electric field E at lattice temperature $T = 4.2\,\text{K}$ for two GaAs/AlGaAs systems with $N_s = 2.5 \times 10^{11}\,\text{cm}^{-2}$ and $\mu_0 = 1.0 \times 10^6\,\text{cm}^2/\text{Vs}$, and $N_s = 3.9 \times 10^{11}\,\text{cm}^{-2}$ and $\mu_0 = 7.9 \times 10^4\,\text{cm}^2/\text{Vs}$, respectively, together with experimental data of μ/μ_0. The crosses are taken from the dark data at 4.2 K of sample 2 of Höpfel and Weimann (1985). The dots are ex-

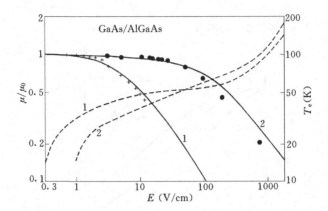

Fig. 2.10 Calculated normalized mobility μ/μ_0 (solid curves) and electron temperature T_e (dashed curves) vs electric field E at lattice temperature $T = 4.2\,\mathrm{K}$ for two systems: (1) $N = 2.5 \times 10^{11}\,\mathrm{cm}^{-2}$, $\mu_0 = 1 \times 10^6\,\mathrm{cm}^2/\mathrm{Vs}$, $s = 25\,\mathrm{nm}$; (2) $N = 3.9 \times 10^{11}\,\mathrm{cm}^{-2}$, $\mu_0 = 7.9 \times 10^4\,\mathrm{cm}^2/\mathrm{Vs}$, $s = 12.5\,\mathrm{nm}$. The experimental data shown in crosses are taken from Höpfel and Weimann (1985) and those shown in dots from Shah et al (1984). From Lei et al (1986).

perimental results of Shah et al (1984) for a multi-quantum-well sample of width 26 nm and layer carrier density $N_s = 3.9 \times 10^{11}\,\mathrm{cm}^{-2}$ at $T = 2\,\mathrm{K}$. The GaAs/AlGaAs material parameters used in the calculations are as follows: density $d_c = 5.31\,\mathrm{g/cm^3}$, effective mass $m = 0.07 m_e$, transverse sound velocity $v_{st} = 2.48 \times 10^3\,\mathrm{m/s}$, longitudinal sound velocity $v_{sl} = 5.29 \times 10^3\,\mathrm{m/s}$, longitudinal-optic-phonon energy $\Omega_{LO} = 35.4\,\mathrm{meV}$, low-frequency dielectric constant $\kappa = 12.9$, optical dielectric constant $\kappa_\infty = 10.8$, acoustic deformation potential $\Xi = 8.5\,\mathrm{eV}$, piezoelectric constant $e_{14} = 1.41 \times 10^9\,\mathrm{V/m}$.

Figure 2.11 is taken from Hirakawa and Sakaki (1988), who compare their experimental data with the prediction of balance-equation approach (i.e. the theoretical curve shown in Fig. 2.10), and with that of electron-temperature model. The measured mobilities are found in excellent agreement with the theoretical results of Lei et al (1986).

Using the balance equation formulation outlined in the preceding and present sections, Xu, Peeters and Devreese (1993) investigate polar optic phonon induced Ohmic mobility in GaAs-based square quantum wells of different well widths, taking account of four lowest subbands. They find strong electro-phonon resonance in the mobility as a function of well width when the energy difference between two subbands is equal to an LO phonon energy Ω_{LO}.

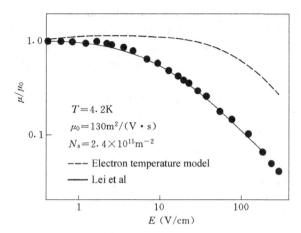

Fig. 2.11 Normalized mobility μ/μ_0 is plotted as a function of the electric field E. Black circles are the measured data of Hirakawa and Sakaki (1988). The dotted line is calculated by the electron temperature model and the solid line is obtained from the balance-equation approach (Lei et al, 1986). After Hirakawa and Sakaki (1988).

It should be noted that taking equations (2.115) and (2.116) as the random phase approximation for density correlation functions of multi-subband systems is only one of possible choices. There may be other recipe for multi-subband RPA used in transport analyses. An example is due to Hasbun (1994). His suggested RPA density correlation function, which looks reasonable theoretically, unfortunately yields some unphysical results in the dynamical resistivity, unless certain matrix elements are discarded. It seems that the problem of multi-band RPA has still been far from being fully solved.

2.7 Transverse Transport in Quantum-Well Superlattices

For quantum-well superlattices, balance-equation analyses of hot carrier transport are carried out by Lei, Horing and Zhang (1986b). The model employed for the superlattice consists of an infinite number of periodically arranged quantum wells of width a and barriers of width b, such that $d = a + b$ is the spatial period, or the distance between two adjacent layer centers. In the effective-mass approximation the electron system can be described also by the Hamiltonian (2.71), with the confining potential $U(z)$ adjusted to reflect the superlattice structure of the quantum wells and barriers. Considering the applied electric field parallel to the layer plane, we again introduce two-dimensional center-of-mass variables and relative electron variables, and separate $H_{\rm cm}$ and $H_{\rm er}$ as in Sec. 2.5 for quasi-two-dimensional systems. We assume that (i) the potential-well depth is deep enough and tunneling is small enough, so that electrons are confined to just one well, and (ii) the potential well is narrow and the electron density is not too high, such that electrons occupy only the lowest subband. The lowest subband wave function of the electron in lth well is taken as (for unit transverse area of the superlattice)

$$\psi_{l\bm{k}_\parallel}(\bm{r},z) = e^{i\bm{k}_\parallel\cdot\bm{r}}\zeta(z-ld) \qquad (l=0,\pm 1,\cdots), \tag{2.130}$$

with

$$\zeta(z) = \begin{cases} (2/a)^{1/2}\sin(\pi z/a) & \text{for } 0 \leqslant z \leqslant a, \\ 0 & \text{for } z < 0 \text{ or } z > a, \end{cases} \tag{2.131}$$

and the corresponding energy $\varepsilon_{\bm{k}_\parallel} = k_\parallel^2/2m$ is degenerate with respect to the layer indices. The Hamiltonian $H_{\rm er}$ for relative electrons can be written in the second quantization notation as

$$H_{\rm er} = \sum_{l,\bm{k}_\parallel,\sigma} \varepsilon_{\bm{k}_\parallel} c_{l\bm{k}_\parallel\sigma}^\dagger c_{l\bm{k}_\parallel\sigma} + \frac{1}{2}\sum_{\substack{l,l',\sigma,\sigma' \\ \bm{k}_\parallel,\bm{k}_\parallel',\bm{q}_\parallel}} V_{ll'}(q_\parallel)\, c_{l\bm{k}_\parallel+\bm{q}_\parallel\sigma}^\dagger c_{l'\bm{k}_\parallel'-\bm{q}_\parallel\sigma'}^\dagger c_{l'\bm{k}_\parallel'\sigma'} c_{l\bm{k}_\parallel\sigma}, \tag{2.132}$$

where $c_{l\bm{k}_\parallel\sigma}^\dagger (c_{l\bm{k}_\parallel\sigma})$ are creation (annihilation) operators for relative electrons with lowest subband wave function $\psi_{l\bm{k}_\parallel}$, and $V_{lm}(q_\parallel)$ is the corresponding matrix element of the Coulomb potential, which can be written as

$$V_{lm}(q_\parallel) = \frac{e^2}{2\epsilon_0\kappa q_\parallel} F_{lm}(q_\parallel). \tag{2.133}$$

In this, κ is the background dielectric constant and the form factor $F_{lm}(q_\parallel)$ has a simple expression if the image charge contribution is negligible:

$$\begin{aligned} F_{lm}(q_\parallel) &= F_{l-m}(q_\parallel) \\ &= \int_{-\infty}^{\infty} dz \int_{-\infty}^{\infty} dz'\, e^{-q_\parallel|z-z'|}\zeta^2(z-ld)\zeta^2(z'-md) \\ &= e^{-q_\parallel|l-m|d}\left[e^{q_\parallel a} I^2(q_\parallel)(1-\delta_{lm}) + \delta_{lm}H(q_\parallel)\right]. \end{aligned} \tag{2.134}$$

Here, δ_{lm} is the Kronecker delta, and the form factors $I(q_\parallel)$ and $H(q_\parallel)$ are determined by the electron wave function within the well ($u \equiv q_\parallel a$):

$$I(q_\parallel) = \int_{-\infty}^{\infty} dz\, e^{-q_\parallel z} \zeta^2(z) = \frac{\pi^2(1-e^{-u})}{u(u^2+4\pi^2)}, \qquad (2.135)$$

$$H(q_\parallel) = \int_{-\infty}^{\infty} dz \int_{-\infty}^{\infty} dz'\, e^{-q_\parallel |z-z'|} \zeta^2(z)\zeta^2(z')$$

$$= 3\frac{1-e^{-u}}{u^2+4\pi^2} + \frac{u}{u^2+4\pi^2}$$

$$-\frac{1-e^{-u}}{(u^2+4\pi^2)^2}(u^2-4\pi^2) + \frac{2}{u}\left(1 - \frac{1-e^{-u}}{u}\right). \qquad (2.136)$$

The charged-impurity–electron interaction takes the form

$$H_{ei} = \sum_{\substack{k_\parallel, q_\parallel \\ l,\sigma,a}} \frac{Z_a e^2}{2\epsilon_0 \kappa q_\parallel} F_l(q_\parallel, z_a)\, e^{i\mathbf{q}_\parallel \cdot (\mathbf{R}-\mathbf{r}_a)} c^\dagger_{lk_\parallel + q_\parallel \sigma} c_{lk_\parallel \sigma}, \qquad (2.137)$$

where $(\mathbf{r}_a, z_a)$ and Z_a are the location and the charge number of the ath impurity and

$$F_l(q_\parallel, z_a) = \int_{-\infty}^{\infty} dz\, e^{-q_\parallel |z-z_a|} \zeta^2(z-ld). \qquad (2.138)$$

Image terms have been neglected again. We consider two kinds of impurities: remote impurities and background impurities. The former are the ionized dopants in AlGaAs regions, which are located in planar sheets at a distance s from the center of each layer with an area density N_r per sheet; the latter are distributed uniformly within the GaAs well regions with an area density N_b per layer. Here, we assume that there is the same impurity number density in each remote sheet (or in each layer). For the expressions of $F_l(q_\parallel, z_a)$ for remote and background impurities we refer the reader to the original paper (Lei, Horing and Zhang, 1986b).

The electron–phonon interaction can be written as

$$H_{ep} = \sum_{q,l,\lambda} M(\mathbf{q},\lambda) I(iq_z)\, e^{i\mathbf{q}_\parallel \cdot \mathbf{R}} (b_{q\lambda} + b^\dagger_{-q\lambda})\, e^{-iq_z ld} \sum_{k_\parallel, \sigma} c^\dagger_{lk_\parallel + q_\parallel \sigma} c_{lk_\parallel \sigma}, \qquad (2.139)$$

in which the form factor $I(iq_z)$ is the same function as (2.135),

$$|I(iq_z)|^2 = \frac{\pi^4 \sin^2 y}{y^2(y^2-\pi^2)^2}, \qquad (2.140)$$

with $y \equiv q_z a/2$.

In the steady state, when the center of mass moves at a constant velocity v_d, the frictional force due to phonons can be shown to be

$$\boldsymbol{f}_\mathrm{p} = 2 \sum_{\boldsymbol{q},\lambda} |M(\boldsymbol{q},\lambda)|^2 |I(\mathrm{i}q_z)|^2 \boldsymbol{q}_\| \sum_{l,l'} e^{-\mathrm{i}q_z(l-l')d} \Pi_2(l,l',\boldsymbol{q}_\|,\Omega_{\boldsymbol{q}\lambda}+\omega_0)$$
$$\times \left[n\left(\frac{\Omega_{\boldsymbol{q}\lambda}}{T}\right) - n\left(\frac{\Omega_{\boldsymbol{q}\lambda}+\omega_0}{T_\mathrm{e}}\right)\right]. \qquad (2.141)$$

Here $\omega_0 \equiv \boldsymbol{q}_\| \cdot \boldsymbol{v}_\mathrm{d}$, and $\Pi_2(l,l',\boldsymbol{q},\Omega)$ is the imaginary part of the Fourier transform of the electron density correlation function of the superlattice:

$$\Pi(l,l',\boldsymbol{q}_\|,t) = -\mathrm{i}\theta(t) \sum_{\boldsymbol{k}_\|,\boldsymbol{k}'_\|,\sigma} \langle [c^\dagger_{l\boldsymbol{k}_\|+\boldsymbol{q}_\|\sigma}(t)c_{l\boldsymbol{k}_\|\sigma}(t), c^\dagger_{l'\boldsymbol{k}'_\|-\boldsymbol{q}_\|\sigma}c_{l'\boldsymbol{k}'_\|\sigma}] \rangle.$$
$$(2.142)$$

In the absence of Coulomb interaction between carriers, the density correlation function has nonzero value only when $l = l'$:

$$\Pi_0(l,l',\boldsymbol{q}_\|,\omega) = \delta_{ll'} \Pi_0(\boldsymbol{q}_\|,\omega), \qquad (2.143)$$

and

$$\Pi_0(\boldsymbol{q}_\|,\omega) = 2 \sum_{\boldsymbol{k}_\|} \frac{f(\varepsilon_{\boldsymbol{k}_\|+\boldsymbol{q}_\|},T_\mathrm{e}) - f(\varepsilon_{\boldsymbol{k}_\|},T_\mathrm{e})}{\omega + \varepsilon_{\boldsymbol{k}_\|+\boldsymbol{q}_\|} - \varepsilon_{\boldsymbol{k}_\|} + \mathrm{i}\delta} \qquad (2.144)$$

is the density correlation function for a single sheet of 2D electrons without Coulomb interaction. The inclusion of both intralayer and interlayer carrier Coulomb interactions for the superlattice can be achieved using the random phase approximation, which leads to the following equation:

$$\Pi(l,l',\boldsymbol{q}_\|,\omega) = \delta_{ll'} \Pi(\boldsymbol{q}_\|,\omega) + \Pi(\boldsymbol{q}_\|,\omega) \sum_m V_{lm}(q_\|)\Pi(m,l',\boldsymbol{q}_\|,\omega). \quad (2.145)$$

Assuming periodic boundary conditions and neglecting image contributions, we have $V_{lm}(q) = V_{l-m}(q_\|)$ and $\Pi(l,l',\boldsymbol{q}_\|,\omega) = \Pi(l-l',\boldsymbol{q}_\|,\omega)$, whence Eq. (2.145) reduces to

$$\Pi(l,\boldsymbol{q}_\|,\omega) = \Pi_0(\boldsymbol{q},\omega)\left[\delta_{l0} + \sum_m V_{l-m}(q_\|)\Pi(m,\boldsymbol{q}_\|,\omega)\right]. \qquad (2.146)$$

By introducing

$$\Pi(q_z,\boldsymbol{q}_\|,\omega) = \sum_l e^{-\mathrm{i}q_z l d} \Pi(l,\boldsymbol{q}_\|,\omega), \qquad (2.147)$$

Eq. (2.146) is easily solved to give

$$\Pi(q_z, \bm{q}_\|, \omega) = \frac{\Pi_0(\bm{q}_\|, \omega)}{1 - V(q_\|, q_z)\Pi_0(\bm{q}_\|, \omega)}, \qquad (2.148)$$

with

$$V(q_\|, q_z) = \sum_l V_l(q) e^{-iq_z ld} = \frac{e^2}{2\epsilon_0 \kappa q_\|} [H(q_\|) + S(q_\|, q_z)], \qquad (2.149)$$

where $S(q_\|, q_z)$ comes from interlayer carrier interactions:

$$S(q_\|, q_z) = \frac{\cos(q_z d) - \exp(-q_\| d)}{\cosh(q_\| d) - \cos(q_x d)} \exp(qa) I^2(q_\|). \qquad (2.150)$$

In terms of the correlation function $\Pi(q_z, \bm{q}_\|, \omega)$, we can write the phonon-induced frictional force $\bm{f}_p$ (per layer) as

$$\bm{f}_p = 2 \sum_{\bm{q}_\|, q_z, \lambda} |M(q, \lambda)|^2 |I(iq_z)|^2 \bm{q}_\| \Pi_2(q_z, \bm{q}_\|, \Omega_{\bm{q},\lambda} + \omega_0)$$

$$\times \left[n\left(\frac{\Omega_{\bm{q}\lambda}}{T}\right) - n\left(\frac{\Omega_{\bm{q}\lambda} + \omega_0}{T_e}\right) \right]. \qquad (2.151)$$

The energy loss of the relative electron system is due to inelastic scattering associated with the electron–phonon coupling. Consideration along these lines yields the energy-transfer rate (per layer) from the electron system to the phonon system as

$$w = 2 \sum_{\bm{q}_\|, q_z, \lambda} |M(\bm{q}, \lambda)|^2 |I(iq_z)|^2 \Omega_{\bm{q}\lambda} \Pi_2(q_z, \bm{q}_\|, \Omega_{\bm{q}\lambda} + \omega_0)$$

$$\times \left[n\left(\frac{\Omega_{\bm{q}\lambda}}{T}\right) - n\left(\frac{\Omega_{\bm{q}\lambda} + \omega_0}{T_e}\right) \right]. \qquad (2.152)$$

Note that although the density correlation function $\Pi_2(q_z, \bm{q}_\|, \omega)$ is a periodic function of q_z, the factor $|M|^2 |I|^2$ is not. The sum over q_z ranges from $-\infty$ to ∞ as determined by the phonon modes:

$$\sum_{q_z} \to \frac{d}{2\pi} \int_{-\infty}^{\infty} dq_z. \qquad (2.153)$$

In deriving the impurity-induced frictional force, we need to deal with the impurity site averaging

$$A = \left\langle \sum_{a,b} e^{i\bm{q}_\| \cdot \bm{r}_{ma} + i\bm{q}'_\| \cdot \bm{r}_{nb}} \right\rangle. \qquad (2.154)$$

If the impurity distribution is random within each sheet (layer) and there is no correlation between different sheets (layers), the quantity in Eq. (2.154) is a two-dimensional extension of the averaging discussed by many authors in the 3D case (Kohn and Luttinger, 1957). Thus, to lowest order in N_i ($N_i = N_r$ or N_b is the impurity sheet density),

$$A = N_i \delta_{\mathbf{q}'_\parallel, -\mathbf{q}_\parallel}, \tag{2.155}$$

independent of the layer indices m and n. The assumption of independent impurity distributions in different sheets generally seems reasonable. Nevertheless, the possibility of correlation between the impurity distributions in different layers could be included in the following way (Lei, Horing and Zhang, 1986a). The periodicity of the superlattice system enables us to assume that the quantity in Eq. (2.154) is a function of $m - n$ after configuration averaging:

$$\left\langle \sum_{a,b} e^{i\mathbf{q}_\parallel \cdot \mathbf{r}_{ma} + i\mathbf{q}'_\parallel \cdot \mathbf{r}_{nb}} \right\rangle = N_i \delta_{\mathbf{q}'_\parallel, -\mathbf{q}_\parallel} g(m-n), \tag{2.156}$$

where the function $g(m)$ ($m = 0, \pm 1, \cdots$), which describes the correlation, may be expressed as a Fourier coefficient of the periodic function $g(q_z)$ with period $2\pi/d$:

$$g(m) = \frac{d}{2\pi} \int_{-\pi/d}^{\pi/d} dq_z\, g(q_z)\, e^{iq_z m d}, \tag{2.157}$$

$$g(q_z) = \sum_m g(m)\, e^{-iq_z m d}. \tag{2.158}$$

The final expression for the frictional force per layer due to remote- and background-impurity scatterings jointly includes a factor $g(q_z)$:

$$\mathbf{f}_i = \left(\frac{e^2}{2\epsilon_0 \kappa}\right)^2 \left(\frac{d}{2\pi}\right) \int_{-\pi/d}^{\pi/d} dq_z \sum_{\mathbf{q}_\parallel} \frac{\mathbf{q}_\parallel}{q_\parallel^2} \widetilde{N}(q_\parallel, q_z) g(q_z) \Pi_2(q_z, \mathbf{q}_\parallel, \omega_0), \tag{2.159}$$

in which $\widetilde{N}(q_\parallel, q_z)$ is an effective combined impurity density,

$$\widetilde{N}(q_\parallel, q_z) = N_r Z_r^2 \left| \frac{\sinh[q_\parallel(d-s)] + \exp(iq_z d)\sinh(q_\parallel s)}{\cosh(q_\parallel d) - \cos(q_z d)} \right|^2 \exp(q_\parallel a) I^2(q_\parallel)$$

$$+ N_b Z_b^2 \left[\frac{\cos(q_z d) - \exp(-q_z d)}{\cosh(q_\parallel d) - \cos(q_z d)} \frac{\exp(q_\parallel a) - 1}{q_\parallel a} I(q_\parallel) + K(q_\parallel) \right]^2, \tag{2.160}$$

Z_r and Z_b are equivalent charge numbers of the remote and background impurities, $I(q_\|)$ is given by (2.135), and ($u \equiv q_\| a$)

$$K(q_\|) = \frac{8\pi^2}{(4\pi^2 + u^2)u}\left(1 + \frac{u^2}{4\pi^2} - \frac{1-e^{-u}}{u}\right). \tag{2.161}$$

If there is no correlation between the impurity distributions of different sheet, $g(m) = 1$, $g(q_z) = 2\pi\delta(q_z d)$ and the impurity-induced frictional force is

$$\boldsymbol{f}_i = \left(\frac{e^2}{2\epsilon_0 \kappa}\right)^2 \sum_{\boldsymbol{q}_\|} \frac{\boldsymbol{q}_\|}{q_\|^2} \widetilde{N}(q_\|, 0) \Pi_2(0, \boldsymbol{q}_\|, \omega_0). \tag{2.162}$$

The force- and energy-balance equations of a superlattice can thus be written in terms of relevant quantities of a single sheet: the electron density N_s, the frictional force $\boldsymbol{f} = \boldsymbol{f}_i + \boldsymbol{f}_p$ and the energy-loss rate w per layer, and have exactly the same forms as those of a quasi-2D system, i.e. Eqs. (2.89) and (2.90).

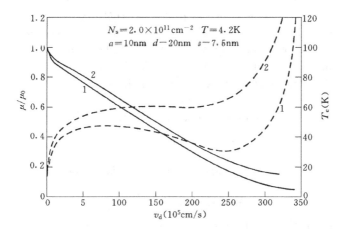

Fig. 2.12 Nonlinear mobility normalized to its Ohmic limit, μ/μ_0 (solid curves), and electron temperature T_e (dashed curves), are shown as functions of drift velocity v_d at lattice temperature $T = 4.2$ K for GaAs-Al$_x$Ga$_{1-x}$As quantum well superlattices with carrier density $N_s = 2.0 \times 10^{11}$ cm^{-2} per sheet but two different values of impurity-limited Ohmic mobilities: 1— 2.5×10^5 cm^2/Vs and 2— 8.0×10^4 cm^2/Vs.

These force- and energy-balance equations are applied to the calculations of Ohmic mobility as a function of temperature, and of nonlinear mobility and electron temperature as functions of drift velocity, by Lei,

Horing and Zhang (1986b) for GaAs/AlGaAs quantum well superlattices, including remote and background impurity scatterings and acoustic and polar-optic phonon scatterings, as well as encompassing the full effects of intralayer and interlayer carrier-carrier interactions within the framework of random phase approximation. A typical result is shown in Fig. 2.12.

2.8 Balance Equations for Transverse Transport in Tunneling Superlattices

In the preceding section we used an ideal model for the superlattice: each electron is confined almost completely within one of the quantum wells and the probability for it to tunnel to the neighboring wells is negligible. The wave functions of electrons in different quantum wells do not overlap each other such that the only coupling between them is the direct Coulomb interaction. Generally, semiconductor superlattice is such a system in which electrons move freely in the x–y plane while along its growth direction z there is a periodic potential $U(z) = U(z+d)$, d being the spatial period of the superlattice. In view of the periodic structure in the z direction, the electron energy spectrum consists of minibands. In this section we assume that electrons occupy only the lowest miniband and neglect effects of higher minibands, thus describe the single electron state of a tunneling superlattice with a wavevector $\bm{k} = (\bm{k}_\parallel, k_z)$, in which $\bm{k}_\parallel = (k_x, k_y)$ and $-\pi/d \leqslant k_z < \pi/d$. For brevity, the spin indices are assumed implicitly included in $\bm{k}$. The wave function is written as (for unit transverse area)

$$\psi_{\bm{k}}(\bm{r}, z) = e^{i\bm{k}_\parallel \cdot \bm{r}} \varphi_{k_z}(z), \qquad (2.163)$$

where $\bm{r} = (x, y)$ is the coordinate vector in the plane, and

$$\varphi_{k_z}(z) = e^{i k_z z} u_{k_z}(z) \qquad (2.164)$$

is the Bloch function, with $u_{k_z}(z)$ the cell periodic function, $u_{k_z}(z+d) = u_{k_z}(z)$. The energy of the $\bm{k}$ state is

$$\varepsilon_{\bm{k}} = k_\parallel^2/2m + \varepsilon(k_z). \qquad (2.165)$$

This is actually an anisotropic three-dimensional system with k_z limited within the Brillouin zone $(-\pi/d, \pi/d)$. For analyzing its transport in the x–y plane, we introduce 2D center-of-mass variables (2.72) and relative electron variables (2.73) and write the relative electron Hamiltonian in the second quantization representation of above single electron state basis as

$$H_{\text{er}} = \sum_{k} \varepsilon_{k} c_{k}^{\dagger} c_{k} + \frac{1}{2} \sum_{k,k',q} V(q, k_z, k_z') c_{k+q}^{\dagger} c_{k'-q}^{\dagger} c_{k'} c_{k}, \quad (2.166)$$

in which $k = (k_{\parallel}, k_z)$, $k' = (k_{\parallel}', k_z')$, $q = (q_{\parallel}, q_z)$, and the summation is performed over $k_{\parallel}$, $k_{\parallel}'$, k_z, k_z', $q_{\parallel}$ and q_z. The ranges of k_z and k_z' are within the Brillouin zone $(-\pi/d \leqslant k_z, k_z' < \pi/d)$, but q_z, which is the difference between two z-direction wavevectors of the electron, goes over the range of $-2\pi/d \leqslant q_z < 2\pi/d$. The Coulomb matrix elements in (2.166) can be expressed as

$$V(q, k_z, k_z') = \frac{e^2}{2\epsilon_0 \kappa q_{\parallel} d} \int_{-d/2}^{d/2} dz_1 \int_{-d/2}^{d/2} dz_2 \, S(z_1 - z_2, q)$$
$$\times \varphi_{k_z+q_z}^{*}(z_1) \varphi_{k_z}(z_1) \varphi_{k_z'-q_z}^{*}(z_2) \varphi_{k_z'}(z_2), \quad (2.167)$$

$$S(z_1 - z_2, q) = e^{-q_{\parallel}|z_1 - z_2|} + \frac{e^{-q_{\parallel}(z_1-z_2)}}{e^{(q_{\parallel}+i q_z)d} - 1} + \frac{e^{q_{\parallel}(z_1-z_2)}}{e^{(q_{\parallel}-i q_z)d} - 1}. \quad (2.168)$$

Assuming that the elastic scatterings are due to charged impurity atoms distributed throughout the system with a charge number Z, we can write the electron–impurity interaction in the form

$$H_{\text{ei}} = \frac{Ze^2 d}{2\epsilon_0 \kappa} \sum_{k,q,a} \frac{e^{i q_{\parallel} \cdot R}}{q_{\parallel}} e^{-i(q_{\parallel} \cdot r_a + q_z n_a d)} u(q, k_z, \bar{z}_a) c_{k+q}^{\dagger} c_{k}, \quad (2.169)$$

in which

$$u(q, k_z, \bar{z}_a) = \frac{1}{d} \int_{-d/2}^{d/2} dz S(z - \bar{z}_a, q) \varphi_{k_z+q_z}^{*}(z) \varphi_{k_z}(z) \quad (2.170)$$

is a form factor related to the longitudinal distribution of impurity atoms. A summation over all impurity atoms should be carried out in (2.169). Here the ath impurity atom having transverse position r_a and longitudinal position z_a, locates in the n_ath layer of the superlattice and at a distance $\bar{z}_a$ from the center of that layer: $z_a = n_a d + \bar{z}_a$. Note that the wavevector $q = (q_{\parallel}, q_z)$ in (2.169) comes from the difference of two electron wavevectors. Its range is determined by those of the electron wavevectors.

The electron–phonon interaction is given by

$$H_{\text{ep}} = \sum_{q,\lambda} e^{i q_{\parallel} \cdot R} M(q, \lambda)(b_{q\lambda} + b_{-q\lambda}^{\dagger}) \sum_{k} g(q_z, k_z) c_{k+q}^{\dagger} c_{k}, \quad (2.171)$$

in which $M(\boldsymbol{q},\lambda)$ is the matrix element of the electron–phonon interaction in the plane-wave representation, $g(q_z,k_z)$ is a form factor determined completely by the electron wave function (2.164) in the z direction:

$$g(q_z,k_z) = \frac{1}{d}\int_{-d/2}^{d/2} dz\, \mathrm{e}^{\mathrm{i}q_z z}\varphi^*_{k_z+q_z}(z)\varphi_{k_z}(z). \qquad (2.172)$$

It should be noted that the wavevector $\boldsymbol{q}=(\boldsymbol{q}_\parallel,q_z)$ in (2.171) is that of the phonon mode which induces the electron transition. Its range is irrelevant to that of the electron wavevector $\boldsymbol{k}=(\boldsymbol{k}_\parallel,k_z)$. Although the electron wave vector k_z is restricted within $(-\pi/d,\pi/d)$, the phonon wavevector q_z not only can go beyond this range but also can go beyond $(-2\pi/d,2\pi/d)$. An electron–phonon scattering process with the phonon wavevector in the range $-2\pi/d < q_z \leqslant 2\pi/d$, is called a normal process or N-process. For q_z outside this range the electron–phonon scattering processes is called an umklapp process or U-process. This situation does not happen for electron–impurity scattering, where the wavevector $\boldsymbol{q}$ represents the difference of two electron wavevectors, thus the q_z range is determined by those of two electron wavevectors, i.e. within $(-2\pi/d,2\pi/d)$. There is no umklapp process in the case of impurity scattering.

Following the derivation outlined in Chapter 1 we can obtain the force- and energy-balance equations for dc steady-state transport in tunneling superlattices similar to those of (2.89) and (2.90):

$$N_\mathrm{s} eE + f(v_\mathrm{d},T_\mathrm{e}) = 0, \qquad (2.173)$$

$$v_\mathrm{d} f(v_\mathrm{d},T_\mathrm{e}) + w(v_\mathrm{d},T_\mathrm{e}) = 0, \qquad (2.174)$$

where v_d is the velocity of the center of mass, i.e. the average drift velocity of the electron system, T_e is the electron temperature, $f(v_\mathrm{d},T_\mathrm{e})$ is the frictional force induced by impurity and phonon scatterings, and $w(v_\mathrm{d},T_\mathrm{e})$ is the energy-transfer rate from the electron system to the phonon system. Since the system is isotropic within the x–y plane, the electric field, the center-of-mass velocity and the frictional force are in the same (or opposite) direction. The equation can then be written in a scalar form.

The derivation of impurity-induced frictional force requires to do an average over the impurity configuration, i.e. to calculate the quantity

$$A = \sum_{a,b} \left\langle \mathrm{e}^{\mathrm{i}\boldsymbol{q}_\parallel\cdot\boldsymbol{r}_a + \mathrm{i}\boldsymbol{q}'_\parallel\cdot\boldsymbol{r}_b} \mathrm{e}^{\mathrm{i}q_z n_a d + \mathrm{i}q'_z n_b d} u(\boldsymbol{q},k_z,\bar{z}_a) u(\boldsymbol{q}',k'_z,\bar{z}'_b) \right\rangle. \qquad (2.175)$$

Let N_i be the impurity sheet density per layer and assume that the correlation of the impurity distribution among different layers is described by a function $g(q_z)$ [see (2.156)–(2.158)], and that the longitudinal impurity distribution within each layer can be described by a probability function $P(\bar{z})$ satisfying

$$\int_{-d/2}^{d/2} d\bar{z} P(\bar{z}) = 1, \qquad (2.176)$$

we have

$$A = \frac{N_i}{d} \int_{-d/2}^{d/2} d\bar{z} P(\bar{z})\, \delta_{q'_\parallel,-q_\parallel}\, \delta_{q'_z,-q_z}\, g(q_z)\, u(\boldsymbol{q}, k_z, \bar{z}) u(-\boldsymbol{q}', k'_z, \bar{z}) \qquad (2.177)$$

after the impurity configuration average. The impurity-induced frictional force is then

$$f_i = N_i d \left(\frac{Ze^2}{2\epsilon_0 \kappa}\right)^2 \int_{-d/2}^{d/2} d\bar{z} P(\bar{z}) \sum_{\boldsymbol{q}} \frac{q_x}{q_\parallel^2}\, \Pi_2^{(u)}(\boldsymbol{q}, \omega_0), \qquad (2.178)$$

where $\omega_0 \equiv q_x v_d$, $\Pi_2^{(u)}(\boldsymbol{q}, \omega)$ is the imaginary part of the Fourier spectrum of the correlation function

$$\Pi^{(u)}(\boldsymbol{q}, t) = -i\,\theta(t) \langle [\gamma_{\boldsymbol{q}}(t), \gamma_{-\boldsymbol{q}}] \rangle_0, \qquad (2.179)$$

with $\gamma_{\boldsymbol{q}}(t) \equiv \exp(i H_{er} t) \gamma_{\boldsymbol{q}} \exp(-i H_{er} t)$ being the interaction representation of a $\bar{z}$-dependent effective electron density operator associated with impurity scattering,

$$\gamma_{\boldsymbol{q}} = \sum_{\boldsymbol{k}} u(\boldsymbol{q}, k_z, \bar{z}) c^\dagger_{\boldsymbol{k}+\boldsymbol{q}} c_{\boldsymbol{k}}. \qquad (2.180)$$

To calculate the phonon-induced frictional force f_p and the energy-transfer rate w from the electron system to the phonon system, it is necessary to take average over the phonon distribution. We finally obtain

$$f_p = 2 \sum_{\boldsymbol{q},\lambda} q_x |M(\boldsymbol{q}, \lambda)|^2\, \Pi_2(\boldsymbol{q}, \Omega_{\boldsymbol{q}\lambda} + \omega_0) \left[n\!\left(\frac{\Omega_{\boldsymbol{q}\lambda}}{T}\right) - n\!\left(\frac{\Omega_{\boldsymbol{q}\lambda} + \omega_0}{T_e}\right) \right], \qquad (2.181)$$

$$w = 2 \sum_{\boldsymbol{q},\lambda} \Omega_{\boldsymbol{q}\lambda} |M(\boldsymbol{q}, \lambda)|^2\, \Pi_2(\boldsymbol{q}, \Omega_{\boldsymbol{q}\lambda} + \omega_0) \left[n\!\left(\frac{\Omega_{\boldsymbol{q}\lambda}}{T}\right) - n\!\left(\frac{\Omega_{\boldsymbol{q}\lambda} + \omega_0}{T_e}\right) \right]. \qquad (2.182)$$

Here $\Pi_2(\boldsymbol{q},\omega)$ is the imaginary part of the Fourier spectrum of the electron density correlation function of the tunneling superlattice,

$$\Pi(\boldsymbol{q},t) = -\mathrm{i}\,\theta(t)\langle[\rho_{\boldsymbol{q}}(t),\rho_{-\boldsymbol{q}}]\rangle_0, \qquad (2.183)$$

where $\rho_{\boldsymbol{q}}(t) \equiv \exp(\mathrm{i}\,H_{\mathrm{er}}t)\rho_{\boldsymbol{q}}\exp(-\mathrm{i}\,H_{\mathrm{er}}t)$ with the electron density operator

$$\rho_{\boldsymbol{q}} = \sum_{\boldsymbol{k}} g(q_z, k_z) c^\dagger_{\boldsymbol{k}+\boldsymbol{q}} c_{\boldsymbol{k}} \qquad (2.184)$$

associated with electron–phonon scattering (2.171).

For an anisotropic many-body system, the matrix element $V(\boldsymbol{q}, k_z, k'_z)$ of the electron Coulomb interaction, the form factor $g(q_z, k_z)$ in the electron density operator $\rho_{\boldsymbol{q}}$ and the form factor $u(\boldsymbol{q}, k_z, \bar{z})$ in the impurity-related electron density operator $\gamma_{\boldsymbol{q}}$, are dependent on the wavevector k_z of the electron state. This makes the RPA equations of correlation functions $\Pi(\boldsymbol{q},\omega)$ and $\Pi^{(u)}(\boldsymbol{q},\omega)$ quite complicated. For weak tunneling superlattices Lu, Xie and Lei (1987) and Yang and Lu (1988) develop a separation technique, with which the above-mentioned correlation functions in the presence of intercarrier Coulomb interaction can be obtained relatively easier in the random phase approximation by using equations in the matrix form.

In the tight-binding representation, the electron wave function of the superlattice, $\varphi_{k_z}(z)$, can be written as

$$\varphi_{k_z}(z) = A_{k_z} \sum_n \mathrm{e}^{\mathrm{i}\,k_z nd}\phi(z-nd), \qquad (2.185)$$

in which A_{k_z} is a normalized coefficient, $\phi(z-nd)$ is the "single-well" wave function of the nth potential well constituting the superlattice, which is chosen to be a real function and normalized as $\int dz\,\phi^2(z) = d$. For weak tunneling systems we consider only the overlap between nearest-neighbor single-well wave functions, and the normalized coefficient is

$$A_{k_z} = [1 + 2\alpha\cos(k_z d)]^{-1/2}, \qquad (2.186)$$

where α is an overlapping integral

$$\alpha = \frac{1}{d}\int_{-d/2}^{d/2} dz\,\phi(z)\phi(z-d). \qquad (2.187)$$

The energy corresponding to the electron state φ_{k_z} is

$$\varepsilon(k_z) = \frac{\Delta}{2}\left[1 - \cos(k_z d)\right], \qquad (2.188)$$

Δ being the miniband width of the superlattice.

Inserting the wave function expression (2.185) into (2.167) and keeping only the nearest-neighbor overlaps, we can write the matrix elements of the Coulomb potential in the form

$$V(\mathbf{q}, k_z, k_z') = A_{k_z+q_z} A_{k_z} A_{k_z'-q_z} A_{k_z'} \widehat{T}(k_z) \tilde{\nu}(\mathbf{q}) \widehat{T}'(k_z'), \quad (2.189)$$

in which

$$\widehat{T}(k_z) \equiv [1, \cos(k_z d), \sin(k_z d)] \quad (2.190)$$

is a one-row matrix, $\widehat{T}'(k_z)$ stands for the transpose of $\widehat{T}(k_z)$; $\tilde{\nu}(\mathbf{q})$ is a 3×3 matrix with elements $(i, j = 1, 2, 3)$

$$\tilde{\nu}_{ij}(\mathbf{q}) = \frac{e^2}{2\epsilon_0 \kappa q_\| d} \int_{-d/2}^{d/2} dz_1 \int_{-d/2}^{d/2} dz_2 \, S(z_1 - z_2, \mathbf{q}) f_i(z_1, q_z) f_j(z_2, -q_z), \quad (2.191)$$

in which $S(z_1 - z_2, \mathbf{q})$ is given by (2.168), and functions $f_i(z, q_z)$ $(i = 1, 2, 3)$ are defined as the following:

$$\begin{aligned}
f_1(z, q_z) &= \phi^2(z) + [\phi^2(z-d) + \phi^2(z+d)] \cos(q_z d) \\
&\quad + i[\phi^2(z+d) - \phi^2(z-d)] \sin(q_z d), \\
f_2(z, q_z) &= \phi(z)[\phi(z+d) + \phi(z-d)][\cos(q_z d) + 1] \\
&\quad + i\phi(z)[\phi(z+d) - \phi(z-d)] \sin(q_z d), \\
f_3(z, q_z) &= i\phi(z)[\phi(z+d) - \phi(z-d)][\cos(q_z d) - 1] \\
&\quad - \phi(z)[\phi(z+d) + \phi(z-d)] \sin(q_z d).
\end{aligned} \quad (2.192)$$

In the case without wave-function overlap, $f_2(z, q_z) = 0 = f_3(z, q_z)$ and $f_1(z, q_z) = \phi^2(z)$, then all the other components of $\tilde{\nu}_{ij}(\mathbf{q})$ vanish except $\tilde{\nu}_{11}(\mathbf{q}) = dV(q_\|, q_z)$ with

$$V(q_\|, q_z) = \frac{e^2}{2\epsilon_0 \kappa q_\|} [H(q_\|) + S(q_\|, q_z)]. \quad (2.193)$$

This is the formula (2.149) in Sec. 2.7.

Under the nearest-neighbor overlap approximation, the form factor (2.170) associated with electron–impurity scattering can be written as

$$\begin{aligned}
u(\mathbf{q}, k_z, \bar{z}) &= A_{k_z+q_z} A_{k_z} \widehat{u}(\mathbf{q}) \widehat{T}'(k_z) \\
&= A_{k_z+q_z} A_{k_z} \widehat{T}(k_z) \widehat{u}'(\mathbf{q}).
\end{aligned} \quad (2.194)$$

Here $\widehat{u}(\boldsymbol{q})$ is a one-row matrix

$$\widehat{u}(\boldsymbol{q}) = [u_1(\boldsymbol{q}), u_2(\boldsymbol{q}), u_3(\boldsymbol{q})] \tag{2.195}$$

with elements ($j = 1, 2, 3$)

$$u_j(\boldsymbol{q}) = \frac{1}{d} \int_{-d/2}^{d/2} dz S(z - \bar{z}, \boldsymbol{q}) f_j(z, q_z), \tag{2.196}$$

and $\widehat{u}'(\boldsymbol{q})$ is the transpose of $\widehat{u}(\boldsymbol{q})$. The form factor (2.172) associated with electron–phonon scattering can be written as

$$g(q_z, k_z) = A_{k_z+q_z} A_{k_z} \widehat{\rho}(q_z) \widehat{T}'(k_z) = A_{k_z+q_z} A_{k_z} \widehat{T}(k_z) \widehat{\rho}'(q_z), \tag{2.197}$$

where $\widehat{\rho}(q_z)$ is a one-row matrix

$$\widehat{\rho}(q_z) = [\rho_1(q_z), \rho_2(q_z), \rho_3(q_z)] \tag{2.198}$$

with elements ($j = 1, 2, 3$)

$$\rho_j(q_z) = \frac{1}{d} \int_{-d/2}^{d/2} dz\, e^{i q_z z} f_j(z, q_z), \tag{2.199}$$

and $\widehat{\rho}'(q_z)$ is the transpose of $\widehat{\rho}(q_z)$. The correlation functions $\Pi_2^{(u)}(\boldsymbol{q}, \omega)$ and $\Pi_2(\boldsymbol{q}, \omega)$ are given, respectively, by the imaginary parts of the multiplications of three matrices:

$$\Pi_2^{(u)}(\boldsymbol{q}, \omega) = \mathrm{Im}\,[\widehat{u}(\boldsymbol{q})\, \widetilde{\pi}(\boldsymbol{q}, \omega)\, \widehat{u}'(-\boldsymbol{q})], \tag{2.200}$$

$$\Pi_2(\boldsymbol{q}, \omega) = \mathrm{Im}\,[\widehat{\rho}(q_z)\, \widetilde{\pi}(\boldsymbol{q}, \omega)\, \widehat{\rho}'(-q_z)]. \tag{2.201}$$

Here $\widetilde{\pi}(\boldsymbol{q}, \omega)$ is a 3×3 matrix density correlation function. It satisfies the following matrix Dyson equation in the random phase approximation:

$$\widetilde{\pi}(\boldsymbol{q}, \omega) = \widetilde{\pi}_0(\boldsymbol{q}, \omega) + \widetilde{\pi}_0(\boldsymbol{q}, \omega)\, \widetilde{\nu}(\boldsymbol{q})\, \widetilde{\pi}(\boldsymbol{q}, \omega), \tag{2.202}$$

in which

$$\widetilde{\pi}_0(\boldsymbol{q}, \omega) = 2 \sum_{\boldsymbol{k}} A_{k_z+q_z}^2 A_{k_z}^2\, \widehat{T}'(k_z + q_z) \widehat{T}(k_z) \frac{f(\varepsilon_{\boldsymbol{k}+\boldsymbol{q}}, T_\mathrm{e}) - f(\varepsilon_{\boldsymbol{k}}, T_\mathrm{e})}{\varepsilon_{\boldsymbol{k}+\boldsymbol{q}} - \varepsilon_{\boldsymbol{k}} + \omega + i\delta} \tag{2.203}$$

is the matrix (3×3) density correlation function in the absence of Coulomb interaction, $f(\varepsilon_{\boldsymbol{k}}, T_\mathrm{e})$ is the Fermi distribution function at temperature T_e. The factor 2 on the right-hand-side of (2.203) comes from the summation over the degenerate spin indices. With the help of the matrix Dyson equation (2.202), the correlation functions $\Pi_2(\boldsymbol{q}, \omega)$ and $\Pi_2^{(u)}(\boldsymbol{q}, \omega)$ can be obtained in the random phase approximation and the transport properties of a tunneling superlattice can be investigated using balance equations (2.173) and (2.174).

2.9 Transport in Quantum Wires

The development of semiconductor technology has made it possible to confine carriers moving within a quite uniform thin quantum wire with its lateral size around a few nanometers, such that their states are quantized along both lateral directions. The transport properties of quantum wires can be conveniently investigated within the framework of the balance equation approach (Wang and Lei, 1993a, 1993b, 1994a).

In this section we are going to discuss electron transport in quantum wires, or quasi-one-dimensional systems.

2.9.1 *Cylindrical quantum wires*

As a simplified model, we consider electrons confined within a cylindrical wire of diameter 2ϱ. The electron state in the wire can be described by lateral quantum numbers (m,l) and a wavevector k_z along its axis in the z direction. For brevity, we assume that the wire has a unit length and use a single index n to denote (m,l): $m = 0, \pm 1, \pm 2, \cdots$ and $l = 0, 1, 2, \cdots, |m|$. The electron wave function is written as

$$\psi_{nk_z}(\boldsymbol{r}, z) = \mathrm{e}^{\mathrm{i}\,k_z z}\varphi_n(\boldsymbol{r}), \tag{2.204}$$

$$\varphi_n(\boldsymbol{r}) = c_l^m \mathrm{J}_m\!\left(\frac{r}{\varrho}x_l^m\right) \mathrm{e}^{\mathrm{i}\,m\phi}. \tag{2.205}$$

Here $\boldsymbol{r} = (r, \phi)$ represents the position vector in the plane perpendicular to the z axis and its polar coordinate, $c_l^m = (\sqrt{\pi}\varrho y_l^m)^{-1}$ is a normalization coefficient, x_l^m is the lth zero of the mth-order Bessel function, i.e. $\mathrm{J}_m(x_l^m) = 0$, and $y_l^m \equiv \mathrm{J}_{m+1}(x_l^m)$. The corresponding eigen energy is

$$\varepsilon_n(k_z) = \varepsilon_n + \frac{k_z^2}{2m^*}, \tag{2.206}$$

$$\varepsilon_n \equiv \varepsilon_l^m = \frac{(x_l^m)^2}{2m^*\varrho^2}, \tag{2.207}$$

m^* being the effective mass of the electron.

When a uniform electric field is applied in the z direction, the electrons will drift along the axis of the quantum wire. Introducing the center-of-mass momentum and coordinate variables P and Z in the z direction, we can express the Hamiltonian of the system as the sum of a center-of-mass part $H_{\rm cm}$, a relative electron part $H_{\rm er}$, and a phonon part $H_{\rm ph}$, together with electron–impurity and electron–phonon interactions, $H_{\rm ei}$ and $H_{\rm ep}$. The center-of-mass part is

$$H_{\rm cm} = \frac{P^2}{2N_1 m^*} - N_1 eEZ, \tag{2.208}$$

N_1 is the total number of electrons in the quantum wire system of unit length, i.e. the electron line density of the wire.

In terms of creation (annihilation) operators $c^\dagger_{nk_z}$ (c_{nk_z}) in the basis of single electron states (2.204) we write the Hamiltonian of the relative-electron system in the form

$$H_{\rm er} = \sum_{n,k_z} \varepsilon_n(k_z)\, c^\dagger_{nk_z} c_{nk_z}$$
$$+ \frac{1}{2} \sum_{\substack{n,n',m',m \\ k_z,k'_z,q_z}} V_{nn'm'm}(q_z)\, c^\dagger_{nk_z+q_z} c^\dagger_{n'k'_z-q_z} c_{m'k'_z} c_{mk_z}. \tag{2.209}$$

Here

$$V_{nn'm'm}(q_z) = \frac{e^2}{2\pi\epsilon_0 \kappa} \iint d\mathbf{r}\, d\mathbf{r}'\, \varphi_n^*(\mathbf{r})\varphi_{n'}^*(\mathbf{r}')\varphi_{m'}(\mathbf{r}')\varphi_m(\mathbf{r}) {\rm K}_0(|q_z(\mathbf{r}-\mathbf{r}')|) \tag{2.210}$$

is the intercarrier Coulomb potential, κ is the dielectric constant of the background material of the quantum wire, ${\rm K}_0(x)$ is the modified Bessel function of zero-th order (Wang and Guo, 1989).

We assume that relevant phonon modes in the quantum wire are the same as the bulk modes of the background material, and the phonon Hamiltonian is given by

$$H_{\rm ph} = \sum_{\mathbf{q},\lambda} \Omega_{\mathbf{q}\lambda} b^\dagger_{\mathbf{q}\lambda} b_{\mathbf{q}\lambda} \tag{2.211}$$

in terms of creation and annihilation operators $b^\dagger_{\mathbf{q}\lambda}$ and $b_{\mathbf{q}\lambda}$ of a phonon with wavevector $\mathbf{q} \equiv (\mathbf{q}_\parallel, q_z) \equiv (q_x, q_y, q_z)$ in banch λ having frequency $\Omega_{\mathbf{q}\lambda}$. The electron–phonon interaction can be written as

$$H_{\rm ep} = \sum_{n,n'} \sum_{\mathbf{q},\lambda} M(n,n',\mathbf{q},\lambda)\, e^{{\rm i}q_z Z}(b_{\mathbf{q}\lambda} + b^\dagger_{-\mathbf{q}\lambda})\, c^\dagger_{nk_z+q_z} c_{n'k_z}, \tag{2.212}$$

where

$$M(n, n', \boldsymbol{q}, \lambda) = M(\boldsymbol{q}, \lambda) F_{nn'}(q_\parallel) \qquad (2.213)$$

with $M(\boldsymbol{q}, \lambda)$ being the conventional electron–phonon matrix element in the 3D Fourier representation,

$$F_{nn'}(q_\parallel) \equiv F_{ll'}^{mm'}(q_\parallel) = 2 \int_0^1 d\xi \, \frac{\xi}{y_l^m y_{l'}^{m'}} \mathrm{J}_m(x_l^m \xi) \mathrm{J}_{m'}(x_{l'}^{m'} \xi) \mathrm{J}_{|m-m'|}(q_\parallel \varrho \xi) \qquad (2.214)$$

is a form factor related to the wave function of the quasi-1D electron system.

The electron–impurity coupling is written as

$$H_{\mathrm{ei}} = \sum_{\substack{n,n' \\ k_z, q_z, a}} U(n, n', q_z, \boldsymbol{r}_a) \, e^{i q_z (Z - z_a)} c_{nk_z + q_z}^\dagger c_{n'k_z}, \qquad (2.215)$$

where

$$U(n, n', q_z, \boldsymbol{r}_a) = \frac{Z_a e^2}{2\pi\epsilon_0 \kappa} \int d\boldsymbol{r} \, \varphi_n^*(\boldsymbol{r}) \varphi_{n'}(\boldsymbol{r}) \, \mathrm{K}_0(|q_z(\boldsymbol{r} - \boldsymbol{r}_a)|) \qquad (2.216)$$

is the potential due to ath impurity locating at $(\boldsymbol{r}_a, z_a)$ having charge number Z_a.

According to balance-equation treatment, we use the z-direction drift velocity v_d and electron temperature T_e to describe the electron transport in the quantum wire under the influence of a uniform electric field E applied along the wire axis. Following the derivation outlined in Chapter 1 we can obtain the force- and energy-balance equations for dc steady-state transport:

$$N_1 e E + f(v_\mathrm{d}, T_\mathrm{e}) = 0, \qquad (2.217)$$

$$v_\mathrm{d} f(v_\mathrm{d}, T_\mathrm{e}) + w(v_\mathrm{d}, T_\mathrm{e}) = 0. \qquad (2.218)$$

Here, the frictional force due to impurity and phonon scatterings (for the quantum wire of unit length) is written as

$$\begin{aligned} f(v_\mathrm{d}, T_\mathrm{e}) &= \sum_{n,n',q_z} |U(n, n', q_z)|^2 q_z \Pi_2(n', n, q_z, \omega_0) \\ &+ 2 \sum_{n,n',\boldsymbol{q},\lambda} |M(n, n', \boldsymbol{q}, \lambda)|^2 q_z \Pi_2(n', n, q_z, \Omega_{\boldsymbol{q}\lambda} + \omega_0) \\ &\quad \times \left[n\left(\frac{\Omega_{\boldsymbol{q}\lambda}}{T}\right) - n\left(\frac{\Omega_{\boldsymbol{q}\lambda} + \omega_0}{T_\mathrm{e}}\right) \right], \qquad (2.219) \end{aligned}$$

in which
$$|U(n,n',q_z)|^2 = \int d\mathbf{r}\, n_\mathrm{i}(\mathbf{r})\, |U(n,n',q_z,\mathbf{r})|^2 \qquad (2.220)$$
is the effective impurity potential (Assuming that the impurity volume density is $n_\mathrm{i}(\mathbf{r})$ at $\mathbf{r}$ and all the impurities have the same charge number). The electron energy-loss rate (for the quantum wire of unit length) is
$$w(v_\mathrm{d}, T_\mathrm{e}) = 2\sum_{n,n',\mathbf{q},\lambda} |M(n,n',\mathbf{q},\lambda)|^2 \Omega_{\mathbf{q}\lambda} \Pi_2(n',n,q_z,\Omega_{\mathbf{q}\lambda}+\omega_0)$$
$$\times \left[n\!\left(\frac{\Omega_{\mathbf{q}\lambda}}{T}\right) - n\!\left(\frac{\Omega_{\mathbf{q}\lambda}+\omega_0}{T_\mathrm{e}}\right) \right]. \qquad (2.221)$$
In Eqs. (2.219) and (2.221) $\omega_0 \equiv q_z v_\mathrm{d}$, $\Pi_2(n,n',q_z,\omega)$ is the imaginary part of the Fourier spectrum $\Pi(n,n',q_z,\omega)$ of the electron density correlation function
$$\Pi(n,n',q_z,t) = -\mathrm{i}\,\theta(t)\sum_{\mathbf{k}} \left\langle \left[c^\dagger_{nk_z+q_z}(t)\, c_{n'k_z}(t),\, c^\dagger_{n'k_z}\, c_{nk_z+q_z} \right] \right\rangle_0. \qquad (2.222)$$
In the absence of electron–electron Coulomb interaction the corresponding density correlation function is
$$\Pi_0(n,n',q_z,\omega) = 2\sum_{k_z} \frac{f(\varepsilon_n(k_z+q_z),T_\mathrm{e}) - f(\varepsilon_{n'}(k_z),T_\mathrm{e})}{\omega + \varepsilon_n(k_z+q_z) - \varepsilon_{n'}(k_z) + \mathrm{i}\delta}, \qquad (2.223)$$
where $f(\varepsilon, T_\mathrm{e}) = 1/[\exp((\varepsilon-\mu)/T_\mathrm{e})+1]$ is the Fermi distribution function at temperature T_e, and μ is the chemical potential to be determined by the electron line density N_l from the relation
$$N_\mathrm{l} = 2\sum_{n,k_z} f(\varepsilon_n(k_z), T_\mathrm{e}). \qquad (2.224)$$

On the basis of the above equations, Wang and Lei (1993a) calculate the transport properties of GaAs-based cylindrical quantum wires having radii ϱ from 5 to 40 nm in the absence of impurity scattering. They take into account effects of the lowest 15 subbands with both intraband and interband transitions due to longitudinal and transverse acoustic phonons and longitudinal optical (LO) phonons. The 3D bulk phonon model is good enough for a quantum wire with radius greater than 5 nm, in which the contribution of surface phonons is generally unimportant (Wang and Lei, 1994b). To evaluate the electron–phonon coupling in quantum wires the form factor $F_{nn'}(q_\parallel)$ has to be calculated numerically. Typical examples of $F_{nn'}(q_\parallel)$ are shown in Fig. 2.13. The intrasubband form factor $F_{nn}(q_\parallel)$ reaches its maximum at $q_\parallel = 0$ with $F_{nn}(0) = 1$, then decreases to about one-half at $q_\parallel = 2/\varrho$, and almost vanishes for $q_\parallel$ larger than $5/\varrho$. Therefore, if the radius of the quantum wire is large, the intrasubband transition contributes to resistivity only from very small $q_\parallel$. The intersubband form factors $F_{nn' \neq n}(q_\parallel)$, on the other hand, are zero at $q_\parallel = 0$, then increase with increasing $q_\parallel$ and arrive at their first maximum around $3.5/\varrho$ (generally several peaks exist). Thus the contribution of intersubband transitions comes mainly from large $q_\parallel$.

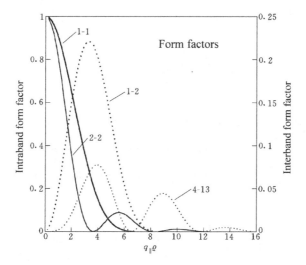

Fig. 2.13 Typical form factors of both intrasubband (solid curves) and intersubband (dotted curves) transitions.

In the case of fixed electron volume density $n = 3.18 \times 10^{17}$ cm^{-3}, the low-temperature ($T = 2$ K) linear mobilities μ_0 for quantum wires having different lateral radii ϱ are calculated and plotted in Fig. 2.14 as the thick solid curve. In the figure all 15 subband bottoms ε_n measured from that of the ground subband ε_1 are also shown as the function of wire radius by the thin dotted curve. The thin solid curve denotes the Fermi energy of the quantum wire having different ϱ at $T = 2$ K. Depending on the position of the Fermi energy relative to the subband bottom, the mobility of a quantum wire may be much higher or lower than that of the corresponding 3D system with same electron volume density. When the wire lateral radius is small, the mobility is very low because of the singularity of the 1D density

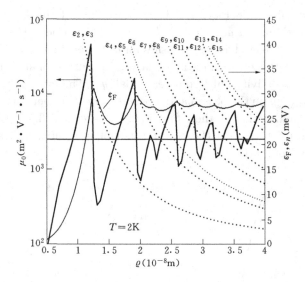

Fig. 2.14 Linear mobility μ_0 (thick solid curve) and Fermi energy ε_F (thin solid curve) as functions of wire radius ϱ at lattice temperature $T = 2\,\mathrm{K}$. The long bar marks the mobility of the 3D system. The dotted curves represent the subband bottoms (thick ones denote degenerate cases). All the quantum wires and the 3D system are assumed to have the same electron volume density $n = 3.18 \times 10^{17}$ cm^{-3}. From Wang and Lei (1993a).

of states and the low Fermi energy. With the increase of lateral radius ϱ and the corresponding rise of Fermi energy, the mobility increases rapidly until the Fermi energy touches the bottom of the second subband where the mobility reaches a maximum and then decreases steeply. The mobility exhibits a deep minimum following a sharp maximum, because electrons can be scattered into the second subband. Moreover, maximum and minimum show up repeatedly once the Fermi energy touches a new subband, exhibiting oscillations with increasing lateral size of the wire. The 3rd and 8th peaks in Fig. 2.14 are somewhat smaller because the corresponding subbands are without degeneration so the changes of the density-of-states are smaller than those of other degenerate subbands.

The amplitude of the oscillation of quantum-wire linear mobility with changing its lateral size decreases at rising lattice temperature. Fig. 2.15 shows the linear mobility versus the wire diameter $d = 2\varrho$ at lattice temperatures $T = 10, 50$ and $100\,\mathrm{K}$ (thin solid curves) for the same 1D system as described in Fig. 2.14. At higher temperatures, electrons with lower energy can also be scattered into the next subband, thus the peaks in the linear mobility move toward smaller diameter where the Fermi energy is lower. At the same time, since the electron distribution becomes more broadened at higher T the intensity of the mobility oscillation weakens.

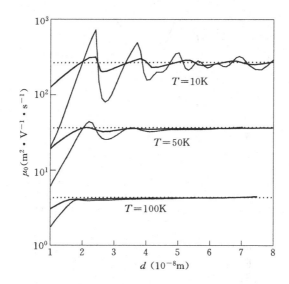

Fig. 2.15 Linear mobility μ_0 of cylindrical quantum wires with diameter $d = 2\varrho$ (thin solid curves), of quantum wells with width d (thick solid curves), and of the 3D bulk system (thin dotted lines) at temperature $T = 10, 50$ and 100 K. All systems are assumed to have same electron volume density $n = 3.18 \times 10^{17}$ cm^{-3}. From Wang and Lei (1993b).

The mobility oscillations with carrier confinement size also appear in quasi-2D systems but weaker than in quasi-1D ones, as shown in Fig. 2.15. There we also plot the linear mobility of a quasi-2D quantum well having width d at $T = 10, 50$ and 100 K (thick solid curves) calculated from the balance equations in Sec. 2.5, keeping the same electron volume density $n = 3.18 \times 10^{17}$ cm^{-3}. Note that oscillations of both 1D and 2D mobilities are around those of the 3D system with the same electron volume density, which are shown as dotted horizontal lines in Fig. 2.15 respectively at the three temperatures. Mobility oscillations gradually weaken when temperature increases. At $T = 100$ K they are almost washed out and both quasi-1D and quasi-2D mobilities approach the bulk value for $d > 20$ nm systems when electrons begin to occupy the second subband in 1D and 2D cases.

For fixed confinement size, the mobility and other transport properties of quantum wires and quantum wells also exhibit oscillation with changing

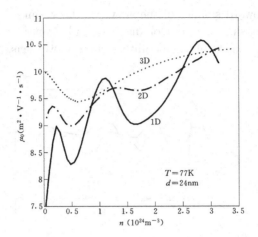

Fig. 2.16 Phonon-limited linear mobilities μ_0 at $T = 77$ K in quantum wires with diameter $d = 24$ nm (1D), in quantum wells with width $d = 24$ nm (2D) and in the 3D bulk, versus the electron volume density n. From Wang and Lei (1994).

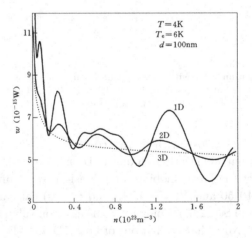

Fig. 2.17 The electron energy-loss rate (per carrier) w in quantum wires with diameter $d = 100$ nm (1D), in quantum wells with width $d = 100$ nm (2D) and in the 3D bulk, versus the electron volume density n. From Wang and Lei (1993b).

electron volume density. Fig. 2.16 illustrates the phonon-limited linear mobilities μ_0 in cylindrical quantum wires of diameter $d = 24$ nm, in quantum wells of width $d = 24$ nm and in 3D bulks, as functions of electron volume density n at temperature $T = 77$ K. Fig. 2.17 shows the energy-loss rates (per electron) w in quantum wires of $d = 100$ nm, in quantum wells of $d = 100$ nm and in the 3D bulk. as functions of n at $T = 4$ K and $T_e = 6$ K.

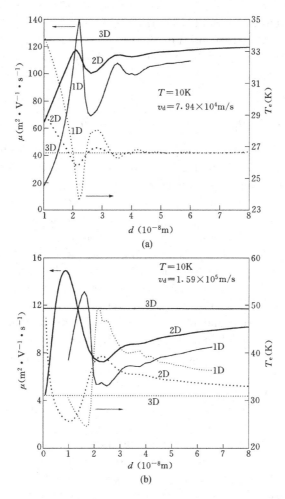

Fig. 2.18 The nonlinear mobility μ and electron temperature T_e of cylindrical quantum wires with diameter $d = 2\varrho$, of quantum wells with width d, and of the 3D bulk system, obtained under the conditions: (a) the drift velocity $v_d = 7.94 \times 10^4$ m/s and (b) $v_d = 1.59 \times 10^5$ m/s. All systems are assumed to have same electron volume density $n = 3.18 \times 10^{17}$ cm^{-3}. From Wang and Lei (1993b).

High-field transport properties also change with changing the dimensionality and exhibit oscillation when varying confinement size. The nonlinear mobility μ and electron temperature T_e for quantum wires, quantum wells and the 3D bulk with carrier volume density $n = 3.18 \times 10^{17}$ cm^{-3}, calculated respectively from corresponding balance equations at lattice temperature $T = 10$ K, are shown as functions of lateral confinement d in Fig. 2.18(a) under the condition of drift velocity $v_\mathrm{d} = 7.94 \times 10^4$ m/s and in Fig. 2.18(b) under the condition of drift velocity $v_\mathrm{d} = 1.59 \times 10^5$ m/s. Strong oscillations in μ and T_e show up in both one-dimensional and two-dimensional systems.

2.9.2 Harmonically confined quantum wires

Although the assumption that the electrons move within a cylindrical quantum wire and scattered by bulk phonons of the background material is good for the demonstration of one-dimensional confinement effect of carriers in transport, it is an academically ideal model. Among several real nanofabrication technique to create a real quasi-one-dimensional electron gas holographic lithography (Katthaus, 1987) starts from a laterally (x–y plane) homogeneous Al$_x$Ga$_{1-x}$As/GaAs heterojunction and confines the existing quasi-2D electron gas in one additional (y) direction by imposing a modulated voltage. The result is a collapse of the electron density in an array consisting of many parallel quasi-1D channels. The selfconsistent electric confining potential (Lai and Das Sarma, 1986; Laux and Stern, 1986) in the direction normal to the interface in each channel can be approximately described by an exactly solvable model of parabolic potential with a natural harmonic frequency of order of a few meV (Berggren et al, 1986; Vasilopoulos et al, 1989). We assume that these parallel channels are separated far enough that it is sufficient to treat the electron motion in just one channel, save for their interaction with phonons (see Chapter 5).

In such a channel the electrons are free to move in the x direction with the effective mass m^*, but subject to a confined potential $U(z)$ (for the formation of the 2D electron gas in the x–y plane) in the z direction and subject to a parabolic potential $\frac{1}{2}m^*\omega_0 y^2$ in the y direction. The electron state is described by the wavevector k_x and a transverse quantum number $n \equiv (l, m)$ ($l = 0, 1, 2, \cdots$ and $m = 0, 1, 2, \cdots$) with the wave function written as (for unit length of the channel)

$$\psi_{nk_x}(x, y, z) = \mathrm{e}^{\mathrm{i} k_x x}\, \varphi_n(y, z) = \mathrm{e}^{\mathrm{i} k_x x}\, \phi_m(y)\, \zeta_l(z), \qquad (2.225)$$

$$\phi_m(y) = \left(\frac{1}{\sqrt{\pi}\, 2^m m!\, l_0}\right)^{1/2} \exp\left(-\frac{y^2}{2 l_0^2}\right) \mathrm{H}_m\!\left(\frac{y}{l_0}\right), \qquad (2.226)$$

where $\zeta_l(z)$ is the envelope function of the lth subband of the heterojunction (Sec. 2.5), $l_0 \equiv (m^*\omega_0)^{-1/2}$, and $H_m(x)$ are Hermite polynomials. The eigen energies are (ε_l is the energy bottom position of the lth subband)

$$\varepsilon_n(k_x) = \varepsilon_l + \left(m + \frac{1}{2}\right)\omega_0 + \frac{k_x^2}{2m^*}. \quad (2.227)$$

If a magnetic field $\boldsymbol{B} = (0, 0, B)$ is applied perpendicular to the interfaces, in the Landau gauge $\boldsymbol{A} = (-By, 0, 0)$ the single electron eigen energies and wave functions can be solved from the schrödinger equation to be (Landau and Lifshitz, 1977)

$$\psi_{nk_x}(x, y, z) = e^{i k_x x} \varphi_n(y, z) = e^{i k_x x} \phi_m(y - \tilde{y}_0)\,\zeta_l(z), \quad (2.228)$$

$$\varepsilon_n(k_x) = \varepsilon_l + \left(m + \frac{1}{2}\right)\tilde{\omega} + \frac{k_x^2}{2\tilde{m}}. \quad (2.229)$$

The natural frequency of the parabolic confined potential, the effective mass, and the Larmor radius are renormalized by the magnetic field:

$$\tilde{\omega} = \sqrt{\omega_c^2 + \omega_0^2}, \quad \tilde{m} = m^*\tilde{\omega}^2/\omega_0^2, \quad \tilde{y}_0 = \tilde{b}\,\tilde{l}_B^2 k_x, \quad \tilde{l}_B^2 = (m^*\tilde{\omega})^{-1},$$

where $\omega_c = |eB|/m$ is the cyclotron frequency and $\tilde{b} = \omega_c/\tilde{\omega}$.

Under influence of an electric field along the x direction, balance equations for electron transport in this quasi-1D channel can be derived as in the preceding subsection (da Cunha Lima, Wang and Lei, 1997):

$$N_l e E + f(v_d, T_e) = 0, \quad (2.230)$$

$$v_d f(v_d, T_e) + w(v_d, T_e) = 0. \quad (2.231)$$

The expressions for the frictional force $f(v_d, T_e)$ and electron energy-loss rate $w(v_d, T_e)$ are similar to (2.219) and (2.221) if the impurities and phonons interacting with electrons are those described in Sec. 2.9.1. The main difference is that the Coulomb matrix elements (2.209), the impurity potential (2.197) and the form factor (2.195), now relate to the transverse wave functions (2.228). For instance, for bulk phonon modes the electron–phonon matrix elements are given by [$n = (m, l)$ and $n' = (m', l')$]

$$|M(n, n', \boldsymbol{q}, \lambda)|^2 = |M(\boldsymbol{q}, \lambda)|^2\,|F_{nn'}(\boldsymbol{q})|^2,$$

$$|F_{nn'}(\boldsymbol{q})|^2 = |I_{ll'}(\mathrm{i}\,q_z)|^2\,C_{mm'}(u), \quad (2.232)$$

in which $I_{ll'}$ is the subband form factor defined in (2.85), and

$$C_{mm'}(u) \equiv \frac{n_2!}{n_1!} u^{n_1-n_2} \mathrm{e}^{-u} \left[\mathrm{L}_{n_2}^{n_1-n_2}(u) \right]^2, \qquad (2.233)$$

where $u \equiv \tilde{l}_{\mathrm{B}}^2(q_y^2 + \tilde{b}^2 q_x^2)/2$, $n_1 = \max(m,m')$, $n_2 = \min(m,m')$, and $\mathrm{L}_m^l(u)$ are associated Laguerre polynomials (Wang and Guo, 1989):

$$\mathrm{L}_m^l(x) = \sum_{s=0}^{m} (-1)^s \frac{(m+l)! x^s}{(l+s)!(m-s)!s!}. \qquad (2.234)$$

The model and formulation presented in this subsection provide a convenient basis for the investigation of nonlinear transport, including the magnetophonon resonance of conductivity, in harmonically confined quantum wires. However, the assumption that phonons are three dimensional neglects any change in the electron–phonon interaction brought by the quasi-one-dimensional confinement of the electron. When the quasi-1D channel is created, e.g. from a GaAs quantum well in between two thick AlAs layers, electrons are scattered mainly by the confined and interface phonons rather than by bulk phonons. This may induce quite different effects. We refer the reader to the original works of da Cunha Lima, Wang and Lei (1997) and Wang, da Cunha Lima and Lei (1998) for the discussion of this issue.

Chapter 3

Time-Dependent and High-Frequency Transport

3.1 Balance Equations for Slowly Time-Dependent Transport

The force-balance equation (1.40) and energy-balance equation (1.55) or (1.60) are also the starting point to investigate the time-dependent transport in the balance-equation approach. When treating the center-of-mass (CM) motion classically and neglecting the velocity fluctuation $\delta \bm{V}$ in (1.19) due to its enormous mass, the $A(\bm{q}, t, t')$ function defined by (1.48) depends only on the time-dependent drift velocity $\bm{v}(t)$ of the electron system:

$$A(\bm{q}, t, t') \approx \exp\left[\mathrm{i}\bm{q} \cdot \int_{t'}^{t} \bm{v}(s)\, ds \right]. \tag{3.1}$$

This form of $A(\bm{q}, t, t')$, however, still renders the time-dependent force- and energy-balance equations (1.40) and (1.55) as complicated coupled integro-differential equations, and significant computational labor is required to solve them numerically. Therefore, approximations are usually made in different cases to simplify the treatment.

If the time variation of the CM velocity is slow, i.e. $\dot{\bm{v}}$ and higher-order derivatives of $\bm{v}$ are small (the drift velocity $\bm{v}$ itself may not be small), one can expand the integrand in (3.1) around t, $\bm{v}(s) \approx \bm{v}(t) + (s-t)\dot{\bm{v}}(t) + \cdots$, to give

$$A(\bm{q}, t, t') \approx \exp[\mathrm{i}\bm{q} \cdot \bm{v}(t)(t-t')] \left[1 - \frac{\mathrm{i}}{2} \bm{q} \cdot \dot{\bm{v}}(t)(t-t')^2 + \cdots \right]. \tag{3.2}$$

Substituting the expansion (3.2) into expressions (1.42), (1.43) and (1.50) for $\tilde{\bm{f}}_\mathrm{i}$, $\tilde{\bm{f}}_\mathrm{p}$ and $\tilde{w}$ and keeping the necessary lowest order terms of $\dot{\bm{v}}(t)$ in equations (1.40) and (1.55), we can write the force- and energy-balance equations for system of unit volume in the form

$$(nm\mathcal{I} + \mathcal{B}) \cdot \frac{d}{dt}\boldsymbol{v}(t) = ne\boldsymbol{E}(t) + \boldsymbol{f}_\mathrm{i} + \boldsymbol{f}_\mathrm{p}, \tag{3.3}$$

$$C_\mathrm{e}\frac{d}{dt}T_\mathrm{e}(t) = -\boldsymbol{v} \cdot (\boldsymbol{f}_\mathrm{i} + \boldsymbol{f}_\mathrm{p}) - w. \tag{3.4}$$

Here n is the electron density, C_e is the specific heat of the electron system, the frictional forces and energy-transfer rate, $\boldsymbol{f}_\mathrm{i}$, $\boldsymbol{f}_\mathrm{p}$ and w, have the same formal expressions as those in the dc steady-state transport in Sec. 2.1:

$$\boldsymbol{f}_\mathrm{i} = n_\mathrm{i} \sum_{\boldsymbol{q}} \boldsymbol{q} |u(\boldsymbol{q})|^2 \Pi_2(\boldsymbol{q}, \omega_0), \tag{3.5}$$

$$\boldsymbol{f}_\mathrm{p} = 2 \sum_{\boldsymbol{q},\lambda} \boldsymbol{q} |M(\boldsymbol{q},\lambda)|^2 \Pi_2(\boldsymbol{q}, \Omega_{\boldsymbol{q}\lambda} + \omega_0) \left[n\left(\frac{\Omega_{\boldsymbol{q}\lambda}}{T}\right) - n\left(\frac{\Omega_{\boldsymbol{q}\lambda} + \omega_0}{T_\mathrm{e}}\right) \right], \tag{3.6}$$

$$w = 2 \sum_{\boldsymbol{q},\lambda} \Omega_{\boldsymbol{q}\lambda} |M(\boldsymbol{q},\lambda)|^2 \Pi_2(\boldsymbol{q}, \Omega_{\boldsymbol{q}\lambda} + \omega_0) \left[n\left(\frac{\Omega_{\boldsymbol{q}\lambda}}{T}\right) - n\left(\frac{\Omega_{\boldsymbol{q}\lambda} + \omega_0}{T_\mathrm{e}}\right) \right], \tag{3.7}$$

with $\omega_0 \equiv \boldsymbol{q} \cdot \boldsymbol{v}(t)$ and $T_\mathrm{e} = T_\mathrm{e}(t)$. They are dependent on time through the time dependence of variables $\boldsymbol{v}(t)$ and $T_\mathrm{e}(t)$. In Eq. (3.3) $\mathcal{I}$ represents the unit tensor. The tensor $\mathcal{B}$, which comes from the contribution of the second term on the right-hand-side of (3.2) to $\tilde{\boldsymbol{f}}_\mathrm{i}$ and $\tilde{\boldsymbol{f}}_\mathrm{p}$ in Eqs. (1.42) and (1.43), gives rise to a drift-velocity dependent effective-mass correction on the center-of-mass motion induced by impurity and phonon scatterings,

$$\mathcal{B} = -n_\mathrm{i} \sum_{\boldsymbol{q}} \frac{\boldsymbol{qq}}{2} |u(\boldsymbol{q})|^2 \frac{\partial^2}{\partial \omega^2} \Pi_1(\boldsymbol{q},\omega)\Big|_{\omega=\omega_0}$$

$$- \sum_{\boldsymbol{q},\lambda} \frac{\boldsymbol{qq}}{2} |M(\boldsymbol{q},\lambda)|^2 \frac{\partial^2}{\partial \omega^2} \Lambda_1(\boldsymbol{q},\lambda,\omega)\Big|_{\omega=\omega_0}, \tag{3.8}$$

where $\Pi_1(\boldsymbol{q},\omega)$ and $\Lambda_1(\boldsymbol{q},\lambda,\omega)$ are respectively the real part of electron density correlation function $\Pi(\boldsymbol{q},\omega)$ and the real part of electron–phonon correlation function $\Lambda(\boldsymbol{q},\lambda,\omega)$. The $\Pi(\boldsymbol{q},\omega)$ function has been discussed in Sec. 2.1. The imaginary part $\Lambda_2(\boldsymbol{q},\lambda,\omega)$ and the real part $\Lambda_1(\boldsymbol{q},\lambda,\omega)$ of the electron–phonon correlation function can be expressed as

$$\Lambda_2(\boldsymbol{q},\lambda,\omega) = \Pi_2(\boldsymbol{q},\omega + \Omega_{\boldsymbol{q}\lambda}) \left[n\left(\frac{\Omega_{\boldsymbol{q}\lambda}}{T}\right) - n\left(\frac{\Omega_{\boldsymbol{q}\lambda} + \omega}{T_\mathrm{e}}\right) \right]$$

$$+ \Pi_2(\boldsymbol{q},\omega - \Omega_{\boldsymbol{q}\lambda}) \left[n\left(\frac{\Omega_{\boldsymbol{q}\lambda}}{T}\right) - n\left(\frac{\Omega_{\boldsymbol{q}\lambda} - \omega}{T_\mathrm{e}}\right) \right], \tag{3.9}$$

$$\Lambda_1(\boldsymbol{q},\lambda,\omega) = [\Pi_1(\boldsymbol{q},\omega+\Omega_{\boldsymbol{q}\lambda}) + \Pi_1(\boldsymbol{q},\omega-\Omega_{\boldsymbol{q}\lambda})]\, n\!\left(\frac{\Omega_{\boldsymbol{q}\lambda}}{T}\right)$$
$$-\frac{1}{\pi}\int_{-\infty}^{\infty} d\omega_1\, \Pi_2(\boldsymbol{q},\omega_1)\, n\!\left(\frac{\omega_1}{T_\mathrm{e}}\right)$$
$$\times \left(\frac{1}{\omega_1+\omega-\Omega_{\boldsymbol{q}\lambda}} + \frac{1}{\omega_1-\omega-\Omega_{\boldsymbol{q}\lambda}}\right), \qquad (3.10)$$

in which the electron density correlation functions $\Pi_2(\boldsymbol{q},\omega)$ and $\Pi_1(\boldsymbol{q},\omega)$ are calculated at electron temperature T_e. Note that the integral in (3.10) implicitly means taking the principal value.

For an isotropic system under the influence of a time-dependent electric field $\boldsymbol{E}(t)$ along a fixed direction (then $\dot{\boldsymbol{v}}$ and $\boldsymbol{v}$ are in the same direction), the effective-mass-correction tensor $\mathcal{B}$ is reduced to a scalar and dependent only on the magnitude of the drift velocity $v(t)$. Eqs. (3.3) and (3.4) retain memory effects to some extent by including $\boldsymbol{v}(t)$ in the effective-mass correction, yet reduce the balance equations (1.40) and (1.55) to a coupled set of nonlinear ordinary differential equations. They are easily integrated for a specified $\boldsymbol{E}(t)$ by starting from given initial values of $\boldsymbol{v}(t)$ and $T_\mathrm{e}(t)$.

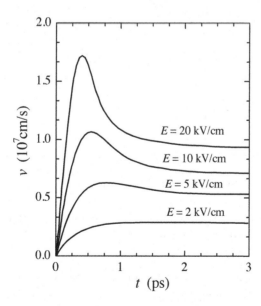

Fig. 3.1 Transient drift velocity $v(t)$ (solid curves) plotted against time when a time-step electric field is applied to a p-type Ge at $T = 300\,\mathrm{K}$, obtained from Eqs. (3.3) and (3.4) without $\mathcal{B}$ term. After Xing and Ting (1987).

Xing and Ting (1987) calculate the transient response of drift velocity $v(t)$ and electron temperature $T_e(t)$ of a p-type Ge at lattice temperature $T = 300\,\text{K}$ to suddenly-impressed constant electric fields of various strengths using balance equations (3.3) and (3.4) without effective-mass correction term. The results are shown in Fig. 3.1. Pronounced overshoot in velocities appear near $t = 0.5\,\text{ps}$ after turn-on of the electric field in the curves of $E = 20$ and $10\,\text{kV/cm}$. The transport is assumed primarily due to heavy holes with effective mass $m = 0.34\,m_e$ (m_e being the free electron mass). Scatterings from longitudinal acoustic phonons and from nonpolar optic phonons are considered with $\Xi = 4.6\,\text{eV}$ and $D = 9 \times 10^8\,\text{eV/cm}$ for acoustic and optic deformation potential parameters, $v_s = 5.4 \times 10^5\,\text{cm/s}$ as the longitudinal sound velocity, and $\Omega_o = 430\,\text{K}$ as the optic phonon frequency.

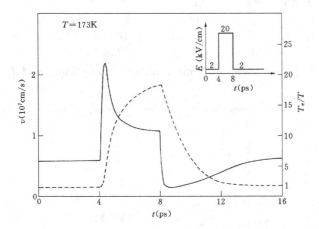

Fig. 3.2 Transient drift velocity $v(t)$ (solid curve) and temperature ratio $T_e(t)/T$ (dashed curve) versus time t when an impulse and a dc electric field are applied simultaneously to a p-type Ge at $T = 173\,\text{K}$. From Xing and Ting (1987).

Fig. 3.2 shows the drift velocity and electron temperature response to an impulse electric field in the same p-type Ge system at $T = 173\,\text{K}$, calculated by Xing and Ting (1987).

The time-dependent balance equations (3.3) and (3.4) are written for 3D systems. Apparently, under the condition of slow time-varying drift velocity, we have equations for quasi-2D or quasi-1D system similar to Eqs. (3.3) and (3.4) except that n should be replaced by the sheet density N_s (2D) or line density N_l (1D), and $\boldsymbol{f}_i$, $\boldsymbol{f}_p$, w and $\mathcal{B}$ should be expressed by the corresponding correlation functions appropriate for low-dimensional systems. They have been discussed in Chapter 2.

Lei, Cui and Horing (1987) analyze the hot carrier transient transport in GaAs-based heterojunctions and superlattices using the corresponding version of time-dependent force- and energy-balance equations (3.3) and (3.4) without $\mathcal{B}$ term. They consider scatterings from impurities (remote and background) and from polar optic phonons (with finite relaxation time τ_p, see Chapter 4) and include full RPA dynamic screening in the calculation. As an example, Fig. 3.3 shows their calculated transient response of drift velocity v and electron temperature T_e to a step electric field of $1.0\,\mathrm{kV/cm}$ turned on at time $t = 0$ in a quantum well superlattice of well width $a = 15\,\mathrm{nm}$, superlattice period $d = 15\,\mathrm{nm}$, carrier sheet density $N_\mathrm{s} = 2.0 \times 10^{11}\,\mathrm{cm}^{-2}$ and zero-temperature mobility $\mu_0 = 3.5 \times 10^5\,\mathrm{cm}^2/\mathrm{Vs}$, at lattice temperature $T = 77\,\mathrm{K}$.

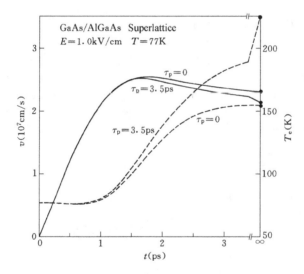

Fig. 3.3 The drift velocity v (full curves) and electron temperature T_e (broken curves) are shown as functions of time t after an electric field of $E = 1.0\,\mathrm{kV/cm}$ is suddenly turned on at $t = 0$ in a quantum-well superlattice at lattice temperature $T = 77\,\mathrm{K}$ for the phonon relaxation time $\tau_\mathrm{p} = 0$ and $\tau_\mathrm{p} = 3.5\,\mathrm{ps}$. From Lei, Cui and Horing (1987).

Transient transport in GaAs/AlGaAs heterostructures has also been studied by Hasbun (1995) at $T = 77\,\mathrm{K}$ from the quasi-2D one-subband version of balance equations (3.3) and (3.4) with impurity and polar optical scatterings without $\mathcal{B}$ term. He presents numerical pictures of the frictional force f and energy-loss rate w as functions of v and T_e, and compares transient $v(t)$ and $T_e(t)$ results obtained with and without screening, and those obtained using the full nonlinear equations (3.3) and (3.4) and using a simplified (for moderate v) version of them (Hasbun and Nee, 1991).

3.2 Memory Effects in Transient Response for the Weak Current Case

Ting and Nee (1986) and Xing and Ting (1987) carry out detailed examinations of memory effects involved in the transient response to a weak electric field in three-dimensional semiconductors. They make use of the fact that the CM velocity v is small under the influence of a weak electric field, such that one can linearize both force- and energy-balance equations by expanding the function $A(\boldsymbol{q}, t, t')$ to the linear order in $\boldsymbol{v}$:

$$A(\boldsymbol{q}, t, t') = \exp\left[\mathrm{i}\boldsymbol{q} \cdot \int_{t'}^{t} \boldsymbol{v}(s)\, ds\right] \approx 1 + \mathrm{i}\boldsymbol{q} \cdot \int_{t'}^{t} \boldsymbol{v}(s)\, ds. \tag{3.11}$$

Substituting this into Eqs. (1.42) and (1.43), one readily finds that the contributions to the frictional forces from the first term of the right hand side of Eq. (3.11) vanish. By exchanging the order of time integrations, the force-balance equation (1.40) (along the field direction) may be written as (Ting and Nee, 1986)

$$\frac{dv(t)}{dt} = \frac{e}{m} E(t) + \int_{-\infty}^{t} K(t-s)\, v(s)\, ds, \tag{3.12}$$

with the kernel given, at $t - s \geqslant 0$, by

$$K(t-s) = \frac{1}{nm} \sum_{\boldsymbol{q}} q_x^2 \int_{-\infty}^{s} dt' \Big[n_{\mathrm{i}}\, |u(\boldsymbol{q})|^2\, \Pi(\boldsymbol{q}, t - t') \\ + \sum_{\lambda} |M(\boldsymbol{q}, \lambda)|^2\, \Lambda(\boldsymbol{q}, \lambda, t - t') \Big]. \tag{3.13}$$

Since equation (3.12) depends only on the value of kernel $K(t)$ at $t \geqslant 0$, it is irrelevant to assign whatever value to $K(t)$ at $t < 0$. For instance, we can consider that $K(t)$ at $t < 0$ is also given by (3.13), or we can set $K(t) = 0$ at $t < 0$. In the following we will choose the latter assignment and the upper limit of the integral in (3.13) is taken to be ∞.

Note that in addition to the explicit t-dependence shown in Eq. (3.13), $K(t - s)$ depends implicitly on t through the t-dependence of the electron temperature $T_e(t)$ in the initial density matrix $\hat{\rho}_0$. However, this dependence of $T_e(t)$ is of higher order in $\boldsymbol{v}$, such that to the lowest order of $\boldsymbol{v}$ we can set $T_e = T$ and neglect this implicit t-dependence of $K(t-s)$.

Assuming that the applied electric field is suddenly turned on at $t = 0$: $E(t) = E\,\theta(t)$, and the CM velocity $v(t) = 0$ at $t \leqslant 0$, such that the lower limit of the integral in (3.12) is zero, we can write Eq. (3.12) as

$$\frac{dv(t)}{dt} = \frac{e}{m} E\,\theta(t) + \int_0^\infty K(t-s)v(s)ds. \tag{3.14}$$

Taking Laplace transformation of Eq. (3.12) and noticing $v(t) = 0$ at $t = 0$, we have

$$v(\varepsilon) = \frac{eE}{m\varepsilon\left[\varepsilon - K(\varepsilon)\right]}, \tag{3.15}$$

where $v(\varepsilon)$ and $K(\varepsilon)$ are the Laplace transforms of $v(t)$ and $K(t)$. The time-dependent drift velocity $v(t)$ may then be obtained by the inverse Laplace transformation of Eq. (3.15):

$$v(t) = \frac{eE}{m}\frac{1}{2\pi\mathrm{i}} \int_{c-\mathrm{i}\infty}^{c+\mathrm{i}\infty} \frac{e^{\varepsilon t}d\varepsilon}{\varepsilon\left[\varepsilon - K(\varepsilon)\right]}. \tag{3.16}$$

The Laplace transform of kernel $K(t)$ is easily calculated from the definition

$$K(\varepsilon) = \int_0^\infty e^{-\varepsilon t} K(t) dt, \tag{3.17}$$

yielding

$$K(\varepsilon) = \frac{n_\mathrm{i}}{nm\varepsilon} \sum_{\boldsymbol{q}} q_x^2 |u(\boldsymbol{q})|^2 \left[\Pi(\boldsymbol{q},0) - \Pi(\boldsymbol{q},\mathrm{i}\varepsilon)\right]$$

$$+ \frac{1}{nm\varepsilon} \sum_{\boldsymbol{q},\lambda} q_x^2 |M(\boldsymbol{q},\lambda)|^2 \left[\Lambda(\boldsymbol{q},\lambda,0) - \Lambda(\boldsymbol{q},\lambda,\mathrm{i}\varepsilon)\right]. \tag{3.18}$$

Here $\Pi(\boldsymbol{q},\omega)$ and $\Lambda(\boldsymbol{q},\lambda,\omega)$ are, respectively, the Fourier spectrum of the electron density correlation function $\Pi(\boldsymbol{q},t)$ and that of the electron–phonon correlation function $\Lambda(\boldsymbol{q},\lambda,t)$ [see (3.9) and (3.10)]. A memory function defined by $M(\omega) \equiv -\mathrm{i}K(-\mathrm{i}\omega)$ is conventionally used in the literature (Götze and Wölfle, 1972; Ting, Ying, and Quinn, 1976b). It consists of contributions from impurity and phonon scatterings:

$$M(\omega) = M^{(\mathrm{i})}(\omega) + M^{(\mathrm{p})}(\omega), \tag{3.19}$$

$$M^{(i)}(\omega) = \frac{n_i}{nm\omega} \sum_{\bm{q}} |u(\bm{q})|^2 q_x^2 [\Pi(\bm{q},0) - \Pi(\bm{q},\omega)], \qquad (3.20)$$

$$M^{(p)}(\omega) = \frac{1}{nm\omega} \sum_{\bm{q},\lambda} |M(\bm{q},\lambda)|^2 q_x^2 [\Lambda(\bm{q},\lambda,0) - \Lambda(\bm{q},\lambda,\omega)]. \qquad (3.21)$$

The drift velocity $v(t)$ expression (3.16) can be rewritten in terms of the memory function as

$$v(t) = -\frac{eE}{2\pi m} \int_{-\infty+ic}^{\infty+ic} \frac{e^{-i\omega t} d\omega}{\omega [\omega + M(\omega)]}, \qquad (3.22)$$

in which the small positive real number c ensures the integral path going away from the pole $\omega = 0$. This integral can be carried out by means of an integration of contour consisting of the real axis and a large curve in the lower half plane. It can also be calculated directly along the real axis. Note that for ω on the real axis, one can define the real part $M_1(\omega)$ and the imaginary part $M_2(\omega)$ of the memory function by

$$M(\omega + i\delta) = M_1(\omega) + iM_2(\omega). \qquad (3.23)$$

$M_1(\omega)$ is an odd function and $M_2(\omega)$ is an even function: $M_1(\omega) = -M_1(-\omega)$ and $M_2(\omega) = M_2(-\omega)$. In terms of $M_1(\omega)$ and $M_2(\omega)$, (3.22) can be written as

$$v(t) = \frac{eE}{m} \left[\frac{1}{2M_2(0)} - \frac{1}{\pi} \int_0^\infty \frac{d\omega}{\omega} \frac{[\omega + M_1(\omega)]\cos\omega t - M_2(\omega)\sin\omega t}{[\omega + M_1(\omega)]^2 + [M_2(\omega)]^2} \right]. \qquad (3.24)$$

The first term on the right-hand-side of Eq. (3.24) comes from the integration of the small semi-circle around the pole $\omega = 0$. If one uses the low-frequency values of the memory function,

$$M_1(0) = 0, \qquad M_2(0) = \frac{1}{\tau_i} + \frac{1}{\tau_p} \equiv \frac{1}{\tau} \qquad (3.25)$$

to replace $M_1(\omega)$ and $M_2(\omega)$ in the integrand of (3.24), the classical Drude formula, as the result of simple relaxation time approximation,

$$v(t) = \frac{eE}{m}\tau \left(1 - e^{-t/\tau}\right) \qquad (3.26)$$

emerges.

For the calculation of transient velocity response $v(t)$ from Eq. (3.24) we need first to acquire the imaginary part and the real part of the memory function. One way to obtain $M_2(\omega)$ and $M_1(\omega)$ is the direct calculation from Eqs. (3.20) and (3.21). Generally, calculation for the imaginary part of a correlation function is much easier than that for its real part. Therefore, one can first obtain the imaginary part of the memory function, $M_2(\omega)$, as a function of ω in the real axis from Eqs. (3.20) and (3.21), and than calculate the real part $M_1(\omega)$ according to the Kramers–Kronig relation

$$M_1(\omega) = \frac{2\omega}{\pi} \int_0^\infty \frac{M_2(\omega_1)}{\omega_1^2 - \omega^2} d\omega_1, \qquad (3.27)$$

which involves a principal-value integral at $\omega_1 = \omega$.

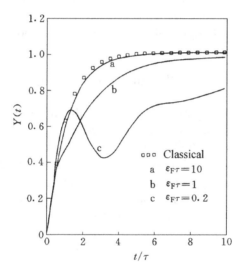

Fig. 3.4 Normalized drift velocity response of a 2D degenerate system to a suddenly turn-on electric field. The Solid curves a, b, and c are calculated results from (3.24) respectively in the cases of $\varepsilon_F\tau = 10, 1$ and 0.2. The small squares represent the Drude formula (3.26). From Ting and Nee (1986).

Ting and Nee (1986) calculate the transient drift velocity response to a suddenly turn-on electric field E from formula (3.24) in two different cases and compare the results with classical Drude formula (3.26). The first case is a degenerate 2D system at low temperature with only impurity scattering. Fig. 3.4 illustrates the normalized velocity response $Y(t) \equiv (m/eE\tau)v(t)$ versus the dimensionless time t/τ. The solid curves a, b and c represent $Y(t)$ determined by (3.24) for systems with $\varepsilon_F\tau = 10, 1$ and 0.2 respectively. The small squares are those from Drude formula (3.26). Though the results from (3.24) almost agree with Drude formula when $\varepsilon_F\tau \gg 1$, significant difference appears when $\varepsilon_F \tau \ll 1$: the transient velocity response of (3.24) exhibits oscillatory behavior. This comes from the fact that, in this system $M_2(\omega)$ can have a maximum and $M_1(\omega)$ can take negative value that $\omega + M_1(\omega)$ becomes zero. The condition $\varepsilon_F \tau \ll 1$ indicates that the de Broglie wavelength of electrons around Fermi level, $\lambda_F = 2\pi/k_F$, is much larger than the electron mean free path $l = k_F \tau / m$ and interference may appear. The second case is a high-temperature nondegenerate (Maxwell–Boltzmann) 3D system with nonpolar optic phonons as the dominant scattering mechanism. They find that, in this system for almost all the cases, even $T\tau < 1$ [the electron mean free path less than the thermal wavelength $1/(3mT)^{1/2}$], $M_2(\omega)$ is monotonous and $M_1(\omega)$ is always positive. The drift velocity response given by (3.24) shows little difference from Drude formula.

3.3 High-Frequency Small-Signal Conductivity

The time-dependent force-balance equation (1.40) and energy-balance equation (1.55) or (1.60) are also the starting point for the investigation of high-frequency transport. We first discuss small-signal high-frequency steady-state conduction (Lei and Zhang, 1986; Lei, Horing and Zhang, 1986a, 1987). Assume that a uniform small-amplitude high-frequency electric field $2\boldsymbol{E}_1 \cos\omega t$ of frequency ω and amplitude $2\boldsymbol{E}_1$, and a uniform constant dc bias electric field $\boldsymbol{E}_0$,

$$\boldsymbol{E}(t) = \boldsymbol{E}_0 + 2\boldsymbol{E}_1 \cos\omega t = \boldsymbol{E}_0 + \boldsymbol{E}_1 \mathrm{e}^{-\mathrm{i}\omega t} + \boldsymbol{E}_1 \mathrm{e}^{\mathrm{i}\omega t}, \qquad (3.28)$$

are applied to the system. Since effects of the associated high-frequency magnetic field are much weaker in comparison with those of the high-frequency electric field for frequency up to terahertz regime, the role of the associated ac magnetic field is generally neglected unless for some special cases. We describe the transport behavior of the system using the time-dependent CM velocity $\boldsymbol{v}(t)$ and electron temperature $T_e(t)$, satisfying the force-balance equation (1.40) and the energy-balance equation (1.55) or (1.60) with a periodical time-dependent electric field $\boldsymbol{E} = \boldsymbol{E}(t)$ given by (3.28). It is obvious that after a transient process the system will reach an oscillatory (ac) steady state in which the center of mass moves at a constant drift velocity $\boldsymbol{v}_\mathrm{d}$ and oscillates with a small amplitude at the single driving frequency ω, if the high-frequency signal field is sufficiently small:

$$\boldsymbol{v}(t) = \boldsymbol{v}_\mathrm{d} + \boldsymbol{v}_1 \mathrm{e}^{-\mathrm{i}\omega t} + \boldsymbol{v}_1^* \mathrm{e}^{\mathrm{i}\omega t}, \qquad (3.29)$$

in which $\boldsymbol{v}_1$ is a small quantity. Substituting (3.29) into Eq. (3.1) for $A(\boldsymbol{q},t,t')$ function and expanding it to the linear order in $\boldsymbol{v}_1$, we obtain

$$\begin{aligned}A(\boldsymbol{q},t,t') &= \exp\left[\mathrm{i}\boldsymbol{q}\cdot\int_{t'}^{t}\boldsymbol{v}(s)ds\right] \\ &= \exp\left\{\mathrm{i}\boldsymbol{q}\cdot\boldsymbol{v}_\mathrm{d}(t-t') - \left[\boldsymbol{q}\cdot\boldsymbol{v}_1\mathrm{e}^{-\mathrm{i}\omega t}(1-\mathrm{e}^{\mathrm{i}\omega(t-t')})/\omega - \mathrm{c.c.}\right]\right\} \\ &\approx \exp\left[\mathrm{i}\boldsymbol{q}\cdot\boldsymbol{v}_\mathrm{d}(t-t')\right] \\ &\quad \times \left\{1 - \left[\boldsymbol{q}\cdot\boldsymbol{v}_1\mathrm{e}^{-\mathrm{i}\omega t}(1-\mathrm{e}^{\mathrm{i}\omega(t-t')})/\omega - \mathrm{c.c.}\right]\right\}.\end{aligned} \quad (3.30)$$

In the ac steady state the electron temperature $T_\mathrm{e}(t)$ will also perform a small amplitude harmonic oscillation of frequency ω about a constant temperature value T_e,

$$T_\mathrm{e}(t) = T_\mathrm{e} + T_1 \mathrm{e}^{-\mathrm{i}\omega t} + T_1^* \mathrm{e}^{\mathrm{i}\omega t}. \quad (3.31)$$

Taken jointly Eq. (3.30) for $A(\boldsymbol{q},t,t')$ and Eq. (3.31) for $T_\mathrm{e}(t)$, provide the frictional force $\tilde{\boldsymbol{f}} = \tilde{\boldsymbol{f}}_\mathrm{i} + \tilde{\boldsymbol{f}}_\mathrm{p}$ [Eqs. (1.42) and (1.43)], and the energy transfer rate $\tilde{w}$ [Eq. (1.50)], in the following forms:

$$\tilde{\boldsymbol{f}} = \boldsymbol{f}_0 + \boldsymbol{f}_0^{(1)}(T_1\mathrm{e}^{-\mathrm{i}\omega t} + T_1^*\mathrm{e}^{\mathrm{i}\omega t}) + (\boldsymbol{f}_1\mathrm{e}^{-\mathrm{i}\omega t} + \boldsymbol{f}_1^*\mathrm{e}^{\mathrm{i}\omega t}), \quad (3.32)$$

$$\tilde{w} = w_0 + w_0^{(1)}(T_1\mathrm{e}^{-\mathrm{i}\omega t} + T_1^*\mathrm{e}^{\mathrm{i}\omega t}) + (w_1\mathrm{e}^{-\mathrm{i}\omega t} + w_1^*\mathrm{e}^{\mathrm{i}\omega t}). \quad (3.33)$$

Here $\boldsymbol{f}_0 \equiv \boldsymbol{f}_0(\boldsymbol{v}_\mathrm{d},T_\mathrm{e})$ and $w_0 \equiv w_0(\boldsymbol{v}_\mathrm{d},T_\mathrm{e})$ are the frictional force and the energy-loss rate of the system in the dc steady-state transport having drift velocity $\boldsymbol{v}_\mathrm{d}$ and electron temperature T_e; $\boldsymbol{f}_0^{(1)} \equiv \partial\boldsymbol{f}_0(\boldsymbol{v}_\mathrm{d},T_\mathrm{e})/\partial T_\mathrm{e}$, $w_0^{(1)} \equiv \partial w_0(\boldsymbol{v}_\mathrm{d},T_\mathrm{e})/\partial T_\mathrm{e}$;

$$\begin{aligned}\boldsymbol{f}_1 &\equiv \boldsymbol{f}_1(\omega,\boldsymbol{v}_\mathrm{d},T_\mathrm{e}) \\ &= \mathrm{i}n_\mathrm{i}\sum_{\boldsymbol{q}}(\boldsymbol{v}_1\cdot\boldsymbol{q})\boldsymbol{q}|u(\boldsymbol{q})|^2\left[\Pi(\boldsymbol{q},\omega_0) - \Pi(\boldsymbol{q},\omega_0+\omega)\right]/\omega \\ &\quad + \mathrm{i}\sum_{\boldsymbol{q},\lambda}(\boldsymbol{v}_1\cdot\boldsymbol{q})\boldsymbol{q}|M(\boldsymbol{q},\lambda)|^2\left[\Lambda(\boldsymbol{q},\lambda,\omega_0) - \Lambda(\boldsymbol{q},\lambda,\omega_0+\omega)\right]/\omega,\end{aligned} \quad (3.34)$$

$$\begin{aligned}w_1 &\equiv w_1(\omega,\boldsymbol{v}_\mathrm{d},T_\mathrm{e}) \\ &= \mathrm{i}\sum_{\boldsymbol{q},\lambda}(\boldsymbol{v}_1\cdot\boldsymbol{q})\Omega_{\boldsymbol{q}\lambda}|M(\boldsymbol{q},\lambda)|^2\left[\Gamma(\boldsymbol{q},\lambda,\omega_0) - \Gamma(\boldsymbol{q},\lambda,\omega_0+\omega)\right]/\omega\end{aligned} \quad (3.35)$$

are frequency-dependent first-order small quantities relevant to memory effects. In expressions (3.34) and (3.35), $\omega_0 \equiv \boldsymbol{q}\cdot\boldsymbol{v}_\mathrm{d}$, $\Pi(\boldsymbol{q},\omega)$ is the electron density correlation function discussed in Chapter 2; $\Lambda(\boldsymbol{q},\lambda,\omega)$ is the electron–phonon correlation function given by (3.9) and (3.10); $\Gamma(\boldsymbol{q},\lambda,\omega)$ is another electron–phonon correlation function introduced in (1.50) related to the electron energy transferred to phonons, with the imaginary part $\Gamma_2(\boldsymbol{q}, \lambda, \omega)$ and real part $\Gamma_1(\boldsymbol{q}, \lambda, \omega)$ given by

$$\Gamma_2(\boldsymbol{q}, \lambda, \omega) = \Pi_2(\boldsymbol{q}, \omega + \Omega_{q\lambda}) \left[n\left(\frac{\Omega_{q\lambda}}{T}\right) - n\left(\frac{\Omega_{q\lambda} + \omega}{T_e}\right) \right]$$
$$- \Pi_2(\boldsymbol{q}, \omega - \Omega_{q\lambda}) \left[n\left(\frac{\Omega_{q\lambda}}{T}\right) - n\left(\frac{\Omega_{q\lambda} - \omega}{T_e}\right) \right], \quad (3.36)$$

$$\Gamma_1(\boldsymbol{q}, \lambda, \omega) = [\Pi_1(\boldsymbol{q}, \omega + \Omega_{q\lambda}) - \Pi_1(\boldsymbol{q}, \omega - \Omega_{q\lambda})]\, n\left(\frac{\Omega_{q\lambda}}{T}\right)$$
$$+ \frac{1}{\pi} \int_{-\infty}^{\infty} d\omega_1\, \Pi_2(\boldsymbol{q}, \omega_1)\, n\left(\frac{\omega_1}{T_e}\right)$$
$$\times \left(\frac{1}{\omega_1 + \omega - \Omega_{q\lambda}} - \frac{1}{\omega_1 - \omega - \Omega_{q\lambda}} \right). \quad (3.37)$$

The force- and energy-balance equations should be held for the zero-th and first orders. The zero-th order equations are just those of the dc steady state:

$$ne\boldsymbol{E}_0 + \boldsymbol{f}_0 = 0, \quad (3.38)$$
$$\boldsymbol{v}_{\rm d} \cdot \boldsymbol{f}_0 + w_0 = 0. \quad (3.39)$$

They determine the drift velocity $\boldsymbol{v}_{\rm d}$ and electron temperature T_e under the dc bias electric field $\boldsymbol{E}_0$. The equations for the linear order are

$$-{\rm i}\,\omega n m \boldsymbol{v}_1 = ne\boldsymbol{E}_1 + \boldsymbol{f}_0^{(1)} T_1 + \boldsymbol{f}_1, \quad (3.40)$$

$${\rm i}\omega C_e T_1 = \boldsymbol{v}_1 \cdot \boldsymbol{f}_0 + \boldsymbol{v}_{\rm d} \cdot \boldsymbol{f}_0^{(1)} T_1 + \boldsymbol{v}_{\rm d} \cdot \boldsymbol{f}_1 + w_0^{(1)} T_1 + w_1. \quad (3.41)$$

Here C_e is the specific heat of the electron system at temperature T_e. Given the ac signal electric field $\boldsymbol{E}_1$, the ac quantities $\boldsymbol{v}_1$ and T_1 can be determined from these two equations.

3.3.1 The case of zero dc bias

In the absence of a dc bias $E_0 = 0$, the zero-order balance equations (3.38) and (3.39) yield $\boldsymbol{f}_0 = 0$, $w_0 = 0$, $\boldsymbol{v}_{\rm d} = 0$, and $T_e = T$. Furthermore it can be easily seen from Eq. (3.35) that $w_1 = 0$ when $\omega_0 = 0$ because the system is isotropic and thus Eq. (3.41) gives $T_1 = 0$. Therefore, when only a small-amplitude signal is applied, the electron temperature remains constant T

to linear order in the signal. Any change of the electron temperature must be of higher order. This reduces the linear order force-balance equation (3.40) to

$$-\mathrm{i}\omega nm\boldsymbol{v}_1 = ne\boldsymbol{E}_1 + \boldsymbol{f}_1. \tag{3.42}$$

Moreover, since $\omega_0 = 0$, $\boldsymbol{f}_1$ is parallel to $\boldsymbol{v}_1$ and can be written as

$$\boldsymbol{f}_1 = \mathrm{i}nm\boldsymbol{v}_1 M(\omega), \tag{3.43}$$

where $M(\omega)$ is the memory function defined by Eqs. (3.19)–(3.21). The dynamic or high-frequency small-signal resistivity $\rho(\omega)$ and conductivity $\sigma(\omega)$ are defined as

$$\rho(\omega) \equiv \frac{1}{\sigma(\omega)} \equiv \frac{E_1}{nev_1}, \tag{3.44}$$

or the high-frequency small-signal mobility $\mu(\omega)$ as

$$\mu(\omega) \equiv \frac{v_1}{E_1}. \tag{3.45}$$

Eqs. (3.42) and (3.43) yield

$$\rho(\omega) = -\mathrm{i}\frac{m}{ne^2}\left[\omega + M(\omega)\right], \tag{3.46}$$

$$\sigma(\omega) = \mathrm{i}\frac{ne^2}{m}\frac{1}{\omega + M(\omega)}, \tag{3.47}$$

or

$$\mu(\omega) = \mathrm{i}\frac{e}{m}\frac{1}{\omega + M(\omega)}. \tag{3.48}$$

$\rho(\omega)$, $\sigma(\omega)$ and $\mu(\omega)$ are generally complex numbers, with their arguments to reflect the phase difference between the ac current and ac electric field. The expression (3.47) for dynamic conductivity was first obtained by Götze and Wölfle (1976) in their memory function approach.

Writing $M(\omega)$ as the sum of its real and imaginary parts, $M(\omega) = M_1(\omega) + \mathrm{i}M_2(\omega)$, we can easily see that in the zero frequency ($\omega \to 0$) limit

$$\rho(0) = \frac{1}{\sigma(0)} = \frac{m}{ne^2}M_2(0). \tag{3.49}$$

Therefore $[M_2(0)]^{-1}$ is the direct measurement of the low-frequency effective relaxation time τ_0 or the low-frequency small-signal mobility μ_0:

$M_2(0) = 1/\tau_0 = e/(m\mu_0)$. Formally, the expression (3.47) for high-frequency small-signal complex conductivity can also be written in a Drude-type form:

$$\sigma(\omega) = i\frac{ne^2}{m^*}\frac{1}{\omega + i/\tau}, \qquad (3.50)$$

The parameters $m^* = m[1 + M_1(\omega)/\omega]$ and $1/\tau = M_2(\omega)/[1 + M_1(\omega)/\omega]$ in it, however, are not constant but depend on ω. The relative change of the effective mass

$$\frac{\Delta m}{m} \equiv \frac{m^* - m}{m} = \frac{M_1(\omega)}{\omega} \qquad (3.51)$$

is generally much less than 1, thus the relaxation time is essentially determined by the imaginary part of the memory function:

$$\tau^{-1} \approx M_2(\omega). \qquad (3.52)$$

3.3.2 The case of finite dc bias

In the presence of a finite dc bias the situation becomes somewhat more complicated. First one has to distinguish different configurations with regard to the relative direction of $\boldsymbol{E}_0$ and $\boldsymbol{E}_1$. Let $\boldsymbol{E}_0$ be along the x direction and $\boldsymbol{E}_1$ in the x–y plane: $\boldsymbol{E}_0 = E_0\hat{\boldsymbol{x}}$, $\boldsymbol{E}_1 = E_{1x}\hat{\boldsymbol{x}} + E_{1y}\hat{\boldsymbol{y}}$. For an isotropic system $\boldsymbol{v}_\text{d}$ and $\boldsymbol{f}_0$ are along the x direction, and $\boldsymbol{v}_1$ and $\boldsymbol{f}_1$ are in the x–y plane. We can write $\boldsymbol{v}_1 = v_{1x}\hat{\boldsymbol{x}} + v_{1y}\hat{\boldsymbol{y}}$, and

$$\boldsymbol{f}_1 = \mathrm{i}nmv_{1x}M_\parallel(\omega, v_\text{d})\hat{\boldsymbol{x}} + \mathrm{i}nmv_{1y}M_\perp(\omega, v_\text{d})\hat{\boldsymbol{y}}, \qquad (3.53)$$

where $M_{\perp(\parallel)}(\omega, v_\text{d})$ are the momentum-related perpendicular (parallel) memory functions in the presence of a dc bias current $\boldsymbol{v}_\text{d} = v_\text{d}\hat{\boldsymbol{x}}$. They consist of contributions $M^{(\text{i})}_{\perp(\parallel)}(\omega, v_\text{d})$ from impurities and $M^{(\text{p})}_{\perp(\parallel)}(\omega, v_\text{d})$ from phonons, having expressions

$$M^{(\text{i})}_{\perp(\parallel)}(\omega, v_\text{d}) = \frac{n_\text{i}}{nm\omega}\sum_{\boldsymbol{q}}|u(\boldsymbol{q})|^2 q^2_{y(x)}\left[\Pi(\boldsymbol{q}, q_x v_\text{d}) - \Pi(\boldsymbol{q}, q_x v_\text{d} + \omega)\right], \qquad (3.54)$$

$$M^{(\text{p})}_{\perp(\parallel)}(\omega, v_\text{d}) = \frac{1}{nm\omega}\sum_{\boldsymbol{q},\lambda}|M(\boldsymbol{q},\lambda)|^2 q^2_{y(x)}\left[\Lambda(\boldsymbol{q},\lambda, q_x v_\text{d}) - \Lambda(\boldsymbol{q},\lambda, q_x v_\text{d} + \omega)\right]. \qquad (3.55)$$

Here $\Pi(\boldsymbol{q},\omega)$ and $\Lambda(\boldsymbol{q},\lambda,\omega)$ are respectively the electron density correlation function and electron–phonon correlation function. The momentum-related memory function will be simply called the memory function. On the other hand, in the presence of a nonzero dc bias the quantity w_1 in Eq. (3.41) is finite, which is proportional to v_{1x} and can be expresses as

$$w_1 = \mathrm{i} n m v_\mathrm{d} v_{1x} N(\omega, v_\mathrm{d}), \tag{3.56}$$

where

$$N(\omega, v_\mathrm{d}) = \frac{1}{n m \omega v_\mathrm{d}} \sum_{\boldsymbol{q},\lambda} q_x \Omega_{\boldsymbol{q}\lambda} |M(\boldsymbol{q}, \lambda)|^2 \left[\Gamma(\boldsymbol{q}, \lambda, q_x v_\mathrm{d}) - \Gamma(\boldsymbol{q}, \lambda, q_x v_\mathrm{d} + \omega) \right] \tag{3.57}$$

is an energy-related memory function (Zou and Lei, 1995a, 1996). The memory functions $M_{\perp(\|)}^{(\mathrm{i})}(\omega, \boldsymbol{v}_\mathrm{d})$, $M_{\perp(\|)}^{(\mathrm{p})}(\omega, \boldsymbol{v}_\mathrm{d})$ and $N(\omega, \boldsymbol{v}_\mathrm{d})$ are Fourier spectra of retarded Green's functions, and the Kramers–Kronig relation (3.27) between the real and imaginary parts is valid for each of them.

From (3.53), the y-component of Eq. (3.40) determines v_{1y} proportional to E_{1y}, giving the small signal resistivity or conductivity in the perpendicular configuration as

$$\rho_\perp(\omega, v_\mathrm{d}) \equiv 1/\sigma_\perp(\omega, v_\mathrm{d}) \equiv \frac{E_{1y}}{n e v_{1y}} = -\mathrm{i}\frac{m}{n e^2}[\omega + M_\perp(\omega, v_\mathrm{d})]. \tag{3.58}$$

On the other hand, the x-components of Eq. (3.40) and Eq. (3.41) determine the relation between v_{1x} and E_{1x}, yielding the small signal resistivity or conductivity in the parallel configuration as

$$\rho_\|(\omega, v_\mathrm{d}) \equiv 1/\sigma_\|(\omega, v_\mathrm{d}) \equiv \frac{E_{1x}}{n e v_{1x}} = -\mathrm{i}\frac{m}{n e^2}[\omega + M_\|(\omega, v_\mathrm{d}) + D(\omega, v_\mathrm{d})], \tag{3.59}$$

in which

$$D(\omega, v_\mathrm{d}) = -\frac{v_\mathrm{d} f_0^{(1)} [M_\|(\omega, v_\mathrm{d}) + N(\omega, v_\mathrm{d}) + \mathrm{i}\tau_0^{-1}]}{v_\mathrm{d} f_0^{(1)} + w_0^{(1)} - \mathrm{i}\omega C_\mathrm{e}}, \tag{3.60}$$

with $\tau_0^{-1} = -f_0/(n m v_\mathrm{d})$ being the inverse scattering time related to nonlinear dc resistivity. In the zero dc bias $\boldsymbol{E}_0 = 0$, the zero-order equation requires $v_\mathrm{d} = 0$ and $T_\mathrm{e} = T$, yielding $D(\omega, 0) = 0$ and $M_\|(\omega, 0) = M_\perp(\omega, 0) = M(\omega)$. The small-signal resistivity in the absence of dc bias is direction independent. In the presence of a finite dc bias, $\rho_\|(\omega, v_\mathrm{d})$ differs from $\rho_\perp(\omega, v_\mathrm{d})$ not only due to the anisotropy induced by the bias [$M_\|(\omega, v_\mathrm{d})$ differs from $M_\perp(\omega, v_\mathrm{d})$], but also due to the additional term $D(\omega, v_\mathrm{d})$ appeared in the parallel configuration. This is because that a small ac current parallel to the dc bias field induces an energy change (thus an electron temperature change) in the linear order, while a small perpendicular ac current can induce the energy (temperature) change only in higher order. The $D(\omega, v_\text{d})$ function, which reflects the effect of the electron temperature oscillation, has the following features. In the zero-frequency limit ($\omega \to 0$), its real part $D_1(\omega, v_\text{d})$ approaches zero and its imaginary part $D_2(\omega, v_\text{d})$ approaches a finite value. Under small dc biases (small v_d), both $D_1(\omega, v_\text{d})$ and $D_2(\omega, v_\text{d})$ are negligibly small in the whole frequency range, and $\rho_\parallel$ and $\rho_\perp$ share essentially the same expression (the difference between $M_\parallel$ and $M_\perp$ also becomes very small at small dc biases). Under strong dc biases, e.g. $v_\text{d} = 0.5\, v_\text{F}$, the $D(\omega, v_\text{d})$-related contribution is important at low frequencies. The small-signal resistivity for the perpendicular (parallel) configuration in the low-frequency limit is

$$\rho_{\perp(\parallel)}(0, v_\text{d}) = \frac{m}{ne^2}\tau_{\perp(\parallel)}, \tag{3.61}$$

in which

$$\tau_\perp = M_{\perp 2}(0, v_\text{d}), \tag{3.62}$$
$$\tau_\parallel = M_{\parallel 2}(0, v_\text{d}) + D_2(0, v_\text{d}). \tag{3.63}$$

To have an idea of orders of magnitude of the related quantities, we show $D_1(\omega, v_\text{d})$ and $D_2(\omega, v_\text{d})$ in Fig. 3.5 as functions of ω/ω_p (ω_p is the plasma frequency of the electron system) in the case of $v_\text{d} = 0.5 v_\text{F}$, together with the real and imaginary parts of the memory function, $M_{\parallel 1}$ and $M_{\parallel 2}$, in the parallel configuration. The figure is obtained for an n-doped GaAs system at lattice temperature $T = 0\,\text{K}$, with the following material parameters: doping (impurity) density $n_\text{i} = 1.4 \times 10^{16}\,\text{cm}^{-3}$, electron density $n = n_\text{i}$, electron effective mass $m = 0.067 m_\text{e}$ (m_e is the free electron mass), static dielectric constant $\kappa = 12.9$, optic dielectric constant $\kappa_\infty = 10.9$. This system has a zero-temperature Fermi level $\varepsilon_\text{F} = 3.14\,\text{meV}$ (counted from the conduction-band bottom) or Fermi temperature $T_\text{F} = 36.4\,\text{K}$, Fermi velocity $v_\text{F} = 1.28 \times 10^5\,\text{m/s}$ and plasma frequency $\omega_\text{p} = 7.79 \times 10^{12}\,\text{s}^{-1}$. Scatterings from impurities, transverse and longitudinal acoustic phonons, and longitudinal optic phonons are taken into account in the calculation. Under the dc bias of $v_\text{d} = 0.5 v_\text{F}$, the electron temperature determined from the zero-order equation is $T_\text{e} = 86\,\text{K}$.

We see from the figure that both $D_1(\omega, v_\text{d})$ and $D_2(\omega, v_\text{d})$ approach zero at high frequency ($\omega > 0.2\,\omega_\text{p}$), reflecting the fact that the electron temperature cannot follow a very rapid oscillation of the driving field. This should be compared with the momentum-related memory functions $M_{\parallel 1}(\omega, v_\text{d})$ and

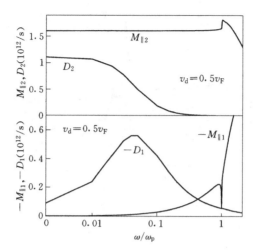

Fig. 3.5 Real and imaginary parts of $D(\omega, v_{\rm d})$ function and memory function $M_\parallel(\omega, v_{\rm d})$ in the parallel configuration for an n-type GaAs. The lattice temperature $T = 0\,\text{K}$ and dc bias velocity $v_{\rm d} = 0.5\,v_{\rm F}$. From Zou and Lei (1995a).

$M_{\parallel 2}(\omega, v_{\rm d})$, which approach zero at much higher frequency ($\omega \gg 2\omega_{\rm p}$). This indicates that the energy relaxation time of the system is much longer than the momentum relaxation time.

It can be seen from the figure that $M_1(\omega, v_{\rm d})$ has a sharp peak and $M_2(\omega, v_{\rm d})$ exhibits a hump at frequency $\omega \approx \omega_{\rm p}$. This comes from the contribution of the plasma excitation induced by the intercarrier Coulomb interactions in the electron density correlation function $\Pi(\boldsymbol{q}, \omega)$. Within random phase approximation the electron density correlation function is given by

$$\Pi(\boldsymbol{q}, \omega) = \frac{\Pi_0(\boldsymbol{q}, \omega)}{1 - \nu_{\rm c}(q)\Pi_0(\boldsymbol{q}, \omega)}$$
$$= \frac{\Pi_{01}(\boldsymbol{q}, \omega) - \nu_{\rm c}(q)|\Pi_0(\boldsymbol{q}, \omega)|^2 + {\rm i}\Pi_{02}(\boldsymbol{q}, \omega)}{[1 - \nu_{\rm c}(q)\Pi_{01}(\boldsymbol{q}, \omega)]^2 + [\nu_{\rm c}(q)\Pi_{02}(\boldsymbol{q}, \omega)]^2}, \qquad (3.64)$$

in which $\Pi_0(\boldsymbol{q}, \omega)$ is the electron density correlation function (2.9) in the absence of electron–electron Coulomb interactions, $\nu_{\rm c}(q)$ is the Coulomb potential. In the small region where both $[1 - \nu_{\rm c}(q)\Pi_{01}(\boldsymbol{q}, \omega)]$ and $\Pi_{02}(\boldsymbol{q}, \omega)$ become small, the real part and the imaginary part of $\Pi(\boldsymbol{q}, \omega)$ can be quite large. This is so called the region of plasma excitation. Though the plasma excitation range is generally small, its integration may yield a large contri-

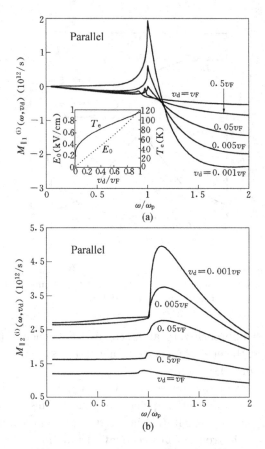

Fig. 3.6 Real part $M_{\|1}^{(i)}(\omega, v_d)$ (a) and imaginary part $M_{\|2}^{(i)}(\omega, v_d)$ (b) of impurity-contributed memory function of parallel configuration under several different dc bias velocities. From Zou and Lei (1995a).

bution to the memory function $M_{\perp(\|)}(\omega, v_d)$. For impurity-induced memory function $M_{\perp(\|)}^{(i)}(\omega, v_d)$, the plasma contribution gives rise to a sharp peak (in the real part) or a hump (in the imaginary part) around $\omega \approx \omega_p$. This feature is salient at low temperature and gradually smeared out with rising electron temperature. Therefore, for given lattice temperature it is most remarkable at zero dc bias since a finite dc bias always increases the electron temperature and thus weakens the plasma effect. These features can be seen quite clearly in Figs. 3.6(a) and 3.6(b), where the impurity-induced memory functions $M_{\|1}^{(i)}(\omega, v_d)$ and $M_{\|2}^{(i)}(\omega, v_d)$ are shown in several different dc biases $v_{\rm d}/v_{\rm F} = 0.001, 0.005, 0.05, 0.5$ and 1.0. The real part and the imaginary part of polar-optic-phonon induced memory function in parallel configuration, $M_{\|1}^{\rm (LO)}(\omega, v_{\rm d})$ and $M_{\|2}^{\rm (LO)}(\omega, v_{\rm d})$, are shown in Figs. 3.7(a) and (b) as functions of the frequency under different dc biases. At small dc bias when the electron temperature is quite low ($T_{\rm e} = 0.6\,{\rm K}$ at $v_{\rm d} = 0.001 v_{\rm F}$), $M_{\|2}^{\rm (LO)}(\omega, v_{\rm d})$ is almost zero at low frequency ($\omega < 6\,\omega_{\rm p}$), and then rises rapidly at frequency around $\omega \sim \Omega_{\rm LO} \approx 7\,\omega_{\rm p}$, reflecting the resonant excitation of LO phonons by the high-frequency electric field. When dc bias increases, electron temperature grows ($T_{\rm e} = 113\,{\rm K}$ at $v_{\rm d} = v_{\rm F}$), the low-frequency value of $M_{\|2}^{\rm (LO)}(\omega, v_{\rm d})$ rises rapidly. Furthermore, we can see that at higher dc biases the rise of $M_{\|2}^{\rm (LO)}(\omega, v_{\rm d})$ with increasing frequency is slower and the position at which $M_{\|2}^{\rm (LO)}(\omega, v_{\rm d})$ reaches the maximum, shifts towards lower frequency. Correspondingly, the peaks showing up in the real part of the LO-phonon-induced memory function, $M_{\|1}^{\rm (LO)}(\omega, v_{\rm d})$ at $\omega \sim \Omega_{\rm LO}$, which are rather sharp at low dc biases, are rounded up significantly at high dc biases. There is no appreciable structure related to the electron plasma excitation in $M_{\|}^{\rm (LO)}(\omega, v_{\rm d})$.

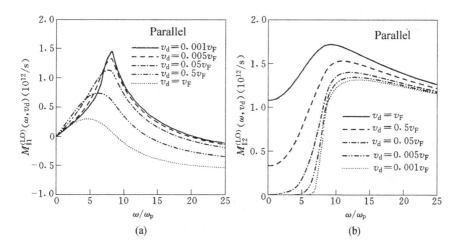

Fig. 3.7 Real part $M_{\|1}^{\rm (LO)}(\omega, v_{\rm d})$ (a) and imaginary part $M_{\|2}^{\rm (LO)}(\omega, v_{\rm d})$ (b) of LO-phonon-contributed memory function of parallel configuration under several different dc bias velocities. From Zou and Lei (1995a).

3.4 Plasmon Contribution to High-Frequency Conductivity in Quasi-2D Systems and Superlattices

Most results derived in the preceding section for 3D semiconductors can, with minor changes, apply to a quasi-2D system or a superlattice, in which a small amplitude high-frequency signal electric field of frequency ω and a dc bias field are applied within the 2D plane. For instance, the expression for the small-signal high-frequency sheet resistivity $R(\omega)$ of a quasi-2D system or a superlattice (per layer) in the zero dc bias can be obtained by changing the carrier volume density n in Eq. (3.46) into the carrier sheet density N_s (per layer):

$$R(\omega) = -\mathrm{i}\frac{m}{N_s e^2}\left[\omega + M(\omega)\right], \qquad (3.65)$$

while the mobility has the same expression as (3.48),

$$\mu(\omega) = \mathrm{i}\frac{e}{m}\frac{1}{\omega + M(\omega)}, \qquad (3.66)$$

in which $M(\omega)$ is the two-dimensional memory function in the zero dc bias.

In the case of finite dc bias the system becomes anisotropic and one has to distinguish the parallel and perpendicular configurations. The small-signal high-frequency sheet resistivity (per layer) can be obtained directly from (3.58) and (3.59) as

$$R_\perp(\omega, v_\mathrm{d}) = -\mathrm{i}\frac{m}{N_s e^2}\left[\omega + M_\perp(\omega, v_\mathrm{d})\right], \qquad (3.67)$$

$$R_\parallel(\omega, v_\mathrm{d}) = -\mathrm{i}\frac{m}{N_s e^2}\left[\omega + M_\parallel(\omega, v_\mathrm{d}) + D(\omega, v_\mathrm{d})\right], \qquad (3.68)$$

in which $D(\omega, v_\mathrm{d})$ is given by (3.60). At zero dc bias, $D(\omega, 0) = 0$ and $M_\perp(\omega, 0) = M_\parallel(\omega, 0) = M(\omega)$. The memory functions consist of impurity and phonon parts: $M_{\perp(\parallel)}(\omega, v_\mathrm{d}) = M^{(\mathrm{i})}_{\perp(\parallel)}(\omega, v_\mathrm{d}) + M^{(\mathrm{p})}_{\perp(\parallel)}(\omega, v_\mathrm{d})$.

For quasi-2D systems discussed in Sec. 2.6,

$$M^{(\mathrm{i})}_{\perp(\parallel)}(\omega, v_\mathrm{d}) = \frac{1}{mN_s\omega}\sum_{n',n,\mathbf{q}_\parallel} q^2_{y(x)}|U_{n'n}(\mathbf{q}_\parallel)|^2$$
$$\times \left[\Pi(n',n,\mathbf{q}_\parallel,\omega_0) - \Pi(n',n,\mathbf{q}_\parallel,\omega_0+\omega)\right], \qquad (3.69)$$

$$M^{(\mathrm{p})}_{\perp(\parallel)}(\omega, v_\mathrm{d}) = \frac{1}{mN_s\omega}\sum_{n',n,\mathbf{q},\lambda} q^2_{y(x)}|M_{n'n}(\mathbf{q},\lambda)|^2$$
$$\times \left[\Lambda(n',n,\mathbf{q},\lambda,\omega_0) - \Lambda(n',n,\mathbf{q},\lambda,\omega_0+\omega)\right], \qquad (3.70)$$

$$N(\omega, v_\mathrm{d}) = \frac{1}{mN_\mathrm{s}\omega v_\mathrm{d}} \sum_{n',n,\boldsymbol{q},\lambda} q_x \Omega_{\boldsymbol{q}\lambda} |M_{n'n}(\boldsymbol{q},\lambda)|^2$$
$$\times [\Gamma(n',n,\boldsymbol{q},\lambda,\omega_0) - \Gamma(n',n,\boldsymbol{q},\lambda,\omega_0+\omega)], \quad (3.71)$$

where $\omega_0 = q_x v_\mathrm{d}$, v_d is the dc bias velocity along the x direction. Correlation function $\Lambda(n',n,\boldsymbol{q},\omega)$ and $\Gamma(n',n,\boldsymbol{q},\omega)$ can be expressed in terms of the electron density correction function $\Pi(n',n,\boldsymbol{q},\omega)$ and phonon distribution function through the relations similar to (3.9) and (3.10), and (3.36) and (3.37).

For superlattices described in Sec. 2.7,

$$M_{\perp(\|)}^{(\mathrm{i})}(\omega,v_\mathrm{d}) = \frac{1}{mN_\mathrm{s}\omega} \left(\frac{e^2}{2\epsilon_0\kappa}\right)^2 \frac{d}{2\pi} \int_{-\pi/d}^{\pi/d} dq_z \sum_{\boldsymbol{q}_\|} \frac{q_{y(x)}^2}{q_\|^2} \widetilde{N}(\boldsymbol{q}_\|,q_z) g(q_z)$$
$$\times [\Pi(q_z,\boldsymbol{q}_\|,\omega_0) - \Pi(q_z,\boldsymbol{q}_\|,\omega_0+\omega)], \quad (3.72)$$

$$M_{\perp(\|)}^{(\mathrm{p})}(\omega,v_\mathrm{d}) = \frac{1}{mN_\mathrm{s}\omega} \sum_{\boldsymbol{q}_\|,q_z,\lambda} q_{y(x)}^2 |M(\boldsymbol{q},\lambda)|^2 |I(\mathrm{i}q_z)|^2$$
$$\times [\Lambda(q_z,\boldsymbol{q}_\|,\lambda,\omega_0) - \Lambda(q_z,\boldsymbol{q}_\|,\lambda,\omega_0+\omega)], \quad (3.73)$$

$$N(\omega,v_\mathrm{d}) = \frac{1}{mN_\mathrm{s}\omega v_\mathrm{d}} \sum_{\boldsymbol{q}_\|,q_z,\lambda} q_x \Omega_{\boldsymbol{q}\lambda} |M(\boldsymbol{q},\lambda)|^2 |I(\mathrm{i}q_z)|^2$$
$$\times [\Gamma(q_z,\boldsymbol{q}_\|,\lambda,\omega_0) - \Gamma(q_z,\boldsymbol{q}_\|,\lambda,\omega_0+\omega)]. \quad (3.74)$$

Here $\Lambda(q_z,\boldsymbol{q}_\|,\lambda,\omega)$ and $\Gamma(q_z,\boldsymbol{q}_\|,\lambda,\omega)$ can be expressed through the relations similar to (3.9) and (3.10), and (3.36) and (3.37) in terms of the electron density correction function of the superlattice $\Pi(q_z,\boldsymbol{q}_\|,\omega)$, which includes intercarrier Coulomb interactions within a layer and between different layers in the system and is given by (2.146) in the random phase approximation. The effective impurity density $\widetilde{N}(\boldsymbol{q}_\|,q_z)$, correlation factor $g(q_z)$ and form factor $I(\mathrm{i}q_z)$ are also those as given in Sec. 2.7.

Calculations of linear dynamic conductivity for two-dimensional electron systems were carried out by many authors (Tzoar, Platzman and Simons, 1976; Ando, 1976; Ting, Ying and Quinn, 1976b; Ganguly and Ting, 1977). Effects due to Coulomb interactions among carriers, especially those of the plasmon modes, on dynamic conductivity have long been recognized in 2D semiconductors (Kennedy et al, 1975, Ting, Ganguly and Lai, 1981). Since in a 2D system the long wavelength plasmon excitation appears at quite a low frequency, its contribution to the dynamic conductivity is expected to be much more important than in its 3D counterpart. Previous estimates

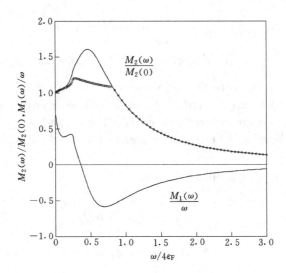

Fig. 3.8 The imaginary part $M_2(\omega)/M_2(0)$ and real part $M_1(\omega)/\omega$ (scaled in units of $mM_2(0)/(2\pi\hbar N_s)$) of zero-temperature memory function versus $\omega/4\varepsilon_F$ (ε_F is the Fermi level) for a GaAs quantum well with width $a = 10$ nm and electron density $N_s = 1.8 \times 10^{11}$ cm^{-1}. The scatterings are entirely due to remote impurities located at a distance $s = 2.5$ nm from the interfaces. The discrete points in the figure are the $M_2(\omega)/M_2(0)$ obtained by neglecting the plasma pole contribution. From Lei and Zhang (1986).

relating to this, however, have been very crude. The balance-equation formulation facilitates accurate evaluation of such dynamic effects of intercarrier Coulomb interactions. Based on Eq. (3.69), Lei and Zhang (1986) carry out a full RPA calculation of the real and imaginary parts of the impurity-limited high-frequency conductivity in a quasi-two-dimensional electron gas at zero temperature without dc bias. The calculated functions of the real part, $M_1(\omega)$, and the imaginary part, $M_2(\omega)$, are shown as solid curves in Fig. 3.8 for a GaAs-based quantum well of width $a = 10$ nm, electron sheet density $N_s = 1.8 \times 10^{11}$ cm^{-2}, assuming scattering only by remote impurities located at a distance $s = 2.5$ nm from the mobile electrons. The discrete points in the figure are the calculated values of $M_2(\omega)/M_2(0)$ obtained by neglecting the plasma contribution, which should correspond to the single-layer result of Tzoar and Zhang (1985). The plasma induced enhancement in $M_2(\omega)$ is significant.

In a superlattice, the contribution of electron collective modes to high-frequency conductivity is even more distinctive. Lei, Horing and Zhang (1986a) calculate $M_1(\omega)$ and $M_2(\omega)$ due to remote and background im-

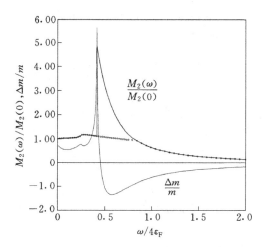

Fig. 3.9 Calculated $M_2(\omega)/M_2(0)$ and $\Delta m/m$ (scaled in units of $mM_2(0)/(2\pi\hbar N_s)$) are shown as functions of $\omega/4\varepsilon_{\mathrm{F}}$ (ε_{F} is the Fermi level) for a quantum well superlattice with $a = 10$ nm, $d = 20$ nm and $N_s = 2.3 \times 10^{11}$ cm^{-2}. The scatterings are due to remote and background impurities. The former are located at a distance $s = 7.5$ nm from the center of each quantum well. The discrete points are $M_2(\omega)/M_2(0)$ calculated by neglecting the plasma pole contribution. From Lei, Horing and Zhang (1986a).

purity scatterings for GaAs/AlGaAs superlattices of different geometrical parameters at zero temperature. An example of their calculated results is shown in Fig. 3.9. For comparison, the values of $M_2(\omega)/M_2(0)$ calculated by neglecting the plasma pole contribution are also shown as discrete points in this figure. The enhancement due to collective modes is very pronounced, resulting in the most striking feature of the frequency dependence of the conductivity: a sudden rise of $M_2(\omega)$ and a sharp peak of $M_1(\omega)$ at the bulk superlattice plasmon frequency $\omega_{\mathrm{p}} = (e^2 N_s/\epsilon_0 \kappa m d)^{1/2}$. These are characteristic of a closely-packed infinitely-repeated layer system, in which the electrons in one layer are scattered by charged impurities located in all different layers, giving rise to a $(qd)^{-1}$ divergence of $\tilde{N}(\boldsymbol{q}_\parallel, 0)$ at small q. This $M_2(\omega)$ jump arises from the long-wavelength ($q \approx 0$ and $q_z \approx 0$) plasmon contribution at $\omega \approx \omega_{\mathrm{p}}$ and is easily derived from the behavior of $\tilde{N}(\boldsymbol{q}_\parallel, 0)$ and $\Pi_2(0, \boldsymbol{q}_\parallel, \omega)$ at small q as

$$\triangle M_2(\omega_{\mathrm{p}}) = \frac{1}{12\pi}\left(\frac{e^2}{\epsilon_0\kappa}\right)^{3/2}\frac{m^{1/2}}{(N_s d)^{3/2}}(N_{\mathrm{r}} Z_{\mathrm{r}}^2 + N_{\mathrm{b}} Z_{\mathrm{b}}^2). \qquad (3.75)$$

This is the result of $T = 0$ K. With increasing lattice temperature the height of $\triangle M_2(\omega_p)$ decreases but the contribution from plasma excitation still exhibits a sharp edge, as shown in Fig. 3.10.

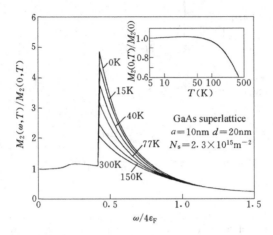

Fig. 3.10 Calculated $M_2(\omega)/M_2(0)$ versus $\omega/4\varepsilon_F$ (ε_F is the Fermi level) at different lattice temperatures $T = 0, 15, 40, 70, 150, 300$ K, for the same superlattice described in Fig. 3.9. The inset shows the temperature variation of $M_2(0)$. From Lei, Horing and Zhang (1987).

It is worth noting that the assumption that the impurity distribution is random within each layer and has no correlation from one layer to another, is relevant to the sharp edge in the plasma resonance of the conductivity for a superlattice. While the existence of certain correlation may round off the sharp edge of resonance, the plasmon contribution will remain distinctive (Lei, Horing and Zhang, 1986a).

3.5 High-Frequency Conductivity of Quantum Wires

The analyses in the preceding two sections can also apply to quasi-1D systems. Zou and Lei (1995b) carry out a high-frequqncy conductivity calculation using the cylindrical model as described in Sec. 2.9 for quantum wires: electrons move freely along the axis in the z direction, but transversely are confined within a cylinder of radius ϱ. The electron wave function and energy in the absence of applied electric field are given by (2.204)–(2.207). The expressions for electron–impurity scattering, electron–phonon scattering and electron–electron interaction are all given in Sec. 2.9. If the quantum wire is thin and the applied electric field is moderate (the electron temperature is not too high), one can assume that carriers occupy only the lowest subband and the influence of all high-lying subbands is negligible. We only need to keep the $n = n' = 0$ and $m = m' = 0$ term in the relevant matrix elements and form factors. For instance, the coupling matrix element between electrons and 3D phonons can be written as $M(\boldsymbol{q}, \lambda)F(q_\parallel)$, where $\boldsymbol{q} = (\boldsymbol{q}_\parallel, q_z)$ is the 3D wavevector, $\boldsymbol{q}_\parallel$ is the transverse wavevector, $M(\boldsymbol{q}, \lambda)$ is the electron–phonon matrix element in the 3D Fourier representation, and $F(q_\parallel) = F_{00}(q_\parallel)$ is the form factor. The matrix element of inter-carrier Coulomb interaction is

$$V(q_z) = V_{0000}(q_z) = \frac{e^2}{2\pi\epsilon_0\kappa} \int d\boldsymbol{r} d\boldsymbol{r}' |\varphi(r)|^2 |\varphi(r')|^2 \mathrm{K}_0(|q_z(\boldsymbol{r} - \boldsymbol{r}')|), \quad (3.76)$$

where $\varphi(r) \equiv \varphi_0(\boldsymbol{r})$ [see (2.205)], $\mathrm{K}_0(x)$ is the modified Bessel function of zero-th order, and κ is the dielectric constant of the wire material. A numerical test shows that, the Coulomb matrix element $V(q_z)$ of (3.76) can be modeled by the following expressions (Gold and Ghazali, 1990) with sufficient accuracy:

$$V(q_z) = \frac{18e^2}{\pi\epsilon_0\kappa(q_z\varrho)^2} \left[\frac{1}{10} - \frac{2}{3(q_z\varrho)^2} + \frac{32}{3(q_z\varrho)^4} - \frac{64}{(q_z\varrho)^4}\mathrm{I}_3(q_z\varrho)\mathrm{K}_3(q_z\varrho)\right]$$
$$(\text{for } |q_z|\varrho > 0.2), \quad (3.77)$$

$$V(q_z) = -\frac{e^2}{2\pi\epsilon_0\kappa} \left\{\ln(\frac{\gamma|q_z|\varrho}{2}) - 0.6083 + (q_z\varrho)^2 \left[\frac{1}{8}\ln(\frac{\gamma|q_z|\varrho}{2}) - 0.1450\right]\right\}$$
$$(\text{for } |q_z|\varrho < 0.2), \quad (3.78)$$

in which $\mathrm{I}_3(x)$ and $\mathrm{K}_3(x)$ are modified Bessel functions of the 3rd order (Wang and Guo, 1989) and $\ln\gamma = 0.577$ is the Eular constant. To save computing time they are used in the numerical calculation.

When a small-amplitude high-frequency electric field and a constant dc bias electric field (3.28) are applied along the axis of the quantum wire, the center of mass of the electrons moves with a constant drift velocity $\boldsymbol{v}_\mathrm{d}$ and slightly oscillates around it at the driving frequency ω as given in (3.29). The dynamic small-signal mobility of the quantum wire is obtained as

$$\mu(\omega, v_\mathrm{d}) = \frac{ie}{m} \frac{1}{\omega + M(\omega, v_\mathrm{d}) + D(\omega, v_\mathrm{d})}, \quad (3.79)$$

where $D(\omega, v_\mathrm{d})$, which relates to the electron temperature oscillation, can still be expressed as (3.60). The momentum-related memory function $M(\omega, v_\mathrm{d})$ consists of contributions due to impurity and phonon scatterings,

$M(\omega,v_{\rm d}) = M^{(\rm i)}(\omega,v_{\rm d}) + M^{(\rm p)}(\omega,v_{\rm d})$, and the energy-related memory function $N(\omega,v_{\rm d})$ is determined by the phonon scattering. Their expressions are, respectively,

$$M^{(\rm i)}(\omega,v_{\rm d}) = \frac{n_{\rm i}}{N_{\rm l}m\omega} \sum_{\bm{q}} |U(\bm{q})|^2 |F(q_\parallel)|^2 q_z^2 \left[\Pi(q_z, q_z v_{\rm d}) - \Pi(q_z, q_z v_{\rm d} + \omega) \right], \tag{3.80}$$

$$M^{(\rm p)}(\omega,v_{\rm d}) = \frac{1}{N_{\rm l}m\omega} \sum_{\bm{q},\lambda} |M(\bm{q},\lambda)|^2 |F(q_\parallel)|^2 q_z^2$$
$$\times \left[\Lambda(\bm{q},\lambda,q_z v_{\rm d}) - \Lambda(\bm{q},\lambda,q_z v_{\rm d} + \omega) \right], \tag{3.81}$$

$$N(\omega,v_{\rm d}) = \frac{1}{N_{\rm l}m\omega v_{\rm d}} \sum_{\bm{q},\lambda} |M(\bm{q},\lambda)|^2 |F(q_\parallel)|^2 q_z \Omega_{\bm{q}\lambda}$$
$$\times \left[\Gamma(\bm{q},\lambda,q_z v_{\rm d}) - \Gamma(\bm{q},\lambda,q_z v_{\rm d} + \omega) \right], \tag{3.82}$$

in which $N_{\rm l}$ is the electron line density, and $\Pi(q_z,\omega)$ is the 1D electron density correlation function including effect of electron–electron Coulomb interaction. The correlation functions $\Lambda(\bm{q},\lambda,\omega)$ and $\Gamma(\bm{q},\lambda,\omega)$ can be expressed through the relations similar to (3.9) and (3.10), and (3.36) and (3.37) in terms of electron density correlation function $\Pi(q_z,\omega)$ and phonon distribution $n(\Omega_{\bm{q}\lambda}/T)$ [replace all $\Pi(\bm{q},\omega)$ by $\Pi(q_z,\omega)$]. Under random phase approximation the 1D electron density correlation function is given by

$$\Pi(q_z,\omega) = \frac{\Pi_0(q_z,\omega)}{1 - V(q_z)\Pi_0(q_z,\omega)}, \tag{3.83}$$

where $\Pi_0(q_z,\omega)$ is the 1D electron density correlation function without Coulomb interaction [(2.223) in the case of $n = n' = 0$],

$$\Pi_0(q_z,\omega) = 2\sum_{k_z} \frac{f(\varepsilon(k_z + q_z)) - f(\varepsilon(k_z))}{\omega + \varepsilon(k_z + q_z) - \varepsilon(k_z) + {\rm i}\delta}, \tag{3.84}$$

in which $f(\varepsilon) = [\exp(\varepsilon - \mu)/T_{\rm e} + 1]^{-1}$ is the Fermi distribution function with μ the chemical potential determined by the electron line density:

$$N_{\rm l} = 2\sum_{k_z} f(\varepsilon(k_z), T_{\rm e}). \tag{3.85}$$

In the zero temperature limit ($T_{\rm e} \to 0$) the real part and the imaginary part of $\Pi_0(q_z,\omega)$ are given by

$$\Pi_{01}(q_z,\omega) = \frac{m}{\pi q_z} \ln \left| \frac{(m\omega/q_z)^2 - (q_z/2 - k_{\rm F})^2}{(m\omega/q_z)^2 - (q_z/2 + k_{\rm F})^2} \right|, \tag{3.86}$$

$$\Pi_{02}(q_z,\omega) = \frac{m}{|q_z|}\left\{\theta\left[k_F^2 - \left(\frac{q_z}{2} + \frac{m\omega}{q_z}\right)^2\right] - \theta\left[k_F^2 - \left(\frac{q_z}{2} - \frac{m\omega}{q_z}\right)^2\right]\right\}.$$
(3.87)

Here $k_F = \pi N_1/2$ is the Fermi wavevector at zero temperature and $\theta[x]$ is the Heaviside unit step function.

The plasma mode of a 1D system starts from zero frequency, having a logarithmic divergence at small wavevectors. These small-wavevector plasma modes may give rise to a significant contribution to the impurity-related memory function and thus affects the high-frequency small-signal response in a quantum wire.

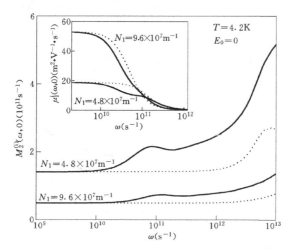

Fig. 3.11 The imaginary part of impurity-induced memory function for a GaAs cylindrical quantum well with $\varrho = 5$ nm at $T = 4.2$ K and zero dc bias, $M_2^{(i)}(\omega, 0)$, is shown as a function of angular frequency ω at two electron wire densities $N_1 = 4.8 \times 10^7$ m^{-1} and $N_1 = 9.6 \times 10^7$ m^{-1}. The inset displays the variation of the real part of the ac mobility of the quantum well, $\mu_1(\omega, 0)$. The solid curves include the contribution from plasma excitations and the dashed curves exclude that. From Zou and Lei (1995b).

As an example we examine two n-type GaAs-based cylindrical quantum wires with same transverse radius $\varrho = 5$ nm and impurity volume density $n_i = 10^{22}$ m^{-3}, but having different conduction electron line densities $N_1 = 9.6 \times 10^7$ m^{-1} and $N_1 = 4.8 \times 10^7$ m^{-1}. The calculated imaginary part of impurity-induced memory function of zero dc bias ($v_d = 0$), $M_2^{(i)}(\omega, 0)$ at lattice temperature $T = 4.2$ K, is plotted as a function of angular frequency

ω of the driving ac field in Fig. 3.11, together with the real part of the ac mobility $\mu_1(\omega,0)$ (inset). The solid curves are obtained by including contributions from the plasma modes and the dashed curves without them. The plasma contribution in these quantum wires starts from relatively low frequency, exhibits a broad hump and is more significant than that in 2D systems. In the case of $N_1 = 9.6 \times 10^7$ m^{-1}, the plasma-induced contribution constitutes 1.4% of $M_2^{(i)}(\omega,0)$ at frequency $\nu \equiv \omega/2\pi = 2$ GHz and increases to the maximum 38.4% at $\nu = 4.4$ THz. In the case of $N_1 = 4.8 \times 10^7$ m^{-1}, the contribution due to plasma excitation constitutes 3.2% of $M_2^{(i)}(\omega,0)$ at $\nu = 2$ GHz and increases to the maximum 57.4% at $\nu = 2.8$ THz. This indicates that the collective excitation plays a more important role with decreasing electron line density. The effect of the plasma excitation on the ac mobility is also quite significant for frequency ν higher than 1 GHz.

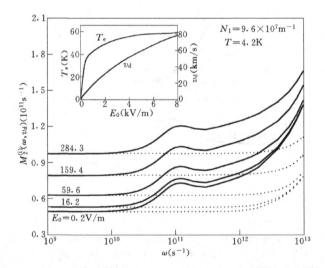

Fig. 3.12 The imaginary part of impurity-induced memory function, $M_2^{(i)}(\omega, v_\mathrm{d})$, is shown against the angular frequency ω for the quantum wire as described in Fig. 3.11 at $T = 4.2$ K and $N_1 = 9.6 \times 10^7$ m^{-1} under different dc biases. Under the bias fields $E_0 = 0.2, 16.2, 59.6, 159.4$ and 284.3 V/m, the system has bias drift velocities $v_\mathrm{d} = 10, 760, 2260, 4550$ and 6550 m/s and electron temperatures $T_\mathrm{e} = 4.2, 5.6, 10.2, 20$ and 30 K. The solid curves include the contribution from plasma excitations and the dashed curves exclude that. The inset indicates the drift velocity v_d and the electron temperature T_e versus the electric field E_0, as determined by zero-order balance equations. From Zou and Lei (1995b).

When a dc bias field E_0 is simultaneously applied, as the situation of the parallel configuration discussed in Sec. 3.2, a small ac current parallel to the dc bias field will induce an energy (thus an electron temperature) change in the linear order in order to satisfy the force and energy balance equations. The ac mobility $\mu(\omega, v_d)$ differs from $\mu(\omega, 0)$ without a dc bias not only due to $M(\omega, v_d)$ differing from $M(\omega, 0)$, but also due to the additional term $D(\omega, v_d)$ induced by the electron temperature oscillation. Fig. 3.12 displays the imaginary part of the impurity-induced memory function $M_2^{(i)}(\omega, v_d)$ versus the angular frequency under different dc bias fields. The solid curves include contributions from the plasma excitation and the dashed curves exclude those. Similar to the case without a dc bias, the plasma excitation-induced contribution to $M_2^{(i)}(\omega, v_d)$ is still significant for frequencies above 2 GHz. This contribution decreases with increasing dc bias field but not as fast as that in a 3D system as shown in Fig. 3.6. Note that in contrast to bulk materials, in a GaAs quantum wire $M_2^{(i)}(\omega, v_d)$ increases with increasing electron temperature (dc bias field) at low frequencies. This owes to the fact that at low temperatures the chemical potential of a quantum wire increases with increasing electron temperature, while the chemical potential of a bulk system decreases with increasing electron temperature.

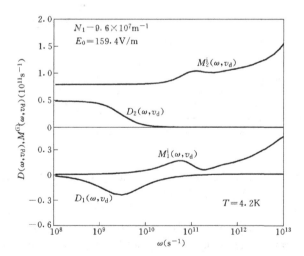

Fig. 3.13 Real and imaginary parts of the $D(\omega, v_d)$ function and the impurity-induced memory function $M^{(i)}(\omega, v_d)$ for the GaAs quantum wire of $N_1 = 9.6 \times 10^7$ m^{-1}, are shown as functions of angular frequency ω under the bias field $E_0 = 159.4$ V/m ($T_e = 20$ K, $v_d = 4.55 \times 10^3$ m/s). From Zou and Lei (1995b).

In Fig. 3.13, we plot the $D(\omega, v_d)$ function as defined in Eq. (3.60), together with $M^{(i)}(\omega, v_d)$, for the case of $E_0 = 159.4\,\text{V/m}$. Similar to the 3D electron system (Fig. 3.5), the $D(\omega, v_d)$-related contribution to mobility in a quantum wire is important at low frequencies. At high frequencies, the electron temperature can not follow the rapid oscillation of the driving field. As a result, both the real part $D_1(\omega, v_d)$ and imaginary part $D_2(\omega, v_d)$ of this function approach zero.

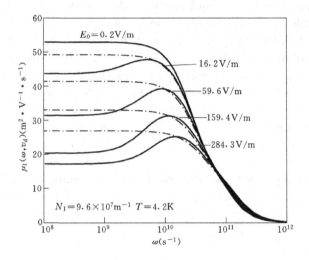

Fig. 3.14 Variation of the real part of the ac mobility $\mu_1(\omega, v_d)$ with angular frequency ω for the GaAs quantum wire of $N_1 = 9.6 \times 10^7\,\text{m}^{-1}$ under dc bias fields $E_0 = 0.2, 16.2, 59.6, 159.4$ and $284.3\,\text{V/m}$. The solid curves include contributions from the plasma excitation and from the electron temperature oscillation, while the dot-dashed curves exclude the latter. From Zou and Lei (1995b).

The ac mobility $\mu(\omega, v_d)$ is obtained from results of $M(\omega, v_d)$ and $D(\omega, v_d)$ functions. Figure 3.14 shows the real part $\mu_1(\omega, v_d)$ of the ac mobility versus the angular frequency ω under different dc bias fields. The solid curves include the contributions from the plasma excitation and from electron temperature oscillation, while the dot-dashed curves exclude the latter $[D(\omega, v_d) = 0]$. As shown by the solid curves, under moderate dc bias fields $\mu_1(\omega, v_d)$ is greatly suppressed by $D(\omega, v_d)$ at low frequencies, leading to the formation of a broad hump in $\mu_1(\omega, v_d)$ before it finally decreases at high frequency. For given carrier concentration and wire radius, $\mu_1(\omega, v_d)$ is generally lower at higher dc bias than at lower dc bias due to the hot-electron nonlinearity.

3.6 Nonlinear High-Frequency Conductivity

3.6.1 *Harmonic expansion method*

We have dealt with linear (i.e. small signal) high-frequency response in Secs. 3.3–3.5. There, though may exist an arbitrarily strong dc bias current, all the quantities which oscillate with time, are assumed to be small and kept to the lowest order. Under this approximation, if the applied ac electric field is a small-amplitude signal of single frequency ω, all the time-dependent parts of transport quantities, such as drift velocity and electron temperature, are small time-oscillating functions with single frequency as given by (3.29) and (3.31), and the $A(\mathbf{q}, t, t')$ function (1.49) can be expanded into a form of (3.30). We then have, except the nonlinear steady-state equations on the dc bias, linear equations regarding the time-dependent small quantities. When the applied ac electric field becomes large, these analyses are no longer valid. Employing time-dependent force- and energy-balance equations equivalent to Eqs. (1.40), (1.55) or (1.60), Cai *et al* (1989) carry out a systematic study of steady-state high-frequency nonlinear transport.

A periodically time-dependent electric field $\mathbf{E}(t)$ applying in the system can always be written as the sum of a dc part $\mathbf{E}_0$ and several harmonic components ($n = \pm 1, \pm 2, \cdots$):

$$\mathbf{E}(t) = \mathbf{E}_0 + \sum_n \mathbf{E}_n e^{-\mathrm{i} n\omega t}, \qquad (3.88)$$

ω being the base frequency. When system reaches a steady time-dependent state, the center-of-mass velocity $\mathbf{v}(t)$, the electron temperature $T_\mathrm{e}(t)$, and other transport quantities can be written as the sum of a dc part and several harmonic components, e.g.

$$\mathbf{v}(t) = \mathbf{v}_\mathrm{d} + \sum_n \mathbf{v}_n e^{-\mathrm{i} n\omega t}, \qquad (3.89)$$

$$T_\mathrm{e}(t) = T_\mathrm{e} + \sum_n T_n e^{-\mathrm{i} n\omega t} \equiv T_\mathrm{e} + \delta T. \qquad (3.90)$$

This type of $\mathbf{v}(t)$ renders $A(\mathbf{q}, t, t')$ function (3.1) of the form

$$\exp\left[\mathrm{i}\mathbf{q} \cdot \int_{t'}^{t} \mathbf{v}(s) ds\right] = \exp[\mathrm{i}\mathbf{q} \cdot \mathbf{v}_\mathrm{d}(t - t')]$$

$$\times \exp\left\{\sum_n \frac{\mathbf{q} \cdot \mathbf{v}_n}{n\omega} e^{-\mathrm{i} n\omega t}\left[e^{\mathrm{i} n\omega(t-t')} - 1\right]\right\}. \qquad (3.91)$$

Expanding the second exponential factor of the right hand side of (3.91) in terms of the quantity inside the curly bracket, as well as all the $T_e(t)$-dependent quantities in the equation to a given order,

$$g(T_e(t)) = g(T_e) + \left[\frac{\partial g(T_e)}{\partial T_e}\right]\delta T + \cdots , \qquad (3.92)$$

one can express the frictional force $\tilde{f}$ and the energy transfer rate $\tilde{w}$ as the sum of harmonic components:

$$\tilde{f} = f_0 + \sum_n f_n \, e^{-i n \omega t}, \qquad (3.93)$$

$$\tilde{w} = w_0 + \sum_n w_n \, e^{-i n \omega t}. \qquad (3.94)$$

With this, Cai *et al* (1989) establish force- and the energy-balance equations for each harmonic component, and, by retaining terms up to the second harmonics ($n = 0, \pm 1, \pm 2$), develop a nonlinear high-frequency transport theory for the nondegenerate limit in particular, to calculate the nonlinear high-frequency response of electrons in the Γ-valley of GaAs.

In principle one can retain harmonic components to any order and such a procedure is valid if the partial sums of the Taylor expansion series of (3.91) and (3.92) can be made sufficiently accurate. However, to keep the approach tractable one has to truncate the expansions to the first few terms [Cai *et al* (1989) cut off them at the third order]. This of course limits the maximum amplitude of the harmonic drift velocity v_n for validity of the procedure. They have tested an ac field with base frequency $\nu = \omega/2\pi = 35\,\text{GHz}$ and maximum amplitude of $1.2\,\text{kV/cm}$. From the point of view of nonlinear conduction, this is only a modest field strength, a little beyond the linear conduction regime in the velocity–field curve of bulk GaAs. Nevertheless, interesting nonlinear results have been obtained by Cai *et al* (1989). We refer the reader to their original paper and the review article of Cai and Lax (1992) for detail.

3.6.2 *Nonlinear calculation without memory effect*

The development of the free electron laser technique, which provides a reliable source of linearly polarized intense electromagnetic waves in the terahertz frequency range, has spurred widespread experimental studies on semiconductor nonlinear dynamics subjected to a strong high-frequency electric field and many interesting phenomena have been observed. Theoretical analysis of this issue, however, is still in a stage far from maturity. The harmonic expansion method discussed in the last subsection is generally not able to deal with highly nonlinear high-frequency response. And there has not yet been a simple and tractable method to include memory effects in large-signal dynamic transport. Though people generally think that memory effect may play some role in the high-frequency transport, there has been no clear understanding as to how large and in what a situation it has to be taken into full account. The balance equations for slowly time-dependent transport, (3.3) and (3.4), which contain memory effects to some extent by including a velocity dependent cenetr-of-mass effective-mass correction, yield, in most cases, a result only slightly different from that obtained by neglecting the effective-mass correction at the expense of much greater computational cost. In this section, as a simplified treatment, we will use simplified equations, i.e. Eqs. (3.3) and (3.4) without the effective-mass correction, or the time-dependent balance equations without memory effect, to calculate the response of a semiconductor system to an intense high-frequency electromagnetic field.

The electron states of a quasi-2D system are characterized by a 2D wavevector $\bm{k}_\| = (k_x, k_y)$ and a subband index n with the energy and wave function given by (2.76) and (2.77). When a dc electric field $\bm{E}_0$ and a sinusoidal ac field of frequency ω and amplitude $\bm{E}_\omega$,

$$\bm{E}(t) = \bm{E}_0 + \bm{E}_\omega \sin \omega t, \qquad (3.95)$$

are applied in the x–y plane, the time-dependent transport state can be described by a time-dependent drift velocity $\bm{v}(t)$ and a time-dependent electron temperature $T_e(t)$. They satisfy the following force- and energy-balance equations (neglecting the effective-mass correction term)

$$N_s m \frac{d}{dt} \bm{v}(t) = N_s e \bm{E}(t) + \bm{f}_i + \bm{f}_p, \qquad (3.96)$$

$$C_e \frac{d}{dt} T_e(t) = -\bm{v}(t) \cdot (\bm{f}_i + \bm{f}_p) - w, \qquad (3.97)$$

These equations are written for a 2D system of unit area. Here the frictional forces $\bm{f}_i$ and $\bm{f}_p$, and the electron energy-loss rate w, have the expressions (2.92)–(2.94) as functions of $\bm{v}(t)$ and $T_e(t)$, $C_e = \partial U/\partial T_e$ is the specific heat, and

$$U = 2 \sum_{n,\bm{k}_\|} \varepsilon_{n\bm{k}_\|} f(\varepsilon_{n\bm{k}_\|}, T_e) \qquad (3.98)$$

is the internal energy of the 2D electron system.

We consider a GaAs quantum well confined by its two interfaces with $Al_{0.3}Ga_{0.7}As$, having well depth 280 meV and well width $a = 12.5$ nm. For carriers of sheet density $N_s = 5.5 \times 10^{15}/m^2$, which occupy only the lowest subband at zero temperature, we take into account the role of two lowest subbands ($n = 0, 1$) separating by an energy distance $\varepsilon_{10} = \varepsilon_1 - \varepsilon_0 = 69$ meV, to partially cover the effects of possible excitation due to transient electron energy enhancement under intense electromagnetic irradiation.

The electron scatterings are assumed due to polar optic phonons (via the Fröhlich coupling with electrons), acoustic phonons (via the deformation potential and the piezoelectric couplings with electrons), and remote charged impurities located at a distance $s = 40$ nm from the center plane of the well having a density to yield the low-temperature (4.2 K) total linear mobility (including phonon contribution) $\mu_0 = 31$ m^2/Vs. The other material parameters are taken to be typical values for GaAs (see Sec. 2.6) in the calculation.

In an isotropic 2D system, if both $\boldsymbol{E}_0$ and $\boldsymbol{E}_\omega$ are along the x direction, the frictional force and the drift velocity are also along the same direction. Under the influence of a time-dependent electric field $\boldsymbol{E}(t)$, the time-dependent response of drift velocity $v(t)$ and electron temperature $T_e(t)$ can be solved from Eqs. (3.96) and (3.97) by starting from initial values of v and T_e. For a periodic driving field as given by Eq. (3.95), after the transient dies out, both $v(t)$ and $T_e(t)$ approach steady time-dependent values, which are independent of the initial values to use. In this steady state both the drift velocity and the electron temperature are periodic functions of time with period $T_\omega = 2\pi/\omega$. Under an intense THz drive, due to the quick rise of the electron temperature, the periodical steady state is generally reached within a time delay of less than ten picoseconds from turning on the field. This time delay exhibits weak dependence on the frequency of the THz field. However, the steady-state behavior of $v(t)$ and $T_e(t)$ varies rather strongly with changing frequency.

Figure 3.15 shows one cycle of the steady time-dependent response of drift velocity $v(t)$ (dashed curves) and electron temperature $T_e(t)$ (solid curves) for three driving THz fields of different frequencies (periods) and amplitudes in the case of zero dc bias $E_0 = 0$. We can see from this figure, that (1) the velocity oscillates at a fundamental frequency ω of the driving electric field, while the electron temperature oscillates at a fundamental frequency of 2ω; (2) the average electron temperature T_{av} is significantly higher than the lattice temperature; (3) to obtain a similar amplitude of the

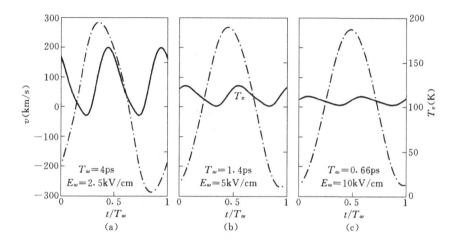

Fig. 3.15 Steady time-dependent response (one cycle) of the drift velocity v (dashed curves) and the electron temperature T_e (solid curves) to sinusoidal electric fields $E(t) = E_\omega \sin(2\pi t/T_\omega)$ of three different periods and amplitudes: (a) $T_\omega = 4.0$ ps and $E_\omega = 2.5$ kV/cm, (b) $T_\omega = 1.4$ ps and $E_\omega = 5.0$ kV/cm, and (c) $T_\omega = 0.66$ ps and $E_\omega = 10.0$ kV/cm. The system is a GaAs-based quantum well described in the text and the lattice temperature is $T = 10$ K. From Lei, Dong, and Chen (1997).

velocity oscillation, much larger strength of the driving ac field is needed for higher frequency (e.g. $T_\omega = 0.66$ ps) than for lower frequency (e.g. $T_\omega = 4$ ps); (4) in the three cases shown in Fig. 3.15, the amplitudes of the velocity oscillations are approximately the same and the average electron temperature $T_{\rm av}$ turns out to be similar; (5) the maximum amplitudes of the temperature oscillation, $\Delta T_M \equiv T_e|_{\rm max} - T_e|_{\rm min}$, however, are strongly dependent on the frequency of the driving field (from 76 K for $T_\omega = 4$ ps in case (a), 23 K for $T_\omega = 1.4$ ps in case (b), to 9.5 K for $T_\omega = 0.66$ ps in case (c). The rapid decrease of ΔT_M with increasing frequency indicates that the electron temperature does not follow the rapid oscillation of the applied field, such that T_e remains to be a relatively slow varying quantity. When the frequency of the driven field becomes high, the electron temperature essentially keeps unchanged at a value determined by the strength and frequency of the driving field.

Figure 3.16 shows the calculated average electron temperature in the steady time-dependent response, $T_{\rm av} \equiv T_\omega^{-1} \int_0^{T_\omega} T_e(t) dt$, as a function of the amplitude E_ω of the driving electric field of three different periods ($T_\omega = 0.66, 1.4$ and 4 ps) in the case of zero dc-field strength $E_0 = 0$ at lattice temperature $T = 10$ K. The theoretical results are in agreement with

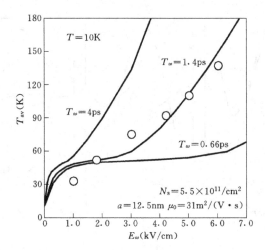

Fig. 3.16 The calculated average electron temperature T_{av} (solid curves) in the steady time-dependent response is shown as a function of the amplitude E_ω of the driving THz electric field of three different periods ($T_\omega = 0.66, 1.4$ and $4\,\text{ps}$) in the case of zero dc field $E_0 = 0$ at lattice temperature $T = 10\,\text{K}$. The small circles indicate the effective electron temperatures derived by Asmar et al (1995) from the photoluminescence measurement for the driving ac field of frequency $\omega/2\pi = 0.68\,\text{THz}$. From Lei, Dong and Chen (1997).

the photoluminescence measurement for different driving ac-field strengths at frequency $\omega/2\pi = 0.68\,\text{THz}$.

The experiments show that, the dc ohmic conductivity of a 2D system is strongly affected by the irradiation of intense high-frequency electromagnetic fields. To analyze this effect we perform numerical calculation of the steady time-dependent response when a sinusoidal THz electric field with period T_ω and amplitude E_ω and a small dc bias electric field E_0 (in the same direction as the ac electric field) are simultaneously applied to the system. We use a value of E_0 in the range around $10\,\text{V/cm}$ to ensure the calculating accuracy and to keep the system close to the ohmic conduction regime when $E_\omega = 0$. At each of the three THz frequencies ($T_\omega = 0.66, 1.4$ and $4\,\text{ps}$) we calculate the average drift velocity in the steady periodic state of the time-dependent response, $v_{av} \equiv T_\omega^{-1} \int_0^{T_\omega} v(t)dt$, for different values of E_ω from 0 to $10\,\text{kV/cm}$ with fixed E_0. This v_{av} essentially proportional to the measured dc current in the experiment. We define a THz-field dependent dc mobility as $\mu_{dc}(E_\omega) \equiv v_{av}/E_0$. $\mu_{dc}(0)$ is the mobility at $E_0 = 0.01\,\text{kV/cm}$ without radiation field. The calculated values of $\mu_{dc}(E_\omega)/\mu_{dc}(0)$ are shown (solid curves) in Fig. 3.17 as a function

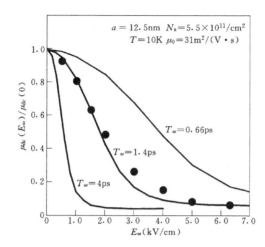

Fig. 3.17 The calculated $\mu_{\rm dc}(E_\omega)/\mu_{\rm dc}(0)$ is shown (solid curves) as a function of THz-field amplitude E_ω at three different periods $T_\omega = 0.66, 1.4$ and $4\,\rm ps$. The dots in the figure are the experimental data taken from Asmar *et al* (1995) in the case of 0.68 THz frequency. From Lei, Dong and Chen (1997).

of THz-field amplitude E_ω for three different frequencies: $T_\omega = 0.66, 1.4$ and $4\,\rm ps$. The dots in the figure are the experimental data of Asmar *et al* (1995) in the case of 0.68 THz. The agreement between theoretical predictions, which are obtained with no adjustable parameter, and experimental measurements is reasonably good.

Chapter 4

Center-of-Mass Velocity Fluctuations, Noise and Diffusion

In Chapter 1, we derived the general force-balance equation (1.40) and energy-balance equation (1.55) or (1.60) with the frictional forces $\tilde{\boldsymbol{f}}_i$, $\tilde{\boldsymbol{f}}_p$ and energy-transfer rate $\tilde{w}$, in which the $A(\boldsymbol{q},t,t')$ function (1.48) is involved. By replacing the difference of the center-of-mass position at different times, $\boldsymbol{R}(t) - \boldsymbol{R}(t')$, with the integral of the average or the drift velocity $\boldsymbol{v}(t)$,

$$A(\boldsymbol{q},t,t') \equiv \exp[\,\mathrm{i}\boldsymbol{q}\cdot(\boldsymbol{R}(t)-\boldsymbol{R}(t'))] \approx \exp\left[\mathrm{i}\boldsymbol{q}\cdot\int_{t'}^{t}\boldsymbol{v}(s)ds\right], \qquad (4.1)$$

simplified balance equations were obtained in different transport situations in Chapters 2 and 3. The approximation (4.1) implies the neglect of the center-of-mass velocity fluctuations $\delta\boldsymbol{V}$ and its effects on the frictional forces and other physical quantities, and thus the derived equations contain only information relating to the drift motion of the system. To take account of the Brownian motion of the center of mass, we must retain the fluctuating part of the center-of-mass velocity and other fluctuating quantities at least to their lowest order and establish Langevin-type equations. This can be achieved by using the Heisenberg equation of motion (Hu and Ting, 1986; Xing and Ting, 1987; Lei and Horing, 1987b; Hu and O'Connell, 1987).

4.1 Langevin-Type Equation

Before we proceed, let us first return to the statistical average of a dynamical variable. Instead of using Eq. (1.27) of the Schrödinger picture, or Eq. (1.31) of the interaction picture, we can formulate the statistical average of a dynamic variable $\mathcal{O}$ in the Heisenberg picture as

$$\langle\mathcal{O}\rangle = \mathrm{tr}\{\hat{\rho}_0\widetilde{\mathcal{O}}(t)\}, \qquad (4.2)$$

where $\widetilde{\mathcal{O}}(t)$ is the Heisenberg operator of the dynamic variable $\mathcal{O}$, and $\hat{\rho}_0$ is the initial ($t = -\infty$) density matrix of the system. The Heisenberg operator $\widetilde{\mathcal{O}}(t)$ is given by

$$\widetilde{\mathcal{O}}(t) = U^\dagger(t, -\infty)\mathcal{O}(t)U(t, -\infty), \qquad (4.3)$$

where $\mathcal{O}(t)$ is the operator of the dynamical variable $\mathcal{O}$ in the interaction picture defined by Eq. (1.32), and

$$U(t, -\infty) \equiv T \exp\left[i \int_{-\infty}^{t} H_{It'}(t')\, dt'\right] \qquad (4.4)$$

is the unitary evolution operator in which T stands for the time-ordering operation.

The operator Eq. (1.20) in Chapter 1 can be regarded as the Heisenberg equation of motion of the center-of-mass momentum $\boldsymbol{P}$ and can be written in the Heisenberg picture, if all the operators in the equation are considered to be Heisenberg operators. Thus we can write (for system of unit volume)

$$nm\frac{d}{dt}\boldsymbol{V}(t) = \dot{\boldsymbol{P}} = ne\boldsymbol{E}(t) + \widetilde{\boldsymbol{F}}(t), \qquad (4.5)$$

with n the electron density. Expanding the evolution operator to lowest order in H_I (Ting and Nee, 1986; Hu and Ting, 1986; Lei and Horing, 1987b) we have

$$\widetilde{\boldsymbol{F}}(t) = \hat{\boldsymbol{F}}(t) - i \int_{-\infty}^{t} \left[\hat{\boldsymbol{F}}(t), H_{It'}(t')\right] dt'. \qquad (4.6)$$

Here

$$\hat{\boldsymbol{F}}(t) = e^{i(H_{er}+H_{ph})t}\hat{\boldsymbol{F}}\, e^{-i(H_{er}+H_{ph})t} \qquad (4.7)$$

is the force operator $\hat{\boldsymbol{F}}$ in the interaction picture. It is a fluctuating (random) force, being of linear order in H_I, and has null contribution to the statistical average

$$\langle \hat{\boldsymbol{F}}(t) \rangle = 0. \qquad (4.8)$$

The second term on the right-hand-side of Eq. (4.6), which is of order of H_I^2, includes a nonzero average part and a fluctuating part. Being of higher order than H_I, this fluctuating part can be neglected in comparison with $\hat{\boldsymbol{F}}(t)$. Thus, keeping both the average part and the fluctuating part to their leading orders, we can take the statistical average for the relative electron and phonon variables over the second term of the right-hand-side of Eq. (4.5), yielding

$$nm\frac{d}{dt}\boldsymbol{V}(t) = ne\boldsymbol{E} + \hat{\boldsymbol{F}}(t) + \tilde{\boldsymbol{f}}. \tag{4.9}$$

Here $\tilde{\boldsymbol{f}} = \tilde{\boldsymbol{f}}_i + \tilde{\boldsymbol{f}}_p$ has the same expressions as Eqs. (1.42) and (1.43), with the function

$$A(\boldsymbol{q}, t, t') \equiv \exp\bigl[i\boldsymbol{q} \cdot (\boldsymbol{R}(t) - \boldsymbol{R}(t'))\bigr] = \exp\left[i\boldsymbol{q} \cdot \int_{t'}^{t} \boldsymbol{V}(s)ds\right] \tag{4.10}$$

in its original form Eq. (1.48) without the elimination of fluctuations by averaging. $\boldsymbol{V}(t)$ now consists of both the average or drift velocity $\boldsymbol{v}(t)$ and a small fluctuation or noise $\delta\boldsymbol{V}(t)$,

$$\boldsymbol{V}(t) = \boldsymbol{v}(t) + \delta\boldsymbol{V}(t) \tag{4.11}$$

with

$$\langle \delta\boldsymbol{V}(t) \rangle = 0. \tag{4.12}$$

Expanding the exponential factor of the right-hand-side of Eq. (4.10) to linear order in δV (Ting and Nee, 1986), we have

$$\exp\left[i\boldsymbol{q} \cdot \int_{t'}^{t} \boldsymbol{V}(s)ds\right] \approx \exp\left[i\boldsymbol{q} \cdot \int_{t'}^{t} \boldsymbol{v}(s)ds\right]\left[1 + i\boldsymbol{q} \cdot \int_{t'}^{t} \delta\boldsymbol{V}(s)ds\right]. \tag{4.13}$$

Substituting this into Eq. (4.9) we can write (Hu and Ting, 1986)

$$nm\frac{d}{dt}\bigl[\boldsymbol{v}(t) + \delta\boldsymbol{V}(t)\bigr] = ne\boldsymbol{E}(t) + \boldsymbol{f} + \hat{\boldsymbol{F}}(t)$$
$$+ nm\int_{-\infty}^{t} \mathcal{K}(t-s) \cdot \delta\boldsymbol{V}(s)\, ds, \tag{4.14}$$

$$\mathcal{K}(t-s) = \frac{1}{nm}\sum_{\boldsymbol{q}} \boldsymbol{q}\boldsymbol{q} \int_{-\infty}^{s} dt' \exp\left[i\boldsymbol{q} \cdot \int_{t'}^{t} \boldsymbol{v}(s)ds\right]$$
$$\times \left[n_i |u(\boldsymbol{q})|^2 \Pi(\boldsymbol{q}, t-t') + \sum_{\lambda} |M(\boldsymbol{q}, \lambda)|^2 \Lambda(\boldsymbol{q}, \lambda, t-t')\right]. \tag{4.15}$$

After the ensemble average to eliminate the fluctuation parts, Eq. (4.14) leaves, with approximation (3.1), the momentum balance equation

$$nm\frac{d}{dt}\boldsymbol{v}(t) = ne\boldsymbol{E}(t) + \boldsymbol{f}. \tag{4.16}$$

This equation, together with the energy balance equation obtained following the same procedure,

$$-C_\mathrm{e}\frac{dT_\mathrm{e}}{dt} = \boldsymbol{v}(t)\cdot\boldsymbol{f} + w, \qquad (4.17)$$

determines $\boldsymbol{v}(t)$ and $T_\mathrm{e}(t)$.

It should be noted that the velocity fluctuation or noise is different in nature from a small alternating signal. An ac signal velocity generally induces an additional change of the electron temperature in linear order in the case of a nonzero dc bias. However, since a noise always consists of various components of the frequency spectrum with random phases, it does not influence the electron temperature in linear order (Lei and Horing 1987b). With this in view, the remaining part of Eq. (4.14),

$$\frac{d}{dt}\delta\boldsymbol{V}(t) = \frac{1}{nm}\hat{\boldsymbol{F}}(t) + \int_{-\infty}^{t} \mathcal{K}(t-s)\cdot\delta\boldsymbol{V}(s)\,ds, \qquad (4.18)$$

becomes a Langevin-type equation relating the random velocity $\delta\boldsymbol{V}(t)$ and the fluctuating force (Ting and Nee, 1986; Hu and Ting, 1986; Lei and Horing, 1987b).

4.2 Thermal Noise and Diffusion: General Relations

Based on the Langevin equation (4.18) we can determine thermal noise temperatures and diffusion coefficients of the system in the presence of a strong dc bias. In the following analysis we assume that the fluctuating force $\hat{\boldsymbol{F}}(t)$ is either parallel to (longitudinal) or perpendicular to (transverse) the dc electric field $\boldsymbol{E}$ along the x direction. For $\hat{\boldsymbol{F}}(t)$ in an arbitrary direction we can resolve it into longitudinal and transverse components and calculate the longitudinal and transverse fluctuating velocities respectively. In either the longitudinal or transverse configuration the random velocity $\delta\boldsymbol{V}$ is along the same direction as the fluctuating force $\hat{\boldsymbol{F}}(t)$, and can be written as

$$\frac{d}{dt}\delta V = \frac{1}{nm}\hat{F}(t) + \int_{-\infty}^{t} K(t-s)\,\delta V(s)\,ds, \qquad (4.19)$$

with $K = \mathcal{K}_{xx}$ for the longitudinal and $K = \mathcal{K}_{yy}$ for the transverse configuration. Eq. (4.19) is readily solved with Fourier transformation, leading to

$$\delta V(\omega) = \frac{1}{n^2}\sigma_\mathrm{n}(\omega) F(\omega). \qquad (4.20)$$

Here, $\delta V(\omega)$ and $F(\omega)$ are the Fourier transforms of random velocity $\delta V(t)$ and fluctuating force $\hat{F}(t)$, and $\sigma_n(\omega)$ is a dynamic conductivity defined by

$$\sigma_n(\omega) = i\frac{ne^2}{m}\frac{1}{\omega + M(\omega)}, \qquad (4.21)$$

which is different from the high-frequency small-signal conductivity discussed in Sec. 3.3 in the case of a nonzero dc bias. It may be identified as a noise conductivity (Lei and Horing, 1987b). The memory function $M(\omega) = M^{(i)}(\omega) + M^{(p)}(\omega)$ in Eq. (4.21) consists of an impurity part $M^{(i)}(\omega)$ and a phonon part $M^{(p)}(\omega)$, with (Hu and Ting, 1986; Lei and Horing, 1987b)

$$M^{(i)}(\omega) = \frac{n_i}{nm\omega}\sum_{\boldsymbol{q}} q_\alpha^2 |u(\boldsymbol{q})|^2 \left[\Pi(\boldsymbol{q},\omega_0) - \Pi(\boldsymbol{q},\omega_0+\omega)\right], \qquad (4.22)$$

$$M^{(p)}(\omega) = \frac{1}{nm\omega}\sum_{\boldsymbol{q},\lambda} q_\alpha^2 |M(\boldsymbol{q},\lambda)|^2 \left[\Lambda(\boldsymbol{q},\lambda,\omega_0) - \Lambda(\boldsymbol{q},\lambda,\omega_0+\omega)\right], \qquad (4.23)$$

in which $\alpha = x$ or y for longitudinal or transverse configuration. The electron density correlation function $\Pi(\boldsymbol{q},\omega)$ and electron–phonon correlation function $\Lambda(\boldsymbol{q},\lambda,\omega)$ are those given by Eqs. (3.64), (3.9) and (3.10).

Defining a generalized diffusion coefficient

$$D_n(\omega) = \frac{N}{4}\int_{-\infty}^{\infty}\langle \delta V(t)\delta V(0) + \delta V(0)\delta V(t)\rangle e^{i\omega t}dt \qquad (4.24)$$

($N = nV$, is the total number of carriers in the real volume V of the system), and noticing that $\delta V(t) = N^{-1}\sum_i \delta v_i(t)$, $\delta v_i(t)$ being the random velocity of the ith particle, we have

$$D_n(\omega) = \frac{1}{4N}\sum_{i,j}\int_{-\infty}^{\infty}\langle \delta v_i(t)\delta v_j(0) + \delta v_i(0)\delta v_j(t)\rangle e^{i\omega t}dt. \qquad (4.25)$$

If the velocities of different particles are uncorrelated, $\langle \delta v_i(t)\delta v_j(0)\rangle = \delta_{ij}\langle \delta v_i(t)\delta v_i(0)\rangle$, then $D_n(\omega)$ is just the noise diffusion coefficient as usually defined for independent particles (Nougier, 1980). However, the definition (4.24) is more generally valid for the interacting particle system under consideration here. Employing Eq. (4.21) one can verify that (Hu and Ting, 1986; Lei and Horing, 1987b)

$$D_n(\omega) = \frac{1}{2n^2e^4}|\sigma_n(\omega)|^2 S(\omega), \qquad (4.26)$$

where $S(\omega)$ is the Fourier transform of the symmetrized fluctuating-force correlation function:

$$S(\omega) = \frac{V}{2n}\int_{-\infty}^{\infty}\langle \hat{F}(t)\hat{F}(0) + \hat{F}(0)\hat{F}(t)\rangle\, e^{i\omega t} dt. \qquad (4.27)$$

On the other hand, this generalized diffusion coefficient is proportional to the power spectrum of the current-density fluctuation, and thus we can define a thermal noise temperature $T_n(\omega)$ by a generalized Einstein relation connecting $D_n(\omega)$ and the real part of the noise conductivity $\sigma_n(\omega)$:

$$D_n(\omega) = \frac{T_n(\omega)}{ne^2}\mathrm{Re}\left[\sigma_n(\omega)\right]. \qquad (4.28)$$

This noise temperature has the physical meaning that the maximum available noise power in the frequency range $\omega \to \omega + d\omega$ for a passive two-terminal network, made of material with noise temperature $T_n(\omega)$, is $T_n(\omega)\, d\omega$ (Hu and Ting, 1987a; Lei and Horing, 1987b). Combining Eqs. (4.26) and (4.28) and using $\sigma_n(\omega)$ given by Eq. (4.21) we obtain a simple expression for $T_n(\omega)$:

$$T_n(\omega) = \frac{S(\omega)}{2mM_2(\omega)}. \qquad (4.29)$$

The fluctuating-force correlation function $S(\omega)$, defined in Eq. (4.27), is also composed of an impurity contribution $S^{(i)}(\omega)$ and a phonon contribution $S^{(p)}(\omega)$:

$$S(\omega) = S^{(i)}(\omega) + S^{(p)}(\omega), \qquad (4.30)$$

which may be determined as (Hu and Ting, 1986; Lei and Horing, 1987b)

$$S^{(i)}(\omega) = -\frac{n_i}{n}\sum_{\mathbf{q}} q_\alpha^2 |u(\mathbf{q})|^2 \Pi_2(\mathbf{q}, \omega - \omega_0)\left[2n\!\left(\frac{\omega - \omega_0}{T_e}\right) + 1\right], \qquad (4.31)$$

$$S^{(p)}(\omega) = -\frac{\omega}{2n}\sum_{\mathbf{q},\lambda} q_\alpha^2 |M(\mathbf{q},\lambda)|^2 \Big\{\Lambda_2(\mathbf{q}, \Omega_{\mathbf{q}\lambda} + \omega - \omega_0)$$

$$\times\left[\coth\!\left(\frac{\Omega_{\mathbf{q}\lambda} + \omega - \omega_0}{2T_e}\right)(2n_{\mathbf{q}\lambda} + 1) - 1\right] + (\omega \to -\omega)\Big\}, \qquad (4.32)$$

where $n_{\mathbf{q}\lambda} = n(\Omega_{\mathbf{q}\lambda}/T)$ is the number of phonons with wavevector $\mathbf{q}$ in branch λ, $\omega_0 \equiv q_x v_\mathrm{d}$, and $\alpha = x$ or y for the longitudinal or transverse configuration.

In the case of zero dc bias, where $\omega_0 \to 0$ and $T_e \to T$, it is easily seen that

$$T_{nl} = T_{nt} = \frac{\omega}{2}\coth\left(\frac{\omega}{2T}\right). \tag{4.33}$$

This is the Nyquist relation (Seeger, 1982). Moreover, the noise conductivity $\sigma_n(\omega)$ (4.21) is equivalent to the small-signal conductivity $\sigma(\omega)$ (3.47) in the zero dc bias. For low frequency, $\text{Re}[\sigma(\omega)]/ne$ is just the linear mobility μ_0, such that Eq. (4.28) reduces to the conventional Einstein relation

$$D = T\frac{\mu_0}{e}, \tag{4.34}$$

where $D = D_n(0)$ is the diffusion coefficient. In the case of nonzero dc bias, we find that at high frequency $\omega \gg qv_d$ the longitudinal and transverse noise temperatures are equal and a Nyquist relation still holds,

$$T_{nl} = T_{nt} = \frac{\omega}{2}\coth\left(\frac{\omega}{2T_e}\right), \tag{4.35}$$

but at the electron temperature T_e.

4.3 Thermal Noise in Three-Dimensional Semiconductors

This section will discuss the low-frequency thermal noise of degenerate electron systems in the case of zero dc bias, warm electron regime or at zero lattice temperature, subject to impurity and acoustic deformation potential scatterings. We will also treat nondegenerate systems with dominant optic phonon deformation potential scatterings (Hu and Ting, 1986; Lei and Horing, 1987b).

4.3.1 3D degenerate systems, impurity and acoustic phonon deformation potential scatterings

We first consider the case of small dc current bias, i.e. Ohmic and warm electron regime, $\alpha_v \equiv v_d/v_s \ll 1$ (v_d is the drift velocity and v_s is the longitudinal sound velocity), and assume $v_s \ll v_F$. After neglecting terms of order of v_s/v_F, the low-frequency limits of the impurity-induced memory function $M_2^{(i)}(\omega)$ and the fluctuating force correlation function $S^{(i)}(\omega)$ are given by

$$M_2^{(i)}(0) = \frac{ne^2}{m}\rho_{i0}, \qquad (4.36)$$

$$S^{(i)}(0) = 2T_e ne^2 \rho_{i0}\left[1 + C\frac{\lambda}{12}\left(\frac{\Theta_F}{T_e}\right)^2 \alpha_v^2\right], \qquad (4.37)$$

where ρ_{i0} is the impurity-induced linear resistivity (2.32), $\Theta_F \equiv 2k_F v_s$,

$$\lambda \equiv \frac{3}{20k_F^2}\frac{\int_0^{2k_F} dq\, |u(q)|^2 q^5}{\int_0^{2k_F} dq\, |u(q)|^2 q^3} \qquad (4.38)$$

is a c-number depending on the impurity potential $u(q)$ ($\lambda = 2/5$ for $u(q) =$ constant), and C is a c-number depending on the configuration ($C = 1$ for the longitudinal configuration and $C = 1/3$ for the transverse configuration).

4.3.1.1 *Finite (nonzero) lattice temperature*

In the case of finite lattice temperature T, to the order of α_v^2 the electron temperature is

$$T_e = T(1 + A\alpha_v^2). \qquad (4.39)$$

The coefficient A depends on the impurity resistivity ρ_{i0}/ρ^* and the reduced lattice temperature $t_F \equiv T/\Theta_F$ [see (2.43)]. And (4.37) can be written as

$$S^{(i)}(0) = 2Tne^2\rho_{i0}(1 + B'_s \alpha_v^2), \qquad (4.40)$$

with

$$B'_s = A + C\frac{\lambda}{12 t_F^2}. \qquad (4.41)$$

The low-frequency limits of the imaginary part of the memory function and the fluctuating force correlation function related to longitudinal acoustic phonon deformation potential scattering, are

$$M_2^{(p)}(0) = \frac{ne^2}{m}\rho_{p0}(1 + B_m\alpha_v^2), \qquad (4.42)$$

$$S^{(p)}(0) = 2Tne^2\rho_{p0}(1 + B_s\alpha_v^2), \qquad (4.43)$$

where ρ_{p0} is phonon-induced linear resistivity (2.36),

$$B_{\rm m} = A\left[4 + \frac{n'(1/t_{\rm F})}{g(t_{\rm F})t_{\rm F}}\right] + C\frac{3}{10}\left[24 + \frac{4n'(1/t_{\rm F})}{g(t_{\rm F})t_{\rm F}} - \frac{n''(1/t_{\rm F})}{g(t_{\rm F})t_{\rm F}^2}\right], \qquad (4.44)$$

$$B_{\rm s} = \frac{A}{2}\left[6 + \frac{n'(1/t_{\rm F})}{g(t_{\rm F})t_{\rm F}}\right] + C\frac{3}{20}\left[30 + \frac{5n'(1/t_{\rm F})}{g(t_{\rm F})t_{\rm F}} - \frac{n''(1/t_{\rm F})}{g(t_{\rm F})t_{\rm F}^2} - \frac{2g_2(t_{\rm F})}{g(t_{\rm F})}\right], \qquad (4.45)$$

$$g_2(t) = \frac{1}{t^3}\int_0^1 \frac{y^7 dy}{(e^{y/t}-1)^2(1-e^{-y/t})^2}. \qquad (4.46)$$

From (4.36), (4.37), (4.42), (4.43) and (4.29), the low-frequency noise temperature is

$$T_{\rm n} \approx T\left\{1 + [x_{\rm i}B'_{\rm s} + x_{\rm p}(B_{\rm s} - B_{\rm m})]\alpha_v^2\right\}, \qquad (4.47)$$

in which

$$x_{\rm i} = \frac{\rho_{\rm i0}}{\rho_{\rm i0} + \rho_{\rm p0}}, \quad x_{\rm p} = \frac{\rho_{\rm p0}}{\rho_{\rm i0} + \rho_{\rm p0}}. \qquad (4.48)$$

In the high-temperature range $t_{\rm F} \gg 1$ of the degenerate case ($T \ll \varepsilon_{\rm F}$), $B_{\rm m} \approx 1$ and $A \approx B'_{\rm s} \approx B_{\rm s} \approx 1/3$. If the impurity contribution is negligible in comparison with that of phonons ($x_{\rm i} \approx 0$, $x_{\rm p} \approx 1$), then

$$T_{\rm n} \approx T\left(1 + \frac{\alpha_v^2}{3}\right) \approx T_{\rm e}, \qquad (4.49)$$

i.e., to the order of α_v^2, the noise temperature is equal to the electron temperature. In the low-temperature range $t_{\rm F} \ll 1$, where phonon-induced resistivity approaches zero and impurity contribution dominates ($x_{\rm i} \approx 1$, $x_{\rm p} \approx 0$), we have

$$T_{\rm n} \approx T\left(1 + B'_{\rm s}\alpha_v^2\right) \approx T_{\rm e}\left(1 + C\frac{\lambda}{12\,t_{\rm F}^2}\alpha_v^2\right), \qquad (4.50)$$

i.e., the noise temperature is higher than the electron temperature.

4.3.1.2 Zero lattice temperature

At $T = 0\,{\rm K}$, the electron temperature is given by (2.50),

$$\left(\frac{T_{\rm e}}{\Theta_{\rm F}}\right)^5 = 0.0134\frac{\rho_{\rm i0}}{\rho^*}\frac{\alpha_v^2}{(1-\alpha_v^2)^2}, \qquad (4.51)$$

and Eqs. (4.36) and (4.37) are still valid. The low-frequency limits of the imaginary part of the memory function and the fluctuating force correlation function associated with phonon scattering are (to the order of α_v^2)

$$M_2^{(p)}(0) = 1.334 \frac{ne^2}{m} \rho_{i0}\, \alpha_v^2, \tag{4.52}$$

$$S^{(p)}(0) = 1.726\, T_e\, ne^2 \rho_{i0}\, \alpha_v^2. \tag{4.53}$$

From (4.36), (4.37), (4.52), (4.53) and (4.51), the noise temperature can be expressed as (to the order of α_v^2)

$$T_n = T_e \left[1 + 0.468\, C\lambda \left(\frac{\rho^*}{\rho_{i0}}\right)^{2/5} \alpha_v^{6/5} - 0.516\, \alpha_v^2\right], \tag{4.54}$$

which is valid when the absolute values of the second and the third terms in the bracket are much less than 1. Depending the parameters, T_n may be larger or lower than T_e.

4.3.2 Nondegenerate 3D system, optic deformation potential scattering

We consider the case of $T_e \gg \varepsilon_F$, assuming that optic phonon deformation potential scatterings dominate (impurity and acoustic phonon scatterings are neglected). Optic phonons essentially have the same frequency, $\Omega_{q\lambda} = \Omega_o$, and the effective matrix elements of optical deformation potential scattering are a constant. The main feature of the dc steady state transport under pure optical deformation potential scattering is that the electron temperature T_e becomes very high and the drift velocity approaches a saturation value. In the case of $T_e/\Omega_o \gg 1$, the energy-balance equation gives

$$\left(\frac{v_d}{v_o}\right)^2 = 3\left[2 - \frac{2T\, K_1'(\Omega_o/2T_e)}{\Omega_o\, K_1(\Omega_o/2T_e)}\right]^{-1} \tanh\left(\frac{\Omega_o}{2T}\right), \tag{4.55}$$

where $K_1(x)$ and $K_1'(x)$ are the modified Bessel function and its derivative, $v_o \equiv (\Omega_o/m)^{1/2}$ is the velocity scale associated with optical phonon scattering defined in Sec. 2.3.2. The saturation value of the drift velocity is obtained by taking $T_e \to \infty$ in the above equation,

$$\left(\frac{v_m}{v_o}\right)^2 = \frac{3}{4} \tanh\left(\frac{\Omega_o}{2T}\right). \tag{4.56}$$

In the range of $T_e/\Omega_o \gg 1$, the dc resistivity ρ, the imaginary part of the memory function, and the low-frequency limit of the fluctuating force correlation function $S(0)$ are, respectively,

$$\rho = \bar{\rho}^* \left(\frac{T_e}{\Omega_o}\right)^{1/2} n\left(\frac{\Omega_o}{T}\right), \tag{4.57}$$

$$M_2(0) = \frac{ne^2}{2m}\bar{\rho}^* \left(\frac{T_e}{\Omega_o}\right)^{1/2} \frac{1}{2}\coth\left(\frac{\Omega_o}{2T}\right), \tag{4.58}$$

$$S(0) = T_e ne^2 \bar{\rho}^* \left(\frac{T_e}{\Omega_o}\right)^{1/2} \coth\left(\frac{\Omega_o}{2T}\right), \tag{4.59}$$

where (d_c is the mass density of the lattice, D is the optical deformation potential, see Sec. 2.3.2)

$$\bar{\rho}^* \equiv \frac{8\sqrt{2}m^2 D^2}{3\pi^{3/2} e^2 n d_c v_o}. \tag{4.60}$$

Therefore both the longitudinal and transverse noise temperatures are equal to the electron temperature $T_{nl} = T_{nt} = T_e$.

4.4 Quasi-2D Systems and Superlattices

The main results obtained in Sections 4.1–4.3 for 3D systems are also valid for carrier transport parallel to the well or interface plane in single and multiple quantum wells. For instance, the noise sheet conductivity per layer can be defined as

$$\sigma_n(\omega) = i\frac{N_s e^2}{m}\frac{1}{\omega + M(\omega)}, \tag{4.61}$$

with N_s the carrier sheet density per layer. The noise temperature is

$$T_n(\omega) = \frac{S(\omega)}{2mM_2(\omega)}. \tag{4.62}$$

The diffusion coefficient is

$$D_n(\omega) = \frac{T_n(\omega)}{m}\operatorname{Re}\left[\frac{i}{\omega + M(\omega)}\right]. \tag{4.63}$$

Here the memory functions $M(\omega)$ of quasi-2D and superlattice systems were given by (3.69)–(3.70) and (3.72)–(3.73) in Chapter 3. The fluctuating force correlation function $S(\omega)$ consists of impurity and phonon contributions:

$S(\omega) = S^{(i)}(\omega) + S^{(p)}(\omega)$. For quasi-2D systems described in Sec. 2.5, the expressions of $S^{(i)}(\omega)$ and $S^{(p)}(\omega)$ are

$$S^{(i)}(\omega) = -\frac{1}{N_s} \sum_{n',n,\boldsymbol{q}_\parallel} q_\alpha^2 |U_{n'n}(\boldsymbol{q}_\parallel)|^2 \Pi_2(n',n,\boldsymbol{q}_\parallel,\omega-\omega_0) \coth\left(\frac{\omega-\omega_0}{2T_e}\right),$$
(4.64)

$$S^{(p)}(\omega) = -\frac{\omega}{2N_s} \sum_{n',n,\boldsymbol{q},\lambda} q_\alpha^2 |M_{n'n}(\boldsymbol{q},\lambda)|^2 \Big\{ \Pi_2(n',n,\boldsymbol{q}_\parallel,\Omega_{\boldsymbol{q}\lambda}+\omega-\omega_0)$$

$$\times \left[\coth\left(\frac{\Omega_{\boldsymbol{q}\lambda}+\omega-\omega_0}{2T_e}\right)(2n_{\boldsymbol{q}\lambda}+1)-1\right] + (\omega \to -\omega)\Big\}. \quad (4.65)$$

For superlattices described in Sec. 2.7, we have

$$S^{(i)}(\omega) = -\frac{n_i}{N_s}\left(\frac{e^2}{2\epsilon_0\kappa}\right)^2 \frac{d}{2\pi} \int_{-\pi/d}^{\pi/d} dq_z \sum_{\boldsymbol{q}_\parallel} \frac{q_\alpha^2}{q_\parallel^2} \tilde{N}(\boldsymbol{q}_\parallel,q_z) g(q_z)$$

$$\times \Pi_2(q_z,\boldsymbol{q}_\parallel,\omega-\omega_0)\coth\left(\frac{\omega-\omega_0}{2T_e}\right), \quad (4.66)$$

$$S^{(p)}(\omega) = -\frac{\omega}{2N_s} \sum_{\boldsymbol{q},q_z,\lambda} q_\alpha^2 |M(\boldsymbol{q},\lambda)|^2 |I(\mathrm{i}\, q_z)|^2 \Big\{ \Pi_2(q_z,\boldsymbol{q}_\parallel,\Omega_{\boldsymbol{q}\lambda}+\omega-\omega_0)$$

$$\times \left[\coth\left(\frac{\Omega_{\boldsymbol{q}\lambda}+\omega-\omega_0}{2T_e}\right)(2n_{\boldsymbol{q}\lambda}+1)-1\right] + (\omega \to -\omega)\Big\}. \quad (4.67)$$

Here $n_{\boldsymbol{q}\lambda} = n(\Omega_{\boldsymbol{q}\lambda}/T)$ is the number of phonons with wavevector $\boldsymbol{q}$ in the λth branch, $\omega_0 \equiv q_x v_d$, $\alpha = x$ for the longitudinal configuration and $\alpha = y$ for the transverse configuration.

Besides, the low-frequency ($\omega \to 0$) Nyquist relation (4.33) and Einstein relation (4.34), and the high-frequency relation (4.35) are also valid for quasi-2D systems and superlattices.

Based on these results Lei and Horing (1987b) carry out careful numerical analyses of noise temperatures in the electron transport parallel to the plane in GaAs/AlGaAs heterojunctions, quantum wells and superlattices, assuming carriers to occupy only the lowest subband but incorporating the Coulomb interaction between carriers by using the full temperature-, wavevector-, and frequency-dependent density correlation function in the random phase approximation. In this, scatterings due to impurities, acoustic phonons and polar optic phonons are taken into account. As an example, Fig. 4.1 presents the calculated longitudinal and transverse noise temperatures $T_{\mathrm{nl}}(\omega)$ and $T_{\mathrm{nt}}(\omega)$ as functions of $\omega/\Omega_{\mathrm{LO}}$

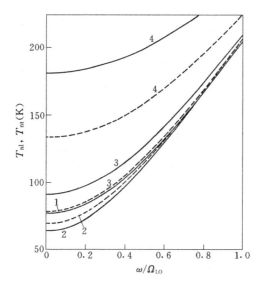

Fig. 4.1 Longitudinal (solid curves) and transverse (dashed curves) noise temperatures, $T_{nl}(\omega)$ and $T_{nt}(\omega)$, are shown as functions of the normalized frequency ω/Ω_{LO} in a GaAs quantum well superlattice at lattice temperature $T = 77\,\text{K}$ for different dc bias fields E: 1— $E \sim 0$, i.e. the Nyquist relation (4.33); 2— $E = 100\,\text{V/cm}$; 3— $E = 400\,\text{V/cm}$; 4— $E = 1\,\text{kV/cm}$.

at lattice temperature $T = 77\,\text{K}$, for a GaAs-based quantum-well superlattice of well width $a = 20\,\text{nm}$, period $d = 30\,\text{nm}$, carrier sheet density $N_s = 2.0 \times 10^{11}\,\text{cm}^{-2}$ per layer, and zero-temperature Ohmic mobility $\mu_0 = 1.0 \times 10^6\,\text{cm}^2/\text{Vs}$. The curve labeled 1 represents the Nyquist relation (4.33) at $T = 77\,\text{K}$. The behavior of the noise temperature is essentially determined by the electron temperature. Generally, at lower dc bias, $T_{nl}(\omega)$ is lower than $T_{nt}(\omega)$, whileas at higher dc bias $T_{nl}(\omega)$ is higher than $T_{nt}(\omega)$. Note that the frequency scale in the figure is the optic-phonon frequency $\Omega_{LO} \sim 5.4 \times 10^{13}\,\text{s}^{-1}$, around 10 THz. Therefore ω/Ω_{LO} is always small even for the millimeter microwave. In Fig. 4.2 we show the zero-frequency values of the longitudinal and transverse noise temperatures $T_{nl} \equiv T_{nl}(0)$ and $T_{nt} \equiv T_{nt}(0)$ versus the bias electric field E, together with the electron temperature T_e. Though T_{nl} and T_{nt} can be higher or lower than T_e, their trend is essentially the same as the electron temperature.

An interesting phenomenon worth mentioning is that $T_{nl}(\omega)$ and $T_{nt}(\omega)$ can be lower than the lattice temperature $T = 77\,\text{K}$, over a quite wide range of dc bias field and over a quite wide range of frequency, as seen

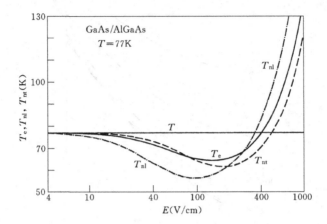

Fig. 4.2 The electron temperature T_e and longitudinal and transverse noise temperatures T_{nl} and T_{nt} at $\omega \sim 0$ are shown as functions of the bias dc field for the same system as described in Fig. 4.1 at lattice temperature $T = 77$ K.

in Figs. 4.1 and 4.2. This is a kind of electron cooling similar to that discussed in Chapter 2, where the electron temperature can be lower than the lattice temperature when acoustic phonon scattering dominates. The electron cooling discussed here in GaAs-based systems, is mainly due to polar optic phonon scattering and shows up as the electron temperature and noise temperatures lowering than the lattice temperature. It should be noted that the electron temperature in the balance-equation theory is not a directly measurable quantity, but the noise temperatures T_{nl} and T_{nt} are directly measurable. Therefore, measuring the noise temperatures provides a way to detect the effect of electron cooling.

In the case of higher carrier density or higher temperature, the lowest subband model is not good enough and we have to consider the effect of higher subbands. Fig 4.3 shows the electron temperature T_e, and longitudinal and transverse noise temperatures T_{nl} and T_{nt} versus the bias electric field, for two GaAs/AlGaAs heterojunctions at lattice temperature $T = 80$ K, calculated using a two subband model (Dong and Lei, 1998b). The system in Fig. 4.3(a) has an electron density $N_s = 1.9 \times 10^{11}\,\mathrm{cm}^{-2}$ and an energy separation $\varepsilon_1 - \varepsilon_0 = 9$ meV between two lowest subbands (derived from variational calculation). The system in Fig. 4.3(b) has an electron density $N_s = 5.7 \times 10^{11}\,\mathrm{cm}^{-2}$ and an energy separation $\varepsilon_1 - \varepsilon_0 = 31$ meV. For comparison the results calculated using the single lowest subband model are also shown in the figures. We can see that, in the case of higher carrier

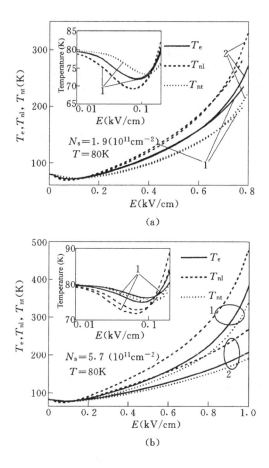

Fig. 4.3 The electron temperature T_e (solid curves) and longitudinal and transverse noise temperatures T_{nl} (dashed curves) and T_{nt} (dotted curves) at $\omega \sim 0$ versus the bias electric field at lattice temperature $T = 80\,\text{K}$, obtained from one-subband and two-subband models for two GaAs heterojunctions as described in the text: (a) electron density $N_s = 1.9 \times 10^{11}\,\text{cm}^{-2}$; (b) electron density $N_s = 5.7 \times 10^{11}\,\text{cm}^{-2}$. From Dong and Lei (1998b).

density ($N_s = 5.7 \times 10^{11}\,\text{cm}^{-2}$), the electron temperature and noise temperatures in the high bias field range are much lower and effects of electron cooling are much weaker obtained from two subband model than those from one subband model. In the case of lower carrier density ($N_s = 1.9 \times 10^{11}\,\text{cm}^{-2}$), the results from two-band model and from one-band model exhibit only minor difference.

4.5 Effects of Velocity Fluctuation on the Drift Motion

The expansion of $A(\mathbf{q}, t, t')$ function to linear order of the velocity fluctuation $\delta \mathbf{V}$ in Sec. 4.1, formula (4.13), gives the impression that the velocity fluctuation does not affect the drift motion. This may not be true, as can be seen by continuing the expansion to higher order of $\delta \mathbf{V}$. In fact, in deriving the force- and energy-balance equations, we may take the average over the whole exponential factor without expanding it, to obtain the following result from the cumulate approximation:

$$\left\langle \exp\left[i\mathbf{q} \cdot \int_{t'}^{t} (\mathbf{v}(s) + \delta \mathbf{V}(s))\, ds\right] \right\rangle \approx$$
$$\exp\left[i\mathbf{q} \cdot \int_{t'}^{t} \mathbf{v}(s)\, ds - \frac{q^2}{6} \int_{t'}^{t} du \int_{t'}^{t} ds\, \langle \delta \mathbf{V}(u) \cdot \delta \mathbf{V}(s) \rangle \right]. \quad (4.68)$$

Thus, the correlation of velocity fluctuations enters the frictional force $\mathbf{f}$ and the energy transfer rate w. Furthermore, in view of $\langle (\mathbf{v}+\delta\mathbf{V})^2 \rangle = v^2 + \langle \delta V^2 \rangle$ there may appear an extra term

$$nm \frac{d}{dt} \langle \delta V(t)^2 \rangle \quad (4.69)$$

in the energy-balance equation.

In this way the velocity fluctuation affects the drift motion through the velocity-fluctuation autocorrelation function

$$\gamma(u - s) \equiv \langle \delta \mathbf{V}(u) \cdot \delta \mathbf{V}(s) \rangle. \quad (4.70)$$

Assuming τ_v to be the characteristic correlation time of this correlation function, such that

$$\gamma(u - s) \approx \begin{cases} \gamma(0) & \text{if } |u - s| \ll \tau_v, \\ 0 & \text{if } |u - s| \gg \tau_v, \end{cases} \quad (4.71)$$

we have

$$B \equiv \frac{q^2}{6} \int_{t'}^{t} du \int_{t'}^{t} ds \langle \delta \mathbf{V}(u) \cdot \delta \mathbf{V}(s) \rangle$$

$$\approx \begin{cases} \frac{1}{6} q^2 \gamma(0)(t - t')^2 & \text{for } |t - t'| \ll \tau_v, \quad (4.72) \\ \frac{1}{3N} q^2 D_n(0)|t - t'| & \text{for } |t - t'| \gg \tau_v, \quad (4.73) \end{cases}$$

where
$$D_n(0) = \frac{N}{2} \int_{-\infty}^{\infty} \gamma(\tau) d\tau \qquad (4.74)$$

is the generalized diffusion coefficient (4.24) defined by the autocorrelation function of center-of-mass velocity fluctuations (Lei and Horing, 1987b).

Two different approximations are used in examining this. Xing and Ting (1987) estimate the autocorrelation function of velocity fluctuations using the high-temperature result obtained from the Langevin equation with zero dc bias (Ting and Nee, 1986)
$$\langle \delta \mathbf{V}(u) \cdot \delta \mathbf{V}(s) \rangle = \frac{3T_e}{Nm} \exp\left(-\frac{u-s}{\tau}\right), \qquad (4.75)$$

which involves a characteristic correlation time $\tau_v = \tau = -1/M_2(0)$. They assume that τ is much larger than the correlation time τ_c of the density correlation functions $\Pi(\mathbf{q}, t-t')$ and $\Lambda(\mathbf{q}, \lambda, t-t')$ and use the approximation (4.72) for B:
$$B = \frac{q^2 T_e}{2Nm}(t-t')^2. \qquad (4.76)$$

Here N is the total number of particles. Since the effective $t-t'$ range of integration is limited by τ_c, we can use $t-t' = \tau_c = 10^{-13}$ s in estimating B. Assuming $T_e = 3 \times 10^3$ K, $q = 1 \times 10^{10}$ m^{-1} and $m = 9 \times 10^{-31}$ kg, we find that only when the total number of particles N is as low as 10^5 can the velocity-fluctuation autocorrelations be appreciable.

Hu and O'Connell (1988a), on the other hand, use the approximation (4.73). This carries the implication that $\tau_c \gg \tau_v$, such that the major contributions to the frictional force and the energy transfer rate come from the range $t-t' \gg \tau_v$. With
$$B = \frac{D_n}{3N} q^2 (t-t'), \qquad (4.77)$$

we obtain the same force- and energy-balance equations as those without velocity fluctuation, save for the replacement
$$\omega_0 \to \omega_0 + i\frac{D_n}{3N} q^2, \qquad (4.78)$$

leading to an energy broadening. Its effects have been discussed extensively by Hu and O'Connell (1988a,d,e; 1989a,b) and we refer the reader to their original papers for further detail. For ordinary macroscopic systems in which the total number of particles is large, this velocity fluctuation autocorrelation effect is always negligible. For small systems, e.g. in microstructures, where N may be as low as 10^5, velocity-fluctuation autocorrelations become appreciable.

It is worth noting that, although choosing Eq. (1.55) or Eq. (1.60) as the energy balance equation does not make any difference up to the linear order in δV, it may have some difference in the terms of order of δV^2. This problem merits further investigation.

Chapter 5

Effects of Nonequilibrium and Confined Phonons

5.1 The Energy-Loss Rate of Electron Systems, Nonequilibrium Phonons

When a dc electric field applies to or a high-frequency electromagnetic wave irradiates on a semiconductor, the carriers in it generally gain energy and deviate from the equilibrium state. These nonequilibrium carriers will transfer energy to the lattice through their coupling with phonons, and return back to equilibrium after the applied field or irradiation is turned off, or approach a steady transport state under the persistent influence of the dc or high-frequency field. The energy that carriers transfer to the lattice in a unit time interval, i.e. the energy-loss rate of carriers, or the energy-transfer rate from the carrier system to the phonon system, is a key physical quantity in the process of nonequilibrium carrier relaxation and in the transport driven by intense electric field.

We have dealt with the electron energy-transfer rate w many times in the energy-balance equations for electron transport driven by an electric field. In the preceding chapters, the electron energy-transfer rate w was derived based on the assumptions that (i) the internal thermalization time of the carrier system is sufficiently short that it can be described by an electron temperature T_e, and (ii) the phonon system is always in an equilibrium state during the whole process of carrier relaxation or transport. With these assumptions, the energy-transfer rate w of a transported electron system can be expressed as formula (2.6). Without considering effect of carrier drifting (there is no carrier drifting in the relaxation process of photon-excited hot carriers), the electron energy-loss rate induced by the coupling between carriers and longitudinal optic (LO) phonons of frequency Ω_{LO}, can be written as

$$\begin{aligned} w_K &= \frac{\Omega_{\rm LO}}{\tau_0}\left[n\left(\frac{\Omega_{\rm LO}}{T_e}\right) - n\left(\frac{\Omega_{\rm LO}}{T}\right)\right] \\ &\approx \frac{\Omega_{\rm LO}}{\tau_0}\left[\exp\left(-\frac{\Omega_{\rm LO}}{T_e}\right) - \exp\left(-\frac{\Omega_{\rm LO}}{T}\right)\right]. \end{aligned} \quad (5.1)$$

The last approximate equality is valid when both T_e and T are much less than $\Omega_{\rm LO}$. (5.1) is the well-known Kogan (1963) formula, where

$$1/\tau_0 = -2\sum_{\boldsymbol{q}} |M(\boldsymbol{q},{\rm LO})|^2 \Pi_2(\boldsymbol{q},\Omega_{\rm LO}) \quad (5.2)$$

is the inverse of the relaxation time associated with electron–LO-phonon scattering. It is easily obtained from the electron–LO-phonon scattering matrix element $M(\boldsymbol{q},{\rm LO})$ and the density correlation function $\Pi_2(\boldsymbol{q},\Omega)$ of the carrier system. Under the influence of an electric field the transported electron system has an integrative drift motion, and the energy-tranfer rate from the electron system to phonon system (the electron energy-loss rate) is given by (2.6). The major difference between Eq. (2.6) and the Kogan formula (5.1) lies in a Dopler shift in the phonon frequency associated with electron temperature: $\Omega_{\rm LO}/T_e \to (\Omega_{\rm LO} - \boldsymbol{q}\cdot\boldsymbol{v})/T_e$, where $\boldsymbol{v}$ is the drift velocity of the electron system. In the case of nonlinear transport driven by an intense electric field, this Dopler frequency shift can not only induce strongly anisotropic phonon emission, but also give rise to a total electron energy loss from Eq. (2.6) which has several times difference from that of Eq. (5.1). Nevertheless, to calculate the energy-loss rate from expression (2.6) one has to first obtain the relation between v and T_e from the balance equations or from other transport analysis. The Kogan formula (5.1), which neglects the effect of Dopler shift and directly gives the energy-loss rate as a function of T_e and τ_0 without need to solve any equation, still serves as a qualitative estimation of the energy loss of transported electron system subject to an electric field.

In the middle 1980s, Shah et al (1985) observed carrier energy-loss rates in multilayer GaAs-based quantum wells of high carrier density which are much lower than (only $1/8 \sim 1/10$ of) that theoretically predicted by Eq. (2.6) and Eq. (5.1). They attributed this to the nonequilibrium population of optic phonons during the carrier relaxation or transport process, so call nonequlibirium phonon effect or hot phonon effect. This discovery stimulated substantial studies on the interplay between hot electrons and nonequilibrium phonons (Price, 1985; Kocevar, 1985; Cai, Marchetti and

Lax, 1986, 1988; Lei and Horing, 1987a, 1987b; Lei, Cui and Horing, 1987; Lugli and Goodnick, 1987; Ridley, 1989a; Rieger et al, 1989). Despite that there may exist other mechanisms to reduce the electron energy-loss rate in low-dimensional semiconductors, people now generally believe that the occurrence of nonequilibrium or hot phonons is indeed the main reason for this significant reduction of electron energy-loss rate in these quantum well systems. The occurrence of nonequilibrium population of phonons can not only greatly change the electron energy-loss rate but also affect the frictional force and other transport quantities. This is the subject of hot phonon dynamics.

Hot phonon dynamics have been extensively explored by using the Boltzmann equation and the Monte-Carlo technique, as well as with the balance equation approach. It is particularly convenient to investigate the effects of nonequilibrium phonon population on hot electron transport within the framework of balance equation theory. In this chapter we provide a brief description based on the analysis of Lei and Horing (1987a, 1987b), and Lei, Cui and Horing (1987) in dealing with high-field transport. The relaxation of photon-excited nonequilibrium carriers can be treated as a specific case of transport without drift motion. An alternative description of the issue can be found in a review article by Lax and Cai (1992).

5.2 Nonequilibrium Phonons in 3D Bulk Systems

We consider a three-dimensional electron–phonon system with unit volume $V = 1$. To investigate nonequilibrium phonons during the carrier transport process one has to somewhat take the interactions inside the phonon system into account. Under the influence of a uniform electric field $\boldsymbol{E}$, the Hamiltonian of a three-dimensional electron–phonon system consists of a center-of-mass part $H_{\rm cm}$, a relative electron part $H_{\rm er}$ (including electron–electron interaction), a free phonon part $H_{\rm ph}$, together with electron–impurity scattering $H_{\rm ei}$, electron–phonon coupling $H_{\rm ep}$, and phonon–phonon interaction $H_{\rm pp}$:

$$H = H_{\rm cm} + H_{\rm er} + H_{\rm ph} + H_{\rm ei} + H_{\rm ep} + H_{\rm pp}. \qquad (5.3)$$

Here $H_{\rm cm}$, $H_{\rm er}$, $H_{\rm ph}$, $H_{\rm ei}$, and $H_{\rm ep}$ have been given by (1.10)–(1.15). In 3D bulk systems the phonon eigenmodes are lattice waves characterized by $\boldsymbol{q}\lambda$. The energy of a phonon with wavevector $\boldsymbol{q}$ in branch λ is $\Omega_{\boldsymbol{q}\lambda}$. We first consider the time variation of the phonon number operator of $\boldsymbol{q}\lambda$ mode,

$$\hat{N}_{\boldsymbol{q}\lambda} = b^\dagger_{\boldsymbol{q}\lambda} b_{\boldsymbol{q}\lambda}. \tag{5.4}$$

According to Heisenberg equation of motion, the electron–phonon interaction H_{ep} (1.15) gives rise to a rate of change of $\hat{N}_{\boldsymbol{q}\lambda}$,

$$\left(\frac{\partial \hat{N}_{\boldsymbol{q}\lambda}}{\partial t}\right)_{\mathrm{ep}} = -\mathrm{i}\left[\hat{N}_{\boldsymbol{q}\lambda}, H_{\mathrm{ep}}\right]$$
$$= \mathrm{i}M(q,\lambda)\,\mathrm{e}^{\mathrm{i}\boldsymbol{q}\cdot\boldsymbol{R}} b_{\boldsymbol{q}\lambda}\rho_{\boldsymbol{q}} - \mathrm{i}M(-q,\lambda)\,\mathrm{e}^{-\mathrm{i}\boldsymbol{q}\cdot\boldsymbol{R}} b^\dagger_{\boldsymbol{q}\lambda}\rho_{-\boldsymbol{q}}, \tag{5.5}$$

where $\rho_{\boldsymbol{q}}$ is the electron density operator (1.16). The density matrix $\hat{\rho}$ of the relative-electron–phonon system described by the Hamiltonian (5.3) still obeys the Liouville equation (1.28), but now the phonon–phonon interaction H_{pp} is put into the perturbation Hamiltonian H_{I} as a possible scenario for the investigation of nonequilibrium phonon occupation:

$$H_{\mathrm{I}} = H_{\mathrm{ei}} + H_{\mathrm{ep}} + H_{\mathrm{pp}}. \tag{5.6}$$

We explain this treatment as follows. In the presence of electron–phonon interaction the transported electron system driven by an electric field may excite or absorb phonons, tending to drive the phonon system out of equilibrium. The phonon–phonon coupling H_{pp}, on the other hand, tends to relax the whole phonon system towards equilibrium, just as the electron–electron interaction to thermalize the relative electron system. Thus far, we have in fact assumed that phonons relax much faster than the tendency to produce nonequilibrium phonons (mainly due to H_{ep}), such that phonons are always in a thermal equilibrium state with a heat bath at the "lattice temperature" T during the whole electron transport process. In other words, H_{pp} is considered to be included in the phonon part of the Hamiltonian: $\widetilde{H}_{\mathrm{ph}} = H_{\mathrm{ph}} + H_{\mathrm{pp}}$ and only $H_{\mathrm{ei}} + H_{\mathrm{ep}}$ is left as the perturbation. Because of this the initial density matrix is chosen as

$$\hat{\rho}_0 = \frac{1}{Z}\mathrm{e}^{-H_{\mathrm{er}}/T_{\mathrm{e}}}\mathrm{e}^{-\widetilde{H}_{\mathrm{ph}}/T}, \tag{5.7}$$

in which the electron temperature T_{e} is a parameter, which may be time-dependent during a slow transient process. The lattice temperature T, on the other hand, is a given external condition. The main role of H_{pp} in $\widetilde{H}_{\mathrm{ph}}$, as that of electron–electron interaction in H_{er}, is to promote the rapid thermalization of the phonon system. Although in addition to this, H_{pp} may result in a renormalization of the phonon dispersion, since the phonon system is essentially in an equilibrium state, the renormalization-induced changes of phonon dispersion rarely produce important effects on the electron transport. However, if the internal thermalization trend is not strong enough in comparison with the out-of-equilibrium tendency due to electron–phonon interaction, we have to give up the assumption that the phonon system is in an equilibrium state. To deal with effects of nonequilibrium phonons, one should use proper phonon modes, separate $H_{\rm pp}$ from $H_{\rm ph}$ and put it into $H_{\rm I}$ (together with $H_{\rm ei}$ and $H_{\rm ep}$) treated as perturbation as indicated in Eq. (5.6). The initial density matrix of the electron–phonon system can be chosen in the following way. We image to turn off $H_{\rm I}$ and the applied electric field after the system reaches a steady state. The center of mass, the relative electron system, and the phonon system are decoupled from each other and rapidly approach individual equilibrium states. More accurately speaking, due to strong electron–electron interaction the relative electron system rapidly approaches an equilibrium state with temperature $T_{\rm e}$. If the phonon eigenmodes are spatially matched with the electron states, the phonons, excited by electrons in transport due to electron–phonon interaction, are essentially those close to the eigenmodes. Therefore, after turning off $H_{\rm ep}$, phonons should quickly reach a quasi-equilibrium distribution first within each eigenmode. Since $H_{\rm pp}$ is also turned off, different phonon eigenmodes become independent and may have different temperatures $T_{\boldsymbol{q}\lambda}$. Therefore, the density matrix of the decoupled relative-electron–phonon system takes the form

$$\hat{\rho}_0 = \frac{1}{Z}\exp(-H_{\rm er}/T_{\rm e})\exp\bigl(-\sum_{\boldsymbol{q}\lambda} H_{\boldsymbol{q}\lambda}/T_{\boldsymbol{q}\lambda}\bigr), \qquad (5.8)$$

where $H_{\boldsymbol{q}\lambda} \equiv \Omega_{\boldsymbol{q}\lambda}\hat{N}_{\boldsymbol{q}\lambda}$ is the Hamiltonian of the $\boldsymbol{q}\lambda$ phonon mode and $T_{\boldsymbol{q}\lambda}$ is a mode-dependent temperature parameter. We choose (5.8) as the initial density matrix. The density matrix $\hat{\rho}$ of the relative-electron–phonon system under the influence of an electric field should be determined from the Liouville equation (1.28) and the initial condition (5.8). To the linear order of the interaction $H_{\rm I}$, we have

$$\hat{\rho}(t) = \hat{\rho}_0 + {\rm i}\int_{-\infty}^{t} dt'\,\exp\Bigl[{\rm i}\sum_{\boldsymbol{q},\lambda}(1-\alpha_{\boldsymbol{q}\lambda})H_{\boldsymbol{q}\lambda}(t'-t)\Bigr]$$
$$\times\bigl[\hat{\rho}_0, H_{\rm I}(t'-t)\bigr]\exp\Bigl[-{\rm i}\sum_{\boldsymbol{q},\lambda}(1-\alpha_{\boldsymbol{q}\lambda})H_{\boldsymbol{q}\lambda}(t'-t)\Bigr]. \qquad (5.9)$$

Here $\alpha_{q\lambda} \equiv T_e/T_{q\lambda}$, $O(t) \equiv e^{iH_0 t} O e^{-iH_0 t}$ (O stands for H_I or $H_{q\lambda}$, and $H_0 \equiv H_{er} + H_{ph}$ is the Hamiltonian of relative electrons and noninteracting phonons). The statistical average of the operator $(\partial \hat{N}_{q\lambda}/\partial t)_{ep}$ can be evaluated to the lowest nonzero order of H_I in accordance with (1.38), yielding

$$\left[\frac{\partial n_{q\lambda}}{\partial t}\right]_{ep} \equiv \left\langle \left(\frac{\partial \hat{N}_{q\lambda}}{\partial t}\right)_{ep}\right\rangle$$

$$= 2|M(\bm{q},\lambda)|^2 \Pi_2(\bm{q}, \Omega_{q\lambda} - \bm{q}\cdot\bm{v}) \left[n_{q\lambda} - n\left(\frac{\Omega_{q\lambda} - \bm{q}\cdot\bm{v}}{T_e}\right)\right].$$

(5.10)

Here,
$$n_{q\lambda} = \langle \hat{N}_{q\lambda}\rangle = n(\Omega_{q\lambda}/T_{q\lambda}) \qquad (5.11)$$

is the population number of phonons in mode $\bm{q}\lambda$, and $\bm{v}$ is the drift velocity of the electron system in transport under the influence of a uniform electric field. In deriving (5.10), we have used the approximation $A(\bm{q},t,t') \approx \exp[i\bm{q}\cdot\bm{v}(t)(t-t')]$, neglecting the memory effect and assuming the slowly time-varying $\bm{v}(t)$ [(3.2) without $\dot{\bm{v}}(t)$ and higher order terms]. The frictional force experienced by the center of mass of the electron system, $\bm{f} = \bm{f}_i + \bm{f}_p$ ($\bm{f}_i$ is due to impurities and $\bm{f}_p$ due to phonons), and the energy transfer rate from the electron system to the phonon system, w, are readily determined in the same approximation. Since $\bm{f}_i$ is irrelevant to the phonon system, it has the same expression as Eq. (3.5) in Chapter 3,

$$\bm{f}_i = n_i \sum_{\bm{q}} \bm{q} |u(\bm{q})|^2 \Pi_2(\bm{q}, \bm{q}\cdot\bm{v}), \qquad (5.12)$$

while $\bm{f}_p$ and w can be written in the physically transparent forms by means of $[\partial n_{q\lambda}/\partial t]_{ep}$, the rate of change of the phonon number in mode $\bm{q}\lambda$ due to electron–phonon interaction:

$$\bm{f}_p = -\sum_{\bm{q},\lambda} \bm{q} \left[\frac{\partial n_{q\lambda}}{\partial t}\right]_{ep}, \qquad (5.13)$$

$$w = \sum_{\bm{q},\lambda} \Omega_{q\lambda} \left[\frac{\partial n_{q\lambda}}{\partial t}\right]_{ep}, \qquad (5.14)$$

involving $\bm{q}$ and $\Omega_{q\lambda}$, the quanta of phonon momentum and energy.

The relaxation, or the rate of change of phonon occupation number due to phonon–phonon interaction, $[\partial n_{q\lambda}/\partial t]_{pp} = -i\langle[\hat{N}_{q\lambda}, H_{pp}]\rangle$, can not be obtained without a detailed analysis of anharmonic couplings, as well as structure and boundary effects which are sample dependent. We will not treat these issues in detail but simply represent them with an effective phonon relaxation time parameter τ_p to describe the process for the phonon system to approach the equilibrium:

$$\left[\frac{\partial n_{q\lambda}}{\partial t}\right]_{pp} = -\frac{1}{\tau_p}\left[n_{q\lambda} - n\left(\frac{\Omega_{q\lambda}}{T}\right)\right], \tag{5.15}$$

$n(\Omega_{q\lambda}/T)$ is the phonon number in mode $q\lambda$ in the equilibrium state of temperature T. Nonequilibrium phonons may occur if the phonon relaxation is not fast enough. τ_p may be mode dependent. For instance, a LO phonon wavepacket relaxes much slower than that of the acoustic phonon due to its much slower group velocity. Therefore, nonequilibrium phonon effect is generally much more important for LO phonons than for acoustic phonons.

The total rate of change of phonon occupation number can be written for each phonon mode as

$$\frac{d}{dt}n_{q\lambda} = \left[\frac{\partial n_{q\lambda}}{\partial t}\right]_{ep} - \frac{1}{\tau_p}\left[n_{q\lambda} - n\left(\frac{\Omega_{q\lambda}}{T}\right)\right]. \tag{5.16}$$

These equations, taken jointly with the force- and energy-balance equations of the electron system,

$$nm\frac{d\boldsymbol{v}}{dt} = ne\boldsymbol{E}(t) + \boldsymbol{f}_i + \boldsymbol{f}_p, \tag{5.17}$$

$$C_e\frac{dT_e}{dt} = -\boldsymbol{v}\cdot(\boldsymbol{f}_i + \boldsymbol{f}_p) - w, \tag{5.18}$$

together with the expressions (5.10)–(5.14) for $[\partial n_{q\lambda}/\partial t]_{ep}$, $\boldsymbol{f}_i$, $\boldsymbol{f}_p$ and w, constitute a complete set of ordinary differential equations to determine all the parameters involved, $\boldsymbol{v}$, T_e and $n_{q\lambda}$ (or $T_{q\lambda}$), as functions of time t from a given initial state.

For a dc steady state, the occupation number for each phonon mode is constant, $dn_{q\lambda}/dt = 0$, and Eqs. (5.15) and (5.10) lead to

$$n_{q\lambda} = n\left(\frac{\Omega_{q\lambda} - \boldsymbol{q}\cdot\boldsymbol{v}}{T_e}\right) + [1 + \tau_p\Gamma(\boldsymbol{q},\lambda,\Omega_{q\lambda} - \boldsymbol{q}\cdot\boldsymbol{v})]^{-1}$$
$$\times\left[n\left(\frac{\Omega_{q\lambda}}{T}\right) - n\left(\frac{\Omega_{q\lambda} - \boldsymbol{q}\cdot\boldsymbol{v}}{T_e}\right)\right], \tag{5.19}$$

where

$$\Gamma(\boldsymbol{q}, \lambda, \Omega_{q\lambda} - \boldsymbol{q} \cdot \boldsymbol{v}) = -2|M(\boldsymbol{q}, \lambda)|^2 \Pi_2(\boldsymbol{q}, \Omega_{q\lambda} - \boldsymbol{q} \cdot \boldsymbol{v}) \quad (5.20)$$

is a phonon-mode and electron-drift-velocity dependent inverse scattering time, reflecting the coupling strength between electrons and mode-$\boldsymbol{q}\lambda$ phonons. If $\tau_\text{p} = 0$, Eq. (5.19) yields $n_{q\lambda} = n(\Omega_{q\lambda}/T)$, i.e. the equilibrium phonon occupation. For nonzero τ_p, the phonon occupation number deviates from the equilibrium value, such that the mode-dependent phonon temperature $T_{q\lambda}$, which is determined by $n_{q\lambda} \equiv n(\Omega_{q\lambda}/T_{q\lambda})$ and dependent on $\boldsymbol{v}$, can be much higher than the lattice temperature T. This is so-called hot-phonon or "bottle-neck" phenomenon. Note that, in the case of nonzero τ_p, the phonon distribution is strongly affected by the electron drift velocity and no longer isotropic for finite $\boldsymbol{v}$. Its peak in the $\boldsymbol{q}$ space generally shifts to somewhere along the direction of the electron drift velocity, as indicated in the argument $\Omega_{q\lambda} - \boldsymbol{q} \cdot \boldsymbol{v}$ on the right-hand-side of (5.19). This is a kind of electron-drift induced phonon drag.

In a steady transport state, the phonon induced frictional force and the energy transfer rate of the electron system may be written as

$$\boldsymbol{f}_\text{p} = \sum_{\boldsymbol{q},\lambda} \frac{\boldsymbol{q}\Gamma(\boldsymbol{q}, \lambda, \Omega_{q\lambda} + \boldsymbol{q} \cdot \boldsymbol{v})}{1 + \tau_\text{p}\Gamma(\boldsymbol{q}, \lambda, \Omega_{q\lambda} + \boldsymbol{q} \cdot \boldsymbol{v})} \left[n\left(\frac{\Omega_{q\lambda} + \boldsymbol{q} \cdot \boldsymbol{v}}{T_\text{e}}\right) - n\left(\frac{\Omega_{q\lambda}}{T}\right) \right], \quad (5.21)$$

$$w = \sum_{\boldsymbol{q},\lambda} \frac{\Omega_{q\lambda}\Gamma(\boldsymbol{q}, \lambda, \Omega_{q\lambda} + \boldsymbol{q} \cdot \boldsymbol{v})}{1 + \tau_\text{p}\Gamma(\boldsymbol{q}, \lambda, \Omega_{q\lambda} + \boldsymbol{q} \cdot \boldsymbol{v})} \left[n\left(\frac{\Omega_{q\lambda} + \boldsymbol{q} \cdot \boldsymbol{v}}{T_\text{e}}\right) - n\left(\frac{\Omega_{q\lambda}}{T}\right) \right]. \quad (5.22)$$

Note that for this steady-state case, the nonequilibrium phonon occupation number $n_{q\lambda}$ does not appear in the expressions (5.21) and (5.22) for $\boldsymbol{f}_\text{p}$ and w. Therefore, for the determination of steady-state hot-electron transport from the electron force- and energy-balance equations,

$$ne\boldsymbol{E} + \boldsymbol{f}_\text{i} + \boldsymbol{f}_\text{p} = 0, \quad (5.23)$$

$$\boldsymbol{v} \cdot (\boldsymbol{f}_\text{i} + \boldsymbol{f}_\text{p}) + w = 0, \quad (5.24)$$

involves only an extra knowledge of the phonon relaxation time τ_p. If the phonon relaxation is fast enough, $\tau_\text{p}\Gamma \ll 1$, The expressions (5.21) and (5.22) for $\boldsymbol{f}_\text{p}$ and w become (3.6) and (3.7), i.e. those in the case of equilibrium phonon occupation. Otherwise the phonon-induced frictional force and the energy transfer rate for given values of T, T_e and v are evidently reduced.

It can be seen from (5.21) and (5.22) that the importance of the nonequilibrium phonon effect is determined by the parameter $\tau_\text{p}\Gamma$. The Γ function

(5.20) contains the electron density correlation function $\Pi_2(\boldsymbol{q},\omega)$ of unit volume and thus is proportional to the electron volume density. Therefore, the nonequilibrium phonon effect becomes more important for system with larger electron density if τ_p is kept unchanged.

5.3 Semiconductor Superlattices, 3D Phonon Model

A superlattice, which is composed of infinite number of periodically arranged quantum wells, fills the whole three-dimensional space. The electron density, though concentrating in the regions of potential wells, distributes periodically throughout the whole space. The phonons, excited (absorbed) by such periodical electron wavepackets, should still be able to rapidly relax to spatially uniform phonon eigenmodes. Therefore, it is reasonable to describe the phonons in a semiconductor superlattice with the 3D phonon model, i.e. the extended 3D lattice plane waves characterized by a wavevector $\boldsymbol{q}$ and a branch index λ, and assume that the excited nonequilibrium phonons quickly relax to an equilibrium state within each $\boldsymbol{q}\lambda$ mode. The treatment described in the preceding section is thus applicable. We consider a model plane superlattice as that discussed in Sec. 2.7, having transverse area $S = 1$, period d and length L with totally $N_\mathrm{L} = L/d$ layers. The electron–phonon interaction H_ep is given by (2.139). It is easily seen that the expression (5.10) now becomes

$$\left[\frac{\partial n_{\boldsymbol{q}\lambda}}{\partial t}\right]_\mathrm{ep} = \Gamma(\boldsymbol{q},\lambda,\Omega_{\boldsymbol{q}\lambda}-\boldsymbol{q}_\|\cdot\boldsymbol{v})\left[n\left(\frac{\Omega_{\boldsymbol{q}\lambda}-\boldsymbol{q}_\|\cdot\boldsymbol{v}}{T_\mathrm{e}}\right)-n_{\boldsymbol{q}\lambda}\right], \quad (5.25)$$

with the Γ function given by

$$\Gamma(\boldsymbol{q},\lambda,\omega) = -\frac{2}{d}|M(\boldsymbol{q},\lambda)|^2\,|I(\mathrm{i}\,q_z)|^2\,\Pi_2(q_z,\boldsymbol{q}_\|,\omega), \quad (5.26)$$

in which $\Pi_2(q_z,\boldsymbol{q}_\|,\omega)$ is the imaginary part of the (per layer) electron density correlation function defined by (2.148), $I(\mathrm{i}\,q_z)$ is the form factor defined by (2.140). The phonon induced frictional force and energy-loss rate of the electron system per layer can be written as

$$\boldsymbol{f}_\mathrm{p} = -\sum_{\boldsymbol{q}_\|,\lambda}\frac{d}{2\pi}\int_{-\infty}^{\infty}dq_z\,\boldsymbol{q}_\|\left[\frac{\partial n_{\boldsymbol{q}\lambda}}{\partial t}\right]_\mathrm{ep}, \quad (5.27)$$

$$w = \sum_{\boldsymbol{q}_\|,\lambda}\frac{d}{2\pi}\int_{-\infty}^{\infty}dq_z\,\Omega_{\boldsymbol{q}\lambda}\left[\frac{\partial n_{\boldsymbol{q}\lambda}}{\partial t}\right]_\mathrm{ep}. \quad (5.28)$$

The impurity-induced frictional force is given by (2.159). The phonon population satisfies equations similar to (5.15) and (5.16). Equations (5.17) and (5.18) for the time variations of the electron momentum and energy remain valid, as long as $\boldsymbol{f}_\mathrm{i}$, $\boldsymbol{f}_\mathrm{p}$ and w in them are replaced by the per-layer quantities defined by (2.159), (5.27) and (5.28), and C_e replaced by the specific heat per layer and n replaced by the electron density per layer N_s.

In the dc steady state, the phonon occupation number is given by

$$n_{q\lambda} = n\left(\frac{\Omega_{q\lambda} - \boldsymbol{q}_\parallel \cdot \boldsymbol{v}}{T_\mathrm{e}}\right) + \left[1 + \tau_\mathrm{p}\Gamma(\boldsymbol{q},\lambda,\Omega_{q\lambda} - \boldsymbol{q}_\parallel \cdot \boldsymbol{v})\right]^{-1}$$
$$\times \left[n\left(\frac{\Omega_{q\lambda}}{T}\right) - n\left(\frac{\Omega_{q\lambda} - \boldsymbol{q}_\parallel \cdot \boldsymbol{v}}{T_\mathrm{e}}\right)\right]. \quad (5.29)$$

The frictional force and the energy-loss rate of electrons are

$$\boldsymbol{f}_\mathrm{p} = \sum_{\boldsymbol{q}_\parallel,\lambda} \boldsymbol{q}_\parallel K(\boldsymbol{q}_\parallel,\lambda,\Omega_{q\lambda} + \boldsymbol{q}_\parallel \cdot \boldsymbol{v})\left[n\left(\frac{\Omega_{q\lambda} + \boldsymbol{q}_\parallel \cdot \boldsymbol{v}}{T_\mathrm{e}}\right) - n\left(\frac{\Omega_{q\lambda}}{T}\right)\right], \quad (5.30)$$

$$w = \sum_{\boldsymbol{q}_\parallel,\lambda} \Omega_{q\lambda} K(\boldsymbol{q}_\parallel,\lambda,\Omega_{q\lambda} + \boldsymbol{q}_\parallel \cdot \boldsymbol{v})\left[n\left(\frac{\Omega_{q\lambda} + \boldsymbol{q}_\parallel \cdot \boldsymbol{v}}{T_\mathrm{e}}\right) - n\left(\frac{\Omega_{q\lambda}}{T}\right)\right], \quad (5.31)$$

in which

$$K(\boldsymbol{q}_\parallel,\lambda,\Omega_{q\lambda} + \boldsymbol{q}_\parallel \cdot \boldsymbol{v}) = \frac{d}{2\pi}\int_{-\infty}^{\infty} dq_z \frac{\Gamma(\boldsymbol{q},\lambda,\Omega_{q\lambda} + \boldsymbol{q}_\parallel \cdot \boldsymbol{v})}{1 + \tau_\mathrm{p}\Gamma(\boldsymbol{q},\lambda,\Omega_{q\lambda} + \boldsymbol{q}_\parallel \cdot \boldsymbol{v})}. \quad (5.32)$$

Since the electron density correlation function $\Pi_2(q_z,\boldsymbol{q}_\parallel,\omega)$ is proportional to the electron density N_s per layer, the Γ function given by (5.26) is then proportional to $N_\mathrm{s}/d = n$, the equivalent bulk electron density. Therefore, as in 3D bulk systems, in the case of given τ_p, the quantity $\tau_\mathrm{p}\Gamma$, which determines the degree of nonequilibrium phonon effect in superlattices, is proportional to the volume density of electrons.

5.4 Nonequilibrium Phonons in Quasi-Two-Dimensional Systems

The emergence of nonequilibrium phonons during the carrier transport process is due to the electron–phonon interaction. Both the phonon eigenmodes (3D lattice waves) and the electron wave functions or density distributions discussed in Sections 5.1 and 5.2 are extending ones throughout the whole

3D space. The nonequilibrium phonons produced by the electron–phonon interaction are essentially spatially homogeneous and not far from lattice plane waves. These phonons quickly reach a quasi-equilibrium state for each plane wave mode $\boldsymbol{q}\lambda$, such that we have a temperature $T_{\boldsymbol{q}\lambda}$ for each $\boldsymbol{q}\lambda$ mode, before they relax to the equilibrium state for the whole phonon system with a unique lattice temperature T in a finite time of order of $\tau_{\rm p}$. According to this model, the importance of nonequilibrium phonon effect (determined by $\tau_{\rm p}\varGamma$) is proportional to the 3D bulk density of carriers. Such a model, though may be approximately useful for periodically arranged multiple quantum-well structures (treated as superlattices), is not appropriate for a quasi-two-dimensional system such as a single heterojunction or a quantum well. In a quasi-2D system, the carriers are confined in a very thin layer in, e.g. the z direction. Emission and absorption of phonons due to electron–phonon interaction occur mainly in the small region where electrons have substantial density. The spatial distribution of these phonons are, in any case, far from a spatially uniform 3D plane wave. The 3D-phonon model discussed in Secs. 5.1 and 5.2 is not a suitable description in this case, despite that the phonon eigenmodes (without electron–phonon interaction) remain the 3D extended lattice-waves characterized by $\boldsymbol{q}\lambda$. As a matter of fact, if the system volume has a length L in the z direction and the carrier sheet density of the single heterojunction or quantum well is $N_{\rm s}$, the equivalent 3D bulk carrier density of this quasi-2D system equals $N_{\rm s}/L$. Application of the 3D-phonon model to the quasi-2D system immediately leads to a nonphysical conclusion: the degree of hot-phonon effect would depend on the selection of the length L. The problem lies in the assumption that the nonequilibrium phonons excited by the electron–phonon interaction can rapidly reach a quasi-equilibrium state for each plane wave mode. This is inadequate when the electron wave functions greatly mismatch the spatially uniform distribution of phonon $\boldsymbol{q}\lambda$ modes.

Cai, Marchetti and Lax (1986) propose a way to overcome this difficulty, obtaining a physically reasonable description of hot-phonon effect in quasi-2D systems. Lei and Horing (1987a) reformulate it in the framework of balance equation approach. The main physical idea is to adapt, as the first step, the quasi-equilibrium distribution of phonon wavepackets which spatially match the electron wave function in quasi-2D systems.

We consider a quasi-2D system, e.g. a GaAs-based single heterojunction or quantum well, in the x–y plane having an area $S = 1$ and being contained inside a volume with a length L in the z direction. The electron state is characterized by a subband index n and a 2D wavevector $\boldsymbol{k}_{\|} = (k_x, k_y)$, with the energy and wave function given by (2.76) and (2.77). For simplicity, we consider only the lowest electron subband ($n = 0$) and neglect the subband indices hereafter in most cases. At the same time we will deal only with longitudinal optic (LO) phonons and assume them to be dispersionless ($\Omega_{\mathrm{LO}} = $ constant).

We define a set of creation and annihilation operators $A^\dagger(\boldsymbol{q}_\parallel)$ and $A(\boldsymbol{q}_\parallel)$ for the "effective two-dimensional phonons", which are constructed by the weighted sum of creation and annihilation operators $b^\dagger_{\boldsymbol{q}} \equiv b^\dagger_{\boldsymbol{q}_\parallel q_z}$ and $b_{\boldsymbol{q}} \equiv b_{\boldsymbol{q}_\parallel q_z}$ of the 3D plane-wave LO phonon with wavevector $\boldsymbol{q} \equiv (\boldsymbol{q}_\parallel, q_z)$ as

$$A^\dagger(\boldsymbol{q}_\parallel) = \sum_{q_z} g^*(\boldsymbol{q}_\parallel, q_z) b^\dagger_{\boldsymbol{q}_\parallel q_z}, \qquad (5.33)$$

$$A(\boldsymbol{q}_\parallel) = \sum_{q_z} g(\boldsymbol{q}_\parallel, q_z) b_{\boldsymbol{q}_\parallel q_z}. \qquad (5.34)$$

Here the weight factor

$$g(\boldsymbol{q}_\parallel, q_z) = \frac{M(\boldsymbol{q}_\parallel, q_z, \mathrm{LO})}{L^{1/2} M(q_\parallel)} I_{00}(\mathrm{i}\, q_z), \qquad (5.35)$$

$M(\boldsymbol{q}_\parallel, q_z, \mathrm{LO}) \equiv M(\boldsymbol{q}, \mathrm{LO})$ is the matrix element of the electron–LO-phonon interaction in the plane wave representation (2.129), $I(\mathrm{i}\, q_z) \equiv I_{00}(\mathrm{i}\, q_z)$ is the intraband form factor of the $n = 0$ subband (2.85). $M(q_\parallel)$ is a positive real function with its square given by

$$M^2(q_\parallel) = \frac{1}{L} \sum_{q_z} |M(\boldsymbol{q}_\parallel, q_z, \mathrm{LO})|^2 |I(\mathrm{i}\, q_z)|^2 = \frac{e^2 \Omega_{\mathrm{LO}}}{2\epsilon_0} \left(\frac{1}{\kappa_\infty} - \frac{1}{\kappa}\right) \frac{H(q_\parallel)}{2q_\parallel}, \qquad (5.36)$$

in which [see (2.80), neglecting the contribution from images]

$$H(q_\parallel) = H_{0000}(q_\parallel) = \int_{-\infty}^\infty dz_1 \int_{-\infty}^\infty dz_2\, \zeta_0^2(z_1) \zeta_0^2(z_2)\, \mathrm{e}^{-q_\parallel |z_1 - z_2|}, \qquad (5.37)$$

$\zeta_0(z)$ is the envelope function of the lowest subband of the quasi-2D system. For heterojunctions,

$$H(q_\parallel) = \frac{b_0}{8(b_0 + q_\parallel)^3} \left(8b_0^2 + 9 b_0 q_\parallel + 3 q_\parallel^2\right). \qquad (5.38)$$

For quantum wells, $H(q_\parallel)$ is given by (2.136). It is easy to verify that $A(\boldsymbol{q}_\parallel)$ and $A^\dagger(\boldsymbol{q}_\parallel)$ satisfy the commutation relation

$$[A(\boldsymbol{q}_\parallel), A^\dagger(\boldsymbol{q}'_\parallel)] = \delta_{\boldsymbol{q}_\parallel \boldsymbol{q}'_\parallel}. \qquad (5.39)$$

Thus they indeed represent the creation and annihilation operators of a kind of two-dimensional phonons, which are spatially matched with the electron distribution in the quasi-2D system. This 2D phonon has an energy Ω_{LO}, and its number operator is given by

$$\hat{N}(\boldsymbol{q}_\parallel) = A^\dagger(\boldsymbol{q}_\parallel)A(\boldsymbol{q}_\parallel) = \int_{-\infty}^{\infty} dz_1 \int_{-\infty}^{\infty} dz_2 \, \zeta_0(z_1)^2 \zeta_0(z_2)^2 \, b^\dagger_{\boldsymbol{q}_\parallel}(z_1) b_{\boldsymbol{q}_\parallel}(z_2), \tag{5.40}$$

in which $b^\dagger_{\boldsymbol{q}_\parallel}(z)$ and $b_{\boldsymbol{q}_\parallel}(z)$ are creation and annihilation operators of LO phonon having parallel wavevector $\boldsymbol{q}_\parallel$ and perpendicularly located at z:

$$b^\dagger_{\boldsymbol{q}_\parallel}(z) = \sum_{q_z} e^{-i q_z z} b^\dagger_{\boldsymbol{q}_\parallel q_z}, \tag{5.41}$$

$$b_{\boldsymbol{q}_\parallel}(z) = \sum_{q_z} e^{i q_z z} b_{\boldsymbol{q}_\parallel q_z}. \tag{5.42}$$

Expression (5.40) indicates that the contribution to $\hat{N}(\boldsymbol{q}_\parallel)$ comes solely from those phonons, that are localized within the region where the electron density is not zero. The electron–LO-phonon interaction H_{ep} (2.87) can be expressed as

$$H_{\text{ep}} = \sum_{\boldsymbol{q}_\parallel, q_z} M(\boldsymbol{q}_\parallel, q_z, \text{LO}) I(i\, q_z) \, e^{i \boldsymbol{q}_\parallel \cdot \boldsymbol{R}} \rho_{\boldsymbol{q}_\parallel} \left(b_{\boldsymbol{q}_\parallel q_z} + b^\dagger_{-\boldsymbol{q}_\parallel, -q_z} \right)$$

$$= \sum_{\boldsymbol{q}_\parallel} M(q_\parallel) \, e^{i \boldsymbol{q}_\parallel \cdot \boldsymbol{R}} \rho_{\boldsymbol{q}_\parallel} \left[A(\boldsymbol{q}_\parallel) + A^\dagger(-\boldsymbol{q}_\parallel) \right], \tag{5.43}$$

where $\boldsymbol{R}$ is the center-of-mass coordinate of 2D electrons,

$$\rho_{\boldsymbol{q}_\parallel} = \sum_{\boldsymbol{k}_\parallel, \sigma} c^\dagger_{\boldsymbol{k}_\parallel + \boldsymbol{q}_\parallel \sigma} c_{\boldsymbol{k}_\parallel \sigma} \tag{5.44}$$

is the electron density operator of the lowest subband. (5.43) shows that the electron–LO-phonon interaction can be expressed entirely in terms of 2D phonon operators $A(\boldsymbol{q}_\parallel)$ and $A^\dagger(-\boldsymbol{q}_\parallel)$. The frictional force experienced by the electron center of mass and the energy-loss rate of relative electrons due to electron–LO-phonon interaction are given by

$$\boldsymbol{F} = -i\,[\boldsymbol{P}, H_{\text{ep}}] = -i \sum_{\boldsymbol{q}_\parallel} M(q_\parallel) \boldsymbol{q}_\parallel \, e^{i \boldsymbol{q}_\parallel \cdot \boldsymbol{R}} \rho_{\boldsymbol{q}_\parallel} \left[A(\boldsymbol{q}_\parallel) + A^\dagger(-\boldsymbol{q}_\parallel) \right], \tag{5.45}$$

$$W = i\,[H_{\text{er}}, H_{\text{ep}}] = i \sum_{\boldsymbol{q}_\parallel} M(q_\parallel) \, \Omega_{\text{LO}} \, e^{i \boldsymbol{q}_\parallel \cdot \boldsymbol{R}} \rho_{\boldsymbol{q}_\parallel} \left[A(\boldsymbol{q}_\parallel) - A^\dagger(-\boldsymbol{q}_\parallel) \right]. \tag{5.46}$$

The rate of change of the 2D phonon number is

$$\left(\frac{\partial \hat{N}(\boldsymbol{q}_\parallel)}{\partial t}\right)_{\text{ep}} = -\mathrm{i}\left[\hat{N}(\boldsymbol{q}_\parallel), H_{\text{ep}}\right]$$

$$= -\mathrm{i}M(q_\parallel)\left[e^{\mathrm{i}\boldsymbol{q}_\parallel\cdot\boldsymbol{R}}\rho_{\boldsymbol{q}_\parallel}A(\boldsymbol{q}_\parallel) - e^{-\mathrm{i}\boldsymbol{q}_\parallel\cdot\boldsymbol{R}}\rho_{-\boldsymbol{q}_\parallel}A^\dagger(-\boldsymbol{q}_\parallel)\right]. \quad (5.47)$$

The above formulation shows that to deal with the electron transport of quasi-2D systems governed by electron–LO-phonon interactions, it is sufficient to consider the 2D phonons described by $A^\dagger(\boldsymbol{q}_\parallel)$ and $A(\boldsymbol{q}_\parallel)$. Therefore, one can take

$$H^A_{\text{ph}} = \sum_{\boldsymbol{q}_\parallel} \Omega_{\text{LO}} \hat{N}(\boldsymbol{q}_\parallel) = \sum_{\boldsymbol{q}_\parallel} \Omega_{\text{LO}} A^\dagger(\boldsymbol{q}_\parallel) A(\boldsymbol{q}_\parallel) \quad (5.48)$$

as the Hamiltonian of the phonon system interacting with quasi-2D electrons. The phonon part of the initial density matrix $\hat{\rho}_0$ for solving the Liouville equation of the system is chosen as the quasi-equilibrium state determined by H^A_{ph}, having different temperatures $T_{\boldsymbol{q}_\parallel}$ for different $\boldsymbol{q}_\parallel$ modes:

$$\hat{\rho}_0 = \frac{1}{Z}\exp(-H_{\text{er}}/T_{\text{e}})\exp\left(-\sum_{\boldsymbol{q}_\parallel}\Omega_{\text{LO}} A^\dagger(\boldsymbol{q}_\parallel)A(\boldsymbol{q}_\parallel)/T_{\boldsymbol{q}_\parallel}\right). \quad (5.49)$$

The other part of H_{ph} can be put into H_{pp}, playing a role to promote phonons of different modes approaching the final equilibrium. In other words, we expect that, after turning off the electron–phonon interaction and the electric field, the electron–phonon system can approach the final equilibrium state in a shorter time by starting from this initial state than that by starting from the initial state described by (5.8). Solving the Liouville equation to obtain the density matrix $\hat{\rho}$ and calculating the statistical average of the physical quantity to the lowest nonzero order of H_{ep}, we have

$$\left[\frac{\partial N(\boldsymbol{q}_\parallel)}{\partial t}\right]_{\text{ep}} = \left\langle\left[\frac{\partial \hat{N}(\boldsymbol{q}_\parallel)}{\partial t}\right]_{\text{ep}}\right\rangle$$

$$= \Gamma(\boldsymbol{q}_\parallel, \Omega_{\text{LO}} - \boldsymbol{q}_\parallel\cdot\boldsymbol{v})\left[n\left(\frac{\Omega_{\text{LO}} - \boldsymbol{q}_\parallel\cdot\boldsymbol{v}}{T_{\text{e}}}\right) - N(\boldsymbol{q}_\parallel)\right], \quad (5.50)$$

in which

$$N(\boldsymbol{q}_\parallel) = \langle\hat{N}(\boldsymbol{q}_\parallel)\rangle \quad (5.51)$$

is the occupation number of 2D phonons having wavevector $\boldsymbol{q}_\parallel$,

$$\Gamma(\boldsymbol{q}_\parallel, \Omega_{\text{LO}} - \boldsymbol{q}_\parallel\cdot\boldsymbol{v}) = -2M^2(q_\parallel)\Pi_2(\boldsymbol{q}_\parallel, \Omega_{\text{LO}} - \boldsymbol{q}_\parallel\cdot\boldsymbol{v}) \quad (5.52)$$

is the $q_\parallel$-dependent relaxation rate (inverse relaxation time) determined by the electron–phonon interaction, and $\Pi_2(q_\parallel,\omega)$ is the imaginary part of the electron density correlation function of the quasi-2D system (single heterojunction or quantum well) with lowest-subband occupancy only [(2.96) with $n = n' = 0$]. The frictional force and the energy-loss rate of the electron system can be written as

$$f_p = -\sum_{q_\parallel} q_\parallel \left(\frac{\partial N(q_\parallel)}{\partial t}\right)_{ep}, \qquad (5.53)$$

$$w = \sum_{q_\parallel} \Omega_{LO} \left(\frac{\partial N(q_\parallel)}{\partial t}\right)_{ep}. \qquad (5.54)$$

The total rate of change of the occupation number of 2D phonons with wavevector $q_\parallel$ consists of the contribution from electron–phonon interaction $[\partial N(q_\parallel)/\partial t]_{ep}$ and that from phonon–phonon interaction $[\partial N(q_\parallel)/\partial t]_{pp}$. The latter is phenomenologically taken into account with a phonon relaxation time τ_p to give

$$\left[\frac{\partial N(q_\parallel)}{\partial t}\right]_{pp} = -\frac{1}{\tau_p}\left[N(q_\parallel) - n_{eq}\right], \qquad (5.55)$$

where n_{eq} represents the final equilibrium distribution of phonons. Assuming that at equilibrium the whole phonon system shares a unique temperature T with the lattice, we have $n_{eq} = n(\Omega_{LO}/T)$ and then

$$\frac{dN(q_\parallel)}{dt} = \left[\frac{\partial N(q_\parallel)}{\partial t}\right]_{ep} - \frac{1}{\tau_p}\left[N(q_\parallel) - n\left(\frac{\Omega_{LO}}{T}\right)\right]. \qquad (5.56)$$

This equation, plus the momentum- and energy-balance equations of the electron system, (5.17) and (5.18), together with the expressions (5.50)–(5.54) for f_p, w and $[\partial N(q_\parallel)/\partial t]_{ep}$ and that for f_i, constitute a complete set of ordinary differential equations to determine the transport quantities v, T_e and $N(q_\parallel)$.

In the dc steady state, the condition $dN(q_\parallel)/dt = 0$ immediately gives the expression for the occupation number of 2D phonons:

$$N(q_\parallel) = n\left(\frac{\Omega_{LO} - q_\parallel \cdot v}{T_e}\right) + \left[1 + \tau_p \Gamma(q_\parallel, \Omega_{LO} - q_\parallel \cdot v)\right]^{-1}$$
$$\times \left[n\left(\frac{\Omega_{LO}}{T}\right) - n\left(\frac{\Omega_{LO} - q_\parallel \cdot v}{T_e}\right)\right]. \qquad (5.57)$$

The frictional force and the energy-loss rate in the dc steady state are thus

$$f_\mathrm{p} = \sum_{\boldsymbol{q}_\|} \boldsymbol{q}_\| \frac{\Gamma(\boldsymbol{q}_\|, \Omega_\mathrm{LO}+\boldsymbol{q}_\|\cdot\boldsymbol{v})}{1+\tau_\mathrm{p}\Gamma(\boldsymbol{q}_\|, \Omega_\mathrm{LO}+\boldsymbol{q}_\|\cdot\boldsymbol{v})} \left[n\left(\frac{\Omega_\mathrm{LO}+\boldsymbol{q}_\|\cdot\boldsymbol{v}}{T_\mathrm{e}}\right) - n\left(\frac{\Omega_\mathrm{LO}}{T}\right)\right],$$
(5.58)

$$w = \sum_{\boldsymbol{q}_\|} \Omega_\mathrm{LO} \frac{\Gamma(\boldsymbol{q}_\|, \Omega_\mathrm{LO}+\boldsymbol{q}_\|\cdot\boldsymbol{v})}{1+\tau_\mathrm{p}\Gamma(\boldsymbol{q}_\|, \Omega_\mathrm{LO}+\boldsymbol{q}_\|\cdot\boldsymbol{v})} \left[n\left(\frac{\Omega_\mathrm{LO}+\boldsymbol{q}_\|\cdot\boldsymbol{v}}{T_\mathrm{e}}\right) - n\left(\frac{\Omega_\mathrm{LO}}{T}\right)\right].$$
(5.59)

Eqs. (5.57)–(5.59) show that the degree of nonequilibrium phonon effect in quasi-2D systems is still determined by the quantity $\tau_\mathrm{p}\Gamma$. However, in the case of 2D-phonon model discussed in this section, the Γ function (5.52) is proportional to the sheet density N_s of quasi-2D carriers rather than the equivalent 3D bulk density n. Therefore, for given τ_p, the degree of nonequilibrium phonon effect is proportional to the carrier sheet density of the quasi-2D system.

So far we have discussed quasi-2D electron systems by considering only one subband. The case with multiple subbands of electrons can be treated in a formulation similar to that in the next section for superlattices.

5.5 Semiconductor Superlattices, Quasi-2D Phonon Model

For a semiconductor superlattice described in Sec. 2.7 (having period d and length L in the z direction and transverse area S in the x–y plane), We can also introduce quasi-2D phonons matching with the spatial distribution of the electron density as

$$A_l^\dagger(\boldsymbol{q}_\|) = \sum_{q_z} g_l^*(\boldsymbol{q}_\|, q_z) b^\dagger_{\boldsymbol{q}_\| q_z},$$
(5.60)

$$A_l(\boldsymbol{q}_\|) = \sum_{q_z} g_l(\boldsymbol{q}_\|, q_z) b_{\boldsymbol{q}_\| q_z}.$$
(5.61)

Here the weight factor is

$$g_l(\boldsymbol{q}_\|, q_z) = \frac{M(\boldsymbol{q}_\|, q_z, \mathrm{LO})}{L^{1/2} M(q_\|)} I(\mathrm{i}\, q_z) \mathrm{e}^{-\mathrm{i}\, q_z l d},$$
(5.62)

with $M(q_\parallel)$ still defined by (5.36). The commutation relations between $A_l(\boldsymbol{q}_\parallel)$ and $A_{l'}^\dagger(\boldsymbol{q}_\parallel')$ are

$$\left[A_l(\boldsymbol{q}_\parallel), A_{l'}^\dagger(\boldsymbol{q}_\parallel')\right] = \beta_{ll'}(q_\parallel)\delta_{\boldsymbol{q}_\parallel \boldsymbol{q}_\parallel'}, \quad (5.63)$$

with

$$\beta_{ll'}(q_\parallel) = \beta_{l-l'}(q_\parallel)$$
$$= \frac{1}{LM^2(q_\parallel)} \sum_{q_z} |M(\boldsymbol{q}_\parallel, q_z)|^2 |I(\mathrm{i}\,q_z)|^2\, \mathrm{e}^{\mathrm{i}\,(l-l')q_z d}. \quad (5.64)$$

The electron–phonon interaction H_ep (2.139) can be written as

$$H_\mathrm{ep} = \sum_{l,\boldsymbol{q}_\parallel} M(q_\parallel)\,\mathrm{e}^{\mathrm{i}\,\boldsymbol{q}_\parallel\cdot\boldsymbol{R}} \rho_{l\boldsymbol{q}_\parallel}\left[A_l(\boldsymbol{q}_\parallel) + A_l^\dagger(-\boldsymbol{q}_\parallel)\right], \quad (5.65)$$

in which

$$\rho_{l\boldsymbol{q}_\parallel} = \sum_{\boldsymbol{k}_\parallel,\sigma} c_{l\boldsymbol{k}_\parallel+\boldsymbol{q}_\parallel\sigma}^\dagger c_{l\boldsymbol{k}_\parallel\sigma} \quad (5.66)$$

is the density operator of electrons in the lth quantum well. The frictional force experienced by the center-of-mass and the energy-loss rate of relative electrons induced by the electron–phonon interaction are given by

$$\boldsymbol{F}_\mathrm{p} = -\mathrm{i}\sum_{l,\boldsymbol{q}_\parallel} M(q_\parallel)\,\boldsymbol{q}_\parallel\,\mathrm{e}^{\mathrm{i}\,\boldsymbol{q}_\parallel\cdot\boldsymbol{R}}\rho_{l\boldsymbol{q}_\parallel}\left[A_l(\boldsymbol{q}_\parallel) + A_l^\dagger(-\boldsymbol{q}_\parallel)\right], \quad (5.67)$$

$$W = \mathrm{i}\sum_{l,\boldsymbol{q}_\parallel} M(q_\parallel)\,\Omega_\mathrm{LO}\,\mathrm{e}^{\mathrm{i}\,\boldsymbol{q}_\parallel\cdot\boldsymbol{R}}\rho_{l\boldsymbol{q}_\parallel}\left[A_l(\boldsymbol{q}_\parallel) - A_l^\dagger(-\boldsymbol{q}_\parallel)\right]. \quad (5.68)$$

Defining a general occupation number operator of quasi-2D phonons,

$$\hat{N}_{lm}(\boldsymbol{q}_\parallel) = A_l^\dagger(\boldsymbol{q}_\parallel) A_m(\boldsymbol{q}_\parallel), \quad (5.69)$$

we can see that the rate of change of $\hat{N}_{lm}(\boldsymbol{q}_\parallel)$ due to electron–phonon interaction is

$$\left(\frac{d\hat{N}_{lm}(\boldsymbol{q}_\parallel)}{dt}\right)_\mathrm{ep} = -\mathrm{i}\left[\hat{N}_{lm}(\boldsymbol{q}_\parallel), H_\mathrm{ep}\right]$$
$$= \mathrm{i}M(q_\parallel)\sum_s \left[\beta_{sl}(q_\parallel)A_m(\boldsymbol{q}_\parallel)\mathrm{e}^{\mathrm{i}\,\boldsymbol{q}_\parallel\cdot\boldsymbol{R}}\rho_{s\boldsymbol{q}_\parallel}\right.$$
$$\left. - \beta_{ms}(q_\parallel)A_l^\dagger(\boldsymbol{q}_\parallel)\mathrm{e}^{-\mathrm{i}\,\boldsymbol{q}_\parallel\cdot\boldsymbol{R}}\rho_{s,-\boldsymbol{q}_\parallel}\right]. \quad (5.70)$$

This shows that it is the quasi-2D phonon described by the creation and annihilation operators $A_l^\dagger(\mathbf{q}_\parallel)$ and $A_l(\mathbf{q}_\parallel)$, that determines the electron–phonon interaction and electron transport. Therefore, we naturally choose $\hat{N}_{lm}(\mathbf{q}_\parallel)$ as the dynamical variable for the phonon system, and use

$$\hat{\rho}_0 = \frac{1}{Z}\exp\bigl(-H_{\rm er}/T_{\rm e}\bigr)\exp\biggl(-\sum_{l,m,\mathbf{q}_\parallel} F_{lm\mathbf{q}_\parallel}\hat{N}_{lm}(\mathbf{q}_\parallel)\biggr) \qquad (5.71)$$

as the initial density matrix for the Liouville equation. Here $F_{lm\mathbf{q}_\parallel}$ are parameters. Solving the density matrix $\hat{\rho}$ from the Liouville equation and calculating the averages of physical quantities to the lowest nonzero order of $H_{\rm ep}$, we have ($\tilde{\omega} \equiv \Omega_{\rm LO} - \mathbf{q}_\parallel\cdot\mathbf{v}$)

$$\langle \mathbf{F}_{\rm p}\rangle = -2\sum_{l,l',\mathbf{q}_\parallel} M^2(q_\parallel)\,\mathbf{q}_\parallel\, \Pi_2(l',l,\mathbf{q}_\parallel,\tilde{\omega})\left[N_{l'l}(\mathbf{q}_\parallel) - \beta_{l'l}(\mathbf{q}_\parallel)n\!\left(\frac{\tilde{\omega}}{T_{\rm e}}\right)\right], \qquad(5.72)$$

$$\langle W\rangle = 2\sum_{l,l',\mathbf{q}_\parallel} M^2(q_\parallel)\,\Omega_{\rm LO}\,\Pi_2(l',l,\mathbf{q}_\parallel,\tilde{\omega})\left[N_{l'l}(\mathbf{q}_\parallel) - \beta_{l'l}(\mathbf{q}_\parallel)n\!\left(\frac{\tilde{\omega}}{T_{\rm e}}\right)\right], \qquad(5.73)$$

$$\left[\frac{\partial}{\partial t}N_{l'l}(\mathbf{q}_\parallel)\right]_{\rm ep} = i\sum_{s',s} M^2(q_\parallel)\bigl[\beta_{l's'}(\mathbf{q}_\parallel)N_{ls}(\mathbf{q}_\parallel)\Pi(s',s,\mathbf{q}_\parallel,-\tilde{\omega})$$
$$- \beta_{ls'}(\mathbf{q}_\parallel)N_{l's}(\mathbf{q}_\parallel)\Pi(s',s,\mathbf{q}_\parallel,\tilde{\omega})\bigr]$$
$$- 2\sum_{s's} M^2(q_\parallel)\beta_{ls}(\mathbf{q}_\parallel)\beta_{l's'}(\mathbf{q}_\parallel)\Pi_2(s',s,\mathbf{q}_\parallel,\tilde{\omega})n\!\left(\frac{\tilde{\omega}}{T_{\rm e}}\right). \qquad(5.74)$$

In these equations

$$N_{lm}(\mathbf{q}_\parallel) = \langle\hat{N}_{lm}(\mathbf{q}_\parallel)\rangle \qquad(5.75)$$

are generalized occupation numbers of quasi-2D phonons. If the system is the periodic repeat of the investigated superlattice of length L (periodic boundary condition), then both $N_{lm}(\mathbf{q}_\parallel)$, $\beta_{lm}(\mathbf{q}_\parallel)$ and $\Pi(l,m,\mathbf{q}_\parallel,\omega)$ are functions of $l-m$, and it is more convenient to use their Fourier spectra

$$N(\mathbf{q}_\parallel,q_z) = \sum_m N_{lm}(\mathbf{q}_\parallel)\,e^{-i(l-m)q_z d}, \qquad(5.76)$$

$$\beta(\mathbf{q}_\parallel,q_z) = \sum_m \beta_{lm}(\mathbf{q}_\parallel)\,e^{-i(l-m)q_z d}, \qquad(5.77)$$

and the electron density correlation function $\Pi(q_z, \boldsymbol{q}_\parallel, \omega)$ given by (2.148). Apparently, $N(\boldsymbol{q}_\parallel, q_z)$, $\beta(\boldsymbol{q}_\parallel, q_z)$, and $\Pi(q_z, \boldsymbol{q}_\parallel, \omega)$ are periodic functions of q_z having period $2\pi/d$. With the help of these results, (5.72) and (5.73) can be written in terms of the force and energy-loss rate for each layer of the superlattice:

$$\boldsymbol{f}_\mathrm{p} = -\frac{d}{\pi} \sum_{\boldsymbol{q}_\parallel} M^2(q_\parallel) \boldsymbol{q}_\parallel \int_{-\frac{\pi}{d}}^{\frac{\pi}{d}} dq_z \, \Pi_2(q_z, \boldsymbol{q}_\parallel, \tilde{\omega}) \left[N(\boldsymbol{q}_\parallel, q_z) - \beta(\boldsymbol{q}_\parallel, q_z) n\left(\frac{\tilde{\omega}}{T_\mathrm{e}}\right) \right], \tag{5.78}$$

$$w = \frac{d}{\pi} \sum_{\boldsymbol{q}_\parallel} M^2(q_\parallel) \Omega_\mathrm{LO} \int_{-\frac{\pi}{d}}^{\frac{\pi}{d}} dq_z \, \Pi_2(q_z, \boldsymbol{q}_\parallel, \tilde{\omega}) \left[N(\boldsymbol{q}_\parallel, q_z) - \beta(\boldsymbol{q}_\parallel, q_z) n\left(\frac{\tilde{\omega}}{T_\mathrm{e}}\right) \right], \tag{5.79}$$

and (5.74) can be written as

$$\left[\frac{\partial}{\partial t} N(\boldsymbol{q}_\parallel, q_z)\right]_\mathrm{ep} = 2 M^2(q_\parallel) \beta(\boldsymbol{q}_\parallel, q_z) \Pi_2(q_z, \boldsymbol{q}_\parallel, \tilde{\omega})$$
$$\times \left[N(\boldsymbol{q}_\parallel, q_z) - \beta(\boldsymbol{q}_\parallel, q_z) n\left(\frac{\tilde{\omega}}{T_\mathrm{e}}\right) \right]. \tag{5.80}$$

The effects of other couplings than the electron–phonon interaction in the system, on the phonon occupation $N(\boldsymbol{q}_\parallel, q_z)$ are to promote it to approach an equilibrium distribution. We use a relaxation time τ_p to represent their contribution to the rate of change of $N(\boldsymbol{q}_\parallel, q_z)$, to give

$$\frac{d}{dt} N(\boldsymbol{q}_\parallel, q_z) = \left[\frac{\partial}{\partial t} N(\boldsymbol{q}_\parallel, q_z)\right]_\mathrm{ep} - \frac{1}{\tau_\mathrm{p}} \left[N(\boldsymbol{q}_\parallel, q_z) - N_\mathrm{eq}(\boldsymbol{q}_\parallel, q_z) \right]. \tag{5.81}$$

Here $N_\mathrm{eq}(\boldsymbol{q}_\parallel, q_z)$ denotes the equilibrium value of $N(\boldsymbol{q}_\parallel, q_z)$:

$$N_\mathrm{eq}(\boldsymbol{q}_\parallel, q_z) = \beta(q_\parallel, q_z) n\left(\frac{\Omega_\mathrm{LO}}{T}\right), \tag{5.82}$$

in which

$$\beta(q_\parallel, q_z) = \frac{1}{d M^2(q_\parallel)} \sum_n |M(\boldsymbol{q}_\parallel, q_z + 2\pi n/d)|^2 |I(\mathrm{i}\, q_z + \mathrm{i}\, 2\pi n/d)|^2, \tag{5.83}$$

with the summation over all integers $n = 0, \pm 1, \cdots$.

Equation (5.81), plus Eqs. (5.17) and (5.18), together with the expressions (5.78)–(5.80) for $\boldsymbol{f}_\mathrm{p}$, w and $[\partial N(\boldsymbol{q}_\parallel, q_z)/\partial t]_\mathrm{ep}$ and that of $\boldsymbol{f}_\mathrm{i}$, form a complete set of time-dependent ordinary differential equations to determine the transport properties of semiconductor superlattice in the case of nonequilibrium phonon occupation.

In the dc steady-state transport, the condition $dN(\mathbf{q}_\parallel, q_z)/dt = 0$ yields

$$N(\mathbf{q}_\parallel, q_z) = \beta(\mathbf{q}_\parallel, q_z) n\left(\frac{\tilde{\omega}}{T_e}\right) + \frac{\beta(\mathbf{q}_\parallel, q_z)}{1 + \tau_p \Gamma(\mathbf{q}_\parallel, q_z, \tilde{\omega})} \left[n\left(\frac{\Omega_{LO}}{T}\right) - n\left(\frac{\tilde{\omega}}{T_e}\right)\right], \tag{5.84}$$

in which

$$\Gamma(q_z, \mathbf{q}_\parallel, \omega) = -2M^2(q_\parallel)\beta(\mathbf{q}_\parallel, q_z)\Pi_2(q_z, \mathbf{q}_\parallel, \omega)$$
$$= \sum_n \Gamma_n(q_z, \mathbf{q}_\parallel, \omega), \tag{5.85}$$

$$\Gamma_n(q_z, \mathbf{q}_\parallel, \omega) = -\frac{2}{d}M(\mathbf{q}_\parallel, q_z + 2\pi n/d)^2 |I(\mathrm{i}\, q_z + \mathrm{i}\, 2\pi n/d)|^2 \Pi_2(q_z, \mathbf{q}_\parallel, \omega). \tag{5.86}$$

We thus have

$$\mathbf{f}_p = \sum_{\mathbf{q}_\parallel} \mathbf{q}_\parallel K(\mathbf{q}_\parallel, \Omega_{LO} + \mathbf{q}_\parallel \cdot \mathbf{v}) \left[n\left(\frac{\Omega_{LO} + \mathbf{q}_\parallel \cdot \mathbf{v}}{T_e}\right) - n\left(\frac{\Omega_{LO}}{T}\right)\right], \tag{5.87}$$

$$w = \sum_{\mathbf{q}_\parallel} \Omega_{LO} K(\mathbf{q}_\parallel, \Omega_{LO} + \mathbf{q}_\parallel \cdot \mathbf{v}) \left[n\left(\frac{\Omega_{LO} + \mathbf{q}_\parallel \cdot \mathbf{v}}{T_e}\right) - n\left(\frac{\Omega_{LO}}{T}\right)\right], \tag{5.88}$$

where

$$K(\mathbf{q}_\parallel, \Omega_{LO} + \mathbf{q}_\parallel \cdot \mathbf{v}) = \frac{d}{2\pi} \int_{-\frac{\pi}{d}}^{\frac{\pi}{d}} dq_z \frac{\Gamma(q_z, \mathbf{q}_\parallel, \Omega_{LO} + \mathbf{q}_\parallel \cdot \mathbf{v})}{1 + \tau_p \Gamma(q_z, \mathbf{q}_\parallel, \Omega_{LO} + \mathbf{q}_\parallel \cdot \mathbf{v})}. \tag{5.89}$$

Note that, in the case of 3D-phonon model discussed in Sec. 5.3, the $K(\mathbf{q}_\parallel, \mathrm{LO}, \Omega_{LO} + \mathbf{q}_\parallel \cdot \mathbf{v})$ function [(5.32), with $\lambda = \mathrm{LO}$] in the frictional force and energy-loss rate in the dc steady state transport can be written as

$$K(\mathbf{q}_\parallel, \mathrm{LO}, \Omega_{LO} + \mathbf{q}_\parallel \cdot \mathbf{v}) = \frac{d}{2\pi} \int_{-\frac{\pi}{d}}^{\frac{\pi}{d}} dq_z \frac{\Gamma(q_z, \mathbf{q}_\parallel, \Omega_{LO} + \mathbf{q}_\parallel \cdot \mathbf{v})}{1 + \tau_p \Gamma_0(q_z, \mathbf{q}_\parallel, \Omega_{LO} + \mathbf{q}_\parallel \cdot \mathbf{v})}. \tag{5.90}$$

The only difference between (5.90) and (5.89) is that Γ in the denominate of (5.89) is replaced by Γ_0. In the case of $\tau_p = 0$, the results from 3D- and 2D-phonon models are the same, both returning to those of Sec. 2.7.

If the period d, i.e. the distance of neighboring quantum wells in the superlattice, is large, (5.89) becomes

$$K(\boldsymbol{q}_\|, \Omega_{\mathrm{LO}} + \boldsymbol{q}_\| \cdot \boldsymbol{v}) = \frac{\Gamma(0, \boldsymbol{q}_\|, \Omega_{\mathrm{LO}} + \boldsymbol{q}_\| \cdot \boldsymbol{v})}{1 + \tau_{\mathrm{p}} \Gamma(0, \boldsymbol{q}_\|, \Omega_{\mathrm{LO}} + \boldsymbol{q}_\| \cdot \boldsymbol{v})}, \quad (5.91)$$

with

$$\Gamma(0, \boldsymbol{q}_\|, \omega) = -\frac{2}{d} \sum_n M^2(\boldsymbol{q}_\|, 2\pi n/d) |I(\mathrm{i}\, 2\pi n/d)|^2 \Pi_2(0, \boldsymbol{q}_\|, \omega)$$

$$= -2 M^2(q_\|) \Pi_2(\boldsymbol{q}_\|, \omega). \quad (5.92)$$

Note that in writing this expression we have made use of the facts that the summation over n can be replaced by an integral in the case of large d such that the definition (5.36) of $M^2(q_\|)$ can be used in superlattices, and that the $q_z = 0$ density correlation function of the superlattice, $\Pi_2(0, q_\|, \omega)$, is the same as the density correlation function of a single quantum well (lowest subband only), $\Pi_2(\boldsymbol{q}_\|, \omega)$. Apparently, the result of a large d superlattice returns to that of a single quantum well in the case of 2D-phonon model, as it should be.

5.6 Effect of Nonequilibrium Phonons on Hot-Electron Transport

We have, for several electron–phonon coupled systems, discussed the transport behavior of electrons driven by intense electric fields. Nonequilibrium phonons may occur in the system having relatively high carrier bulk or sheet density when the relaxation time τ_{p} of phonon system is finite (nonzero). This may lead to a substantial change of the electron energy-loss rate as well as the frictional force in comparison with those of $\tau_{\mathrm{p}} = 0$ case, and thus greatly affect the steady-state and transient transport of hot electrons. On the basis of formulae presented above, extensive calculations of hot carrier transport properties (transient and steady state, mobility as well as electron and noise temperatures) are carried out in the presence of nonequilibrium phonon occupation for semiconductor heterostructures (Lei and Horing, 1987a,b, 1988; Lei, Cui, and Horing, 1987).

First, the nonequilibrium phonon occupation may have a significant impact on the carrier energy-transfer rate induced by electron–polar-optic-phonon interaction. This effect can be understood from the fact that scatterings with small momentum transfer have a considerable weight in contributing to the electron energy loss, and in the case of a Fröhlich coupling between electrons and polar optic phonons, the effective scattering matrix element $|M(q_\parallel)|^2$ (5.36) and consequently the relaxation rate $\Gamma(\boldsymbol{q}_\parallel, \Omega_{\rm LO} - \boldsymbol{q}_\parallel \cdot \boldsymbol{v}) \sim |M(q_\parallel)|^2 \Pi_2(\boldsymbol{q}_\parallel, \Omega_{\rm LO} - \boldsymbol{q}_\parallel \cdot \boldsymbol{v})$, are large at small $q_\parallel$. The contributions of such large values of Γ are sharply reduced when taken in conjunction with a finite phonon relaxation time $\tau_{\rm p}$. Of course, the phonon-induced frictional force is also modified by the finite $\tau_{\rm p}$. However, this effect is not as pronounced as that in the energy transfer rate because of the relatively small weight of small $q_\parallel$ scattering in the constitution of the force, in contrast to the energy loss.

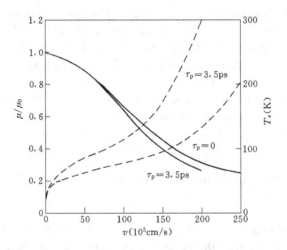

Fig. 5.1 Nonlinear mobility normalized to its Ohmic limit, μ/μ_0 (solid curves), and electron temperatures $T_{\rm e}$ (dashed curves) are shown as functions of drift velocity v for phonon relaxation times $\tau_{\rm p} = 0$ and $\tau_{\rm p} = 3.5\,{\rm ps}$ at lattice temperature $T = 2\,{\rm K}$ for an n-type GaAs heterojunction with carrier sheet density $N_{\rm s} = 3.9 \times 10^{11}\,{\rm cm}^{-2}$ and $\mu_0 = 7.9 \times 10^4\,{\rm cm}^2/{\rm Vs}$. From Lei and Horing (1987a).

Fig. 5.1 shows the nonlinear mobility μ (defined as the ratio of the drift velocity to the electric field, $\mu = v/E$), normalized to the Ohmic mobility μ_0, and the carrier temperature $T_{\rm e}$, determined from steady-state balance equations with $\boldsymbol{f}_{\rm p}$ and w given by (5.58) and (5.59) for an n-type GaAs quantum well of width $a = 25\,{\rm nm}$ at lattice temperature $T = 2\,{\rm K}$, as functions of drift velocity $\boldsymbol{v}$ for two different phonon relaxation times: $\tau_{\rm p} = 0$ and $\tau_{\rm p} = 3.5\,{\rm ps}$. The system has an electron sheet density $N_{\rm s} = 3.9 \times 10^{11}\,{\rm cm}^{-2}$ and a linear mobility $\mu_0 = 7.9 \times 10^4\,{\rm cm}^2$ due to both background and remote impurity scatterings. The well potential is assumed deep enough that $\cos(\pi z/a)$ adequately represents the envelope function $\zeta(z)$. The most striking effect of the finite phonon relaxation time $\tau_{\rm p} = 3.5\,{\rm ps}$ on hot carrier transport is that, for a given drift velocity, the electron temperature is much higher than in the case of $\tau_{\rm p} = 0$, as can be seen from Fig. 5.1. Such an increase of electron temperature tends to increase the frictional force in a measure which outweighs the tendency toward the diminution of it due to finite $\tau_{\rm p}$. As a result, the normalized mobility μ/μ_0 is lower for finite $\tau_{\rm p}$ than for $\tau_{\rm p} = 0$ (Lei and Horing, 1987a).

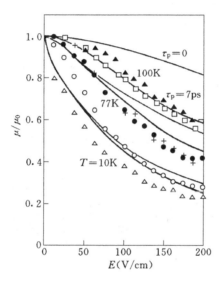

Fig. 5.2 Normalized mobility μ/μ_0 vs electric field E. Solid curves are the theoretical results of the balance-equation approach for phonon relaxation times $\tau_{\rm p} = 0$ and $\tau_{\rm p} = 7\,{\rm ps}$ at lattice temperatures $T = 10, 77$, and $100\,{\rm K}$, respectively. The experimental data are taken from Drummond et al (1981). From Lei and Horing (1988).

This effect is more pronounced at higher lattice temperature than at lower lattice temperature, as is seen in Fig. 5.2, in which the nonlinear mobility μ normalized by the linear mobility μ_0 at temperature T, calculated with $\tau_{\rm p} = 0$ and $\tau_{\rm p} = 7\,{\rm ps}$, is shown as a function of the electric field at lattice temperatures $T = 10, 77$ and $100\,{\rm K}$ for a GaAs/AlGaAs heterojunction having carrier sheet density $4.0 \times 10^{11}\,{\rm cm}^{-2}$ and zero-temperature Ohmic mobility $\mu_0(0) = 2.2 \times 10^5\,{\rm cm}^2/{\rm Vs}$. The experimental data from Drummond et al (1981) are also shown in the figure for comparison. The inclusion of finite phonon relaxation time greatly improves the agreement between the theory and experiment, especially at higher temperatures (Lei and Horing, 1988; Marchetti, Cai and Lax, 1988).

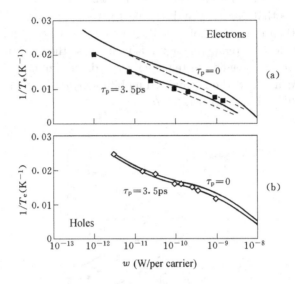

Fig. 5.3 The inverse electron temperature $1/T_e$ versus the average energy loss rate per carrier, w, for phonon relaxation times $\tau_p = 0$ and $\tau_p = 7\,\text{ps}$. (a) Electron conduction at $T = 2\,\text{K}$ in an n-type GaAs quantum well described in Fig. 5.1, with the experimental results (Shah et al, 1986) shown as square dots. (b) Hole conduction at $T = 2\,\text{K}$ in a p-type GaAs quantum well with well width $a = 9.5\,\text{nm}$, carrier density $N_s = 3.5 \times 10^{11}\,\text{cm}^{-2}$ and zero-temperature linear mobility $\mu_0 = 3.4\,\text{m}^2/\text{Vs}$, together with experimental data (Shah et al, 1986) shown as rhombuses. From Lei and Horing (1987a).

To have a concrete impression and to further compare with experiments, we plot in Fig. 5.3(a) the calculated inverse electron temperature $1/T_e$ in the steady state transport of the n-type quantum well described in Fig. 5.1, versus the average energy-loss rate per carrier, w, together with the experimental data of Shah et al (1986). For a given T_e, the electron energy-loss rate due to electron-phonon interaction in the case of $\tau_p = 3.5\,\text{ps}$ is only $1/5 \sim 1/10$ of that for the case $\tau_p = 0$. The results of $\tau_p = 3.5\,\text{ps}$ in the figure agree well with the experiment.

We have also calculated the steady state transport of holes at $T = 2\,\text{K}$ in a p-type GaAs quantum well having well width $a = 9.5\,\text{nm}$, carrier density $N_s = 3.5 \times 10^{11}\,\text{cm}^{-2}$ and zero-temperature linear mobility $\mu_0 = 3.4\,\text{m}^2/\text{Vs}$, for phonon relaxation times $\tau_p = 0$ and $\tau_p = 3.5\,\text{ps}$. The $1/T_e$-vs-w curves are shown in Fig. 5.3(b). In the case of hole conduction, in addition to polar-optic phonon scattering, optic deformation-potential scattering and acoustic phonon scattering also play roles in transport. Contributions from large wave vectors are important in the last two mechanisms, thus reduce the weight of small wave vectors in contributing to the carrier energy loss. As a result, the nonequilibrium phonon effect is much weaker for the same τ_p in the hole case, in agreement with experiment.

It is worth mentioning that to calculate the energy-loss rate of carriers from (5.59) or from (5.88) one needs first to solve the steady-state balance equations to derive the drift velocity v and the electron temperature T_e. The main difference of w expressions (5.59) and (5.88) from Kogan formula (5.1) is that there is a Dopler frequency shift of the phonon frequency in the electron temperature term in the former: $\Omega_{LO} \to \Omega_{LO} - \bm{q}_\parallel \cdot \bm{v}$. In other words, $\bm{v} = 0$ result of (5.59) or (5.88) is equivalent to that of Kogan formula or the electron-temperature model. Since no drift velocity is involved and thus one can directly obtain w–vs–T_e from the latter without need to solve the balance equation, Kogan formula is still widely used to qualitatively analyze the energy-loss rate of carriers in transport. For the system described in Fig. 5.3(a), The $1/T_e$–vs–w results for the cases of $\tau_p = 0$ and $\tau_p = 3.5$ ps, directly obtained from Kogan formula, are shown as dashed curves in the figure. We see that, for a given $1/T_e$, the predicted w by the solid and dashed curves can have a difference as large as a factor of three.

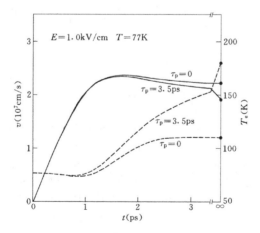

Fig. 5.4 The drift velocity v (full curves) and electron temperature T_e (broken curves), calculated with quasi-2D phonon model, are shown as functions of time t after an electric field of $E = 1.0$ kV/cm is suddenly turned on at $t = 0$ in a GaAs quantum-well at lattice temperature $T = 77$ K in the cases of phonon relaxation time $\tau_p = 0$ and $\tau_p = 3.5$ ps. From Lei, Cui and Horing (1987).

Effects of nonequilibrium phonon occupation on the transient response in quas-2D systems and superlattices has been investigated by Lei, Cui, and Horing (1987). Using the quasi-2D phonon model they calculate the transient response of quasi-2D systems to suddenly-impressed constant electric fields without ($\tau_\mathrm{p} = 0$) and with ($\tau_\mathrm{p} = 3.5\,\mathrm{ps}$) nonequilibrium phonons, by means of time-dependent balance equations (5.17), (5.18) and (5.56) for the momentum, energy and phonon occupation number with the expressions for $\boldsymbol{f}_\mathrm{p}$, w and $M(q_\parallel)$ given in Sec. 5.4. Fig. 5.4 shows the calculated transient response of drift velocity v and electron temperature T_e to a step electric field of $1.0\,\mathrm{kV/cm}$ turned on at time $t=0$ in a single GaAs-based quantum-well system of well width $a = 15\,\mathrm{nm}$, carrier sheet density $N_\mathrm{s} = 2.0 \times 10^{11}\,\mathrm{cm}^{-2}$ and zero-temperature linear mobility $\mu_0 = 50\,\mathrm{m^2/Vs}$. In addition, the transient response of a GaAs-based quantum-well superlattice with period $d = 15\,\mathrm{nm}$, well width $a = 10\,\mathrm{nm}$, carrier sheet density $N_\mathrm{s} = 2.0 \times 10^{11}\,\mathrm{cm}^{-2}$ and zero-temperature linear mobility $\mu_0 = 35\,\mathrm{m^2/Vs}$ is also calculated at lattice temperature $T = 77\,\mathrm{K}$ by using the 3D phonon model discussed in Sec. 5.3. The result has shown in Fig. 3.3 (Chapter 3) for both $\tau_\mathrm{p} = 0$ and $\tau_\mathrm{p} = 3.5\,\mathrm{ps}$. We can see from these two figures that, at the initial stage ($t \leqslant 1\,\mathrm{ps}$), the responses for $\tau_\mathrm{p} = 0$ and $\tau_\mathrm{p} = 3.5\,\mathrm{ps}$ are similar, thus a finite τ_p has only a small influence on the velocity overshoot. However, the final electron temperature T_e is much higher and the final velocity is lower, especially the time to approach the final steady state is much longer in the case of $\tau_\mathrm{p} = 3.5\,\mathrm{ps}$ than in the case of $\tau_\mathrm{p} = 0$.

Effects of nonequilibrium phonon occupation on high-frequency response, noise and diffusion can be investigated based on the formulation in the preceding sections (Lei and Horing, 1987b). In fact, the expressions of the high-frequency small-signal resistivity in terms of $M(\omega, v_\mathrm{d})$, $N(\omega, v_\mathrm{d})$ and $D(\omega, v_\mathrm{d})$ functions (Chapter 3) and the expressions of the noise temperature in terms of $M(\omega)$ and $S(\omega)$ functions (Chapter 4), remain valid in the case of finite τ_p in the 3D-phonon model. The occurrence of nonequilibrium phonons affects only the phonon occupation number $n_{\boldsymbol{q}\lambda}$ of mode $\boldsymbol{q}\lambda$, which was, in Chapters 3 and 4, taken to be $n(\Omega_{\boldsymbol{q}\lambda}/T)$ under the assumption of phonons being in equilibrium, where T is the lattice temperature. In the case of nonequilibrium phonon occupation, the $n_{\boldsymbol{q}\lambda}$ expressions derived in the present chapter should be used instead. For instance, in the case of 3D-phonon model, $n_{\boldsymbol{q}\lambda}$ is given by (5.19) for 3D bulk systems or (5.29) for semiconductor superlattices. Quasi-2D phonon model was not directly discussed in Chapters 3 and 4 since there is no difference between 3D and quasi-2D phonon model if phonons are in equilibrium. The relevant expressions in the case of quasi-2D phonon model can be obtained from those of 3D phonon model by taking the following replacements: for quasi-2D systems,

$$\sum_{\mathbf{q}} |M(\mathbf{q}, \text{LO})|^2 |I(\mathrm{i}\, q_z)|^2 \to \sum_{\mathbf{q}_\|} M^2(\mathbf{q}_\|), \tag{5.93}$$

$$n_{\mathbf{q}\lambda} \to N(\mathbf{q}_\|); \tag{5.94}$$

for semiconductor superlattices,

$$\sum_{\mathbf{q}_\|} |M(\mathbf{q}, \text{LO})|^2 |I(\mathrm{i}\, q_z)|^2 \to \sum_{\mathbf{q}_\|} \frac{d}{2\pi} \int_{-\pi/d}^{\pi/d} dq_z M^2(\mathbf{q}_\|), \tag{5.95}$$

$$n_{\mathbf{q}\lambda} \to N(\mathbf{q}_\|, q_z). \tag{5.96}$$

The occupation number of quasi-2D phonons $N(\mathbf{q}_\|)$ and $N(\mathbf{q}_\|, q_z)$ are given in Secs. 5.4 and 5.5.

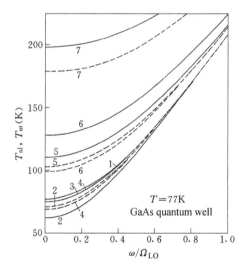

Fig. 5.5 Longitudinal (solid curves) and transverse (dashed curves) noise temperatures $T_{\text{nl}}(\omega)$ and $T_{\text{nt}}(\omega)$ are shown as functions of the normalized frequency ω/Ω_0 in a GaAs quantum well described in the text at lattice temperature $T = 77$ K for different dc bias fields at $\tau_{\text{p}} = 0$ or $\tau_{\text{p}} = 3.5$ ps. 1: $E \approx 0$, i.e., Nyquist relation; 2: $E = 100$ V/cm, $\tau_{\text{p}} = 0$; 3: $E = 100$ V/cm, $\tau_{\text{p}} = 3.5$ ps; 4: $E = 400$ V/cm, $\tau_{\text{p}} = 0$; 5: $E = 400$ V/cm, $\tau_{\text{p}} = 3.5$ ps; 6: $E = 1$ kV/cm, $\tau_{\text{p}} = 0$; 7: $E = 1$ kV/cm, $\tau_{\text{p}} = 3.5$ ps.

Fig. 5.5 demonstrates the calculated longitudinal and transverse noise temperatures $T_{\text{nl}}(\omega)$ and $T_{\text{nt}}(\omega)$ as functions of the normalized frequency

$\omega/\Omega_{\rm LO}$ for a single GaAs quantum well having well width $a = 25$ nm, carrier sheet density $N_{\rm s} = 2.0 \times 10^{11}\,{\rm cm}^{-2}$ and zero-temperature linear mobility $\mu_0(0) = 100\,{\rm m}^2/{\rm Vs}$ at lattice temperature $T = 77$ K, obtained with quasi-2D phonon model in the cases of $\tau_{\rm p} = 0$ and $\tau_{\rm p} = 3.5$ ps. The curve 1 represents the Nyquist relation at $T = 77$ K. Fig. 5.6 shows the zero-frequency values of longitudinal and transverse noise temperatures $T_{\rm nl} \equiv T_{\rm nl}(0)$ and $T_{\rm nt} \equiv T_{\rm nt}(0)$, as well as the electron temperature $T_{\rm e}$, versus the dc bias electric field E in the cases of $\tau_{\rm p} = 0$ (a) and $\tau_{\rm p} = 3.5$ ps (b). We can see that there is appreciable electron cooling: both $T_{\rm nl}(\omega)$ and $T_{\rm nt}(\omega)$ are lower than $T = 77$ within quite wide ranges of bias electric field and frequency. The occurrence of nonequilibrium phonons, however, always suppresses the electron cooling. When $\tau_{\rm p} = 3.5$ ps, the electron cooling disappears completely.

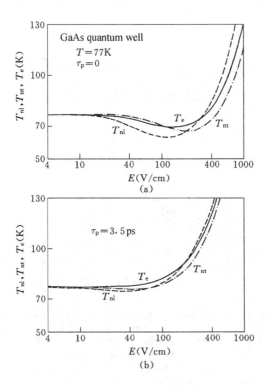

Fig. 5.6 The electron temperature $T_{\rm e}$ and longitudinal and transverse noise temperatures $T_{\rm nl}$ and $T_{\rm nt}$ at $\omega \sim 0$, are shown as functions of the bias dc field for the same GaAs quantum well as discussed in Fig. 5.5 at lattice temperature $T = 77$ K for optic-phonon relaxation times (a) $\tau_p = 0$ and $\tau_p = 3.5$ ps. From Lei and Horing (1987b).

Note that, acoustic phonons, though giving a much smaller contributions to the force correlation function $S(\omega)$ and memory function $M_2(\omega)$ than optic phonons, play an important role in determining the bias points. They are included in the above calculation (assuming equilibrium distribution).

5.7 Confined Phonons in Quasi-2D Semiconductors

Semiconductor microstructures, such as heterjunctions, quantum wells and quantum wires, are spatially inhomogeneous materials, in which always exist interfaces between different semiconductors or between a semiconductor and the vacuum. The eigenmodes, especially the optic modes of the lattice vibration in these systems, may be different from those in the homogeneous bulk system consisting of a single semiconductor. The early theoretical study on this matter can trace back to middle 1960s (Fuchs and Kliewer, 1965). It was till the middle of 1980s when the confined bulk modes and interface modes of optic phonons were experimentally observed in GaAs-based quantum-well superlattices (Sood et al, 1985; Klein, 1986), the issue of phonon confinement in semiconductor microstructures begun to arouse widespread attention. Macroscopic theoretical models, based on dielectric continuum theory, including slab modes (Fuchs and Kliewer, 1965) and guided models (Sood et al, 1985; Babiker, 1986; Ridley, 1989b), and microscopic models (Huang and Zhu, 1988a,b) were developed to describe these confined optic phonons. Quite a deep understanding on the characteristics of confined phonon modes in different kind semiconductor microstructures and their interactions with electrons have been reached (Nash, 1992; Xia and Zhu, 1995).

Unlike light spectrum, transport properties depend on the totality of scatterings from all possible phonons, and may not be very sensitive to the detailed structure of the phonon modes. In fact, theoretical analysis (Nash, 1992) and concrete calculations for quasi-2D (Tsuchiya and Ando, 1992) and quasi-1D (Wang and Lei, 1994) systems show that the details of phonon eigenmodes have no significant influence on the properties of electron transport if phonons are in equilibrium and contributions from all the phonon modes are included. However, situation may become different if phonons are not in equilibrium. In quasi-2D and quasi-1D systems, the nonequilibrium occupation of these confined phonon modes in quantum well superlattices may lead to substantial modifications of the electron power loss rate and other transport properties due to the symmetry related selection rules at low temperature (Jain and Das Sarma, 1989).

We can discuss effects of confined and interface phonon modes on hot electron transport in a quasi-2D system with the method presented in the preceding sections in this chapter, including nonequilibrium phonon occupations. Since the confined phonon modes do not spatially mismatch the electron density distribution, the assumption that each eigenmode is in a quasi-equilibrium state before the whole phonon system relaxes (in time τ_p) to thermal equilibrium, is reasonable. With this, we can immediately write down the corresponding expressions from those in Sec. 5.1.

For a quasi-2D quantum well, the electron state in the well can be described by a 2D wavevector $\boldsymbol{k}_\parallel$ and a subband index n, while the confined and interface phonons in the well region are described by a 2D wavevector $\boldsymbol{q}_\parallel$ and a branch index j. All the results can be obtained from the relevant formulae in Sec. 2.5 for the coupled system of quasi-2D electrons and 3D bulk phonons [represented by a 3D wavevector $\boldsymbol{q} = (\boldsymbol{q}_\parallel, q_z)$ and a branch index j]. For instance, the electron–phonon interaction can be obtained by rewriting (2.87) to be

$$H_\mathrm{ep} = \sum_{\boldsymbol{q}_\parallel, j, n', n} M_{n'n}(\boldsymbol{q}_\parallel, j) e^{i\boldsymbol{q}_\parallel \cdot \boldsymbol{R}} \left(b_{\boldsymbol{q}_\parallel j} + b^\dagger_{-\boldsymbol{q}_\parallel j}\right) \sum_{\boldsymbol{k}_\parallel, \sigma} c^\dagger_{n'\boldsymbol{k}_\parallel + \boldsymbol{q}_\parallel \sigma} c_{n\boldsymbol{k}_\parallel \sigma}, \quad (5.97)$$

$M_{n'n}(\boldsymbol{q}_\parallel, j)$ being the scattering matrix elements between electrons (subbands n, n') and the confined phonon $(\boldsymbol{q}_\parallel, j)$. The rate of change of the phonon occupation number $n_{\boldsymbol{q}_\parallel j}$ due to electron–phonon interaction is

$$\left[\frac{dn_{\boldsymbol{q}_\parallel j}}{dt}\right]_\mathrm{ep} = \Gamma(\boldsymbol{q}_\parallel, j, \Omega_{\boldsymbol{q}_\parallel j} - \boldsymbol{q}_\parallel \cdot \boldsymbol{v}) \left[n\left(\frac{\Omega_{\boldsymbol{q}_\parallel j} - \boldsymbol{q}_\parallel \cdot \boldsymbol{v}}{T_\mathrm{e}}\right) - n_{\boldsymbol{q}_\parallel j}\right], \quad (5.98)$$

in which

$$\Gamma(\boldsymbol{q}_\parallel, j, \Omega) = -2 \sum_{n', n} |M_{n'n}(\boldsymbol{q}_\parallel, j)|^2 \Pi_2(n', n, \boldsymbol{q}_\parallel, \Omega). \quad (5.99)$$

The H_ep-induced frictional force and energy-loss rate of electrons are

$$\boldsymbol{f}_\mathrm{p} = -\sum_{\boldsymbol{q}_\parallel, j} \boldsymbol{q}_\parallel \left[\frac{\partial n_{\boldsymbol{q}_\parallel j}}{\partial t}\right]_\mathrm{ep}, \quad (5.100)$$

$$w = \sum_{\boldsymbol{q}_\parallel, j} \Omega_{\boldsymbol{q}_\parallel j} \left[\frac{\partial n_{\boldsymbol{q}_\parallel j}}{\partial t}\right]_\mathrm{ep}. \quad (5.101)$$

The time-dependent equation for the phonon occupation number is written in terms of the relaxation time τ_p of the phonon system as

$$\frac{dn_{\boldsymbol{q}_\parallel j}}{dt} = \left[\frac{\partial n_{\boldsymbol{q}_\parallel j}}{\partial t}\right]_\mathrm{ep} - \frac{1}{\tau_\mathrm{p}}\left[n_{\boldsymbol{q}_\parallel j} - n\left(\frac{\Omega_{\boldsymbol{q}_\parallel j}}{T}\right)\right]. \quad (5.102)$$

In the steady state ($dn_{\boldsymbol{q}_\parallel j}/dt = 0$), this equation gives

$$n_{\boldsymbol{q}_\parallel j} = n\left(\frac{\Omega_{\boldsymbol{q}_\parallel j} - \boldsymbol{q}_\parallel \cdot \boldsymbol{v}}{T_\mathrm{e}}\right) + \left[1 + \tau_\mathrm{p}\Gamma(\boldsymbol{q}_\parallel, j, \Omega_{\boldsymbol{q}_\parallel j} - \boldsymbol{q}_\parallel \cdot \boldsymbol{v})\right]^{-1}$$
$$\times \left[n\left(\frac{\Omega_{\boldsymbol{q}_\parallel j}}{T}\right) - n\left(\frac{\Omega_{\boldsymbol{q}_\parallel j} - \boldsymbol{q}_\parallel \cdot \boldsymbol{v}}{T_\mathrm{e}}\right)\right], \quad (5.103)$$

$$\boldsymbol{f}_\mathrm{p} = \sum_{\boldsymbol{q}_\parallel, j} \frac{\boldsymbol{q}_\parallel \Gamma(\boldsymbol{q}_\parallel, j, \Omega_{\boldsymbol{q}_\parallel j} + \boldsymbol{q}_\parallel \cdot \boldsymbol{v})}{1 + \tau_\mathrm{p}\Gamma(\boldsymbol{q}_\parallel, j, \Omega_{\boldsymbol{q}_\parallel j} + \boldsymbol{q}_\parallel \cdot \boldsymbol{v})} \left[n\left(\frac{\Omega_{\boldsymbol{q}_\parallel j} + \boldsymbol{q}_\parallel \cdot \boldsymbol{v}}{T_\mathrm{e}}\right) - n\left(\frac{\Omega_{\boldsymbol{q}_\parallel j}}{T}\right)\right],$$
$$(5.104)$$

$$w = \sum_{\boldsymbol{q}_\parallel, j} \frac{\Omega_{\boldsymbol{q}_\parallel j}\Gamma(\boldsymbol{q}_\parallel, j, \Omega_{\boldsymbol{q}_\parallel j} + \boldsymbol{q}_\parallel \cdot \boldsymbol{v})}{1 + \tau_\mathrm{p}\Gamma(\boldsymbol{q}_\parallel, j, \Omega_{\boldsymbol{q}_\parallel j} + \boldsymbol{q}_\parallel \cdot \boldsymbol{v})} \left[n\left(\frac{\Omega_{\boldsymbol{q}_\parallel j} + \boldsymbol{q}_\parallel \cdot \boldsymbol{v}}{T_\mathrm{e}}\right) - n\left(\frac{\Omega_{\boldsymbol{q}_\parallel j}}{T}\right)\right].$$
$$(5.105)$$

Applying the above formulation to a quantum-well LO phonon model ($\Omega_{\boldsymbol{q}_\parallel j} = \Omega_\mathrm{LO}$), Guillimot et al (1990) calculate high-field electron transport in GaAs/AlGaAs multiple quantum well samples in the presence of nonequilibrium phonons and compare theoretical results with their experimental measurements. Fig. 5.7(a) exhibits the electric field dependent drift velocity v obtained in the cases of $\tau_\mathrm{p} = 0$ and $\tau_\mathrm{p} = 5\,\mathrm{ps}$ for a quantum well having width $a = 16.1\,\mathrm{nm}$, carrier sheet density $N_\mathrm{s} = 1.62 \times 10^{11}\,\mathrm{cm}^{-2}$ and low-temperature linear mobility $\mu_0 = 13\,\mathrm{m}^2/\mathrm{Vs}$, together with experimental data (small circles) at Helium temperature. Fig 5.7(b) plots the inverse electron temperature $1/T_\mathrm{e}$ versus the input power (Watt per electron). They reach the same conclusion as that obtained in the preceding section: a finite τ_p greatly reduces the electron energy-loss rate and increases the electron temperature. As a result, the drift velocity is also reduced at high electric fields, in spite of the momentum reabsorption on the forward peaked phonon distribution, as shown in Fig. 5.8. The theoretical results of $\tau_\mathrm{p} = 5\,\mathrm{ps}$ are in relatively good agreement with the experimental data of this sample.

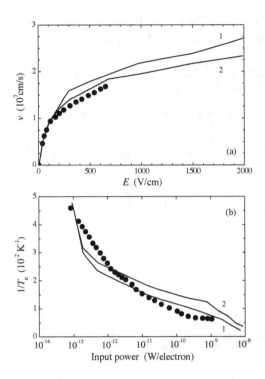

Fig. 5.7 The electron drift velocity v is shown as a function of the electric field E (a) and the inverse electron temperature $1/T_e$ versus the input power per electron (b). The system is a GaAs-based quantum well having width $a = 16.1\,\text{nm}$, carrier sheet density $N_s = 1.62 \times 10^{11}\,\text{cm}^{-2}$ and low-temperature linear mobility $\mu_0 = 13\,\text{m}^2/\text{Vs}$. The solid curves are theoretical results for $\tau_p = 0$ (line 1) and $\tau_p = 5\,\text{ps}$ (line 2). the black dots are experimental data. After Guillemot et al (1990).

According to (5.103), phonons are in equilibrium distribution $n_{\bm{q}_\parallel j} = n(\Omega_{\bm{q}_\parallel j}/T)$ for $\tau_p = 0$. Nonequilibrium phonons occur when $\tau_p \neq 0$. In addition to the general $\bm{q}_\parallel$ dependence in $\Omega_{\bm{q}_\parallel j}$ and $\Gamma(\bm{q}_\parallel, j, \omega)$, the wavevector $\bm{q}_\parallel$ distribution of nonequilibrium phonons under the condition of high-field driven transport is mainly affected by the Dopler shift $(\Omega_{\bm{q}_\parallel j} - \bm{q}_\parallel \cdot \bm{v})$ of the phonon frequency. This may push nonequilibrium phonons distributing to and peaking on the forward direction of the electron drift velocity $\bm{v}$ in the wavevetor space. The peak of the nonequilibrium phonon population appears roughly around the wavevector $\bm{q}_\parallel$ determined by $\Omega_{\text{LO}} - \bm{q}_\parallel \cdot \bm{v} \approx 0$. In the case of large drift velocity v, the peak may be quite remarkable. In this case the electron gas interacts mainly with those LO phonons, whose phase velocities are near the drift velocity of the electrons. Fig. 5.8 presents the LO phonon population distribution in the wavevector space along the direction of the electron drift velocity v, at several different input powers (different electron temperatures).

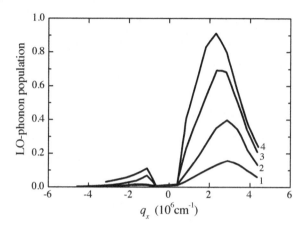

Fig. 5.8 The distribution of nonequilibrium LO phonon population in the wavevector space along the direction of the electron drift velocity v: $q_x = \bm{q}_\parallel \cdot \bm{v}/v$. The phonon relaxation time $\tau_p = 5\,\mathrm{ps}$. The curves 1, 2, 3 and 4 are respectively at electron temperatures $T_e = 61, 96, 137$ and $194\,\mathrm{K}$. After Guillemot et al (1990).

Chapter 6

Systems with Several Species of Carriers

6.1 Balance Equations for Type-II Superlattices

A type-II superlattice is a material composed of alternating layers of different semiconductor constituents, like an InAs-GaSb system in which the band discontinuities at the heterojunction interfaces are so large that the bottom of the conduction band of InAs lies below the top of the valence band of GaSb. There is a transfer of electrons from GaSb layers to the conduction band of InAs, producing a periodic structure with spatially separated electrons and holes in adjacent layers. Such a structure forms one of the simplest systems having two species of carriers: electrons and holes. The total number of electrons and that of holes are assumed fixed. The Coulomb interaction between them plays an important role in determining the transport properties of the system.

We consider a model system consisting of two sets of periodically arranged quantum wells of width a_1 and a_2 respectively, but with the same spatial period d. The central positions of quantum wells in the first set are at $z = ld$ ($l = 0, \pm 1, \cdots$), and the carriers in them are electrons with charge $-e$, effective mass m_e and sheet density N_e per layer. The central positions of quantum wells in the second set are at $z = ld+b$ ($l = 0, \pm 1, \cdots$), and the carriers in them are holes with charge e, effective mass m_h and per-layer density N_h. For simplicity we assume that electrons and holes populate only their respective first (lowest) subbands and that the wave functions of different wells do not overlap, irrespective of type. The lowest subband wave function for electrons in the lth InAs well is given by [$(\bm{r}_1, z_1)$ is the electron position variable]

$$\psi^{(1)}_{l\bm{k}_\|}(\bm{r}_1, z_1) = e^{i\bm{k}_\| \cdot \bm{r}_1} \zeta_1(z_1 - ld), \tag{6.1}$$

$$\zeta_1(z) = \begin{cases} \sqrt{2/a_1}\cos(\pi z/a_1), & -a_1/2 < z < a_1/2; \\ 0, & \text{otherwise.} \end{cases} \quad (6.2)$$

The lowest subband wave function for holes in the lth GaSb well is taken as [$(\boldsymbol{r}_2, z_2)$ is the hole position variable]

$$\psi_{l\boldsymbol{k}_\|}^{(2)}(\boldsymbol{r}_2, z_2) = e^{i\boldsymbol{k}_\| \cdot \boldsymbol{r}_2}\zeta_2(z_2 - ld - b), \quad (6.3)$$

$$\zeta_2(z) = \begin{cases} \sqrt{2/a_2}\cos(\pi z/a_2), & -a_2/2 < z < a_2/2; \\ 0, & \text{otherwise.} \end{cases} \quad (6.4)$$

The electrons and holes are subject to scatterings by impurities and by phonons just as in their respective type-I superlattices. In order to investigate transport parallel to the layer plane we introduce 2D center-of-mass coordinates $\boldsymbol{R}_1 = (X_1, Y_1)$ and $\boldsymbol{R}_2 = (X_2, Y_2)$ for electrons and holes respectively. Under the influence of a uniform electric field $\boldsymbol{E}$ parallel to the plane, the Hamiltonian can be written as

$$H = H_{\text{cm}} + H_{\text{er}} + H_{\text{hr}} + H_{\text{I}} + H_{\text{ph}}, \quad (6.5)$$

in which H_{ph} is the phonon Hamiltonian; H_{cm} is the center-of-mass (CM) Hamiltonian, including the kinetic energies of the electron center of mass and the hole center of mass and their potential energies in the electric field $\boldsymbol{E}$; H_{er} and H_{hr} are respectively the Hamiltonian of relative electrons and that of relative holes; H_{I} contains, in addition to electron–impurity (H_{ei}), electron–phonon (H_{ep}), hole–impurity (H_{hi}) and hole–phonon (H_{hp}) couplings, the Coulomb interaction between electrons and holes,

$$H_{12} = \sum_{\boldsymbol{q}_\|} e^{i\boldsymbol{q}_\| \cdot (\boldsymbol{R}_1 - \boldsymbol{R}_2)} \sum_{l,m} V_{lm}^{12}(\boldsymbol{q}_\|)\rho_{l\boldsymbol{q}_\|}^{(1)}\rho_{m,-\boldsymbol{q}_\|}^{(2)}, \quad (6.6)$$

where $\rho_{l\boldsymbol{q}}^{(1)}$ and $\rho_{m,-\boldsymbol{q}_\|}^{(2)}$ are respectively the density operator of electrons in the lth well and that of holes in the mth well,

$$V_{lm}^{12}(q_\|) = -\frac{e^2}{2\epsilon_0 \kappa q} \int\int dz\,dz'\, e^{-q_\| |z-z'|}\zeta_1(z-ld)^2\zeta_2(z'-md-b)^2 \quad (6.7)$$

is the matrix element of the electron–hole Coulomb interaction. We take the initial density matrix to be

$$\hat{\rho}\Big|_{t=-\infty} = \hat{\rho}_0 = \frac{1}{Z}e^{-H_{\text{ph}}/T - H_{\text{er}}/T_{\text{e}} - H_{\text{hr}}/T_{\text{h}}}, \quad (6.8)$$

where T, T_e, and T_h are, respectively, the lattice temperature, electron temperature, and hole temperature, and treat H_I as a perturbation. By calculating the statistical averages of the rates of changes of the CM momenta of electrons and holes, and the rates of changes of the internal energies of relative electrons and relative holes to the lowest order in H_I following the procedure described in Chapter 1, we obtain the following force and energy balance equations for nonlinear transport in a type-II superlattice in the presence of an electric field $\boldsymbol{E} = (E, 0, 0)$ along the x direction (Lei, 1987a; Lei, Cui and Horing, 1988):

$$N_\mathrm{e} m_\mathrm{e} \frac{dv_1}{dt} = -N_\mathrm{e} eE + f_1(v_1) + f_{12}(v_1 - v_2), \tag{6.9}$$

$$N_\mathrm{h} m_\mathrm{h} \frac{dv_2}{dt} = N_\mathrm{h} eE + f_2(v_2) + f_{21}(v_2 - v_1), \tag{6.10}$$

$$-\frac{du_\mathrm{e}}{dt} = v_1 f_1(v_1) + w_1(v_1) + w_{12}(v_1 - v_2), \tag{6.11}$$

$$-\frac{du_\mathrm{h}}{dt} = v_2 f_2(v_2) + w_2(v_2) + w_{21}(v_2 - v_1). \tag{6.12}$$

Equations (6.9) and (6.10) are actually the equations of motion of the electron center of mass and the hole center of mass. In the above equations, N_e and N_h are, respectively, the electron sheet density and the hole sheet density per layer, u_e and u_h are, respectively, the relative electron energy and relative hole energy per layer, v_1 and v_2 are, respectively, the velocity of the electron center of mass and that of the hole center of mass, i.e. the average drift velocities of the electron and hole systems. The total current (per layer) is due to both electron and hole contributions:

$$J = -N_\mathrm{e} e v_1 + N_\mathrm{h} e v_2. \tag{6.13}$$

In Eqs. (6.9)–(6.12), $f_j(v_j)$ are the frictional forces (per layer) exerted on the electron center of mass ($j = 1$) and on the hole center of mass ($j = 2$), due to impurity and phonon scatterings; $w_j(v_j)$ are the energy transfer rates (per layer) for electron system ($j = 1$) and hole system ($j = 2$) due to carrier–phonon interactions. Their expressions are exactly the same as those in Sec. 2.7 for separate electron and hole superlattices. Finally, $f_{ij}(v_i - v_j)$ and $w_{ij}(v_i - v_j)$ ($j = 1, 2$) are the forces and energy transfer rates due to electron–hole interactions. They satisfy

$$f_{21}(v_2 - v_1) = -f_{12}(v_1 - v_2), \tag{6.14}$$

$$w_{21}(v_2 - v_1) = (v_1 - v_2) f_{12}(v_1 - v_2) - w_{12}(v_1 - v_2). \tag{6.15}$$

The expressions of $f_{12}(v_1 - v_2)$ and $w_{12}(v_1 - v_2)$ are as follows:

$$f_{12}(v_1 - v_2) = \left(\frac{e^2}{2\epsilon_0 \kappa}\right)^2 \frac{d}{\pi^2} \int_0^{\pi/d} dq_z \sum_{\mathbf{q}_\parallel} \frac{q_x}{q_\parallel^2} \int_{-\infty}^{\infty} d\omega$$
$$\times \left[n\left(\frac{\omega}{T_e}\right) - n\left(\frac{\omega - \omega_{12}}{T_h}\right)\right] I_1^2(q_\parallel) I_2^2(q_\parallel) S^{12}(\mathbf{q}_\parallel, q_z)$$
$$\times \Pi_2^{(1)}(q_z, \mathbf{q}_\parallel, \omega) \Pi_2^{(2)}(q_z, -\mathbf{q}_\parallel, \omega - \omega_{12}), \quad (6.16)$$

$$w_{12}(v_1 - v_2) = \left(\frac{e^2}{2\epsilon_0 \kappa}\right)^2 \frac{d}{\pi^2} \int_0^{\pi/d} dq_z \sum_{\mathbf{q}_\parallel} \frac{1}{q_\parallel^2} \int_{-\infty}^{\infty} \omega \, d\omega$$
$$\times \left[n\left(\frac{\omega}{T_e}\right) - n\left(\frac{\omega - \omega_{12}}{T_h}\right)\right] I_1^2(q_\parallel) I_2^2(q_\parallel) S^{12}(\mathbf{q}_\parallel, q_z)$$
$$\times \Pi_2^{(1)}(q_z, \mathbf{q}_\parallel, \omega) \Pi_2^{(2)}(q_z, -\mathbf{q}_\parallel, \omega - \omega_{12}), \quad (6.17)$$

in which the form factors $I_j(q_\parallel)$ ($j = 1, 2$) are given by (2.135), the density correlation functions $\Pi_2^{(j)}(q_z, \mathbf{q}_\parallel, \omega)$ are given by (2.148) for electron ($j = 1$) and hole ($j = 2$) subsuperlattices, respectively, and finally

$$S^{12}(\mathbf{q}_\parallel, q_z) \equiv \frac{\cosh(q_\parallel d)\cos(q_z d) - 1}{[1 - \cosh(q_\parallel d)\cos(q_z d)]^2 + [\sinh(q_\parallel d)\sin(q_z d)]^2}$$
$$+ \frac{\cosh[q_\parallel (d - 2b + a_1 - a_2)]}{\cosh(q_\parallel d) - \cos(q_z d)}, \quad (6.18)$$

$$\omega_{12} \equiv q_x(v_1 - v_2). \quad (6.19)$$

Transport quantities v_1, v_2, T_e and T_h are involved in equations (6.9)–(6.12) as variables. For a given electric field E, this set of ordinary differential equations completely determines the time variation of these variables starting from their initial values. In the steady state under a constant E, these equations reduce to

$$-N_e eE + f_1(v_1) + f_{12}(v_1 - v_2) = 0, \quad (6.20)$$
$$N_h eE + f_2(v_2) + f_{21}(v_2 - v_1) = 0, \quad (6.21)$$
$$v_1 f_1(v_1) + w_1(v_1) + w_{12}(v_1 - v_2) = 0, \quad (6.22)$$
$$v_2 f_2(v_2) + w_2(v_2) + w_{21}(v_2 - v_1) = 0. \quad (6.23)$$

Apparently, all the equations presented in this section for a type-II superlattice remain useful in discussing a system consisting of two sublattices having the same kind of carriers. For instance, if carriers in both sublattices are electrons, what we need to do is to change the hole charge e into $-e$ in Eqs. (6.13) and (6.21), while keep everything else unchanged.

6.2 DC Steady State Transport

The nonlinear resistivity function is defined as $R(v_1, v_2) \equiv E/J$, J being the per-layer current given by (6.13). One immediately obtains from Eqs. (6.20) and (6.21) of the dc steady state,

$$R(v_1, v_2) = \frac{f_1(v_1) + f_{12}(v_1 - v_2)}{N_e e^2 (N_h v_2 - N_e v_1)} = -\frac{f_2(v_2) - f_{12}(v_1 - v_2)}{N_h e^2 (N_h v_2 - N_e v_1)}. \tag{6.24}$$

We first discuss the weak field limit. For small v_1 and v_2, the energy-balance equations (6.22) and (6.23) yield $T_e = T_h = T$ up to the linear order in the drift velocities. The force-balance equations give rise to

$$N_h [f_1(v_1) + f_{12}(v_1 - v_2)] + N_e [f_2(v_2) - f_{12}(v_1 - v_2)] = 0. \tag{6.25}$$

Expanding this equation to the linear order in v_1 and v_2, we have

$$\frac{v_2}{v_1} = -\frac{N_h [f_1'(0) + f_{12}'(0)] - N_e f_{12}'(0)}{N_e [f_2'(0) + f_{12}'(0)] - N_h f_{12}'(0)}, \tag{6.26}$$

where $f_j'(0) \equiv [\partial f_j(v)/\partial v]_{v=0}$ $(j = 1, 2)$,

$$f_{12}'(0) \equiv \left[\frac{\partial}{\partial v} f_{12}(v)\right]_{v=0}$$
$$= \left(\frac{e^2}{2\epsilon_0 \kappa}\right)^2 \frac{d}{\pi^2} \int_0^{\pi/d} dq_z \sum_{\mathbf{q}_\parallel} \frac{q_x}{q_\parallel^2} I_1^2(q_\parallel) I_2^2(q_\parallel) S^{12}(\mathbf{q}_\parallel, q_z)$$
$$\times \int_{-\infty}^\infty d\omega \left[-n'\left(\frac{\omega}{T}\right)\right] \Pi_2^{(1)}(q_z, \mathbf{q}_\parallel, \omega) \Pi_2^{(2)}(q_z, -\mathbf{q}_\parallel, -\omega). \tag{6.27}$$

The weak-field resistivity (linear resistivity) is derived by expanding (6.24) in terms of small v_1 and v_2 and making use of (6.26):

$$R_0 \equiv R(v_1 \to 0, v_2 \to 0)$$
$$= -e^{-2} \frac{f_1'(0) f_2'(0) + f_{12}'(0) [f_1'(0) + f_2'(0)]}{N_h^2 f_1'(0) + N_e^2 f_2'(0) + (N_e - N_h)^2 f_{12}'(0)}. \tag{6.28}$$

For an intrinsic system, $N_e = N_h = N_s$, (6.28) reduces to

$$R_0 = R_{e-h} + R_e R_h / (R_e + R_h). \tag{6.29}$$

In this, $R_\mathrm{e} = -f_1'(0)/(N_\mathrm{s}^2 e^2)$ and $R_\mathrm{h} = -f_2'(0)/(N_\mathrm{s}^2 e^2)$ are linear resistivities associated separately with electron and hole systems, $R_\mathrm{e-h} = -f_{12}'(0)/(N_\mathrm{s}^2 e)$ is a linear resistivity arising from the electron–hole interaction. Equation (6.29) indicates that the Ohmic resistivity of a type-II superlattice with equal electron and hole concentrations can be thought of two parallel connected resistances R_e and R_h of the electron and hole superlattices, taken jointly with a series-connected resistance $R_\mathrm{e-h}$ due to electron–hole coupling.

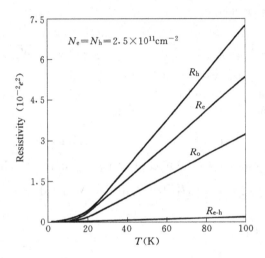

Fig. 6.1 Linear resistivities R_0, R_e, R_h and $R_\mathrm{e-h}$ of an InAs/GaSb-based type-II superlattice, are shown as functions of lattice temperature. The system has equal electron and hole densities $N_\mathrm{e} = N_\mathrm{h} = N_\mathrm{s} = 2.5 \times 10^{11}$ cm^{-2}. From Lei, Cui and Horing (1988).

On the basis of the formulation detailed above, Lei, Cui and Horing (1988) carry out numerical studies of nonlinear dc transport in an InAs-GaSb type-II superlattice with electrons in InAs layers having effective mass $m_\mathrm{e} = 0.026 m_0$ and holes in GaSb layers having effective mass $m_\mathrm{h} = 0.03 m_0$ (m_0 is the free electron mass). The geometric parameters of the superlattice are $a_1 = 10$ nm, $a_2 = 15$ nm, $b = 20$ nm, and $d = 50$ nm. The system is assumed intrinsic that electrons and holes have equal densities, $N_\mathrm{e} = N_\mathrm{h} = N_\mathrm{s}$, and effects of impurity scattering can be negligible. In addition to electron–hole interaction, the main scattering mechanisms considered include electron–polar-optic-phonon scattering with matrix element (2.129), electron–acoustic-phonon deformation potential scattering with matrix element (2.125), hole–polar-optic-phonon scattering with matrix element (2.129), hole–acoustic-phonon deformation potential scattering with matrix element (2.125), and hole–optic-phonon deformation potential scattering with constant matrix element (2.66). The material and coupling parameters are: n-type InAs optical dielectric constant 11.74, static dielectric constant 14.54, LO-phonon frequency 29.0 meV, average sound velocity 3.09×10^5 cm/s, acoustic deformation potential 5.8 eV; p-type GaSb optical dielectric constant 14.44, static dielectric constant 15.69, LO-phonon frequency 29.8 meV, average sound velocity 3.22×10^5 cm/s, acoustic deformation potential 8.3 eV, optic deformation potential 9.0×10^8 eV/cm; lattice mass density 5.61 g/cm^3.

Figure 6.1 plots the linear resistivity R_e of the electron subsystem, the linear resistivity R_h of the hole subsystem, the linear resistivity $R_\text{e-h}$ due to electron–hole interaction, and the total linear resistivity R_0 determined by Eq. (6.29) for an InAs/GaSb-based type-II superlattice having equal electron and hole densities $N_\text{e} = N_\text{h} = N_\text{s} = 2.5 \times 10^{11}$ cm^{-2}.

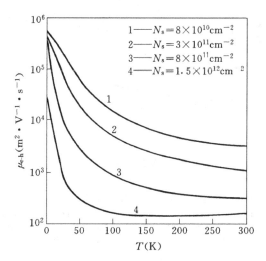

Fig. 6.2 The linear mobility due to electron–hole interaction, $\mu_\text{e-h} = N_\text{s} e R_\text{e-h}$, versus the lattice temperature for type-II superlattices described in the text having different carrier densities. From Lei, Cui and Horing (1988).

The effect of electron–hole interaction depends strongly on the carrier density and temperature. Fig. 6.2 shows the linear dc mobility due to electron–hole interaction alone, $\mu_\text{e-h} = N_\text{s} e R_\text{e-h}$, as a function of lattice temperature, in type-II superlattices of equal electron and hole densities ($N_e = N_h = N_s$) having four different carrier concentrations $N_s = 8 \times 10^{10}, 3 \times 10^{11}, 8 \times 10^{11}$, and $1.5 \times 10^{12}\,\text{cm}^{-2}$.

Nonlinear hot-carrier transport of type-II superlattices can be determined from balance equations (6.20)–(6.23). Fig. 6.3 demonstrates the electron and hole temperatures T_e and T_h, the electron drift velocity v_1, and the hole drift velocity v_2, as functions of the applied electric field E, obtained for a type-II superlattice having $N_e = N_h = N_s = 2.5 \times 10^{11}\,\text{cm}^{-2}$ at lattice temperature $T = 4.2\,\text{K}$. We can see that holes drift slower and tend to stay cooler than electrons under a given electric field. This is because that they have a larger effective mass than electrons, $m_h > m_e$, and hole–phonon couplings are more effective than electron–phonon couplings.

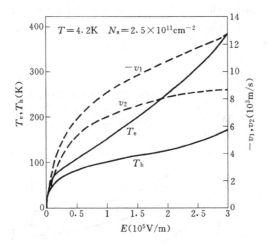

Fig. 6.3 The electron temperature T_e, the hole temperature T_h, the minus electron drift velocity $-v_1$, and the hole drift velocity v_2, versus the applied electric field E. The system is an InAs/GaSb-based type-II superlattice having $N_e = N_h = N_s = 2.5 \times 10^{11}\,\text{cm}^{-2}$ at lattice temperature $T = 4.2\,\text{K}$. From Lei, Cui and Horing (1988).

We note that in a type-II superlattice the electron–hole interaction generally makes only a small contribution to the total dc resistivity as can be seen from Fig. 6.1. The main reason for this is the spatial separation of electrons and holes ($b = 20\,\text{nm}$ in the discussed system). If both types of carriers are confined in the same layer, the attraction between them can be much stronger to cause the minority carriers to drift along the direction opposite to the electric field acceleration. This will be discussed in more detail in the following section.

6.3 Negative Minority-Electron Mobility in Electron–Hole Plasma

Höpfel et al (1986) reported their interesting experimental observation of negative minority-carrier mobility in GaAs/AlGaAs quantum well structures and attributed it physically to the electron–hole attraction. This immediately stimulated further theoretical investigations (Cui, Lei and Horing, 1988a; Cai, Zheng and Lax, 1988).

The two-component plasma having electrons and holes in a single quantum well can be regarded as a special case of long period (large d) type-II superlattices composed of overlapping electron and hole sublattices ($b = 0$). The general balance equation formulation derived in Sections 6.1 and 6.2 for type-II superlattices is directly usable in this case. The theoretical examination of Cui, Lei and Horing (1988) on carrier transport in a two-component plasma consisting of minority electrons and majority holes ($N_e \ll N_h$) is based on these equations. Similar analyses are separately performed by Cai, Zheng and Lax (1988) directly for the electron–hole plasma in a semiconductor quantum well.

For given electric field E and lattice temperature T, the coupled force- and energy-balance equations (6.20)–(6.23) can be solved to give v_1, v_2, T_e and T_h for the dc steady state. The electron and hole (nonlinear) mobilities are defined as

$$\mu_e = -v_1/E, \qquad \mu_h = v_2/E, \tag{6.30}$$

in which the minus sign in the μ_e definition keeps it positive. The possibility of μ_e becoming negative can be readily recognized from the linear limit of the balance equations. In the case of $v_1 \to 0$ and $v_2 \to 0$, the energy-balance equations require $T_e = T_h = T$. From the expressions of force-balance equations (6.20) and (6.21) in the small v_1 and v_2 limit,

$$-N_e e E + v_1 f_1'(0) + (v_1 - v_2) f_{12}'(0) = 0 \tag{6.31}$$

and

$$N_h e E + v_2 f_2'(0) - (v_1 - v_2) f_{12}'(0) = 0, \tag{6.32}$$

one immediately obtains the expression for the electron linear mobility:

$$\mu_e = -\frac{v_1}{E} = \frac{e\,[f_{12}'(0)(N_h - N_e) - f_2'(0)N_e]}{f_1'(0)f_2'(0) + f_2'(0)^2 + f_1'(0)f_{12}'(0)}. \tag{6.33}$$

Without Coulomb interaction ($f'_{12}(0) = 0$) μ_e is positive. In the case of $N_h \gg N_e$, the term $-f'_{12}(0)N_h$ can be of the same order of magnitude as $-f'_2(0)N_e$. Since $-f'_2(0)$, which is mainly due to phonon scattering, decreases faster than Coulomb-interaction induced $-f'_{12}(0)$ when temperature drops, μ_e can change sign and become negative at low temperatures when $-f'_{12}(0)N_h$ dominates over $-f'_2(0)N_e$.

Numerical calculations are carried out for a GaAS/AlGaAs quantum well with structure parameters pertinent to the experiments (well width $a = 11.2$ nm, electron density $N_e = 3 \times 10^{10}$ cm^{-2}, hole density $N_h = 1.5 \times 10^{11}$ cm^{-2}), assuming that the electron effective mass $m_e = 0.07 m_0$ (m_0 is the free electron mass) and hole effective mass $m_h = 0.4\, m_0$. Polar-optical phonon coupling and acoustic deformation potential coupling with electrons and optical and acoustic deformation potential couplings with holes, are considered as scattering mechanisms with typical material and coupling parameters for GaAs. The density correlation functions in the forces and energy-transfer rates are calculated by taking full account of temperature-dependent RPA dynamic screening.

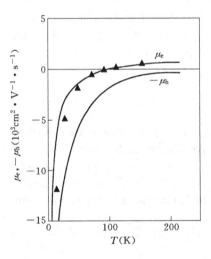

Fig. 6.4 Electron linear mobility μ_e and minus hole linear mobility $-\mu_h$ versus temperature T. Solid curves are theoretical results. Solid triangles are experimental data of electron mobility taken from Höpfel et al (1986). From Cui, Lei and Horing (1988a).

Numerical results for electron (minority carrier) linear mobility μ_e and hole (majority carrier) linear mobility μ_h multiplied by -1, are shown in Fig. 6.4 as functions of temperature T, together with the experimental data

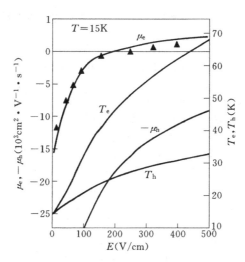

Fig. 6.5 Electron mobility μ_e, minus hole mobility $-\mu_h$, electron temperature T_e and hole temperature T_h are shown as functions of the applied electric field E at lattice temperature $T = 15$ K. Solid triangles are experimental values of μ_e taken from Höpfel et al (1986). From Cui, Lei and Horing (1988a).

of electron ohmic mobility from Höpfel et al (1986). At high temperatures μ_e is positive. When T goes down below 100 K the electron mobility becomes negative, while the hole mobility keeps positive through the whole temperature range. Theoretical results are in good agreement with the experiment. Fig. 6.5 illustrates the nonlinear electron mobility μ_e, hole mobility $\mu_{h,}$, electron temperature T_e and hole temperature T_h, calculated at $T = 15$ K, as functions of the applied electric field. We see that T_e is always much higher than T_h and the hole mobility μ_h is always positive. At high electric field when both T_e and and T_h are comparatively high, μ_e is positive. With decreasing field strength, however, T_e and T_h are dropping and μ_e becomes negative.

6.4 Coulomb Drag between Two Semiconductor Layers, Transfer Resistivity

Carrier drag in semiconductors was known to lead to interesting physical phenomena (McLean and Paige, 1960), especially its direct influence on transport (Pogrebinskii, 1977; Price, 1983). The direct experimental observation of mutual drag between carriers in two separated semiconductor system (Solomon et al, 1989) stimulated a great deal of research activity. Although there may be other mechanisms to induce carrier drag in different situations, Coulomb interaction between charged carriers is usually the most important one in the typical experimental setup: two spatially separated pools of carriers, such as two-dimensional electron (hole) gases in parallel placed quantum wells. Two quantum wells are in close proximity so that the inter-well Coulomb scattering is effective in imparting momentum (and/or energy) of a carrier in one well to that in the other well. But their separation is still large enough that the tunneling between two wells is negligible to keep the numbers of carriers independent and fixed in each well. Thus when there is a current of sheet density J_1 flowing in a well, a current can be induced in the other well; or, if the second well is electrically disconnected, an electric field E_2 would emerge to cancel the drag force induced by the current in the first well. The transfer sheet resistivity ρ_t is defined as the ratio of the induced electric field E_2 to the sheet current density J_1:

$$\rho_t = \frac{E_2}{J_1}. \tag{6.34}$$

This kind of carrier drag can be conveniently analyzed based on the balance equations derived in Sec. 6.1 for nonlinear transport in type-II superlattices. If carriers occupy only the lowest minibands in respective quantum wells, the corresponding equations for two parallel placed quantum wells are immediately obtained from those in Sec. 6.1 for type-II superlattices by taking $d \to \infty$. For instance, the balance equations for the dc steady state, (6.20)–(6.23), are now written as (E_1 and E_2 are, respectively, the electric fields in the first and in the second quantum wells)

$$-N_e e E_1 + f_1(v_1) + f_{12}(v_1 - v_2) = 0, \tag{6.35}$$

$$N_h e E_2 + f_2(v_2) - f_{12}(v_1 - v_2) = 0, \tag{6.36}$$

$$v_1 f_1(v_1) + w_1(v_1) + w_{12}(v_1 - v_2) = 0, \tag{6.37}$$

$$v_2 f_2(v_2) + w_2(v_2) + (v_1 - v_2) f_{12}(v_1 - v_2) - w_{12}(v_1 - v_2) = 0. \tag{6.38}$$

The expressions for $f_1(v_1)$, $f_2(v_2)$, $w_1(v_1)$ and $w_2(v_2)$ in these equations can be obtained from the corresponding expressions for type-I superlattices in Sec. 2.7 by taking $d \to \infty$ limit. They can be also obtained from the corresponding expressions for quasi-2D system with multiple subbands by taking $n' = n = 0$. The expressions for $f_{12}(v_1 - v_2)$ and $w_{12}(v_1 - v_2)$ are obtained from Eqs. (6.16) and (6.17) under $d \to \infty$ limit. Noting that

$$\lim_{d \to \infty} S^{12}(\boldsymbol{q}_\parallel, q_z) = e^{-2bq_\parallel} \qquad (6.39)$$

(b is the distance between two quantum-well centers), we can write

$$f_{12}(v_1 - v_2) = \sum_{\boldsymbol{q}_\parallel} q_x V_c^2(q_\parallel) I_1(q_\parallel)^2 I_2(q_\parallel)^2 e^{-2bq_\parallel} \int_{-\infty}^{\infty} \frac{d\omega}{\pi}$$
$$\times \left[n\left(\frac{\omega}{T_e}\right) - n\left(\frac{\omega - \omega_{12}}{T_h}\right) \right] \Pi_2^{(1)}(\boldsymbol{q}_\parallel, \omega) \Pi_2^{(2)}(-\boldsymbol{q}_\parallel, \omega - \omega_{12}), \qquad (6.40)$$

$$w_{12}(v_1 - v_2) = \sum_{\boldsymbol{q}_\parallel} V_c^2(q_\parallel) I_1(q_\parallel)^2 I_2(q_\parallel)^2 e^{-2bq_\parallel} \int_{-\infty}^{\infty} \omega \frac{d\omega}{\pi}$$
$$\times \left[n\left(\frac{\omega}{T_e}\right) - n\left(\frac{\omega - \omega_{12}}{T_h}\right) \right] \Pi_2^{(1)}(\boldsymbol{q}_\parallel, \omega) \Pi_2^{(2)}(-\boldsymbol{q}_\parallel, \omega - \omega_{12}). \qquad (6.41)$$

Here $V_c(q_\parallel) \equiv e^2/(2\epsilon_0 \kappa q_\parallel)$, $\Pi_2^j(\boldsymbol{q}_\parallel, \omega)$ is the imaginary part of the single-well density correlation function $\Pi^j(\boldsymbol{q}_\parallel, \omega) = \Pi^j(0, \boldsymbol{q}_\parallel, \omega)$ [given by (2.148)] for electrons ($j = 1$) or holes ($j = 2$) alone. Under random phase approximation

$$\Pi^j(\boldsymbol{q}_\parallel, \omega) = \frac{\Pi_0^j(\boldsymbol{q}_\parallel, \omega)}{1 - V^j(q_\parallel) \Pi_0^j(\boldsymbol{q}_\parallel, \omega)}, \qquad (6.42)$$

in which

$$V^j(q_\parallel) = \frac{e^2}{2\epsilon_0 \kappa q_\parallel} H^j(q_\parallel) \qquad (6.43)$$

is the effective Coulomb potential in the jth well, κ is the static dielectric constant, $H^j(q_\parallel)$ is the form factor of the jth well given by (2.136), $\Pi_0^j(\boldsymbol{q}_\parallel, \omega)$ is the density correlation function for noninteracting electrons ($j = 1$) or holes ($j = 2$), given by (2.144):

$$\Pi_0^j(\boldsymbol{q}_\parallel, \omega) = 2 \sum_{\boldsymbol{k}_\parallel} \frac{f((\varepsilon_{\boldsymbol{k}_\parallel + \boldsymbol{q}_\parallel}^{(j)} - \mu^{(j)})/T_j) - f((\varepsilon_{\boldsymbol{k}_\parallel}^{(j)} - \mu^{(j)})/T_j)}{\omega + \varepsilon_{\boldsymbol{k}_\parallel + \boldsymbol{q}_\parallel}^{(j)} - \varepsilon_{\boldsymbol{k}_\parallel}^{(j)} + i\delta}, \qquad (6.44)$$

where $\varepsilon_{\boldsymbol{k}_\parallel}^{(j)} = k_\parallel^2/2m_j$ is the electron ($m_1 \equiv m_e$) or hole ($m_2 \equiv m_h$) energy, $T_1 \equiv T_e$ and $T_2 \equiv T_h$ are the electron and hole temperatures, $f(x) =$

$(e^x + 1)^{-1}$ is the Fermi distribution function. The chemical potential of the electron system $\mu^{(1)}$ and that of the hole system $\mu^{(2)}$ are determined respectively by the electron density $(N_1 \equiv N_e)$ and hole density $(N_2 \equiv N_h)$ through the relation

$$N_j = 2 \sum_{\bm{k}_\|} f\big((\varepsilon^{(j)}_{\bm{k}_\|} - \mu^{(j)})/T_j\big). \tag{6.45}$$

In terms of the Fermi energy $\varepsilon^{(j)}_F = \pi N_j/m_j$ at $T_j = 0$, the chemical potential $\mu^{(j)}$ can be written as

$$\mu^{(j)} = T_j \ln\big[\exp(\varepsilon^{(j)}_F/T_j) - 1\big]. \tag{6.46}$$

We now specialize to the situation when the current flowing through only one of the quantum well (say, the electron well, $v_1 \neq 0$), and the other well keeps "open circuit" $(v_2 = 0)$. Because $f_2(0) = 0$, Eq. (6.36) gives

$$E_2 = \frac{f_{12}(v_1)}{N_h e}. \tag{6.47}$$

On the other hand, the current density carried by electrons in the first quantum well is

$$J_1 = -N_e e v_1, \tag{6.48}$$

yielding the transfer resistivity

$$\rho_t(v_1) = -\frac{f_{12}(v_1)}{N_e N_h e^2 v_1}. \tag{6.49}$$

In the linear case, $v_1 \to 0$ and $T_e = T_h = T$. From the f_{12} expression (6.40), we have

$$\begin{aligned}
\rho_t(0) &= -\frac{f'_{12}(0)}{N_e N_h e^2} \\
&= \frac{1}{4\pi^2 N_e N_h e^2 T} \int_0^\infty dq_\| q_\|^3 V_c^2(q_\|) I_1^2(q_\|) I_2^2(q_\|) e^{-2bq_\|} \\
&\quad \times \int_{-\infty}^\infty d\omega \left[-n'\left(\frac{\omega}{T}\right)\right] \Pi_2^{(1)}(\bm{q}_\|, \omega) \Pi_2^{(2)}(-\bm{q}_\|, -\omega),
\end{aligned} \tag{6.50}$$

where $f'_{12}(0) \equiv [d f_{12}(v)/dv]_{v=0}$, $n'(x) \equiv d n(x)/dx$.

Based on Eqs. (6.49) and (6.50), Cui, Lei and Horing (1993) numerically calculate the transfer resistivity of the electron–hole system in a GaAs/AlGaAs/GaAs double quantum well structure in both linear and nonlinear regimes, taking electron effective mass $m_e = 0.07m_0$, hole effective mass $m_h = 0.45m_0$. Both well widths are assumed to be zero that $I_1(q_\parallel) = I_2(q_\parallel) = H(q_\parallel) = 1$, and the well separation is $b = 20$ nm. The full RPA dynamic density correlation functions as given by (6.42) are used in the calculation. Figure 6.6 shows the linear transfer resistivity $\rho_t(0)$ normalized by ρ_0 as a function of temperature at three different electron densities with fixed hole concentration $N_h = 5 \times 10^{10}\,\mathrm{cm}^{-2}$. Here $\rho_0 \equiv 4m_e^2 e^2/(\pi^2 \kappa^2 N_h)$ is a reference resistivity. We can see that the transfer resistivity increases rapidly with increasing temperature, and its value is very sensitive to the carrier density. These are in qualitative agreement with experiments.

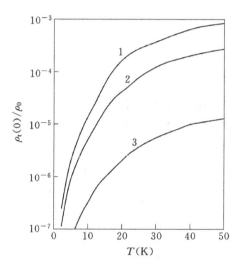

Fig. 6.6 The normalized transfer resistivity $\rho_t(0)/\rho_0$ due to Coulomb drag of electrons and holes in two parallel placed quantum wells separated by $b = 20$ nm, is shown as a function of temperature at three different electron densities: (1) $N_e = 5 \times 10^{10}\,\mathrm{cm}^{-2}$; (2) $N_e = 1 \times 10^{11}\,\mathrm{cm}^{-2}$; (3) $N_e = 2 \times 10^{11}\,\mathrm{cm}^{-2}$. The hole density is fixed with $N_h = 5 \times 10^{10}\,\mathrm{cm}^{-2}$. From Cui, Lei and Horing (1993).

6.5 High-Frequency Small-Signal Transport in Type-II Superlattices

High-frequency small-signal transport in type-II superlattices can be investigated using the same method as that in Sec. 3.3 for 3D systems (Lei, 1987;

Lei, Cui and Horing, 1988). When a small ac field of frequency ω,

$$E(t) = E_\omega e^{-i\omega t} + E_\omega e^{i\omega t}, \tag{6.51}$$

is applied along the x direction, the centers of mass of electrons and holes, after the transient process, will execute simple harmonic oscillations in the field direction at the frequency of the ac field:

$$v_j(t) = v_{j\omega} e^{-i\omega t} + v_{j\omega}^* e^{i\omega t} \tag{6.52}$$

($j = 1, 2$). The current density of the system can be written as

$$J(t) = -N_e e v_1(t) + N_h e v_2(t) = J_\omega e^{-i\omega t} + J_\omega^* e^{i\omega t}, \tag{6.53}$$

with

$$J_\omega = -N_e e v_{1\omega} + N_h e v_{2\omega}. \tag{6.54}$$

To the linear order of the signal field the electron temperature and hole temperature remain unchanged, $T_e = T_h = T$. The equations of motion for the centers of mass of the electrons and holes yield

$$-i\omega v_{1\omega} = -\frac{e}{m_e} E_\omega + i v_{1\omega} M_e(\omega) + i(v_{1\omega} - v_{2\omega}) M_a(\omega), \tag{6.55}$$

$$-i\omega v_{2\omega} = \frac{e}{m_h} E_\omega + i v_{2\omega} M_h(\omega) + i(v_{1\omega} - v_{2\omega}) M_b(\omega), \tag{6.56}$$

Here the memory functions $M_e(\omega)$ and $M_h(\omega)$ of the respective electron and hole subsystems, are given by (3.20) and (3.21) due to impurity and phonon scatterings; $M_a(\omega) = -i F_{12}(\omega)/(N_e m_e \omega)$ and $M_b(\omega) = -i F_{12}(\omega)/(N_h m_h \omega)$, with function $F_{12}(\omega)$ associated with electron-hole interaction:

$$F_{12}(\omega) = -\left(\frac{e^2}{2\epsilon_0 \kappa}\right)^2 \frac{d}{\pi} \int_0^{\pi/d} dq_z \sum_{\mathbf{q}_\parallel} \frac{q_x^2}{q_\parallel^2} I_1(q)^2 I_2(q)^2 S^{12}(\mathbf{q}_\parallel, q_z)$$

$$\times \left[\chi(q_z, \mathbf{q}_\parallel, 0) - \chi(q_z, \mathbf{q}_\parallel, \omega)\right], \tag{6.57}$$

in which

$$\chi(q_z, \mathbf{q}_\parallel, \omega) = \frac{1}{\pi} \int_{-\infty}^{\infty} d\omega' \, n\left(\frac{\omega'}{T}\right) \left[\Pi_2^{(1)}(q_z, \mathbf{q}_\parallel, \omega) \Pi^{(2)}(-q_z, -\mathbf{q}_\parallel, \omega - \omega')\right.$$

$$\left. - \Pi^{(1)}(q_z, \mathbf{q}_\parallel, \omega + \omega') \Pi_2^{(2)}(-q_z, -\mathbf{q}_\parallel, -\omega')\right]. \tag{6.58}$$

Eqs. (6.55) and (6.56) determine the two complex drift velocities $v_{1\omega}$ and $v_{2\omega}$, which are linear in E_ω. The complex resistivity is defined as

$$R(\omega) = \frac{E_\omega}{J_\omega} = -\frac{\mathrm{i}\,m_\mathrm{e}}{e^2}\frac{A}{B}, \qquad (6.59)$$

$$A = m_\mathrm{h}\{[\omega + M_\mathrm{e}(\omega)]\,[\omega + M_\mathrm{h}(\omega)] \\ + [\omega + M_\mathrm{e}(\omega)]\,M_\mathrm{b}(\omega) + [\omega + M_\mathrm{h}(\omega)]\,M_\mathrm{a}(\omega)\},$$
$$B = N_\mathrm{e}\{m_\mathrm{h}\,[\omega + M_\mathrm{h}(\omega) + M_\mathrm{b}(\omega)] - m_\mathrm{e}M_\mathrm{a}(\omega)\} \\ + N_\mathrm{h}\{m_\mathrm{e}\,[\omega + M_\mathrm{e}(\omega) + M_\mathrm{a}(\omega)] - m_\mathrm{h}M_\mathrm{b}(\omega)\}.$$

For an intrinsic system having equal electron and hole densities ($N_\mathrm{e} = N_\mathrm{h} = N_\mathrm{s}$), (6.59) reduces to

$$R(\omega) = -\mathrm{i}\frac{m_\mathrm{r}}{e^2 N_\mathrm{s}}\,[\omega + M(\omega)], \qquad (6.60)$$

$$M(\omega) = \frac{(m_\mathrm{e} + m_\mathrm{h})M_\mathrm{e}(\omega)M_\mathrm{h}(\omega) + \omega\,[m_\mathrm{e}M_\mathrm{h}(\omega) + m_\mathrm{h}M_\mathrm{e}(\omega)]}{(m_\mathrm{e} + m_\mathrm{h})\omega + m_\mathrm{e}M_\mathrm{e}(\omega) + m_\mathrm{h}M_\mathrm{h}(\omega)} + M_{12}(\omega), \qquad (6.61)$$

$$M_{12}(\omega) = M_\mathrm{a}(\omega) + M_\mathrm{b}(\omega) = -\frac{\mathrm{i}\,F_{12}(\omega)}{m_\mathrm{r}N_\mathrm{s}\omega}, \qquad (6.62)$$

with $m_\mathrm{r} = m_\mathrm{e}m_\mathrm{h}/(m_\mathrm{e} + m_\mathrm{h})$. If we define

$$R_{1(2)}(\omega) = -\mathrm{i}\frac{m_\mathrm{r}}{e^2 N_\mathrm{s}}\,[\omega + M_\mathrm{e(h)}(\omega)], \qquad (6.63)$$

$$R_{12}(\omega) = -\mathrm{i}\frac{m_\mathrm{r}}{e^2 N_\mathrm{s}}M_{12}(\omega), \qquad (6.64)$$

the expression (6.60) for $R(\omega)$ can be written as

$$R(\omega) = R_{12}(\omega) + \frac{R_1(\omega)R_2(\omega)}{R_1(\omega) + R_2(\omega)}. \qquad (6.65)$$

It indicates that the total small-signal high-frequency resistivity $R(\omega)$ of a type-II superlattice can also be regarded as two parallel-connected resistances $R_1(\omega)$ and $R_2(\omega)$ of electron and hole subsuperlattices, taken jointly with a series-connected resistance $R_{12}(\omega)$ due to coupling between electrons and holes.

For an ideal intrinsic system, the hole density is equal to the electron density $N_\mathrm{e} = N_\mathrm{h} = N_\mathrm{s}$, and impurity and other elastic scatterings do not exist. When temperature is low enough that phonon scattering is also negligible, the dominant contribution to the resistivity comes from electron–hole interaction. In this case we have

$$R(\omega) = -\frac{\mathrm{i}\, m_\mathrm{r}}{e^2 N_\mathrm{s}}[\omega + M_{12}(\omega)]. \qquad (6.66)$$

It is worthy noting that the plasmon excitation in the electron and hole subsuperlattices can contribute to the memory function and thus to the high-frequency resistivity of the composed type-II superlattice. The approach in Chapter 3 can directly apply here to calculate relevant quantities.

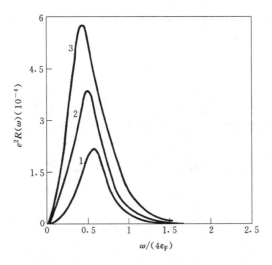

Fig. 6.7 Linear ac resistivity $R(\omega)$ contributed by electron–hole scattering in a type-II superlattice at temperature $T = 0\,\mathrm{K}$, versus frequency ω. The superlattice parameters are: $a_1 = 10\,\mathrm{nm}$, $a_2 = 15\,\mathrm{nm}$, $b = 20\,\mathrm{nm}$, $N_e = N_h = 2.5 \times 10^{11}\,\mathrm{cm}^{-2}$. Three curves represent different ratios of hole to electron effective mass: (1) $m_\mathrm{h}/m_e = 1.15$; (2) $m_\mathrm{h}/m_e = 5.0$; (3) $m_\mathrm{h}/m_e = 10$. $\varepsilon_\mathrm{F} = \pi N_e/m_e$ is the Fermi level of 2D electrons. From Lei, Cui and Horing (1988).

Numerical calculations of linear high frequency transport are carried out for InAs-GaSb type-II superlattices (Lei, Cui and Horing, 1988), focusing on the electron–hole scattering mechanism. Plasmon contributions are taken into consideration in the calculation. The electron effective mass is taken to be $m_e = 0.026 m_0$, but the hole effective mass is assumed to have different values. The calculated linear resistivity $R(\omega)$ contributed by electron-hole scattering in a type-II superlattice at temperature $T = 0\,\mathrm{K}$, is shown in Fig. 6.7 as a function of frequency. The high-frequency resistivity depends strongly not only on the separation between the electrons and holes, but also on the hole to electron mass ratio m_h/m_e. At high values of m_h/m_e, electron–hole scattering resembles electron–ionized impurity scattering, which is more effective in limiting electron mobility than easily mobile holes.

6.6 Multivalley Semiconductors, Systems of Several Carrier Species with Particle Interchange

The $\boldsymbol{k}$-space band structures of conventional semiconductors (such as Ge, Si and GaAs) contain multiple energy minima at some symmetrical points in the Brillouin zone, forming an energy valley around each of these minima. We take n-type GaAs as an example. The lowest energy minimum of its conduction band is at the center of Brillouin zone, the Γ point, around which forms a local band structure called Γ-valley, having almost spherically symmetrical iso-energy surface with effective mass $m_\Gamma \approx 0.067 m_0$ (m_0 is the free electron mass). The next lowest energy minimum is at four L points having an energy $\varepsilon_g \approx 0.3$ eV higher than that of the Γ point, around which form four equivalent L-valleys: each one is almost spherical around the valley bottom $\boldsymbol{k}_L$, with effective mass $m_L \approx 0.23 m_0$. Even if another three equivalent X-valleys at higher energy are ignored, these two nonequivalent sets of isotropic valleys with different effective masses have to be taken into account in analyzing high-field transport in GaAs, i.e., one has to treat an n-type GaAs at least as a system composed of two species of carriers with particle interchange (Lei et al, 1987; Xing, Liu and Ting, 1988).

To give general expressions, we consider a model consisting of s species of carriers of the same type (with charge e) having effective mass m_α and particle number N_α for the αth species of carriers. The particles of different species are exchangeable thus N_α ($\alpha = 1, \cdots, s$) are not fixed quantities, but the total number of carriers N keeps constant:

$$\sum_\alpha N_\alpha = N. \tag{6.67}$$

We introduce the center-of-mass coordinates $\boldsymbol{R}_\alpha$ ($\alpha = 1, \cdots, s$) for each species of carriers (subsystem) and the total center-of-mass coordinate $\boldsymbol{R}$ for the whole carrier system

$$\boldsymbol{R} = \frac{1}{N} \sum_\alpha \sum_i \boldsymbol{r}_{\alpha i} = \frac{1}{N} \sum_\alpha N_\alpha \boldsymbol{R}_\alpha \tag{6.68}$$

($r_{\alpha i}$ represents the coordinate of the ith carrier of the αth subsystem); introduce the total momentum P_α and velocity V_α for each species of carriers ($\alpha = 1, \cdots, s$)

$$P_\alpha = N_\alpha m_\alpha V_\alpha = \sum_i p_{\alpha i} \qquad (6.69)$$

($p_{\alpha i}$ represents the momentum of the ith carrier of the αth subsystem); and define the relative electron coordinates and momenta of the αth subsystem

$$r'_{\alpha i} = r_{\alpha i} - R_\alpha, \qquad (6.70)$$
$$p'_{\alpha i} = p_{\alpha i} - m_\alpha V_\alpha. \qquad (6.71)$$

From these definitions, we have $[N_\alpha R_{\alpha\mu}, P_{\alpha\nu}] = \mathrm{i} N_\alpha \delta_{\mu\nu}$ ($\mu, \nu = x, y, z$). Therefore, R_α and P_α can be thought of canonical conjugate variables, satisfying the commutation relations

$$[R_{\alpha\mu}, P_{\alpha\nu}] = \mathrm{i}\delta_{\mu\nu}. \qquad (6.72)$$

Relative electron variables commute with center-of-mass variables. The particle number of the αth subsystem, N_α, which is a relative electron variable, should also commute with the center-of-mass variable. In addition, the relative coordinate and momentum variables of the αth subsystem satisfy

$$[r'_{\alpha i \mu}, p'_{\alpha j \nu}] = \mathrm{i}\delta_{\mu\nu}\left[\delta_{ij} + o\left(\frac{1}{N_\alpha}\right)\right], \qquad (6.73)$$

and the variables of different subsystems are commutable with each other.

With the help of these variables, the total Hamiltonian of the system in the presence of a uniform electric field E can be written as

$$H = H_\mathrm{E} + H_\mathrm{T} + \sum_\alpha H^\alpha_\mathrm{er} + H_\mathrm{ph} + H_\mathrm{I}, \qquad (6.74)$$

$$H_\mathrm{I} = \sum_\alpha H^\alpha_\mathrm{ei} + \sum_\alpha H^\alpha_\mathrm{ep} + \sum_{\alpha \neq \beta} H^{\alpha\beta}_\mathrm{ep} + \sum_{\alpha \neq \beta} H_{\alpha\beta}, \qquad (6.75)$$

where

$$H_\mathrm{E} = -NeE \cdot R \qquad (6.76)$$

is the potential of the electric field,

$$H_\mathrm{T} = \frac{1}{2}\sum_\alpha m_\alpha N_\alpha V_\alpha^2 \qquad (6.77)$$

is the sum of kinetic energies of individual subsystems. The particle number N_α of the αth subsystem can be written in the second quantization representation of the subsystem as

$$N_\alpha = \sum_{k,\sigma} c^\dagger_{\alpha k \sigma} c_{\alpha k \sigma}, \qquad (6.78)$$

where $c^\dagger_{\alpha k \sigma}$ ($c_{\alpha k \sigma}$) is the creation (annihilation) operator of the relative carrier of the αth subsystem. H^α_{er} in H of (6.74) is the Hamiltonian of the αth relative carrier system, including intravalley Coulomb interaction:

$$H^\alpha_{\text{er}} = \sum_{k,\sigma} \varepsilon_{\alpha k} c^\dagger_{\alpha k \sigma} c_{\alpha k \sigma} + \frac{1}{2} \sum_{k,k',q,\sigma,\sigma'} \nu_c(q) c^\dagger_{\alpha k+q\sigma} c^\dagger_{\alpha k'-q\sigma'} c_{\alpha k'\sigma'} c_{\alpha k\sigma},$$
$$(6.79)$$

in which $\nu_c(q) = e^2/(\epsilon_0 \kappa q^2)$ is the Coulomb potential,

$$\varepsilon_{\alpha k} = \frac{(k - k_\alpha)^2}{2m_\alpha} + \varepsilon_\alpha \qquad (6.80)$$

is the energy spectrum of the αth subsystem, with the energy bottom ε_α located at k_α. The fourth term on the right-hand-side of (6.74), H_{ph}, represents the Hamiltonian of the phonon system. H^α_{ei} and H^α_{ep} in H_I (6.75), the interaction part of the Hamiltonian, are electron–impurity and electron–phonon couplings in the αth subsystem:

$$H^\alpha_{\text{ei}} = \sum_{q,a} u_\alpha(q)\, e^{i q \cdot (R_\alpha - r_a)} \rho_{\alpha q}, \qquad (6.81)$$

$$H^\alpha_{\text{ep}} = \sum_{q,\lambda} M_\alpha(q,\lambda) \phi_{q\lambda}\, e^{i q \cdot R_\alpha} \rho_{\alpha q}, \qquad (6.82)$$

where

$$\rho_{\alpha q} = \sum_{k,\sigma} c^\dagger_{\alpha k+q\sigma} c_{\alpha k\sigma} \qquad (6.83)$$

is the density operator of the αth subsystem, r_a represents the impurity position, $u_\alpha(q)$ and $M_\alpha(q,\lambda)$ are the electron–impurity potential and electron–phonon matrix element of the αth subsystem, $\phi_{q\lambda} \equiv b_{q\lambda} + b^\dagger_{-q\lambda}$ is the phonon field operator. The third term on the right-hand-side of (6.75) represents the electron–phonon interaction associated with different subsystems ($\alpha \neq \beta$, i.e. nonequivalent valleys) given by

$$H^{\alpha\beta}_{\text{ep}} = \sum_{q,\lambda} M_{\alpha\beta}(q,\lambda) \phi_{q\lambda}\, e^{i q \cdot R_\alpha} \sum_{k,\sigma} e^{i k \cdot (R_\alpha - R_\beta)} c^\dagger_{\alpha k+q\sigma} c_{\beta k\sigma}, \qquad (6.84)$$

where $M_{\alpha\beta}(\boldsymbol{q},\lambda) = M_{\beta\alpha}^*(-\boldsymbol{q},\lambda)$ is the matrix element of the electron–phonon interaction associated different subsystems. The role of this interaction is to absorb or emit a phonon while creating an electron in the αth subsystem and annihilating an electron in the βth subsystem. Therefore, $H_{\text{ep}}^{\alpha\beta}$ does not commute with particle number operators N_α nor N_β, i.e., $H_{\text{ep}}^{\alpha\beta}$ induces particle exchange between αth and βth subsystems. The last term on the right-hand-side of (6.75) represents the Coulomb interaction between different subsystems, taken as

$$H_{\alpha\beta} = \sum_{\boldsymbol{q}} v_{\text{c}}(q)\, e^{i\boldsymbol{q}\cdot(\boldsymbol{R}_\alpha-\boldsymbol{R}_\beta)} \rho_{\alpha\boldsymbol{q}}\rho_{\beta,-\boldsymbol{q}}, \qquad (6.85)$$

with the same Coulomb potential $v_{\text{c}}(q) = e^2/(\epsilon_0\kappa q^2)$ as in the intravalley interaction in (6.79). In fact, this is only a part of the inter-subsystem Coulomb interaction, which does not induce particle exchange between different subsystems. The other part of the inter-subsystem Coulomb interaction, which may yield particle exchange between different subsystems, is neglected in the present treatment. Besides, the electron–impurity scattering associated nonequivalent valleys is also assumed negligible.

The rates of changes of the particle number N_α, the center-of-mass momentum $\boldsymbol{P}_\alpha$, and the relative electron energy H_{er}^α of the αth subsystem, as well as the rate of change of the phonon energy H_{ph}, can be obtained from the general Heisenberg equation $\dot{\mathcal{O}} = -\mathrm{i}[\mathcal{O}, H]$ for an arbitrary operator $\mathcal{O}$ and the Hamiltonian H of (6.74). We have the following operator equations $(\alpha = 1, \cdots, s)$

$$\dot{N}_\alpha = -\mathrm{i} \sum_{\beta(\neq\alpha)} \left(H_{\text{ep}}^{\alpha\beta} - H_{\text{ep}}^{\beta\alpha} \right), \qquad (6.86)$$

$$\begin{aligned}
\dot{\boldsymbol{P}}_\alpha &= -\mathrm{i}[\boldsymbol{P}_\alpha, H] \\
&= N_\alpha e\boldsymbol{E} - \mathrm{i}\sum_{\boldsymbol{q},a} u_\alpha(\boldsymbol{q})\,\boldsymbol{q}\, e^{\mathrm{i}\boldsymbol{q}\cdot(\boldsymbol{R}_\alpha-\boldsymbol{r}_a)} \rho_{\alpha\boldsymbol{q}} - \mathrm{i}\sum_{\boldsymbol{q},\lambda} M_\alpha(\boldsymbol{q},\lambda)\,\boldsymbol{q}\,\phi_{\boldsymbol{q}\lambda} e^{\mathrm{i}\boldsymbol{q}\cdot\boldsymbol{R}_\alpha}\rho_{\alpha\boldsymbol{q}} \\
&\quad - \left[\mathrm{i}\sum_{\substack{\boldsymbol{q},\lambda,\boldsymbol{k},\sigma \\ \beta(\neq\alpha)}} M_{\alpha\beta}(\boldsymbol{q},\lambda)\phi_{\boldsymbol{q}\lambda} e^{\mathrm{i}\boldsymbol{q}\cdot\boldsymbol{R}_\alpha} e^{\mathrm{i}\boldsymbol{k}\cdot(\boldsymbol{R}_\alpha-\boldsymbol{R}_\beta)}(\boldsymbol{k}+\boldsymbol{q}) c_{\alpha\boldsymbol{k}+\boldsymbol{q}\sigma}^\dagger c_{\beta\boldsymbol{k}\sigma} + \text{c.c.}\right] \\
&\quad - \mathrm{i}\sum_{\substack{\boldsymbol{q} \\ \beta(\neq\alpha)}} v_{\text{c}}(q)\,\boldsymbol{q}\, e^{\mathrm{i}\boldsymbol{q}\cdot(\boldsymbol{R}_\alpha-\boldsymbol{R}_\beta)} \rho_{\alpha\boldsymbol{q}}\rho_{\beta,-\boldsymbol{q}}, \qquad (6.87)
\end{aligned}$$

$$\begin{aligned}
\dot{H}_{\text{er}}^\alpha &= -\mathrm{i}\big[H_{\text{er}}^\alpha, H\big] \\
&= -\mathrm{i} \sum_{q,a} u_\alpha(q)\,\mathrm{e}^{\mathrm{i}\boldsymbol{q}\cdot(\boldsymbol{R}_\alpha-\boldsymbol{r}_a)} \sum_{\boldsymbol{k},\sigma} (\varepsilon_{\alpha\boldsymbol{k}+\boldsymbol{q}} - \varepsilon_{\alpha\boldsymbol{k}})\, c^\dagger_{\alpha\boldsymbol{k}+\boldsymbol{q}\sigma} c_{\alpha\boldsymbol{k}\sigma} \\
&\quad - \mathrm{i}\sum_{\boldsymbol{q},\lambda} M_\alpha(\boldsymbol{q},\lambda)\phi_{\boldsymbol{q}\lambda} \mathrm{e}^{\mathrm{i}\boldsymbol{q}\cdot\boldsymbol{R}_\alpha} \sum_{\boldsymbol{k},\sigma}(\varepsilon_{\alpha\boldsymbol{k}+\boldsymbol{q}}-\varepsilon_{\alpha\boldsymbol{k}})\, c^\dagger_{\alpha\boldsymbol{k}+\boldsymbol{q}\sigma} c_{\alpha\boldsymbol{k}\sigma} \\
&\quad - \bigg[\mathrm{i}\sum_{\substack{\boldsymbol{q},\lambda \\ \beta(\neq\alpha)}} M_{\alpha\beta}(\boldsymbol{q},\lambda)\,\phi_{\boldsymbol{q}\lambda}\mathrm{e}^{\mathrm{i}\boldsymbol{q}\cdot\boldsymbol{R}_\alpha}\sum_{\boldsymbol{k},\sigma}\mathrm{e}^{\mathrm{i}\boldsymbol{k}\cdot(\boldsymbol{R}_\alpha-\boldsymbol{R}_\beta)}\varepsilon_{\alpha\boldsymbol{k}+\boldsymbol{q}} c^\dagger_{\alpha\boldsymbol{k}+\boldsymbol{q}\sigma} c_{\beta\boldsymbol{k}\sigma} + \text{c.c.}\bigg] \\
&\quad - \mathrm{i}\sum_{\substack{\boldsymbol{q} \\ \beta(\neq\alpha)}} \nu_{\text{c}}(q)\,\mathrm{e}^{\mathrm{i}\boldsymbol{q}\cdot(\boldsymbol{R}_\alpha-\boldsymbol{R}_\beta)} \sum_{\boldsymbol{k},\sigma}(\varepsilon_{\alpha\boldsymbol{k}+\boldsymbol{q}}-\varepsilon_{\alpha\boldsymbol{k}})\,c^\dagger_{\alpha\boldsymbol{k}+\boldsymbol{q}\sigma} c_{\alpha\boldsymbol{k}\sigma}\rho_{\beta,-\boldsymbol{q}}. \quad (6.88)
\end{aligned}$$

6.7 Balance Equations and Langevin Equations for Systems of Several Carrier Species

The derivation of balance equations and Langevin equations for systems of multiple carrier species proceeds in a manner similar to those of Chapter 1 and Chapter 4, i.e. treating the center-of-mass variables classically and treating the interaction H_I perturbatively to the lowest order. The coordinates and velocities of the centers of mass enter the density matrix only through their time-dependent statistical averages. It should be noted, however, that there is an important difference in the present case of multiple carrier species with particle interchange from those discussed previously. The separation of centers of mass from the relative degrees of freedom is incomplete in that H_T includes the relative electron variables N_α ($\alpha = 1,\cdots,s$) which do not commute with the interactions associated with particle exchange between different subsystems. Therefore the Liouville equation for the density matrix $\hat{\rho}$ of the relative electron and phonon systems takes the form (Lei et al, 1987)

$$\mathrm{i}\frac{\partial\hat{\rho}}{\partial t} = \Big[H_\text{T} + \sum_\alpha H_\text{er}^\alpha + H_\text{ph} + H_\text{I}, \hat{\rho}\Big]. \tag{6.89}$$

This requires to define the interaction picture of an operator $\mathcal{O}$ as

$$\mathcal{O}(t) = \exp(\mathrm{i}\,H_0 t)\mathcal{O}\exp(-\mathrm{i}\,H_0 t), \tag{6.90}$$

in which

$$H_0 = H_\text{T} + \sum_\alpha H_\text{er}^\alpha + H_\text{ph}, \qquad (6.91)$$

including H_T. The initial density matrix can be chosen in the same way as in Chapter 1. We image to turn off all electron–impurity and electron–phonon couplings, as well as inter-subsystem electron–electron Coulomb interactions, together with the applied electric field at time t. The subsystems of different carrier species, which are then decoupled from each other without particle exchange, will approach equilibrium states separately, having individual carrier temperatures T_α and chemical potentials μ_α ($\alpha = 1, \cdots, s$). They, of course, depend on the transport state of the system at time t. The phonon system decoupled from electron systems is assumed to approach an equilibrium state with lattice temperature T without hot-phonon effect for simplicity. The initial density matrix for the Liouville equation (6.89) is chosen as

$$\hat{\rho}_0 = \frac{1}{Z}\exp(-H_\text{ph}/T)\exp\left[-\sum_\alpha (H_\text{er}^{(\alpha)} - \mu_\alpha N_\alpha)/T_\alpha\right]. \qquad (6.92)$$

We proceed in the same way as in Sec. 4.1 to derive the Langevin equation for the system of one carrier species. Transferring Eqs. (6.86)–(6.88) to Heisenberg picture and expanding the time-evolution operator (4.4) to the lowest order in H_I such that all the Heisenberg operators are written in a form similar to that of Eq. (4.6), and then neglecting the fluctuation parts in the terms corresponding to the second term on the right-hand-side of Eq. (4.6), which is of order of H_I^2, we obtain (Xing, Liu and Ting, 1988, Xing *et al*, 1988)

$$\frac{d}{dt}N_\alpha = B_\alpha(t) + b_\alpha, \qquad (6.93)$$

$$m_\alpha \frac{d}{dt}[N_\alpha(t)\boldsymbol{V}_\alpha(t)] = N_\alpha e\boldsymbol{E} + \boldsymbol{F}_\alpha(t) + \boldsymbol{f}_\alpha, \qquad (6.94)$$

$$\frac{d}{dt}H_\text{er}^{(\alpha)} = W_\alpha(t) + w_\alpha, \qquad (6.95)$$

($\alpha = 1, \cdots, s$). In these equations, $B_\alpha(t)$, $F_\alpha(t)$ and $W_\alpha(t)$ are the interaction-picture representations and b_α, $\boldsymbol{f}_\alpha$ and w_α are the ensemble averages in the relative-electron–phonon space, of the operators (excluding

$N_\alpha e\boldsymbol{E}$, if any), respectively on the right-hand-sides of Eqs. (6.86), (6.87) and (6.88). b_α and $\boldsymbol{f}_\alpha$ are given by

$$b_\alpha = -\mathrm{i}\sum_{\beta(\neq\alpha)}\sum_{\boldsymbol{k},\boldsymbol{q},\lambda}|M_{\alpha\beta}(\boldsymbol{q},\lambda)|^2\int_{-\infty}^{\infty}dt'\,A_{\alpha\beta}(\boldsymbol{k},\boldsymbol{q},t,t')\Lambda^{\alpha\beta}(\boldsymbol{k},\boldsymbol{q},\lambda,t-t'),$$
(6.96)

$$\boldsymbol{f}_\alpha = -\mathrm{i}\,n_\mathrm{i}\sum_{\boldsymbol{k},\boldsymbol{q}}|u_\alpha(\boldsymbol{q})|^2\boldsymbol{q}\int_{-\infty}^{\infty}dt'\,A_{\alpha\alpha}(\boldsymbol{k},\boldsymbol{q},t,t')\Pi^\alpha(\boldsymbol{k},\boldsymbol{q},t-t')$$

$$-\mathrm{i}\sum_{\boldsymbol{k},\boldsymbol{q},\lambda}|M_\alpha(\boldsymbol{q},\lambda)|^2\boldsymbol{q}\int_{-\infty}^{\infty}dt'\,A_{\alpha\alpha}(\boldsymbol{k},\boldsymbol{q},t,t')\Lambda^\alpha(\boldsymbol{k},\boldsymbol{q},\lambda,t-t')$$

$$-\mathrm{i}\sum_{\substack{\boldsymbol{k},\boldsymbol{q},\lambda\\\beta(\neq\alpha)}}|M_{\alpha\beta}(\boldsymbol{q},\lambda)|^2(\boldsymbol{k}+\boldsymbol{q})\int_{-\infty}^{\infty}dt'\,A_{\alpha\beta}(\boldsymbol{k},\boldsymbol{q},t,t')\Lambda^{\alpha\beta}(\boldsymbol{k},\boldsymbol{q},\lambda,t-t')$$

$$-\mathrm{i}\sum_{\substack{\boldsymbol{k},\boldsymbol{q}\\\beta(\neq\alpha)}}\nu_\mathrm{c}(q)\boldsymbol{q}\int_{-\infty}^{\infty}dt'\,A_{\alpha\beta}(\boldsymbol{q},0,t,t')\Xi^{\alpha\beta}(\boldsymbol{k},\boldsymbol{q},t-t').$$
(6.97)

The expression for w_α can be obtained from (6.97) by changing $\boldsymbol{q}$ into $\varepsilon_{\alpha\boldsymbol{k}+\boldsymbol{q}} - \varepsilon_{\alpha\boldsymbol{k}}$, and changing $\boldsymbol{k}+\boldsymbol{q}$ into $\varepsilon_{\alpha\boldsymbol{k}+\boldsymbol{q}}$, but the first term on the right-hand-side of (6.97) (impurity scattering related term) gives no contribution to w_α. In these equations,

$$A_{\alpha\beta}(\boldsymbol{k},\boldsymbol{q},t-t') \equiv \exp\left[\mathrm{i}\,(\boldsymbol{k}+\boldsymbol{q})\cdot\int_{t'}^{t}\boldsymbol{v}_\alpha(s)ds - \mathrm{i}\,\boldsymbol{k}\cdot\int_{t'}^{t}\boldsymbol{v}_\beta(s)ds\right],$$
(6.98)

and the relevant correlation functions are defined as

$$\Pi^\alpha(\boldsymbol{k},\boldsymbol{q},t-t') = -\mathrm{i}\,\theta(t-t')\sum_\sigma\left\langle\left[c^\dagger_{\alpha\boldsymbol{k}+\boldsymbol{q}\sigma}(t)c_{\alpha\boldsymbol{k}\sigma}(t),c^\dagger_{\alpha\boldsymbol{k}\sigma}(t')c_{\alpha\boldsymbol{k}+\boldsymbol{q}\sigma}(t')\right]\right\rangle_0,$$
(6.99)

$$\Lambda^\alpha(\boldsymbol{k},\boldsymbol{q},\lambda,t-t') = -\mathrm{i}\,\theta(t-t')$$
$$\times\sum_\sigma\left\langle\left[\phi_{\boldsymbol{q}\lambda}(t)c^\dagger_{\alpha\boldsymbol{k}+\boldsymbol{q}\sigma}(t)c_{\alpha\boldsymbol{k}\sigma}(t),\phi_{-\boldsymbol{q}\lambda}(t')c^\dagger_{\alpha\boldsymbol{k}\sigma}(t')c_{\alpha\boldsymbol{k}+\boldsymbol{q}\sigma}(t')\right]\right\rangle_0,\quad(6.100)$$

$$\Lambda^{\alpha\beta}(\boldsymbol{k},\boldsymbol{q},\lambda,t-t') = -\mathrm{i}\,\theta(t-t')$$
$$\times\sum_\sigma\left\langle\left[\phi_{\boldsymbol{q}\lambda}(t)c^\dagger_{\alpha\boldsymbol{k}+\boldsymbol{q}\sigma}(t)c_{\beta\boldsymbol{k}\sigma}(t),\phi_{-\boldsymbol{q}\lambda}(t')c^\dagger_{\beta\boldsymbol{k}\sigma}(t')c_{\alpha\boldsymbol{k}+\boldsymbol{q}\sigma}(t)\right]\right\rangle_0,\quad(6.101)$$

$$\Xi^{\alpha\beta}(\boldsymbol{k},\boldsymbol{q},t-t') = -\mathrm{i}\theta(t-t')$$
$$\times \sum_\sigma \left\langle \left[c^\dagger_{\alpha\boldsymbol{k}+\boldsymbol{q}\sigma}(t)c_{\alpha\boldsymbol{k}\sigma}(t)\rho_{\beta,-\boldsymbol{q}}(t), c^\dagger_{\alpha\boldsymbol{k}\sigma}(t')c_{\alpha\boldsymbol{k}+\boldsymbol{q}\sigma}(t')\rho_{\beta\boldsymbol{q}}(t') \right] \right\rangle_0. \quad (6.102)$$

$B_\alpha(t)$, $\boldsymbol{F}_\alpha(t)$ and $W_\alpha(t)$ in Eqs. (6.93)–(6.95) are respectively the fluctuating rates of changes of the carrier number, momentum and energy of the αth subsystem, with vanishing averages: $\langle B_\alpha(t)\rangle = 0$, $\langle \boldsymbol{F}_\alpha(t)\rangle = 0$, $\langle E_\alpha(t)\rangle = 0$. Taking the ensemble average of Eqs. (6.93)–(6.95) with $V_\alpha(t)$ and $N_\alpha(t)$ replaced by the averaged drift velocity $\boldsymbol{v}_\alpha$ and averaged carrier number n_α, we obtain the balance equations for n_α, $\boldsymbol{v}_\alpha$ and T_α as

$$\frac{d}{dt}n_\alpha = b_\alpha, \quad (6.103)$$

$$m_\alpha \frac{d}{dt}(n_\alpha \boldsymbol{v}_\alpha) = n_\alpha e\boldsymbol{E} + \boldsymbol{f}_\alpha, \quad (6.104)$$

$$\frac{d}{dt}U_\alpha = w_\alpha. \quad (6.105)$$

Here b_α, $\boldsymbol{f}_\alpha$ and w_α are still given by (6.96) and (6.97), with V_α and N_α replaced by $\boldsymbol{v}_\alpha$ and n_α. For the αth subsystem, the average carrier number n_α relates to the chemical potential μ_α and carrier temperature T_α through the equation

$$n_\alpha = \sum_{\boldsymbol{k},\sigma} f((\varepsilon_{\alpha\boldsymbol{k}} - \mu_\alpha)/T_\alpha). \quad (6.106)$$

The average energy of the αth subsystem is given by

$$U_\alpha = \sum_{\boldsymbol{k},\sigma} \varepsilon_{\alpha\boldsymbol{k}} f((\varepsilon_{\alpha\boldsymbol{k}} - \mu_\alpha)/T_\alpha). \quad (6.107)$$

Note that the summations over $\boldsymbol{k}$ and $\boldsymbol{q}$ in all the equations are considered to go through the entire Brillouin zone, thus include contributions from all the equivalent valleys of each subsystem.

When neglecting the memory effect, the integral $\int_{t'}^{t} \boldsymbol{v}_\alpha(s)\,ds$ in $A_{\alpha\beta}(\boldsymbol{k},\boldsymbol{q},t-t')$ function (6.98) can be replaced by $\boldsymbol{v}_\alpha(t)(t-t')$, and the balance equations for the drift motion are reduced to ($\alpha = 1,\cdots,s$)

$$\frac{d}{dt}n_\alpha = \sum_{\beta(\neq\alpha)} \chi_\mathrm{p}^{\alpha\beta}(\boldsymbol{v}_\alpha,\boldsymbol{v}_\beta), \quad (6.108)$$

$$\frac{d}{dt}(n_\alpha \boldsymbol{v}_\alpha) = n_\alpha e\boldsymbol{E} + \boldsymbol{f}^\alpha(\boldsymbol{v}_\alpha) + \sum_\beta \boldsymbol{f}_\mathrm{p}^{\alpha\beta}(\boldsymbol{v}_\alpha,\boldsymbol{v}_\beta) + \sum_{\beta(\neq\alpha)} \boldsymbol{f}_{\alpha\beta}(\boldsymbol{v}_\alpha - \boldsymbol{v}_\beta), \quad (6.109)$$

$$\frac{d}{dt}U_\alpha = \boldsymbol{v}_\alpha \cdot \boldsymbol{f}^\alpha(\boldsymbol{v}_\alpha) + w^\alpha(\boldsymbol{v}_\alpha) + \sum_\beta w_{\mathrm{p}}^{\alpha\beta}(\boldsymbol{v}_\alpha, \boldsymbol{v}_\beta) + \sum_{\beta(\neq\alpha)} w_{\alpha\beta}(\boldsymbol{v}_\alpha - \boldsymbol{v}_\beta). \tag{6.110}$$

In these equations, $\boldsymbol{f}^\alpha(\boldsymbol{v}_\alpha)$ is the frictional force of the αth subsystem due to intravalley impurity and phonon scatterings and $w^\alpha(\boldsymbol{v}_\alpha)$ is the energy transfer rate from the αth subsystem to the lattice due to intravalley phonon scattering:

$$\boldsymbol{f}^\alpha(\boldsymbol{v}_\alpha) = n_i d_\alpha \sum_{\boldsymbol{q}} \boldsymbol{q} |u_\alpha(\boldsymbol{q})|^2 \Pi_2^\alpha(\boldsymbol{q}, \omega_\alpha) + 2d_\alpha \sum_{\boldsymbol{q},\lambda} \boldsymbol{q} |M_\alpha(\boldsymbol{q}, \lambda)|^2$$
$$\times \Pi_2^\alpha(\boldsymbol{q}, \Omega_{\boldsymbol{q}\lambda} + \omega_\alpha) \left[n\left(\frac{\Omega_{\boldsymbol{q}\lambda}}{T}\right) - n\left(\frac{\Omega_{\boldsymbol{q}\lambda} + \omega_\alpha}{T_\alpha}\right) \right], \tag{6.111}$$

$$w^\alpha(\boldsymbol{v}_\alpha) = 2d_\alpha \sum_{\boldsymbol{q},\lambda} \Omega_{\boldsymbol{q}\lambda} |M_\alpha(\boldsymbol{q}, \lambda)|^2 \Pi_2^\alpha(\boldsymbol{q}, \Omega_{\boldsymbol{q}\lambda} + \omega_\alpha)$$
$$\times \left[n\left(\frac{\Omega_{\boldsymbol{q}\lambda}}{T}\right) - n\left(\frac{\Omega_{\boldsymbol{q}\lambda} + \omega_\alpha}{T_\alpha}\right) \right]. \tag{6.112}$$

Here $\omega_\alpha \equiv \boldsymbol{q} \cdot \boldsymbol{v}_\alpha$, $\Pi_2^{(\alpha)}(\boldsymbol{q}, \omega)$ is the imaginary part of the density correlation function of electrons in a valley of the αth subsystem as given by (2.16), d_α is the number of equivalent valleys of the αth subsystem.

In Eqs. (6.109) and (6.110), $\boldsymbol{f}_{\mathrm{p}}^{\alpha\alpha}(\boldsymbol{v}_\alpha, \boldsymbol{v}_\alpha)$ and $w_{\mathrm{p}}^{\alpha\alpha}(\boldsymbol{v}_\alpha, \boldsymbol{v}_\alpha)$ are the frictional force and the energy-loss rate of the αth subsystem due to electron–phonon scatterings among equivalent valleys, $f_{\mathrm{p}}^{\alpha\beta}(\boldsymbol{v}_\alpha, \boldsymbol{v}_\beta)$ and $w_{\mathrm{p}}^{\alpha\beta}(\boldsymbol{v}_\alpha, \boldsymbol{v}_\beta)$ ($\beta \neq \alpha$) are the frictional force and the energy-loss rate of the αth subsystem due to intervalley electron–phonon interaction between the α and β subsystems. For both $\alpha = \beta$ and $\alpha \neq \beta$ cases, they can be written as

$$\boldsymbol{f}_{\mathrm{p}}^{\alpha\beta}(\boldsymbol{v}_\alpha, \boldsymbol{v}_\beta) = 4\pi \sum_{\mathrm{pair}\ \boldsymbol{k},\boldsymbol{q},\lambda} |M_{\alpha\beta}(\boldsymbol{q}, \lambda)|^2 (-\boldsymbol{k}) \left[f\left(\frac{\xi_{\alpha\boldsymbol{k}}}{T_\alpha}\right) - f\left(\frac{\xi_{\beta\boldsymbol{k}+\boldsymbol{q}}}{T_\beta}\right) \right]$$
$$\times \left\{ \left[n\left(\frac{\Omega_{\boldsymbol{q}\lambda}}{T}\right) - n\left(\frac{\xi_{\alpha\boldsymbol{k}}}{T_\alpha} - \frac{\xi_{\beta\boldsymbol{k}+\boldsymbol{q}}}{T_\beta}\right) \right] \delta(E_{\beta\boldsymbol{k}+\boldsymbol{q}} - E_{\alpha\boldsymbol{k}} + \Omega_{\boldsymbol{q}\lambda}) \right.$$
$$\left. + \left[n\left(\frac{\Omega_{\boldsymbol{q}\lambda}}{T}\right) - n\left(\frac{\xi_{\beta\boldsymbol{k}+\boldsymbol{q}}}{T_\beta} - \frac{\xi_{\alpha\boldsymbol{k}}}{T_\alpha}\right) \right] \delta(E_{\beta\boldsymbol{k}+\boldsymbol{q}} - E_{\alpha\boldsymbol{k}} - \Omega_{\boldsymbol{q}\lambda}) \right\}, \tag{6.113}$$

$$w_{\rm p}^{\alpha\beta}(\bm{v}_\alpha,\bm{v}_\beta) = 4\pi \sum_{\text{pair}}\sum_{\bm{k},\bm{q},\lambda} |M_{\alpha\beta}(\bm{q},\lambda)|^2 \varepsilon_{\alpha\bm{k}} \left[f\left(\frac{\xi_{\alpha\bm{k}}}{T_\alpha}\right) - f\left(\frac{\xi_{\beta\bm{k}+\bm{q}}}{T_\beta}\right)\right]$$
$$\times \left\{\left[n\left(\frac{\Omega_{q\lambda}}{T}\right) - n\left(\frac{\xi_{\alpha\bm{k}}}{T_\alpha} - \frac{\xi_{\beta\bm{k}+\bm{q}}}{T_\beta}\right)\right]\delta(E_{\beta\bm{k}+\bm{q}} - E_{\alpha\bm{k}} + \Omega_{q\lambda})\right.$$
$$\left. + \left[n\left(\frac{\Omega_{q\lambda}}{T}\right) - n\left(\frac{\xi_{\beta\bm{k}+\bm{q}}}{T_\beta} - \frac{\xi_{\alpha\bm{k}}}{T_\alpha}\right)\right]\delta(E_{\beta\bm{k}+\bm{q}} - E_{\alpha\bm{k}} - \Omega_{q\lambda})\right\}.$$
(6.114)

Here we have used notations $\xi_{\alpha\bm{k}} \equiv \varepsilon_{\alpha\bm{k}} - \mu_\alpha$ and $E_{\alpha\bm{k}} \equiv \varepsilon_{\alpha\bm{k}} + \frac{1}{2}m_\alpha v_\alpha^2 + \bm{k}\cdot\bm{v}_\alpha$. The summation means first to go over wavevectors $\bm{k}$ and $\bm{q}$ in the regions giving rise to a pair of the intervalley scattering, and then to sum over all possible pairs. If contributions from all the scattering pairs are equal, then $\sum_{\text{pair}} \to d_\alpha d_\beta$.

In Eq. (6.108), $\chi_{\rm p}^{\alpha\beta}(\bm{v}_\alpha,\bm{v}_\beta)$ is the increase rate of the particle number of the αth subsystem due to intervalley electron–phonon scattering between the αth and the βth subsystems:

$$\chi_{\rm p}^{\alpha\beta}(\bm{v}_\alpha,\bm{v}_\beta) = 4\pi \sum_{\text{pair}}\sum_{\bm{k},\bm{q},\lambda} |M_{\alpha\beta}(\bm{q},\lambda)|^2 \left[f\left(\frac{\xi_{\alpha\bm{k}}}{T_\alpha}\right) - f\left(\frac{\xi_{\beta\bm{k}+\bm{q}}}{T_\beta}\right)\right]$$
$$\times \left\{\left[n\left(\frac{\Omega_{q\lambda}}{T}\right) - n\left(\frac{\xi_{\alpha\bm{k}}}{T_\alpha} - \frac{\xi_{\beta\bm{k}+\bm{q}}}{T_\beta}\right)\right]\delta(E_{\beta\bm{k}+\bm{q}} - E_{\alpha\bm{k}} + \Omega_{q\lambda})\right.$$
$$\left. + \left[n\left(\frac{\Omega_{q\lambda}}{T}\right) - n\left(\frac{\xi_{\beta\bm{k}+\bm{q}}}{T_\beta} - \frac{\xi_{\alpha\bm{k}}}{T_\alpha}\right)\right]\delta(E_{\beta\bm{k}+\bm{q}} - E_{\alpha\bm{k}} - \Omega_{q\lambda})\right\}.$$
(6.115)

Apparently $\chi_{\rm p}^{\alpha\alpha} = 0$.

At last, in Eqs. (6.109) and (6.110), $\bm{f}_{\alpha\beta}(\bm{v}_\alpha-\bm{v}_\beta)$ and $w_{\alpha\beta}(\bm{v}_\alpha-\bm{v}_\beta)$ ($\alpha \neq \beta$) are the frictional force and energy-loss rate of the αth subsystem due to intervalley Coulomb interaction between the αth and the βth subsystems:

$$\bm{f}_{\alpha\beta}(\bm{v}_\alpha - \bm{v}_\beta) = d_\alpha d_\beta \sum_{\bm{q}} |v_{\rm c}(q)|^2 \bm{q} \int_{-\infty}^{\infty} \frac{d\omega}{\pi} \left[n\left(\frac{\omega}{T_\alpha}\right) - n\left(\frac{\omega - \omega_{\alpha\beta}}{T_\beta}\right)\right]$$
$$\times \Pi_2^{(\alpha)}(\bm{q},\omega)\Pi_2^{(\beta)}(\bm{q},\omega-\omega_{\alpha\beta}), \qquad (6.116)$$

$$w_{\alpha\beta}(\bm{v}_\alpha - \bm{v}_\beta) = d_\alpha d_\beta \sum_{\bm{q}} |v_{\rm c}(q)|^2 \int_{-\infty}^{\infty} \frac{d\omega}{\pi} \omega \left[n\left(\frac{\omega}{T_\alpha}\right) - n\left(\frac{\omega - \omega_{\alpha\beta}}{T_\beta}\right)\right]$$
$$\times \Pi_2^{(\alpha)}(\bm{q},\omega)\Pi_2^{(\beta)}(\bm{q},\omega-\omega_{\alpha\beta}), \qquad (6.117)$$

in which $\omega_{\alpha\beta} \equiv \boldsymbol{q}\cdot(\boldsymbol{v}_\alpha-\boldsymbol{v}_\beta) = \omega_\alpha - \omega_\beta$, and $\Pi_2^\alpha(\boldsymbol{q},\omega)$ is the imaginary part of the density correlation function of electrons in a valley of the αth subsystem.

Balance equations (6.103)–(6.105), or the corresponding Eqs. (6.108)–(6.110) neglecting memory effects, contain $\boldsymbol{v}_\alpha$, T_α, n_α, μ_α and U_α as parameters and constitute a complete set of equations, together with Eqs. (6.106) and (6.107), to determine these ensemble averaged transport quantities when the electric field is given.

6.8 DC Steady-State Transport in GaAs

As an example, we first apply the balance equations derived in the last section for several carrier species systems with particle interchange to discuss the dc steady-state transport in an n-type GaAs system.

The conduction band of GaAs is generally considered to consist of valleys at Γ, L and X points in the k-space. The central Γ valley ($d_1 = 1$) has the lowest energy with its bottom at the center of the Brillouin zone, $\boldsymbol{k}_\Gamma = (0,0,0)$, having roughly spherical iso-energy surface and an effective mass $m_1 = m_\Gamma = 0.067 m_0$ in its vicinity, i.e. the energy spectrum $\varepsilon_{1\boldsymbol{k}} = k^2/2m_1$ (the bottom of the Γ-vallay is taken as the energy zero). With the lowest energy $\varepsilon_{\mathrm{L}\Gamma} = \varepsilon_\mathrm{g} \approx 0.3$ eV higher than the Γ-valley bottom, there are four equivalent valleys ($d_2 = 4$) at the L points, $\boldsymbol{k}_\mathrm{L} = (\pm 1, \pm 1, \pm 1)$, on the boundary of the Brillouin zone, having approximately spherical iso-energy surface and effective mass $m_2 = m_\mathrm{L} = 0.58 m_0$ in the vicinities around these bottoms, or the energy spectrum $\varepsilon_{2\boldsymbol{k}} = (\boldsymbol{k}-\boldsymbol{k}_\mathrm{L})^2/2m_2 + \varepsilon_{\mathrm{L}\Gamma}$. At even higher energy $\varepsilon_{\mathrm{X}\Gamma} \approx 0.49$ eV from the Γ-valley bottom, there are three ($d_3 = 3$) equivalent valleys, with bottoms located at the X points $\boldsymbol{k}_\mathrm{X} = (\pm 1, 0, 0), (0, \pm 1, 0), (0, 0, \pm 1)$ on the boundary of the Brillouin zone, having approximately spherical iso-energy surface in the vicinities around these bottoms, or the energy spectrum $\varepsilon_{3\boldsymbol{k}} = (\boldsymbol{k}-\boldsymbol{k}_\mathrm{X})^2/2m_3 + \varepsilon_{\mathrm{X}\Gamma}$. Since the energy spectrum around the bottom of individual valley is approximately spherical, the four L valleys are equivalent each other and three X valleys are equivalent each other no matter in what direction the electric field is applied, thus the conduction of the system is isotropic. One can describe the transport in a GaAs system using a two carrier species (Γ–L) model (Lei et al, 1987; Xing et al, 1988), or using a three carrier species (Γ–L–X) model (Xing, Liu and Ting, 1988; Liu, Xing and Ting, 1989).

As an example, in the following we will analyze the high-field electron transport of n-type GaAs systems using the Γ–L–X model. The Γ-valley constitutes the first subsystem (the 1st species of carriers), 4 equivalent L-valleys constitute the second subsystem (the 2nd species of carriers), and 3 equivalent X-valleys constitute the third subsystem (the 3rd species of carriers). Thus, the number density n_1 of the 1st species of carriers is the number density of carriers in the Γ valley and the intra-1st-subsyem scattering is the scattering intra-Γ-valley. The number density n_2 of the second species of carriers equals d_2 ($=4$) times of the number density of carriers in a L-valley, and the intra-2nd-subsystem scatterings include all the intra-L-valley scatterings and all inter-valley scattering between two different L valleys. The number density n_3 of the third species of carriers equals d_3 ($=3$) times of the number density of carriers in a X-valley, and the intra-3rd-subsystem scatterings include all the intra-X-valley scatterings and all inter-valley scattering between two different X valleys. The total drift velocity of carriers is

$$\bm{v}_\mathrm{d} = \frac{1}{n}(n_1\bm{v}_1 + n_2\bm{v}_2 + n_3\bm{v}_3) = rn_1\bm{v}_1 + rn_2\bm{v}_2 + rn_3\bm{v}_3, \qquad (6.118)$$

where $\bm{v}_1$, $\bm{v}_2$ and $\bm{v}_3$ are respectively the average drift velocities of carriers of Γ, L and X valley, $rn_1 = n_1/n$, $rn_2 = n_2/n$ and $rn_3 = n_3/n$, n being the total number density of carriers. The carrier density n_α of the αth subsystem relates to its chemical potential μ_α by ($\alpha = 1, 2, 3$)

$$n_\alpha = 2d_\alpha \left(\frac{m_\alpha T_\alpha}{2\pi}\right)^{3/2} F_{\frac{1}{2}}\left(\frac{\mu_\alpha}{T_\alpha}\right), \qquad (6.119)$$

$F_{\frac{1}{2}}(x)$ being the Fermi integral of $1/2$ order defined as

$$F_{\frac{1}{2}}(x) \equiv \frac{2}{\sqrt{\pi}} \int_0^\infty dy \frac{y^{1/2}}{\exp(y-x)+1}. \qquad (6.120)$$

We consider the following scattering mechanisms.

Γ valley (subsystem 1): (i) intravalley acoustic-phonon deformation potential scattering, with matrix element

$$|M_1(\bm{q}, \lambda_1)|^2 = \frac{E_1^2 q}{2d_c v_s}, \qquad (6.121)$$

having GaAs lattice mass density $d_c = 5.36$ g/cm^3, longitudinal sound velocity $v_s = 5.2 \times 10^5$ cm/s, and Γ-valley acoustic deformation potential

$E_1 = 7.0$ eV; (ii) polar optic-phonon scattering with matrix element

$$|M_1(\boldsymbol{q}, \lambda_2)|^2 = \frac{e^2 \Omega_{\text{LO}}}{2\epsilon_0 q^2} \left(\frac{1}{\kappa_\infty} - \frac{1}{\kappa} \right), \quad (6.122)$$

having static dielectric constant $\kappa = 12.9$, optical dielectric constant $\kappa_\infty = 10.9$, and longitudinal optical phonon energy $\Omega_{\text{LO}} = 36.2$ meV.

4 equivalent L valleys (subsystem 2): (i) intravalley acoustic-phonon deformation potential scattering, with matrix element $|M_2(\boldsymbol{q}, \lambda_1)|^2$ similar to (6.121) (E_1 replaced by E_2), having L-valley acoustic deformation $E_2 = 7.0$ eV; (ii) inter-valley optic-phonon deformation potential scattering between different L valleys, with matrix element

$$|M_2(\boldsymbol{q}, \lambda_2)|^2 = \frac{D_{\text{LL}}^2}{2d_c \Omega_o}, \quad (6.123)$$

having optical deformation potential $D_{\text{LL}} = 0.5 \times 10^9$ eV/cm and optic-phonon energy $\Omega_o = 29.9$ meV.

3 equivalent X valleys (subsystem 3): (i) intravalley acoustic-phonon deformation potential scattering, with matrix element $|M_3(\boldsymbol{q}, \lambda_1)|^2$ similar to (6.121) (E_1 replaced by E_3), having L-valley acoustic deformation $E_3 = 7.0$ eV; (ii) inter-valley optic-phonon deformation potential scattering between different X valleys, with matrix element

$$|M_3(\boldsymbol{q}, \lambda_3)|^2 = \frac{D_{\text{XX}}^2}{2d_c \Omega_o}, \quad (6.124)$$

having optical deformation potential $D_{\text{XX}} = 1 \times 10^9$ eV/cm.

In addition, we also need to consider the inter-subsystem optic deformation potential scattering, i.e. the optic-phonon deformation potential induced scattering of carriers between inequivalent valleys, which has the matrix element of form

$$|M_{\alpha\beta}(\boldsymbol{q}, \lambda)|^2 = \frac{D_{\alpha\beta}^2}{2d_c \Omega_o}, \quad (6.125)$$

with optical deformation potentials $D_{\Gamma\text{L}} = 0.285 \times 10^9$ eV/cm, $D_{\Gamma\text{X}} = 1 \times 10^9$ eV/cm, and $D_{\text{LX}} = 0.316 \times 10^9$ eV/cm in GaAs.

Using the three-carrier-species model and taking the above-mentioned scattering mechanisms into account but neglecting effect of inter-subsystem Coulomb interaction ($f_{\alpha\beta} = 0$, $w_{\alpha\beta} = 0$), Xing, Liu and Ting (1988) calculate the dc steady-state transport of GaAs. Fig. 6.8 illustrates the total drift velocity v_d as a function of electric field, obtained from the steady-state form of equations (6.108)–(6.110) at $T = 300\,\text{K}$, together with experimental dada. If changing $D_{\text{L}\Gamma}$ value to 0.36×10^9 eV/cm (other parameters remain), an even better agreement between theory (dashed curve) and experiment can be achieved.

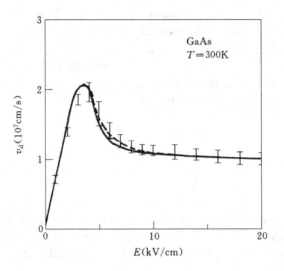

Fig. 6.8 The average drift velocity v_d versus the electric field E at room temperature $T = 300$ K. The theoretical curve (solid) is obtained based on Γ–L–X three-carrier-species model with scattering mechanisms and parameters given in the text. The dash curve is the theoretical result using $D_{\text{L}\Gamma} = 0.36 \times 10^9$ eV/cm (other parameters remain the same as in the text). From Xing, Liu and Ting (1988).

Starting from the time-dependent equations (6.108)–(6.110) and using the same model, same scattering mechanisms and parameters, Liu, Xing and Ting (1989) calculated the transient transport of GaAs. The time-dependent response of the drift velocity $v_\text{d}(t)$ to step electric fields suddenly impressed at $t = 0$ of several different strengths at lattice temperature $T = 300$ K, is shown in Fig. 6.9. The inset figure compares a part of curve A with the results of a Monte-Carlo calculation (black dots).

The three-carrier-species Γ–L–X model can also apply to $\text{Al}_x\text{Ga}_{1-x}\text{As}$ systems. Utilizing the known band parameters of $\text{Al}_x\text{Ga}_{1-x}\text{As}$ at different Aluminum content x (Sadachi, 1985) and the $D_{\alpha\beta}$ values for GaAs, Cao and Lei (1996) calculate the dc steady state transport of $\text{Al}_{0.2}\text{Ga}_{0.8}\text{As}$. As an example, the total average drift velocity v_d, the average drift velocities

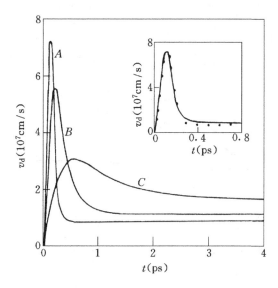

Fig. 6.9 Transient response of drift velocity $v_{\rm d}(t)$ to step electric fields suddenly impressed at $t = 0$. Curve A: $E = 40\,{\rm kV/cm}$; Curve B: $E = 15\,{\rm kV/cm}$; Curve C: $E = 5\,{\rm kV/cm}$. The temperature is $T = 300\,{\rm K}$. The inset shows a part of curve A and the results (dots) of a Monte-Carlo calculation (in Reggiani, 1985). From Liu, Xing and Ting (1989).

v_Γ, $v_{\rm L}$ and $v_{\rm X}$, the electron temperature T_Γ, $T_{\rm L}$ and $T_{\rm X}$, and the fractional electron occupations, in Γ, L and X valleys, are shown in Fig. 6.10.

The most salient feature of the velocity–field curve of dc transport in n-type GaAs ($\text{Al}_x\text{Ga}_{1-x}\text{As}$) systems is that, after reaching a maximum, the drift velocity $v_{\rm d}$ decreases with further increase of the electric field. The main reason for this negative differential conductance is the simultaneous existence of higher mobility valley located at lower energy (γ valley) and lower mobility valleys at higher energy (L and X valleys). At low fields, the conduction is mainly due to electrons in the Γ-vally. When the applied electric field increases, the electron temperature rises and more and more electrons transfer from the Γ valley of higher mobility to the L and X valleys having lower mobility, such that the total average mobility decreases. As long as such a physics of electron transfer is included the Γ–L two-carrier-species model can also produce a $v_{\rm d}$–E curve exhibiting negative differential mobility (Lei et al, 1987; Xing et al, 1988).

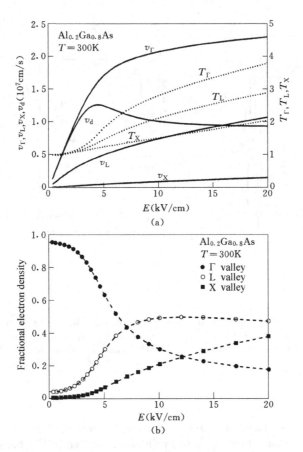

Fig. 6.10 Transport properties of an n-type $Al_{0.2}Ga_{0.8}As$ calculated by Γ–L–X three-carrier-species model, including the total average drift velocity v_d, the average drift velocities v_Γ, v_L and v_X, the electron temperature T_Γ, T_L and T_X, and the fractional electron occupations, in Γ, L and X valleys. From Cao and Lei (1996).

The electron transfer from the Γ valley to the L and X valleys discussed above in the bulk GaAs system can be thought to happen in the k space, so called k-space transfer. It can be described as carriers transferring from certain states in a region of the k space to states in other regions of the k space. In Sec. 6.10, we will see that the rise of the electron temperature due to enhancing electric field strength can induce carrier transfer between those states which have quite different density distribution in real space, so called real-space transfer.

6.9 Quasi-Two-Dimensional Multiple Subband Systems

6.9.1 *Multisubband system treated as composed of multiple species of carriers*

In discussing the electron transport of quasi-two-dimensional systems, it is often required to consider the carrier occupation of several subbands due to the enhanced carrier density or the elevated electron temperature. The subband energy spectra of a quasi-2D system generally have the form

$$\varepsilon_{\alpha \bm{k}_\parallel} = \varepsilon_\alpha + k_\parallel^2/2m, \qquad (6.126)$$

where m is the effective mass, $\bm{k}_\parallel$ is the 2D wavevector, α is the subband index ($\alpha = 0, 1, \cdots, s-1$, assuming there are s subbands), and ε_α is the energy bottom of the αth subband. These spectra have the same shape only that the positions of their energy bottoms may be different for different subbands. Therefore, it is natural to treat the whole system as composed of the same species of carriers, and the occupation of different subbands is referred solely to the carrier statistical distribution based on their energy levels. This is certainly reasonable if the system is in a thermal equilibrium state. Things, however, may become different when system is not in thermal equilibrium. In dealing with carrier transport of a quasi-2D system driven by an electric field in Sections 2.5 and 2.6, we describe the transport state of the whole 2D electron system with a one-carrier-species model, i.e. in terms of a single electron temperature T_e, a single chemical potential μ, and a single average drift velocity $\bm{v}_d$. This requires that intercarrier interactions, both intersubband and intrasubband ones, are strong enough to produce frequent energy, momentum and particle exchanges among carriers, that all carriers of the system, whatever subband they dwell in, can share a common electron temperature T_e, a common chemical potential μ, and have the same drift velocity $\bm{v}_d$. In a realistic system, however, in view of the difference of the spatial density distribution and the wave-function behavior of carriers in different subbands, the effective intersubband Coulomb interaction between carriers is often much weaker than the intrasubband one. Therefore, even if the carriers in each individual subband can be well described by an electron temperature, a chemical potential and a drift velocity, these temperatures, chemical potentials and drift velocities for different subbands may be different. In other words, generally, we should treat a multiple subband 2D semiconductor as a system composed of multiple species of carriers, having repective drift velocity, electron temperature and chemical potential for each subband. Compared with that of a multivalley semiconductor, a multisubband 2D system has the following features. First, in view of the possible energy overlap of neighboring subbands, impurities and acoustic phonons contribute not only to intrasubband scattering but also to intersubband scattering. Therefore, the interaction Hamiltonian H_I (6.75) should include an intersubband impurity scattering term

$$H_\mathrm{ei}^{\alpha\beta} = \sum_{\boldsymbol{q}_\parallel,a} U_{\alpha\beta}(q_\parallel, z_a) e^{i\boldsymbol{q}_\parallel\cdot(\boldsymbol{R}_\alpha-\boldsymbol{r}_a)} \sum_{\boldsymbol{k}_\parallel,\sigma} e^{i\boldsymbol{k}_\parallel\cdot(\boldsymbol{R}_\alpha-\boldsymbol{R}_\beta)} c^\dagger_{\alpha\boldsymbol{k}_\parallel+\boldsymbol{q}_\parallel\sigma} c_{\beta\boldsymbol{k}_\parallel\sigma}, \tag{6.127}$$

in addition to the intersubband electron–phonon scattering

$$H_\mathrm{ep}^{\alpha\beta} = \sum_{\boldsymbol{q},\lambda} M_{\alpha\beta}(\boldsymbol{q},\lambda)\phi_{\boldsymbol{q}\lambda} e^{i\boldsymbol{q}_\parallel\cdot\boldsymbol{R}_\alpha} \sum_{\boldsymbol{k}_\parallel,\sigma} e^{i\boldsymbol{k}_\parallel\cdot(\boldsymbol{R}_\alpha-\boldsymbol{R}_\beta)} c^\dagger_{\alpha\boldsymbol{k}_\parallel+\boldsymbol{q}_\parallel\sigma} c_{\beta\boldsymbol{k}_\parallel\sigma}. \tag{6.128}$$

Here $U_{\alpha\beta}(q,z_a)$ and $M_{\alpha\beta}(\boldsymbol{q},\lambda)$ are intersubband ($\alpha \neq \beta$) impurity potential (2.83) and electron–phonon matrix element (2.88). Secondly, because the subband wave function of a quasi-2D system is far from a plane wave, the form factors play more important roles in the electron–impurity, electron–phonon, and electron–electron scattering. In a quasi-2D multisubband system, the electron–electron Coulomb interaction can be written as

$$\frac{1}{2}\sum_{\substack{\alpha',\alpha,\beta',\beta \\ \boldsymbol{k}_\parallel,\boldsymbol{k}'_\parallel,\boldsymbol{q}_\parallel \\ \sigma,\sigma'}} V_{\alpha'\alpha\beta'\beta}(q_\parallel) c^\dagger_{\alpha'\boldsymbol{k}_\parallel+\boldsymbol{q}_\parallel\sigma} c^\dagger_{\beta'\boldsymbol{k}'_\parallel-\boldsymbol{q}_\parallel\sigma'} c_{\beta\boldsymbol{k}'_\parallel\sigma'} c_{\alpha\boldsymbol{k}_\parallel\sigma}, \tag{6.129}$$

in which

$$V_{\alpha'\alpha\beta'\beta}(q_\parallel) = \frac{e^2}{2\epsilon_0\kappa q_\parallel} H_{\alpha'\alpha\beta'\beta}(q_\parallel), \tag{6.130}$$

$$H_{\alpha'\alpha\beta'\beta}(q_\parallel) = \iint dz_1 dz_2\, \zeta^*_{\alpha'}(z_1)\zeta_\alpha(z_1)\zeta^*_{\beta'}(z_2)\zeta_\beta(z_2)(e^{-q_\parallel|z_1-z_2|} + \text{images}). \tag{6.131}$$

The terms with subscript $(\alpha',\alpha;\beta',\beta)$ in (6.129) represent such parts of Coulomb interaction, which scatter an electron in subband α and an electron in subband β into one in subband α' and one in subband β'. Among them the intrasubband parts, i.e. the terms with subscript $(\alpha,\alpha;\alpha,\alpha)$, have the role to establish an electron temperature in the relative electron system of subband α. The intersubband parts $(\alpha,\alpha;\beta,\beta)$ $[\alpha\neq\beta]$, have the effect to exchange momentum and energy between these two subbands without the exchange of their particles. Therefore, they intend to equalize the center-of-mass velocities and electron temperatures of these two subbands. The other types of intersubband collision terms, such as $(\alpha, \beta; \beta, \alpha)$ and $(\alpha, \beta; \alpha, \beta)$ $[\alpha \neq \beta]$, which induce particle exchange between α and β subbands (the former keep the particle number conservation of each subband, while the latter transfer two electrons from subband α to subband β), have vanishing matrix elements at $q_\parallel = 0$ and thus much smaller effects in comparison with those of $(\alpha, \alpha; \beta, \beta)$ terms. They are generally neglected.

As explained in Sec. 6.6, the intrasubband terms $(\alpha, \alpha; \alpha, \alpha)$ of the Coulomb interaction have been included in the relative electron Hamiltonian (6.79) of the αth subband; the terms of $(\alpha, \alpha; \beta, \beta)$ type $(\alpha \neq \beta)$ will be treated as perturbations. They can be written, in terms of center-of-mass and relative electron variables of subbands, as

$$H_{\alpha\beta} = \sum_{\boldsymbol{k}_\parallel, \boldsymbol{k}'_\parallel, \boldsymbol{q}_\parallel, \sigma, \sigma'} V_{\alpha\alpha\beta\beta}(q_\parallel) e^{i\boldsymbol{q}_\parallel \cdot (\boldsymbol{R}_\alpha - \boldsymbol{R}_\beta)} c^\dagger_{\alpha \boldsymbol{k}_\parallel + \boldsymbol{q}_\parallel \sigma} c^\dagger_{\beta \boldsymbol{k}'_\parallel - \boldsymbol{q}_\parallel \sigma'} c_{\beta \boldsymbol{k}'_\parallel \sigma'} c_{\alpha \boldsymbol{k}_\parallel \sigma}.$$
(6.132)

Assuming a drift velocity (i.e. average velocity of the center-of-mass) $\boldsymbol{v}_\alpha$, an electron temperature T_α, and a chemical potential μ_α for each subband and repeating the derivation carried out in Sec. 6.6, we obtain the steady state transport balance equations for a quasi-2D multisubband system driven by a dc electric field $\boldsymbol{E}$ as

$$\sum_{\beta(\neq\alpha)} \chi^{\alpha\beta}(\boldsymbol{v}_\alpha, \boldsymbol{v}_\beta) = 0, \tag{6.133}$$

$$N_\alpha e\boldsymbol{E} + \boldsymbol{f}^\alpha(\boldsymbol{v}_\alpha) + \sum_{\beta(\neq\alpha)} \boldsymbol{f}^{\alpha\beta}(\boldsymbol{v}_\alpha, \boldsymbol{v}_\beta) + \sum_{\beta(\neq\alpha)} \boldsymbol{f}_{\alpha\beta}(\boldsymbol{v}_\alpha - \boldsymbol{v}_\beta) = 0, \tag{6.134}$$

$$\boldsymbol{v}_\alpha \cdot \boldsymbol{f}^\alpha(\boldsymbol{v}_\alpha) + w^\alpha(\boldsymbol{v}_\alpha) + \sum_{\beta(\neq\alpha)} w^{\alpha\beta}(\boldsymbol{v}_\alpha, \boldsymbol{v}_\beta) + \sum_{\beta(\neq\alpha)} w_{\alpha\beta}(\boldsymbol{v}_\alpha - \boldsymbol{v}_\beta) = 0, \tag{6.135}$$

$\alpha = 0, 1, \cdots, s-1$. In these equations, N_α is the electron sheet density of the αth subband,

$$\boldsymbol{f}^\alpha = \boldsymbol{f}^\alpha_i + \boldsymbol{f}^\alpha_p,$$

$$\chi^{\alpha\beta} = \chi^{\alpha\beta}_i + \chi^{\alpha\beta}_p,$$

$$\boldsymbol{f}^{\alpha\beta} = \boldsymbol{f}^{\alpha\beta}_i + \boldsymbol{f}^{\alpha\beta}_p,$$

$$w^{\alpha\beta} = w_{\rm p}^{\alpha\beta},$$

where $\chi_{\rm i}^{\alpha\beta}$ and $\boldsymbol{f}_{\rm i}^{\alpha\beta}$ are rates of changes of electron sheet density and frictional force (for system of unity area) of the αth subband induced by intersubband impurity scattering between the αth and βth subbands:

$$\chi_{\rm i}^{\alpha\beta} = 2\pi \sum_{\boldsymbol{k}_\|,\boldsymbol{q}_\|} |U_{\alpha\beta}(q_\|)|^2 \left[f\left(\frac{\xi_{\alpha\boldsymbol{k}_\|}}{T_\alpha}\right) - f\left(\frac{\xi_{\beta\boldsymbol{k}_\|+\boldsymbol{q}_\|}}{T_\beta}\right) \right] \delta(E_{\alpha\boldsymbol{k}_\|} - E_{\beta\boldsymbol{k}_\|+\boldsymbol{q}_\|}), \tag{6.136}$$

$$\boldsymbol{f}_{\rm i}^{\alpha\beta} = 2\pi \sum_{\boldsymbol{k}_\|,\boldsymbol{q}_\|} |U_{\alpha\beta}(q_\|)|^2 (-\boldsymbol{k}_\|) \left[f\left(\frac{\xi_{\alpha\boldsymbol{k}_\|}}{T_\alpha}\right) - f\left(\frac{\xi_{\beta\boldsymbol{k}_\|+\boldsymbol{q}_\|}}{T_\beta}\right) \right]$$
$$\times \delta(E_{\alpha\boldsymbol{k}_\|} - E_{\beta\boldsymbol{k}_\|+\boldsymbol{q}_\|}). \tag{6.137}$$

In this, $U_{\alpha\beta}(q_\|)$ is the effective average impurity potential (2.95), and

$$E_{\alpha\boldsymbol{k}_\|} = \varepsilon_{\alpha\boldsymbol{k}_\|} + \frac{1}{2}mv_\alpha^2 + \boldsymbol{k}_\| \cdot \boldsymbol{v}_\alpha. \tag{6.138}$$

The other functions in (6.133)–(6.135), $\boldsymbol{f}_{\rm p}^\alpha$, w^α, $\chi_{\rm p}^{\alpha\beta}$, $\boldsymbol{f}_{\rm p}^{\alpha\beta}$, $w_{\rm p}^{\alpha\beta}$, $\boldsymbol{f}_{\alpha\beta}$ and $w_{\alpha\beta}$, can be obtained from the corresponding expressions in (6.111)–(6.117) through the following transformation: changing all the wavevector $\boldsymbol{k}$ into $\boldsymbol{k}_\|$; changing all $\boldsymbol{q}$ into $\boldsymbol{q}_\|$ except those in $\Omega_{\boldsymbol{q}\lambda}$, $M_\alpha(\boldsymbol{q},\lambda)$ and $M_{\alpha\beta}(\boldsymbol{q},\lambda)$; replacing $\Pi^{(\alpha)}(\boldsymbol{q},\omega)$ by $\Pi(\alpha,\alpha,\boldsymbol{q}_\|,\omega)$; replacing $\nu_{\rm c}(q)$ in (6.116) and (6.117) by $V_{\alpha\alpha\beta\beta}(q_\|)$, where d_α is the degeneracy of the αth subband. Besides, the frictional force due to intraband impurity scattering in the αth subband is written as

$$\boldsymbol{f}_{\rm i}^\alpha = \sum_{\boldsymbol{k}_\|,\boldsymbol{q}_\|} \boldsymbol{q}_\| |U_{\alpha\alpha}(q_\|)|^2 \Pi_2(\alpha,\alpha,\boldsymbol{q}_\|,\omega_\alpha). \tag{6.139}$$

Note that, since the total number of particles $\sum_\alpha N_\alpha = N$ is fixed, there are only $s - 1$ independent equations in the s ($\alpha = 0, 1, \cdots, s-1$) particle-number balance equations. Therefore we have totally $3s - 1$ equations, which, together with s relations between the particle number and the chemical potential for each subband,

$$N_\alpha = 2 \sum_{\boldsymbol{k}_\|} f(\xi_{\alpha\boldsymbol{k}_\|}/T_\alpha), \tag{6.140}$$

determine all the steady-state transport parameters $\boldsymbol{v}_\alpha$, T_α, μ_α (or N_α) when the electric field $\boldsymbol{E}$ is given. Here we treat $\boldsymbol{v}_\alpha$ for each subband as one variable, and the related force balance equation as one equation.

6.9.2 Chemical potential difference considered for different subbands

In analyzing high-field transport in GaAs quantum wells, Guillemot, Clérot and Regreny (1992) utilize a somewhat simpler model for the electron system. They assume that intersubband Coulomb interactions of $(\alpha,\alpha;\beta,\beta)$ type are sufficiently strong to yield a unique drift velocity $\boldsymbol{v}_d$ and a single electron temperature T_e for all the subbands. But intersubband Coulomb interactions relevant to the particle exchange between different subbands, which generally involve form factors that are quite small, may not be strong enough that the chemical potentials should be considered different for different subbands. In this case, one needs only to deal with the total momentum and energy balances for the whole system, i.e. to sum up separately the force balance equations (6.134) and energy balance equations (6.135) for all the subbands. In view of the cancellation of all $\boldsymbol{f}_{\alpha\beta}$ and $w_{\alpha\beta}$ terms, the following simpler equations result:

$$Ne\boldsymbol{E} + \sum_{\alpha}\boldsymbol{f}^{\alpha}(\boldsymbol{v}_\alpha) + \sum_{\substack{\alpha,\beta \\ \alpha\neq\beta}}\boldsymbol{f}^{\alpha\beta}(\boldsymbol{v}_\alpha,\boldsymbol{v}_\beta) = 0, \qquad (6.141)$$

$$\sum_{\alpha}\boldsymbol{v}_\alpha \cdot \boldsymbol{f}^{\alpha}(\boldsymbol{v}_\alpha) + \sum_{\alpha}w^{\alpha}(\boldsymbol{v}_\alpha) + \sum_{\substack{\alpha,\beta \\ \alpha\neq\beta}}w^{\alpha\beta}(\boldsymbol{v}_\alpha,\boldsymbol{v}_\beta) = 0. \qquad (6.142)$$

The particle-number balance, however, has to be written down for each subband,

$$\sum_{\beta(\neq\alpha)}\chi^{\alpha\beta}(\boldsymbol{v}_\alpha,\boldsymbol{v}_\beta) = 0, \qquad (6.143)$$

having totally $s-1$ independent equations. These equations, together with the relation between the particle number and the chemical potential of each subband, determine the steady-state transport parameters $\boldsymbol{v}_d$, T_e and μ_α (or N_α) for given electric field $\boldsymbol{E}$.

Note that, for a multiple subband system in which all subbands share the same center-of-mass velocity $\boldsymbol{v}_d$, we have

$$\boldsymbol{f}_i^{\alpha\beta} + \boldsymbol{f}_i^{\beta\alpha} = 2\pi \sum_{\boldsymbol{k}_\|,\boldsymbol{q}_\|}|U_{\alpha\beta}(q_\|)|^2 \boldsymbol{q}_\| \left[f\left(\frac{\xi_{\alpha\boldsymbol{k}_\|}}{T_\alpha}\right) - f\left(\frac{\xi_{\beta\boldsymbol{k}_\|+\boldsymbol{q}_\|}}{T_\beta}\right)\right]$$
$$\times \delta(\varepsilon_{\beta\boldsymbol{k}_\|+\boldsymbol{q}_\|} - \varepsilon_{\alpha\boldsymbol{k}_\|} + \boldsymbol{q}_\| \cdot \boldsymbol{v}_d), \qquad (6.144)$$

$$f_p^{\alpha\beta} + f_p^{\beta\alpha} = 4\pi \sum_{k_{\|},q_{\|},\lambda} |M_{\alpha\beta}(q,\lambda)|^2 q_{\|} \left[f\left(\frac{\xi_{\alpha k_{\|}}}{T_\alpha}\right) - f\left(\frac{\xi_{\beta k_{\|}+q_{\|}}}{T_\beta}\right) \right]$$

$$\times \left\{ \left[n\left(\frac{\Omega_{q\lambda}}{T}\right) - n\left(\frac{\xi_{\alpha k_{\|}}}{T_\alpha} - \frac{\xi_{\beta k_{\|}+q_{\|}}}{T_\beta}\right) \right] \delta(\varepsilon_{\beta k_{\|}+q_{\|}} - \varepsilon_{\alpha k_{\|}} + q_{\|} \cdot v_d + \Omega_{q\lambda}) \right.$$

$$\left. + \left[n\left(\frac{\xi_{\alpha k_{\|}}}{T_\alpha} - \frac{\xi_{\beta k_{\|}+q_{\|}}}{T_\beta}\right) - n\left(\frac{\Omega_{q\lambda}}{T}\right) \right] \delta(\varepsilon_{\beta k_{\|}+q_{\|}} - \varepsilon_{\alpha k_{\|}} + q_{\|} \cdot v_d - \Omega_{q\lambda}) \right\},$$

(6.145)

$$w^{\alpha\beta} + w^{\beta\alpha} - v_d \cdot \left(f_p^{\alpha\beta} + f_p^{\beta\alpha}\right) = \sum_{q,\lambda} \Omega_{q\lambda} \left(\frac{\partial n_{q\lambda}}{\partial t}\right)_{\text{ep}}^{\alpha\beta}, \tag{6.146}$$

where

$$\left(\frac{\partial n_{q\lambda}}{\partial t}\right)_{\text{ep}}^{\alpha\beta} = 4\pi \sum_{k_{\|}} |M_{\alpha\beta}(q,\lambda)|^2 \left[f\left(\frac{\xi_{\alpha k_{\|}}}{T_\alpha}\right) - f\left(\frac{\xi_{\beta k_{\|}+q_{\|}}}{T_\beta}\right) \right]$$

$$\times \left[n\left(\frac{\xi_{\alpha k_{\|}}}{T_\alpha} - \frac{\xi_{\beta k_{\|}+q_{\|}}}{T_\beta}\right) - n\left(\frac{\Omega_{q\lambda}}{T}\right) \right]$$

$$\times \delta(\varepsilon_{\beta k_{\|}+q_{\|}} - \varepsilon_{\alpha k_{\|}} + q_{\|} \cdot v_d - \Omega_{q\lambda}) \tag{6.147}$$

is the rate of change of the phonon number of mode $q\lambda$ due to interband electron–phonon interaction between αth and βth subbands. The total rate of change of the $q\lambda$-mode phonon number due to electron–phonon interaction includes all possible intraband and interband contributions:

$$\left(\frac{\partial n_{q\lambda}}{\partial t}\right)_{\text{ep}} = \sum_\alpha \left(\frac{\partial n_{q\lambda}}{\partial t}\right)_{\text{ep}}^\alpha + \sum_{\alpha>\beta} \left(\frac{\partial n_{q\lambda}}{\partial t}\right)_{\text{ep}}^{\alpha\beta}. \tag{6.148}$$

The rate of change of the $q\lambda$-mode phonon number induced by intra-αth-subband electron–phonon interaction is given by

$$\left(\frac{\partial n_{q\lambda}}{\partial t}\right)_{\text{ep}}^\alpha = 2|M_\alpha(q,\lambda)|^2 \Pi_2^\alpha(q_{\|}, \Omega_{q\lambda} - q_{\|} \cdot v_d)$$

$$\times \left[n\left(\frac{\Omega_{q\lambda}}{T}\right) - n\left(\frac{\Omega_{q\lambda} - q_{\|} \cdot v_d}{T_\alpha}\right) \right]. \tag{6.149}$$

Eq. (6.148) can be obtained directly from the Hamiltonian by calculating the rate of change of the $q\lambda$-phonon number operator,

$$\left(\frac{\partial \hat{N}_{q\lambda}}{\partial t}\right)_{\text{ep}} = -\mathrm{i}\,[\hat{N}_{q\lambda}, H_{\text{ep}}], \tag{6.150}$$

and taking the statistical average. Furthermore, though all the equations presented so far in this subsection are written under the condition of equilibrium phonon occupation, the effects of nonequilibrium phonon occupation can be easily included following the prescription in Chapter 5: replacing all the equilibrium phonon occupation number $n(\Omega_{q\lambda}/T)$ in the equation by the nonequilibrium phonon occupation number $n_{q\lambda}$ satisfying the steady state relaxation equation for the phonon system:

$$0 = \left(\frac{\partial n_{q\lambda}}{\partial t}\right)_{\text{ep}} + \left[n_{q\lambda} - n\left(\frac{\Omega_{q\lambda}}{T}\right)\right]\frac{1}{\tau_{\text{p}}}. \tag{6.151}$$

Nonequilibrium phonon number $n_{q\lambda}$ is immediately solved from Eqs. (6.148) and (6.151), and then all the transport quantities in the balance equations are obtained, containing a phonon relaxation time τ_{p}.

Based on these equations Guillemot, Clérot and Regreny (1992) calculate the electron transport of a GaAs quantum well having well width $a = 27.4\,\text{nm}$, carrier sheet density $N_{\text{s}} = 3.51\times 10^{11}\,\text{cm}^{-2}$ (equivalent volume density $1.3 \times 10^{17}\,\text{cm}^{-3}$) and linear mobility $3.3\,\text{m}^2/\text{Vs}$, at lattice temperature $T = 4.2\,\text{K}$. They take into account two lowest subbands ($\alpha, \beta = 0, 1$) with impurity and longitudinal optical phonon scatterings, and consider some many-particle properties (renormalization of energy and polarization) of the two-subband electron gas within the framework of finite-temperature random phase approximation. The energy broadening of the electron states is included by the electron self-energy.

Figure 6.11 shows the calculated chemical potentials and the renormalized subband energies versus the electron temperature. At low electric fields (low T_{e}), the chemical potentials of two subbands are equal. With increasing field strength the splitting between μ_1 and μ_0 appears. The chemical potential of the upper band rises above that of the lower band until $T_{\text{e}} \approx 80\,\text{K}$. Then the upper-band chemical potential becomes lower than the lower-band one with continuing increase of the electric field. A careful inspection shows that the LO-phonon scattering mainly transfers electrons from the upper subband to the ground subband, while the impurity scattering plays a reverse role. At low temperatures, impurity scattering dominates, raising the Fermi level of the upper subband. At higher

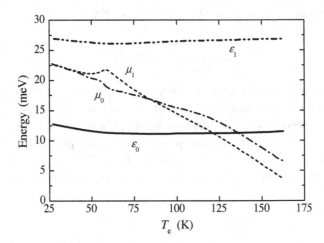

Fig. 6.11 The renormalized energies ε_0, ε_1 and chemical potentials μ_0, μ_1 of two subbands versus the electron temperature T_e. All the energies are measured from the GaAs conduction band bottom. The phonon relaxation time τ_p is set to be zero. After Guillemot, Clérot and Regreny (1992).

temperatures, LO-phonon scattering becomes dominant, leading to a reverse effect. The splitting of the subband chemical potentials accounts for the balance on the exchange of particles between two subbands.

On the other hand, the rise of the electron temperature T_e above the lattice temperature T in the present system mainly results from the energy balance. Fig. 6.12 shows the calculated electron temperature versus the strength of the electric field. In view of the existence of impurity (as well as phonon) scatterings electrons can absorb energy from the electric field. This energy must be transferred to the lattice through the phonon emission to keep the energy balance in a steady transport state. We see that with increasing electric field the electron temperature increases steeply in the first step. This is because that the impurity-scattering induced energy absorption grows rapidly with electric field (proportional to E^2) while the electron energy-loss rate (mainly due to the LO-phonon emission) remains low at low lattice temperatures T and low electric field (therefore low T_e). This trend of electron-temperature increase is lessened because of the transfer of carriers induced by intersubband impurity scattering, which raises the upper-subband Fermi level above the ground subband one: for the same average energy the electron temperature is lower than what it would be if the

Fermi levels were equal. The slowing down in the growth speed of the electron temperature is also due to the increasing efficiency of the LO-phonon emission by growing electron temperature and drift velocity at higher electric fields [as reflected in the Bose factor $n((\Omega_{\rm LO} - \boldsymbol{q}_\| \cdot \boldsymbol{v}_\alpha)/T_{\rm e})$]. After a gentle-slope range of increase, however, $T_{\rm e}$ rises again more rapidly when the electric field goes even higher. For drift velocities above 1.5×10^7 cm/s, the range of wavevector where $\Omega_{\rm LO} - \boldsymbol{q}_\| \cdot \boldsymbol{v}_\alpha$ is negative, becomes noticeable ($q_\| \geqslant 3 \times 10^6$ cm^{-1}) and continues to extend with increasing drift velocity. This leads to a rapid decrease of the LO-phonon emission efficiency and thus an accelerated increase of electron temperature.

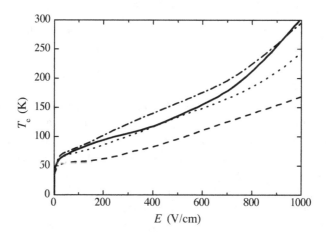

Fig. 6.12 The electron temperature $T_{\rm e}$ of a two-subband GaAs quantum well versus the electric field E. The solid curve is from the photoluminasence measurement. The other three curves are theoretical results with $\tau_{\rm p} = 0$ (dash curve), $\tau_{\rm p} = 3$ ps (dot-dash curve) and $\tau_{\rm p} = 7$ ps (dot curve) at lattice temperature $T = 4.2$ K. After Guillemot, Clérot and Regreny (1992).

6.9.3 *Effect of intersubband Coulomb interaction on transport*

The effective strength of interband Coulomb interaction is determined by the carrier density. For a more quantitative understanding on the effect of intersubband Coulomb interaction Dong and Lei (1998) perform a numerical investigation based on the multisubband formulation as described in Sec. 6.9.1, taking the ground ($\alpha = 0$) and the next lowest ($\alpha = 1$) subbands of a GaAs-based quantum-well having well width $a = 50\,\text{nm}$ to serve as a model two-subband system. They calculate the hot-carrier transport properties of the system assuming that the 0th subband and the 1st subband have individual drift velocity, electron temperature and chemical potential: v_0, T_0, μ_0 and v_1, T_1, μ_1. Then, the particle number, force, and energy balance equations under the influence of a dc electric field E have the following form in the steady state:

$$\chi^{01}(v_0, v_1) = 0, \qquad (6.152)$$
$$N_s^0 eE + f^0(v_0) + f^{01}(v_0, v_1) + f_{01}(v_0 - v_1) = 0, \qquad (6.153)$$
$$N_s^1 eE + f^1(v_1) + f^{10}(v_1, v_0) - f_{01}(v_0 - v_1) = 0, \qquad (6.154)$$
$$v_0 f^0(v_0) + w^0(v_0) + w^{01}(v_0, v_1) + w_{01}(v_0 - v_1) = 0, \qquad (6.155)$$
$$v_1 f^1(v_1) + (v_0 - v_1) f_{01}(v_0 - v_1) + w^1(v_1) + $$
$$\qquad\qquad w^{10}(v_1, v_0) - w_{01}(v_0 - v_1) = 0, \qquad (6.156)$$

in which N_s^0 and N_s^1 are the carrier numbers (sheet densities) of the 0th and the 1st subbands, $N_s^0 + N_s^1 = N_s$, with N_s the total carrier sheet density. The calculation is carried out at two different values of carrier densities, $N_s = 1 \times 10^{10}\,\text{cm}^{-2}$ (equivalent volume density $2 \times 10^{15}\,\text{cm}^{-3}$) and $N_s = 5 \times 10^{11}\,\text{cm}^{-2}$ (equivalent volume density $1 \times 10^{17}\,\text{cm}^{-3}$). The electron temperatures T_0 and T_1 and drift velocities v_0 and v_1 for each subband as well as the total average drift velocity $v_d = (N_s^0 v_0 + N_s^1 v_1)/N_s$, obtained from solving the above equations at lattice temperature $T = 80\,\text{K}$ by considering intrasubband electron–impurity and electron–phonon scatterings and intersubband electron–phonon and electron–electron Coulomb scatterings, are shown in Fig. 6.13.

It is seen that at low carrier density ($N_s = 1 \times 10^{10}\,\text{cm}^{-2}$) there exist relatively large differences in drift velocities, electron temperatures and other transport properties between two subbands. At high carrier density ($N_s = 5 \times 10^{11}\,\text{cm}^{-2}$), these differences become so small that results for this two-subband system from the above model (individual drift velocities, electron temperatures and chemical potentials) are essentially the same as those obtained by using the model with a single drift velocity, single electron temperature and single chemical potential for both subbands.

Note that, the electron temperature T_e in the present case ($T = 80\,\text{K}$) exhibits a quite different behavior from that of $T = 4.2\,\text{K}$ (Fig. 6.12). At such a high lattice temperature (thus high electron temperatures), the LO-phonon-emission induced electron energy-loss rate is large and increases with drift velocity more rapidly than the electron energy-absorption from the electric field, that the electron cooling, i.e. T_e lower than the lattice temperature T, appears in the field range lower than $E \approx 0.3\,\text{kV/cm}$, as clearly seen in Fig. 6.13.

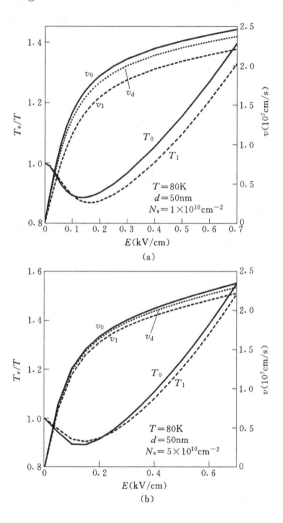

Fig. 6.13 Drift velocities v_0, v_1, and v_d, and electron temperatures T_0 and T_1 of a two-subband GaAs quantum well are shown as functions of the electric field E. The quantum well has a width of $a = 50\,\text{nm}$ and the lattice temperature is $T = 80\,\text{K}$. The carrier sheet density is (a) $N_s = 1 \times 10^{10}\,\text{cm}^{-2}$ or (b) $N_s = 5 \times 10^{11}\,\text{cm}^{-2}$. From Dong and Lei (1998).

6.9.4 δ-doped structures, electric-field induced real-space transfer

Modern epitaxial growth techniques allow fabrication of semiconductor materials, in which the dopant atoms are concentrated in very thin (a few nanometers) periodically arranged multilalyers in an otherwise undoped sample. This is so-called planar doping or δ-doping. In δ-doped systems the sheets of ionized dopant atoms produce spike-like narrow electrostatic potential wells which can bind the emitted carriers (i.e. electrons in the case of n-type δ-doping, and holes in the case of p-type δ-doping) to form a few quantized levels. These bound carriers, which move freely in the plane, are quasi-two-dimensional (2D) in nature. Above the potential barriers there are three-dimensional (3D) extended states. The potential, subband structure and electron density distribution from a selfconsistant calculation (Kostial et al, 1993) for a δ-doped GaAs system with sheet-doping concentration 5×10^{11} cm^{-2} and doping layer spacing $d = 100$ nm (sample A) are shown in Fig. 6.14. Electrons in the ground subband are bound to single narrow V-type wells centered at doping layers and thus quasi-two-dimensional, whereas electrons in the uppermost subbands, whose density distributes in the regions between neighboring doping layers and forms broad minibands, are in fact three-dimensional.

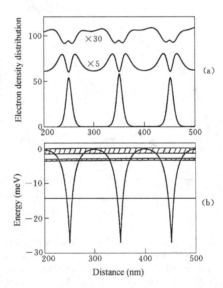

Fig. 6.14 The potential and conduction-band edge (b) and electron densities (a) of the lowest three subbands of the δ-doped sample A. From Kostial et al (1993).

Kostial et al (1993) measure the differential conductivity dj/dE of a δ-doped system in the plane at lattice temperature $T = 4.2\,\mathrm{K}$. Their measured result of sample A is plot in Fig. 6.16(a), showing that dj/dE increases with increasing electric field E from the beginning and reaches a peak at around $E \approx 10\,\mathrm{V/cm}$, then decreases and approaches approximately a constant value with continuing increase of the electric field. This kind of behavior is referred to electric field-induced real space transfer of carriers: in the weak-field region most electrons are in the quasi-2D states inside wells and experience strong scattering by the impurities located in the center planes. When the electric field increases more electrons transfer to upper quasi-3D states due to the rise of electron temperature and experience weakened scatterings because of more weight of electron density distributing to the barrier region away from doping layers.

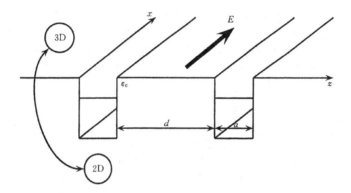

Fig. 6.15 A simple model for the investigation of real space transfer.

To avoid self-consistent and field-dependent calculation for the potential of δ-doping layer and concentrate on the effect of impurity scattering on real-space transfer, Kostial et al (1993) employ a rigid square-well superlattice with well width a and period d, as illustrated in Fig. 6.15, to simulate the δ-doped structure. The energy level is approximated by one quasi-2D subband in the well,

$$\varepsilon^{(1)} = \varepsilon^{(1)}_{\boldsymbol{k}_\parallel} = \frac{1}{2m}(k_x^2 + k_y^2), \tag{6.157}$$

and one almost-free 3D band close to the barrier edge,

$$\varepsilon^{(2)} = \varepsilon^{(2)}_{\boldsymbol{k}} = \varepsilon_c + \frac{1}{2m}(k_x^2 + k_y^2 + k_z^2), \tag{6.158}$$

in which ε_c is the distance between the 3D and quasi-2D band bottoms. The wave function of the 2D state is given by $[\mathbf{r} \equiv (x,y)]$

$$\psi^{(1)}_{l\mathbf{k}_\parallel}(\mathbf{r}, z) = \mathrm{e}^{\mathrm{i}\mathbf{k}_\parallel \cdot \mathbf{r}} \phi^{(1)}_l(z), \qquad (6.159)$$

where l is the well index, $\phi^{(1)}_l(z) = \zeta(z - ld)$ is the normalized single-well wave function of the lth well, taken as the ground-state wave function of an infinite square well of width a:

$$\zeta(z) = \begin{cases} (2/a)^{1/2} \sin(\pi z/a), & \text{when } 0 \leqslant z \leqslant a; \\ 0, & \text{when } z < 0 \text{ or } z > a. \end{cases} \qquad (6.160)$$

The wave function of the 3D state is just assumed to be a plane wave,

$$\psi^{(2)}_\mathbf{k}(\mathbf{r}, z) = \mathrm{e}^{\mathrm{i}(\mathbf{k}_\parallel \cdot \mathbf{r} + k_z z)}. \qquad (6.161)$$

Such a system can be regarded as a two-carrier-species system consisting of electrons in the 2D band (density $n^{(2D)}$) and electrons in the 3D band (density $n^{(3D)}$). Neglecting effects induced by the momentum and energy exchange between two subsystems, the steady-state momentum and energy balance equations read

$$n_j eE + f^{(j)}(v_j, T_j) = 0, \qquad (6.162)$$
$$v_j f^{(j)}(v_j, T_j) + w^{(j)}(v_j, T_j) = 0, \qquad (6.163)$$

where $j = 1, 2$ refer to quasi-2D and quasi-3D subsystems respectively. v_j and T_j are the drift velocity and electron temperature of the subsysetm j. The frictional force $f^{(j)}$ and the energy-transfer rate $w^{(j)}$ result from impurity, acoustic deformation potential, piezoelectric and polar optic-phonon scatterings:

$$f^{(j)} = f^{(j)}_\mathrm{i} + f^{(j)}_\mathrm{ac} + f^{(j)}_\mathrm{op}, \qquad (6.164)$$
$$w^{(j)} = w^{(j)}_\mathrm{ac} + w^{(j)}_\mathrm{op}. \qquad (6.165)$$

Their expressions were given in Secs. 2.1 and 2.9. As a preliminary approximation, a common chemical potential μ is assumed for both subsystems and the occupation numbers of 2D and 3D states are determined by the Fermi distribution. This gives densities of 2D and 3D electrons as

$$n^{(2D)} = \frac{1}{2\pi^2 d} \int d^2 k_\parallel \left\{ \exp\left[\left(\varepsilon^{(1)}_{\mathbf{k}_\parallel} - \mu\right)/T^{(1)}_\mathrm{e}\right] + 1 \right\}^{-1}, \qquad (6.166)$$

$$n^{(3D)} = \frac{1}{4\pi^3} \int d^3 k \left\{ \exp\left[\left(\varepsilon^{(2)}_\mathbf{k} - \mu\right)/T^{(2)}_\mathrm{e}\right] + 1 \right\}^{-1}, \qquad (6.167)$$

and the total electron density is

$$n = n^{(2D)} + n^{(3D)}. \tag{6.168}$$

For given n and E, Eqs. (6.162), (6.163) and (6.168) completely determine $v^{(2D)}$, $T_e^{(2D)}$, $v^{(3D)}$, $T_e^{(3D)}$, and μ, and then $n^{(2D)}$ and $n^{(3D)}$ are obtained from (6.166) and (6.167). The current density of the system is the sum of 2D and 3D contributions,

$$J = e\, n^{(2D)} v^{(2D)} + e\, n^{(3D)} v^{(3D)}. \tag{6.169}$$

By using the frictional force and energy-tranfer rate expressions with RPA dynamical screening for 2D and 3D systems, Kostial et al (1993) calculate transport properties of the model GaAs superlattice with $a = 10$ nm, $d = 100$ nm, $n = 5 \times 10^{16}$ cm^{-3} and $\varepsilon_c = 18$ meV, assuming a residual bulk doping of $n_i = 10^{15}$ cm^{-3}, which is treated as an adjustable parameter to get the measured mobility from cyclotron resonance. At zero electric field $T_e^{(2D)} = T_e^{(3D)} = T = 4.2$ K, almost all electrons are in 2D states: $n^{(2D)} \simeq n$ and $n^{(3D)} \simeq 0$. With increasing electric field strength, 2D and 3D electron temperatures, especially $T_e^{(3D)}$, rapidly increase at relatively weak fields and so does $n^{(3D)}$. When electron temperature rises to the range around 40 K or $T_e^{(3D)} \approx 0.1\Omega_{LO}$, the strong emission of optical phonons produces a very efficient electron energy-transfer to the lattice that balances the further increase of electron energy gained from the electric field. Therefore, in a broad field range the electron temperature exhibits almost no increase with continuing increase of the electric field. On the other hand, since 3D electrons suffer much weaker impurity scattering than 2D electrons, the drift velocity of 3D electrons is much higher than that of 2D electrons. The total current density increases rapidly with increasing electric field at the beginning due to the rapid increase of the 3D electron density induced by the rising electron temperature. When $T_e^{(3D)}$ exhibits saturation at high electric field, however, the total current density becomes sublinearly slow increase with electric field. Thus, there must be a point of inflection in the current-voltage dependence which gives a maximum in the differential conductivity. These are in qualitatively agreement with experiments as shown in Fig. 6.16.

The above model assumes that 2D and 3D subsystems have separated drift velocities and separated electron temperatures, but share a common chemical potential. Usually, this assumption can be valid only when electric field is weak that the electron system is not driven too far out of equilibrium,

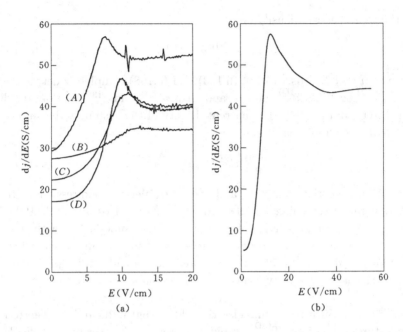

Fig. 6.16 Differential conductivity dj/dE versus the electric field at $T = 4.2\,\text{K}$. (a) The measured results for several δ-doped systems. (b) Model calculation for system A. from Kostial et al (1993).

or there exists very frequent particle exchange between two subsystems due to strong intersubband scatterings. Generally, one should allow different chemical potentials μ_1 and μ_2 for two different subbands. This, for the determination of the additional parameter, requires to consider intersubband scattering induced particle exchange between two subbands (Kleinert and Asche, 1994). Another drawback of the above model is to take isotropic plane waves (6.161) as the wave functions for the 3D states. In view of δ-doping induced potential existing in the z direction, the plane wave is not an eigen function of the system and thus not orthogonal with 2D wave function (6.159). When intersubband scattering is not relevant in solving the problem, this kind of wave function inaccuracy yields only weak influence since the determinative intrasubband matrix elements are not sensitive to the orthogonality of wave functions in different subbands. This is the reason why a qualitative correct result is still obtained even a simple plane wave is used as the 3D wave function in the above model. However, when intersubband scatterings are needed to take into account, the mutual orthogonality of wave functions becomes important: the correct selection rule may be violated and intersubband matrix elements may diverge if the wave functions are not orthogonal.

To overcome this difficulty, Kubrak and Kleinert (1998) employ the orthogonalized plane waves as the wave functions of the 3D states:

$$\psi_{\mathbf{k}}^{(2)}(\mathbf{r}, z) = e^{i \mathbf{k}_\| \cdot \mathbf{r}} \phi_{k_z}^{(2)}(z), \tag{6.170}$$

$$\phi_{k_z}^{(2)}(z) = \Gamma(k_z) \left[e^{i k_z z} - \sum_l e^{i k_z l d} I^{(12)}(k_z) \phi_l^{(1)}(z) \right], \tag{6.171}$$

in which

$$I^{(12)}(k_z) = \frac{2\pi \sqrt{2a}}{\pi^2 - k_z^2 a^2} \cos(k_z a/2) \tag{6.172}$$

is the orthogonal coefficient,

$$\Gamma(k_z) = \left[\frac{d}{d - |I^{(12)}(k_z)|^2} \right]^{1/2} \tag{6.173}$$

is a normalized factor. The wave function (6.170) is orthogonalized to the wave function (6.159) of the 2D state.

The electron-phonon and electron-impurity scattering related intersubband Hamiltonian can be written in the second quantization representation in the basis of wave functions (6.159) and (6.170) as

$$H_{\text{ep}}^{12} = \sum_{\substack{\mathbf{k}, \mathbf{q}, \lambda \\ l, \sigma}} M(\mathbf{q}, \lambda) J^{(12)}(k_z, q_z) \phi_{\mathbf{q}\lambda} e^{i \mathbf{k}_\| \cdot (\mathbf{R}_1 - \mathbf{R}_2)} e^{i(k_z + q_z)ld} c_{1\mathbf{k}_\| + \mathbf{q}_\| l\sigma}^\dagger c_{2\mathbf{k}_\| k_z \sigma}, \tag{6.174}$$

$$H_{\text{ei}}^{12} = \sum_{\substack{\mathbf{k}_\|, \mathbf{q}_\|, a \\ k_z, l, \sigma}} F_{lk_z}^{(12)}(q_\|, z_a) e^{i \mathbf{q}_\| \cdot (\mathbf{R}_1 - \mathbf{r}_a) + i \mathbf{k}_\| \cdot (\mathbf{R}_1 - \mathbf{R}_2)} e^{i k_z l d} c_{1\mathbf{k}_\| + \mathbf{q}_\| l\sigma}^\dagger c_{2\mathbf{k}_\| k_z \sigma}, \tag{6.175}$$

where $\mathbf{R}_1$ and $\mathbf{R}_2$ are respectively the center-of-mass coordinates (in the plane) of the 2D and 3D systems, $(\mathbf{r}_a, z_a)$ denotes the impurity position, $c_{1\mathbf{k}_\| l\sigma}^\dagger$ is the creation operator of a 2D electron in the lth well having transverse wavevector $\mathbf{k}_\|$ and spin σ, $c_{2\mathbf{k}_\| k_z \sigma}$ is the annihilation operator of a 3D electron having transverse wavevector $\mathbf{k}_\|$, longitudinal wavevector k_z, and spin σ, $\phi_{\mathbf{q}\lambda} = b_{\mathbf{q}\lambda} + b_{-\mathbf{q}\lambda}^\dagger$ is the field operator of a bulk phonon having wavevector $\mathbf{q}$ in branch λ, $M(\mathbf{q}, \lambda)$ is the electron–phonon matrix element in the plane wave representation,

$$J^{(12)}(k_z, q_z) = \int dz\, e^{i q_z z} \phi_0^{(1)}(z) \phi_{k_z}^{(2)}(z) \qquad (6.176)$$

is a form factor related to 2D and 3D wave functions (Kubrak and Kleinert, 1998a),

$$F_{lk_z}^{(12)} = \left(\frac{Ze^2}{2\epsilon_0 \kappa q_\|}\right) \int dz\, e^{-q_\| |z + ld - z_a|} \phi_0^{(1)}(z) \phi_{k_z}^{(2)}(z) \qquad (6.177)$$

is the effective intersubband impurity scattering potential (κ is the background dielectric constant and Z is the charge number of the impurity). The particle-number transfer rates from the 2D subsystem to the 3D subsystem induced by intersubband electron–phonon and electron–impurity scatterings, χ_{ep}^{21} and χ_{ei}^{21}, can be obtained based on the method introduced in Secs. 6.7 and 2.9:

$$\chi_{\mathrm{ep}}^{21} = \frac{4\pi}{d} \sum_{k,q,\lambda} |M(q,\lambda)|^2 |J^{(12)}(k_z, q_z)|^2 \left[f\!\left(\frac{\xi_k^{(2)}}{T_2}\right) - f\!\left(\frac{\xi_{k_\|+q_\|}^{(1)}}{T_1}\right) \right]$$

$$\times \left\{ \left[n\!\left(\frac{\Omega_{q\lambda}}{T}\right) - n\!\left(\frac{\xi_{k_\|+q_\|}^{(1)}}{T_1} - \frac{\xi_k^{(2)}}{T_2}\right) \right] \delta(E_{k_\|+q_\|}^{(1)} - E_k^{(2)} - \Omega_{q\lambda}) \right.$$

$$\left. + \left[n\!\left(\frac{\Omega_{q\lambda}}{T}\right) - n\!\left(\frac{\xi_k^{(2)}}{T_2} - \frac{\xi_{k_\|+q_\|}^{(1)}}{T_1}\right) \right] \delta(E_{k_\|+q_\|}^{(1)} - E_k^{(2)} + \Omega_{q\lambda}) \right\},$$

$$(6.178)$$

$$\chi_{\mathrm{ei}}^{21} = \frac{4\pi}{d} \sum_{k_\|, k_z, q_\|} \left(\frac{Ze^2}{2\epsilon_0 \kappa q_\|}\right)^2 N_{\mathrm{i}}^{(12)}(q_\|, k_z) \left[f\!\left(\frac{\xi_k^{(2)}}{T_2}\right) - f\!\left(\frac{\xi_{k_\|+q_\|}^{(1)}}{T_1}\right) \right]$$

$$\times \delta(E_{k_\|+q_\|}^{(1)} - E_k^{(2)}). \qquad (6.179)$$

Here $n(x) \equiv (e^x - 1)^{-1}$ is the Bose function, $f(x) \equiv (e^x + 1)^{-1}$ is the Fermi function, $\xi_{k_\|}^{(1)} = \varepsilon_{k_\|}^{(1)} - \mu_1$, $E_{k_\|}^{(1)} = \varepsilon_{k_\|}^{(1)} + k_x v_1 + \tfrac{1}{2} m v_1^2$, $\xi_k^{(2)} = \varepsilon_k^{(2)} - \mu_2$, $E_k^{(2)} = \varepsilon_k^{(2)} + k_x v_2 + \tfrac{1}{2} m v_2^2$, and $N_{\mathrm{i}}^{(12)}(q_\|, k_z)$ is an equivalent averaged impurity density having different expressions for randomly distributed background impurities, remote impurities or planar impurities in the well centers which are easily derived by the method of Sec. 2.9.

As pointed out by Kubrak, Kleinert and Asche (1998), in view of the orthogonality of $\phi_l^{(1)}(z)$ and $\phi_{k_z}^{(2)}(z)$ functions, the effective intersubband impurity potential and effective impurity density vanish for electron transition without change of transverse wavevector ($q_\parallel = 0$): $F_{lk_z}^{(12)}(q_\parallel = 0, z_a) = 0$ and $N_i^{(12)}(q_\parallel, k_z) = 0$. This indicates that the main contribution to intersubband transition comes from the range of large wavevectors. If the plane wave (6.161) is taken as the 3D wave function, the corresponding $F_{lk_z}^{(12)}$ and $N_i^{(12)}$ do not have this property and χ_{ei}^{21} calculated from (6.179) diverges. By using a 3D state orthogonalized to the 2D state, this unphysical small-$q_\parallel$ divergence of intersubband Coulomb matrix element is eliminated. Because the dominant contribution comes from large wavevectors where screening plays a less important role, the screening is generally negligible for the evaluation of intersubband matrix elements. Whether or not using the orthogonalized wave function for 3D states is also relevant to the form factor of electron–phonon scattering. Since $J^{(12)}(k_z, q_z=0) = 0$ when $\phi_l^{(1)}(z)$ and $\phi_{k_z}^{(2)}(z)$ are orthogonal, the orthogonalization of wave functions suppresses the contribution from small-q_z phonons. However, since there is no divergence in the case of phonon scattering, the influence of wavefunction orthogonalization is not so strong as in the case of impurity scattering.

Neglecting effects of intersubband scatterings on the friction force and energy-transfer rate of each subsystem, one needs only a particle-number balance equation

$$\chi_{\text{ep}}^{21} + \chi_{\text{ei}}^{21} = 0, \tag{6.180}$$

in addition to the force-balance equation (6.162) and energy-balance equation (6.163) for each subsystem, for the determination of transport parameters.

Employing the improved model and starting from Eqs. (6.162), (6.163) and (6.180), Kubrak, Kleinert and Asche (1998) recalculate the transport properties related to real-space transfer in the δ-doped GaAs sample A described in Fig. 6.14, arriving at a result in better agreement with experiment as shown in Fig. 6.17.

6.9.5 *Quasi-one-dimensional multiple subband systems*

In dealing with quasi-one-dimensional multiple subband systems one first faces with the same problem as in the case of quasi-two-dimensional subband systems: treating these carriers in different subbands as one species

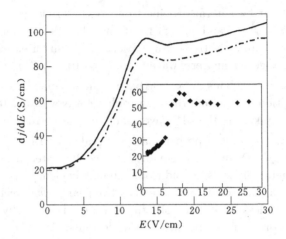

Fig. 6.17 Calculated differential conductivity dj/dE versus the electric field for sample A (solid curve) and another sample. The inset shows experimental results. From Kubrak, Kleinert and Asche (1998).

or several species. In discussing electric-field driven transport in quantum wires in Sec. 2.11, we treat all carriers occupying different subbands as the same species of carriers described by a common drift velocity, a common electron temperature and a common chemical potential. This apparently requires strong interactions between carriers in different subbands. To examine the applicability of this simple description in quantum wires and analyze the influence of carrier density and intersubband Coulomb interaction on transport, we deal with nonlinear conduction in a GaAs-based n-type quantum wire using the several-carrier-species model.

The single-electron eigenenergy and wavefunction and the many-particle Hamiltonian of a quantum wire were discussed in detail in Sec. 2.11. As a simple model, we consider a system consisting of three lowest subbands of a cylindrical quantum wire, i.e. the ground subband ($m = 0, l = 0$ with wave function φ_0) and two lowest degenerate excitation subbands ($m = 0, l = 1$ and $m = 0, l = -1$ with wave functions φ_1 and φ_{-1}). It is in fact a system composed of two species of carriers, since when the electric field is applied along the axis of the cylindrical quantum wire (z-direction) the behaviors of two degenerate subbands φ_1 and φ_{-1} are identical. The transport properties of this system can be described with the following parameters: the drift velocity v_0, electron temperature T_0, and chemical potential μ_0 of the ground subband φ_0, and the drift velocity v_1, electron temperature T_1, and chemical potential μ_1 of φ_1 and φ_{-1} subbands. The steady-state balance equations are

$$n^{01}(v_0, v_1) = 0, \qquad (6.181)$$

$$N_1^0 eE + f^0(v_0) + 2f^{01}(v_0, v_1) + 2f_{01}(v_0 - v_1) = 0, \qquad (6.182)$$

$$N_1^1 eE + f^1(v_1) + f^{10}(v_1, v_0) - f_{01}(v_0 - v_1) = 0, \qquad (6.183)$$

$$v_0 f^0(v_0) + w^0(v_0) + 2w^{01}(v_0, v_1) + 2w_{01}(v_0 - v_1) = 0, \qquad (6.184)$$

$$v_1 f^1(v_1) + (v_0 - v_1) f_{01}(v_0 - v_1) + w^1(v_1)$$
$$\qquad + w^{10}(v_1, v_0) - w_{01}(v_0 - v_1) = 0, \qquad (6.185)$$

where N_1^0 is the line density of electrons in subband φ_0, N_1^1 and $N_1^{-1} = N_1^1$ are the line densities of electrons in subbands φ_1 and φ_{-1}. The total electron density is $N_1 = N_1^0 + 2N_1^1$.

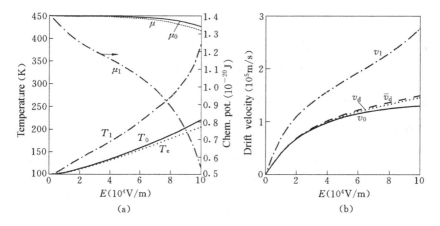

Fig. 6.18 Chemical potentials, electron temperatures and drift velocities of a two-subband GaAs quantum wire versus the electric field E. The system has an electron line density $N_1 = 2 \times 10^5 \, \text{cm}^{-1}$ and the lattice temperature is $T = 100\,\text{K}$. From Wu, Yu and Lei (1994).

By means of this set of equations Wu, Yu and Lei (1994) investigate the hot carrier transport of a cylindrical quantum wire having transverse radius $\varrho = 9\,\text{nm}$ and electron line density $N_1 = 2 \times 10^5 \, \text{cm}^{-1}$ at lattice temperature $T = 100\,\text{K}$, considering three subbands with all intrasubband and intersubband polar-optic phonon and acoustic phonon scatterings. The calculated subband chemical potentials μ_0 and μ_1, electron temperatures T_0 and T_1 and drift velocities v_0 and v_1 by using two-carrier-species model are shown in Fig. 18(a) and (b), together with the total average drift velocity $\bar{v}_{\rm d} = (N_1^0 v_0 + 2N_1^1 v_1)/N_1$. For comparison, the transport properties of this system obtained with single-carrier-species model (i.e. the method described in Sec. 2.11) by assuming a common drift velocity $v_{\rm d}$, a single electron temperature $T_{\rm e}$ and a unique chemical potential μ for all electrons in three subbands, are also shown in the figure (dotted curves). We can see that for this system, μ_0 and μ_1, T_0 and T_1, and v_0 and v_1 in two-carrier-species model are quite different at finite applied electric fields, but the total average drift velocity $\bar{v}_{\rm d}$ is rather close to $v_{\rm d}$, the drift velocity obtained in the single-carrier-species model. Calculation using two-carrier-species model for systems of higher carrier density (Dong and Lei, 1998) shows that large differences between T_0 and T_1 and between v_0 and v_1 still exist at $N_1 = 2 \times 10^6 \, {\rm cm}^{-1}$. Only when the carrier density increases to the order of $N_1 = 5 \times 10^6 \, {\rm cm}^{-1}$, the differences of drift velocities and electron temperatures of the two subbands become small.

6.10 Fluctuations and Diffusion Coefficients in Systems of Several Carrier Species with Particle Exchange

6.10.1 *Velocity and particle-number fluctuations*

To investigate issues related to fluctuations in systems of several carrier species with particle exchange, such as the noise or diffusion, one needs to start from the set of equations (6.93)–(6.95). Xing, Liu and Ting (1988) carry out rather careful investigations on them. We will briefly introduce their methods in this subsection.

Both the center-of-mass velocity $\boldsymbol{V}_\alpha(t)$ and particle number $N_\alpha(t)$ of the αth subsystem can be considered to consist of drifting or average parts $\boldsymbol{v}_\alpha(t)$ and $n_\alpha(t)$, and fluctuating or random parts $\delta\boldsymbol{V}_\alpha(t)$ and $\delta N_\alpha(t)$:

$$\boldsymbol{V}_\alpha(t) = \boldsymbol{v}_\alpha(t) + \delta\boldsymbol{V}_\alpha(t), \qquad (6.186)$$
$$N_\alpha(t) = n_\alpha(t) + \delta N_\alpha(t), \qquad (6.187)$$

$\delta\boldsymbol{V}_\alpha(t)$ and $\delta N_\alpha(t)$ being small quantities, and $\langle \delta V_\alpha(t) \rangle = 0$ and $\langle \delta N_\alpha(t) \rangle = 0$. But the electron temperature T_α is treated differently from $\boldsymbol{v}_\alpha$ or N_α. We do not consider its fluctuation for the time being. Putting (6.186) and (6.187) into Eqs. (6.93) and (6.94) [also into $A_{\alpha\beta}$ function (6.98)], expanding the various parts of the equations to the linear order in $\delta\boldsymbol{V}_\alpha$ and δN_α, and making use of balance equations (6.103) and (6.104) to eliminate the average parts, we obtain the linear set of Langevin equations for $\delta \boldsymbol{V}_\alpha(t)$ and $\delta N_\alpha(t)$ (Xing, Liu and Ting, 1988).

The equations for velocity fluctuations are ($\alpha = 1, 2, \cdots, s$)

$$\frac{d}{dt}\delta \boldsymbol{V}_\alpha(t) = \frac{1}{n_\alpha m_\alpha}\boldsymbol{F}_\alpha(t) + \sum_{\beta=1}^{s}\int_{-\infty}^{t} d\tau \mathcal{K}^{\alpha\beta}(t-\tau)\cdot \delta \boldsymbol{V}_\beta(\tau), \qquad (6.188)$$

in which

$$\mathcal{K}^{\alpha\beta}(t-s) = \frac{1}{n_\alpha m_\alpha}\int_{-\infty}^{s} dt' S_R^{\alpha\beta}(t-t'), \qquad (6.189)$$

$$S_R^{\alpha\beta}(t-t') = -i\,\theta(t-t')\langle[\boldsymbol{F}_\alpha(t),\boldsymbol{F}_\beta(t')]\rangle. \qquad (6.190)$$

In the following we assume that both the bias electric field $\boldsymbol{E}$ and the drift velocity $\boldsymbol{v}_\alpha$ are along the x direction, and the fluctuating force $\boldsymbol{F}_\alpha(t)$ is either parallel (longitudinal configuration) or perpendicular (transverse configuration) to the electric field. A general fluctuating force $\boldsymbol{F}_\alpha(t)$ in arbitrary direction can be decomposed into a parallel and a perpendicular component, from which the longitudinal and transverse velocity fluctuations can be calculated respectively. Since the iso-energy surface of each carrier species is spherical, the tensor $\mathcal{K}^{\alpha\beta}$ has only nonzero diagonal elements and thus there is no mixture between the transverse and longitudinal components. Therefore Eq. (6.188) can be written in a scalar form for parallel and perpendicular configurations ($\alpha = 1, \cdots, s$):

$$\frac{d}{dt}\delta V_\alpha(t) = \frac{1}{n_\alpha m_\alpha}F_\alpha(t) + \sum_{\beta=1}^{s}\int_{-\infty}^{t} d\tau\, \mathcal{K}_{\mu\mu}^{\alpha\beta}(t-\tau)\delta V_\beta(\tau), \qquad (6.191)$$

where $\mathcal{K}_{\mu\mu}^{\alpha\beta} = \mathcal{K}_{xx}^{\alpha\beta}$ for longitudinal configuration, and $\mathcal{K}_{\mu\mu}^{\alpha\beta} = \mathcal{K}_{yy}^{\alpha\beta}$ for transverse configuration. Fourier transforming Eq. (6.191) we get

$$\sum_{\beta=1}^{s}[\omega\delta_{\alpha\beta} + M_{\alpha\beta}(\omega)]\,\delta V_\beta(\omega) = i\,\frac{F_\alpha(\omega)}{n_\alpha m_\alpha}, \qquad (6.192)$$

in which

$$M_{\alpha\beta}(\omega) = \frac{S_R^{\alpha\beta}(0) - S_R^{\alpha\beta}(\omega)}{n_\alpha m_\alpha \omega}, \qquad (6.193)$$

and $S_R^{\alpha\beta}(\omega)$ is the Fourier transform of $S_R^{\alpha\beta}(t-t')$:

$$S_R^{\alpha\beta}(\omega) = \int_{-\infty}^{\infty} dt\, e^{i\omega t} S_R^{\alpha\beta}(t). \qquad (6.194)$$

The solution of Eq. (6.192) can be written as

$$\delta V_\alpha(\omega) = \sum_{\beta=1}^{s} X_{\alpha\beta}(\omega) F_\beta(\omega), \qquad (6.195)$$

with $X_{\alpha\beta}(\omega)$ expressed in terms of $M_{\alpha'\beta'}(\omega)$ (see Xing et al, 1988).

The equations for particle-number fluctuations are ($\alpha = 1, 2, \cdots, s$)

$$\frac{d}{dt}\delta N_\alpha(t) = B_\alpha(t) - \delta N_\alpha(t) \sum_{\beta(\neq\alpha)} \frac{1}{\tau_{\alpha\beta}} + \sum_{\beta(\neq\alpha)} \frac{\delta N_\beta(t)}{\tau_{\alpha\beta}}. \qquad (6.196)$$

$1/\tau_{\alpha\beta}$ relates to scatterings between α and β subsystems (Xing, Liu and Ting, 1988). Note that since the total particle number is fixed without fluctuation, only $s - 1$ independent variables in all δN_α ($\alpha = 1, 2, \cdots, s$). Equation set (6.196) is easily solved by Fourier transformation, yielding $\delta N_\alpha(\omega)$ in terms of $B_{\alpha'}(\omega)$ and $1/\tau_{\alpha'\beta}$ (Xing, Liu and Ting, 1988).

6.10.2 *Diffusion coefficients*

The displacement diffusion coefficient is defined as the average of the square of the particle displacement fluctuation in unit time (Reggiani, 1985). For a system consisting of N particles the displacement diffusion coefficient can be defined as (we consider only the displacement along one direction, thus using the scalar notation for simplicity)

$$D^{(d)}(t) = \frac{N}{2} \frac{d}{dt} \langle [\delta R(t)]^2 \rangle, \qquad (6.197)$$

in which $\delta R(t) = R(t) - \langle R(t) \rangle$ is the fluctuation of the displacement of the center of mass of N particles, which can be expressed in terms of the fluctuation of center-of-mass velocity $\delta V(t)$ as

$$\delta R(t) = \int_0^t d\tau\, \delta V(\tau) + \delta R(0). \qquad (6.198)$$

Substituting (6.198) into (6.197), we have

$$D^{(d)}(t) = \frac{N}{2} \int_0^t d\tau \langle \delta V(\tau)\delta V(0) + \delta V(0)\delta V(\tau) \rangle. \qquad (6.199)$$

It can be written as

$$D^{(d)}(t) = \int_{-\infty}^{\infty} \frac{d\omega}{\pi} \frac{\sin\omega t}{\omega} D_n(\omega), \qquad (6.200)$$

where $D_n(\omega)$ is the noise diffusion coefficient

$$D_n(\omega) = \frac{N}{4} \int_{-\infty}^{\infty} dt\, e^{i\omega t} \langle \delta V(t)\delta V(0) + \delta V(0)\delta V(t)\rangle. \tag{6.201}$$

Apparently,

$$\lim_{t\to\infty} D^{(d)}(t) = D_n(\omega)\big|_{\omega=0}. \tag{6.202}$$

For several-carrier-species system with particle exchange, the total center-of-mass velocity is (taking 3-carrier-species as an example)

$$V = \frac{1}{N}\left(N_1 V_1 + N_2 V_2 + N_3 V_3\right). \tag{6.203}$$

Therefore (noting that $\delta N_3 = -\delta N_1 - \delta N_2$)

$$\delta V = \frac{1}{N}\left[N_1 \delta V_1 + N_2 \delta V_2 + N_3 \delta V_3 + (v_1 - v_3)\delta N_1 + (v_2 - v_3)\delta N_2\right]. \tag{6.204}$$

Putting (6.204) into (6.201), $D_n(\omega)$ can be expressed as

$$D_n(\omega) = D_{\text{th}}(\omega) + D_{\text{tr}}(\omega) + D_{\text{cr}}(\omega), \tag{6.205}$$

$$D_{\text{th}}(\omega) = \frac{1}{N}\left[N_1 D_1(\omega) + N_2 D_2(\omega) + N_3 D_3(\omega)\right], \tag{6.206}$$

$$D_\alpha(\omega) = \frac{N_\alpha}{2}\Gamma_{vv}^{\alpha\alpha}(\omega) \quad (\alpha = 1,2,3), \tag{6.207}$$

$$D_{\text{tr}}(\omega) = -\frac{N}{2}\left[(v_1 - v_2)^2 \Gamma_{nn}^{12}(\omega) + (v_1 - v_3)^2 \Gamma_{nn}^{13}(\omega) + (v_2 - v_3)^2 \Gamma_{nn}^{23}(\omega)\right], \tag{6.208}$$

$$D_{\text{cr}}(\omega) = \frac{1}{N}\left[\sum_{\alpha=1,\beta>\alpha}^{3} N_\alpha N_\beta \Gamma_{vv}^{\alpha\beta}(\omega) + (v_1 - v_3)\sum_{\alpha=1}^{3} N_\alpha \Gamma_{vn}^{\alpha 1}(\omega) \right.$$

$$\left. + (v_2 - v_3)\sum_{\alpha=1}^{3} N_\alpha \Gamma_{vn}^{\alpha 2}(\omega)\right], \tag{6.209}$$

$$\Gamma_{vv}^{\alpha\beta}(\omega) = \int_{-\infty}^{\infty} dt\, e^{i\omega t}\langle \delta V_\alpha(t)\delta V_\beta(0) + \delta V_\beta(0)\delta V_\alpha(t)\rangle, \tag{6.210}$$

$$\Gamma_{nn}^{\alpha\beta}(\omega) = \int_{-\infty}^{\infty} dt\, e^{i\omega t}\langle \delta N_\alpha(t)\delta N_\beta(0) + \delta N_\beta(0)\delta N_\alpha(t)\rangle, \tag{6.211}$$

$$\Gamma_{vn}^{\alpha\beta}(\omega) = \int_{-\infty}^{\infty} dt\, e^{i\omega t}\langle \delta V_\alpha(t)\delta N_\beta(0) + \delta N_\beta(0)\delta V_\alpha(t)\rangle. \tag{6.212}$$

According to above definitions, the thermal diffusion coefficient D_{th} equals the average of all thermal diffusion coefficients of separated subsystems weighted by their particle numbers and are related to the correlation functions of velocity-fluctuations of individual subsystems. D_{tr} is called the carrier-transfer diffusion coefficient, which is related to the particle exchange between different subsystems and determined by the correlation functions of particle-number-fluctuations between subsystems. D_{cr} is determined by the velocity-fluctuation correlation functions and velocity-fluctuation–particle-number-fluctuation correlation functions between different subsystems, and is called the cross diffusion coefficient or the off-diagonal diffusion coefficient (Xing, Liu and Ting, 1988).

The correlation functions (6.210)–(6.212) can be calculated based on $\delta V_\alpha(\omega)$ and $\delta N_\alpha(\omega)$ solved in Sec. 6.10.1, and the diffusion coefficient $D_{\text{n}}(\omega)$ and noise spectrum (see Sec. 4.2) are obtained. In the case of hot electron transport, in view of the anisotropy resulting from the finite drift velocity $\boldsymbol{v}_{\text{d}}$, the correlation functions and diffusion coefficients for velocity fluctuation parallel to $\boldsymbol{v}_{\text{d}}$ are different from those for velocity fluctuation perpendicular to $\boldsymbol{v}_{\text{d}}$. The former is called the longitudinal diffusion coefficient and the latter is called the transverse diffusion coefficient

The room temperature diffusion coefficients of an n-type GaAs are calculated by Xing, Liu and Ting (1988) using Γ–L two-carrier-species model and by Liu, Xing and Ting (1989) using Γ–L–X three-carrier-species model, with parameters described in Sec. 6.8. Fig. 6.19 shows the longitudinal diffusion coefficients obtained by them.

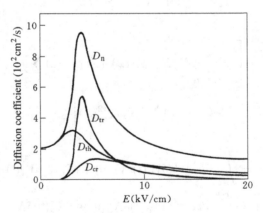

Fig. 6.19 Zero-frequency ($\omega = 0$) longitudinal diffusion coefficients D_{th}, D_{tr}, D_{cr} and D_{n} as functions of the electric field, obtained by using Γ–L–X three-species-of-carriers model at lattice temperature $T = 300\,\text{K}$. From Liu, Xing and Ting (1989).

Chapter 7

Balance Equation Transport Theory and Electron Correlation

We analyzed the main physical consideration in establishing the balance equation transport model in Chapter 1, emphasizing the significance of choosing the initial state for solving the Liouville equation of the density matrix as one in which the electrons exhibit a thermal equilibrium distribution with respect to their center of mass. This choice enables us to establish a succinct high-field transport theory as described in the preceding chapters for systems with strong intercarrier couplings. We will explore this further jointly with other issues related to electron correlation in the balance equation theory.

7.1 Full Quantum-Mechanical Treatment in the Laboratory Reference Frame

In Chapter 1 we started our discussion with writing Hamiltonian (1.1) in terms of center-of-mass variables and relative electron variables defined by Eqs. (1.2) and (1.3). They satisfy the commutation relations (1.4), (1.5) and (1.7) and the constraint (1.6). We have made two assumptions: (a) the center of mass can be treated as a classical particle; (b) for large N system the $1/N$ terms in the exact noncanonical commutation relation (1.7) can be neglected, such that relative electron momentum $\bm{p}'_j$ and coordinate $\bm{r}'_j$ are regarded as canonically conjugate variables. Although these assumptions are valid in obtaining correct results to the lowest orders in the impurity and phonon scatterings in the presence of strong electron–electron (e–e) interactions, they should not be used indiscriminately. For instance, by summing Eq. (1.7) over j, both the left hand side (LHS) and the right hand side (RHS) vanish. This consistency, LHS=RHS, would be seriously violated if one were to neglect the $1/N$ terms in Eq. (1.7). In this context of summing over j, the neglect of the $1/N$ terms incurs an error of order $N^0 = 1$ rather than $1/N$. This fact shows that the relative electron momenta, which satisfy constraint (1.6), do not provide the requisite translation generating qualities and that using them as translation generators may lead to unphysical results (Argyres, 1989).

An alternative discussion has been given by Lei (1990a) using the complete Hamiltonian (1.1), with a full quantum mechanical treatment. We rewrite the Hamiltonian here for convenience:

$$H = H_e + H_E + H_I + H_{\text{ph}}, \tag{7.1}$$

$$H_e = \sum_i \frac{p_i^2}{2m} + H_{ee} = \frac{P^2}{2Nm} + H_{er}, \tag{7.2}$$

$$H_{er} = \sum_i \frac{p_i'^2}{2m} + H_{ee}, \tag{7.3}$$

$$H_{ee} = \frac{e^2}{8\pi\epsilon_0} \sum_{i \neq j} \frac{1}{|r_i - r_j|} = \frac{e^2}{8\pi\epsilon_0} \sum_{i \neq j} \frac{1}{|r_i' - r_j'|}, \tag{7.4}$$

$$H_E = -Ne\boldsymbol{E} \cdot \boldsymbol{R}, \tag{7.5}$$

$$H_I = H_{ei} + H_{ep}, \tag{7.6}$$

$$H_{ei} = \sum_{q,a} u(q) e^{-i q \cdot r_a} \rho_q = \sum_{q,a} u(q) e^{i q \cdot (R - r_a)} \bar{\rho}_q, \tag{7.7}$$

$$H_{ep} = \sum_{q,\lambda} M(q,\lambda) \phi_{q\lambda} \rho_q = \sum_{q,\lambda} M(q,\lambda) e^{i q \cdot R} \phi_{q\lambda} \bar{\rho}_q, \tag{7.8}$$

$$H_{\text{ph}} = \sum_{q,\lambda} \Omega_{q\lambda} b_{q\lambda}^\dagger b_{q\lambda}. \tag{7.9}$$

In this, $\boldsymbol{p}_i$, $\boldsymbol{r}_i$, $\boldsymbol{p}_i'$, $\boldsymbol{r}_i'$, $\boldsymbol{P}$, $\boldsymbol{R}$ and other notations have the same meaning as given in Sec. 1.1, except that

$$\rho_q = \sum_j e^{i q \cdot r_j} \tag{7.10}$$

now denotes the electron density operator in the laboratory reference frame, and

$$\bar{\rho}_q = \sum_j e^{i q \cdot r_j'} = e^{-i q \cdot R} \rho_q \tag{7.11}$$

represents the density operator for relative electrons.

According to quantum statistical mechanics the density matrix of the system, $\hat{\rho}$, obeys the Liouville equation

$$i\frac{d\hat{\rho}}{dt} = [H, \hat{\rho}], \tag{7.12}$$

with the complete Hamiltonian H.

To proceed we assume that both H_I and H_E are turned on at time $t = 0$. Thus for $t < 0$, the Hamiltonian of the system is H_e with electron–electron (e–e) interactions, plus $H_{\rm ph}$ of the independent phonons. As pointed out in Chapter 1 (Lei, 1990b), the balance equation approach is characterized by an initial state selected to be one in which electrons exhibit a thermal equilibrium distribution at the electron temperature with respect to their center of mass, which is moving at a velocity equal to the average drift velocity in the presence of an electric field (i.e. drifted thermal equilibrium state), plus a thermal equilibrium phonon system. The density matrix of this initial state, given in Eq. (1.30), can be written as

$$\hat{\rho}_0 = \frac{1}{Z} e^{-H_{\rm er}/T_e} e^{-H_{\rm ph}/T} |\boldsymbol{P}_{\rm d}\rangle\langle\boldsymbol{P}_{\rm d}|, \tag{7.13}$$

explicitly showing that the center of mass, as a quantum particle, is in a momentum eigenstate $|\boldsymbol{P}_{\rm d}\rangle$ with eigenvalue $Nm\boldsymbol{v}_{\rm d}$. Here $\boldsymbol{v}_{\rm d}$ is the drift velocity of the final state, which, like the electron temperature T_e, is also a parameter to be determined by the resulting balance equations. Since the final steady state bears the same average velocity $\boldsymbol{v}_{\rm d}$ as the initial state (7.13), the evolution from the initial state to the final state is solely a thermalization process without the need for momentum relaxation. The time required for this evolution, $\tau_{\rm th}$, is the thermalization time of the relative electron system, and is closely related to intercarrier interactions $H_{\rm ee}$. The Liouville equation for the density matrix together with the initial condition $\hat{\rho}(t=0) = \hat{\rho}_0$, can be written in a form of integral equation:

$$\hat{\rho}(t) = \hat{\rho}_0 - i \int_0^t dt_1 [H_I(t_1) + H_E(t_1), \hat{\rho}(t_1)]. \tag{7.14}$$

The statistical expectation of any dynamical variable $\mathcal{O}$ should be calculated according to

$$\langle \mathcal{O} \rangle = {\rm tr}\{\hat{\rho}(t)\mathcal{O}(t)\}. \tag{7.15}$$

Here, the interaction-picture operator is defined as

$$\mathcal{O}(t) = e^{i(H_e + H_{\rm ph})t} \mathcal{O} e^{-i(H_e + H_{\rm ph})t}. \tag{7.16}$$

Our purpose is to calculate statistical expectation values $\boldsymbol{f} = \mathrm{tr}\{\hat{\rho}(t)\boldsymbol{F}(t)\}$ and $w = \mathrm{tr}\{\hat{\rho}(t)W(t)\}$, of the force operator

$$\boldsymbol{F} = -\mathrm{i}\sum_{\boldsymbol{q},a} u(\boldsymbol{q})\,\boldsymbol{q}\,\mathrm{e}^{-\mathrm{i}\boldsymbol{q}\cdot\boldsymbol{r}_a}\rho_{\boldsymbol{q}} - \mathrm{i}\sum_{\boldsymbol{q},\lambda} M(\boldsymbol{q},\lambda)\,\boldsymbol{q}\,\phi_{\boldsymbol{q}\lambda}\rho_{\boldsymbol{q}} \qquad (7.17)$$

and the energy transfer rate operator

$$W = -\sum_{\boldsymbol{q},\lambda} M(\boldsymbol{q},\lambda)\,\dot{\phi}_{\boldsymbol{q}\lambda}\rho_{\boldsymbol{q}} \qquad (7.18)$$

to derive the balance equations for the steady state:

$$Ne\boldsymbol{E} + \boldsymbol{f} = 0, \qquad (7.19)$$

$$\boldsymbol{v}_\mathrm{d} \cdot \boldsymbol{f} + w = 0. \qquad (7.20)$$

Of course this requires a knowledge of the density matrix.

7.2 Perturbation Expansion and Steady-State Balance Equations

Unfortunately, we cannot solve Eq. (7.12) or Eq. (7.14) to obtain $\hat{\rho}(t)$ exactly, and thus take recourse to a perturbation expansion as usual. A formal iteration of Eq. (7.14) yields

$$\hat{\rho}(t) = \hat{\rho}_0 + \sum_{n=1} (-\mathrm{i})^n$$
$$\times \int_0^t dt_1 \int_0^{t_1} dt_2 \cdots \int_0^{t_{n-1}} dt_n [H_\mathrm{IE}(t_1), [H_\mathrm{IE}(t_2), \cdots [H_\mathrm{IE}(t_n), \hat{\rho}_0]\cdots]], \qquad (7.21)$$

where $H_\mathrm{IE} \equiv H_\mathrm{I} + H_\mathrm{E}$. With this $\hat{\rho}(t)$, we can write the statistical average of any operator as an expansion in powers of both the electric field H_E and the scattering potential H_I jointly. For instance, the frictional force is given by (Lei, 1990b)

$$\boldsymbol{f}(t) = \sum_{n=1}^{\infty} \mathrm{i}^n \int_0^t dt_1 \int_0^{t_1} dt_2 \cdots \int_0^{t_{n-1}} dt_n$$
$$\times \mathrm{tr}\{\hat{\rho}_0[H_\mathrm{IE}(t_{n-1}), \cdots [H_\mathrm{IE}(t_1), \boldsymbol{F}(t)]\cdots]\}. \qquad (7.22)$$

We can regroup the terms on the right-hand side of Eq.(7.22) according to the respective powers of $H_{\rm E}$ and $H_{\rm I}$:

$$\boldsymbol{f}(t) = \sum_{n=0}^{\infty} \sum_{m=1}^{\infty} \boldsymbol{f}_n^{(2m)}(t). \tag{7.23}$$

Here, $\boldsymbol{f}_n^{(2m)}(t)$ stands for all the terms with n factors of $H_{\rm E}$ and $(2m-1)$ factors of $H_{\rm I}$ in the perturbative expansion of the force.

A serious problem associated with this double expansion is that except for a few lowest order terms, the higher order terms are divergent when $t \to \infty$ for nonzero $H_{\rm I}$ and $H_{\rm E}$, no matter how small they are (Lei and Horing, 1989; Lei, 1990b). Thus the expansion series is divergent at large t. This divergence, however, does not invalidate the balance equation approach. Since the system, starting from the initial state $\hat{\rho}_0$, certainly reaches a unique steady state after the thermalization time $\tau_{\rm th}$. The divergence of the perturbation series is only an indication of the uselessness of the perturbation series itself for describing the evolution of the system at large t, and is irrelevant to the behavior of transport. Such perturbative series may still be useful within the finite time ranges in which they converge.

We can easily determine the expansion parameters for these perturbation series. In the frictional force series (7.23), the expansion parameter with respect to $H_{\rm I}$ is seen to be $t/\tau_{\rm m}$, and that with respect to $H_{\rm E}$ is $t/t_{\rm E}$. Here $\tau_{\rm m}$ is estimated to be of order of the average momentum relaxation time due to impurity and phonon scatterings, and $t_{\rm E}$ is a characteristic time associated with the electric field: $t_{\rm E} \sim (m/e\bar{q}E)^{1/2}$, with $\bar{q}$ being an average wavevector contributing in the momentum relaxation integration. Thus, up to time $t < t_{\rm E}$, the perturbation expansion for the frictional force $\boldsymbol{f}(t)$, Eq. (7.23), is convergent and thus useful in describing the system transport. Since at time $t \sim \tau_{\rm th}$ the system has essentially reached the steady state, the perturbative series Eq. (7.23) for the frictional force $\boldsymbol{f}(t)$ can be used for the determination of steady state properties of transport, if the conditions

$$\tau_{\rm th} \ll \tau_{\rm m} \tag{7.24}$$

and

$$\tau_{\rm th} \ll t_{\rm E} \tag{7.25}$$

are satisfied. For $t < \tau_{\text{th}}$, the leading term in the force series is

$$\begin{aligned}
\boldsymbol{f}_0^{(2)}(t) = &- n_{\text{i}} \sum_{\boldsymbol{q}} |u(\boldsymbol{q})|^2 \boldsymbol{q} \int_0^t dt_1 \text{tr}\{\hat{\rho}_0[\rho_{-\boldsymbol{q}}(t_1), \rho_{\boldsymbol{q}}(t)]\} \\
&- \sum_{\boldsymbol{q},\lambda} |M(\boldsymbol{q},\lambda)|^2 \boldsymbol{q} \int_0^t dt_1 \text{tr}\{\hat{\rho}_0[\phi_{-\boldsymbol{q}}(t_1)\rho_{-\boldsymbol{q}}(t_1), \phi_{\boldsymbol{q}}(t)\rho_{\boldsymbol{q}}(t)]\}. \quad (7.26)
\end{aligned}$$

Direct evaluations from the definition of $\rho_{\boldsymbol{q}}$ show that

$$\rho_{-\boldsymbol{q}}(t_1)\rho_{\boldsymbol{q}}(t) = \exp[\mathrm{i}\boldsymbol{q}\cdot\boldsymbol{P}(t-t_1)/Nm]\bar{\rho}_{-\boldsymbol{q}}(t_1)\bar{\rho}_{\boldsymbol{q}}(t), \quad (7.27)$$

$$[\rho_{-\boldsymbol{q}}(t_1), \rho_{\boldsymbol{q}}(t)] = \exp[\mathrm{i}\boldsymbol{q}\cdot\boldsymbol{P}(t-t_1)/Nm][\bar{\rho}_{-\boldsymbol{q}}(t_1), \bar{\rho}_{\boldsymbol{q}}(t)]. \quad (7.28)$$

Here $\bar{\rho}_{\boldsymbol{q}}(t) \equiv \exp(\mathrm{i} H_{\text{er}} t)\bar{\rho}_{\boldsymbol{q}}\exp(-\mathrm{i} H_{\text{er}} t)$ is the density operator of relative electrons (7.11) in the interaction picture, and the commutation relation $[\bar{\rho}_{-\boldsymbol{q}}(t_1), \bar{\rho}_{\boldsymbol{q}}(t)]$ should be evaluated as if $\boldsymbol{r}'_i, \boldsymbol{p}'_j$ satisfy an exactly canonical commutation relation $[r'_{i\alpha}, p'_{j\beta}] = \mathrm{i}\delta_{ij}\delta_{\alpha\beta}$. Therefore we can write $\boldsymbol{f}_0^{(2)}(t)$ as

$$\begin{aligned}
\boldsymbol{f}_0^{(2)}(t) = &\; n_{\text{i}} \sum_{\boldsymbol{q}} |u(\boldsymbol{q})|^2 \boldsymbol{q} \int_0^t ds \exp(\mathrm{i}\omega_0 s)\langle[\bar{\rho}_{-\boldsymbol{q}}, \bar{\rho}_{\boldsymbol{q}}(s)]\rangle \\
&+ \sum_{\boldsymbol{q},\lambda} |M(\boldsymbol{q},\lambda)|^2 \boldsymbol{q} \int_0^t ds \exp(\mathrm{i}\omega_0 s) \cos(\Omega_{\boldsymbol{q}\lambda} s)\langle[\bar{\rho}_{-\boldsymbol{q}}, \bar{\rho}_{\boldsymbol{q}}(s)]\rangle 2n\left(\frac{\Omega_{\boldsymbol{q}\lambda}}{T}\right) \\
&+ \sum_{\boldsymbol{q},\lambda} |M(\boldsymbol{q},\lambda)|^2 \boldsymbol{q} \int_{-t}^t ds \exp[\mathrm{i}(\omega_0 - \Omega_{\boldsymbol{q}\lambda})s]\langle\bar{\rho}_{-\boldsymbol{q}}\bar{\rho}_{\boldsymbol{q}}(s)\rangle, \quad (7.29)
\end{aligned}$$

in which $\omega_0 \equiv \boldsymbol{q}\cdot\boldsymbol{v}_{\text{d}}$, $n(x) \equiv (\mathrm{e}^x - 1)^{-1}$, and the correlation functions of the density operators are calculated with respect to a canonical ensemble of Hamiltonian H_{er} at the electron temperature T_{e}. The density correlation function of the system, $\langle\rho_{\boldsymbol{q}}(t)\rho_{-\boldsymbol{q}}\rangle$, generally differs appreciably from zero only in a limited time range $-\tau_{\text{cr}} < t < \tau_{\text{cr}}$, where τ_{cr} is the correlation time. When time t is large than the correlation time τ_{cr}, $\boldsymbol{f}_0^{(2)}$ approaches a finite limit value

$$\boldsymbol{f}_0^{(2)} = n_{\text{i}} \sum_{\boldsymbol{q}} |u(\boldsymbol{q})|^2 \boldsymbol{q} \Pi_2(\boldsymbol{q}, \omega_0) + $$

$$\sum_{\boldsymbol{q},\lambda} |M(\boldsymbol{q},\lambda)|^2 \boldsymbol{q} \Pi_2(\boldsymbol{q}, \Omega_{\boldsymbol{q}\lambda} + \omega_0) \left[n\left(\frac{\Omega_{\boldsymbol{q}\lambda}}{T}\right) - n\left(\frac{\Omega_{\boldsymbol{q}\lambda} + \omega_0}{T_{\text{e}}}\right)\right].$$

$$(7.30)$$

Therefore, the force balance equation, valid at time t of the order of, or larger than, τ_{th} and τ_{cr}, takes the form

$$Ne\boldsymbol{E} + \boldsymbol{f}_0^{(2)} = 0. \tag{7.31}$$

This is exactly the force balance equation obtained in Sec. 2.1 for the steady state.

The energy transfer rate $w(t)$ can also be expanded in powers of H_{I} and H_{E}:

$$w(t) = \sum_{n=0}^{\infty}\sum_{m=0}^{\infty} w_n^{(2m)}(t). \tag{7.32}$$

The lowest-order nonzero term $w_0^{(2)}(t)$ can be calculated in a similar fashion with the steady-state value

$$w_0^{(2)} = 2\sum_{\boldsymbol{q},\lambda}|M(\boldsymbol{q},\lambda)|^2 \Omega_{\boldsymbol{q}\lambda}\Pi_2(\boldsymbol{q},\Omega_{\boldsymbol{q}\lambda}+\omega_0)\left[n\left(\frac{\Omega_{\boldsymbol{q}\lambda}}{T}\right) - n\left(\frac{\Omega_{\boldsymbol{q}\lambda}+\omega_0}{T_{\text{e}}}\right)\right]. \tag{7.33}$$

The energy balance equation, which may be valid when $t \geqslant \tau_{\text{th}}$ and τ_{cr}, is

$$\boldsymbol{v}_{\text{d}} \cdot \boldsymbol{f}_0^{(2)} + w_0^{(2)} = 0. \tag{7.34}$$

The correlation time τ_{cr} can be estimated roughly from the frequency range $\triangle\omega$ in which $\Pi_2(\boldsymbol{q},\omega)$ differs appreciably from zero:

$$\tau_{\text{cr}} \approx 1/\triangle\omega. \tag{7.35}$$

As a conservative estimate we use the zero-temperature expression Eq. (2.11) for the density correlation function in the absence of electron–electron interactions, yielding $1/\tau_{\text{cr}} \sim [q(2k_{\text{F}}+q)/m]$, where k_{F} is the Fermi wavevector. In the degenerate case the dominant contribution comes from wavevector transfer q of order of $2k_{\text{F}}$, such that $\tau_{\text{cr}} \sim (16\varepsilon_{\text{F}})^{-1}$. For a typical degenerate semiconductor of $\varepsilon_{\text{F}} \sim 200\,\text{K}$, $\tau_{\text{cr}} \sim 10^{-14}\,\text{sec}$. In the presence of electron–electron interactions or at finite temperatures, we anticipate an even larger $\triangle\omega$ and thus a shorter τ_{cr}.

The condition (7.25) imposes a restriction on the strength of the applied electric field $\boldsymbol{E}$. Fortunately, this restriction can be released on the proviso that the electric field is not treated as a perturbation, but is included in the unperturbed Hamiltonian (Lei and Horing, 1989). Then, effects due to the acceleration by the electric field during the correlation time τ_{cr} interval (intracollisional field effect) will show up. We will discuss this matter in Chapter 9.

The thermalization among relative electrons due to effective e–e couplings can be represented mathematically by a damping factor in the single relative-electron retarded Green's function

$$G_r(\bm{k}, t) \sim e^{-i\varepsilon_k t - |t|/\tau_e}, \qquad (7.36)$$

where the damping time τ_e, or the imaginary part of the electron self-energy or the lifetime of the quasi particle, is related to the effective electron–electron scattering and is relevant to electron thermalization in the center-of-mass reference frame. With this, and the application of Feynman diagrammatic technique jointly with closed-time-path Green's functions, Chen, Ting and Horing (1989, 1990) show that the corresponding perturbation series is devoid of long-time divergences and that the higher-order terms are negligible in the case $\tau_e \ll \tau_m$, thus establishing the validity of the balance equation approach.

The thermalization time τ_{th} is a collective feature of all kinds of effective carrier–carrier interactions: Coulombic, phonon-mediated, etc. It depends on the carrier density and temperature, and varies over a wide range for various systems. Wingreen, Stanton and Wilkins (1986) phenomenologically introduce in the Boltzmann equation a relaxation time τ_{ee} to represent electron–electron collisions, the effect of which is to drive the distribution toward a displaced Maxwellian. This τ_{ee} somewhat resembles our τ_{th}. For an n-GaAs system with the electron density of the order of $10^{17}\,\text{cm}^{-3}$, they estimate an effective contribution of about $4 \times 10^{-13}\,\text{sec}$ from Coulomb scattering to τ_{ee} at electron temperature $T_e = 100\,\text{K}$. Ensemble Monte-Carlo simulation of the relaxation kinetics reckons a 50 fs thermalization time from Coulomb scattering for a 2D electron gas of density $0.96 \times 10^{12}\,\text{cm}^{-2}$ with a final electron temperature 236 K (Haug and Jahho, 1998). Experimentally, thermalization of nonequilibrium electrons by electron–electron interactions in n-type GaAs quantum wells is observed to occur in less than 10 fs (Knox *et al*, 1988). A short thermalization time, or a strong effective electron–electron interaction which quickly strives towards thermal equilibrium of carriers in the center-of-mass reference frame, plays an important role in the balance-equation transport theory.

On the other hand, there are quite a few investigations in the literature to explore what happens in the balance-equation theory if the e–e interaction is weak such that the thermalization time τ_{th} is of the same order of magnitude as τ_m. In this case one has to deal with time scales of order of or greater than the momentum relaxation time, $t > \tau_m$, and the perturbation series is divergent. Thus far the problem has been approached by summing the divergent series. Unfortunately, the sum of a divergent series depends on how to truncate it and how to regroup its terms. If one retains in the frictional force expansion (7.23) only terms of zero and linear orders in E and formally performs the divergent partial sums $\bm{f}_0 = \sum_n \bm{f}_0^{(2n)}$ and $\bm{f}_1 = \sum_n \bm{f}_1^{(2n)}$, the resulting force balance equation for $t \to \infty$ yields a resistivity of the form (7.41) (Lei, 1990b). However, terms of higher order in E are even more strongly divergent in the limit of $t \to \infty$. Therefore one may question the validity of using the balance equation without $\bm{f}_n (n \geqslant 2)$ and taking $t \to \infty$ limit to obtain the resistivity (7.41). On the other hand, if one first performs the partial summations over all orders of E but with fixed power of H_I, $\bm{f}^{2m} = \sum_n \bm{f}_n^{(2m)}$, we find that all $\bm{f}^{2m}$ terms approach zero when $t \to \infty$, such that one might expect a zero resistivity.

The situation becomes even more complicated if one performs the calculation in terms of center-of-mass and relative electron variables. The noncommutability of $\bm{P}$ and $\bm{R}$ as quantum mechanical variables and the noncanonical $1/N$ terms in the commutation relation (1.7), produce large numbers of new terms, which are also divergent when $t \to \infty$. Although each of these terms may be of order of $1/N$, the number of these terms is of order of N, such that we have to do a careful counting of them to avoid missing important contributions. In the literature, Fishman and Mahan (1989) and Fishman (1989) obtain Eq. (7.41) by treating relative electron variables as exactly canonical variables, but taking only partial account of the effects of the quantum mechanical aspect of the center of mass. Chen, Ting and Horing (1990) point out that if the full quantum mechanical noncommutability of $\bm{P}$ and $\bm{R}$ is taken into account, contributions from the additional infinite set of divergent terms fully cancel those picked up by Fishman and Mahan (1989). Terms arising solely from the noncanonical relative-electron commutation relation are also discussed by them and shown to produce resistivity of form (7.41). However, these complicated consideration may be spurious. Lei (1990a) points out that the contribution from the noncanonical $1/N$ terms in the commutation relation (1.7) of the relative electron variables is exactly cancelled by that arising from the noncommutability of the center-of-mass coordinates $\bm{R}(t)$ at different times. Physically this should be the case. The separation of the center-of-mass variables from the relative electron variables is only a mathematical convenience, introducing no new physics. Returning to the original electron variables is of course equivalent to taking both effects [noncommutability of P and R and noncanonical $1/N$ terms in (1.7)] into full consideration, just as we have done in this section: no trace of the divergent terms of both effects emerges, but the long-time divergence of the perturbation series (7.24) remains.

Summing the divergent series to obtain a finite resistivity in the absence of e–e interaction remains an open problem in the force balance theory, as well as in the Kubo linear response theory (Lei and Cai, 1991).

7.3 Isothermal and Adiabatic Resistivities

Theoretically, there have been two different types of approaches to transport. The one is force-balance theory, which expresses the electric field as a function of the current density, i.e. to calculate the electric field at a given current density. The other is dynamic response theory, which calculates the current density as the response to an applied electric field. In principle, these approaches are equivalent (Su and Sakita, 1986). The realization in practical calculations, however, usually relies on perturbation methods, such that the different choices of the initial state, which determines the time required to reach the steady state, make these two approaches dramatically different (Lei, 1990b).

The balance-equation transport theory belongs to the first type of approaches. As discussed above, this theory, which starts from the initial state given by the density matrix (7.13) and features force- and energy-balance equations (7.31) and (7.34) taken to the leading order in the scattering interaction H_I, describes isothermal transport in that it requires a rapid thermalization of the relative electron system due to strong electron–electron interactions. In the balance-equation theory the impurity-induced Ohmic (linear) resistivity as given in Eq. (2.28), and the corresponding conductivity σ, can be written in the form

$$\rho = \frac{m}{ne^2}\left\langle \frac{1}{\tau_\mathrm{i}(\boldsymbol{k})} \right\rangle, \quad \sigma = \frac{ne^2}{m}\left\langle \frac{1}{\tau_\mathrm{i}(\boldsymbol{k})} \right\rangle^{-1}, \quad (7.37)$$

in which

$$\frac{1}{\tau_\mathrm{i}(\boldsymbol{k})} = \frac{n_\mathrm{i} m}{4\pi k^3} \int_0^{2k} dq\, |\bar{u}(q)|^2 q^3 \quad (7.38)$$

is a wavevector-dependent inverse relaxation time for impurity scattering, and the bracket $\langle \ \rangle$ represents an average defined by

$$\langle A(\boldsymbol{k}) \rangle \equiv \frac{4}{3n} \sum_{\boldsymbol{k}} \left(-\frac{\partial f(\varepsilon_{\boldsymbol{k}}, T)}{\partial \varepsilon_{\boldsymbol{k}}} \right) \varepsilon_{\boldsymbol{k}} A(\boldsymbol{k}). \tag{7.39}$$

The impurity scattering here has a generalized meaning, i.e. the elastic scattering, including impurity scattering and elastically treated acoustic phonon scattering. $\tau_i(\boldsymbol{k})$ implies the relaxation time from all kinds of elastic scatterings, not necessary given by (7.38). The resistivity ρ or conductivity σ related to the average of the inverse relaxation time as expressed by (7.37), which is called the resistivity formula or the isothermal resistivity, is expected valid for systems with strong e–e interactions.

The situation is quite different (Lei, 1990b) in the case of Kubo linear-response theory (Kubo, 1959), which is a typical approach of the second type. The initial state of the Kubo linear response theory is a thermal equilibrium state in the fixed reference frame without drift motion:

$$\hat{\rho}_0 = \frac{1}{Z} e^{-H_e/T} e^{-H_{ph}/T}. \tag{7.40}$$

The total momentum of the initial state is zero. In order for the system to evolve from this initial state to the final transport state having drift motion, a momentum relaxation mechanism is necessary. The electron–electron interaction, which induces no momentum relaxation, has no primary importance in this process. The time required for the system to go from this initial state to the final drifting state is mainly determined by τ_m, the momentum relaxation time, and is essentially not affected by the inter-carrier interaction (save for shielding). When the strength of the impurity (and phonon) scattering gets weaker, this time becomes longer. This is the physical reason why one has to sum an infinite series of divergent terms in order to obtain the conductivity even in the weak scattering limit in Kubo linear response theory. The conductivity (resistivity), thus derived in the absence of e–e interaction, has the form

$$\sigma = \frac{ne^2}{m} \langle \tau_i(\boldsymbol{k}) \rangle, \quad \rho = \frac{m}{ne^2} \frac{1}{\langle \tau_i(\boldsymbol{k}) \rangle}. \tag{7.41}$$

This formula, known as the conductivity formula or the adiabatic resistivity, can also be obtained from Boltzmann equation without e–e scattering.

There are many other resistivity theories in the literature. In all these theories the expressions for the linear impurity resistivity in the weak scattering limit are one of the above two: either the resistivity formula (7.37) or the conductivity formula (7.41). The results from these two formulas, though identical at zero temperature, are different at finite temperature as long as $\tau(\boldsymbol{k})$ is not a constant [e.g. $\tau(\boldsymbol{k})$ changing with $\varepsilon_{\boldsymbol{k}}$]. The deviation increases with increasing temperature and can be as large as a factor of 3 when temperature is higher than the energy scale of $\varepsilon_{\boldsymbol{k}}$, within which $\tau(\boldsymbol{k})$ varies appreciably.

The conflict of the resistivity and conductivity formulas (7.37) and (7.41), so-called $\langle 1/\tau \rangle$–vs–$1/\langle \tau \rangle$ dispute, has been in the literature for many years. Edwards (1965), Kendre and Dresdon (1972), and Rousseau, Stoddart and March (1973) proposed to write the linear resistivity in the form of force correlation function and expanded it in powers of scattering strength:

$$\rho = \rho_0 + \rho_1 + \rho_2 + \cdots. \tag{7.42}$$

In the case of weak impurity scattering, $\langle 1/\tau \rangle$ formula (7.37) results if only the lowest order term $\rho = \rho_0$ is retained. This theory was later criticized by Argyres and Sigel (1973) and by Huberman and Chester (1975), who pointed out that all other terms $\rho_1, \rho_2, \cdots$ in (7.42) are divergent and can not be neglected no matter how weak the scattering potential is. Formally summing up all the divergent terms in (7.42), they found that the result turned out to be in agreement with (7.41) of the Kubo linear response theory. Thus, the reason for the force correlation function theory to get formula (7.37) was referred to the incorrect disregard of the higher order divergent terms. From then on the conductivity formula (7.41) was thought by most people to be the only correct formula for the linear resistivity and the force correlation function theory and its $\langle 1/\tau \rangle$ resistivity formula were almost negated in quite a period of time.

Looking back at this dispute carefully, we notice that both the proposer and criticizer of the force correlation function theory investigated an ideal electron system having impurity scattering but without electron–electron interaction. In such a system the electron relaxation is entirely due to impurity scattering. When the impurity scattering becomes very weak, the time required for this system to relax will be very long. This is the physical reason for the divergence of $\rho_1, \rho_2, \cdots$ terms in the expansion (7.42). However, when there exists direct or indirect electron–electron interaction in addition to the impurity scattering, the higher-order terms of the resistivity perturbation expansion in powers of impurity scattering strength, will not be divergent and becomes negligible in the case of weak impurity scattering and strong electron–electron scattering (Lei, 1990b; Chen, Ting and

Horing, 1990). The previous criticism against the resistivity formula (7.37) will no longer be valid. In fact, besides the balance-equation theory and the force correlation function theory mentioned above, the variational solution of Boltzmann equation (Ziman, 1960), the energy-loss method (Gerlach, 1986), the generalized Drude method (Sernelius, 1989), etc, all go to formula (7.37) for weak impurity scattering induced linear resistivity.

Most people now agree (Sernelius and Söderström, 1991a,b, Hu and Flensberg, 1996) that these two types of formulas for weak impurity scattering induced linear resistivity (conductivity), $\langle 1/\tau \rangle$ formula (7.37) and $1/\langle \tau \rangle$ formula (7.41), represent two extreme cases of electron–electron interaction strength in the system: when e–e scattering is very weak, the adiabatic resistivity $1/\langle \tau \rangle$ formula is correct; when e–e scattering is very strong, the isothermal resistivity $\langle 1/\tau \rangle$ formula (7.37) is correct; if the thermalization time τ_{th} or the e–e scattering rate τ_{ee} is of the same order of magnitude as the momentum relaxation time τ_{m}, the correct result lies somewhere in between (7.37) and (7.41).

Similar conflict also exists for phonon-induced linear resistivities, so-called $\frac{3}{2}(k_{\text{B}}T/\hbar\omega_{\text{LO}})$ problem in polaron mobility theory (see Peeters and Devreese, 1984, and the references therein).

7.4 Resistivity of Two-Subband System versus Interband Coulomb Interaction

For the purpose of further understanding of the difference between isothermal and adiabatic resistivities, we will examine the effect of intercarrier Coulomb interaction on the total resistivity of a system having two energy bands, by starting from balance equations for several-carrier-species sysyems introduced in Chapter 6. As a simplest example we consider a model system consisting of two subbands with carriers of charge e and effective mass m, having densities n_1 and n_2 and drift velocities v_1 and v_2 for subband 1 and subband 2.

If the direct Coulomb interaction between carriers is the only intersubband coupling, the steady-state force-balance equations for this two-subband system are given by

$$n_1 eE + f_1(v_1) + f_{12}(v_1 - v_2) = 0, \qquad (7.43)$$

$$n_2 eE + f_2(v_2) - f_{12}(v_1 - v_2) = 0. \qquad (7.44)$$

Here f_1 and f_2 are respectively the frictional forces (including impurity and phonon effects) experienced by the centers of mass of carrriers in subbands 1 and 2 and f_{12} is the mutual drag force of two centers of mass due to Coulomb interactions of carriers between two subbands. In the case of weak electric field, the energy balance requires the electron temperature equal to the lattice temperature and the above force-balance equations are written as

$$n_1 eE + v_1 f_1' + (v_1 - v_2) f_{12}' = 0, \qquad (7.45)$$

$$n_2 eE + v_2 f_2' - (v_1 - v_2) f_{12}' = 0, \qquad (7.46)$$

where $f_1' \equiv f_1'(0)$, $f_2' \equiv f_2'(0)$, and $f_{12}' \equiv f_{12}'(0)$. Eqs. (7.45) and (7.46) yield

$$v_1 = eE \frac{n - f_2' A}{f_1' + f_2'}, \qquad (7.47)$$

$$v_2 = eE \frac{n + f_1' A}{f_1' + f_2'}, \qquad (7.48)$$

in which $n = n_1 + n_2$, and

$$A = \frac{n_1 f_1' - n_2 f_2'}{f_1' f_2' + (f_1' + f_2') f_{12}'}. \qquad (7.49)$$

The total average drift velocity of this two-subband system, $\bar{v} = (n_1 v_1 + n_2 v_2)/n$, can be expressed as

$$\bar{v} = -eE \frac{n_1^2 f_2' + n_2^2 f_1' + n^2 f_{12}'}{n \left[f_1' f_2' + (f_1' + f_2') f_{12}' \right]}. \qquad (7.50)$$

In the absence of intersubband Coulomb interaction, two subbands are independent. The force-balance equation $n_1 eE + v_1 f_1' = 0$ of subband 1 gives rise to the conductivity of the 1st subband, $\sigma_1 = (n_1 e/m)\tau_1$, with the scattering time τ_1 of the 1st subband given by

$$\frac{1}{\tau_1} = -\frac{f_1'}{n_1 m}. \qquad (7.51)$$

Likewise, the conductivity of the 2nd subband is $\sigma_2 = (n_2 e/m)\tau_2$, with the scattering time τ_2 of the 2nd subband given by

$$\frac{1}{\tau_2} = -\frac{f_2'}{n_2 m}. \qquad (7.52)$$

We may also define a scattering time τ_{es} related to intersubband Coulomb interaction:

$$\frac{1}{\tau_{es}} = -\frac{f'_{12}}{nm}. \tag{7.53}$$

The total conductivity σ of this two-subband system is obtained from the total average drift velocity $\bar{v}$ of (7.50) to be

$$\sigma = \frac{ne\bar{v}}{E} = \frac{ne^2}{m}\tau, \tag{7.54}$$

with the total scattering time τ given by

$$\frac{1}{\tau} = \frac{\frac{r_1 r_2}{\tau_1 \tau_2} + \left(\frac{r_1}{\tau_1} + \frac{r_2}{\tau_2}\right)\frac{1}{\tau_{es}}}{r_1 r_2 \left(\frac{r_1}{\tau_1} + \frac{r_2}{\tau_2}\right) + \frac{1}{\tau_{es}}}, \tag{7.55}$$

where $r_1 \equiv n_1/n$ and $r_2 \equiv n_2/n$. If the Coulomb interactions between two subbands are very weak ($\tau_{es} \gg \tau_1, \tau_2$), the terms with $1/\tau_{es}$ in (7.55) are negligible and we have

$$\frac{1}{\tau} = \frac{1}{r_1\tau_1 + r_2\tau_2} \quad \text{or} \quad \tau = r_1\tau_1 + r_2\tau_2 \equiv \tau_a, \tag{7.56}$$

i.e. two subbands are independent carrier systems, with the total conductivity equal to the sum of individual subband conductivities: $\sigma = \sigma_1 + \sigma_2$. This result corresponds to the adiabatic conductivity of this two subband system. If the Coulomb interactions between two subbands are very strong ($\tau_{es} \ll \tau_1, \tau_2$), only terms with τ_{es} in (7.55) are needed to retain and we have

$$\frac{1}{\tau} = \left(\frac{r_1}{\tau_1} + \frac{r_2}{\tau_2}\right) \equiv \frac{1}{\tau_t}. \tag{7.57}$$

This means that the scattering rates of two subbands are additive. Because of strong Coulomb interaction two subbands are linked together with equal drift velocities. In fact, the sum of Eqs. (7.45) and (7.46) immediately gives (7.57) if $v_1 = v_2 = v$ results due to strong interband Coulomb interaction. Eq. (7.57) corresponds to the isothermal resistivity of this two subband system. The general formula (7.55) contains intersubband Coulomb interaction related scattering time τ_{es}. Changing τ_{es} one can see how τ varies from adiabatic resistivity (τ_a) to isothermal resistivity (τ_t), as schematically shown in Fig. 7.1.

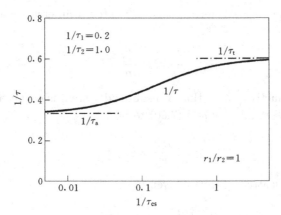

Fig. 7.1 Variation of the scattering time τ of a two-subband system with the intersubband Coulomb scattering time $\tau_{\rm es}$.

7.5 Boltzmann Equation with Electron–Electron Scattering

In a parabolic band semiconductor where the umklapp processes are negligible, the momentum-conserved electron–electron (e–e) scattering does not give a direct contribution to classical linear transport properties (save for pure quantum effects, such as weak localization). However, the e–e interaction affects the carrier distribution and occupation numbers of electron states and thereby indirectly modifies the transport behavior of the system. In principle, e–e scatterings are contained in the collision term of the Boltzmann equation, but so far there has been no simple tractable method to accurately solve the Boltzmann equation with impurity, phonon and e–e scatterings involved in its collision term. For a long time, the solution of Boltzmann equation has been obtained in the absence of e–e scattering, leading to the adiabatic formula (7.41) for the weak impurity resistivity.

Sernelius and Söderström (1990a,b) investigate effects of electron–electron scattering of varying strength in the Boltzmann equation on the impurity resistivity, employing a simple relaxation time approximation. They consider a semiconductor system with an isotropic parabolic energy spectrum $\varepsilon_{\bm{k}} = k^2/2m$ under the influence of a weak electric field $\bm{E}$. The deviation of the system distribution function $f(\bm{k})$ from its thermal equilibrium distribution $f^0(\bm{k})$ is a small quantity of linear order of the electric field $\bm{E}$. The steady state Boltzmann equation with e–e scattering can be written to this order as

$$-e\boldsymbol{E} \cdot \frac{\partial \varepsilon_{\boldsymbol{k}}}{\partial \boldsymbol{k}} \frac{\partial f^0(\boldsymbol{k})}{\partial \varepsilon_{\boldsymbol{k}}} = \left(\frac{\partial f(\boldsymbol{k})}{\partial t}\right)_{\text{imp}} + \left(\frac{\partial f(\boldsymbol{k})}{\partial t}\right)_{\text{e-e}}. \quad (7.58)$$

For a screened charge impurity potential $V(q) = Ze^2/[q^2\epsilon_0\kappa\epsilon(q)]$, where Z is the charge number of the impurity, κ is the static dielectric constant of the semiconductor, and $\epsilon(q)$ is the static dielectric function of electrons, the impurity-scattering related collision terms can be written as

$$\left(\frac{\partial f(\boldsymbol{k})}{\partial t}\right)_{\text{imp}} = -\frac{Z^2 e^4 m n_{\text{i}}}{2\pi^2 \epsilon_0^2 \kappa^2} \int d^3q \frac{[f(\boldsymbol{k}) - f(\boldsymbol{k}+\boldsymbol{q})]}{q^4 \epsilon^2(q)} \delta(q^2 + 2kq\cos(\boldsymbol{k},\boldsymbol{q})), \quad (7.59)$$

in which $(\boldsymbol{k},\boldsymbol{q})$ denotes the angle between $\boldsymbol{k}$ and $\boldsymbol{q}$, n_{i} is the impurity density. If we define an effective transport relaxation time $\tau(\boldsymbol{k})$ by

$$f(\boldsymbol{k}) = f^0(\boldsymbol{k} + e\tau(\boldsymbol{k})\boldsymbol{E}), \quad (7.60)$$

where $f^0(\boldsymbol{k}) = f(\varepsilon_{\boldsymbol{k}}, T)$ is the Fermi distribution function, the linear resistivity can be written in the form

$$\rho = \frac{m}{ne^2} \frac{1}{\langle \tau(\boldsymbol{k}) \rangle}, \quad (7.61)$$

with the average defined by (7.39). In this, $\tau(\boldsymbol{k})$ not only comes from impurity scattering, but also includes effects of e-e scattering. Putting (7.60) into (7.59) and keeping terms to the linear order in $\boldsymbol{E}$, we have

$$\left(\frac{\partial f(\boldsymbol{k})}{\partial t}\right)_{\text{imp}} = -e\boldsymbol{E} \cdot \frac{\partial \varepsilon_{\boldsymbol{k}}}{\partial \boldsymbol{k}} \left(\frac{\partial f^0(\boldsymbol{k})}{\partial \varepsilon_{\boldsymbol{k}}}\right) \frac{\tau(\boldsymbol{k})}{\tau_{\text{i}}(\boldsymbol{k})}, \quad (7.62)$$

where

$$\frac{1}{\tau_{\text{i}}(\boldsymbol{k})} \equiv \frac{Z^2 e^4 m n_{\text{i}}}{2\pi^2 \epsilon_0^2 \kappa^2} \int d^3q \frac{[1 - \cos(\boldsymbol{k}, \boldsymbol{k}+\boldsymbol{q})]}{q^4 \epsilon^2(q)} \delta(q^2 + 2kq\cos(\boldsymbol{k},\boldsymbol{q})) \quad (7.63)$$

is the transport relaxation time when only impurity scattering exists. The role of e-e scattering is to promote thermalization of carriers about the drifted state. If we approximately use a relaxation time τ_{ee} to represent the effect of e-e scattering, the collision term related to e-e scattering can be written as

$$\left(\frac{\partial f(\boldsymbol{k})}{\partial t}\right)_{\text{e-e}} = -\frac{f(\boldsymbol{k}) - f(\boldsymbol{k}+e\boldsymbol{E}\langle\tau(\boldsymbol{k})\rangle)}{\tau_{\text{ee}}}$$

$$= -\frac{\tau(\boldsymbol{k}) - \langle\tau(\boldsymbol{k})\rangle}{\tau_{\text{ee}}} e\boldsymbol{E} \cdot \frac{\partial \varepsilon_{\boldsymbol{k}}}{\partial \boldsymbol{k}} \left(\frac{\partial f^0(\boldsymbol{k})}{\partial \varepsilon_{\boldsymbol{k}}}\right). \quad (7.64)$$

$\langle \tau(\boldsymbol{k}) \rangle$ is the average of $\tau(\boldsymbol{k})$ defined by (7.60). Substituting (7.62) and (7.64) into the linearized steady-state Boltzmann equation (7.58), we have

$$\tau(\boldsymbol{k}) = \frac{\tau_{\mathrm{i}}(\boldsymbol{k})[\tau_{\mathrm{ee}} + \langle \tau(\boldsymbol{k}) \rangle]}{\tau_{\mathrm{ee}} + \tau_{\mathrm{i}}(\boldsymbol{k})}. \tag{7.65}$$

The average of the above equation yields

$$\langle \tau(\boldsymbol{k}) \rangle = \left\langle \frac{\tau_{\mathrm{i}}(\boldsymbol{k})}{1 + \tau_{\mathrm{i}}(\boldsymbol{k})/\tau_{\mathrm{ee}}} \right\rangle \Big/ \left\langle \frac{1}{1 + \tau_{\mathrm{i}}(\boldsymbol{k})/\tau_{\mathrm{ee}}} \right\rangle. \tag{7.66}$$

This result shows that if e–e scattering is very weak in comparison with impurity scattering that $\tau_{\mathrm{i}}(\boldsymbol{k})/\tau_{\mathrm{ee}} \ll 1$, we have $\langle \tau(\boldsymbol{k}) \rangle = \langle \tau_{\mathrm{i}}(\boldsymbol{k}) \rangle$ and (7.61) goes back to the adiabatic resistivity formula (7.41). On the contrary, if e–e scattering is very strong that $\tau_{\mathrm{i}}(\boldsymbol{k})/\tau_{\mathrm{ee}} \gg 1$, we have $\langle \tau(\boldsymbol{k}) \rangle = 1/\langle 1/\tau_{\mathrm{i}}(\boldsymbol{k}) \rangle$ and (7.61) goes to the isothermal resistivity formula (7.37).

Hu and Flensberg (1996) investigate the effect of changing ratio of elastic to e–e scattering strengths on the electron mobility. They numerically solve the accurate linear Boltzmann equation (7.58) with e–e scattering for a two-dimensional system formed by electrons confined in a 10 nm-wide square GaAs quantum well with infinite barriers at temperature $T = 10\,\mathrm{K}$. The electron sheet density is fixed at $N_{\mathrm{s}} = 1.5 \times 10^{11}\,\mathrm{cm}^{-2}$. The elastic scatterings come from remote charged impurities δ-doped a distance d away from the well center having the same density as electrons in the well, and from acoustic phonons (their scattering treated as elastic). By changing the distance d, the impurity scattering strength, and thus the strength ratio of elastic scattering to e–e scattering can be changed (temperature and electron density are fixed). The finite-temperature dynamically screened Coulomb matrix elements are used for calculating the scattering probability in the collision terms. Fig. 7.2 shows their calculated mobility (dots) as a function of ionized impurity distance d from the center of the quantum well. The solid curve is the mobility in the absence of e–e scattering, $\mu_0 = e\langle \tau_{\mathrm{i}} \rangle / m$ (τ_{i} is the transport relaxation time due to elastic scattering), corresponding to that of (7.41). The dashed curve is the mobility at infinitely large e–e scattering, $\mu_\infty = e/(m\langle \tau_{\mathrm{i}}^{-1} \rangle)$, corresponding to that of (7.37). μ_0 is generally large than μ_∞ because e–e scattering tends to scatter large velocity electrons back into lower velocity states. μ_0 can be almost twice μ_∞, as seen in the inset, where μ_∞/μ_0 and $(\mu - \mu_\infty)/(\mu_0 - \mu_\infty)$ are plotted versus d. Since τ_{ee} is essentially fixed, small d means weak e–e scattering, $\tau_{\mathrm{i}} \ll \tau_{\mathrm{ee}}$, and large d means strong e–e scattering, $\tau_{\mathrm{i}} \gg \tau_{\mathrm{ee}}$. The change from weak to strong e–e scattering limit is clearly seen with increasing d in the figure. The crossover occurs in the range having mobility around $3 \times 10^4 \, \text{cm}^2\text{V}^{-1}\text{s}^{-1}$, where the transport scattering rate τ_i^{-1} is around the same as e–e scattering rate τ_ee^{-1}.

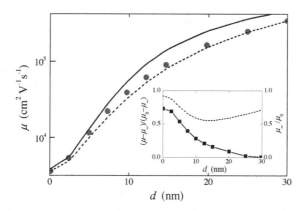

Fig. 7.2 Calculated mobility μ, μ_∞/μ_0 and $(\mu - \mu_\infty)/(\mu_0 - \mu_\infty)$ versus the ionized impurity distance d from the center of the quantum well. After Hu and Flensberg (1996).

7.6 Distribution Function in the Balance Equation Theory

A basic starting point of the balance-equation transport theory is to treat the impurity and phonon scattering Hamiltonian H_I as a small quantity and to proceed with perturbation expansions of the Liouville equation for the density matrix and the statistical averages of dynamical variables (Sec. 1.4). Almost all the equations derived and all the transport quantities calculated in the preceding sections are limited to the lowest nonzero order of the scattering potential H_I. In this sense, the balance-equation theory is also a kind of lowest-order response theory comparable with the Kubo linear response theory for conductivity, but different from the latter in that the expansion quantity of the balance-equation theory is the scattering potential H_I, rather than the electric field (Lei and Ting, 1987). As in the perturbation theory in quantum mechanics, where one only needs the zero-order or unperturbed wave function for calculating the average of a dynamical variable to the first order in the perturbation, in the balance-equation transport theory we only need the initial density matrix ρ_0, or the distribution function of the unperturbed state, to calculate most important transport quantities (such as mobility). This is a thermal equilibrium distribution function of electrons in the center-of-mass frame at temperature T_e (due to electron–electron interaction, this distribution function generally somewhat differs from an ordinary Fermi-Dirac function). Although there is no need to explicitly determine the electron distribution function in the calculation of major transport properties to leading nonzero order in the scattering interactions in the balance-equation theory, such information about electron distribution is necessary in making connection with more detailed experimental measurements. For instance, in the balance-equation theory, the electron temperature T_e, which is defined as the thermodynamic temperature of the relative electron system after turning-off the scattering interactions (Lei and Ting, 1985a) and serves as one of the major parameters in high-field transport, is not a readily experimentally measurable quantity (Lei and Horing, 1987b). Photoluminescence experiments usually can extract an effective temperature T_{eff} by fitting the line shape of luminescence intensity at high energy with an exponential form $\sim \exp(-\omega/T_{\text{eff}})$. A meaningful comparison and the justification of identifying T_{eff} as the electron temperature T_e, however, need the knowledge of the distribution function of electrons under high-field transport conditions. Furthermore, the calculation of nonequilibrium correction of the distribution function, which requires to go at least an order higher in the perturbative expansion of the scattering potential, enables us to check the validity of the balance-equation treatment, i.e. the limit of very strong e–e interaction, in various real situations.

The momentum (wavevector) distribution function of the relative electrons is defined in the balance-equation transport theory as (Lei and Ting, 1987)

$$n_{\bm{k}} = \text{tr}\{\hat{\rho}\hat{n}_{\bm{k}}\}, \tag{7.67}$$

where $\hat{n}_{\bm{k}} \equiv \sum_\sigma c^\dagger_{\bm{k}\sigma} c_{\bm{k}\sigma}$ is the particle number operator in state $\bm{k}$, $\hat{\rho}$ is the density matrix of the transport system satisfying the Liouville equation (1.28). (7.67) can be written as the sum of an equilibrium distribution function at the electron temperature T_e,

$$n_0(\bm{k}) = \text{tr}\{\hat{\rho}_0 \hat{n}_{\bm{k}}\} \tag{7.68}$$

and a nonequilibrium deviation $\delta n_{\bm{k}}$ from it:

$$n_{\bm{k}} = n_0(\bm{k}) + \delta n_{\bm{k}}. \tag{7.69}$$

Note that the equilibrium part $n_0(\bm{k})$, which includes effects of Coulomb interaction between carriers, is generally different from a Fermi-Dirac function. The nonequilibrium part $\delta n_{\bm{k}}$ can be written in the interaction picture as [see (1.31)]

$$\delta n_{\bm{k}} = \mathrm{tr}\{\hat{\rho}_1(t)\hat{n}_{\bm{k}}(t)\}, \tag{7.70}$$

with $\hat{\rho}_1(t) \equiv \hat{\rho}(t) - \hat{\rho}_0$.

The distribution function correction $\delta n_{\bm{k}}$ should obey the sum rule

$$\sum_{\bm{k}} \delta n_{\bm{k}} = 0, \tag{7.71}$$

required by the particle-number conservation.

From the point of view of nonequilibrium many-particle physics, the calculation of $\delta n_{\bm{k}}$ is a complicated problem of Coulomb relaxation kinetics of a dense electron gas. A perturbative treatment for $\delta n_{\bm{k}}$ is possible for a system with relatively strong carrier–carrier interactions. Calculation can proceed using the iterative expansion of the density matrix $\hat{\rho}_1(t)$ (Sec. 1.4), or using the Feynman diagrammatic technique in the closed-time-path-Green's function formulation (Sec. 9.2). Apparently the linear order correction in the scattering potential H_I is zero, and therefore the lowest contribution to $\delta n_{\bm{k}}$ comes from the second-order terms. Guillemot and Clérot (1993) carry out a calculation of the distribution function for an interacting relative electron gas to leading nonvanishing order in the scattering interactions. Their result can be expressed as (μ is the chemical potential)

$$n_0(\bm{k}) = \int d\varepsilon \frac{A(\bm{k}, \varepsilon)}{\exp\left[(\varepsilon - \mu)/T_\mathrm{e}\right] + 1}, \tag{7.72}$$

$$\delta n_{\bm{k}} = -\left[\pi \int d\varepsilon A(\bm{k}, \varepsilon)^2\right] \left\{ n_\mathrm{i} \sum_{\bm{q}} |u(\bm{q})|^2 \,\mathrm{Im}\, T(\bm{k}, \bm{q}, \bm{q} \cdot \bm{v}_\mathrm{d}) \right.$$
$$+ \sum_{\bm{q}\lambda} |M(\bm{q}, \lambda)|^2 \left[n\left(\frac{\Omega_{\bm{q}\lambda}}{T}\right) - n\left(\frac{\Omega_{\bm{q}\lambda} - \bm{q} \cdot \bm{v}_\mathrm{d}}{T_\mathrm{e}}\right) \right]$$
$$\left. \times \left[\mathrm{Im}\, T(\bm{k}, \bm{q}, \bm{q} \cdot \bm{v}_\mathrm{d} - \Omega_{\bm{q}\lambda}) + \mathrm{Im}\, T(\bm{k}, \bm{q}, -\bm{q} \cdot \bm{v}_\mathrm{d} + \Omega_{\bm{q}\lambda})\right] \right\}. \tag{7.73}$$

Here $A(\bm{k}, \varepsilon)$ is the spectral density function of the interacting electron system. It measures the probability of a particle with momentum $\bm{k}$ to be at the many-body eigenstate of energy ε. The expression (7.72) indicates that the equilibrium distribution function $n_0(\bm{k})$ is not the ordinary Fermi–Dirac function, and (7.73) shows that each external scattering mechanism, such as impurities, acoustic and optic phonons, contributes a correction to the equilibrium distribution function. We can express the impurity and phonon terms on the right-hand side of (7.73) in terms of inverse relaxation times $1/\tau_{\bm{k}}^{(i)}$ and $1/\tau_{\bm{k}}^{(p)}$, respectively related to the impurity and phonon scattering strength. With a Coulomb lifetime defined as

$$\tau_{\bm{k}}^{ee} = \pi \int d\varepsilon A^2(\bm{k}, \varepsilon), \tag{7.74}$$

we write $\delta n_{\bm{k}}$ in the form

$$\delta n_{\bm{k}} = \frac{\tau_{\bm{k}}^{ee}}{\tau_{\bm{k}}^{(i)}} + \frac{\tau_{\bm{k}}^{ee}}{\tau_{\bm{k}}^{(p)}} + \cdots . \tag{7.75}$$

The contribution of each external scattering mechanism to the distribution function correction equals the ratio of the Coulomb lifetime to the relaxation time related to that scattering mechanism. This result is physically easily understood: the electron–electron interaction promotes the relative electron gas to the thermal equilibrium distribution $n_0(\bm{k})$, while the electron–impurity and electron–phonon scatterings drive the relative electron system deviating from equilibrium. The validity of the balance-equation approach as a linear response theory requires that the distribution function correction $\delta n_{\bm{k}}$, thus each term of the right-hand side of (7.75) is small.

In the quasiparticle picture, the spectral density function is approximated by a Lorentzian form ($\xi_{\bm{k}} = \varepsilon_{\bm{k}} - \mu$ and $\varepsilon_{\bm{k}} = k^2/2m$):

$$A(\bm{k}, \varepsilon) \approx \frac{1}{\pi} \frac{\Gamma_{\bm{k}}}{(\varepsilon - \xi_{\bm{k}})^2 + \Gamma_{\bm{k}}^2}, \tag{7.76}$$

where

$$\Gamma_{\bm{k}} = -\operatorname{Im} \Sigma(\bm{k}, \xi_{\bm{k}}), \tag{7.77}$$

$\Sigma(\bm{k}, \xi_{\bm{k}})$ being the self-energy of the electron gas in the thermal equilibrium state of temperature T_e. With this form of the spectral function $A(\bm{k}, \varepsilon)$, the Coulomb lifetime is given by

$$\tau_{\bm{k}}^{ee} = \pi \int d\varepsilon A^2(\bm{k}, \varepsilon) = \frac{1}{2\Gamma_{\bm{k}}}. \tag{7.78}$$

This is the standard electron–electron scattering limited lifetime of the quasiparticle at state $\bm{k}$, which is usually calculated from the self-energy to the lowest order in the screened Coulomb interaction. The quasiparticle picture, which is restricted to excitations close to the Fermi level and to temperatures low as compared to the Fermi energy, has been traditionally used in the context of metals. In semiconductors, where the carrier density is much smaller and the Fermi level may be much lower than the electron temperature, the excitation and other relevant energies may be of the same order of, or larger than the Fermi energy, the quasiparticle picture is generally not a good description. A more sophisticated expression for the spectral density function is desirable for semiconductors.

Guillemot and Clérot (1993) carefully investigate the spectral density function $A(\boldsymbol{k}, \varepsilon)$ and Coulomb lifetime (7.47) in typical semiconductor-based 2D electron gases beyond the quasiparticle approximation. The readers are referred to their original paper for detail. An important result is that at zero temperature, where the quasiparticle theory is well studied, the Coulomb damping rate $1/\tau_{\boldsymbol{k}}^{ee}$ calculated from the full spectral density function is more than 10 times larger than that from the quasiparticle picture for any $\boldsymbol{k} \neq \boldsymbol{k}_{\mathrm{F}}$.

The three-particle Green's function $T(\boldsymbol{k}, \boldsymbol{q}, \omega)$ in Eq. (7.73) can be calculated within the frame work of random phase approximation (Guillemot and Clérot, 1993)

$$T(\boldsymbol{k}, \boldsymbol{q}, \omega) = T_0(\boldsymbol{k}, \boldsymbol{q}, \omega)/|\epsilon(\boldsymbol{q}, \omega)|^2, \qquad (7.79)$$

where $\epsilon(\boldsymbol{q}, \omega) = 1 - V(q)\Pi_0(\boldsymbol{q}, \omega)$ is the electron dielectric function in RPA, $V(q)$ is the effective 2D Coulomb potential and $\Pi_0(\boldsymbol{q}, \omega)$ is the density correlation function or polarizability function of 2D electrons in the absence of Coulomb interaction (Sec. 2.7, single subband case). $T_0(\boldsymbol{k}, \boldsymbol{q}, \omega)$ is the three-particle Green's function for noninteracting 2D electron gas:

$$T_0(\boldsymbol{k}, \boldsymbol{q}, \omega) = \frac{f_{\boldsymbol{k}+\boldsymbol{q}}(1 - f_{\boldsymbol{k}})}{\varepsilon_{\boldsymbol{k}+\boldsymbol{q}} - \varepsilon_{\boldsymbol{k}} + \omega + \mathrm{i}\delta} - \frac{f_{\boldsymbol{k}}(1 - f_{\boldsymbol{k}-\boldsymbol{q}})}{\varepsilon_{\boldsymbol{k}} - \varepsilon_{\boldsymbol{k}-\boldsymbol{q}} + \omega + \mathrm{i}\delta} + f_{\boldsymbol{k}}\Pi_0(\boldsymbol{q}, \omega), \quad (7.80)$$

with $f_{\boldsymbol{k}}$ the Fermi-Dirac function at temperature T_{e}.

Based on these results Guillemot and Clérot (1993) calculate the nonequilibrium correction to the distribution function, $\delta n_{\boldsymbol{k}}$, induced by electron–polar-optic-phonon scattering and by electron–impurity scattering, in a GaAs-based 2D electron system having electron sheet density $1.0 \times 10^{15}\,\mathrm{m}^{-2}$ and linear mobility $12\,\mathrm{m}^2\mathrm{V}^{-1}\mathrm{s}^{-1}$ at lattice temperature $T = 2\,\mathrm{K}$. The calculation is performed under the condition that an electric field of $E = 850\,\mathrm{V/cm}$ is applied in the 2D plane, with the electron temper-

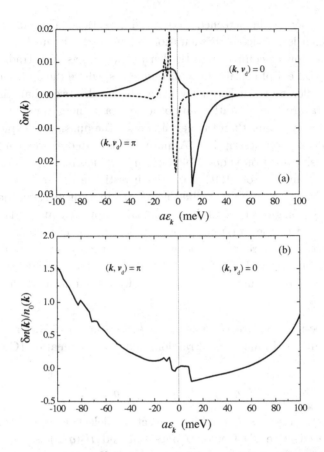

Fig. 7.3 (a) The distribution function correction $\delta n_{\bm{k}}$ induced by electron–LO-phonon scattering (solid line) and by electron–impurity scattering (dashed line) in a GaAs-based interacting 2D electron system subjected to an in-plane electric field $E = 850\,\text{V/cm}$, is plotted along the directions of wavevector $\bm{k}$ parallel ($a = 1$) and antiparallel ($a = -1$) to the drift velocity $\bm{v}_\text{d}$, as a function of $a\varepsilon_{\bm{k}}$ [$a \equiv \sin(\bm{k}, \bm{v}_\text{d})$]. (b) The relative correction $\delta n_{\bm{k}}/n_0(\bm{k})$ of distribution function against $a\varepsilon_{\bm{k}}$. After Guillemot and Clérot (1993).

ature $T_\text{e} = 116.5\,\text{K}$ and the drift velocity $v_\text{d} = 2.04 \times 10^7\,\text{cm/s}$ derived from the force- and energy-balance equations. The calculated $\delta n_{\bm{k}}$ and $\delta n_{\bm{k}}/n_0(\bm{k})$ for wavevector $\bm{k}$ in the directions parallel and antiparallel to the drift velocity $\bm{v}_\text{d}$, are shown in Fig. 7.3 versus $\varepsilon_{\bm{k}} = k^2/2m$. The total distribution function $n_{\bm{k}} = n_0(\bm{k}) + \delta n_{\bm{k}}$ is shown in Fig. 7.4 in logarithmic scale. It can be seen that the differences between the nonequilibrium distribution function of the transport state, the equilibrium distribution function of the Coulomb gas and the Fermi–Dirac distribution function, are significant.

The Coulomb lifetime $\tau_{\bm{k}}^{\text{ee}}$ evaluated by Guillemot and Clérot (1993) ranges between 50 and 150 fs for carrier density of order of $10^{15}\,\text{m}^{-2}$ at temperature around 100 K. The LO-phonon induced $\tau_{\bm{k}}^{(\text{p})}$, though somewhat larger, are of the same order of magnitude as $\tau_{\bm{k}}^{\text{ee}}$. Therefore, a perturbative linear handling of the electron–LO-phonon interaction as compared to the Coulomb interaction may not be good enough.

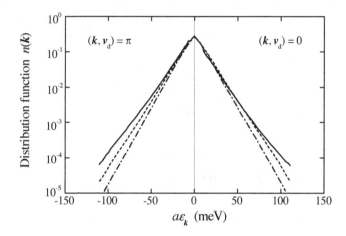

Fig. 7.4 The momentum distribution function for wavevectors $\bm{k}$ parallel ($a = 1$) and antiparallel ($a = -1$) to the drift velocity v_{d} direction, versus $a\varepsilon_{\bm{k}}$ [$a \equiv \sin(\bm{k}, \bm{v}_{\text{d}})$]. Solid line: interaction electron gas with electron–phonon and electron–impurity interactions. Dashed line: Coulomb electron gas without external scattering. Dashed-dotted line: noninteracting electron gas (Fermi-Dirac function). After Guillemot and Clérot (1993).

7.7 Coupled Force-Balance and Particle-Occupation Rate Equations

Balance-equation transport theory treating the external (impurity and phonon) scattering interaction H_{I} as a small quantity, is a kind of perturbation theory. The smallness of H_{I} is in comparison with the internal interaction H_{ee} between carriers, and the perturbative expansion is essentially that in powers of the ratio of these two interactions. Most results presented in the preceding chapters are obtained by truncating the perturbation series at the lowest nonzero order, which correspond to the limit of rapid thermalization, or strong e–e interaction. The calculation of the distribution function to the first order of the scattering interaction in the preceding section indicates that in a polar semiconductor, the e–e interaction may not be strong enough in comparison with LO-phonon scattering that it is desirable to go beyond the strong e–e scattering limit. A natural and simple way along this direction is to put the electron–impurity, electron–phonon and electron–electron scatterings on an equal footing (Lei, 2008).

We start with the same system of N electrons with impurities and phonons under the influence of a uniform electric field $\bm{E}$, and write its full Hamiltonian H in terms of the center-of-mass variables and the relative-electron variables in the form

$$H = H_{\text{cm}} + H_{\text{e0}} + H_{\text{ee}} + H_{\text{ph}} + H_{\text{ei}} + H_{\text{ep}}, \tag{7.81}$$

$$H_{\text{cm}} = \frac{P^2}{2Nm} - N\bm{E}\cdot\bm{R}, \tag{7.82}$$

$$H_{\text{e0}} = \sum_{\bm{k},\sigma} \varepsilon_{\bm{k}} c^\dagger_{\bm{k}\sigma} c_{\bm{k}\sigma} \tag{7.83}$$

$$H_{\text{ee}} = \sum_{\bm{q}} \tfrac{1}{2} \nu_c(q)\left(\rho_{\bm{q}}\rho_{-\bm{q}} - N\right) \tag{7.84}$$

$$H_{\text{ph}} = \sum_{\bm{q},\lambda} \Omega_{\bm{q}\lambda} b^\dagger_{\bm{q}\lambda} b_{\bm{q}\lambda}, \tag{7.85}$$

$$H_{\text{ei}} = \sum_{\bm{q},a} u(\bm{q})\, e^{i\bm{q}\cdot(\bm{R}-\bm{r}_a)} \rho_{\bm{q}}, \tag{7.86}$$

$$H_{\text{ep}} = \sum_{\bm{q},\lambda} M(\bm{q},\lambda)\,(b_{\bm{q}\lambda} + b^\dagger_{-\bm{q}\lambda}) e^{i\bm{q}\cdot\bm{R}} \rho_{\bm{q}}. \tag{7.87}$$

In this, we have neglected the noncanonical $1/N$ terms in the exact commutation relation (1.7) for large N system, i.e. assume that the relative electron system consists of N independent particles characterized by the set of canonical conjugate variables $\bm{p}'_i$ and $\bm{r}'_i$ ($i = 1,\cdots,N$), and thus can be described by means of the creation and annihilation operators $c^\dagger_{\bm{k}\sigma}$ and $c_{\bm{k}\sigma}$ with wave vector $\bm{k}$, spin σ, and energy $\varepsilon_{\bm{k}} = k^2/2m$. In these equations, $\nu_c(q) = e^2/\epsilon_0 \kappa q^2$ is the Coulomb potential (κ is the background dielectric constant), $\rho_{\bm{q}} = \sum_{\bm{k},\sigma} c^\dagger_{\bm{k}+\bm{q}\sigma} c_{\bm{k}\sigma}$ is the density operator of relative electrons, $b^\dagger_{\bm{q}\lambda}$ and $b_{\bm{q}\lambda}$ are creation and annihilation operators of a phonon with wavevector $\bm{q}$ in branch λ having frequency $\Omega_{\bm{q}\lambda}$, $u(\bm{q})$ and $M(\bm{q},\lambda)$ are, respectively, the electron–impurity potential and the electron–phonon matrix element satisfying $u(\bm{q}) = u^*(-\bm{q})$ and $M(\bm{q},\lambda) = M^*(-\bm{q},\lambda)$, and $\bm{r}_a$ stands for the impurity position.

We proceed from the Heisenberg equations for the rates of changes of center-of-mass momentum $\boldsymbol{P}$, the relative electron energy H_{e0}, and the relative electron occupation number $\hat{n}_{\boldsymbol{k}} \equiv \sum_\sigma c_{\boldsymbol{k}\sigma}^\dagger c_{\boldsymbol{k}\sigma}$:

$$\dot{\boldsymbol{P}} = -\mathrm{i}\,[\boldsymbol{P}, H], \tag{7.88}$$

$$\dot{H}_{e0} = -\mathrm{i}\,[H_{e0}, H], \tag{7.89}$$

$$\dot{\hat{n}}_{\boldsymbol{k}} = -\mathrm{i}\,[\hat{n}_{\boldsymbol{k}}, H]. \tag{7.90}$$

The center-of-mass motion will be treated classically and its fluctuation neglected that in the dc steady state $\boldsymbol{R}(t) - \boldsymbol{R}(t') = \boldsymbol{v}\,(t-t')$, $\boldsymbol{v}$ being the drift velocity. For the statistical average of relative electrons and phonons we still follow the formula similar to (1.38) in the interaction picture to the lowest nonzero order in the perturbative scattering potential $H_\mathrm{I} = H_\mathrm{ei} + H_\mathrm{ep} + H_\mathrm{ee}$, but, instead of using an isotropic initial density matrix (1.30), we treat $\hat{\rho}_0$ as more general one having the property that the statistical averages of multiple (relative-electron and phonon) single-particle operators with respect to it are reducible to the product of pair averages. The typical pair averages are the occupation number or the distribution function of relative electrons,

$$n_{\boldsymbol{k}} = \sum_\sigma \langle c_{\boldsymbol{k}\sigma}^\dagger c_{\boldsymbol{k}\sigma} \rangle, \tag{7.91}$$

and that of phonons,

$$N_{\boldsymbol{q}\lambda} = \sum_\sigma \langle b_{\boldsymbol{q}\lambda}^\dagger b_{\boldsymbol{q}\lambda} \rangle. \tag{7.92}$$

The statistical average of operator equation (7.88) yields the force-balance equation:

$$\dot{\boldsymbol{P}} \equiv Nm\frac{d\boldsymbol{v}}{dt} = Ne\boldsymbol{E} + \boldsymbol{f}_\mathrm{i} + \boldsymbol{f}_\mathrm{p}, \tag{7.93}$$

with the frictional forces exerted on the center-of-mass, $\boldsymbol{f}_\mathrm{i}$ and $\boldsymbol{f}_\mathrm{p}$, given by ($\omega_0 \equiv \boldsymbol{q}\cdot\boldsymbol{v}$)

$$\boldsymbol{f}_\mathrm{i} = 2\pi n_\mathrm{i} \sum_{\boldsymbol{k},\boldsymbol{q}} |u(\boldsymbol{q})|^2\,\boldsymbol{q}\,(n_{\boldsymbol{k}} - n_{\boldsymbol{k}+\boldsymbol{q}})\,\delta(\varepsilon_{\boldsymbol{k}+\boldsymbol{q}} - \varepsilon_{\boldsymbol{k}} + \omega_0), \tag{7.94}$$

$$\boldsymbol{f}_\mathrm{p} = 4\pi \sum_{\boldsymbol{k},\boldsymbol{q},\lambda} |M(\boldsymbol{q},\lambda)|^2\,\boldsymbol{q}\,[(N_{\boldsymbol{q}\lambda}(n_{\boldsymbol{k}} - n_{\boldsymbol{k}+\boldsymbol{q}}) - n_{\boldsymbol{k}+\boldsymbol{q}}(1 - n_{\boldsymbol{k}})]$$
$$\times \delta(\varepsilon_{\boldsymbol{k}+\boldsymbol{q}} - \varepsilon_{\boldsymbol{k}} + \omega_0 - \Omega_{\boldsymbol{q}\lambda}). \tag{7.95}$$

The statistical average of Eq. (7.89) yields the energy-balance equation:

$$\langle \dot{H}_{e0} \rangle \equiv \frac{dU_e}{dt} = -\boldsymbol{v} \cdot (\boldsymbol{f}_i + \boldsymbol{f}_p) - w, \qquad (7.96)$$

where $w = \langle \dot{H}_{ph} \rangle = -i\langle [H_{ph}, H] \rangle$ is the energy-transfer rate from the electron system to the phonon system and given by

$$w = 4\pi \sum_{\boldsymbol{k},\boldsymbol{q},\lambda} |M(\boldsymbol{q},\lambda)|^2 \Omega_{\boldsymbol{q}\lambda} [(N_{\boldsymbol{q}\lambda}(n_{\boldsymbol{k}+\boldsymbol{q}} - n_{\boldsymbol{k}}) + n_{\boldsymbol{k}+\boldsymbol{q}}(1 - n_{\boldsymbol{k}})]$$
$$\times \delta(\varepsilon_{\boldsymbol{k}+\boldsymbol{q}} - \varepsilon_{\boldsymbol{k}} + \omega_0 - \Omega_{\boldsymbol{q}\lambda}). \qquad (7.97)$$

Eq. (7.96) states that the energy increase of the relative electron system equals the work done by the center of mass, $-\boldsymbol{v} \cdot (\boldsymbol{f}_i + \boldsymbol{f}_p)$, minus the the energy transferred to the lattice, w.

Equations (7.93) and (7.95) are not enough for the determination of the unknown parameters $\boldsymbol{v}$ and $n_{\boldsymbol{k}}$, we therefore need to take the statistical average of Eq. (7.90), giving rise to the rate equations for the occupation number of relative electrons:

$$\frac{dn_{\boldsymbol{k}}}{dt} = \left[\frac{\partial n_{\boldsymbol{k}}}{\partial t}\right]_{ei} + \left[\frac{\partial n_{\boldsymbol{k}}}{\partial t}\right]_{ep} + \left[\frac{\partial n_{\boldsymbol{k}}}{\partial t}\right]_{ee}, \qquad (7.98)$$

with

$$\left[\frac{\partial n_{\boldsymbol{k}}}{\partial t}\right]_{ei} = 4\pi n_i \sum_{\boldsymbol{q}} |u(\boldsymbol{q})|^2 (n_{\boldsymbol{k}+\boldsymbol{q}} - n_{\boldsymbol{k}}) \delta(\varepsilon_{\boldsymbol{k}+\boldsymbol{q}} - \varepsilon_{\boldsymbol{k}} + \omega_0), \qquad (7.99)$$

$$\left[\frac{\partial n_{\boldsymbol{k}}}{\partial t}\right]_{ep} = 4\pi \sum_{\boldsymbol{q},\lambda} |M(\boldsymbol{q},\lambda)|^2$$
$$\times \{[(N_{\boldsymbol{q}\lambda}(n_{\boldsymbol{k}+\boldsymbol{q}} - n_{\boldsymbol{k}}) + n_{\boldsymbol{k}+\boldsymbol{q}}(1 - n_{\boldsymbol{k}})] \delta(\varepsilon_{\boldsymbol{k}+\boldsymbol{q}} - \varepsilon_{\boldsymbol{k}} + \omega_0 - \Omega_{\boldsymbol{q}\lambda})$$
$$+ [(N_{\boldsymbol{q}\lambda}(n_{\boldsymbol{k}+\boldsymbol{q}} - n_{\boldsymbol{k}}) - n_{\boldsymbol{k}}(1 - n_{\boldsymbol{k}+\boldsymbol{q}})] \delta(\varepsilon_{\boldsymbol{k}+\boldsymbol{q}} - \varepsilon_{\boldsymbol{k}} + \omega_0 + \Omega_{\boldsymbol{q}\lambda})\},$$
$$(7.100)$$

$$\left[\frac{\partial n_{\boldsymbol{k}}}{\partial t}\right]_{ee} = 2\pi n_i \sum_{\boldsymbol{q}} |\nu_c(\boldsymbol{q})|^2 [(1 - n_{\boldsymbol{k}}) n_{\boldsymbol{k}-\boldsymbol{q}} (1 - n_{\boldsymbol{k'}}) n_{\boldsymbol{k'}+\boldsymbol{q}}$$
$$- n_{\boldsymbol{k}}(1 - n_{\boldsymbol{k}-\boldsymbol{q}}) n_{\boldsymbol{k'}}(1 - n_{\boldsymbol{k'}+\boldsymbol{q}})] \delta(\varepsilon_{\boldsymbol{k}} + \varepsilon_{\boldsymbol{k'}} - \varepsilon_{\boldsymbol{k}-\boldsymbol{q}} - \varepsilon_{\boldsymbol{k'}+\boldsymbol{q}}). \qquad (7.101)$$

It is easily seen that Eq. (7.98) obeys the sum rule to conserve the total number of relative electrons:

$$\sum_{\boldsymbol{k}} \frac{dn_{\boldsymbol{k}}}{dt} = 0. \qquad (7.102)$$

Multiplying both sides of Eq. (7.98) by $\varepsilon_{\bm{k}}$ and then summing over $\bm{k}$, one gets[1]

$$\sum_{\bm{k}} \varepsilon_{\bm{k}} \frac{dn_{\bm{k}}}{dt} = -\bm{v}\cdot(\bm{f}_\mathrm{i}+\bm{f}_\mathrm{p}) - w, \qquad (7.103)$$

rederiving the energy-balance equation (7.96). This indicates that the work done by the center-of-mass on the relative electrons, and the energy transfer from the relative electron system to the phonon system are already contained in the rate equation (7.98) of the relative-electron occupation number, and that the energy-balance equation (7.96) is no longer needed as an independent equation.

Note that the total momentum of the relative electrons, which should be zero from the original definition, is not a physically meaningful quantity in the present scheme, in which the $1/N$ term in the exact noncanonical commutation relation (1.7) is ignored. Neverhteless, this does not invalidate the calculation of other useful transport quantities under the assumption that relative electrons constitute a system of N independent particles.

Eq. (7.98) is similar to the Boltzmann equation for the single particle distribution function $n_{\bm{k}}$ of a nondrifting electron system but contains an extra unknown parameter $\bm{v}$. By combining equations (7.93) with (7.98) one can obtain a complete set of equations to determine all the unknown parameters $n_{\bm{k}}$ and $\bm{v}$, and then the transport properties of the system. The coupled set of equations (7.93) and (7.98) is equivalent to the coupled force and Boltzmann equations given by Huang et al (2004) in the absence of high-frequency field.

The appearance of the drift velocity $\bm{v}$ in $[\partial n_{\bm{k}}/\partial t]_\mathrm{ei}$ and $[\partial n_{\bm{k}}/\partial t]_\mathrm{ep}$ may result in anisotropy of the $n_{\bm{k}}$ distribution. The $[\partial n_{\bm{k}}/\partial t]_\mathrm{ee}$ part which contains no $\bm{v}$, always promotes towards an isotropic distribution and in the strong e–e scattering limit the results of the preceding chapters recover. To go beyond this limit one must proceed using the complete set of equations (7.93) with (7.98) with all H_ei, H_ep and H_ee scatterings included.

Unfortunately, the solution to the coupled equations (7.93) and (7.98) is a very hard job in that (i) Eq. (7.98) is a complicated integro-differential equation, and (ii) the $[\partial n_{\bm{k}}/\partial t]_\mathrm{ee}$ term in (7.98) is difficult to handle, just as the Coulomb collisions in the Boltzmann equation.

To illustrate the usage of the coupled set of force-balance and particle-ocupation rate equations (7.93) and (7.98) in analyzing nonlinear transport,

---

[1] Note that $\sum_{\bm{k}} \varepsilon_{\bm{k}}[\partial n_{\bm{k}}/\partial t]_\mathrm{ee} = 0$, i.e. e–e interaction does not change the relative electron energy, as it should be.

Huang et al (2004, 2005) neglect the e–e scattering term $[\partial n_{\bm{k}}/\partial t]_{ee}$ in (7.98) and consider a GaAs system of high electron density under the influence of modest electric fields that the Dopler shift $\omega_0 = \bm{q}\cdot\bm{v}$ and phonon energy and $\Omega_{q\lambda}$ are smaller in comparison with the dominant relative electron kinetic energy $\varepsilon_{\bm{k}}$ which is around the Fermi level $\varepsilon_{\rm F}$. In view of δ-functions in Eqs. (7.99) and (7.100) [as well as in Eqs. (7.74) and (7.75)] $\varepsilon_{\bm{k}+\bm{q}} - \varepsilon_{\bm{k}}$ is small and $n_{\bm{k}+\bm{q}}$ only slightly deviates from $n_{\bm{k}}$, then one can expand $n_{\bm{k}+\bm{q}}$ as

$$n_{\bm{k}+\bm{q}} = n_{\bm{k}} + \frac{\partial n_{\bm{k}}}{\partial \varepsilon_{\bm{k}}}(\varepsilon_{\bm{k}+\bm{q}} - \varepsilon_{\bm{k}}) + \frac{\partial^2 n_{\bm{k}}}{\partial \varepsilon_{\bm{k}}^2}(\varepsilon_{\bm{k}+\bm{q}} - \varepsilon_{\bm{k}})^2. \qquad (7.104)$$

With this approximation in (7.99) and (7.100) the particle-occupation rate equations (7.98) are simplified to pure differential equations for $n_{\bm{k}}$, a type of Fokker-Planck equation, and become much easier to solve. Huang et al (2004, 2005) perform numerical calculations within this scheme, obtaining $n_{\bm{k}}$ which exhibits anisotropic distribution in the $\bm{k}$ space in respect to the direction of the drift velocity $\bm{v}$, as expected. It seems, however, if the e–e scattering terms can also be taken into account, the anisotropy of $n_{\bm{k}}$ distribution would be weakened.

Chapter 8

Balance Equation Approach to Magnetotransport

Balane-equation theory for hot-electron transport introduced in the preceding chapters can be extended to the case of simultaneous presence of electric and magnetic fields. The equations attained have been widely applied to semiclassical and quantum hot-carrier magnetotransport of various configurations (Lei, Cai and Ting, 1985; Cai, Lei and Ting, 1985; Cui, Lei and Horing, 1988). This chapter will give a brief introduction to some basic issues and recent development of this important and interesting field.

8.1 Balance Equations of Hot-Electron Transport in the Presence of a Magnetic field

We first consider the isotropic parabolic band system as described in Chapter 1 under the influence of a uniform electric field $\bm{E}$ and a uniform magnetic field $\bm{B}$. Without losing any generality we assume that the magnetic field is along the z direction, $\bm{B} = (0, 0, B)$. In terms of center-of-mass variables $\bm{P}$, $\bm{R}$ and relative-electron variables $\bm{p}'_i$, $\bm{r}'_i$, the Hamiltonian of this system can be written as

$$H = H_{\rm cm} + H_{\rm er} + H_{\rm ph} + H_{\rm ei} + H_{\rm ep}, \qquad (8.1)$$

but now the magnetic field enters in both the center-of-mass and relative electron parts of Hamiltonian:

$$H_{\rm cm} = \frac{1}{2Nm}(\bm{P} - Ne\bm{A}(\bm{R}))^2 - Ne\bm{E}\cdot\bm{R}, \qquad (8.2)$$

$$H_{\rm er} = \sum_i \frac{1}{2m}(\bm{p}'_i - e\bm{A}(\bm{r}'_i))^2 + g\mu_{\rm B} B \sum_i \sigma_i + \sum_{i<j}\frac{e^2}{4\pi\epsilon_0|\bm{r}'_i - \bm{r}'_j|}. \qquad (8.3)$$

Here $\boldsymbol{A}(\boldsymbol{r})$ is the vector potential of the constant uniform magnetic field $\boldsymbol{B}$. The second term on the right-hand-side of Eq. (8.3) represents the energy splitting of electron spins in the magnetic field with σ_i the spin of the ith electron ($\sigma_i = \pm 1$), μ_B the Bohr magneton and g the effective g-factor of electrons in the system. This Zeeman splitting of electron energy, which may play an important role in many situations, is generally neglected in the following discussions in this chapter where we focus our attention on issues related to Landau quantization of the electron orbital motion in a magnetic field. The expressions for H_{ph}, H_{ei} and H_{ep} in (8.1) are the same as Eqs. (1.13)–(1.15), in which ρ_q is the Fourier representation of the relative electron density operator given by (1.16). It can be seen from Eqs. (8.2) and (8.3) that though the separation between the center-of-mass and relative-electron variables is still realized in the presence of a magnetic field, the magnetic field plays a role different from that of the electric filed. While a spatially uniform electric field acts only on the center of mass and the relative electrons do not feel its existence directly, the magnetic field appears not only in H_{cm} but also in H_{er}, i.e. it applies both on the center of mass and on relative electrons.

In accordance with Hamiltonian (8.1), the center-of-mass velocity and acceleration operators are

$$\boldsymbol{V} = -\mathrm{i}\,[\boldsymbol{R}, H] = \frac{1}{Nm}(\boldsymbol{P} - Ne\boldsymbol{A}(\boldsymbol{R})), \tag{8.4}$$

$$\dot{\boldsymbol{V}} = -\mathrm{i}\,[\boldsymbol{V}, H] = \frac{e}{m}(\boldsymbol{V} \times \boldsymbol{B}) + \frac{e}{m}\boldsymbol{E} + \frac{1}{Nm}\hat{\boldsymbol{F}}_{\mathrm{i}} + \frac{1}{Nm}\hat{\boldsymbol{F}}_{\mathrm{p}}. \tag{8.5}$$

Here $\hat{\boldsymbol{F}}_{\mathrm{i}}$, $\hat{\boldsymbol{F}}_{\mathrm{p}}$, and the rate of change of the phonon energy, $\hat{W} \equiv -\mathrm{i}[H_{\mathrm{ph}}, H]$, have the same expressions as given in Eqs. (1.21), (1.22) and (1.23). The determination of the statistical averages of these quantities to lowest order in the scattering interaction, $H_{\mathrm{I}} = H_{\mathrm{ei}} + H_{\mathrm{ep}}$, follows the same procedures as in Chapter 1 using the initial density matrix $\hat{\rho}_0$ given by (1.30). By identifying $\langle \boldsymbol{V} \rangle = \boldsymbol{v}$ as the average drift velocity of the electron system in the presence of electric and magnetic fields, and $\langle \dot{\boldsymbol{V}} \rangle = d\boldsymbol{v}/dt$ as the time change rate of the drift velocity, we are able to obtain the time-dependent balance equations for statistical average quantities and the Langevin equations for small fluctuating or random quantities. The expressions for frictional forces, energy-loss rate, memory functions, fluctuating-force correlation functions, noise temperatures and diffusion coefficients, when they are written in the plane-wave representation, formally are the same as those without a magnetic field. In comparison with the case without magnetic field, an apparent feature characterizing the existence of the magnetic field is an additional Lorentz force term, $ne(\boldsymbol{v} \times \boldsymbol{B})$, appearing in the equation of motion of the center of mass (i.e. the force-balance equation):

$$Nm\frac{d\boldsymbol{v}}{dt} = Ne\boldsymbol{E} + Ne\boldsymbol{v} \times \boldsymbol{B} + \tilde{\boldsymbol{f}}_{\text{i}} + \tilde{\boldsymbol{f}}_{\text{p}}. \tag{8.6}$$

The energy-balance equation is exactly the same as (1.56),

$$-\frac{dU}{dt} = \boldsymbol{v} \cdot (\tilde{\boldsymbol{f}}_{\text{i}} + \tilde{\boldsymbol{f}}_{\text{p}}) + \tilde{w}, \tag{8.7}$$

where $U = \langle H_{\text{er}} \rangle$ is the average energy of relative electrons. The expressions for the frictional forces $\tilde{\boldsymbol{f}}_{\text{i}}$ and $\tilde{\boldsymbol{f}}_{\text{p}}$, and that for the energy-loss rate $\tilde{w}$, formally are the same as (1.42), (1.43) and (1.50) given in Chapter 1,

$$\tilde{\boldsymbol{f}}_{\text{i}} = -\text{i} n_{\text{i}} \sum_{\boldsymbol{q}} \boldsymbol{q} |u(\boldsymbol{q})|^2 \int_{-\infty}^{\infty} dt' A(\boldsymbol{q},t,t') \Pi(\boldsymbol{q},t-t'), \tag{8.8}$$

$$\tilde{\boldsymbol{f}}_{\text{p}} = -\text{i} \sum_{\boldsymbol{q},\lambda} \boldsymbol{q} |M(\boldsymbol{q},\lambda)|^2 \int_{-\infty}^{\infty} dt' A(\boldsymbol{q},t,t') \Lambda(\boldsymbol{q},\lambda,t-t'), \tag{8.9}$$

$$\tilde{w} = -\sum_{\boldsymbol{q}\lambda} |M(\boldsymbol{q},\lambda)|^2 \int_{-\infty}^{\infty} dt' \Lambda(\boldsymbol{q},t,t') \Gamma(\boldsymbol{q},\lambda,t-t'), \tag{8.10}$$

containing correlation functions $\Pi(\boldsymbol{q},t-t')$, $\Lambda(\boldsymbol{q},\lambda,t-t')$ and $\Gamma(\boldsymbol{q},\lambda,t-t')$ defined by (1.44), (1.45) and (1.51), and the $A(\boldsymbol{q},t,t')$ function defined in (1.49):

$$A(\boldsymbol{q},t,t') \equiv \exp\left[\text{i}\boldsymbol{q}\cdot(\boldsymbol{R}(t) - \boldsymbol{R}(t'))\right] \simeq \exp\left[\text{i}\boldsymbol{q}\cdot\int_{t'}^{t} \boldsymbol{v}(s)ds\right]. \tag{8.11}$$

They depend on the center-of-mass velocity $\boldsymbol{v}(t)$ and electron temperature $T_{\text{e}}(t)$. If the time variation of the CM velocity $\boldsymbol{v}(t)$ is slow, we can use the approximate expansion (3.2) for $A(\boldsymbol{q},t,t')$ and obtain the following force- and energy-balance equations written for systems of unit volume:

$$(nm\mathcal{I} + \mathcal{B}) \cdot \frac{d}{dt}\boldsymbol{v}(t) = ne\boldsymbol{E}(t) + ne\boldsymbol{v} \times \boldsymbol{B} + \boldsymbol{f}_{\text{i}} + \boldsymbol{f}_{\text{p}}, \tag{8.12}$$

$$-C_{\text{e}}\frac{d}{dt}T_{\text{e}}(t) = \boldsymbol{v} \cdot (\boldsymbol{f}_{\text{i}} + \boldsymbol{f}_{\text{p}}) + w. \tag{8.13}$$

Here n is the electron density, C_{e} is the specific heat, $\boldsymbol{f}_{\text{i}}$, $\boldsymbol{f}_{\text{p}}$ and w have exactly the same formal expressions as those in the case without a magnetic field in Chapter 3:

$$f_\mathrm{i} = n_\mathrm{i} \sum_{q} q\, |u(q)|^2 \Pi_2(q,\omega_0), \tag{8.14}$$

$$f_\mathrm{p} = 2 \sum_{q,\lambda} q\, |M(q,\lambda)|^2 \Pi_2(q, \Omega_{q\lambda} + \omega_0) \left[n\!\left(\frac{\Omega_{q\lambda}}{T}\right) - n\!\left(\frac{\Omega_{q\lambda} + \omega_0}{T_\mathrm{e}}\right) \right], \tag{8.15}$$

$$w = 2 \sum_{q,\lambda} \Omega_{q\lambda}\, |M(q,\lambda)|^2 \Pi_2(q, \Omega_{q\lambda} + \omega_0) \left[n\!\left(\frac{\Omega_{q\lambda}}{T}\right) - n\!\left(\frac{\Omega_{q\lambda} + \omega_0}{T_\mathrm{e}}\right) \right], \tag{8.16}$$

in which $\omega_0 \equiv \boldsymbol{q} \cdot \boldsymbol{v}(t)$ and $T_\mathrm{e} = T_\mathrm{e}(t)$. They may depend on time through the time dependence of variables $\boldsymbol{v}(t)$ and $T_\mathrm{e}(t)$. In equation (8.12), $\mathcal{I}$ represents the unit tensor, and the tensor $\mathcal{B}$, which has the same expression as given in Eq. (3.8), gives rise to a drift-velocity dependent effective-mass correction on the center-of-mass motion induced by impurity and phonon scatterings.

The $\Pi_2(\boldsymbol{q},\omega)$ function in the expressions (8.14)–(8.16) for $\boldsymbol{f}_\mathrm{i}$, $\boldsymbol{f}_\mathrm{p}$ and w and the $\Pi_1(\boldsymbol{q},\omega)$ function in the expression (3.8) for tensor $\mathcal{B}$, are respectively the imaginary part and real part of the Fourier transform of the relative-electron density correlation function, $\Pi(\boldsymbol{q},\omega)$, defined by (1.44), in which the statistical average $\langle \cdots \rangle_0$ is taken with reference to the initial density matrix given by the same form as (1.30) but H_er representing Hamiltonian (8.3), i.e. the Hamiltonian of relative electron system with interparticle Coulomb interactions and in the presence of the magnetic field (without electric field). As in the case without the magnetic field, the effect of electron–electron Coulomb interaction is taken into account in the random phase approximation (RPA), that

$$\Pi(\boldsymbol{q},\omega) = \frac{\Pi_0(\boldsymbol{q},\omega)}{\epsilon(\boldsymbol{q},\omega)}. \tag{8.17}$$

Here $\epsilon(\boldsymbol{q},\omega) = 1 - \nu_\mathrm{c}(q) \Pi_0(\boldsymbol{q},\omega)$ is the RPA dielectric function of the electron system, $\nu_\mathrm{c} = e^2/(\epsilon_0 \kappa q^2)$ is the Coulomb potential (κ is the dielectric constant of the background material), $\Pi_0(\boldsymbol{q},\omega)$ is the Fourier transform of density correlation function of noninteracting relative electrons in the presence of the magnetic field.

Therefore, in comparison with balance equations (3.3) and (3.4) in the case without magnetic field, balance equations (8.12) and (8.13) have two features characterizing the existence of a magnetic field: the appearance of a classical Lorentz force term $Ne\boldsymbol{v} \times \boldsymbol{B}$ in the force-balance equation for the center of mass, and the installation of the proper quantized magnetic field dependence into the structure of the electron density correlation function $\Pi_0(\boldsymbol{q},\omega)$.

For calculating the density correlation function $\Pi_0(\boldsymbol{q},\omega)$ of the noninteracting electron system in the presence of a magnetic field, we proceed in the Landau representation. By choosing the vector potential as $\boldsymbol{A}(\boldsymbol{r}) = (-By, 0, 0)$, the eigenfunctions of the single-particle Hamiltonian of noninteracting relative electron system in the presence of magnetic field $\boldsymbol{B} = (0, 0, B)$,

$$h_e^{(0)} = \frac{1}{2m}\left[(p_x + eBy)^2 + p_y^2 + p_z^2\right] \tag{8.18}$$

(the spin-related part and the "′" denoting the relative-electron variables are omitted for simplicity), are the Landau wave functions of the form (assuming a system of unit volume, $V = L_x L_y L_z = 1$):

$$\psi_{nk_x k_z}(\boldsymbol{r}) = e^{ik_x x + ik_z z} \chi_{nk_x}(y), \tag{8.19}$$

$$\chi_{nk_x}(y) = \left(\frac{1}{\sqrt{\pi} 2^n n! l_B}\right)^{1/2} \exp\left[-\frac{(y-y_0)^2}{2l_B^2}\right] H_n\left(\frac{y-y_0}{l_B}\right), \tag{8.20}$$

in which $y_0 \equiv -k_x/(eB)$, $l_B \equiv |eB|^{-1/2}$ is the magnetic length, $H_n(x)$ are Hermite polynomials. The eigen energies form the Landau levels:

$$\varepsilon_n(k_z) = \left(n + \frac{1}{2}\right)\omega_c + \frac{k_z^2}{2m} \tag{8.21}$$

($n = 0, 1, 2, \cdots$), where $\omega_c = |eB|/m$ is the cyclotron frequency. The Landau level is degenerate with respect to k_x and to the spin index (due to the neglect of the spin splitting). Since the electron motion in x–y plane is restricted in the area $S = L_x L_y$, the number of available k_x points is $|eB|S/(2\pi)$, proportional to the strength of the magnetic field. In the z direction the electron is restricted within the length L_z, therefore the number of states in the Δk_z interval is $|eB|\Delta k_z/(2\pi)^2$ for each Landau level (Landau and Lifshitz, 1977).

The density operator of the electron system, $\rho_{\boldsymbol{q}} = \sum_j e^{i\boldsymbol{q}\cdot\boldsymbol{r}_j}$, can be expressed in the Landau representation as

$$\rho_{\boldsymbol{q}} = \sum_{\substack{n',n \\ k_x, k_z, \sigma}} J_{n'n}(q_y, k_x + q_x, k_x)\, c^\dagger_{n'k_x+q_x, k_z+q_z, \sigma} c_{nk_x k_z \sigma}, \tag{8.22}$$

in which $c^\dagger_{nk_x k_z \sigma}$ ($c_{nk_x k_z \sigma}$) are creation (annihilation) operators in the Landau basis (8.19) and the form factor

$$J_{n'n}(q_y, k_x + q_x, k_x) = \int dy\, e^{iq_y y} \chi^*_{n' k_x + q_x}(y) \chi_{n k_x}(y). \tag{8.23}$$

The expression of $\Pi_0(\boldsymbol{q}, \omega)$ function in the Landau representation is attained when expanding the density operators $\rho_{\boldsymbol{q}}$ and $\rho_{-\boldsymbol{q}}$ inside the correlation function (1.44) in the Landau representation. Noticing that the product $J_{nn'}(q_y, k_x + q_x, k_x) J_{n'n}(-q_y, k_x, k_x + q_x)$ depends on n, n', q_x and q_y, but is independent of k_x, one gets (Mermin and Canal, 1964)

$$\Pi_0(\boldsymbol{q}, \omega) = \frac{1}{2\pi l_B^2} \sum_{n,n'} C_{nn'}(l_B^2 q_\parallel^2 / 2) \Pi_0(n, n', q_z, \omega). \tag{8.24}$$

Here $q_\parallel^2 \equiv q_x^2 + q_y^2$,

$$C_{nn'}(x) \equiv \frac{n_2!}{n_1!} x^{n_1 - n_2} e^{-x} [L^{n_1 - n_2}_{n_2}(x)]^2, \tag{8.25}$$

in which $n_1 = \max(n, n')$, $n_2 = \min(n, n')$, and $L^l_m(x)$ are associated Laguerre polynomials (Wang and Guo, 1989):

$$L^l_m(x) = \sum_{s=0}^{m} (-1)^s \frac{(m+l)! x^s}{(l+s)!(m-s)!s!}. \tag{8.26}$$

In Eq. (8.24) $\Pi_0(n, n', q_z, \omega)$ is a correlation function in the Landau representation. Its real part $\Pi_{01}(n, n', q_z, \omega)$ and imaginary part $\Pi_{02}(n, n', q_z, \omega)$ are given respectively by (without spin splitting and collision broadening)

$$\Pi_{01}(n, n', q_z, \omega) = 2 \sum_{k_z} \frac{f(\varepsilon(n', k_z + q_z), T_e) - f(\varepsilon(n, k_z), T_e)}{\omega + \varepsilon(n', k_z + q_z) - \varepsilon(n, k_z)}, \tag{8.27}$$

$$\Pi_{02}(n, n', q_z, \omega) = 2\pi \sum_{k_z} [f(\varepsilon(n, k_z), T_e) - f(\varepsilon(n', k_z + q_z), T_e)]$$
$$\times \delta(\omega + \varepsilon(n', k_z + q_z) - \varepsilon(n, k_z)). \tag{8.28}$$

Carrying out the summation over k_z in (8.28) we can write

$$\Pi_{02}(n,n',q_z,\omega) = \frac{m}{|q_z|}$$
$$\times \left\{ f\left(\left(n+\frac{1}{2}\right)\omega_c + \frac{m}{2q_z^2}\left(\frac{q_z^2}{2m} + (n'-n)\omega_c + \omega\right)^2, T_e\right)\right.$$
$$\left. - f\left(\left(n'+\frac{1}{2}\right)\omega_c + \frac{m}{2q_z^2}\left(\frac{q_z^2}{2m} - (n'-n)\omega_c - \omega\right)^2, T_e\right)\right\}. \quad (8.29)$$

The real and imaginary parts $\Pi_{01}(\boldsymbol{q},\Omega)$ and $\Pi_{02}(\boldsymbol{q},\Omega)$ of the noninteracting electron density correlation function $\Pi_0(\boldsymbol{q},\Omega)$ in the magnetic field can also be expressed in a closed-form integral representation introduced by Horing (1965) in terms of elementary functions:

$$\Pi_{01}(\boldsymbol{q},\Omega) = \mathcal{P}\int \frac{d\omega}{2\pi}\int \frac{d\omega'}{2\pi}\frac{\omega'}{\Omega^2-\omega'^2} f(\omega,T_e) R(\omega,\omega',\boldsymbol{q}), \quad (8.30)$$

$$\Pi_{02}(\boldsymbol{q},\Omega) = -\frac{1}{2}\int \frac{d\omega}{2\pi} f(\omega,T_e) R(\omega,\Omega,\boldsymbol{q}), \quad (8.31)$$

where $\mathcal{P}$ denotes taking the principal value of the integral, and

$$R(\omega,\Omega,\boldsymbol{q}) = \left(\frac{m}{2\pi}\right)^{3/2}\omega_c \int_{-\infty}^{\infty} dy\, e^{-i\Omega y/2}\int_{-\infty}^{\infty}\frac{dx}{(ix)^{1/2}}\frac{\sin(\Omega x/2)}{\sin(\omega_c x/2)} e^{i\omega x}$$
$$\times \exp\left(-i\frac{q_z^2}{2m}\frac{x^2-y^2}{4x}\right)\exp\left(-i\frac{q_\parallel^2}{2m}\frac{\cos(\omega_c y/2)-\cos(\omega_c x/2)}{\omega_c \sin(\omega_c x/2)}\right). \quad (8.32)$$

In Eqs. (8.27)–(8.31), $f(\omega,T_e) = 1/[1+\exp(\omega-\mu)/T_e]$ is the Fermi function at temperature T_e with μ the chemical potential of the system in the magnetic field. In the nondegenerate case, $f(\omega,T_e) \sim e^{\mu/T_e} e^{-\omega/T_e}$, performing the integration over ω in (8.31), we have

$$\Pi_{02}(\boldsymbol{q},\Omega) = -\left(\frac{m}{2\pi}\right)^{3/2}\omega_c T_e^{1/2}\exp\left[-\frac{q_z^2}{8mT_e} - \frac{q_\parallel^2}{2m\omega_c}\coth\left(\frac{\omega_c}{2T_e}\right)\right]$$
$$\times e^{\mu/T_e}\frac{\sinh(\Omega/2T_e)}{\sinh(\omega_c/2T_e)}\int_0^\infty dy \cos\left(\frac{\Omega y}{2}\right)$$
$$\times \exp\left[-\frac{T_e}{8m}q_z^2 y^2 + \frac{q_\parallel^2 \cos(\omega_c y/2)}{2m\omega_c \sinh(\omega_c/2T_e)}\right]. \quad (8.33)$$

Eqs. (8.31) and (8.33) are the results without spin splitting. The chemical potential μ, which should be determined by the carrier density through the equation (N is the total number of carriers)

$$N = \sum_{n,k_z,\sigma} f(\varepsilon(n,k_z), T_e),$$

is dependent on the magnetic field. For nondegenerate systems without spin splitting,

$$e^{\mu/T_e} = \sqrt{2}\,n \left(\frac{\pi}{mT_e}\right)^{3/2} \left(\frac{2T_e}{\omega_c}\right) \sinh\left(\frac{\omega_c}{2T_e}\right), \qquad (8.34)$$

in which n is the carrier density.

The above two different expressions for $\Pi_0(\boldsymbol{q},\omega)$ are equivalent and valid for arbitrary strength of the magnetic field. Generally, the Landau series representation (8.24)–(8.29) may be more convenient in dealing with issues at strong magnetic field and the Horing integral representation (8.30)–(8.33) may be more convenient for weak magnetic field expansions (Horing, 1965).

8.2 General Results of Steady-State Magnetotransport in Isotropic Bulk Systems

The force and energy balance equations (8.12) and (8.13) with arbitrary electric and magnetic fields can be conveniently used to deal with hot electron magnetotransport in the longitudinal and transverse configurations. For steady transport state these equations reduce to (Lei, Cai and Ting, 1985)

$$ne\boldsymbol{E} + ne\boldsymbol{v}\times\boldsymbol{B} + \boldsymbol{f}(\boldsymbol{v}) = 0, \qquad (8.35)$$

$$\boldsymbol{v}\cdot\boldsymbol{f}(\boldsymbol{v}) + w(\boldsymbol{v}) = 0, \qquad (8.36)$$

with the frictional force and energy-dissipation rate $\boldsymbol{f}(\boldsymbol{v}) \equiv \boldsymbol{f}_i + \boldsymbol{f}_p$ and $w(\boldsymbol{v}) \equiv w$ given by (8.14)–(8.16). For either the longitudinal ($\boldsymbol{E} \parallel \boldsymbol{B}$) or the transverse ($\boldsymbol{E}\perp\boldsymbol{B}$) configuration, the magnitudes of the frictional force and energy-loss rate are functions of $v = |\boldsymbol{v}|$ only, and $\boldsymbol{f}(\boldsymbol{v})$ is in the opposite direction of $\boldsymbol{v}$. Letting $\boldsymbol{f}(\boldsymbol{v}) = f(v)\boldsymbol{v}/v$ or $\boldsymbol{v}\cdot\boldsymbol{f}(\boldsymbol{v}) = vf(v)$, and $w(\boldsymbol{v}) = w(v)$, we can define a resistivity function

$$\rho(v) \equiv -\frac{f(v)}{n^2 e^2 v} \qquad (8.37)$$

and an energy-dissipation resistivity function

$$\rho_E(v) \equiv \frac{w(v)}{n^2 e^2 v^2}. \tag{8.38}$$

The energy-balance equation (8.36) can be written as

$$\rho(v) - \rho_E(v) = 0. \tag{8.39}$$

For given magnetic field strength B and lattice temperature T, Eq. (8.39) determines the electron temperature T_e as a function of drift velocity v. The force-balance equation (8.35) depends explicitly on the relative direction between the magnetic field $\boldsymbol{B}$ and the electric field $\boldsymbol{E}$.

In the case of longitudinal configuration with $\boldsymbol{E} = E\hat{\boldsymbol{z}} \| \boldsymbol{B}$, $\boldsymbol{v}$ is also along the z direction. The force-balance equation (8.35) reduces to (Horing, Cui and Lei, 1992)

$$neE + f(v) = 0,$$

formally the same as that without magnetic field, and the resistivity is given by

$$\rho_{zz} \equiv \frac{E}{nev} = -\frac{f(v)}{n^2 e^2 v} = \rho(v). \tag{8.40}$$

Here $\rho(v)$ should be calculated at electron temperature T_e, i.e., $\rho(v)$ is just the resistivity function satisfying the energy-balance equation (8.39).

In the case of transverse configuration with $\boldsymbol{B} \| \hat{\boldsymbol{z}}$ and $\boldsymbol{E}$ in the x–y plane, $\boldsymbol{v}$ is also in the x–y plane. We can write the force-balance equation (8.35) using the following two different schemes (Lei, Cai and Ting, 1985). Taking the x-axis along the $\boldsymbol{v}$ direction, $\boldsymbol{v} = (v, 0, 0)$, $\boldsymbol{E} = (E_x, E_y, 0)$, we can write Eq. (8.35) as

$$neE_x + f(v) = 0,$$
$$neE_y - neBv = 0,$$

and the current density is along the x direction,

$$j_x = nev, \quad j_y = 0. \tag{8.41}$$

Therefore the nonlinear longitudinal (diagonal) and transverse (Hall) resistivities are immediately obtained:

$$\rho_{xx} \equiv E_x/j_x = \rho(v), \tag{8.42}$$
$$\rho_{yx} \equiv E_y/j_x = B/ne. \tag{8.43}$$

On the other hand, if we choose the x-axis along the $\boldsymbol{E}$ direction, $\boldsymbol{E} = (E,0,0)$, and let φ be the angle between $\boldsymbol{v}$ and $\boldsymbol{E}$, $\boldsymbol{v} = (v\cos\varphi, v\sin\varphi, 0)$, Eq. (8.35) becomes

$$neE + neBv\sin\varphi + f(v)\cos\varphi = 0,$$
$$-neBv\cos\varphi + f(v)\sin\varphi = 0.$$

This yields

$$E = [(nev\rho(v))^2 + B^2 v^2]^{1/2}, \quad (8.44)$$
$$\cot\varphi = -\frac{ne}{B}\rho(v). \quad (8.45)$$

Now the current density is

$$j_x = nev\cos\varphi, \quad j_y = nev\sin\varphi. \quad (8.46)$$

Then the nonlinear longitudinal (diagonal) and transverse (Hall) conductivities are obtained as

$$\sigma_{xx} \equiv \frac{j_x}{E} = \frac{\rho(v)}{\rho(v)^2 + (B/ne)^2}, \quad (8.47)$$
$$\sigma_{yx} \equiv \frac{j_y}{E} = \frac{-B/ne}{\rho(v)^2 + (B/ne)^2}. \quad (8.48)$$

The rest components of the resistivity and conductivity tensors are easily determined by the symmetrical consideration:

$$\rho_{yy} = \rho_{xx}, \quad \rho_{xy} = -\rho_{yx}, \quad (8.49)$$
$$\sigma_{yy} = \sigma_{xx}, \quad \sigma_{xy} = -\sigma_{yx}. \quad (8.50)$$

The nonlinear resistivity tensor and nonlinear conductivity tensor, both defined by the relations between current density vector $\boldsymbol{j}$ and electric field vector $\boldsymbol{E}$, are the inverse tensor of each other:

$$\begin{pmatrix} \rho_{xx} & \rho_{xy} \\ \rho_{yx} & \rho_{yy} \end{pmatrix} = \begin{pmatrix} \sigma_{xx} & \sigma_{xy} \\ \sigma_{yx} & \sigma_{yy} \end{pmatrix}^{-1}. \quad (8.51)$$

Note that, the nonlinear transverse (Hall) resistivity given by (8.43), or the hot-electron Hall coefficient

$$R_{\rm H} = 1/ne, \quad (8.52)$$

is exactly the same as in the linear transport for single parabolic band system (for fixed electron density n). Furthermore, the diagonal resistivity in the transverse configuration, $\rho_{xx} = \rho_{yy} = \rho(v)$, and the resistivity in the longitudinal configuration, $\rho_{zz} = \rho(v)$, are also formally the same. However, the $\Pi_2(\bm{q}, \bm{q}\cdot\bm{v})$ function in the expressions (8.14) and (8.15) for $\bm{f}_\mathrm{i}$ and $\bm{f}_\mathrm{p}$ depends on the direction of the magnetic field relative to the drift velocity $\bm{v}$. Thus the values of resistivity function $\rho(v)$ are different at the longitudinal and transverse configurations. We will discuss these in the next sections.

8.3 Longitudinal Configuration

In the longitudinal configuration, with the electric field parallel to the magnetic field in the z direction, $\bm{E}\|\bm{B}\|\hat{\bm{z}}$, we have $\rho_{zz} = \rho_{zz}^\mathrm{i} + \rho_{zz}^\mathrm{p} = \rho_\mathrm{i}(v) + \rho_\mathrm{p}(v)$,

$$\rho_{zz}^\mathrm{i} = \rho_\mathrm{i}(v) = -\frac{n_\mathrm{i}}{n^2 e^2 v}\sum_{\bm{q}}|u(\bm{q})|^2 q_z \Pi_2(\bm{q}, q_z v), \tag{8.53}$$

$$\rho_{zz}^\mathrm{p} = \rho_\mathrm{p}(v) = -\frac{2}{n^2 e^2 v}\sum_{\bm{q},\lambda}|M(\bm{q},\lambda)|^2 q_z \Pi_2(\bm{q}, \Omega_{\bm{q}\lambda} + q_z v)$$
$$\times\left[n\left(\frac{\Omega_{\bm{q}\lambda}}{T}\right) - n\left(\frac{\Omega_{\bm{q}\lambda} + q_z v}{T_\mathrm{e}}\right)\right]. \tag{8.54}$$

The expression for $\rho_\mathrm{E}(v)$ can be obtained from that of $\rho_\mathrm{p}(v)$ by multiplying $-\Omega_{\bm{q}\lambda}q_z^{-1}v^{-1}$ inside the summation sign. In the small v (low electric field) limit, the energy-balance equation requires $T_\mathrm{e} = T$, and the linearized ρ_{zz}^i and ρ_{zz}^p are given by

$$\rho_{zz}^\mathrm{i}(v\to 0) = -\frac{n_\mathrm{i}}{n^2 e^2}\sum_{\bm{q}}|\bar{u}(\bm{q})|^2 q_z^2 \frac{\partial}{\partial\omega}\Pi_{02}(\bm{q},\omega)\Big|_{\omega=0}, \tag{8.55}$$

$$\rho_{zz}^\mathrm{p}(v\to 0) = -\frac{2}{n^2 e^2 T}\sum_{\bm{q},\lambda}|\bar{M}(\bm{q},\lambda)|^2 q_z^2 \Pi_{02}(\bm{q}, \Omega_{\bm{q}\lambda})\frac{e^{\Omega_{\bm{q}\lambda}/T}}{(e^{\Omega_{\bm{q}\lambda}/T} - 1)^2}, \tag{8.56}$$

in which $|\bar{u}(\bm{q})|^2 \equiv |u(\bm{q})/\epsilon(\bm{q},0)|^2$, $|\bar{M}(\bm{q},\lambda)|^2 \equiv |M(\bm{q},\lambda)/\epsilon(\bm{q},\Omega_{\bm{q}\lambda})|^2$, and $\epsilon(\bm{q},\omega) = 1 - v_\mathrm{c}(q)\Pi_0(\bm{q},\omega)$ is the RPA dielectric function of the electron system.

8.3.1 Linear resistivity due to impurity scattering

Taking the short-range screened impurity potential $|\bar{u}(q)|^2$ as constant u^2, the impurity-induced linear resistivity can be written as (Horing, Cui, and

Lei, 1992)

$$\rho_{zz}^{i}(v \to 0) = -\frac{n_{\mathrm{i}} m^4 u^2 \omega_{\mathrm{c}}^2}{\pi^3 n^2 e^2} \sum_{n,n'} \int_0^\infty d\varepsilon\, f'(\varepsilon) \left[\frac{\varepsilon - (n+\frac{1}{2})\omega_{\mathrm{c}}}{\varepsilon - (n'+\frac{1}{2})\omega_{\mathrm{c}}}\right]^{1/2}$$

$$\times \theta\!\left(\varepsilon - \left(n+\frac{1}{2}\right)\omega_{\mathrm{c}}\right) \theta\!\left(\varepsilon - \left(n'+\frac{1}{2}\right)\omega_{\mathrm{c}}\right), \quad (8.57)$$

where $\theta(x)$ is the unit step function and $f(\varepsilon) \equiv f(\varepsilon, T)$ is the Fermi function at lattice temperature T.

As in the case without magnetic field, the expression (8.57) for linear resistivity applies for systems with strong electron–electron interaction (isothermal resistivity). Though it is in precise agreement with the adiabatic (for systems without e–e interaction) magnetotransport results (Argyres, 1956; Roth and Argyres, 1966) at zero temperature in all regimes of magnetic field strength (low, intermediate and quantum strong fields), they are different at finite temperatures. In the nondegenerate regime $[f(\varepsilon) \sim e^{-\varepsilon/T}]$ the linear longitudinal resistivity expression (8.57) for arbitrary magnetic field strength is ($x \equiv \omega_{\mathrm{c}}/T$)

$$\frac{\rho_{zz}^{\mathrm{i}}}{\rho_0^{\mathrm{i}}} = \frac{3}{8}(1-e^{-x}) \sum_{n,n'} \int_0^\infty ds\, e^{-s} \left(\frac{s-nx}{s-n'x}\right)^{1/2} \theta(s-nx)\,\theta(s-n'x), \quad (8.58)$$

ρ_0^{i} being the zero field resistivity. The corresponding adiabatic linear longitudinal nondegenerate conductivity of Argyres can be written as

$$\frac{\sigma_{zz}^{\mathrm{i}}}{\sigma_0^{\mathrm{i}}} = 3\left(\frac{1-e^x}{x}\right) \int_0^\infty ds\, e^{-s} \frac{\sum_n (s-nx)^{1/2}}{\sum_n (s-nx)^{-1/2}}, \quad (8.59)$$

σ_0^{i} being the zero field conductivity. In the quantum strong magnetic field limit ($\omega_{\mathrm{c}} \gg T$), where only the lowest Landau eigenstate is occupied, $\rho_{zz}^{\mathrm{i}}/\rho_0^{\mathrm{i}} = 3\omega_{\mathrm{c}}/8T$ in contrast to $\sigma_{zz}^{\mathrm{i}}/\sigma_0^{\mathrm{i}} = 3T/\omega_{\mathrm{c}}$. In the zero magnetic field the ratio of resistivities predicted by isothermal and adiabatic theories is $\rho_0^{\mathrm{i}}/(\sigma_0^{\mathrm{i}})^{-1} = 32/9\pi$ (see Sec. 7.3). In the strong magnetic field limit the ratio of two resistivities is $\rho_{zz}^{\mathrm{i}}/(\sigma_{zz}^{\mathrm{i}})^{-1} = 4/\pi$.

8.3.2 Resistivity due to optic phonon scattering, magnetophonon resonance

Optic phonons are almost dispersionless ($\Omega_{q\lambda} \approx$ constant). The main feature of optic-phonon scattering induced resistivity as a function of magnetic field is the appearance of magnetophonon resonance structure. In nonpolar semiconductors the major mechanism is played by the optical deformation potential scattering, having an electron–phonon matrix element $|M(q,\lambda)|^2 = D^2/(2d_c\Omega_o)$, with Ω_o the optic phonon frequency, D the optic deformation potential and d_c the mass density of the lattice. Without screening, the linear resistivity in the nondegenerate limit is

$$\rho_{zz}^{\text{p}}(v \to 0) = \frac{D^2 m^{5/2}\omega_c}{\sqrt{2}\pi^{3/2}ne^2 d_c \Omega_o T^{3/2}} n\left(\frac{\Omega_o}{T}\right) \sum_{n=0}^{\infty}\sum_{\pm} \exp\left(\frac{\Omega_o - n\omega_c}{2T}\right)$$
$$\times |\Omega_o - (\pm)n\omega_c|\, \text{K}_1\left(\frac{|\Omega_o - (\pm)n\omega_c|}{2T}\right), \qquad (8.60)$$

where $\text{K}_1(x)$ is the modified Bessel function (Wang and Guo, 1989). The resistivity ρ_{zz}^{p} itself has no singularity, but its second derivative with respect to the magnetic field exhibits sharp peaks at $n\omega_c = \Omega_o$ ($n = 1, 2, \cdots$), clearly reflecting the structure of magnetophonon resonance.

In the nonlinear regime, the q_z integral in the expressions of optic-phonon induced $\rho(v)$ and $\rho_{\text{E}}(v)$ can not be done analytically. In the nondegenerate limit after carrying out the $q_\parallel$ integration, we have

$$\rho(v) = \frac{D^2 m^{3/2}\omega_c}{2^{5/2}\pi^{3/2}ne^2 d_c \Omega_o T_e^{1/2} v} n\left(\frac{\Omega_o}{T}\right) \sum_{n=0}^{\infty}\sum_{\pm} \exp\left(-\frac{n\omega_c}{2T_e}\right) \int_0^{\infty} dq_z$$
$$\times \exp\left(-\frac{q_z^2}{8mT_e}\right) \left\{\exp\left(-\frac{m(\Omega_o + q_z v - (\pm)n\omega_c)^2}{2T_e q_z^2}\right)\right.$$
$$\left.\times \left[\exp\left(\frac{\Omega_o + q_z v}{2T_e}\right) - \exp\left(\frac{\Omega_o}{T}\right)\exp\left(-\frac{\Omega_o + q_z v}{2T_e}\right)\right] - (v \to -v)\right\}.$$
$$(8.61)$$

The expression of ρ_{E} is obatined from the right-hand-side of (8.61) by multiplying the integrand with a factor $-\Omega_o q_z^{-1} v^{-1}$ and changing the term $-(v \to -v)$ into $+(v \to -v)$.

For polar semiconductors, the main mechanism is the Fröhlish scattering related to longitudinal (polar) optic phonons of frequency Ω_{LO}, having electron–phonon matrix element $|M(\mathbf{q},\lambda)|^2 = M^2/(q_\parallel^2 + q_z^2)$, M^2 being a constant [see (2.129)] (Mahan, 1972). In the nondegenerate limit, polar-optic phonon induced $\rho(v)$ reads

$$\rho(v) = \frac{M^2 m^{1/2}}{(2\pi)^{3/2} n e^2 T_e^{1/2} v} \left(\frac{1 - e^{\omega_c/T_e}}{e^{\Omega_{LO}/T} - 1}\right) \sum_{n_1, n_2} \int_0^\infty dq_z \int_0^\infty dq_\|$$

$$\times \frac{q_\|}{q_\|^2 + q_z^2} C_{n_1 n_2}\left(\frac{q_\|^2}{2m\omega_c}\right) \exp\left(-\frac{(n_1 + n_2)\omega_c}{2T_e}\right) \exp\left(-\frac{q_z^2}{8mT_e}\right)$$

$$\times \left\{ \exp\left(-\frac{m[\Omega_{LO} + q_z v + (n_2 - n_1)\omega_c]^2}{2T_e q_z^2}\right) \left[\exp\left(\frac{\Omega_{LO} + q_z v}{2T_e}\right)\right.\right.$$

$$\left.\left. - \exp\left(\frac{\Omega_{LO}}{T}\right) \exp\left(-\frac{\Omega_{LO} + q_z v}{2T_e}\right)\right] - (v \to -v) \right\}. \quad (8.62)$$

The expression for $\rho_E(v)$ can be obtained from the right-hand-side of (8.62) by multiplying the integrand with a factor $-\Omega_{LO} q_z^{-1} v^{-1}$ and changing the term $-(v \to -v)$ into $+(v \to -v)$.

Generally, the integrals over $q_\|$ and q_z can not be done analytically. To have a rough upper bound estimation, we take an approximate electron–phonon matrix element $|M(q,\lambda)|^2 \approx M^2/q_z^2$ to carry out the $q_\|$ and q_z integrations, giving the following expressions for the linear resistivity induced by polar-optic phonon scattering:

$$\rho^p_{zz}(v \to 0) = \frac{M^2 m^{3/2} \omega_c}{\sqrt{2}\pi^{3/2} n e^2 T^{3/2}} n\left(\frac{\Omega_{LO}}{T}\right) \sum_{n=0}^\infty \sum_\pm \exp\left(\frac{\Omega_{LO} - n\omega_c}{2T}\right)$$

$$\times K_0\left(\frac{|\Omega_{LO} - (\pm)n\omega_c|}{2T}\right). \quad (8.63)$$

It exhibits explicit magnetophonon resonance structure: $\rho^p_{zz}(v \to 0)$ diverges logarithmically when $n\omega_c \to \Omega_{LO}$. This log-divergence is relevant with approximating the matrix element as M^2/q_z^2. Polar optic phonon induced $\rho^p_{zz}(v \to 0)$ is finite at resonant peaks when keeping the original matrix element (see below).

The nonlinear longitudinal resistivity can be directly calculated by solving the energy balance equation (8.39) based on the expression (8.62) of $\rho(v)$ and that of $\rho_E(v)$ from (8.62). The numerically obtained ρ_{zz}/ρ_0 (ρ_0 is linear resistivity in the absence of the magnetic field) at lattice temperature $T = 0.4\Omega_{LO}$ for bias drift velocities $v = 0$, $0.1 v_{LO}$, and $0.2 v_{LO}$ [$v_{LO} \equiv (\Omega_{LO}/m)^{1/2}$], is shown in Fig. 8.1 as a function of the magnetic field in terms of ω_c/Ω_{LO}. Magnetophonon resonance shows up clearly as broad humps at $n\omega_c = \Omega_{LO}$ ($n = 1, 2, 3$), with only relative small amplitude change.

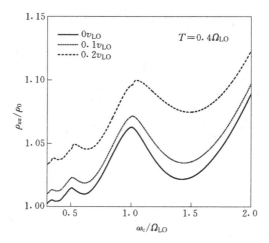

Fig. 8.1 ρ_{zz}/ρ_0 of longitudinal configuration versus $\omega_c/\Omega_{\rm LO}$ at lattice temperature $T = 0.4\Omega_{\rm LO}$ for three different bias drift velocities $v = 0, 0.1v_{\rm LO}$, and $0.2v_{\rm LO}$, in a 3D semiconductor with dominant polar optic phonon Fröhlich scattering. From Cui, Lei and Horing (1988).

Note that the energy-dissipation resistivity function $\rho_{\rm E}$ from (8.62) may have log-divergence at resonant points $\Omega_{\rm LO} + (n_2 - n_1)\omega_c = 0$ unless $v = 0$ and $T_{\rm e} = T$. But it has been found that when polar optic phonon scattering dominates the energy-balance equation (8.39) essentially has a solution of $T_{\rm e}$ very close to T for drift velocity v up to $0.2v_{\rm LO}$, indicating that this log-divergence is numerically tolerable for moderate bias current. Otherwise, the broadening or other mechanism has to be taken into account to eliminate the divergence.

8.4 Transverse Configuration

In the transverse configuration, i.e. the electric field perpendicular to the magnetic field, we choose $\boldsymbol{B} = (0, 0, B)$ and $\boldsymbol{E} = (E_x, E_y, 0)$. Taking x axis along the $\boldsymbol{v}$ direction, $\boldsymbol{v} = (v, 0, 0)$, we have $\rho_{xx} = \rho_{xx}^{\rm i} + \rho_{xx}^{\rm p} = \rho(v)$ with the impurity and phonon terms expressed as

$$\rho_{xx}^{\rm i} = -\frac{n_{\rm i}}{n^2 e^2 v} \sum_{\boldsymbol{q}} |u(\boldsymbol{q})|^2 q_x \Pi_2(\boldsymbol{q}, q_x v), \qquad (8.64)$$

$$\rho_{xx}^{\mathrm{p}} = -\frac{2}{n^2 e^2 v} \sum_{\bm{q},\lambda} |M(\bm{q},\lambda)|^2 q_x \Pi_2(\bm{q}, \Omega_{\bm{q}\lambda} + q_x v)$$
$$\times \left[n\left(\frac{\Omega_{\bm{q}\lambda}}{T}\right) - n\left(\frac{\Omega_{\bm{q}\lambda} + q_x v}{T_{\mathrm{e}}}\right) \right]. \tag{8.65}$$

The linear resistivities are

$$\rho_{xx}^{\mathrm{i}}(v \to 0) = -\frac{n_{\mathrm{i}}}{n^2 e^2} \sum_{\bm{q}} |\bar{u}(\bm{q})|^2 q_x^2 \frac{\partial}{\partial \omega} \Pi_{02}(\bm{q},\omega)\big|_{\omega=0}, \tag{8.66}$$

$$\rho_{xx}^{\mathrm{p}}(v \to 0) = -\frac{2}{n^2 e^2 T} \sum_{\bm{q},\lambda} |\bar{M}(\bm{q},\lambda)|^2 q_x^2 \, \Pi_{02}(\bm{q},\Omega_{\bm{q}\lambda}) \frac{\mathrm{e}^{\Omega_{\bm{q}\lambda}/T}}{(\mathrm{e}^{\Omega_{\bm{q}\lambda}/T} - 1)^2}, \tag{8.67}$$

where $|\bar{u}(\bm{q})|^2 \equiv |u(\bm{q})/\epsilon(\bm{q},0)|^2$, $|\bar{M}(\bm{q},\lambda)|^2 \equiv |M(\bm{q},\lambda)/\epsilon(\bm{q},\Omega_{\bm{q}\lambda})|^2$, and $\epsilon(\bm{q},\omega)$ is the RPA dielectric function.

8.4.1 The logarithmic divergence of linear impurity resistivity and its nonlinear elimination

Though the resistivity expressions (8.66) and (8.67) for the transverse configuration look somewhat similar to expressions (8.55) and (8.56) of longitudinal configuration, they have quite different behaviors. This is because that when there is a magnetic field in the z direction the ω-derivative of the imaginary part of the electron density correlation function, $\Pi_{02}(\bm{q},\omega)$, has $|q_z|^{-1}$ divergence (when $q_z \to 0$) at frequency $\omega = 0$. In the case of longitudinal configuration, i.e. Eqs. (8.55) and (8.56), this divergence is eliminated by the q_z^2 factor in the integrand and thus both linear resistivities $\rho_{zz}^{\mathrm{i}}(v \to 0)$ and $\rho_{zz}^{\mathrm{p}}(v \to 0)$ are finite. In the transverse configuration, where the corresponding factor is q_x^2 rather than q_z^2, the $|q_z|^{-1}$ divergence in (8.66) and (8.67) is not cancelled. This results in the logarithmic divergence in the resistivity after the q_z integration at finite temperature. $T = 0\,\mathrm{K}$ is an exception, where the contribution from $q_z = 0$ is excluded by the Fermi distribution.

This logarithmic divergence was first noted by Davydov and Pomeranchuk (1958) in a calculation of linear magnetoconductivity for systems subject to Landau quantization and may be understood to have its origin in the overlap of two state-density factors which coincide for the identical Landau levels [see below, e.g. (8.68)]. Several cures for this divergence of the transverse linear impurity resistivity were reviewed by Kubo, Hashitsume and Miyake (1965) and by Roth and Argyres (1966), within the framework of linear response theory. These include (1) collision broadening, (2) non-Born scattering, (3) inelastic collision, and (4) classical cutoff.

On the other hand, the nonlinear ($v \neq 0$) resistivities are not divergent (Doran and Gerlach, 1976; Horing, Cui and Lei, 1987). One may verify this from (8.64) and (8,65) even in the absence of screening ($\Pi_2 \to \Pi_{02}$). For instance, assuming a short-range (point δ-function) impurity potential ($|u(q)|^2 \to u^2$, a constant) in (8.64) we have

$$\rho_{xx}^i(v) = \frac{n_i u^2 m^2 \omega_c}{2\pi^2 n^2 e^2 v} \sum_{n_1, n_2} \sum_{\mathbf{q}_\parallel} q_x C_{n_1 n_2}\left(\frac{q_\parallel^2}{2m\omega_c}\right)$$
$$\times \int_0^\infty d\varepsilon \frac{f(\varepsilon, T_e) - f(\varepsilon + q_x v, T_e)}{\left[\varepsilon - (n_1 + \tfrac{1}{2})\omega_c\right]^{1/2} \left[q_x v + \varepsilon - (n_2 + \tfrac{1}{2})\omega_c\right]^{1/2}}. \quad (8.68)$$

Because of a relative shift, $q_x v$, existing in the two inverse-square-root factor in the integrand, the ε integration converges for finite v. This unscreened $\rho_{xx}^i(v)$ exhibits strong oscillations as a function of v and ω_c. The inclusion of dynamic screening or using a non-short-range impurity potential, may somewhat moderate and smooth its oscillatory behavior (Horing, Cui and Lei, 1987).

8.4.2 Optic phonon scattering, magnetophonon resonance

In the case of transverse configuration, magnetophonon resonance structures show up even in the phonon-induced linear resistivity (8.67) itself. For instance, in the nondegenerate limit the polar-optic-phonon (frequency $\Omega_{\mathbf{q}\lambda} = \Omega_{\text{LO}}$) Flöhlich scattering [$|M(q, \lambda)|^2 = M^2/(q_\parallel^2 + q_z^2)$] induced linear resistivity can be written as (Cui, Lei, Horing, 1988)

$$\rho(v \to 0) = \frac{M^2 \omega_c}{(2\pi)^{3/2} n e^2} \left(\frac{m}{T}\right)^{3/2} \left(\frac{1 - e^{-\omega_c/T}}{e^{\Omega_{\text{LO}}/T} - 1}\right)$$
$$\times \sum_{n_1, n_2} \exp\left[\frac{\Omega_{\text{LO}} - (n_2 + n_1)\omega_c}{2T}\right] K_0\left(\frac{|(n_2 - n_1)\omega_c - \Omega_{\text{LO}}|}{2T}\right),$$
$$(8.69)$$

which has peaks of log-divergence when $(n_2 - n_1)\omega_c = \Omega_{\text{LO}}$ due to that of $K_0(x)$ function at $x = 0$.

In the nonlinear case (finite v), the divergence disappears. Given v, the solution to the energy-balance equation (8.39) direct determines the electron temperature T_e and the resistivity function $\rho(v)$, thus the resistivity $\rho_{xx} = \rho(v)$. In the case of dominant polar-optic phonon Fröhlich scattering, numerical calculations based on the $\Pi_{02}(\boldsymbol{q},\omega)$ expressions (8.24) and (8.28) in the Landau representation have been carried out at lattice temperature $T = 0.4\Omega_{\rm LO}$ for bias drift velocities $v = 0.03, 0.1$ and $0.2v_{\rm LO}$ [$v_{\rm LO} = (\Omega_{\rm LO}/m)^{1/2}$]. The obtained resistivity ρ_{xx}/ρ_0 (ρ_0 is the linear resistivity at zero magnetic field) and electron temperature T_e/T are shown in Fig. 8.2 as functions of $\omega_{\rm c}/\Omega_{\rm LO}$. The magnetophonon resonance structures show up markedly in the resistivity curve of the lowest dc bias velocity, but quickly weaken at higher drift velocities.

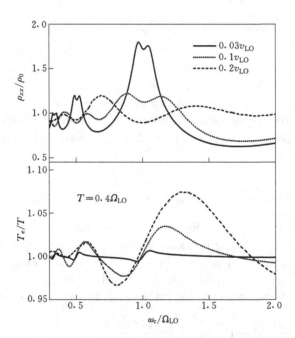

Fig. 8.2 ρ_{zz}/ρ_0 and T_e/T of transverse configuration versus $\omega_{\rm c}/\Omega_{\rm LO}$ at lattice temperature $T = 0.4\Omega_{\rm LO}$ for three different bias drift velocities $v = 0.03, 0.1$ and $0.2v_{\rm LO}$ under dominant polar optic-phonon Fröhlich scattering. From Cui, Lei and Horing (1988).

For nonpolar semiconductors, when optic phonon (frequency $\Omega_{\boldsymbol{q}\lambda} = \Omega_{\rm o}$) deformation potential scattering $[\,|M(\boldsymbol{q},\lambda)|^2 = D^2/(2d_{\rm c}\Omega_{\rm o})$, see Sec. 2.4.2] dominates, the resistivity function $\rho(v)$ and energy-dissipation resistivity function $\rho_{\rm E}(v)$ can be written in the nondegenerate case (using the zero-spin-splitting $\Pi_{02}(\boldsymbol{q},\omega)$ expression (8.33) and neglecting the screening) as

(Cui, Lei and Horing, 1988)

$$\rho(v) = \rho_o^* \frac{\omega_c^3}{8\Omega_o^2 T_e} \frac{\sinh(\omega_c/2T_e)}{\sinh(\Omega_o/2T)} e^{\mu/T_e}$$

$$\times \int_0^\infty \frac{dx}{\sqrt{1+x^2}} \frac{\exp(\xi(1-x^2))}{[\cosh(\omega_c/2T_e) - \cos(\omega_c x/2T_e)]^2}$$

$$\times \left\{ \exp\left(\frac{\Omega_o}{2T} - \frac{\Omega_o}{2T_e}\right) \left[\cos\left(\frac{\Omega_o}{2T_e}x - 2\xi x\right) + x\sin\left(\frac{\Omega_o}{2T_e}x - 2\xi x\right)\right] \right.$$

$$\left. + \exp\left(\frac{\Omega_o}{2T_e} - \frac{\Omega_o}{2T}\right) \left[\cos\left(\frac{\Omega_o}{2T_e}x + 2\xi x\right) - x\sin\left(\frac{\Omega_o}{2T_e}x + 2\xi x\right)\right] \right\}, \tag{8.70}$$

$$\rho_E(v) = -\rho_o^* \frac{\omega_c^2}{2\Omega_o^2} \frac{v_o^2}{v^2} \frac{1}{\sinh(\Omega_o/2T)} e^{\mu/T_e}$$

$$\times \int_0^\infty \frac{dx}{\sqrt{1+x^2}} \frac{\exp(\xi(1-x^2))}{\cosh(\omega_c/2T_e) - \cos(\omega_c x/2T_e)}$$

$$\times \left[\exp\left(\frac{\Omega_o}{2T} - \frac{\Omega_o}{2T_e}\right) \cos\left(\frac{\Omega_o}{2T_e}x - 2\xi x\right) \right.$$

$$\left. - \exp\left(\frac{\Omega_o}{2T_e} - \frac{\Omega_o}{2T}\right) \cos\left(\frac{\Omega_o}{2T_e}x + 2\xi x\right) \right], \tag{8.71}$$

in which $\rho_o^* \equiv m^4 D^2 \Omega_o/(8\pi^3 d_c n^2 e^2)$, $v_o^2 = \Omega_o/m$,

$$\xi = \frac{\omega_c \Omega_o v^2}{8T_e^2 v_o^2} \frac{\sinh(\omega_c/2T_e)}{\cosh(\omega_c/2T_e) - \cos(\omega_c x/2T_e)}. \tag{8.72}$$

and e^{μ/T_e} is given by (8.34) for systems of fixed carrier density n.

We have known in Sec. 2.4.2 that, in the absence of magnetic field, the most striking feature of optic-phonon deformation-potential scattering dominant hot-electron transport is the saturation of the drift velocity at high electric field, with the saturation value v_m determined by (2.70). From the asymtotic behavior of the energy-balance equation (8.36) at $T_e \to \infty$ it is easily seen that this feature remains true in the presence of a magnetic field and the saturation value of drift velocity is still determined by (2.70), independent of the magnetic field.

Nonlinear longitudinal resistivity ρ_{xx} and electron temperature T_e under dominant nonpolar optic-phonon deformation-potential scattering are numerically calculated by Cui, Lei and Horing (1988) at lattice temperature $T = 0.4\Omega_o$ for several bias drift velocities from balance equation (8.39) using the $\rho(v)$ and $\rho_E(v)$ functions of (8.70) and (8.71). The calculated T_e/T and ρ_{xx}/ρ_0 (ρ_0 is the linear resistivity in the absence of the magnetic field) are shown in Fig. 8.3 as functions of ω_c/Ω_o. The magnetophonon resonance clearly shows up in the resistivity and electron temperature. Besides, the magnetophonon resonance structures may also appear in other transport quantities, such as noise temperature (Cui, Lei and Horing, 1988).

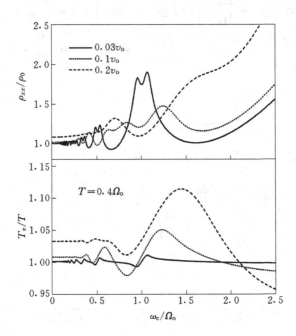

Fig. 8.3 ρ_{zz}/ρ_0 and T_e/T of transverse configuration versus ω_c/Ω_o at lattice temperature $T = 0.4\Omega_o$ for three bias drift velocities, $v = 0.03, 0.1$ and $0.2v_o$, under dominant optic-phonon deformation potential scattering. From Cui, Lei and Horing (1988).

8.5 Small-Signal Transport in a Magnetic Field, Cyclotron Resonance

8.5.1 *Balance equations for high-frequency small-signal transport in a magnetic field*

The small-signal high-frequency transport of semiconductor carriers in a magnetic field can be studied based on the force and energy balance equations (8.6) and (8.7) using the expressions (8.8), (8.9), and (8.10) for the frictional forces $\tilde{\bm{f}}_i$, $\tilde{\bm{f}}_p$, and energy dissipation rate $\tilde{w}$ with memory effect included. As the simplest example, we deal with the case of zero dc bias.

Consider a uniform constant magnetic field $\bm{B} = (0, 0, B)$ and a uniform small ac electric field of frequency ω,

$$\bm{E}(t) = \bm{E}_1 e^{-i\omega t} + \bm{E}_1^* e^{i\omega t}, \tag{8.73}$$

applying in an isotropic semiconductor. After transient dies out the system approaches an ac steady state, in which all the time-dependent transport quantities are ω-frequency oscillating with small amplitudes. The carrier drift velocity $\bm{v}$ and frictional force $\tilde{\bm{f}} = \tilde{\bm{f}}_i + \tilde{\bm{f}}_p$ can be written as

$$\bm{v} = \bm{v}_1 e^{-i\omega t} + \bm{v}_1^* e^{i\omega t}, \tag{8.74}$$

$$\tilde{\bm{f}} = \bm{f}_1 e^{-i\omega t} + \bm{f}_1^* e^{i\omega t}, \tag{8.75}$$

and the electron temperature is given by $T_e = T + T_1 e^{-i\omega t} + T_1^* e^{i\omega t}$, where $\bm{v}_1$, $\bm{f}_1$ and T_1 are small quantities.

To linearize balance equations (8.6) and (8.7), we insert (8.73)–(8.75) into expressions (8.8), (8.9), and (8.10) of $\tilde{\bm{f}}_i$, $\tilde{\bm{f}}_p$, and $\tilde{w}$ and expand them in powers of small quantities, with $A(\bm{q}, t, t')$ function simplified as [see (3.30)]

$$A(\bm{q}, t, t') \approx 1 - \left\{ \bm{q} \cdot \bm{v}_1 e^{-i\omega t} \left[1 - e^{i\omega(t-t')} \right] / \omega - \text{c.c.} \right\}. \tag{8.76}$$

To the linear order of small quantities the energy-balance equation requires $T_1 = 0$, or $T_e = T$, and the force-balance equation becomes

$$-i\omega n m \bm{v}_1 = ne\bm{E}_1 + ne\bm{v}_1 \times \bm{B} + \bm{f}_1. \tag{8.77}$$

We consider either the longitudinal configuration $\bm{E} \parallel \bm{B}$ or the transverse configuration $\bm{E} \perp \bm{B}$. In either case $\bm{f}_1$ is along (opposite to) the $\bm{v}_1$ direction and can be expressed as

$$\bm{f}_1 = i n m \bm{v}_1 M(\omega), \tag{8.78}$$

$M(\omega)$ being the memory function in the presence of the magnetic field. Different from the case without magnetic field, $M(\omega)$ is now configuration dependent. Symbols $M_\parallel(\omega)$ and $M_\perp(\omega)$ will be used to distinguish the memory functions in longitudinal and transverse configurations when needed. In the case without confusion, however, we still simply use $M(\omega)$ to represent either one of them. In the presence of both impurity and phonon scatterings $M(\omega)$ consists of impurity and phonon contributions,

$$M(\omega) = M^{\mathrm{i}}(\omega) + M^{\mathrm{p}}(\omega). \tag{8.79}$$

The impurity part $M^{\mathrm{i}}(\omega)$ and phonon part $M^{\mathrm{p}}(\omega)$ share the same expressions as those without magnetic field, (3.16) and (3.17), given by

$$M^{\mathrm{i}}_{\perp(\parallel)}(\omega) = \frac{n_{\mathrm{i}}}{nm\omega}\sum_{\bm{q}}|u(\bm{q})|^2 q^2_{x(z)}\left[\Pi(\bm{q},0)-\Pi(\bm{q},\omega)\right], \tag{8.80}$$

$$M^{\mathrm{p}}_{\perp(\parallel)}(\omega) = \frac{1}{nm\omega}\sum_{\bm{q},\lambda}|M(\bm{q},\lambda)|^2 q^2_{x(z)}\left[\Lambda(\bm{q},\lambda,0)-\Lambda(\bm{q},\lambda,\omega)\right]. \tag{8.81}$$

The imaginary part $\Lambda_2(\bm{q},\lambda,\omega)$ and the real part $\Lambda_1(\bm{q},\lambda,\omega)$ of the electron–phonon correlation function $\Lambda(\bm{q},\lambda,\omega)$ are given by (3.9) and (3.10), where $\Pi_2(\bm{q},\omega)$ and $\Pi_1(\bm{q},\omega)$ are the imaginary part and the real part of the electron density correlation function in the presence of the magnetic field along the z direction, $\Pi(\bm{q},\omega)$ [the same function in (8.80)], given by (8.17).

8.5.2 High-frequency complex resistivity and complex conductivity in a magnetic field

The drift velocity $\bm{v}$ (8.74) or the current density $(\bm{j}_1 = ne\bm{v}_1)$

$$\bm{j}(t) = \bm{j}_1 \mathrm{e}^{-\mathrm{i}\omega t} + \bm{j}_1^* \mathrm{e}^{\mathrm{i}\omega t} \tag{8.82}$$

under a high-frequency small signal electric field (8.73) can be immediately solved from the linearized force-balance equation (8.77).

In the longitudinal configuration, both $\bm{E}_1$ and $\bm{v}_1$ are along the z direction. Eq. (8.77) gives

$$\bm{E}_1 = -\mathrm{i}\frac{m}{e}\left[\omega + M_{\parallel}(\omega)\right]\bm{v}_1, \tag{8.83}$$

and the high-frequency small signal complex resistivity

$$\rho_{zz}(\omega) \equiv \frac{E_1}{nev_1} = -\mathrm{i}\frac{m}{ne^2}\left[\omega + M_{\parallel}(\omega)\right]. \tag{8.84}$$

The complex conductivity is $\sigma_{zz}(\omega) = 1/\rho_{zz}(\omega)$.

In the transverse configuration, $\bm{E}$ and $\bm{v}$ are within the x–y plane. By choosing x-axis along the $\bm{v}$ direction: $\bm{v}_1 = (v_1,0,0)$, $\bm{E}_1 = (E_{1x},E_{1y},0)$,

Eq. (8.77) gives

$$E_{1x} = -i\frac{m}{e}\left[\omega + M_{\perp}(\omega)\right]v_1,$$
$$E_{1y} = Bv_1,$$

and the high-frequency small signal complex resistivity

$$\rho_{xx}(\omega) \equiv \frac{E_{1x}}{nev_1} = -i\frac{m}{ne^2}\left[\omega + M_{\perp}(\omega)\right], \tag{8.85}$$

$$\rho_{yx}(\omega) \equiv \frac{E_{1y}}{nev_1} = \frac{B}{ne} = \frac{m}{ne^2}\omega_c, \tag{8.86}$$

with carrier charge e and $\omega_c \equiv eB/m$. On the other hand, taking x-axis along the $\boldsymbol{E}$ direction, $\boldsymbol{E}_1 = (E_1, 0, 0)$ and $\boldsymbol{v}_1 = (v_{1x}, v_{1y}, 0)$, we obtain the following high-frequency small signal complex conductivity from Eq. (8.77):

$$\sigma_{xx}(\omega) \equiv \frac{nev_{1x}}{E_1} = i\frac{ne^2}{m}\frac{\omega + M_{\perp}(\omega)}{\left[\omega + M_{\perp}(\omega)\right]^2 - \omega_c^2}, \tag{8.87}$$

$$\sigma_{yx}(\omega) \equiv \frac{nev_{1y}}{E_1} = \frac{ne^2}{m}\frac{\omega_c}{\left[\omega + M_{\perp}(\omega)\right]^2 - \omega_c^2}. \tag{8.88}$$

Restricted to the linear response to high-frequency small signals, the above results for the longitudinal and transverse configurations are sufficient for us to write down the current density for a system having a magnetic field in the z direction when a small high-frequency electric field applies in an arbitrary direction:

$$\boldsymbol{j}_1 = \mathcal{G}\cdot\boldsymbol{E}_1, \tag{8.89}$$

where $\mathcal{G}$ is the conductivity tensor of the form

$$\mathcal{G} = \begin{pmatrix} \sigma_{xx} & \sigma_{xy} & 0 \\ \sigma_{yx} & \sigma_{yy} & 0 \\ 0 & 0 & \sigma_{zz} \end{pmatrix} = \begin{pmatrix} \rho_{xx} & \rho_{xy} & 0 \\ \rho_{yx} & \rho_{yy} & 0 \\ 0 & 0 & \rho_{zz} \end{pmatrix}^{-1}, \tag{8.90}$$

having symmetrical relations between its components: $\sigma_{xx} = \sigma_{yy}$, $\sigma_{xy} = -\sigma_{yx}$, $\rho_{xx} = \rho_{yy}$, $\rho_{xy} = -\rho_{yx}$.

With the general linear tensor relation (8.89) between the current density and the electric field vectors, we can study the system response to a circularly polarized small electromagnetic radiation. A circularly polarized electromagnetic wave propagating along the z direction can be considered as the superposition of two electromagnetic waves linearly polarized respectively along the x and y directions having $\pi/2$ phase difference. An electric field of circular polarization can also be expressed in the form of (8.73) if

$$\boldsymbol{E}_1 = E_{1x}(\hat{\boldsymbol{x}} \pm \mathrm{i}\hat{\boldsymbol{y}}). \tag{8.91}$$

Here $\hat{\boldsymbol{x}}$ and $\hat{\boldsymbol{y}}$ are unit vectors in the x and y directions, E_{1x} is the amplitude of the linearly polarized electric field along the x direction, and the plus and minus signs represent the left and right circular polarizations. From Eq. (8.89), the system current density $\boldsymbol{j}_1$ in (8.82) is obtained as

$$\boldsymbol{j}_1 = j_{1x}(\hat{\boldsymbol{x}} \pm \mathrm{i}\hat{\boldsymbol{y}}), \tag{8.92}$$

where

$$j_{1x} = (\sigma_{xx} \pm \mathrm{i}\sigma_{xy})E_{1x} = \sigma_{\pm}E_{1x}. \tag{8.93}$$

Thus, we can use

$$\sigma_{\pm}(\omega) \equiv \sigma_{xx}(\omega) \pm \mathrm{i}\,\sigma_{xy}(\omega) \tag{8.94}$$

to describe the system response to an electromagnetic radiation of circular polarization. According to (8.87) and (8.88),

$$\sigma_{\pm}(\omega) = \mathrm{i}\frac{ne^2}{m}\frac{1}{\omega \pm \omega_{\mathrm{c}} + M_{\perp}(\omega)}. \tag{8.95}$$

In the same vein,

$$\rho_{\pm}(\omega) \equiv \rho_{xx}(\omega) \pm \mathrm{i}\,\rho_{xy}(\omega) \tag{8.96}$$

can be considered as the complex resistivity related to circularly polarized radiation,

$$\rho_{\pm}(\omega) = -\mathrm{i}\frac{m}{ne^2}[\omega \pm \omega_{\mathrm{c}} + M_{\perp}(\omega)] = 1/\sigma_{\pm}(\omega). \tag{8.97}$$

8.5.3 *Cyclotron resonance*

In terms of the expressions (8.73) and (8.82) of the electric field and current density, the power absorbed by the carriers of the system from the high-frequency field is

$$S_{\mathrm{p}} = (\boldsymbol{j}_1 \cdot \boldsymbol{E}_1^* + \boldsymbol{j}_1^* \cdot \boldsymbol{E}_1) = 2\mathrm{Re}(\boldsymbol{j}_1 \cdot \boldsymbol{E}_1^*). \tag{8.98}$$

For circularly polarized fields with current density (8.93),

$$S_{\mathrm{p}} = 4\,\mathrm{Re}[\sigma_{\pm}(\omega)]|E_{1x}|^2, \tag{8.99}$$

which is proportional to the energy-flow density along the propagating z direction. Therefore the absorption coefficients for circularly polarized electromagnetic fields are

$$\alpha_\pm \sim \mathrm{Re}[\sigma_\pm(\omega)]. \qquad (8.100)$$

For a linearly polarized electromagnetic field, the absorption coefficient is

$$\alpha_x \sim \mathrm{Re}[\sigma_{xx}(\omega)]. \qquad (8.101)$$

Separating the memory function $M_\perp(\omega)$ into its real and imaginary parts, $M_\perp(\omega) = M_1(\omega) + iM_2(\omega)$, and defining

$$m^* \equiv m\left[1 + M_1(\omega)/\omega\right], \qquad (8.102)$$
$$1/\tau \equiv M_2(\omega)/\left[1 + M_1(\omega)/\omega\right], \qquad (8.103)$$
$$\omega_c^* \equiv eB/m^*, \qquad (8.104)$$

we can write $\sigma_\pm(\omega)$ and $\sigma_{xx}(\omega)$ in the form

$$\sigma_\pm(\omega) = \frac{\sigma^*}{1 - i(\omega \pm \omega_c^*)\tau}, \qquad (8.105)$$

$$\sigma_{xx}(\omega) - \sigma^* \frac{1 - i\omega\tau}{(1 - i\omega\tau)^2 + (\omega_c^*\tau)^2}, \qquad (8.106)$$

in which $\sigma^* \equiv ne^2\tau/m^*$. The absorption coefficients of circularly polarized electromagnetic waves, (8.100), are then

$$\alpha_\pm \sim \frac{\sigma^*}{1 + (\omega \pm \omega_c^*)^2\tau^2}. \qquad (8.107)$$

Eq. (8.107) indicates that, the absorption coefficients $\alpha_\pm$ reach a maximum when $\omega \pm \omega_c^* = 0$, as long as there is no drastic change in σ^* in the vicinity of ω_c^*. This implies that, depending on the sign of the carrier charge e (electron or hole), the absorption coefficient of one of the two (left and right) circularly polarized high-frequency fields will exhibit a resonant peak around $\omega = |\omega_c^*|$. See Fig. 8.4(a). This is the cyclotron resonance.

According to (8.106), the real part of $\sigma_{xx}(\omega)/\sigma^*$ is

$$\mathrm{Re}[\sigma_{xx}(\omega)/\sigma^*] = \frac{1 + (\omega^2 + \omega_c^{*2})\tau^2}{1 + 2(\omega^2 + \omega_c^{*2})\tau^2 + (\omega^2 - \omega_c^{*2})^2\tau^4}. \qquad (8.108)$$

The absorption coefficient of a linearly polarized high-frequency field also exhibits cyclotron resonance. Fig. 8.4(b) plots $\mathrm{Re}[\sigma_{xx}(\omega)/\sigma^*]$ versus $|\omega_c^*|/\omega$ for several fixed values of $\omega\tau$. The curve begins to exhibit a peak at ω

somewhat smaller than $|\omega_c^*|$ when $\omega\tau > 1/\sqrt{3}$. For larger $\omega\tau$ value, the peak becomes sharper and its position is closer to $\omega = |\omega_c^*|$. Therefore, when the cyclotron resonance is clearly identified, the magnetic field strength B and the frequency ω of the radiation field at resonance provide a good measurement of the carrier effective mass $m^* = |eB|/\omega$.

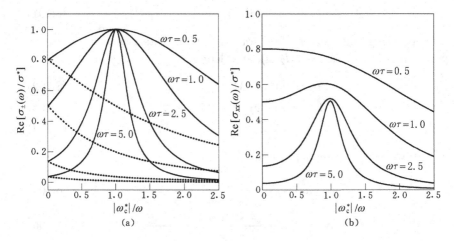

Fig. 8.4 $\text{Re}[\sigma_\pm(\omega)/\sigma^*]$ (a) and $\text{Re}[\sigma_{xx}(\omega)/\sigma^*]$ (b), are shown as functions of $|\omega_c^*|/\omega$. The curves are obtained under the condition of fixed $\omega\tau$ value. For positive ω_c^*, the solid curves in (a) represent $\text{Re}[\sigma_-(\omega)/\sigma^*]$, and the dot curves represent $\text{Re}[\sigma_+(\omega)/\sigma^*]$. For negative ω_c^*, the solid and dot curves represent the contrary. Four dot curves are respectively with $\omega\tau = 0.5, 1.0, 2.5$ and 5.0 (from top to bottom).

The above results are obtained for systems with isotropic parabolic energy bands. Even in this simplest case the effective mass m^* and effective scattering time τ given by (8.102) and (8.103) may vary with frequency and magnetic field. The measured absorption coefficients are hardly presented as fixed $\omega\tau$ curves of the Fig. 8.4 type. In a real semiconductor the cyclotron resonance structure is generally much more complicated than a single resonance peak described above. Depending on the magnetic field direction relative to the crystal axis, two or more cyclotron resonance peaks can appear. Nevertheless, all these complexity can be analyzed based on the real energy band structure. Cyclotron resonance has now been a useful tool for studying the band structure and measuring the carrier effective mass of semiconductors.

Formulas (8.87), (8.88) and (8.95) expressing the high-frequency small-signal complex conductivities in a magnetic field in terms of memory functions, were derived by Ting, Ying and Quinn (1977) from Kubo linear response theory.

8.6 Magnetotransport in Quasi-Two-Dimensional Systems

Many results obtained in the preceding sections for magnetotransport in 3D bulk systems, including hot-electron and high-frequency small-signal transport, are easily extended to quasi-2D systems outside the quantum Hall regime. Especially when the magnetic field is perpendicular to the 2D plane and the electric field lies in the plane (Faraday configuration), almost all the formulas of 2D magnetotransport can be directly written down from the corresponding 3D formulas. However, many magnetotransport properties of 2D systems are qualitatively different from those of 3D, due to the intrinsic difference in their Landau states. To take Landau level broadening into account is generally the major issues in 2D quantum magnetotransport.

8.6.1 *Nonlinear resistivity tensor of 2D systems in Faraday configuration*

In Chapter 2 we derived the momentum and energy balance equations for carrier dc steady state transport in the layer plane of quasi-two-dimensional systems (single heterojunction, quantum well or a layer of superlattice). These equations are easily extended to the case in the presence of a magnetic field in the z direction, $\boldsymbol{B} = (0, 0, B)$, i.e. the Faraday configuration. Formally, we only need to add a Lorentz force term to Eq. (2.89) to attain the momentum balance equation in a magnetic field in Faraday configuration, and the energy balance equation is exactly the same as (2.90):

$$N_s e \boldsymbol{E} + N_s e \boldsymbol{v} \times \boldsymbol{B} + \boldsymbol{f}(\boldsymbol{v}) = 0, \tag{8.109}$$

$$\boldsymbol{v} \cdot \boldsymbol{f}(\boldsymbol{v}) + w(\boldsymbol{v}) = 0. \tag{8.110}$$

Here $\boldsymbol{v}$ is the carrier drift velocity in the plane, $\boldsymbol{f}(\boldsymbol{v}) = \boldsymbol{f}_\mathrm{i} + \boldsymbol{f}_\mathrm{p}$ is the (per layer) frictional force and $w(\boldsymbol{v}) = w$ is the (per layer) energy-dissipation rate of carriers in unit area of the quasi-2D system. Since force and energy balance equations (8.109) and (8.110) have the same structure as Eqs. (8.35) and (8.36), we can introduce nonlinear resistivity functions and define the nonlinear resistivity and conductivity for quasi-2D systems in exactly the same way as in Sec. 8.2 for 3D systems. Denoting $\boldsymbol{v} \cdot \boldsymbol{f}(\boldsymbol{v}) = v f(v)$ and

$w(\boldsymbol{v}) = w(v)$ for systems isotropic in the plane, and defining a resistivity function

$$R(v) \equiv -\frac{f(v)}{N_s^2 e^2 v}, \tag{8.111}$$

and an energy-dissipation resistivity function

$$R_E(v) \equiv \frac{w(v)}{N_s^2 e^2 v^2}, \tag{8.112}$$

we can write the energy-balance equation in the form

$$R(v) - R_E(v) = 0. \tag{8.113}$$

This equation determines the electron temperature T_e for a given drift velocity v.

Choosing the x-axis along the $\boldsymbol{v}$ direction that the sheet current density $J = J_x = N_s e v$, we immediately get E_x and E_y from force-balance equation (8.109) and then the nonlinear diagonal (longitudinal) and nondiagonal (transverse) sheet resistivities:

$$R_{xx} \equiv \frac{E_x}{J_x} = R(v), \tag{8.114}$$

$$R_{yx} \equiv \frac{E_y}{J_x} = \frac{B}{N_s e}. \tag{8.115}$$

The other components of the resistivity tensor are $R_{yy} = R_{xx}$, $R_{xy} = -R_{yx}$.

The nonlinear sheet conductivity tensor can be obtained by taking the inverse of the sheet resistivity tensor:

$$\begin{pmatrix} \sigma_{xx} & \sigma_{xy} \\ \sigma_{yx} & \sigma_{yy} \end{pmatrix} = \begin{pmatrix} R_{xx} & R_{xy} \\ R_{yx} & R_{yy} \end{pmatrix}^{-1}. \tag{8.116}$$

Alternatively, choosing the x-axis along the $\boldsymbol{E}$ direction, $\boldsymbol{E} = (E, 0, 0)$, and solving for $J_x = N_s e v_x$ and $J_y = N_s e v_y$ from force-balance equation (8.109), we get

$$\sigma_{xx} \equiv \frac{R(v)}{R(v)^2 + (B/N_s e)^2}, \tag{8.117}$$

$$\sigma_{yx} \equiv -\frac{B/N_s e}{R(v)^2 + (B/N_s e)^2}. \tag{8.118}$$

The (per layer) frictional forces $\boldsymbol{f}_i$ and $\boldsymbol{f}_p$ due to impurity and phonon scatterings and the energy-dissipation rate $w(\boldsymbol{v})$ due to phonon scattering in (8.109) and (8.110) have the same formal expressions as those in Sec. 2.5 for single quasi-2D systems and Sec. 2.7 for superlattices in terms of their electron density correlation functions in the presence of the magnetic field. In the case of superlattices the relevant density correlation function with $\boldsymbol{f}_\mathrm{i}$, $\boldsymbol{f}_\mathrm{p}$, and w [(2.159) and (2.162), (2.151), and (2.152)] is $\Pi(q_z,\boldsymbol{q}_\parallel,\omega)$, which, in the random phase approximation, can be expressed by the noninteracting electron density correlation function $\Pi_0(\boldsymbol{q}_\parallel,\omega)$ of a pure 2D system in the presence of the magnetic field:

$$\Pi(q_z,\boldsymbol{q}_\parallel,\omega) = \frac{\Pi_0(\boldsymbol{q}_\parallel,\omega)}{1 - V(q_\parallel,q_z)\Pi_0(\boldsymbol{q}_\parallel,\omega)} = \frac{\Pi_0(\boldsymbol{q}_\parallel,\omega)}{\epsilon(\boldsymbol{q}_\parallel,q_z,\omega)}, \qquad (8.119)$$

where $V(q_\parallel,q_z)$ is given by (2.149) and (2.150). In the case of a single heterojunction or quantum well the relevant density correlation function is $\Pi(n',n,\boldsymbol{q}_\parallel,\omega)$ (Sec. 2.5). To avoid the complicated expression in multisubband RPA treatment (Sec. 2.6) we consider here only the lowest subband, i.e. $\Pi(0,0,\boldsymbol{q}_\parallel,\omega) \to \Pi(\boldsymbol{q}_\parallel,\omega)$. In the random phase approximation this quasi-2D (a heterojunction or quantum well) electron density correlation function is given by

$$\Pi(\boldsymbol{q}_\parallel,\omega) = \frac{\Pi_0(\boldsymbol{q}_\parallel,\omega)}{1 - V(q_\parallel)\Pi_0(\boldsymbol{q}_\parallel,\omega)} = \frac{\Pi_0(\boldsymbol{q}_\parallel,\omega)}{\epsilon(\boldsymbol{q}_\parallel,\omega)}, \qquad (8.120)$$

where [see (2.79), $H_{0000}(q_\parallel) \to H(q_\parallel)$, or (2.136)]

$$V(q_\parallel) = \frac{e^2}{2\epsilon_0 \kappa q_\parallel} H(q_\parallel), \qquad (8.121)$$

and $\Pi_0(\boldsymbol{q}_\parallel,\omega)$ is the noninteracting electron density correlation function of a pure 2D system in the presence of the magnetic field.

Thus, the key to analyze the nonlinear magnetotransport of different kinds of quasi-2D systems is the calculation of a single pure two-dimensional density correlation function $\Pi_0(\boldsymbol{q}_\parallel,\omega)$ of noninteracting electrons in the magnetic field.

8.6.2 Density correlation function of 2D electron systems in a magnetic field, Landau level broadening

In the presence of a magnetic field $\boldsymbol{B} = (0,0,B)$ in the z direction, the single particle eigenstates of a noninteracting 2D system in the x–y plane [$\boldsymbol{r} \equiv (x,y)$] are Landau wave functions, which, when choosing the vector potential as $\boldsymbol{A}(\boldsymbol{r}) = (-By, 0, 0)$, can be written as (for system of unit area)

$$\psi_{nk_x}(\boldsymbol{r}) = e^{ik_x x} \chi_{nk_x}(y), \tag{8.122}$$

$$\chi_{nk_x}(y) = \left(\frac{1}{\sqrt{\pi} 2^n n! l_B}\right)^{1/2} \exp\left[-\frac{(y-y_0)^2}{2l_B^2}\right] H_n\left(\frac{y-y_0}{l_B}\right), \tag{8.123}$$

where $y_0 \equiv -k_x/(eB)$, $l_B \equiv |eB|^{-1/2}$, and $H_n(x)$ are the Hermite polynomials. The eigen energies form the Landau levels of equidistance:

$$\varepsilon_n = \left(n + \frac{1}{2}\right)\omega_c \tag{8.124}$$

($n = 0, 1, 2, \cdots$), where $\omega_c = |eB|/m$ is the cyclotron frequency. The Landau level is degenerate with respect to k_x and to the spin index (due to the neglect of the spin splitting). The number of degenerate states in each Landau level is $|eB|/\pi$.

The noninteracting electron density correlation function $\Pi_0(\boldsymbol{q}_\parallel, \omega)$ in (8.119) and (8.120) of a pure 2D system in a magnetic field can be expressed as weighted sums of the Landau representation correlation function $\Pi_0(n, n', \omega)$ over all the Landau levels (Ting, Ying and Quinn, 1977):

$$\Pi_0(\boldsymbol{q}_\parallel, \omega) = \frac{1}{2\pi l_B^2} \sum_{n,n'} C_{nn'}(l_B^2 q_\parallel^2/2) \Pi_0(n, n', \omega), \tag{8.125}$$

where $C_{nn'}(x)$ are functions given by (8.25). The correlation function $\Pi_0(n, n', \omega)$ can be constructed using the retarded Green's function $G_n(\omega)$. Its real part and imaginary part are given by (Ting, Ying and Quinn, 1977)

$$\Pi_{01}(n, n', \omega) = -\frac{2}{\pi} \int_{-\infty}^{\infty} d\varepsilon f(\varepsilon) \big[\operatorname{Re} G_n(\varepsilon + \omega) \operatorname{Im} G_{n'}(\varepsilon)$$
$$+ \operatorname{Re} G_{n'}(\varepsilon - \omega) \operatorname{Im} G_n(\varepsilon)\big], \tag{8.126}$$

$$\Pi_{02}(n, n', \omega) = -\frac{2}{\pi} \int_{-\infty}^{\infty} d\varepsilon \big[f(\varepsilon) - f(\varepsilon + \omega)\big] \operatorname{Im} G_n(\varepsilon + \omega) \operatorname{Im} G_{n'}(\varepsilon). \tag{8.127}$$

Here $f(\varepsilon) = f(\varepsilon, T_e)$ is the Fermi distribution function at electron temperature T_e, $\operatorname{Re} G_n(\omega)$ and $\operatorname{Im} G_n(\omega)$ are, respectively, the real part and the imaginary part of the single-particle retarded Green's function of the nth Landau level. The latter is proportional to the density-of-states $D_n(\varepsilon)$ of the nth Landau level:

$$D_n(\varepsilon) = -\frac{1}{\pi^2 l_B^2} \operatorname{Im} G_n(\varepsilon). \tag{8.128}$$

The electron density N_s of the 2D system equals the number of all the occupied states of the Landau levels, i.e.

$$N_s = -\frac{1}{\pi^2 l_B^2} \sum_{n=0}^{\infty} \int_{-\infty}^{\infty} d\varepsilon f(\varepsilon) \mathrm{Im} G_n(\varepsilon). \tag{8.129}$$

This equation determines the chemical potential.

In the absence of scattering, the single-particle retarded Green's function of the nth Landau level is

$$G_n(\varepsilon) = \frac{1}{\varepsilon - \varepsilon_n + \mathrm{i}\delta}, \tag{8.130}$$

and the correlation function is

$$\Pi_0(n, n', \omega) = 2 \frac{f(\varepsilon_{n'}) - f(\varepsilon_n)}{\omega + \varepsilon_{n'} - \varepsilon_n + \mathrm{i}\delta}. \tag{8.131}$$

The density-of-states given by (8.130) has a δ-function peak at each Landau-level position, and the correlation $\Pi_0(n, n', \omega)$ of (8.131) diverges whenever ω equals $\varepsilon_n - \varepsilon_{n'} \equiv (n - n')\omega_c$ if $f(\varepsilon_{n'}) - f(\varepsilon_n) \neq 0$. This divergence of the density-of-states and the correlation function, which would lead to more seriously divergent resistivity in 2D systems, arises from the complete neglect of the scattering in the relative electron system in a magnetic field. In the presence of scatterings the lifetime of the single-particle level becomes finite, or, the Landau level is broadened. The calculation of such level broadening requires to renormalize the single-particle Green's function $G_n(\varepsilon)$ by electron–impurity, electron–phonon and electron–electron scatterings.[1] Several different prescriptions have been proposed to renormalize the single-particle Green's function, and the behavior of the Landau level broadening depends not only on the method of renormalization but strongly on the nature of the scattering, especially the effective range of scattering potentials.

One of these prescriptions is the self-consistent Born approximation, in which the direct contributions of scatterings to observable quantities are taken into account in the lowest Born approximation, while those of level broadening are considered in a selfconsistent way within the Green's functions formulation. With the single-level self-consistent Born approximation for short-range impurity scattering, Ando and Uemura (1974) derive a semielliptic form for the density-of-states of each Landau level in strong

---

[1] The electron–electron interaction so far has been included only in the RPA screening. Its role in contributing to the level broadening has not yet been considered.

magnetic fields:

$$\operatorname{Im} G_n(\varepsilon) = \begin{cases} -2\Gamma_n^{-2}\left[\Gamma_n^2 - (\varepsilon - \varepsilon_n)^2\right]^{1/2} & (\varepsilon - \varepsilon_n)^2 \leqslant \Gamma_n^2, \\ 0 & (\varepsilon - \varepsilon_n)^2 > \Gamma_n^2, \end{cases} \quad (8.132)$$

where ε_n is the level center (8.124) and Γ_n is the half-width of the nth Landau level. This approximation gives a good description in the central range of each Landau level but has an unphysical sharp cutoff of the density-of-states near the spectral edges.

Using a method of cumulant expansion (or path-integral method) Gerhardts (1975) derives a Gaussian form for the density-of-states of the Landau levels:

$$\operatorname{Im} G_n(\varepsilon) = -\frac{\sqrt{2\pi}}{\Gamma_n} \exp\left[-\frac{2(\varepsilon - \varepsilon_n)^2}{\Gamma_n^2}\right]. \quad (8.133)$$

It gives a rapidly but gradually descending spectrum towards the edge. The density-of-states drops down to less than $1/7$ of the center value at $|\varepsilon - \varepsilon_n| = \Gamma_n$. Γ_n can be considered as the half-width of the nth Landau level.

A Lorentz form of the density-of-states is also used in the literature for Landau level broadening:

$$\operatorname{Im} G_n(\varepsilon) = -\frac{\Gamma_n}{\Gamma_n^2 + (\varepsilon - \varepsilon_n)^2}. \quad (8.134)$$

In this case the density-of-states drops down to half of the center value when $|\varepsilon - \varepsilon_n| = \Gamma_n$. Γ_n is also called the half-width of the level.

In addition to these, other forms of Landau level density-of-states are also proposed and used in the literature in dealing with different issues. For instance, a Gaussian lineshape admixed with a flat background is shown to give a good agreement with electronic specific heat measurement (Gornik et al, 1985). Many more numerical calculations are carried out by taking into account effects such as carrier screening, inter-Landau-level coupling, and phonon scattering (Das Sarma, 1981; Cai and Ting, 1986; Murayama and Ando, 1987; Das Sarma and Xie, 1988; Tanatar, Singh and MacDonald, 1991). Most of these investigations concentrate on the high magnetic field regime.

The half-width Γ_n of the nth level may strongly vary with n at the lowest few Landau indices ($n = 0, 1, 2$), but only weakly depends on it at larger n (Cai and Ting, 1986). In the case of impurity scattering, when replacing

Γ_n by a Landau-index independent constant Γ, the self-consistent Born approximation yields (Ando and Uemura, 1974; Ting, Ying and Quinn, 1977)

$$\Gamma^2 = \frac{2\omega_c}{\pi\tau_s}, \qquad (8.135)$$

where τ_s is the single-particle lifetime in the absence of the magnetic field. There are two points we should pay attention to. (1) The expression (8.135), which is derived in the case of high magnetic fields, predicts a $B^{1/2}$-dependent Landau level broadening. (2) Depending on the effective range of scattering potentials, the single-particle lifetime τ_s may be quite different from the transport scattering time $\tau_{\rm tr}$ (or $\tau_{\rm m}$), which is directly related to an easily measurable quantity, the linear mobility $\mu_0 = e\tau_{\rm tr}/m$ in the absence of the magnetic field. The ratio $\alpha = \tau_{\rm tr}/\tau_s$ may change from $\alpha = 1$ for δ-type short-range potential to more than $\alpha = 100$ for long-range scattering potential (Das Sarma and Stern, 1985). Therefore, at same linear mobilities the Landau levels of the system with dominant long-range impurity scattering are much broader than those with short-range scattering (Cai and Ting, 1986; Das Sarma and Xie, 1988). In realistic cases where different ranges of impurity and phonon scatterings may exist simultaneously together with the electron–electron interaction which, though has no direct influence on $\tau_{\rm tr}$, also contributes to τ_s, it is even more difficult to determine the coefficient α. We therefore treat α as a phenomenological parameter and use the following Γ to describe the half-width of the Landau levels for not too low level indices (Lei and Liu, 2003c, 2005b):

$$\Gamma = \left(\frac{8\alpha e \omega_c}{\pi m \mu_0}\right)^{1/2}. \qquad (8.136)$$

Note that the models for Landau level broadening discussed in this section do not consider the effect of the localized states in the tails of the Landau levels on the density correlation function, thus they are unable to deal with the quantum Hall plateaus and related issues.

8.6.3 Linear magnetoresistivity at low temperatures, Shubnikov de-Haas oscillation

At low temperatures when only impurity scattering plays the role, the linear diagonal resistivity $R_{xx}^0 = R(v \to 0) = -f'(0)/N_s^2 e^2$ for a single 2D sheet (a heterojunction or quantum well) in a magnetic field is given by

$$R_{xx}^0 = \frac{1}{N_s^2 e^2} \sum_{\bm{q}_\|} |\bar{U}(\bm{q}_\|)|^2 q_x^2 \frac{\partial}{\partial \omega} \Pi_{02}(\bm{q}_\|, \omega)\Big|_{\omega=0}$$

$$= \frac{1}{\pi^2 N_s^2 e^2 l_B^2} \sum_{\bm{q}_\|} |\bar{U}(\bm{q}_\|)|^2 q_x^2 \sum_{n,n'} C_{nn'}(l_B^2 q_\|^2/2)$$

$$\times \int_{-\infty}^{\infty} d\varepsilon \left[-\frac{\partial f(\varepsilon)}{\partial \varepsilon}\right] \mathrm{Im} G_n(\varepsilon) \mathrm{Im} G_{n'}(\varepsilon). \quad (8.137)$$

Here $|\bar{U}(\bm{q}_\|)|^2 \equiv |U(\bm{q}_\|)/\epsilon(\bm{q}_\|,0)|^2$ is the statically screened effective impurity potential, and $f(\varepsilon) = f(\varepsilon, T)$ is the Fermi distribution function at lattice temperature T.

The most salient feature of R_{xx}^0 is immediately seen from the above expression. There, the main contribution comes from terms of $n = n'$: $D(\varepsilon) \equiv \sum_n C_{nn} \mathrm{Im} G_n(\varepsilon) \mathrm{Im} G_n(\varepsilon)$, which is a function of ε peaking at all Landau-level centers $\varepsilon_n = n\omega_c$ ($n = 0, 1, 2, ...$). Since the effective factor $-\partial f(\varepsilon)/\partial \varepsilon$ in the integrand is a δ-type function around the Fermi level ε_F, the continuing change of the ε_F position relative to ε_n resulting from changing magnetic field strength or changing electron density, will lead to a periodic and sharp variation of R_{xx}^0. This is the Shubnikov–de-Haas (SdH) oscillation (Shubnikov and de Haas, 1930). An example is shown in Fig. 8.5, plotting R_{xx}^0 as the inverse magnetic field $1/B$ at a fixed electron density.

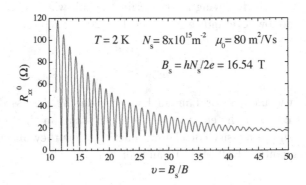

Fig. 8.5 An example of Shubnikov de-Haas (SdH) oscillation of linear diagonal magnetoresisivity R_{xx}^0 is shown as a function of the inverse magnetic field $1/B$ in terms of the filling factor ν at fixed electron density.

At fixed electron density, the SdH oscillation of R_{xx}^0 is periodic in inverse magnetic field strength, $1/B$, and the period is $\Delta(1/B) = \delta_{\text{sdh}} = 2e/\hbar N_{\text{s}}$. In other words, for a given system (fixed N_{s}) the SdH oscillation of R_{xx}^0 is periodic in the filling factor $\nu = B_{\text{s}}/B$ (with $B_{\text{s}} \equiv hN_{\text{s}}/2e$) of the Landau levels, and the period is $\Delta\nu = 1$. The SdH oscillation depends on the sharpness of the Fermi distribution, $-\partial f(\varepsilon)/\partial\varepsilon$, and shows up strongly only at temperature $T \ll \omega_{\text{c}}$. When temperature increases to $T \simeq \omega_{\text{c}}$, the oscillation essentially disappears. Due to this temperature dependence, also due to the effective density-of-states function $D_2(\varepsilon)$ peaking more sharply at higher magnetic field, the SdH oscillation is stronger at higher magnetic field and gradually weakened with decreasing strength of the magnetic field. These features of SdH oscillations remain unchanged irrespective of the form of the Landau level broadening and the behavior of the scattering potential $|\bar{U}(\boldsymbol{q}_\parallel)|^2$. On the other hand, since the total weight of the effective density-of-states function, $\int D(\varepsilon)d\varepsilon$, changes little with changing magnetic field, (8.137) predicts a sublinear dependence of the "averaged" R_{xx}^0 with increasing B at relatively high magnetic field range. In the low-field side, however, R_{xx}^0 generally approaches its zero-field value somewhat faster.

8.6.4 *Magnetoresistance oscillations induced by direct current*

Despite possible oscillation of magnetoresistance as a function of current density passing through the system were anticipated long ago in nonlinear magnetotransport in semiconductors (Elesin, 1969; Magarill and Sarvinkh, 1970; Agaeva, Askerov and Gashimzade (1976); Horing, Cui and Lei, 1987), it did not attract much attention until a few years ago Yang et al (2002) observed experimentally that a relatively weak direct current can induce substantial magnetoresistance oscillations in two-dimensional electron systems. The oscillation shows up saliently in the differential resistance in high-mobility 2D specimens subject to a magnetic field. It is almost periodic in inverse magnetic field and the period is tunable by the current density flowing through the 2D system. These experimental findings, which arouse new interest in nonlinear dc magnetotransport, can be well explained in the magnetotransport model presented in this section.

The nonlinear (diagonal) magnetoresistivity R_{xx} of a quasi-2D system is given by (8.114),

$$R_{xx} = -\frac{f(v)}{N_{\text{s}}^2 e^2 v}, \qquad (8.138)$$

and the differential resistivity at a bias drift velocity v is

$$r_{xx} = -\frac{f'(v)}{N_s^2 e^2}. \tag{8.139}$$

To capture the main physics of the magnetoresistance oscillation directly induced by the direct current we temporarily neglect the dynamical screening and consider only the impurity-induced frictional-force function for a single 2D sheet that

$$f(v) = \sum_{\bm{q}_\parallel} |U(\bm{q}_\parallel)|^2 q_x\, \Pi_{02}(\bm{q}_\parallel, q_x v)$$

$$= \frac{1}{\pi^2 l_B^2} \sum_{\bm{q}_\parallel} |U(\bm{q}_\parallel)|^2 q_x \sum_{n,n'} C_{nn'}(l_B^2 q_\parallel^2/2)$$

$$\times \int_{-\infty}^{\infty} d\varepsilon\, [f(\varepsilon + q_x v) - f(\varepsilon)] \operatorname{Im} G_n(\varepsilon + q_x v) \operatorname{Im} G_{n'}(\varepsilon). \tag{8.140}$$

Here $U(\bm{q}_\parallel)$ is the effective impurity potential and $f(\varepsilon) = f(\varepsilon, T_e)$ is the Fermi distribution function at electron temperature T_e, which should been determined by the energy balance equation $vf(v) + w(v) = 0$, with the energy-dissipation function given by

$$w(v) = 2\sum_{\bm{q}} |M(\bm{q},\lambda)|^2 \Omega_{\bm{q}\lambda} \Pi_{02}(\bm{q}_\parallel, \Omega_{\bm{q},\lambda} + q_x v) \left[n\!\left(\frac{\Omega_{\bm{q}\lambda}}{T}\right) - n\!\left(\frac{\Omega_{\bm{q}\lambda} + q_x v}{T_e}\right) \right]. \tag{8.141}$$

The drift velocity v or the current density $J_x = N_s e v$, affects R_{xx} and r_{xx} through the argument $q_x v$ in the $\Pi_{02}(\bm{q}_\parallel, q_x v)$ function in the frictional force. Physically, since an extra energy $q_x v$ is provided by the moving center-of-mass during the process, a relative electron can be scattered by impurities from Landau level n to n' ($=$ or $\neq n$). The transition rate is proportional to the overlap of two corresponding density-of-states factors $\operatorname{Im} G_n(\varepsilon + q_x v) \operatorname{Im} G_{n'}(\varepsilon)$ shifted by $q_x v$. A change of drift velocity v or magnetic field B would alter the degree of their overlap, giving rise to a change of the longitudinal resistivity. In the case of low electron temperature ($T_e \ll \epsilon_F$, the Fermi level) and many Landau-level occupation, $\Pi_{02}(\bm{q}_\parallel, \omega)$ is essentially a periodical function, i.e. $\Pi_{02}(\bm{q}_\parallel, \omega + \omega_c) \approx \Pi_{02}(\bm{q}_\parallel, \omega)$. Since at low temperatures the contribution to the wavevector integration in Eq. (8.140) generally peaks around $q_\parallel \sim 2k_F$ due to the volume weight of the phase space, the periodicity of Π_{02} function implies that when the drift velocity obtains an increment Δv satisfying $2k_F \Delta v = \omega_c$, the frictional force $f(v)$ at the drift velocity $v + \Delta v$ returns to that at drift velocity v, if the electron temperature remains unchanged. Therefore, the impurity-induced resistivity R_{xx} would exhibit periodical oscillations when changing drift velocity v or changing magnetic field $1/B$. We introduce a frequency-dimension quantity $\omega_j \equiv 2k_\mathrm{F} v$ to trace the change of the drift velocity v or the current $J_x = N_s e v$, and use the dimensionless ratio

$$\frac{\omega_j}{\omega_c} = \frac{2m k_\mathrm{F} v}{eB} = \sqrt{\frac{8\pi}{N_s}\frac{m}{e^2}}\frac{J_x}{B} \qquad (8.142)$$

as the control parameter to demonstrate this oscillation, which exhibits an approximate periodicity $\Delta(\omega_j/\omega_c) \sim 1$.

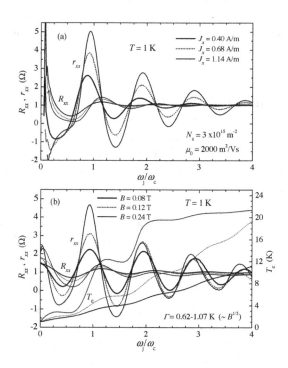

Fig. 8.6 (a) Resistivity R_{xx} and differential resistivity r_{xx} versus ω_j/ω_c at fixed current density $J_x = 0.40, 0.68$ or $1.14\,\mathrm{A/m}$, and (b) R_{xx}, r_{xx} and electron temperature T_e versus ω_j/ω_c at fixed magnetic field $B = 0.08, 0.12$ or $0.24\,\mathrm{T}$, for a GaAs-based 2D system having electron density $N_e = 3.0 \times 10^{15}\,\mathrm{m}^{-2}$, linear mobility $\mu_0 = 2000\,\mathrm{m^2/Vs}$, $\alpha = 40$ at $T = 1\,\mathrm{K}$, assuming short-range impurity scattering. From Lei, 2007a.

Lei (2007a) performs a calculation of nonlinear magnetoresistivity R_{xx} resulting from impurity scattering at lattice temperature $T = 1\,\mathrm{K}$ for an

n-type GaAs-based 2D system with electron density $N_e = 3.0 \times 10^{15}\,\text{m}^{-2}$ and linear mobility $\mu_0 = 2000\,\text{m}^2/\text{Vs}$, considering scatterings from bulk longitudinal and transverse acoustic phonons, as well as from polar optical phonons to contribute to the energy-dissipation rate $w(v)$. Figure 8.6(a) presents the calculated resistivity R_{xx} and differential resistivity r_{xx} versus the inverse magnetic field $1/B$ in terms of ω_j/ω_c subject to three different bias current densities $J_x = 0.40, 0.68$, and $1.14\,\text{A/m}$, which correspond to $\omega_j/2\pi = 32.3, 62.2$ and $103.6\,\text{GHz}$ respectively. The impurity scattering is assumed to be short-ranged, and the Gaussian form (8.133) is used for the Landau level broadening with a half-width Γ of (8.136) and parameter $\alpha = 10$, i.e. a $B^{1/2}$-dependent Γ, $\simeq 0.62\,\text{K}$ at $B = 0.08\,\text{T}$. Oscillations in resistivity R_{xx}, especially in differential resistivity r_{xx} show up remarkably, having an approximate period $\Delta(\omega_j/\omega_c) \sim 1$. The oscillation amplitude decays with increasing ω_j/ω_c (reducing B field, due to increasing overlap of LLs) at fixed bias current, but increases with increasing bias current density within the J_x range shown. The maxima (minima) of the differential resistivity r_{xx} locate quite close to (but somewhat lower than) the integers (half integers) of ω_j/ω_c, while the maxima (minima) of the total resistivity R_{xx} are shifted around a quarter period higher. These features are in good agreement with the experimental finding (Zhang et al, 2007a). Note that, the electron temperature T_e (not shown) exhibits only a weak variation with changing B-field at each fixed current density.

In Fig. 8.6(b) we plot R_{xx}, r_{xx} and electron temperature T_e versus the current density in terms of ω_j/ω_c at fixed magnetic fields $B = 0.08, 0.12$, and $0.24\,\text{T}$ for the same system. Remarkable R_{xx} and r_{xx} oscillations with approximate period $\Delta(\omega_j/\omega_c) \sim 1$ and maxima (minima) positions similar to those in Fig. 8.6(a) can be seen in this current-sweeping figure, but here the oscillation decays with increasing ω_j/ω_c due to the increase in the electron temperature.[2] At lower ω_j/ω_c range, the oscillation amplitude of the higher B-field case is apparent larger than the lower B-field case when the electron temperature T_e is still in the range less than or around $10\,\text{K}$. However, in the case of $B = 0.24\,\text{T}$ the oscillation amplitude decays rapidly with increasing ω_j/ω_c due to the rapid increase of the electron temperature, which rises up to $20\,\text{K}$ range around $\omega_j/\omega_c = 2$. In comparison, the amplitude decay is much slower in the case of $B = 0.08\,\text{T}$ because of the slow T_e rise. Furthermore, the periods of R_{xx} and r_{xx} oscillations are also somewhat shrunk by the rise of T_e, as can be seen at higher orders in the

---

[2] The early small peaks in the $B = 0.24\,\text{T}$ case are relevant to SdH oscillation, when R_{xx} locates below the average at zero dc bias.

$B = 0.24$ T case.

Despite the temperature-dependence exhibiting above, the current-induced magnetoresistance oscillations survive at T_e to above 20 K, much larger than the temperature corresponding to ω_c in the B-field range discussed, where the SdH disappears long before. The amplitude and other detailed behavior of the oscillation depend on the form of the scattering potential thus on the screening. Nevertheless, a detailed analysis of the initial suppression of the magnetoresistivity with increasing current density indicates that the width of the half zero-bias peak of R_{xx}–J_x curve quite accurately measures the width of the Landau levels (Lei, 2007a).

8.6.5 Linear magnetoresistance of GaAs heterojunctions in wide ranges of temperatures and magnetic fields

Leadly et al (1993) carry out a careful measurement on the linear magnetoresistance of high-mobility GaAs/Al$_{0.3}$Ga$_{0.7}$As heterojunctions over the temperature range 1.5–300 K subjected to a perpendicular magnetic field up to 30 T and theoretically analyze the magnetotransport of these quasi-2D systems in wide ranges of magnetic field and temperature within the balance equation formulation.

The linear longitudinal magnetoresistance $R_{xx}^0 \equiv 1/N_s e \mu$ of a single quasi-2D sheet is given by (2.111) and (2.112) in terms of linear mobilities limited by impurity and phonon scatterings with or without a magnetic field. Considering only the lowest subband and neglecting the screening, one can write the inverse mobilities due to impurity (IMP), longitudinal acoustic phonon ($\Omega_q = v_{sl} q$, treated as quasi-elastic) deformation potential (DPA), and polar optic phonon (Ω_{LO} constant) scatterings as

$$\frac{1}{\mu_{\text{IMP}}} = -\frac{1}{N_s e} \sum_{\mathbf{q}_\parallel} |U_{00}(\mathbf{q}_\parallel)|^2 q_x^2 \frac{\partial}{\partial \omega} \Pi_{02}(\mathbf{q}_\parallel, \omega)\Big|_{\omega=0}, \qquad (8.143)$$

$$\frac{1}{\mu_{\text{DPA}}} = -\frac{2T}{N_s e} \sum_{\mathbf{q}} \frac{|M_{00}(\mathbf{q},1)|^2}{v_{sl} q} q_x^2 \frac{\partial}{\partial \omega} \Pi_{02}(\mathbf{q}_\parallel, \omega)\Big|_{\omega=0}, \qquad (8.144)$$

$$\frac{1}{\mu_{\text{LO}}} = -\frac{2}{N_s e T} \sum_{\mathbf{q}} |M_{00}(\mathbf{q}, \text{LO})|^2 q_x^2 \, \Pi_{02}(\mathbf{q}, \Omega_{LO}) \frac{e^{\Omega_{LO}/T}}{(e^{\Omega_{LO}/T} - 1)^2}. \qquad (8.145)$$

Here the $\Pi_{02}(\mathbf{q}, \omega)$ function is given by (2.110) in the absence of the magnetic field, and by (8.125) and (8.127) in the presence of the magnetic field.

The total inverse mobility can be written as the sum of the above three parts:

$$\frac{1}{\mu} = \frac{1}{\mu_{\text{IMP}}} + \frac{1}{\mu_{\text{DPA}}} + \frac{1}{\mu_{\text{LO}}}. \tag{8.146}$$

This expression, however, is not identical to a Matthiessen's rule, since contributions to the total resistivity from different scattering mechanisms are not independent. In the presence of a magnetic field even the linear resistivity due to any individual scattering mechanism is relevant to the broadening of the Landau levels. The latter is related to the interactions of electrons with all scattering mechanisms. Therefore, in the simultaneous presence of several scattering mechanisms, the calculation of magnetotranport should be carried out by including all the scattering processes selfconsistently in the evaluation of both the density-of-states and the resistivities.

In the presence of a magnetic field, the Green's function for the nth Landau level can be written as

$$G_n(\varepsilon) = \frac{1}{\varepsilon - \varepsilon_n - \Sigma_n(\varepsilon)}, \tag{8.147}$$

where $\varepsilon_n = (n + \tfrac{1}{2})\omega_c$ is the center of the nth Ladau level, and

$$\Sigma_n(\varepsilon) = \Delta_n(\varepsilon) - \mathrm{i}\,\Gamma_n(\varepsilon) \tag{8.148}$$

is the self-energy with real part $\Delta_n(\varepsilon)$ and imaginary part $-\Gamma_n(\varepsilon)$.

Leadley et al (1993) calculate the self-energies for systems corresponding to the high-mobility GaAs/Al$_{0.3}$Gas$_{0.7}$As heterojunctions in their experiments, considering electron scattering with impurities, DPA phonons and LO phonons. They use the lowest self-consistent Born approximation for elastic impurity scattering and DPA scattering and the one-phonon self-energy approximation for LO-phonon scattering. This leads to the following equation for the self-energy, in which the effective Born matrix elements for short-range background impurity scattering and for quasi-elastic DPA scattering have the same functional form as in the expressions for their mobilities:

$$\Sigma_n(\varepsilon) = \sum_{n'} \left\{ \left[W^{\text{IMP}} + W^{\text{DPA}} \right] G_{n'}(\varepsilon) \right.$$
$$+ W^{\text{LO}}_{n'n} \left[(n_{\text{LO}} + f_{n'}(\varepsilon)) G_{n'}(\varepsilon + \Omega_{\text{LO}}) \right.$$
$$\left. \left. + (n_{\text{LO}} + 1 - f_{n'}(\varepsilon)) G_{n'}(\varepsilon - \Omega_{\text{LO}}) \right] \right\}. \tag{8.149}$$

where $n_{\text{LO}} = 1/[\exp(\Omega_{\text{LO}}/T) - 1]$ and $f_n(\varepsilon) = 1/\{\exp[(\varepsilon - \varepsilon_n)/T] + 1\}$. W^{DPA} and $W^{\text{LO}}_{n'n}$ are known from the electron–phonon matrix elements of the GaAs material and the form factor of the heterojunction structure, while W^{IMP}, which depends on the type and density of impurities, is just $\Gamma_{\mathrm{sb}}^2/4$, Γ_{sb} being the half-width of the broadened Landau levels obtained in the self-consistent Born approximation when impurities are assumed to be the only scatterers and the inter-level couplings are neglected (Ando and Uemura, 1974). Since Γ_{sb} can be thought to relate to the zero-temperature linear mobility μ_0 by (8.136), Leadly et al (1993) take

$$W^{\mathrm{DPA}} = \frac{e\omega_{\mathrm{c}}}{2\pi m}\frac{\gamma}{\mu_0}, \tag{8.150}$$

with γ as a fitting parameter.

Equations (8.147) and (8.149) constitute a set of self-consistent equations to determine the self-energy function $\Sigma_n(\varepsilon)$ for given lattice temperature and magnetic field strength. The self-energy thus determined depends on all the scattering mechanisms. In their paper, Leadley et al (1993) calculate only the value of the self-energy function at the center $\varepsilon = \varepsilon_n$ of each Landau level n through the self-consistent equations to obtain $\Gamma_n \equiv \Gamma_n(\varepsilon_n)$ and $\Delta_n \equiv \Delta_n(\varepsilon_n)$, and then use the constant $\Delta_n - i\Gamma_n$ to replace $\Sigma_n(\varepsilon)$ in (8.147) to obtain $G_n(\varepsilon)$ for each Landau level. The imaginary part of the Green's function is then Lorentzian:

$$\mathrm{Im}\, G_n(\varepsilon) = -\frac{\Gamma_n}{(\varepsilon - \varepsilon_n - \Delta_n)^2 + \Gamma_n^2}, \tag{8.151}$$

but its half-width Γ_n has different temperature and magnetic field dependence for each n. With the density-of-states function (8.151) thus obtained, the mobilities $1/\mu_{\mathrm{IMP}}$, $1/\mu_{\mathrm{DPA}}$, $1/\mu_{\mathrm{LO}}$ and the total mobility $1/\mu$ can be calculated.

Following the above procedure Leadley (1993) calculate the total linear mobility μ, and the separate contributions from impurity (IMP), DPA-phonon and LO-phonon scatterings, μ_{IMP}, μ_{DPA} and μ_{LO}, for their sample G650 at $B = 0$ and $B = 9\,\mathrm{T}$. The results are presented in Fig. 8.7(a) and 8.7(b). One can see the strong influence of the magnetic field on μ_{IMP}, μ_{DPA} and μ_{LO}. Fig. 8.7(c) shows the calculated result for another sample G137 at $B = 9\,\mathrm{T}$. G137 has a much higher impurity density than G650, that its zero-field low-temperature (4 K) mobility is only 1/10 that of G650. It is worth noting that, despite that the phonon scattering strengths and phonon-limited zero-field mobilities are essentially the same for these two GaAs-based samples, their phonon limited mobilities at high magnetic field are quite different due to their different impurity densities. The theoretical calculation is in good agreement with the experimental measurement.

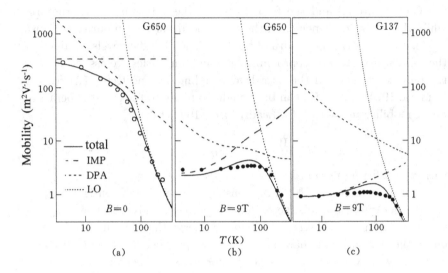

Fig. 8.7 The total linear mobility μ (solid curve) and the separate contributions from impurity (IMP), DPA phonon and LO phonon scatterings, μ_{IMP}, μ_{DPA} and μ_{LO}. (a) Sample G650, zero magnetic field; (b) Sample G650, $B = 9\,\text{T}$; (c) Sample G137, $B = 9\,\text{T}$. The small circles and dots are experimental data. From Leadley at el, 1993.

8.7 Balance Equations of Two Species of Carriers in a Magnetic Field

8.7.1 Type-II superlattices, electron–hole plasma

A Type-II superlattice is a system consisting of two species of carriers (electrons and holes) without carrier number exchange. We have derived the balance equations (6.9)–(6.12) for electric-field driven transport within the superlattice layer plane in this system in the absence of magnetic field. When a magnetic field $\bm{B} = (0,0,B)$ is applied along the the superlattice growth axis the balance equations can be formally obtained from Eqs. (6.9)–(6.12) by adding terms of Lorentz forces acting on the electron and hole centers of mass, provided that the relative electron and hole density correlation functions appearing in the expressions of forces and energy-dissipation rates, are understood to be those in the presence of the magnetic field. In view of the existence of Lorentz forces, the drift velocities $\bm{v}_1$ and $\bm{v}_1$ of electrons and holes, though still within the superlattice plane, may not be in the same direction as the electric field. For convenience we write the momentum and energy balance equations for electrons and holes of a type-II superlattice in a magnetic field as (see Sec. 6.1 for notation explanation):

$$N_e m_e \frac{d\bm{v}_1}{dt} = -N_e e(\bm{E} + \bm{v}_1 \times \bm{B}) + \bm{f}_1(\bm{v}_1) + \bm{f}_{12}(\bm{v}_1 - \bm{v}_2), \tag{8.152}$$

$$N_h m_h \frac{d\bm{v}_2}{dt} = N_h e(\bm{E} + \bm{v}_2 \times \bm{B}) + \bm{f}_2(\bm{v}_2) - \bm{f}_{12}(\bm{v}_1 - \bm{v}_2), \tag{8.153}$$

$$-\frac{du_e}{dt} = \bm{v}_1 \cdot \bm{f}_1(\bm{v}_1) + w_1(\bm{v}_1) + w_{12}(\bm{v}_1 - \bm{v}_2), \tag{8.154}$$

$$-\frac{du_h}{dt} = \bm{v}_2 \cdot \bm{f}_2(\bm{v}_2) + w_2(\bm{v}_2) - w_{12}(\bm{v}_1 - \bm{v}_2) + (\bm{v}_1 - \bm{v}_2) \cdot \bm{f}_{12}(\bm{v}_1 - \bm{v}_2). \tag{8.155}$$

For systems isotropic within the plane, the magnitudes of the frictional forces $\bm{f}_1$, $\bm{f}_2$ and $\bm{f}_{12}$ depend only on the magnitudes of the argument velocities, and their directions are in the directions of the argument velocities, that one can define $\bm{f}_1(\bm{v}_1) = f_1(v_1)(\bm{v}_1/v_1)$, $\bm{f}_2(\bm{v}_2) = f_2(v_2)(\bm{v}_2/v_2)$, and $\bm{f}_{12}(\bm{v}_1 - \bm{v}_2) = f_{12}(v_{12})(\bm{v}_1 - \bm{v}_2)/v_{12}$ ($v_{12} \equiv |\bm{v}_1 - \bm{v}_2|$). The energy-dissipation rates w_1, w_2 and w_{12} depend only on the magnitudes of the argument velocities: $w_1(\bm{v}_1) = w_1(v_1)$, $w_2(\bm{v}_2) = w_2(v_2)$ and $w_{12}(\bm{v}_1 - \bm{v}_2) = w_{12}(v_{12})$. Note that both $\bm{f}_1$, $\bm{f}_2$, $\bm{f}_{12}$ and w_1, w_2, w_{12} are quantities in the presence of the magnetic field. Their expressions formally are the same as those without the magnetic field, except that the density correlation functions are those of electrons and holes in the presence of the magnetic field.

The linear magnetotransport of an electron–hole two-component system is easily studied from these equations. Taking the electric field along the x direction, $\bm{E} = (E, 0, 0)$, we define the absolute mobilities for electrons and holes as

$$\mu_{xx}^e = -v_{1x}/E, \quad \mu_{yx}^e = -v_{1y}/E; \tag{8.156}$$

$$\mu_{xx}^h = v_{2x}/E, \quad \mu_{yx}^h = v_{2y}/E. \tag{8.157}$$

In the limit of small $\bm{v}_1$ and $\bm{v}_2$, $\bm{f}_j(\bm{v}_j) \approx \bm{v}_j f_j'(0)$ ($j = 1, 2$), $\bm{f}_{12}(\bm{v}_1 - \bm{v}_2) \approx (\bm{v}_1 - \bm{v}_2) f_{12}'(0)$, one can easily obtain (Cui and Horing, 1989)

$$\mu_{xx}^e = \frac{\mu_0^e}{D}\left[(\omega_{c2}\tau_2)^2 + \left(1 + \frac{\tau_2}{\tau_{21}} - \frac{m_e \tau_2}{m_h \tau_{12}}\right)\left(1 + \frac{\tau_1}{\tau_{12}} + \frac{\tau_2}{\tau_{21}}\right)\right], \tag{8.158}$$

$$\mu_{yx}^e = \frac{\mu_0^e}{D}(\omega_{c2}\tau_2)\left[(\omega_{c1}\tau_1)(\omega_{c2}\tau_2) + \frac{m_e}{m_h}\left(\frac{\tau_2}{\tau_{12}} + \frac{\tau_1 \tau_2}{\tau_{12}^2}\right)\right.$$
$$\left. + \frac{m_h \tau_1}{m_e \tau_2}\left(1 + \frac{\tau_2}{\tau_{21}}\right)^2 - \left(\frac{\tau_1}{\tau_{12}} + \frac{2\tau_1 \tau_2}{\tau_{12}\tau_{21}}\right)\right], \tag{8.159}$$

$$\mu_{xx}^{\rm h} = \frac{\mu_0^{\rm h}}{\mathcal{D}}\left[(\omega_{c1}\tau_1)^2 + \left(1 + \frac{\tau_1}{\tau_{12}} - \frac{m_{\rm h}\tau_1}{m_{\rm e}\tau_{21}}\right)\left(1 + \frac{\tau_2}{\tau_{21}} + \frac{\tau_1}{\tau_{12}}\right)\right], \tag{8.160}$$

$$\mu_{yx}^{\rm h} = -\frac{\mu_0^{\rm h}}{\mathcal{D}}(\omega_{c1}\tau_1)\left[(\omega_{c1}\tau_1)(\omega_{c2}\tau_2) + \frac{m_{\rm h}}{m_{\rm e}}\left(\frac{\tau_1}{\tau_{21}} + \frac{\tau_1\tau_2}{\tau_{21}^2}\right)\right.$$
$$\left. + \frac{m_{\rm e}\tau_2}{m_{\rm h}\tau_1}\left(1 + \frac{\tau_1}{\tau_{12}}\right)^2 - \left(\frac{\tau_2}{\tau_{21}} + \frac{2\tau_1\tau_2}{\tau_{12}\tau_{21}}\right)\right]. \tag{8.161}$$

In this, $\omega_{c1} = eB/m_{\rm e}$ and $\omega_{c2} = eB/m_{\rm h}$ are the cyclotron frequencies of electrons and holes; $\mu_0^{\rm e} = e\tau_1/m_{\rm e}$ and $\mu_0^{\rm h} = e\tau_2/m_{\rm h}$;

$$\mathcal{D} = \left[(\omega_{c1}\tau_1)(\omega_{c2}\tau_2) + \left(1 + \frac{\tau_1}{\tau_{12}} + \frac{\tau_2}{\tau_{21}}\right)\right]^2$$
$$+ \left[(\omega_{c1}\tau_1)\left(1 + \frac{\tau_2}{\tau_{21}}\right) - (\omega_{c2}\tau_2)\left(1 + \frac{\tau_1}{\tau_{12}}\right)\right]^2, \tag{8.162}$$

with the relaxation times defined by

$$\tau_1 = -N_{\rm e}m_{\rm e}/f_1'(0), \tag{8.163}$$
$$\tau_2 = -N_{\rm h}m_{\rm h}/f_2'(0), \tag{8.164}$$
$$\tau_{12} = -N_{\rm e}m_{\rm e}/f_{12}'(0), \tag{8.165}$$
$$\tau_{21} = -N_{\rm h}m_{\rm h}/f_{12}'(0). \tag{8.166}$$

It is interesting to compare the above results with those of electron–hole systems without magnetic field (Sec. 6.3). One can see from the expression (6.33) of $\mu_{\rm e}$, that in the absence of magnetic field, the condition for electron mobility to become negative is (noting that $N_{\rm h} \gg N_{\rm e}$) $f_{12}'(0)N_{\rm h} - f_2'(0)N_{\rm e} < 0$, i.e. $\tau_{12}/m_{\rm e} < \tau_2/m_{\rm h}$. In the presence of a magnetic field the condition for negative $\mu_{xx}^{\rm e}$ is (assuming $\tau_{12} \gg \tau_1$, $\tau_{21} \gg \tau_2$)

$$(\tau_{12}/m_{\rm e})[1 + (\omega_{c2}\tau_2)^2] < \tau_2/m_{\rm h}. \tag{8.167}$$

It is easy to identify two competing effects of the magnetic field on the left-hand-side of the above inequality. The factor τ_{12} decreases with increasing magnetic field because of the enhancement of the electron–hole scattering associated with their closer proximity due to the bending of the carrier orbits by the magnetic field, while the factor $1 + (\omega_{c2}\tau_2)^2$ obviously grows in connection with the general increase of the magnetoresistance and this effect usually dominates. On the other hand, τ_2 exhibits only moderate change in the range of weak magnetic field. The combined result is that the magnitude of negative electron mobility $\mu^{\rm e}_{xx}$ is reduced (i.e. toward the positive direction) as the magnetic field increases.

The quantitative examination needs the relaxation times defined by (8.163)–(8.166). For the calculation of these quantities one has to derive the density correlation functions of quasi-2D systems in the magnetic field with pertinent Landau-level broadening. This is generally a difficult job, requiring a self-consistent calculation. However, if the magnetic field is relatively weak that the level quantization has not yet played an important role, we can use the weak-field expansion of the 2D density correlation function in an approximate analysis for the effect of magnetic-field induced bending of carrier orbits on the electron–hole scattering. An example of this kind of calculation by Cui and Horing (1986) will be briefly introduced in the next subsection.

8.7.2 Weak magnetic-field expansion of two-dimensional density correlation function

The weak B-field expansion of the magnetic-field-dependent 2D density correlation function $\Pi_0(\boldsymbol{q}_\parallel, \Omega) = \Pi_{01}(\boldsymbol{q}_\parallel, \Omega) + i\Pi_{02}(\boldsymbol{q}_\parallel, \Omega)$ can be carried out by starting from the integral representation of Horing and Yildiz (1976):

$$\Pi_{01}(\boldsymbol{q}_\parallel, \Omega) = \mathcal{P}\int \frac{d\omega}{2\pi} \int \frac{d\omega'}{2\pi} \frac{\omega'}{\Omega^2 - \omega'^2} f(\omega, T_{\rm e}) R(\omega, \omega', \boldsymbol{q}_\parallel), \quad (8.168)$$

$$\Pi_{02}(\boldsymbol{q}_\parallel, \Omega) = -\frac{1}{2} \int \frac{d\omega}{2\pi} f(\omega, T_{\rm e}) R(\omega, \Omega, \boldsymbol{q}_\parallel), \quad (8.169)$$

in which

$$R(\omega, \Omega, \boldsymbol{q}_\parallel) = \frac{m\omega_{\rm c}}{2\pi} \int_{-\infty}^{\infty} dy\, e^{-i\Omega y/2} \int_{-\infty}^{\infty} dx\, e^{i\omega x} \frac{\sin(\Omega x/2)}{\sin(\omega_{\rm c} x/2)}$$
$$\times \exp\left(-i\frac{q_\parallel^2}{2m\omega_{\rm c}} \frac{\cos(\omega_{\rm c} y/2) - \cos(\omega_{\rm c} x/2)}{\sin(\omega_{\rm c} x/2)}\right). \quad (8.170)$$

This expression is written for the case of zero spin splitting and without effect of Landau level broadening. For weak magnetic fields, the function $R(\omega, \Omega, \boldsymbol{q}_\parallel)$ can be expanded in a power series of $\omega_{\rm c} = |eB|/m$, which, to the lowest order of the magnetic field (order of $\omega_{\rm c}^2$), can be simplified by performing the y integration. Substituting this $R(\omega, \Omega, \boldsymbol{q}_\parallel)$ expansion into (8.168) and (8.169), we obtain the following weak-magnetic-field expansion for the 2D density correlation function (Cui and Horing, 1986):

$$\Pi_0(\boldsymbol{q}_\parallel,\omega) \approx \left(1 + h_1\frac{d}{d\zeta} + \frac{h_2}{4}\frac{d^2}{d\zeta^2} + \frac{h_3}{24}\frac{d^3}{d\zeta^3}\right)P(\boldsymbol{q}_\parallel,\omega), \tag{8.171}$$

in which $\zeta \equiv \mu/T_e$ (μ is the chemical potential, T_e is the electron temperature),

$$h_1 \equiv \frac{m\omega_c^2}{8q_\parallel^2 T_e}, \tag{8.172}$$

$$h_2 \equiv \frac{\omega_c^2}{T_e^2}\left(\frac{5}{12} - \frac{m^2\omega^2}{q_\parallel^4}\right), \tag{8.173}$$

$$h_3 \equiv \frac{\omega_c^3}{T_e^2}\left(\frac{q_\parallel^2}{16mT_e} + \frac{m^3\omega^4}{T_e q_\parallel^6} - \frac{m\omega^2}{2T_e q_\parallel^2}\right), \tag{8.174}$$

and $P(\boldsymbol{q}_\parallel,\omega)$ is the density correlation function of noninteracting 2D electrons in the absence of the magnetic field, i.e. the function given by (2.144). The conditions for this expansion to be valid are: (i) $\omega_c/\omega \ll 1$, (ii) $\omega_c/T_e \ll 1$ and $m\omega_c^2/(q_\parallel^2 T_e) \ll 1$ (for nondegenerate systems), or $\omega_c/\mu \ll 1$ and $m\omega_c^2/(q_\parallel^2 \mu) \ll 1$ (for degenerate systems).

Using the weak-field expansion (8.171) for 2D magnetic-field-dependent density correlation function, Cui and Horing (1986) calculate the relaxation times τ_1, τ_2, τ_{12} and τ_{21} defined in (8.163)–(8.166) and analyse the temperature variation of μ_{xx}^e, μ_{xx}^h, μ_{yx}^e and μ_{yx}^h subject to magnetic fields of different strengths. We refer the readers to their original paper for detail.

8.7.3 Transport behavior of two-subband systems and intersubband Coulomb interaction

As one of the simplest examples to investigate the effect of Coulomb interaction on the transport behavior of a multi-subband system in the presence of a magnetic field, we consider a model quasi-2D system consisting of two subbands. The carriers with number densities N_1 and N_2 in two subbands are assumed to have the same charge e and the same effective mass m. Under the condition of linear transport (weak electric field) the electron temperatures and chemical potentials of both subbands are equal, and their drift velocities $\boldsymbol{v}_1$ and $\boldsymbol{v}_2$ can be determined by the force-balance equations:

$$N_1 e\boldsymbol{E} + N_1 e\boldsymbol{v}_1 \times \boldsymbol{B} + \boldsymbol{v}_1 f_1'(0) + (\boldsymbol{v}_1 - \boldsymbol{v}_2)f_{12}'(0) = 0, \tag{8.175}$$

$$N_2 e\boldsymbol{E} + N_2 e\boldsymbol{v}_2 \times \boldsymbol{B} + \boldsymbol{v}_2 f_2'(0) - (\boldsymbol{v}_1 - \boldsymbol{v}_2)f_{12}'(0) = 0. \tag{8.176}$$

The total drift velocity of the system is

$$\bar{v} = (N_1 v_1 + N_2 v_2)/N_s = r_1 v_1 + r_2 v_2, \tag{8.177}$$

where $r_1 \equiv N_1/N_s$ and $r_2 \equiv N_2/N_s$, with $N_s = N_1 + N_2$ being the total carrier density. Taking the x-axis along the $\boldsymbol{E}$ direction and solving v_1 and v_2 from the above equations, we obtain the conductivities $\sigma_{xx} \equiv N_s e \bar{v}_x/E$ and $\sigma_{yx} \equiv N_s e \bar{v}_y/E$ for this two-subband system as follows:

$$\sigma_{xx} = \frac{N_s e^2}{m} \frac{\tau_1 \tau_2}{D \tau_{\rm es}} \left[\frac{2}{r_1 r_2} + \frac{(\tau_2 - \tau_1)^2}{\tau_1 \tau_2} + \left(\frac{r_1}{\tau_2} + \frac{r_2}{\tau_1} \right) \tau_{\rm es} + \frac{r_1 \tau_2 + r_2 \tau_1}{(r_1 r_2)^2 \tau_{\rm es}} \right.$$
$$\left. + \omega_c^2 (r_1 \tau_2 + r_2 \tau_1) \right], \tag{8.178}$$

$$\sigma_{yx} = -\frac{N_s e^2}{m} \frac{\omega_c}{D} \left(\frac{\tau_1 \tau_2}{\tau_{\rm es}} \right)^2 \left[\frac{1}{r_1 r_2} + r_1 \left(\frac{\tau_{\rm es}}{\tau_2} \right)^2 + r_2 \left(\frac{\tau_{\rm es}}{\tau_1} \right)^2 \right.$$
$$\left. + 2\tau_{\rm es} \left(\frac{1}{r_1 \tau_1} + \frac{1}{r_2 \tau_2} \right) + \omega_c^2 \tau_{\rm es}^2 \right], \tag{8.179}$$

$$D = \left[1 + \left(\frac{\tau_1}{r_1} + \frac{\tau_2}{r_2} \right) \frac{1}{\tau_{\rm es}} - \omega_c^2 \tau_1 \tau_2 \right]^2 + \omega_c^2 \left(\tau_1 + \tau_2 + \frac{\tau_1 \tau_2}{r_1 r_2 \tau_{\rm es}} \right)^2, \tag{8.180}$$

in which $\omega_c = |eB|/m$, and the relaxation times are defined by

$$\tau_1 = -N_1 m/f_1'(0), \tag{8.181}$$
$$\tau_2 = -N_2 m/f_2'(0), \tag{8.182}$$
$$\tau_{\rm es} = -N_s m/f_{12}'(0). \tag{8.183}$$

If the intersubband Coulomb scatterings are sufficiently strong that the related relaxation time $\tau_{\rm es}$ is much shorter than the relaxation times τ_1 and τ_2 related to intraband scatterings ($\tau_{\rm es} \ll \tau_1, \tau_2$), the general expressions (8.178) and (8.179) reduce to

$$\sigma_{xx} = \frac{N_s e^2}{mS} \left(\frac{r_1}{\tau_1} + \frac{r_2}{\tau_2} \right), \tag{8.184}$$

$$\sigma_{yx} = -\frac{N_s e^2}{mS} \left[1 + (r_1 r_2 \tau_{\rm es})^2 \omega_c^2 \right], \tag{8.185}$$

$$S \equiv \left(\frac{r_1}{\tau_1} + \frac{r_2}{\tau_2} \right)^2 + \omega_c^2 \left[1 + (r_1 r_2 \tau_{\rm es})^2 \omega_c^2 \right]. \tag{8.186}$$

The longitudinal and transverse resistivities ρ_{xx} and ρ_{yx} can be obtained from the relation (8.51) between the resistivity tensor and conductivity tensor. For not too strong magnetic fields ($\omega_c^2 \tau_{es}^2 \ll 1$), we have

$$\rho_{xx} = \frac{m}{N_s e^2} \left(\frac{r_1}{\tau_1} + \frac{r_2}{\tau_2} \right) = \frac{m}{N_s e^2} \frac{1}{\tau_t}, \qquad (8.187)$$

$$\rho_{yx} = \frac{B}{N_s e}. \qquad (8.188)$$

This means that under the condition of strong intersubband Coulomb scattering, the longitudinal resistivity ρ_{xx} of this coupled two-subband system is determined by an effective scattering rate $1/\tau_t$, which equals the weighted sum of the scattering rates $1/\tau_1$ and $1/\tau_2$ of two individual subbands independent of each other:

$$\frac{1}{\tau_t} = \frac{r_1}{\tau_1} + \frac{r_2}{\tau_2}. \qquad (8.189)$$

The Hall resistivity of this coupled two-subband system, ρ_{yx}, equals that of a single parabolic band system having carrier density N_s.

On the other hand, if the intersubband Coulomb scatterings are weak that $\tau_{es} \gg \tau_1, \tau_2$, the general expressions (8.178) and (8.179) become

$$\sigma_{xx} = \frac{e^2}{m} \left(\frac{N_1 \tau_1}{1 + \omega_c^2 \tau_1^2} + \frac{N_2 \tau_2}{1 + \omega_c^2 \tau_2^2} \right), \qquad (8.190)$$

$$\sigma_{yx} = -\frac{e^2}{m} \omega_c \left(\frac{N_1 \tau_1^2}{1 + \omega_c^2 \tau_1^2} + \frac{N_2 \tau_2^2}{1 + \omega_c^2 \tau_2^2} \right). \qquad (8.191)$$

Both the longitudinal conductivity σ_{xx} and the transverse conductivity σ_{yx} of the system equal the sum of longitudinal conductivities $\sigma_{xx}^{(1)}$ and $\sigma_{xx}^{(2)}$ and the sum of the transverse conductivities $\sigma_{yx}^{(1)}$ and $\sigma_{yx}^{(2)}$ of two independent subsystems.

For the convenience of seeing how the longitudinal resistivity ρ_{xx} and Hall resistivity ρ_{yx} vary with intersubband relaxation time τ_{es}, we rewrite the resistivities ρ_{xx} and ρ_{yx} derived from the general σ_{xx} and σ_{xy} expressions (8.178) and (8.179) in the following form:

$$\rho_{xx} = \frac{m}{N_s e^2} \frac{1}{\tau}, \qquad (8.192)$$

$$\rho_{yx} = \frac{B}{N_s e} \gamma. \qquad (8.193)$$

The effective relaxation time τ and the dimensionless coefficient γ so defined depend on τ_1, τ_2, τ_{es} and ω_c. In the strong Coulomb scattering limit $\tau = \tau_t$ and $\gamma = 1$. Fig. 8.6 illustrates how the scattering rate $1/\tau$ and the coefficient γ vary with ω_c and τ_{es} for a sample two-subband system. One can see that with increasing $1/\tau_{es}$, the longitudinal and transverse resistivities (conductivities) change gradually from the behavior described by (8.190) and (8.191) to that described by (8.187) and (8.188).

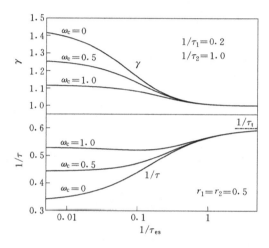

Fig. 8.8 The effective scattering time τ and γ coefficient versus the intersubband scattering rate $1/\tau_{es}$, for a Coulomb-coupled two-subband system at different values of ω_c. All the inverse scatttering times and ω_c are in the same (arbitrary) units.

Chapter 9

Higher Order Scatterings and Alternative Formulations of Balance Equation Theory

9.1 General Forms of Balance Equations, Intracollisional Field Effect

The role of the electric field in electron acceleration during the collision process, known as "intracollisional field effect" or the "interference effect", was first proposed by Argyres (1960) and later explored by Levinson (1970) and by Barker (1973). Much attention was given to it in the literature and a preliminary understanding of its influences on linear and nonlinear transport has already been achieved through many years of investigations (Horing and Argyres, 1962; Argyres, 1963; Argyres and Kelley, 1964; Barker, 1978; Thornber, 1978; Barker and Ferry, 1979; Calecki and Pottier, 1981; Arora, 1983; Sarker, 1986; Khan, Davis, and Wilkins, 1987; Krieger and Iafrate, 1987; Abdolsalami and Khan, 1990; Lipavský et al, 1991; Rossi and Jacoboni, 1991).

The appearance of the intracollisional field effect in the balance equation transport theory was pointed out by Xing and Ting (1987). In fact, carrier acceleration by the electric field during a collision is naturally incorporated when one treats the electric field in the unperturbed Hamiltonian (Lei and Horing, 1989; Lei, 1990c). In this section we provide a brief discussion of this matter in the process to express the force and energy balance equations in a more general form, equally valid for crystalline, amorphous and liquid systems (Lei, 1990c).

9.1.1 The frictional force and energy-loss rate in an electron–atom system, intracollisonal field effect

To write the balance equations in a more general form we consider N interacting electrons moving in a medium composed of atoms (ions) under the influence of a uniform electric field $\boldsymbol{E}$. The Hamiltonian of this electron–atom system consists of an atom part H_A, an electron part H_e, and electron–atom interaction $H_{\text{e-A}}$. The electron Hamiltonian H_e can be separated into a center-of-mass part and a relative electron part, $H_e = H_{\text{cm}} + H_{\text{er}}$, as did in Chapter 1. The atom Hamiltonian depends on the momenta and positions of all atoms but the effects of the electric field on the atom system are neglected. The electron–atom interaction

$$H_{\text{e-A}} = \sum_{i,l} v_l(\boldsymbol{r}_i - \boldsymbol{R}_l), \tag{9.1}$$

where $v_l(\boldsymbol{r}_i - \boldsymbol{R}_l)$ represents the potential experienced by the ith electron at position $\boldsymbol{r}_i$ and produced by the lth atom at position $\boldsymbol{R}_l$, can be written in the form

$$H_{\text{e-A}} = \sum_{\boldsymbol{q}} \rho_{\boldsymbol{q}} \gamma_{-\boldsymbol{q}} \mathrm{e}^{\mathrm{i}\boldsymbol{q}\cdot\boldsymbol{R}}, \tag{9.2}$$

where

$$\rho_{\boldsymbol{q}} = \sum_j \mathrm{e}^{\mathrm{i}\boldsymbol{q}\cdot\boldsymbol{r}'_j} \tag{9.3}$$

is the electron density operator in the relative frame, and

$$\gamma_{\boldsymbol{q}} = \sum_l v_l(\boldsymbol{q}) \mathrm{e}^{\mathrm{i}\boldsymbol{q}\cdot\boldsymbol{R}_l} \tag{9.4}$$

is the atom scattering operator with

$$v_l(\boldsymbol{q}) = \int \mathrm{e}^{\mathrm{i}\boldsymbol{q}\cdot\boldsymbol{r}} v_l(\boldsymbol{r}) \, d\boldsymbol{r} \tag{9.5}$$

being the Fourier transform of the electron–atom potential. In accordance with this Hamiltonian, the force operator $\boldsymbol{F}$ due to electron–atom interaction takes the form

$$\boldsymbol{F} = -\mathrm{i} \sum_{\boldsymbol{q}} \boldsymbol{q} \, \mathrm{e}^{\mathrm{i}\boldsymbol{q}\cdot\boldsymbol{R}} \rho_{\boldsymbol{q}} \gamma_{-\boldsymbol{q}}, \tag{9.6}$$

and the energy loss rate W of the electron system is equal to the energy increase rate of the atom system due to electron–atom interaction:

$$W = -\mathrm{i}\,[H_\mathrm{A}, H] = -\sum_{\boldsymbol{q}} \mathrm{e}^{\mathrm{i}\boldsymbol{q}\cdot\boldsymbol{R}} \rho_{\boldsymbol{q}} \dot{\gamma}_{-\boldsymbol{q}}, \tag{9.7}$$

where we have defined

$$\dot{\gamma}_{\boldsymbol{q}} \equiv -\mathrm{i}\,[\gamma_{\boldsymbol{q}},\ H_\mathrm{A}]. \tag{9.8}$$

It has been pointed out in Sec. 7.2 that the perturbation expansion with respect to the electric field is convergent and useful for a steady state calculation if the strength of the electric field is limited by the condition $\tau_\mathrm{th} \ll t_\mathrm{E}$. This condition can be relaxed if the electric field is not treated as a perturbation but is included in the unperturbed Hamiltonian (Lei and Horing, 1989). In other words, we assume that electron–impurity and electron–phonon interactions are absent before time $t = 0$, wherein the relative variables of the electron are decoupled from the center of mass. On the other hand, the electric field is considered to already be present, so that the system is subject to the Hamiltonian $H_\mathrm{cm} + H_\mathrm{er} + H_\mathrm{A}$ at $t \leqslant 0$. Therefore the initial density matrix is chosen as

$$\hat{\rho}_0 = \frac{1}{Z} \mathrm{e}^{-H_\mathrm{er}/T_\mathrm{e}} \mathrm{e}^{-H_\mathrm{A}/T} |\Psi\rangle \langle\Psi| \tag{9.9}$$

where $|\Psi\rangle$ is the traveling eigenstate of the center-of-mass Hamiltonian H_cm with average momentum of $\langle\Psi|\,\boldsymbol{P}\,|\Psi\rangle = Nm\boldsymbol{v}$, $\boldsymbol{v}$ being the drift velocity, and T is the lattice atom temperature, which is kept unchanged by assuming the existence of an energy dissipation mechanism which quickly drains energy out of the atom system. The Liouville equation for the evolution of the density matrix may be written in a form of an integral equation:

$$\hat{\rho}(t) = \hat{\rho}_0 - \int_0^t dt_1 [H_\text{e-A}(t),\ \hat{\rho}(t_1)], \tag{9.10}$$

in which an interaction representation has been defined for any operator $\mathcal{O}$ (in the Schrödinger picture) as

$$\mathcal{O}(t) = \mathrm{e}^{\mathrm{i}(H_\mathrm{er}+H_\mathrm{A})t} \mathcal{O} \mathrm{e}^{-\mathrm{i}(H_\mathrm{er}+H_\mathrm{A})t}. \tag{9.11}$$

The steady state calculation of the statistical average of the frictional force $\boldsymbol{f} = \mathrm{tr}\{\hat{\rho}(t)\boldsymbol{F}(t)\}$ and the electron energy loss rate $w = \mathrm{tr}\{\hat{\rho}(t)W(t)\}$ becomes easy for a system with strong electron–electron interactions. In such a system the time to approach the steady state from the initial state

(9.9), τ_{th}, is determined mainly by the intercarrier coupling and is short in comparison with the relaxation time due to electron–atom scattering. The perturbative expansions of the frictional force $\bm{f}$ and the energy loss rate w in powers of the electron–atom scattering potential, obtained by using the formal iterative series of Eq. (9.10), converge up to time $t \sim \tau_{\text{th}}$. In the case of a weak electron–atom potential it is a good approximation to keep only leading terms, yielding (Lei, 1990c)

$$\bm{f} = -\sum_{\bm{q}} \int_0^t ds \, \langle [\rho_{\bm{q}}(s)\gamma_{-\bm{q}}(s), \, \rho_{-\bm{q}}\gamma_{\bm{q}}] \rangle \exp\left[i\bm{q} \cdot \bm{v}s + \frac{ie}{2m}\bm{q} \cdot \bm{E}s^2\right], \quad (9.12)$$

$$w = i\sum_{\bm{q}} \int_0^t ds \, \langle [\rho_{\bm{q}}(s)\dot{\gamma}_{-\bm{q}}(s), \, \rho_{-\bm{q}}\gamma_{\bm{q}}] \rangle \exp\left[i\bm{q} \cdot \bm{v}s + \frac{ie}{2m}\bm{q} \cdot \bm{E}s^2\right]. \quad (9.13)$$

Here, the average should be taken with respect to the initial density matrix (9.9). The s^2 term in the argument of the exponential function, which is due to the acceleration of the center-of-mass by the electric field,

$$\bm{R}(t) = \bm{R} + \frac{\bm{P}}{Nm}t + \frac{e\bm{E}}{2m}t^2, \quad (9.14)$$

gives rise to the intracollisional field effect. The dominant contributions to the integrations in Eqs. (9.12) and (9.13) come from the time range of s less than the characteristic time τ_{cr} of the correlation function. When time t is larger than the correlation time the integrations in Eqs. (9.12) and (9.13) will approach steady values and the upper limits of the time integrals can be extended to ∞.

The correlation time may be identified with the duration of a collision. The s^2 term is expected relevant only when $|e|\bm{q} \cdot \bm{E}\tau_{\text{cr}}^2/2m \geqslant 1$. If the electric field strength E is much smaller than the field

$$E_{\text{cr}} = \frac{2m}{|e|q\tau_{\text{cr}}^2}, \quad (9.15)$$

the intracollisional field term is obviously negligible. Employing the estimation $\tau_{\text{cr}} \approx (4\varepsilon_{\text{F}})^{-1}$ we have from Eq. (9.15) $E_{\text{cr}} \approx 1 \times 10^6\,\text{V/cm}$ for $m = 1.8 \times 10^{-31}\,\text{kg}$ and $\varepsilon_{\text{F}} = 1000\,\text{K}$; and $E_{\text{cr}} \approx 1 \times 10^4\,\text{V/cm}$ if $m = 0.6 \times 10^{-31}\,\text{kg}$ and $\varepsilon_{\text{F}} = 70\,\text{K}$.

9.1.2 General forms of balance equations without intracollisional field effect

Without the intracollisional field terms, the $\boldsymbol{f}$ and w expressions (9.12) and (9.13) become ($\omega_0 \equiv \boldsymbol{q} \cdot \boldsymbol{v}$)

$$\boldsymbol{f} = -2 \sum_{\boldsymbol{q}} \int \frac{d\omega}{2\pi} \boldsymbol{q} \Pi_2(\boldsymbol{q}, \omega + \omega_0) \Gamma(\boldsymbol{q}, \omega) n\left(\frac{\omega + \omega_0}{T_e}\right), \qquad (9.16)$$

$$w = -2 \sum_{\boldsymbol{q}} \int \frac{d\omega}{2\pi} \omega \Pi_2(\boldsymbol{q}, \omega + \omega_0) \Gamma(\boldsymbol{q}, \omega) n\left(\frac{\omega + \omega_0}{T_e}\right). \qquad (9.17)$$

Here $n(x) \equiv 1/(e^x - 1)$ is the Bose function, and $\Gamma(\boldsymbol{q}, \omega)$ is the dynamic scattering factor, i.e. the spectral function of the time correlation function $\Gamma(\boldsymbol{q}, t)$ of the atom system interacting with electrons:

$$\Gamma(\boldsymbol{q}, \omega) = \int dt\, e^{i\omega t} \Gamma(\boldsymbol{q}, t), \qquad (9.18)$$

$$\Gamma(\boldsymbol{q}, t) = \langle \gamma_{-\boldsymbol{q}}(t)\, \gamma_{\boldsymbol{q}} \rangle, \qquad (9.19)$$

where the average is taken with respect to the initial state (9.9), which, as far as the atom system is concerned, is a thermal equilibrium state at temperature T. Because of this, the spectral function has the general property

$$\Gamma(-\boldsymbol{q}, -\omega) = e^{-\omega/T} \Gamma(\boldsymbol{q}, \omega). \qquad (9.20)$$

Furthermore, in deriving Eq. (9.17) we have used the following relations:

$$\langle \dot{\gamma}_{-\boldsymbol{q}}(t)\, \gamma_{\boldsymbol{q}} \rangle = -\langle \gamma_{-\boldsymbol{q}}(t)\, \dot{\gamma}_{\boldsymbol{q}} \rangle = \frac{\partial}{\partial t} \Gamma(\boldsymbol{q}, t). \qquad (9.21)$$

Note that in Eqs. (9.16) and (9.17) the Bose factor $n((\omega + \omega_0)/T_e)$ can be replaced by

$$\left[\frac{e^{-\omega/T}}{1 - e^{(\omega+\omega_0)/T_e}} - 1\right] \qquad (9.22)$$

or by

$$\frac{1}{2}\left[\frac{1 - e^{-\omega/T}}{1 - e^{-(\omega+\omega_0)/T_e}} - 1\right]. \qquad (9.23)$$

The force and the energy balance equations can be written, in terms of $\boldsymbol{f}$ and w given by (9.16) and (9.17), as

$$ne\boldsymbol{E} + \boldsymbol{f} = 0, \qquad (9.24)$$

$$\boldsymbol{v} \cdot \boldsymbol{f} + w = 0, \qquad (9.25)$$

and the resistivity is directly obtained from the frictional force:

$$\rho = -\frac{\boldsymbol{f} \cdot \boldsymbol{v}}{n^2 e^2 v^2}. \qquad (9.26)$$

In the weak current limit we can expand Eq. (9.16) to linear order in $\boldsymbol{v}$ and make use of the approximate expression Eq. (2.35) for $\Pi_2(\boldsymbol{q}, \omega)$ in the degenerate case, obtaining the Ohmic resistivity from Eq. (9.26) as (Lei, 1990c)

$$\rho = \frac{12\pi}{e^2 v_F^2} \int_0^1 d\left(\frac{q}{2k_F}\right) \left(\frac{q}{2k_F}\right)^3 \int_{-\infty}^{\infty} \frac{d\omega}{2\pi} \frac{\omega/T}{e^{\omega/T} - 1} \Gamma(\boldsymbol{q}, \omega). \qquad (9.27)$$

This is the generalized Ziman formula for the linear resistivity of liquid and amorphous metals (Ziman 1961, 1967), as well as for crystalline systems if the elastic Bragg part due to the long-range ordered atoms of the lattice is substracted from the dynamic scattering factor $\Gamma(\boldsymbol{q}, \omega)$ (Lei, 1980). For solids it is convenient to write the atom position $\boldsymbol{R}_l$ as the sum of its static position and a small oscillatory deviation $\boldsymbol{u}_l$ from it. The dynamic scattering factor is then divided into an elastic part $2\pi \Gamma_0(\boldsymbol{q}) \delta(\omega)$ and an inelastic part $\Gamma_{\text{in}}(\boldsymbol{q}, \omega)$:

$$\Gamma(\boldsymbol{q}, \omega) = 2\pi \Gamma_0(\boldsymbol{q}) \delta(\omega) + \Gamma_{\text{in}}(\boldsymbol{q}, \omega). \qquad (9.28)$$

The latter includes all multiphonon effects. Both the elastic and the inelastic parts contribute to the frictional force, but only the inelastic part contributes to the electron energy-loss rate. For crystalline solids only randomly distributed impurities contribute to the elastic part of the dynamic scattering factor, $\Gamma_0(\boldsymbol{q}) = n_i |u(\boldsymbol{q})|^2$, yielding the frictional force from Eq. (9.16)

$$\boldsymbol{f}_i = n_i \sum_{\boldsymbol{q}} |u(\boldsymbol{q})|^2 \boldsymbol{q} \Pi_2(\boldsymbol{q}, \omega_0) \qquad (9.29)$$

due to elastic scattering. Here n_i is the impurity density and $u(\boldsymbol{q})$ is the difference of an impurity potential and a regular lattice atom potential $v(\boldsymbol{q})$. Retaining only the dynamic single-phonon process in the inelastic part of the scattering factor, we obtain the phonon-induced frictional force $\boldsymbol{f}_\mathrm{p}$ and the energy loss rate w of the form

$$\boldsymbol{f}_\mathrm{p} = 2 \sum_{\boldsymbol{q},\lambda} |M(\boldsymbol{q},\lambda)|^2 \boldsymbol{q}\, \Pi_2(\boldsymbol{q},\Omega_{\boldsymbol{q}\lambda}+\omega_0)\left[n\left(\frac{\Omega_{\boldsymbol{q}\lambda}}{T}\right) - n\left(\frac{\Omega_{\boldsymbol{q}\lambda}+\omega_0}{T_\mathrm{e}}\right)\right], \tag{9.30}$$

$$w = 2 \sum_{\boldsymbol{q},\lambda} |M(\boldsymbol{q},\lambda)|^2 \Omega_{\boldsymbol{q}\lambda}\, \Pi_2(\boldsymbol{q},\Omega_{\boldsymbol{q}\lambda}+\omega_0)\left[n\left(\frac{\Omega_{\boldsymbol{q}\lambda}}{T}\right) - n\left(\frac{\Omega_{\boldsymbol{q}\lambda}+\omega_0}{T_\mathrm{e}}\right)\right]. \tag{9.31}$$

In this

$$|M(\boldsymbol{q},\lambda)|^2 = \frac{N_\mathrm{L}(\boldsymbol{q}\cdot\boldsymbol{e}_{\boldsymbol{q}\lambda})^2}{2M_\mathrm{L}\Omega_{\boldsymbol{q}\lambda}}|v(q)|^2, \tag{9.32}$$

where M_L and N_L are the mass and number density of the atom, and $\Omega_{\boldsymbol{q}\lambda}$ and $\boldsymbol{e}_{\boldsymbol{q}\lambda}$ are the phonon frequency and polarization vector.

9.1.3 *Evaluation of the intracollisional field effect*

Although it has been shown that the intracollisional field terms in Eqs. (9.12) and (9.13) are negligible when the applied electric field $E \ll E_\mathrm{cr}$, it is still not clear how important they are when $E \geqslant E_\mathrm{cr}$. To evaluate the effect of these intracollisional field terms on the high-field mobility for $E \geqslant E_\mathrm{cr}$, we rewrite Eqs. (9.12) and (9.13) in the form (Lei and Song, 1992)

$$\boldsymbol{f} = 2\sum_{\boldsymbol{q}} \boldsymbol{q} \int\!\!\int_{-\infty}^{\infty}\frac{d\omega d\omega'}{(2\pi)^2} D(\omega_0-\omega-\omega',\alpha)\Pi_2(\boldsymbol{q},\omega)\Gamma(\boldsymbol{q},\omega')\frac{\mathrm{e}^{\omega/T_\mathrm{e}}-\mathrm{e}^{-\omega'/T}}{\mathrm{e}^{\omega/T_\mathrm{e}}-1}, \tag{9.33}$$

$$w = 2\sum_{\boldsymbol{q}} \int\!\!\int_{-\infty}^{\infty}\frac{d\omega d\omega'}{(2\pi)^2} \omega' D(\omega_0-\omega-\omega',\alpha)\Pi_2(\boldsymbol{q},\omega)\Gamma(\boldsymbol{q},\omega')\frac{\mathrm{e}^{\omega/T_\mathrm{e}}-\mathrm{e}^{-\omega'/T}}{\mathrm{e}^{\omega/T_\mathrm{e}}-1}, \tag{9.34}$$

where

$$\alpha \equiv \frac{e}{2m}\boldsymbol{q}\cdot\boldsymbol{E} \tag{9.35}$$

and the function $D(\beta, \alpha)$ is defined by

$$D(\beta, \alpha) \equiv \int_0^\infty \cos(\beta t + \alpha t^2) dt. \tag{9.36}$$

Retaining contributions from the elastic part and the single phonon part of the dynamic scattering factor $\Gamma(\mathbf{q}, \omega)$, we have the frictional forces $\mathbf{f}_i$ and $\mathbf{f}_p$, due to elastic and single phonon processes respectively, as

$$\mathbf{f}_i = \sum_{\mathbf{q}} \mathbf{q} \Gamma_0(\mathbf{q}) \int \frac{d\omega}{\pi} D(\omega_0 - \omega, \alpha) \Pi_2(\mathbf{q}, \omega), \tag{9.37}$$

$$\mathbf{f}_p = 2 \sum_{\mathbf{q}, \lambda} \mathbf{q} |M(\mathbf{q}, \lambda)|^2 \int \frac{d\omega}{\pi} \Pi_2(\mathbf{q}, \omega) \left[n\left(\frac{\Omega_{\mathbf{q}\lambda}}{T}\right) - n\left(\frac{\omega}{T_e}\right) \right]$$
$$\times D(\omega_0 - \omega + \Omega_{\mathbf{q}\lambda}, \alpha), \tag{9.38}$$

and the energy transfer rate w, due to the single phonon process, as

$$w = 2 \sum_{\mathbf{q}, \lambda} \Omega_{\mathbf{q}\lambda} |M(\mathbf{q}, \lambda)|^2 \int \frac{d\omega}{\pi} \Pi_2(\mathbf{q}, \omega) \left[n\left(\frac{\Omega_{\mathbf{q}\lambda}}{T}\right) - n\left(\frac{\omega}{T_e}\right) \right]$$
$$\times D(\omega_0 - \omega + \Omega_{\mathbf{q}\lambda}, \alpha). \tag{9.39}$$

The function $D(\beta, \alpha)$ defined in Eq. (9.36), which was discussed by Thornber (1978) and by Barker (1979), can be written in terms of Fresnel integrals (Abramowitz and Stegun, 1965)

$$D(\beta, \alpha) = \left(\frac{\pi}{2|\alpha|}\right)^{1/2} \left\{ \cos\left(\frac{\beta^2}{4|\alpha|}\right) \left[\frac{1}{2} - C\left(\frac{\alpha\beta}{2|\alpha|^{3/2}}\right)\right] \right.$$
$$\left. + \sin\left(\frac{\beta^2}{4|\alpha|}\right) \left[\frac{1}{2} - S\left(\frac{\alpha\beta}{2|\alpha|^{3/2}}\right)\right] \right\}. \tag{9.40}$$

with

$$C(x) = \left(\frac{2}{\pi}\right)^{1/2} \int_0^x \cos y^2 dy, \tag{9.41}$$

$$S(x) = \left(\frac{2}{\pi}\right)^{1/2} \int_0^x \sin y^2 dy. \tag{9.42}$$

It has a delta-function behavior when $\alpha \to 0$:

$$D(\beta, \alpha) \to \pi \delta(\beta). \tag{9.43}$$

Thus, in the limit of $E \ll E_{\text{cr}}$, Eqs. (9.37)–(9.39) reduce to the corresponding expressions without intracollisional field terms in the preceding subsection. The effect of high electric field resides in the oscillatory function $D(\beta, \alpha)$. The complicated oscillatory behavior of the integrands in Eqs. (9.41)–(9.42) makes numerical calculation very difficult. In most previous investigations of intracollisional field effect, the field-dependent oscillatory functions were replaced by non-oscillatory functions to facilitate calculation. Lei and Song (1992) perform a numerical calculation directly from these equations by taking advantage of the Fresnel integral

$$C(x) = \frac{1}{2} + f(x)\sin\left(\frac{\pi}{2}x^2\right) - g(x)\cos\left(\frac{\pi}{2}x^2\right), \qquad (9.44)$$

$$S(x) = \frac{1}{2} - f(x)\cos\left(\frac{\pi}{2}x^2\right) - g(x)\sin\left(\frac{\pi}{2}x^2\right), \qquad (9.45)$$

expressed in terms of slowly varying auxiliary functions $f(x)$ and $g(x)$, which are tabulated (Abramowitz and Stegun, 1965). Using (9.44) and (9.45), one only needs to treat oscillatory integrations with the simple factor $\sin y^2$. A special interpolation is employed to carry out these oscillatory integrations semi-analytically. In this way Lei and Song (1992) are able to solve the force- and energy-balance equations for a Ge system including the intracollisional field effect correction to the first order, with the help of a microcomputer. They find that the intracollisional field effect is unimportant in the steady-state transport even for very high electric field, because it is always suppressed by the field-induced increase of the carrier temperature which leads to a broadening of the density correlation function. However, in the transient process of transport when the electron temperature has not been able to significantly rise after the high electric field turned on, the role of intracollisional field effect may show up. This result is in conformity with most previous studies.

9.2 Balance Equations in the CTPG Representation, Effects of Higher Order Scatterings

Despite that the force- and energy-balance equations are generally treated only to the lowest order in the scattering interactions in the framework of balance equation theory, effects of higher order scatterings are needed to be taken into account from time to time. We have already known the case of magnetotransport in 2D electron systems in Chapter 8, where the effects of higher order scattering have to be included properly to account for the broadening of the Landau levels. There are more balance-equation calculations beyond the lowest order treatment. Most of them are based on Feynman diagrammatic techniques involving close-time-path Green's functions (CTPG) (Lei and Cai, 1990b; Lei and Wu, 1992, 1993). A few examples will be briefly discussed in this section.

9.2.1 Balance equations in the CTPG function representation

The two important quantities in the balance-equation theory are the frictional force experienced by the center of mass, $\boldsymbol{f}$, and the energy-transfer rate from the electron system to the phonon system, w. They are statistical averages of the frictional force and the energy-transfer rate operators, $\hat{\boldsymbol{F}} = \hat{\boldsymbol{F}}_\mathrm{i} + \hat{\boldsymbol{F}}_\mathrm{p}$ and $\hat{W}$, as given in (1.21)–(1.23). The force- and energy-balance equations (1.40) and (1.55),

$$nm\frac{d\boldsymbol{v}}{dt} = ne\boldsymbol{E} + \boldsymbol{f}, \tag{9.46}$$

$$-\frac{dU}{dt} = \boldsymbol{v}\cdot\boldsymbol{f} + w, \tag{9.47}$$

have extensive validity beyond the lowest order in the scattering interaction, provided that the frictional force $\boldsymbol{f}$ and the energy transfer rate w are calculated in general forms:

$$\boldsymbol{f} = \mathrm{tr}\{\hat{\rho}(t)\hat{\boldsymbol{F}}\} = \mathrm{tr}\left\{\hat{\rho}_0\widetilde{\boldsymbol{F}}(t)\right\}, \tag{9.48}$$

$$w = \mathrm{tr}\{\hat{\rho}(t)\hat{W}\} = \mathrm{tr}\left\{\hat{\rho}_0\widetilde{W}(t)\right\}. \tag{9.49}$$

Here $\hat{\rho}(t)$ is the density matrix satisfying the Liouville equation (1.28) with the initial condition $\hat{\rho}(t_0) = \hat{\rho}_0$ as given by (1.30) at time $t = t_0 = \infty$, $\widetilde{\boldsymbol{F}}(t)$ and $\widetilde{W}(t)$ are the frictional force and energy-transfer rate operators in the Heisenberg picture with respect to the Hamiltonian $H = H_\mathrm{er} + H_\mathrm{ph} + H_\mathrm{ei} + H_\mathrm{ep}$ having $\widetilde{\boldsymbol{F}}(t_0) = \hat{\boldsymbol{F}}$ and $\widetilde{W}(t_0) = \hat{W}$. One can determine these statistical averages by treating $H_\mathrm{I} = H_\mathrm{ei} + H_\mathrm{ep}$ as a perturbation and writing Eqs. (9.48) and (9.49) in terms of the Keldysh closed-time-path integration (Chou et al, 1985; Rammer and Smith, 1986),

$$\boldsymbol{f} = \mathrm{tr}\{T_\mathrm{p}\left[\hat{\rho}_0\hat{\boldsymbol{F}}(t)L_\mathrm{p}\right]\} \tag{9.50}$$

with

$$L_\mathrm{p} = \exp\left[-\mathrm{i}\int_p H_\mathrm{I}(t_1)dt_1\right]. \tag{9.51}$$

Here $\hat{\boldsymbol{F}}(t)$ and $H_{\rm I}(t)$ are the interaction picture representations of operators $\hat{\boldsymbol{F}}$ and $H_{\rm I}$. The time path p is composed of a "+" branch (from $-\infty$ to $+\infty$) and a "$-$" branch (from $+\infty$ to $-\infty$), and $T_{\rm p}$ is the time-ordering operator on the path p. Note that the random impurity site average is implicitly included in the trace operations. The expression (9.48) can be recast into the form

$$\boldsymbol{f} = {\rm tr}\left\{T_{\rm p}\left[\hat{\rho}_0 \hat{\boldsymbol{F}} \exp\left(-{\rm i}\int_p H_{\rm I}(t_1-t)dt_1\right)\right]\right\}. \tag{9.52}$$

which is manifestly time independent. By expanding the exponential part inside the trace of Eq. (9.52) we obtain a perturbative expansion. After averaging over the impurity sites, there remains only the configurations of electron and phonon momentum variables that are paired in each term of the perturbation. Despite that the trace operation with respect to the relative electron variables and phonon variables is in reference to an initial density matrix $\hat{\rho}_0$ having different temperatures $T_{\rm e}$ and T for electrons and phonons, it can still be calculated by the use of the diagrammatic technique based on the Feynman-Keldysh rules as follows (Lei and Cai, 1990b). There is a Keldysh path branch index ($\alpha = +$ or $-$) with each vertex. The solid and wavy lines represent the components of the electron and phonon closed-time-path Green's functions, and dashed lines represent the impurity average pairings. All these lines are directional and carry wavevectors and frequencies. Furthermore, with each impurity (dashed) or phonon (wavy) line of wavevector $\boldsymbol{q}$, there is an additional line of the same direction, which carries only frequency $\omega_{\boldsymbol{q}} = -\boldsymbol{q}\cdot\boldsymbol{v}$. Wavevectors and frequencies are conserved at each vertex, i.e., the total wavevectors and total frequencies in are equal to those out. The only new feature is that all the Wick contractions for relative electron operators are taken at the electron temperature $T_{\rm e}$, and those for phonon operators are taken at the phonon temperature T (Lei and Wu, 1992). According to these rules we can easily construct the diagrams for the frictional force $\boldsymbol{f}$ and the energy loss rate w.

In each frictional-force diagram we must add a factor of wavevector $\boldsymbol{q}$ at the primary vertex, which is indicated by a dot. The lowest order diagram for the impurity-induced frictional force is shown in Fig. 9.1(a), and it may be evaluated as

$$\boldsymbol{f}_{\rm i} = n_{\rm i} u^2 \sum_{\boldsymbol{q},\boldsymbol{k}} \int \frac{d\omega}{2\pi}\, \boldsymbol{q}\, \xi_\alpha \eta_{\alpha'} G_{\alpha\alpha'}(\boldsymbol{k}-\boldsymbol{q},\omega+\boldsymbol{q}\cdot\boldsymbol{v})\, G_{\alpha'\alpha}(\boldsymbol{k},\omega). \tag{9.53}$$

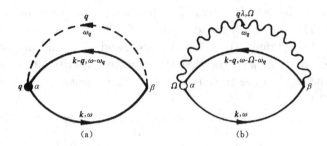

Fig. 9.1 Fig.12.1 Feynman diagrams for (a) the frictional force due to impurities and (b) the electron energy-loss rate due to phonons.

Here two symbols ξ_α and η_β ($\alpha, \beta = +, -$) are defined by $\xi_+ = \xi_- = 1$, $\eta_+ = 1$ and $\eta_- = -1$, and the Einstein summation rule is implied for the repeated Greek-letter scripts. In writing Eq. (9.53) we have assumed that the impurity potential is short-ranged, such that $|u(\bm{q})|^2 = u^2$. Employing the relation between the components of the closed-time-path Green's function $G_{\alpha\beta}$ and the retarded, advanced and correlation Green's functions $G^{\rm r}, G^{\rm a}$ and $G^{\rm c}$,

$$G_{\alpha\beta} = \frac{1}{2}(\xi_\alpha \eta_\beta \, G^{\rm r} + \eta_\alpha \xi_\beta \, G^{\rm a} + \xi_\alpha \xi_\beta \, G^{\rm c}), \tag{9.54}$$

we can rewrite (9.53) as

$$\bm{f}_{\rm i} = -n_{\rm i} u^2 \int \frac{d\omega}{2\pi} \sum_{\bm{q},\bm{k}} \bm{q} \left[G^{\rm r}(\bm{k}-\bm{q}, \omega + \bm{q}\cdot\bm{v})\, G^{\rm c}(\bm{k},\omega) + G^{\rm c}(\bm{k}-\bm{q}, \omega + \bm{q}\cdot\bm{v})\, G^{\rm a}(\bm{k},\omega) \right]. \tag{9.55}$$

These Green's functions are defined with respect to the initial density matrix $\hat{\rho}_0$, Eq. (1.30), i.e. a thermal equilibrium ensemble with Hamiltonian $H_{\rm er}$ at the electron temperature $T_{\rm e}$. Therefore

$$G^{\rm c}(\bm{k},\omega) = \left[1 - 2f(\omega)\right]\left[G^{\rm r}(\bm{k},\omega) - G^{\rm a}(\bm{k},\omega)\right], \tag{9.56}$$

where $f(\omega)$ is the Fermi function of the electron system at electron temperature $T_{\rm e}$.

The effects of elastic (impurity) and inelastic (electron–electron and electron–phonon) scatterings on the self-energy can be taken into account by introducing a life-time in the retarded and advanced electron Green's functions:

$$G^{r(a)}(\mathbf{k},\omega) = \frac{1}{\omega - \varepsilon_{\mathbf{k}} \pm i/\tau_e}. \qquad (9.57)$$

Both elastic scattering and inelastic scattering contribute to $1/\tau_e$, such that

$$\frac{1}{\tau_e} = \frac{1}{\tau} + \frac{1}{\tau_{in}}, \qquad (9.58)$$

where τ and τ_{in} are, respectively, elastic and inelastic scattering times. Substituting $G^{r(a)}$ expression (9.57) into Eq. (9.55) and noticing that terms with $G^r G^r$ and $G^a G^a$ vanish after ω-integration, we are left with

$$\mathbf{f}_i = -i n_i u^2 \sum_{\mathbf{q}} \mathbf{q}\, \Pi(\mathbf{q}, \mathbf{q} \cdot \mathbf{v}) = n_i u^2 \sum_{\mathbf{q}} \mathbf{q}\, \Pi_2(\mathbf{q}, \mathbf{q} \cdot \mathbf{v}). \qquad (9.59)$$

Here the correlation function

$$\Pi(\mathbf{q}, \omega_1) = -i \int \frac{d\omega}{\pi} \sum_{\mathbf{k}} [f(\omega) - f(\omega + \omega_1)] G^r(\mathbf{k} - \mathbf{q}, \omega + \omega_1) G^a(\mathbf{k}, \omega) \qquad (9.60)$$

satisfies $\Pi(\mathbf{q}, \omega) = \Pi(-\mathbf{q}, -\omega)^*$, and in the case of small $1/\tau_e$ its imaginary part can be written as

$$\Pi_2(\mathbf{q}, \omega_1) = 2\pi \sum_{\mathbf{k}} [f(\varepsilon_{\mathbf{k}}) - f(\varepsilon_{\mathbf{k}+\mathbf{q}})] \delta(\varepsilon_{\mathbf{k}+\mathbf{q}} - \varepsilon_{\mathbf{k}} + \omega_1). \qquad (9.61)$$

Eq. (9.59) is the expression for impurity-induced frictional force given in Sec. 2.4.

The diagrams for the electron energy-loss rate can be constructed in the same way. The time derivative of the phonon field operator $\dot{\phi}_{\mathbf{q}\lambda}$ results in an additional frequency factor at the primary vertex in the ω-representation, which is indicated by a circle. The lowest order diagram for the electron energy-loss rate is shown in Fig. 9.1(b) and it gives

$$w = i \int \frac{d\omega d\Omega}{(2\pi)^2} \sum_{\mathbf{k},\mathbf{q},\lambda} |M(\mathbf{q},\lambda)|^2 \Omega\, \xi_\alpha \eta_{\alpha'}$$
$$G_{\alpha\alpha'}(\mathbf{k} - \mathbf{q}, \omega - \Omega + \mathbf{q} \cdot \mathbf{v}) G_{\alpha'\alpha}(\mathbf{k},\omega) D_{\alpha\alpha'}(\mathbf{q}\lambda, \Omega). \qquad (9.62)$$

Here the components of the phonon closed-time-path Green's function $D_{\alpha\beta}$ are connected with the retarded, advanced and correlation functions of phonons as (9.54):

$$D_{\alpha\beta} = \frac{1}{2}\left(\xi_\alpha \eta_\beta D^r + \eta_\alpha \xi_\beta D^a + \xi_\alpha \xi_\beta D^c\right). \qquad (9.63)$$

In the case of infinite phonon life-time

$$D^{\mathrm{r}}(\bm{q}\lambda,\Omega) = (\Omega - \Omega_{q\lambda} + \mathrm{i}\delta)^{-1} - (\Omega + \Omega_{q\lambda} + \mathrm{i}\delta)^{-1},$$
$$D^{\mathrm{a}}(\bm{q}\lambda,\Omega) = -(\Omega - \Omega_{q\lambda} - \mathrm{i}\delta)^{-1} + (\Omega + \Omega_{q\lambda} - \mathrm{i}\delta)^{-1}, \qquad (9.64)$$
$$D^{\mathrm{c}}(\bm{q}\lambda,\Omega) = -\mathrm{i}2\pi(1 + 2n(\Omega_{q\lambda}/T))\left[\delta(\Omega - \Omega_{q\lambda}) + \delta(\Omega + \Omega_{q\lambda})\right],$$

where $n(x) = 1/(\mathrm{e}^x - 1)$ is the Bose function and $\delta = 0^+$. We can calculate the expression (9.62) for w by reducing it to

$$\begin{aligned}
w = \frac{\mathrm{i}}{2}\int \frac{d\omega d\Omega}{(2\pi)^2} &\sum_{\bm{k},\bm{q},\lambda} |M(\bm{q},\lambda)|^2 \Omega \\
\times \Big\{ &\left[G^{\mathrm{r}}(\bm{k}-\bm{q},\omega-\Omega+\bm{q}\cdot\bm{v})\, G^{\mathrm{c}}(\bm{k},\omega) \right. \\
&+ G^{\mathrm{c}}(\bm{k}-\bm{q},\omega-\Omega+\bm{q}\cdot\bm{v})\, G^{\mathrm{a}}(\bm{k},\omega) \Big] D^{\mathrm{c}}(\bm{q}\lambda,\Omega) \\
&+ \left[G^{\mathrm{r}}(\bm{k}-\bm{q},\omega-\Omega+\bm{q}\cdot\bm{v})\, G^{\mathrm{a}}(\bm{k},\omega) \right. \\
&+ G^{\mathrm{c}}(\bm{k}-\bm{q},\omega-\Omega+\bm{q}\cdot\bm{v})\, G^{\mathrm{c}}(\bm{k},\omega) \Big] D^{\mathrm{r}}(\bm{q}\lambda,\Omega) \Big\}. \qquad (9.65)
\end{aligned}$$

The final result is

$$w = 2\sum_{\bm{q},\lambda} |M(\bm{q},\lambda)|^2 \Omega_{q\lambda} \Pi_2(\bm{q},\Omega_{q\lambda}+\bm{q}\cdot\bm{v})\left[n\!\left(\frac{\Omega_{q\lambda}}{T}\right) - n\!\left(\frac{\Omega_{q\lambda}+\bm{q}\cdot\bm{v}}{T_{\mathrm{e}}}\right)\right]. \qquad (9.66)$$

This is the energy-transfer rate given in Sec. 2.1

Note that the phonon contribution to the frictional force $\bm{f}_{\mathrm{p}}$ can also be represented by a diagram shown in Fig. 9.1(b) if the circle vertex Ω is replaced by a dot vertex $\bm{q}$. The expression for $\bm{f}_{\mathrm{p}}$ is obtained if the factor $\Omega_{q\lambda}$ in Eq. (9.66) is replaced by $\bm{q}$.

9.2.2 Effect of an electric field on weak localization

We assume that the impurity scattering dominates the momentum relaxation such that the phonon contribution to the frictional force is negligible. Phonons, however, still play an important role in dissipating energy from the electron system. It is well known that an important quantum correction to the conductivity arises from the maximally crossed diagrams for elastic scattering. In the balance equation theory the contribution of the nth order maximally crossed diagram to the frictional force is shown in Fig. 9.2, and is given by

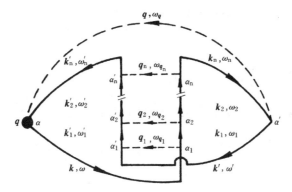

Fig. 9.2 Contribution of the nth-order maximally crossed diagram to the frictional force.

$$\delta \bm{f}_n = (n_\text{i} u^2)^{n+1} \sum_{\bm{q},\bm{k},\bm{q}_1\cdots\bm{q}_n} \bm{q} \int \frac{d\omega}{2\pi} \xi_\alpha \eta_{\alpha'} \eta_{\alpha_1} \eta_{\alpha'_1} \cdots \eta_{\alpha_n} \eta_{\alpha'_n}$$
$$\times G_{\alpha\alpha'_n}(\bm{k}'_n,\omega'_n) G_{\alpha'_n \alpha'_{n-1}}(\bm{k}'_{n-1},\omega'_{n-1}) \cdots G_{\alpha'_1 \alpha'}(\bm{k}',\omega') G_{\alpha'\alpha_n}(\bm{k}_n,\omega_n)$$
$$\times G_{\alpha_n \alpha_{n-1}}(\bm{k}_{n-1},\omega_{n-1}) \cdots G_{\alpha_1 \alpha}(\bm{k},\omega), \tag{9.67}$$

where $\bm{k}_i = \bm{k} - \bm{q}_1 - \bm{q}_2 \cdots - \bm{q}_i$, $\omega_i = \omega + (\bm{q}_1 + \bm{q}_2 \cdots + \bm{q}_i)\cdot\bm{v}$, $\bm{k}'_i = \bm{k} - \bm{q}_{i+1} \cdots - \bm{q}_n - \bm{q}$, $\omega'_i = \omega + (\bm{q}_{i+1} \cdots + \bm{q}_n + \bm{q})\cdot\bm{v}$ ($i = 1,\cdots n$) and $\bm{k}_n = \bm{k}' + \bm{q}$, $\omega_n = \omega' - \bm{q}\cdot\bm{v}$, $\bm{k}'_n = \bm{k} - \bm{q}$, $\omega'_n = \omega + \bm{q}\cdot\bm{v}$. Using relations (9.54) and (9.56) we can express the above combinations of the components of the closed-time-path Green's functions as the sum of terms composed of various numbers of retarded and advanced Green's function factors. The only terms that survive the summation over $\bm{q}_1, \bm{q}_2 \cdots \bm{q}_n$ are those with equal number of retarded and advanced Green's functions:

$$2\left[f(\omega_n) - f(\omega')\right] G^\text{r}(\bm{k}'_n,\omega'_n) G^\text{r}(\bm{k}'_{n-1},\omega'_{n-1}) \cdots$$

$$G^\text{r}(\bm{k}',\omega') G^\text{a}(\bm{k}_n,\omega_n) G^\text{a}(\bm{k}_{n-1},\omega_{n-1}) G^\text{a}(\bm{k},\omega)$$

and their calculation is straightforward (Lei and Cai, 1990b).

In the case of $v \ll v_\text{F}$ and $1/\tau_\text{e} \ll \varepsilon_\text{F}$, the sum of all orders of maximally crossed diagrams yields a total correction to the frictional force

$$\delta f = \sum_{n=1}^\infty \delta f_n,$$

and thus an addition to the steady state resistivity:

$$\begin{aligned}
\delta\rho &= -\frac{\delta f}{n^2 e^2 v} \\
&= -\frac{n_i u^2}{n^2 e^2 v^2} \sum_{q,k,k'} q \cdot v \int \frac{d\omega}{\pi} \left[f(\omega' - q \cdot v) - f(\omega') \right] C(K_v) \\
&\quad \times G^{\mathrm{r}}(k-q, \omega + q \cdot v)\, G^{\mathrm{r}}(k', \omega')\, G^{\mathrm{a}}(k+q, \omega' - q \cdot v)\, G^{\mathrm{a}}(k, \omega).
\end{aligned} \quad (9.68)$$

where $\omega' = \omega + (k - k') \cdot v$,

$$C(K_v) = \frac{n_i u^2 \tau_e^{-1}}{1/\tau_{\mathrm{in}} + (\tau_e/\tau)^2 D K_v^2}, \quad (9.69)$$

with $K_v = |k + k' + 2mv|$, and $D = v_{\mathrm{F}}^2 \tau / d$ is the diffusion coefficient due to impurity scattering in a system of d dimensions. If $\tau_{\mathrm{in}} \gg \tau$, or $\tau_e \simeq \tau$, which is the case for weak localization, the $C(K_v)$ function can be approximately written as

$$C(K_v) = \frac{1}{2\pi N(0)\tau^2} \frac{1}{DK_v^2 + 1/\tau_{\mathrm{in}}}, \quad (9.70)$$

where we have used $\tau^{-1} = 2\pi N(0) n_i u^2$, $N(0)$ being the density of states for single spin. This is the well-known expression for the particle–particle diffusion propagator (Lee and Ramakrishnan, 1985). The existence of the inverse inelastic scattering time $1/\tau_{\mathrm{in}}$ in Eq. (9.70) implies a cutoff of the quantum interference effect beyond the Thouless length $L_{\mathrm{Th}} = (D\tau_{\mathrm{in}})^{1/2}$ (Thouless, 1977). Such a cutoff was phenomenologically introduced previously (Anderson et al, 1979). The present derivation suggests a simple way to achieve this inelastic-scattering cutoff of the localization effect. Furthermore, this expression for $C(K_v)$ in the presence of a finite drift velocity ($v \neq 0$) indicates that a dc electric field produces no direct cutoff related to it. This is in agreement with Altshuler et al (1981, 1982), who point out that a dc electric field does not break the time-reversal invariance and should not affect the quantum interference. However, the drift velocity v, hence the dc electric field, does enter in the other factors of Eq. (9.68) and affects the quantum correction to the resistivity contributed by maximally crossed diagrams.

To calculate $\delta\rho$ from Eq. (9.68) we proceed by changing the summation variables k and k' to $k_v = k' - k$ and $K_v = k + k' + 2mv$. The expression (9.70) of $C(K_v)$ indicates that the dominant contribution to the resistivity correction $\delta\rho$ comes from the small K_v region, such that in other factors of the integrand, we can neglect K_v in comparison with k_v, for which the dominant contribution region is of order of k_F. With these considerations, the summation over k_v can be carried out, yielding at low temperature T_e

$$\delta\rho = \frac{2m^2}{\pi n^2 e^2 \tau^2} S_d \, g_d(2k_F v\tau), \tag{9.71}$$

where

$$S_d = \sum_{K_v} \frac{1}{K_v^2 + L_{\text{Th}}^{-2}} \tag{9.72}$$

is the conventional dimension-dependent scale factor. Note that all the above formulations are written per unit volume and we assume the system size to be much larger than L_{Th}, which is the length cutoff for the interference. The upper cutoff for K_v is generally taken to be ℓ^{-1}, $\ell = v_F\tau$ being the electron mean free path related to elastic scattering. We have

$$S_3 = \frac{1}{2\pi^2}\left(\frac{1}{\ell} - \frac{\pi}{2L_{\text{Th}}}\right), \tag{9.73}$$

$$S_2 = \frac{1}{2\pi}\ln\left(\frac{L_{\text{Th}}}{\ell}\right), \tag{9.74}$$

$$S_1 = \frac{1}{2}L_{\text{Th}}. \tag{9.75}$$

These factors depend strongly on the electron temperature T_e because of the temperature dependence of the inelastic scattering time τ_{in}. The direct effect of a dc electric field is included in the function $g_d(2k_F v\tau)$:

$$g_3(x) = \frac{6}{x^2}\left[1 - \frac{1}{x}\arctan x + \frac{1}{x^2}\ln(1+x^2)\right] \simeq \begin{cases} 1 & x \ll 1 \\ 6/x^2 & x \gg 1 \end{cases} \tag{9.76}$$

$$g_2(x) = \frac{8}{\pi x^2}\int_0^1 \frac{d\xi}{\sqrt{1-\xi^2}}\left(1 - \frac{1}{\sqrt{1+x^2\xi^2}}\right) \simeq \begin{cases} 1 & x \ll 1 \\ 4/x^2 & x \gg 1 \end{cases} \tag{9.77}$$

$$g_1(x) = \frac{1}{1+x^2}. \tag{9.78}$$

The magnitude of $g_d(x)$ decreases with increasing x, indicating that localization effects are reduced when the electric field increases (Lei and Cai, 1990b).

9.2.3 The phonon–plasma coupled mode and electron energy-loss rate

As an important physical quantity, the carrier energy-loss rate in semiconductor, i.e. the energy transferred per unit time from the electron system to the lattice, has repeatedly appeared in various analyses in the preceding chapters, especially those related to nonequilibrium phonons in Chapter 5. In polar semiconductors, it is generally believed that hot electrons lose their energy by first emitting longitudinal optic (LO) phonons due to Fröhlich interaction, except at very low lattice temperatures ($\leqslant 15\,\text{K}$), where acoustic phonons are directly responsible for the carrier energy dissipation to the lattice. Then these energies relax through the coupling between the LO and acoustic phonons. Treating both the electron system and the phonon system, which are assumed weakly coupled, as in equilibrium separately at electron temperature T_e and at lattice temperature T, Kogan (1963) derived the formula (5.1) for the energy-loss rate w_K of electrons to the lattice by directly using the Fermi golden rule for one-phonon process without the effect of electron drifting. The Kogan formula, extended to include hot phonon effects accounts for the experimental electron energy-loss data reasonably well in the temperature region of 40–150 K for wide ranges of carrier density in bulk and low dimensional systems. The situation is different, however, in the temperature range 15–40 K, where the experimental energy-loss rate per carrier shows a trend of significant enhancement over the Kogan prediction (Shah, 1986; Ridley, 1991). This induced a debate on the dominant energy-loss mechanism and promoted our understanding of the issue (Jain, Jalabert and Das Sarma, 1988; Das Sarma, Jain and Jalabert, 1990; Dharma-wardana, 1991; Das Sarma and Korenman, 1991; Lei and Wu, 1993; Tao, Ting and Singh, 1993). We will give a brief analysis on this issue in the framework of balance equation theory.

We consider a model system of electrons and LO phonons with the Hamiltonian $H = H_{\text{er}} + H_{\text{ph}} + H_{\text{ep}} + H_{\text{B}}$, where H_{er} and $H_{\text{ph}} = \sum_{q} H_q$ stand for the electron and LO-phonon Hamiltonian respectively, H_{ep} is the electron–LO-phonon interaction. H_{B} represents the rest part of the lattice vibrations and the heat bath, and their coupling with LO phonons. Physically, it is clear that the concept of the electron energy loss to LO phonons, or the energy flow from the electron system to phonon system, is meaningful only when the coupling H_{ep} between them is weak that the identification of the electron system and the phonon system is possible during the entire process of relaxation. Otherwise, they become an indistinguishable coupled entity, rather than separate electron and phonon systems and the energy flow between them is meaningless. Whether H_{ep} being weak or strong is in comparison with the internal interactions, or the internal thermalization trends of both electron and LO-phonon systems. If the internal thermalization (relaxation) in each subsystem (electron and phonon) is quick in comparison with the relaxation time due to the coupling between them, H_{ep}, we can identify the electron system and the phonon system, consider them to have respective temperatures, and treat H_{ep} as a perturbation. This is the two-temperature model, which have been used from the beginning of this book and contained in the initial density matrix. To include the effect of nonequilibrium phonon occupation (Chapter 5) we take the initial density matrix as

$$\hat{\rho}_0 = \frac{1}{Z} \exp\left(-H_{er}/T_e\right) \exp\left(-\sum_q H_q/T_q\right), \tag{9.79}$$

i.e., different wavevector modes of LO phonons may have different temperatures T_q.

The internal thermalization comes from scatterings due to electron–electron interaction, which can be included by a finite imaginary part i/τ_e in the single electron retarded and advanced Green's functions, as (9.57). Therefore, the life-time τ_e in the single electron Green's functions can be considered as the internal thermalization time of the electron system. Likewise, the ability of internal thermalization of LO phonons can also be represented by a finite life-time τ_{pp} in the phonon retarded and advanced Green's functions:

$$D^r(q,\Omega) = \frac{1}{\Omega - \Omega_q + i/\tau_{pp}} - \frac{1}{\Omega + \Omega_q + i/\tau_{pp}}, \tag{9.80}$$

$$D^a(q,\Omega) = \frac{1}{\Omega + \Omega_q - i/\tau_{pp}} - \frac{1}{\Omega - \Omega_q - i/\tau_{pp}}, \tag{9.81}$$

$$D^c(q,\Omega) = \left[2n\left(\frac{\Omega}{T_q}\right) + 1\right] \left[D^r(q,\Omega) - D^a(q,\Omega)\right]. \tag{9.82}$$

The quasi-equilibrium distribution of mode-q phonons, $n_q \equiv n(\Omega_q/T_q)$, can be determined using the method of Sec. 5.2. The role of electron–phonon interaction is to drive the LO phonon system out of equilibrium, inducing a rate of change of the mode-q phonon population:

$$\left[\frac{\partial n_q}{\partial t}\right]_{ep} = 2M_q^2 \Pi_2(q,\Omega_q) \left[n_q - n\left(\frac{\Omega_q}{T_e}\right)\right]. \tag{9.83}$$

Here M_q is the matrix element of the electron–LO-phonon scattering, $\Pi_2(\bm{q},\omega)$ is the imaginary part of the electron density correlation function $\Pi(\bm{q},\omega)$. On the other hand, there must be another interaction in the lattice, which promotes the whole lattice (all phonons of different modes) towards the final equilibrium at the unique temperature T. Using a single relaxation time τ_p to describe this effect, we write

$$\left[\frac{\partial n_q}{\partial t}\right]_\mathrm{ep} - \frac{1}{\tau_\mathrm{p}}\left[n_q - n\left(\frac{\Omega_q}{T}\right)\right] = 0 \tag{9.84}$$

for the steady state. This gives

$$n\left(\frac{\Omega_q}{T_q}\right) = n\left(\frac{\Omega_q}{T_\mathrm{e}}\right) + \left[1 + \frac{\tau_\mathrm{p}}{\tau(\bm{q},\Omega_q)}\right]^{-1}\left[n\left(\frac{\Omega_q}{T}\right) - n\left(\frac{\Omega_q}{T_\mathrm{e}}\right)\right], \tag{9.85}$$

in which we have defined

$$1/\tau(\bm{q},\Omega_q) \equiv -2M_q^2\,\Pi_2(\bm{q},\Omega_q). \tag{9.86}$$

The time for all phonons of different modes to approach the final equilibrium, τ_p, is considered longer than the time needed for a part of them (e.g. the LO phonons of mode $\bm{q}$) to approach equilibrium, τ_pp.

Using the CTPG representation (Sec. 9.2.1), the electron energy-loss rate including all orders of electron–phonon interaction, is given by

$$w = \mathrm{tr}\{T_\mathrm{p}\left[\hat{\rho}_0 \hat{W}(t) L_\mathrm{p}\right]\}, \tag{9.87}$$

in which

$$L_\mathrm{p} = \exp\left[-\mathrm{i}\int_p H_\mathrm{ep}(t_1)\mathrm{d}t_1\right] \tag{9.88}$$

is an integral operator along the Keldysh path p, T_p is the time ordering operator along the path p, $\hat{W}(t)$ and $H_\mathrm{ep}(t)$ are the operators in the interaction picture of the energy-loss rate $\hat{W}$ (Sec. 1.2) and the electron–phonon interaction H_ep. The lowest order contribution of electron–phonon interaction to the electron energy-loss rate, w_1, is depicted in the Feynmann diagram Fig. 9.1(b). Expanding it according to the Feynmann–Keldysh rule and taking summation in the random phase approximation, we have

$$w_1 = -\sum_{\bm{q}}\int_0^\infty \frac{\mathrm{d}\Omega}{\pi}\frac{\Omega}{\tau(\bm{q},\Omega) + \tau_\mathrm{p}}\left[n\left(\frac{\Omega}{T}\right) - n\left(\frac{\Omega}{T_\mathrm{e}}\right)\right]\mathrm{Im}D^\mathrm{r}(\bm{q},\Omega). \tag{9.89}$$

This is a generalized Kogan formula, including effects of nonequilibrium phonon occupation (finite phonon relaxation time $\tau_p \neq 0$) and finite LO-phonon life-time $(1/\tau_{pp} \neq 0)$. For GaAs-based two-dimensional semiconductors, in which τ_p (around picoseconds) is of the same order of magnitude as $\tau(q, \Omega)$, the nonequilibrium phonon effect is negligible. In Eq. (9.89)

$$\mathrm{Im} D^r(q, \Omega) = \frac{4\Omega\Omega_q \tau_{pp}^{-1}}{(\Omega^2 - \Omega_q^2 - \tau_{pp}^{-2})^2 + 4\Omega^2 \tau_{pp}^{-2}} \quad (9.90)$$

is a spectral density function having finite double peaks. It becomes a double-δ function when $\tau_{pp}^{-1} \to 0$,

$$\lim_{\tau_{pp} \to \infty} \mathrm{Im} D^r(q, \Omega) = -\pi \left[\delta(\Omega - \Omega_q) - \delta(\Omega + \Omega_q) \right], \quad (9.91)$$

and the energy-loss rate to the lowest order of electron–phonon scattering reduces to that of (9.66) at $v = 0$, i.e. the conventional Kogan formula with nonequilibrium phonon effect. In fact, the effect of a finite width of the spectral peaks on the energy-loss rate is minor. There is almost no appreciable change in w_1 when τ_{pp}^{-1} varies from zero to several ps^{-1}. Therefore, when only the lowest order of electron–phonon interaction is taken into account, the finite life-time of phonons is not important and usually disregarded. Thus, the low-temperature deviation of experimental energy-loss rate from the Kogan formula can not be explained within the lowest order of electron–phonon interaction.

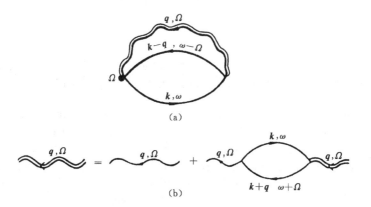

Fig. 9.3 Fig.12.1 Feynman diagrams for (a) the frictional force due to impurities and (b) the electron energy-loss rate due to phonons.

A class of higher order H_{ep} terms which may contribute to the electron energy-loss rate is known as phonon–plasma coupled mode. The Feynman diagram of the electron energy-loss rate with phonon–plasma coupled mode can be obtained by changing the single wavy line in the diagram of Fig. 9.1(b) to a double wavy line, as shown in Fig. 9.3(a). Here the double wavy line represents a matrix propagator in the Keldysh space, i.e. the phonon-coupled mode. It obeys the Dyson equation as depicted in Fig. 9.3(b). Writing Fig. 9.3(a) out we obtain an electron energy-loss rate, which includes the lowest order and all higher orders of the electron–phonon interaction:

$$w = -\sum_{\bm{q}} \int_0^\infty \frac{d\Omega}{\pi} \frac{\Omega}{\tau(\bm{q},\Omega)+\tau_p} \left[n\left(\frac{\Omega}{T}\right) - n\left(\frac{\Omega}{T_e}\right) \right] R(\bm{q},\Omega), \quad (9.92)$$

where

$$R(\bm{q},\Omega) = \frac{-4\Omega\Omega_q \tau_{pp}^{-1}}{\left[\Omega^2 - \Omega_q^2 - \tau_{pp}^{-2} + \Omega_q/\tau_1(\bm{q},\Omega)\right]^2 + \left[2\Omega\tau_{pp}^{-1} + \Omega_q/\tau(\bm{q},\Omega)\right]^2}, \quad (9.93)$$

$$1/\tau_1(\bm{q},\Omega) = -2M_q^2 \Pi_1(\bm{q},\Omega). \quad (9.94)$$

In the limit of weak electron–phonon coupling, one can take $\Omega_q/\tau_1 \to 0$ and $\Omega_q/\tau \to 0$ in (9.93) while keeping τ_{pp}^{-1} intact, and (9.92) becomes the exact expression (9.89) for the lowest order energy-loss rate w_1. Then, by taking the limit $\tau_{pp}^{-1} \to 0$, it reduces to the modified Kogan formula with nonequilibrium phonon effect:

$$w_K = -\sum_{\bm{q}} \frac{\Omega_q}{\tau(\bm{q},\Omega_q)+\tau_p} \left[n\left(\frac{\Omega_q}{T}\right) - n\left(\frac{\Omega_q}{T_e}\right) \right]. \quad (9.95)$$

Such an order of limiting process is reasonable if the internal thermalization trends within the LO-phonon system and within the electron system are stronger than that caused by the electron–phonon interaction. This is exactly what is required in the present model. The electron energy-loss rate formula (9.92), which represents a sum of the lowest and all the higher order terms, can be applied to the case of stronger electron–phonon coupling as long as the series shown in Fig. 9.3(b) remains convergent, and this formula should give a better description for the energy-loss rate than the lowest-order formula. On the other hand, if one takes the limit $1/\tau_{pp} \to 0$ first, the energy-loss rate given by Eq. (9.92) vanishes. It will not return to

Kogan formula for any strength of the electron–phonon interaction, whatever small. In this case the higher order terms in the series are divergent, and the formal summation results in a renormalization, which turns the electron and phonon system into a unity. We can not distinguish individual electron system and phonon system, and the concept of the electron energy-loss to the LO phonon system will no longer be meaningful.

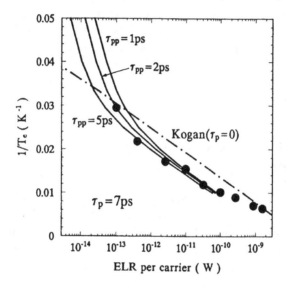

Fig. 9.4 Inverse electron temperature $1/T_e$ vs energy-loss rate per carrier for a two-dimensional GaAs quantum-well system at lattice temperature $T = 1.8$ K. The solid curves are calculated from Eq.(28) and (29) with $\tau_p = 7$ ps and $\tau_{pp} = 1$, 2 and 5 ps respectively. The chain line is obtained from Kogan formula (22) with $\tau_p = 0$. The closed circles are the experimental data of Shah *et al* (1986). From Lei and Wu (1993).

The hot-electron energy-loss rate has been calculated as a function of electron temperature T_e at lattice temperature $T = 1.8$ K from Eq. (9.92) for a two-dimensional GaAs quantum-well system with well width $a = 26$ nm, electron sheet density $N_s = 3.9 \times 10^{15}/\mathrm{m}^2$. The material parameters used in the calculation are: electron effective mass $m = 0.07 m_e$ (m_e is the free electron mass), LO phonon energy $\Omega_q = \Omega_{LO} = 35.4$ meV, static dielectric constant $\kappa = 12.9$, and optic dielectric constant $\kappa_\infty = 10.8$. Fig. 9.4 plots the inverse electron temperature $1/T_e$ versus the electron energy-loss rate per carrier for three values of the LO-phonon internal thermalization time $\tau_{pp} = 1$, 2 and 5 ps. The hot-phonon relaxation time is assumed to be

$\tau_p = 7\,\text{ps}$ for all three cases. The energy-loss rate per carrier calculated from the lowest order H_{ep} contribution, Eq. (9.89) (generalized Kogan formula), is also shown for the case of $\tau_p = 0$. It behaves almost like a straight line. Change in τ_{pp} value has little effect on the curve. Finite value of τ_p essentially shifts the whole $\tau_p = 0$ Kogan curve down rigidly. The main effect of higher order contributions of H_{ep} is the significant low-temperature enhancement of the electron energy-loss rate over the straight-line Kogan behavior as shown in the figure. This enhancement is more pronounced at smaller τ_{pp}. In the case $\tau_{\text{pp}} = 5\,\text{ps}$ it begins around $T_e \simeq 40\,\text{K}$. For larger τ_{pp} the enhancement appears at lower electron temperature. The predicted low-temperature enhancement of electron energy-loss rate is in agreement with the experimental data of Shah et al (1986), which we show in the figure as closed circles.

9.3 Derivation of Force Balance Equation from Dielectric Response

One can gain insight into the physical processes associated with the balance equation description of transport dynamics through a rederivation of the resistive force on the center of mass (CM) due to impurity scattering from the dielectric response (Horing, Lei and Cui, 1986). The impurity scattering potential energy in Eq. (1.2) can be written in terms of CM and relative electron coordinates $\boldsymbol{R}$ and $\boldsymbol{r}'_i$ as

$$H_{\text{ei}} = \sum_{i,a} u(\boldsymbol{r}'_i + \boldsymbol{R} - \boldsymbol{r}_a), \tag{9.96}$$

where $\boldsymbol{r}_a$ represents the position of the ath impurity. Considering that H_{ei} is the only mechanism which couples the CM and relative electron variables, one may take the view that the moving CM projects the single-particle potential energy $u(2)$ into the space and time point $2 \equiv (\boldsymbol{r}'_2, t_2)$:

$$u(2) \equiv u(\boldsymbol{r}'_2, t_2) = \sum_a u(\boldsymbol{r}'_2 + \boldsymbol{R}(t_2) - \boldsymbol{r}_a), \tag{9.97}$$

which polarizes the relative electrons, resulting in an effective total electric potential $\phi(1)$ at $1 \equiv (\boldsymbol{r}'_1, t_1)$ by $(d2 \equiv d\boldsymbol{r}_2 dt_2)$

$$\phi(1) = \int d2\, K(1,2) u(2)/e, \tag{9.98}$$

where e is the charge of an electron, $K(1,2)$ is the inverse dielectric function of the relative electron gas, representing its dynamic, nonlocal (and inhomogeneous) screening action in response to $u(2)/e$ impressed by the CM. The associated electric field acting on the relative electrons is

$$\boldsymbol{E}(1) = -\boldsymbol{\nabla}_1 \phi(1). \tag{9.99}$$

The charge density perturbation of the relative electron gas may be identified at position $1 \equiv (\boldsymbol{r}_1', t_1)$ as

$$\rho(1) = -\epsilon_0 \nabla_1^2 (\phi(1) - u(1)/e), \tag{9.100}$$

where we have separated the unperturbed $[u(1)/e$- or impurity-related] charge density. The force experienced by the relative electrons per unit volume at that point is $\rho(1)\boldsymbol{E}(1)$. The total force experienced by the relative electrons is obtained by integrating over the relative electron gas, and by the law of action and reaction, the resistive force on the CM is given by

$$\boldsymbol{F} = -\int d\boldsymbol{r}_1' \boldsymbol{E}(1)\rho(1). \tag{9.101}$$

The integrand term involving the first term in (9.100) of $\rho(1)$ is proportional to $\nabla_1^2 \phi(1) \boldsymbol{\nabla}_1 \phi(1) = \boldsymbol{\nabla}_1 \cdot (\boldsymbol{\nabla}_1 \phi(1) \boldsymbol{\nabla}_1 \phi(1))/2$, which vanishes after integration. We have

$$\boldsymbol{F} = \frac{\epsilon_0}{e^2} \int d\boldsymbol{r}_1' \int d2 \left[\boldsymbol{\nabla}_1 K(1,2) u(2)\right] \nabla_1^2 u(1). \tag{9.102}$$

To leading order in the impurity scattering potential, this $\boldsymbol{F}$ is explicitly of order of u^2, and $K(1,2)$ can be evaluated in the absence of u, and it is therefore decoupled from the electric field as well as from impurity scattering. Such an evaluation of the inverse dielectric function nevertheless involves electron–electron interactions which can be treated within the framework of the random phase approximation and can proceed using an equilibrium ensemble for the relative electrons with the lattice temperature by assuming that there is a bath in contact with the system which dissipates the energy quickly enough while contributes no resistive force. With this, $K(1,2)$ is translationally invariant in time and space for a homogeneous bulk medium, and it is convenient to use its Fourier transform: $K(1-2) \to K(p) \equiv K(\boldsymbol{p}, \omega)$. Note that in the steady state when the CM coordinate $\boldsymbol{R}(t_2) = \boldsymbol{v} t_2$ ($\boldsymbol{v}$ is the drift velocity), the resistive force per unit volume after average over randomized impurity sites, may be written as

$$\boldsymbol{f} = n_\mathrm{i} \frac{\epsilon_0}{e^2} \int \frac{d\boldsymbol{p}}{(2\pi)^3} \boldsymbol{p}\, p^2 K_2(\boldsymbol{p}, \boldsymbol{p}\cdot\boldsymbol{v}) |u(\boldsymbol{p})|^2. \tag{9.103}$$

Here n_i is the impurity density, and $u(\boldsymbol{p})$ is the impurity scattering potential in the Fourier representation. $K_2(\boldsymbol{p}, \omega)$ is the imaginary part of $K(\boldsymbol{p}, \omega) = K_1(\boldsymbol{p}, \omega) + \mathrm{i} K_2(\boldsymbol{p}, \omega)$, which is the inverse of the dielectric function (2.17) in the random phase approximation, $K(\boldsymbol{p}, \omega) = 1/\epsilon(\boldsymbol{p}, \omega)$, and its relation to the density correlation function $\Pi(\boldsymbol{p}, \omega)$ is given by

$$K_2(\boldsymbol{p}, \omega) = \frac{e^2}{\epsilon_0 p^2} \Pi_2(\boldsymbol{p}, \omega). \tag{9.104}$$

Thus, (9.103) is exactly the impurity-induced frictional force given by Eq. (2.4).

9.4 Nonequilibrium Statistical Operator Method

The nonequlibrium statistical operator method, developed by Zubarev (1961, 1974), McLennan (1961) and Peletminskii and Yatsenko (1967), is based on Bogolyubov's idea (Bogolyubov, 1962) that a physical system has a hierarchy of distinct relaxation times. If one considers the evolution of the system on a time scale which is larger than the characteristic time, τ_0, of micro-relaxation in the system, it is possible to use a set of macroscopic parameters to describe the system such that the time-dependence of these variables determines the time evolution of the statistical ensemble.

We assume that the system Hamiltonian can be broken up into two parts $H = H_0 + H_\mathrm{I}$, where H_0 is the main Hamiltonian and H_I is a relatively weak interaction. A short micro-relaxation time τ_0 inherent in H_0 makes it possible to describe the system in a time scale larger than τ_0 by a set of macroscopic parameters $\gamma = \{\gamma_m\}$, which is the average value of the set of dynamic operators $\hat{\gamma} = \{\hat{\gamma}_m\}$, satisfying the commutation relations: $[H_0, \hat{\gamma}_m] = \sum_n \alpha_{mn} \hat{\gamma}_n$, α_{mn} being c-numbers. The density matrix, or the nonequilibrium statistical operator of the system depends on t only via the time dependence of the parameters $\gamma_m(t)$: $\hat{\rho} = \hat{\rho}(t) = \hat{\rho}(\gamma)$. To determine such a density matrix satisfying the Liouville equation

$$\frac{d\hat{\rho}}{dt} = \mathrm{i}\,[\hat{\rho}, H], \tag{9.105}$$

a set of time-dependent parameters $\{X_m\}$ (which can be understood as the thermodynamic variables conjugate to $\{\gamma_m\}$), and a quasi-equilibrium statistical operator $\hat{\rho}^0 = \hat{\rho}^0(t) = \hat{\rho}^0(\gamma)$,

$$\hat{\rho}^0 = \frac{1}{Z}\exp\left(-\sum_m X_m \hat{\gamma}_m\right), \tag{9.106}$$

are introduced in such a way that the equality

$$\gamma_m \equiv \text{tr}\{\hat{\rho}\hat{\gamma}_m\} = \text{tr}\{\rho^0 \hat{\gamma}_m\} \tag{9.107}$$

is satisfied. Eq. (9.107) gives the conditions for the determination of the parameters X_m. $\hat{\rho}^0(t)$ has the following meaning: if one turns off the interaction H_I at time t, the system, starting from $\hat{\rho}(t)$ and following the Liouville equation of Hamiltonian H_0 for a time longer than τ_0, would approach $\hat{\rho}^0(t)$. With these conditions the general form of the nonequilibrium statistical operator $\hat{\rho}(t)$ can be derived (Peletmiskii and Yatsenko, 1967; Zubarev, 1974). To linear order in H_I, it can be written as

$$\hat{\rho} = \hat{\rho}^0 + \text{i}\int_{-\infty}^{0} d\tau\, \text{e}^{\epsilon\tau}[\hat{\rho}^0, H_\text{I}(\tau)] + \text{i}\int_{-\infty}^{0} d\tau\, \text{e}^{\epsilon\tau}\sum_m \frac{\partial \hat{\rho}^0}{\partial \gamma_m}\text{tr}\{\hat{\rho}^0[\hat{\gamma}_m, H_\text{I}(\tau)]\}, \tag{9.108}$$

where $\epsilon \to 0^+$ and $H_\text{I}(\tau) \equiv \text{e}^{\text{i}H_0\tau}H_\text{I}\text{e}^{-\text{i}H_0\tau}$. If $[\hat{\gamma}_m, \hat{\gamma}_n] = 0$, the last term of the right-hand-side of the above equation vanishes, leading to

$$\hat{\rho} = \hat{\rho}^0 + \text{i}\int_{-\infty}^{0} d\tau\, \text{e}^{\epsilon\tau}[\hat{\rho}^0, H_\text{I}(\tau)]. \tag{9.109}$$

With the help of these explicit expressions for the nonequilibrium statistical operator, the expectation value of any physical quantity can be calculated on a time scale larger than τ_0 under the nonequilibrium conditions subject to the Hamiltonian $H_0 + H_\text{I}$.

The nonequilibrium statistical operator method was first applied by Kalashnikov (1970), and then used by other authors (Niez and Ferry, 1983) in the study of warm-electron transport. Xing, Hu and Ting (1987) employ this method to discuss steady-state hot-electron transport in a strong electric field. Choosing the total electron momentum $\boldsymbol{P}_\text{e} = \sum_{\boldsymbol{k}\sigma} \boldsymbol{k}\, c_{\boldsymbol{k}\sigma}^\dagger c_{\boldsymbol{k}\sigma}$, the electron energy $H_\text{e} = \sum_{\boldsymbol{k}\sigma} \varepsilon_{\boldsymbol{k}} c_{\boldsymbol{k}\sigma}^\dagger c_{\boldsymbol{k}\sigma}$, the electron number $N = \sum_{\boldsymbol{k}\sigma} c_{\boldsymbol{k}\sigma}^\dagger c_{\boldsymbol{k}\sigma}$, and the phonon energy H_ph to form the parameter set, they derive nonlinear force and energy balance equations to the lowest order in the electron-impurity and electron-phonon interactions. These equations are exactly the same as the balance equations of Lei and Ting (1984; 1985b) as presented in Chapter 2. Liu et al (1988) apply the nonequilibrium statistical operator method to steady state dc transport in many-valley semiconductors, obtaining balance equations identical to those discussed in Sec. 6.7 with an appropriate selection of the parameter set. We refer the readers to the review article by Xing and Liu (1992) for a systematic discussion. The nonequilibrium statistical operator method has also been applied to issues related to nonequilibrium phonons (Cai, Marchetti and Lax, 1986). Further detail may be found in their original papers.

9.5 Generalized Quantum Langevin Equation Approach

Starting from the separation of the center-of-mass variables from relative electron variables and the Hamiltonian described in Eqs. (1.9)–(1.15), Hu and O'Connell (1987, 1988a, 1989a) develop a generalized quantum Langevin equation (GLE) approach to transport with the operator rates of change, $\dot{P}$, $\dot{H}_{\rm ph}$ and $\dot{H}_{\rm er}$ given by Eqs. (1.20), (1.23) and (1.24), direct treated in the Heisenberg representation. For instance, they rewrite the operator Eq. (1.20) as

$$\dot{P} = NeE - i\sum_{q} qU_q(R)\rho_q - i\sum_{q\lambda} qM_{q\lambda}(R)O_{q\lambda}, \quad (9.110)$$

with the notations

$$U_q(R) \equiv \sum_{a} u(q)\, e^{i q \cdot (R - r_a)}, \quad (9.111)$$

$$M_{q\lambda}(R) \equiv M(q,\lambda)\, e^{i q \cdot R}, \quad (9.112)$$

$$\rho_q \equiv \sum_{k} \rho_{kq}, \quad \rho_{kq} \equiv \sum_{\sigma} c^{\dagger}_{k+q\sigma} c_{k\sigma}, \quad (9.113)$$

$$O_{q\lambda} \equiv (b_{q\lambda} + b^{\dagger}_{-q\lambda})\rho_q \equiv O^A_{q\lambda} + O^E_{q\lambda}, \quad (9.114)$$

$$O^A_{q\lambda} \equiv \sum_{k} O^A_{kq\lambda}, \quad O^A_{kq\lambda} \equiv b_{q\lambda}\rho_{kq}, \quad (9.115)$$

$$O^E_{q\lambda} \equiv \sum_{k} O^E_{kq\lambda}, \quad O^E_{kq\lambda} \equiv b^{\dagger}_{-q\lambda}\rho_{kq}. \quad (9.116)$$

Treating Eq. (9.110) as the Heisenberg equation of motion, i.e. all the operators in it, P, R, ρ_q and $O_{q\lambda}$, as time-dependent Heisenberg operators (thus $\dot{P} \equiv dP/dt$, $\dot{R} \equiv dR/dt$), Hu and O'connell are able to obtain explicit expressions for these Heisenberg operators by formally solving the Heisenberg equations of motion for ρ_{kq}, $O^A_{kq\lambda}$ and $O^E_{kq\lambda}$. For instance, the equations $\dot{\rho}_{kq} = -i[\rho_{kq}, H_{\rm B} + H_{\rm I}]$ and $\ddot{\rho}_{kq} = -i[\dot{\rho}_{kq}, H_{\rm B} + H_{\rm I}]$ ($H_{\rm B} = H_{\rm cm} + H_{\rm er} + H_{\rm ph}$,

$H_{\rm I} \equiv H_{\rm ei} + H_{\rm ep}$) yield the following second order differential equation:

$$\ddot{\rho}_{\bm{kq}} + \omega_{\bm{kq}}^2 \rho_{\bm{kq}} = -\bigl[[\rho_{\bm{kq}}, H_{\rm I}], H_{\rm B} + H_{\rm I}\bigr], \qquad (9.117)$$

where $\omega_{\bm{kq}} \equiv \varepsilon_{\bm{k+q}} - \varepsilon_{\bm{k}} = \bm{k}\cdot\bm{q}/m$. Assuming that as $t \to -\infty$, $H_{\rm I} = 0$ and the initial state of the relative electron system is an isotropic free electron gas (which is in agreement with the initial density matrix in Sec. 1.3), they obtain the solution of the differential equation (9.117) as (retaining only the lowest order terms in electron–impurity and electron–phonon interactions and neglecting electron–electron interactions)

$$\rho_{\bm{kq}}(t) = {\rm e}^{-{\rm i}\omega_{\bm{kq}}t}\rho_{\bm{kq}}^0 - {\rm i}\int_{-\infty}^{t} dt'\, {\rm e}^{-{\rm i}\omega_{\bm{kq}}(t-t')}(n_{\bm{k+q}} - n_{\bm{k}})$$

$$\times \left[U_q(\bm{R}) + \sum_{\lambda} M_{q\lambda}(\bm{R})(b_{\bm{q}\lambda} + b^{\dagger}_{-\bm{q}\lambda}) \right]. \qquad (9.118)$$

Here $\rho_{\bm{kq}}$ denotes the relative electron density operator in the absence of $H_{\rm I}$. Similar expressions are also obtained for $O_{\bm{kq}\lambda}^A(t)$ and $O_{\bm{kq}\lambda}^E(t)$. With these results Hu and O'Connell write Eq. (9.110) in the form

$$M\ddot{\bm{R}} = Ne\bm{E}(t) + \hat{\bm{F}}(t) - \int_{-\infty}^{t} dt'\, \mu(\dot{\bm{R}}; t, t') M\dot{\bm{R}}, \qquad (9.119)$$

which serves as momentum-GLE, in which $M = Nm$ and

$$\hat{\bm{F}}(t) = -{\rm i}\sum_{\bm{k},\bm{q}} \bm{q}\, U_q(\bm{R})\, {\rm e}^{-{\rm i}\omega_{\bm{kq}}t}\rho_{\bm{kq}}^0$$

$$-{\rm i}\sum_{\bm{k},\bm{q}} \bm{q} M_{q\lambda}(\bm{R}) \left[{\rm e}^{-{\rm i}(\omega_{\bm{kq}}+\Omega_{\bm{q}\lambda})t} O_{\bm{kq}\lambda}^{0A} + {\rm e}^{-{\rm i}(\omega_{\bm{kq}}-\Omega_{\bm{q}\lambda})t} O_{\bm{kq}\lambda}^{0E} \right] \qquad (9.120)$$

is the random force. The last term of the right-hand-side of (9.119) is the frictional force, which contains a memory function (Hu and O'Connell, 1988a, 1989a)

$$\mu(\dot{\bm{R}}; t, t') \equiv \sum_{\bm{k},\bm{q},s} d_{\bm{kq}}^s\, {\rm e}^{{\rm i}\bm{q}\cdot[\bm{R}(t)-\bm{R}(t')] - {\rm i}\omega_{\bm{kq}}^s(t-t')}. \qquad (9.121)$$

The definitions of $d_{\bm{kq}}^s$ and $\omega_{\bm{kq}}^s$ are detailed in their paper. It is easily seen that the random force $\hat{\bm{F}}(t)$, Eq. (9.120), is equivalent to the fluctuation force operator given in Eq. (4.7). Furthermore, in obtaining Eq. (9.121) they have used the approximation

$$ {\rm e}^{{\rm i}\bm{q}\cdot\bm{R}(t)} {\rm e}^{-{\rm i}\bm{q}\cdot\bm{R}(t')} \approx {\rm e}^{-{\rm i}\bm{q}\cdot[\bm{R}(t)-\bm{R}(t')]}, \qquad (9.122)$$

neglecting the noncommutivity of the Heisenberg operators $\boldsymbol{R}(t)$ at differing times. This is reasonable for large N, and is obviously equivalent to treating the center-of-mass as a classical particle (Sec. 1.2).

The energy-GLE of Hu and O'Connell (1988a) is obtained from

$$\frac{d}{dt}\left[\frac{1}{2}Nm\dot{\boldsymbol{R}}(t)^2\right] + \dot{H}_{\mathrm{er}}(t) + \dot{H}_{\mathrm{ph}}(t) = Ne\boldsymbol{E}\cdot\dot{\boldsymbol{R}}(t), \qquad (9.123)$$

written in a form similar to Eq. (9.110). This equation is equivalent to the operator equation (1.26) in Sec. (1.2).

It is worth noting that the GLE's of Hu and O'connell, which correspond to the operator equations in the Heisenberg picture in the balance-equation description in Chapter 1, do not incorporate statistical information. To establish balance equations and Langevin equations for transport, one must average them over the (initial) statistical ensemble. Although Hu and O'Connell do not clearly state what an ensemble to use for the average, they adopt the idea of the electron temperature T_{e} in relation to the average $\langle \dot{H}_{\mathrm{er}} \rangle$, indicating that the same initial density matrix $\hat{\rho}_0$ as (1.30) is employed. Furthermore, the establishment of an energy balance equation involves an identification of the kind

$$\frac{d}{dt}\langle H_{\mathrm{er}} \rangle = \langle \dot{H}_{\mathrm{er}} \rangle. \qquad (9.124)$$

By this means the electron temperature enters the theory as a time-dependent parameter.

Hu and O'Connell (1988b-e, 1989b) have used these equations in many investigations, particularly in problems related to CM velocity fluctuations. We refer the readers to their original papers for detail.

9.6 Drifted Electron-Temperature Model

There are many similarities between the balance equation formulation and generalized drifted electron temperature model. Marchetti and Cai (1987) show that the force- and energy-balance equations (3.3) (without $\mathcal{B}$ term) and (3.4) for hot-electron transport in the random phase approximation, can be obtained under certain conditions from the Boltzmann equation for the electron distribution function $f(\boldsymbol{k})$:

$$\frac{\partial f(\boldsymbol{k})}{\partial t} - e\boldsymbol{E}\cdot\frac{\partial f(\boldsymbol{k})}{\partial \boldsymbol{k}} = \left[\frac{\partial f(\boldsymbol{k})}{\partial t}\right]_{\mathrm{ep}} + \left[\frac{\partial f(\boldsymbol{k})}{\partial t}\right]_{\mathrm{ei}} + \left[\frac{\partial f(\boldsymbol{k})}{\partial t}\right]_{\mathrm{ee}}. \qquad (9.125)$$

In this, $\boldsymbol{E}$ is the applied electric field and the right-hand side comprises the electron–phonon(ep), electron–impurity(ei) and electron–electron(ee) collision terms. The electron–phonon collision term, for instance, is given by

$$\left[\frac{\partial f(\boldsymbol{k})}{\partial t}\right]_{\text{ep}} = -\sum_{\boldsymbol{k}_1}\{W_{\text{ep}}(\boldsymbol{k}_1,\boldsymbol{k})f(\boldsymbol{k})[1-f(\boldsymbol{k}_1)] - W_{\text{ep}}(\boldsymbol{k},\boldsymbol{k}_1)f(\boldsymbol{k}_1)[1-f(\boldsymbol{k})]\}. \tag{9.126}$$

The transition probability of an electron from state $\boldsymbol{k}$ to $\boldsymbol{k}_1$ due to electron–phonon interaction, $W_{\text{ep}}(\boldsymbol{k}_1,\boldsymbol{k})$, takes the form

$$W_{\text{ep}}(\boldsymbol{k}_1,\boldsymbol{k}) = 2\pi\sum_{\boldsymbol{q},\lambda}|M(\boldsymbol{q},\lambda)|^2\sum_{\sigma=\pm 1}\delta_{\boldsymbol{k}_1,\boldsymbol{k}+\boldsymbol{q}}\,\delta(\varepsilon_{\boldsymbol{k}}-\varepsilon_{\boldsymbol{k}_1}-\sigma\Omega_{\boldsymbol{q}\lambda})$$
$$\times\left[n\left(\frac{\Omega_{\boldsymbol{q}\lambda}}{T}\right)+\frac{1}{2}+\frac{\sigma}{2}\right]. \tag{9.127}$$

The electron–electron collision term $[\partial f(\boldsymbol{k})/\partial t]_{\text{ee}}$ is assumed to play the following two roles: (a) to make the distribution function $f(\boldsymbol{k})$ taking the form of a displaced Fermi-Dirac function at electron temperature T_{e} with the wavevector shift $\boldsymbol{k}_{\text{d}} = m\boldsymbol{v}$, $\boldsymbol{v}$ being the drift velocity of the system,

$$f(\boldsymbol{k}) \to f(\varepsilon_{\boldsymbol{k}-\boldsymbol{k}_{\text{d}}},T_{\text{e}}) \equiv \frac{1}{\exp[(\varepsilon_{\boldsymbol{k}-\boldsymbol{k}_{\text{d}}}-\mu)/T_{\text{e}}]+1}, \tag{9.128}$$

and (b) to provide a dynamic nonlocal screening of the electron–phonon matrix element and the electron–impurity potential, such that $M(\boldsymbol{q},\lambda)$ in (9.127) should be replaced by

$$\bar{M}(\boldsymbol{q},\lambda) = M(\boldsymbol{q},\lambda)/\epsilon_{\text{R}}(\boldsymbol{q},\Omega_{\boldsymbol{q}\lambda}) \tag{9.129}$$

and $u(\boldsymbol{q})$ by

$$\bar{u}(\boldsymbol{q}) = u(\boldsymbol{q})/\epsilon_{\text{R}}(\boldsymbol{q},0). \tag{9.130}$$

Here, $\Omega_{\boldsymbol{q}\lambda}$ is the frequency of mode $\boldsymbol{q}\lambda$ phonon, and $\epsilon_{\text{R}}(\boldsymbol{q},\omega)$ is the "RPA" dielectric function defined by

$$\epsilon_{\text{R}}(\boldsymbol{q},\omega) = 1 - 2\nu_c(q)\sum_{\boldsymbol{k}}\frac{f(\varepsilon_{\boldsymbol{k}+\boldsymbol{q}-\boldsymbol{k}_{\text{d}}},T_{\text{e}}) - f(\varepsilon_{\boldsymbol{k}-\boldsymbol{k}_{\text{d}}},T_{\text{e}})}{\varepsilon_{\boldsymbol{k}+\boldsymbol{q}}-\varepsilon_{\boldsymbol{k}}+\omega+i\delta} = \epsilon(\boldsymbol{q},\omega+\boldsymbol{q}\cdot\boldsymbol{v}), \tag{9.131}$$

$\epsilon(\boldsymbol{q},\omega)$ being the dielectric function given in Eq. (2.17). Multiplying Eq. (9.125) (without $[\partial f(\boldsymbol{k}/\partial t]_{\text{ee}}$ term) by $\boldsymbol{k}$, and by $(\boldsymbol{k}-\boldsymbol{k}_{\text{d}})^2/2m$ respectively, summing over $\bm{k}$ and identifying

$$\bm{v} = \frac{2}{n}\sum_{\bm{k}}\frac{\bm{k}}{m}f(\bm{k}) \qquad (9.132)$$

$$E_e = 2\sum_{\bm{k}}\frac{(\bm{k}-\bm{k}_d)^2}{2m}f(\bm{k}) \qquad (9.133)$$

as the average drift velocity and electron internal energy and with the replacements (9.129) and (9.130), Marchetti and Cai (1987) obtain two coupled nonlinear equations:

$$\frac{\partial}{\partial t}mn\bm{v} + ne\bm{E} = \bm{f}(\bm{v},T_e) \qquad (9.134)$$

$$\frac{\partial}{\partial t}E_e = -\bm{v}\cdot\bm{f}(\bm{v},T_e) - w(\bm{v},T_e), \qquad (9.135)$$

with $\bm{f} = \bm{f}_i + \bm{f}_p$ and w having the same expressions as given by Eqs. (2.4)-(2.6).

Note that since the displaced Fermi-Dirac function is not a distribution function satisfying the dissipative Boltzmann equation in the presence of impurity and (or) phonon scattering(s), the above derivation by Marchetti and Cai (1987) gives no indication about the equivalence between the balance equation approach and the Boltzmann equation method (Lei and Ting, 1987). On the other hand, a displaced equilibrium (Fermi-Dirac) distribution function is not the distribution function of the balance equation theory neither (Sec 7.6). In principle, a system in transport is in a nonequilibrium state and can not be described by an equilibrium distribution function of any kind. To leading order of scattering interaction, $H_I = H_{ei} + H_{ep}$, the balance equation approach is a kind of linear response theory. The linear response of a system in the scattering interaction is determined by the properties of the system in the absence of the scattering interaction. Consequently, one only needs the distribution function of the initial state (or unperturbed state) before turning on the scattering, i.e. the displaced Fermi-Dirac function, in order to obtain major transport properties to the leading order of H_I. This is one of the advantages of the balance-equation approach. For the evaluation of the nonequilibrium distribution function to the next order in H_I in the balance equation theory, we refer the readers to Sec. 7.6 in Chapter 7 of this book.

Chapter 10

Weakly Nonuniform Systems, Hydrodynamic Balance Equations

Up to now we have dealt with spatially homogeneous case involving only a uniform applied electric field. In real systems, especially in semiconductor devices, the spatial inhomogeneity is not only widely existent, but, in many cases, is purposely designed for achieving certain function. The investigation of carrier transport in spatially inhomogeneous systems has developed into an independent field of discipline: semiconductor device physics, which is beyond the scope of this book. This chapter and Chapter 13 will give a brief introduction to several basic concepts related to device modeling within the framework of balance-equation theory. In this chapter we first extend the balance-equation approach to weakly nonuniform systems to establish hydrodynamic balance equations (Lei, Cai and Xie, 1988), and apply them directly to the analyses of thermoelectric power and effect of phonon drag. We will examine the validity of the general Onsager relation within the framework of hydrodynamic balance equation theory before proceeding to the description of carrier transport in semiconductor devices.

We first focus systems made of compositionally uniform material, in which the energy band structure can be assumed the same throughout the whole space volume considered, despite that the spatial variations of the carrier density, drift velocity and electron temperature may occur due to some external reasons (such as spatially selected doping, nonuniform applied electric field, or externally introduced temperature gradient). The extension to compositionally nonuniform (with spatial variation of band structure) semiconductors can be made without difficulty.

10.1 Hamiltonian of Small Fluid Elements

We consider an interacting electron system under the influence of an inhomogeneous external electric field $\boldsymbol{E}_e(\boldsymbol{r}) = -\nabla \phi_e(\boldsymbol{r})$, with the Hamiltonian

$$H = \sum_i \left[\frac{p_i^2}{2m} + \sum_{j \neq i} \frac{e^2}{8\pi\epsilon_0} \frac{1}{|\boldsymbol{r}_i - \boldsymbol{r}_j|} + e\phi_e(\boldsymbol{r}_i) + \Phi(\boldsymbol{r}_i) \right]. \quad (10.1)$$

Here $\boldsymbol{p}_i$ and $\boldsymbol{r}_i$ are the momentum and coordinate of the ith electron, and

$$\Phi(\boldsymbol{r}_i) = \sum_a u(\boldsymbol{r}_i - \boldsymbol{r}_a) + \sum_l v_l(\boldsymbol{r}_i - \boldsymbol{R}_l) \quad (10.2)$$

is the scattering potential due to randomly distributed impurities ($\boldsymbol{r}_a$ stands for the position of the ath impurity) and lattice vibrations ($\boldsymbol{R}_l$ stands for the lattice site). The whole system can be divided into many macroscopically small subsystems. A subsystem, which is termed a fluid element, or simply an element, is composed of electrons in a small volume $d\tau$ about a position $\boldsymbol{R}$. The Hamiltonian of the element $d\tau$ can be written as

$$\delta H = \sum_{i \in d\tau} \left[\frac{p_i^2}{2m} + \frac{e^2}{8\pi\epsilon_0} \sum_{j \neq i} \frac{1}{|\boldsymbol{r}_i - \boldsymbol{r}_j|} + e\phi_e(\boldsymbol{r}_i) + \Phi(\boldsymbol{r}_i) \right]. \quad (10.3)$$

Here, the second term in parentheses is the Coulomb interaction potential of the ith electron with all other electrons. We divide it into two parts: contribution from electrons inside $d\tau$ and those from outside, and use a mean-field treatment for the latter. For small $d\tau$ it becomes the macroscopic induction potential $\phi_i(\boldsymbol{r}_i)$ caused by the electron charge distribution (plus positive charge background), such that

$$\delta H = \sum_{i \in d\tau} \left[\frac{p_i^2}{2m} + \sum_{\substack{j \neq i \\ j \in d\tau}} \frac{e^2}{8\pi\epsilon_0} \frac{1}{|\boldsymbol{r}_i - \boldsymbol{r}_j|} + e\phi(\boldsymbol{r}_i) + \Phi(\boldsymbol{r}_i) \right], \quad (10.4)$$

with $\phi(\boldsymbol{r}) = \phi_e(\boldsymbol{r}) + \phi_i(\boldsymbol{r})$ being the total potential. In this, $\sum_{i \in d\tau}$ indicates the restricted sum over the particles inside $d\tau$ around $\boldsymbol{R}$, which, in the limit of small $d\tau$, can be conveniently represented by a δ-function,

$$\sum_{i \in d\tau} \sim d\tau \sum_i \delta(\boldsymbol{r}_i - \boldsymbol{R}), \quad (10.5)$$

such that we can write

$$\delta H = d\tau H(\mathbf{R}) \quad \text{and} \quad H = \int d\tau H(\mathbf{R}). \tag{10.6}$$

Here

$$H(\mathbf{R}) = H_e(\mathbf{R}) + \sum_i [e\phi(\mathbf{r}_i) + \Phi(\mathbf{r}_i)]\,\delta(\mathbf{r}_i - \mathbf{R}), \tag{10.7}$$

$$H_e(\mathbf{R}) = \sum_i \left[\frac{p_i^2}{2m} + \frac{e^2}{8\pi\epsilon_0} \sum_{\substack{j \neq i \\ j \in d\tau}} \frac{1}{|\mathbf{r}_i - \mathbf{r}_j|} \right] \delta(\mathbf{r}_i - \mathbf{R}), \tag{10.8}$$

and $H_e(\mathbf{R})\,d\tau$ represents the kinetic and Coulombic energies of electrons in the element $d\tau$. Note that, under the present approximation, different elements consist of different sets of electrons, and the Coulomb coupling between electrons in different elements has been included in the total potential $\phi(\mathbf{r})$, which is a parametrized variable to be determined self-consistently by the resulting equations. Therefore, different subparts of H are independent and commutative. The number of electrons in the element $d\tau$ is $\delta N = d\tau N(\mathbf{R})$ and

$$N(\mathbf{R}) = \sum_i \delta(\mathbf{r}_i - \mathbf{R}) \tag{10.9}$$

is the number density of electrons around $\mathbf{R}$. It is convenient to identity $\mathbf{R}$ as the coordinate of the center of mass of the electrons in $d\tau$:

$$\mathbf{R} = \sum_{i \in d\tau} \mathbf{r}_i / \delta N. \tag{10.10}$$

The total momentum of the electrons in the element $d\tau$ is $\delta \mathbf{P} = \sum_{i \in d\tau} \mathbf{p}_i = d\tau \mathbf{P}(\mathbf{R})$,

$$\mathbf{P}(\mathbf{R}) = \sum_i \mathbf{p}_i\,\delta(\mathbf{r}_i - \mathbf{R}) \tag{10.11}$$

being the momentum density around $\mathbf{R}$. Letting $\mathbf{v}(\mathbf{R})$ be the average velocity of the electrons around $\mathbf{R}$, which is a parameter to be determined self-consistently by the resulting balance equations, we can write the statistical expectation of the momentum density as

$$\langle \mathbf{P}(\mathbf{R}) \rangle = mn(\mathbf{R})\,\mathbf{v}(\mathbf{R}), \tag{10.12}$$

where $n(\mathbf{R}) = \langle N(\mathbf{R}) \rangle$ is the statistical average of the electron number density around $\mathbf{R}$. It is convenient to use the relative electron variables $\mathbf{p}'_i$ and $\mathbf{r}'_i$, defined by

$$\mathbf{p}'_i = \mathbf{p} - m\mathbf{v}(\mathbf{R}), \qquad \mathbf{r}'_i = \mathbf{r}_i - \mathbf{R}. \tag{10.13}$$

They are momentum and coordinate of the ith electron relative to the center of mass of a small element around $\mathbf{R}$. With these relative electron variables, the statistical average of $H_e(\mathbf{R})$ can be written as

$$h_e(\mathbf{R}) = \langle H_e(\mathbf{R}) \rangle = u(\mathbf{R}) + \frac{1}{2}mn(\mathbf{R})v^2(\mathbf{R}), \tag{10.14}$$

where

$$u(\mathbf{R}) = \left\langle \sum_i \frac{p_i'^2}{2m} \delta(\mathbf{r}_i - \mathbf{R}) \right\rangle \tag{10.15}$$

is the average kinetic energy density of the relative electrons around $\mathbf{R}$, and $u(\mathbf{R})d\tau$ is their kinetic energy in the element $d\tau$. $h_e(\mathbf{R})$ is the total kinetic energy density of electrons, including that of the center of mass. The interaction potential energy between electrons inside $d\tau$, which is a higher-order quantity in $d\tau$, has been neglected.

$n(\mathbf{R})$, $\mathbf{v}(\mathbf{R})$ and $u(\mathbf{R})$ are fundamental field variables, with which all other field quantities can be directly expressed or through some relation linked. For instance, the density of particle current can be defined as

$$\mathbf{J}(\mathbf{R}) = \sum_i \frac{\mathbf{p}_i}{m} \delta(\mathbf{r}_i - \mathbf{R}) = \frac{1}{m}\mathbf{P}(\mathbf{R}), \tag{10.16}$$

with the statistical expectation

$$\mathbf{j}_n(\mathbf{R}) = \langle \mathbf{J}(\mathbf{R}) \rangle = n(\mathbf{R})\mathbf{v}(\mathbf{R}). \tag{10.17}$$

The density of energy flux defined as

$$\mathbf{J}_H(\mathbf{R}) = \sum_i \frac{p_i^2}{2m} \frac{\mathbf{p}_i}{m} \delta(\mathbf{r}_i - \mathbf{R}), \tag{10.18}$$

has the statistical expectation

$$\mathbf{j}_H(\mathbf{R}) = \langle \mathbf{J}_H(\mathbf{R}) \rangle = \frac{5}{3}u(\mathbf{R})\mathbf{v}(\mathbf{R}) + \frac{m}{2}n(\mathbf{R})v^2(\mathbf{R})\mathbf{v}(\mathbf{R}). \tag{10.19}$$

10.2 Rates of Change of Particle Number, Momentum and Energy

We now consider the rates of change of particle number, momentum and energy. First, the rate of change of particle number density $\dot{N}(\boldsymbol{R}) = -\mathrm{i}\,[N(\boldsymbol{R}), H]$ can be easily calculated and this results (after taking statistical average) in the continuity equation:

$$\frac{\partial n}{\partial t} + \boldsymbol{\nabla}\cdot(\boldsymbol{v}n) = 0. \qquad (10.20)$$

Secondly, the rate of change of the momentum density $\langle \boldsymbol{P}(\boldsymbol{R})\rangle$ can be calculated from Eq. (10.12) to give

$$\frac{\partial}{\partial t}\langle \boldsymbol{P}(\boldsymbol{R})\rangle = m\frac{\partial n}{\partial t}\boldsymbol{v} + mn\frac{\partial \boldsymbol{v}}{\partial t}. \qquad (10.21)$$

On the other hand, according to $\dot{\boldsymbol{P}}(\boldsymbol{R}) = -\mathrm{i}\,[\boldsymbol{P}(\boldsymbol{R}), H]$,

$$\dot{\boldsymbol{P}}(\boldsymbol{R}) = -\boldsymbol{\nabla}_{\boldsymbol{R}}\cdot\left[\sum_i \frac{\boldsymbol{p}_i\boldsymbol{p}_i}{m}\delta(\boldsymbol{r}_i - \boldsymbol{R})\right] - \sum_i \left[\boldsymbol{\nabla}\Phi(\boldsymbol{r}_i) + e\boldsymbol{\nabla}\phi(\boldsymbol{r}_i)\right]\delta(\boldsymbol{r}_i - \boldsymbol{R}). \qquad (10.22)$$

Changing to the relative electron variables and taking the statistical average, we have

$$\langle \dot{\boldsymbol{P}}(\boldsymbol{R})\rangle = -\boldsymbol{\nabla}\left[\tfrac{2}{3}u(\boldsymbol{R})\right] - \boldsymbol{\nabla}\cdot\left[mn(\boldsymbol{R})\,\boldsymbol{v}(\boldsymbol{R})\,\boldsymbol{v}(\boldsymbol{R})\right] + en(\boldsymbol{R})\,\boldsymbol{E}(\boldsymbol{R}) + \boldsymbol{f}(\boldsymbol{R}), \qquad (10.23)$$

where

$$\boldsymbol{f}(\boldsymbol{R}) = -\left\langle \sum_i \boldsymbol{\nabla}\Phi(\boldsymbol{r}'_i + \boldsymbol{R})\delta(\boldsymbol{r}'_i)\right\rangle, \qquad (10.24)$$

$\boldsymbol{f}(\boldsymbol{R})d\tau$ being the resistive force experienced by the fluid element $d\tau$ around $\boldsymbol{R}$ due to impurity and phonon scatterings. Equating Eq. (10.23) to Eq. (10.21) and making use of the continuity equation (10.20), we obtain an Euler-type momentum balance equation:

$$\frac{\partial \boldsymbol{v}}{\partial t} + \boldsymbol{v}(\boldsymbol{\nabla}\cdot\boldsymbol{v}) = -\frac{2}{3}\frac{\boldsymbol{\nabla}u}{mn} + \frac{e}{m}\boldsymbol{E} + \frac{1}{mn}\boldsymbol{f}. \qquad (10.25)$$

Thirdly, the rate of change of $\langle H_\mathrm{e}(\boldsymbol{R})\rangle$ is calculated from Eq. (10.14) to be

$$\frac{d}{dt}\langle H_\mathrm{e}(\boldsymbol{R})\rangle = \frac{du}{dt} + \frac{m}{2}\frac{dn}{dt}v^2 + mn\boldsymbol{v}\cdot\frac{d\boldsymbol{v}}{dt}, \qquad (10.26)$$

which should be equated to that obtained from averaging the operator rate of change $\dot{H}_e(\bm{R}) = -\mathrm{i}\,[H_e(\bm{R}), H]$:

$$\langle \dot{H}_e(\bm{R}) \rangle = -\bm{\nabla} \cdot \bm{v}(\bm{R}) \frac{5}{3} u(\bm{R}) + \frac{m}{2} n(\bm{R}) v^2(\bm{R}) + e n(\bm{R}) \bm{v}(\bm{R}) \cdot \bm{E}(\bm{R}) - w(\bm{R}), \tag{10.27}$$

where

$$w(\bm{R}) = \left\langle \sum_i \frac{\bm{p}_i'}{m} \cdot \bm{\nabla} \Phi(\bm{r}_i' + \bm{R}) \delta(\bm{r}_i') \right\rangle - \bm{v}(\bm{R}) \cdot \bm{f}(\bm{R}), \tag{10.28}$$

$w(\bm{R}) d\tau$ being the energy-loss rate of electrons in the element $d\tau$ around $\bm{R}$. Combining Eqs. (10.26) and (10.27) and considering Eqs. (10.20) and (10.25) we obtain the energy balance equation:

$$\frac{\partial u}{\partial t} + \bm{v} \cdot \bm{\nabla} u = -\frac{5}{3} u (\bm{\nabla} \cdot \bm{v}) - w - \bm{v} \cdot \bm{f}. \tag{10.29}$$

According to expressions (10.17) and (10.19) for $\bm{j}_\mathrm{n}$ and $\bm{j}_\mathrm{H}$, the momentum balance equation (10.25) can be written in terms of the rate of change of the particle current density:

$$\frac{\partial}{\partial t} \bm{j}_\mathrm{n} + \bm{\nabla} \cdot (\bm{j}_\mathrm{n} \bm{v}) = -\frac{2}{3} \frac{\bm{\nabla} u}{m} + \frac{e n \bm{E}}{m} + \frac{\bm{f}}{m}. \tag{10.30}$$

With the expression (10.14) for the electron energy density h_e, the energy balance equation (10.29) can be written as

$$\frac{\partial h_e}{\partial t} + \bm{\nabla} \cdot \bm{j}_\mathrm{H} = e n \bm{v} \cdot \bm{E} - w. \tag{10.31}$$

10.3 Hydrodynamic Balance Equations

To calculate the resistive force density $\bm{f}$, the energy-loss rate density w and other average local quantities, we need the density matrix $\hat{\rho}$, which can be obtained by solving the Liouville equation with the appropriate initial condition. In the present scheme the interactions between different fluid elements are included approximately in the total potential with a mean-field treatment. Therefore different fluid elements, though geometrically correlative, are dynamically independent, and thus evolve separately from their own initial states.

In accordance with the precepts of the balance-equation approach, the unperturbed state is chosen to be described by a density matrix $\hat{\rho}_0$ of the local equilibrium form,

$$\hat{\rho}_0 = \frac{1}{Z} \exp\left\{-\int d\tau \left[H_e(\boldsymbol{R}) - \boldsymbol{v}(\boldsymbol{R}) \cdot \boldsymbol{P}(\boldsymbol{R}) - \mu(\boldsymbol{R}) N(\boldsymbol{R})\right]/T_e(\boldsymbol{R})\right\}$$
$$\times \exp(-H_{\rm ph}/T), \qquad (10.32)$$

where $H_e(\boldsymbol{R})$, $\boldsymbol{P}(\boldsymbol{R})$ and $N(\boldsymbol{R})$ are operators of local energy, momentum and particle-number densities of the electron system given by (10.8), (10.11) and (10.9), and the velocity field $\boldsymbol{v}(\boldsymbol{R})$, electron-temperature field $T_e(\boldsymbol{R})$ and chemical potential field $\mu(\boldsymbol{R})$ are parameters to be determined selfconsistently from the resulting balance equations. The phonon system described by $H_{\rm ph}$ is assumed to be in an equilibrium state with lattice temperature T.

The processes involved in deriving the perturbed density matrix $\hat{\rho} = \hat{\rho}_0 + \hat{\rho}_1$ to linear order in the scattering interactions and in calculating the statistical expectations follow those of Chapter 1. To the lowest nonzero order, the average local kinetic energy density of the relative electrons is determined by $\hat{\rho}_0$ only:

$$u(\boldsymbol{R}) = 2 \sum_{\boldsymbol{k}} \varepsilon_{\boldsymbol{k}} f((\varepsilon_{\boldsymbol{k}} - \mu(\boldsymbol{R}))/T_e(\boldsymbol{R})), \qquad (10.33)$$

and the local chemical potential $\mu(\boldsymbol{R})$ is related to the local electron density $n(\boldsymbol{R})$ via the relation

$$n(\boldsymbol{R}) = 2 \sum_{\boldsymbol{k}} f((\varepsilon_{\boldsymbol{k}} - \mu(\boldsymbol{R}))/T_e(\boldsymbol{R})). \qquad (10.34)$$

Here $\varepsilon_{\boldsymbol{k}} = k^2/2m$ and $f(x) \equiv (e^x + 1)^{-1}$. The resistive force density $\boldsymbol{f}(\boldsymbol{R})$ and the energy loss rate density $w(\boldsymbol{R})$ are found to be

$$\boldsymbol{f} = \boldsymbol{f}(n(\boldsymbol{R}), T_e(\boldsymbol{R}), \boldsymbol{v}(\boldsymbol{R})) = n_{\rm i} \int \frac{d^3q}{(2\pi)^3} \boldsymbol{q} \, |u(\boldsymbol{q})|^2 \Pi_2(\boldsymbol{q}, \omega_0)$$
$$+ 2 \int \frac{d^3q}{(2\pi)^3} \boldsymbol{q} \sum_{\lambda} |M(\boldsymbol{q}, \lambda)|^2 \Pi_2(\boldsymbol{q}, \omega_0 + \Omega_{\boldsymbol{q}\lambda}) \left[n\left(\frac{\Omega_{\boldsymbol{q}\lambda}}{T}\right) - n\left(\frac{\Omega_{\boldsymbol{q}\lambda} + \omega_0}{T_e}\right)\right],$$
$$(10.35)$$

$$w = w(n(\boldsymbol{R}), T_e(\boldsymbol{R}), \boldsymbol{v}(\boldsymbol{R})) = 2 \int \frac{d^3q}{(2\pi)^3} \sum_{\lambda} |M(\boldsymbol{q}, \lambda)|^2 \Omega_{\boldsymbol{q}\lambda}$$
$$\times \Pi_2(\boldsymbol{q}, \omega_0 + \Omega_{\boldsymbol{q}\lambda}) \left[n\left(\frac{\Omega_{\boldsymbol{q}\lambda}}{T}\right) - n\left(\frac{\Omega_{\boldsymbol{q}\lambda} + \omega_0}{T_e}\right)\right]. \qquad (10.36)$$

Here $\omega_0 \equiv \boldsymbol{q} \cdot \boldsymbol{v}(\boldsymbol{R})$, $n(x) \equiv (e^x - 1)^{-1}$, n_i is the impurity density around $\boldsymbol{R}$, $\Omega_{q\lambda}$ is the phonon frequency, and $\Pi_2(\boldsymbol{q}, \omega)$ is the density correlation function of relative electrons in a unit volume with density $n(\boldsymbol{R})$ at temperature $T_\text{e}(\boldsymbol{R})$, including the effect of Coulomb interaction between carriers, which can be treated in RPA or beyond. We see that although the intercarrier Coulomb coupling inside the element gives a vanishing contribution to the average internal energy of the electron system, it plays an important role in screening. It can also be seen that all the above expressions for the local particle-number density n, energy density u, frictional force density $\boldsymbol{f}$, energy-loss density w, and electron density correlation function Π_2, are the same as for a spatially uniform system.

There are six parametric field variables: $\boldsymbol{v}(\boldsymbol{R})$, $T_\text{e}(\boldsymbol{R})$, $u(\boldsymbol{R})$, $n(\boldsymbol{R})$, $\mu(\boldsymbol{R})$, as well as the total electric potential $\phi(\boldsymbol{R})$ or field $\boldsymbol{E}(\boldsymbol{R})$. These variables may be time dependent on a macroscopic time scale for time-dependent processes. We have three hydrodynamic-type equations: continuity equation (10.20), momentum-balance equation (10.25), and energy-balance equation (10.29), supplemented by two relations (10.33) and (10.34) plus the Poisson equation relating electric charge density and potential:

$$\nabla^2 \phi = -\frac{e}{\epsilon_0 \kappa}[n(\boldsymbol{R}) - n_\text{D}], \qquad (10.37)$$

in which n_D denotes the density of the positive charge background and κ is the dielectric constant of the background material. All the space-time field variables can be determined for given initial and boundary conditions.

10.4 Thermoelectric Power

One of the direct applications of the hydrodynamic balance equations for weakly nonuniform systems derived in the preceding section is the evaluation of thermoelectric power (Lei, 1994a).

We consider electron transport under the influence of an electric field $\boldsymbol{E}$ and in the presence of a small lattice temperature gradient ∇T. The occurrence of lattice temperature gradient makes the transport problem an inhomogeneous one. Since temperature gradient is small we are dealing with a weakly nonuniform case and its transport properties are described by the hydrodynamic balance equations (10.20), (10.25) and (10.29). Without loss of generality we assume that the small lattice temperature gradient is in the x direction: $\nabla T = (\nabla_x T, 0, 0)$. There may be a small drift velocity (current density) and a small electric field in the x direction in addition to the drift velocity (current) and the electric field along the y-direction: $\boldsymbol{v} = (v_x, v_y, 0)$ and $\boldsymbol{E} = (E_x, E_y, 0)$. Here v_x and E_x are small, and the spatial variations of all the field quantities are along the x direction only. We will treat the steady state transport and consider the particle number, force and energy balance equations to the first order in the small quantities. For instance, the gradient operator $\nabla_x \equiv \partial/\partial x$ is a first order small quantity and v_x is also a first order small quantity, thus $\nabla_x v_x$ is a higher order small quantity and can be neglected. With these in mind we can write the momentum-balance equation (10.25) as

$$0 = -\frac{2}{3}\nabla_x u + neE_x + f_x \qquad (10.38)$$

in the x direction, and

$$0 = neE_y + f_y \qquad (10.39)$$

in the y direction. The energy-balance equation (10.29) is reduced to

$$w + v_y f_y = 0. \qquad (10.40)$$

For small v_x, f_x is proportional to v_x, and

$$\rho = -\frac{f_x}{n^2 e^2 v_x} \qquad (10.41)$$

is the resistivity along the x direction in the presence of a drift velocity v_y in the y direction. Eq. (10.38) can then be written in terms of ρ and the current density $j_x = nev_x$ along the x direction:

$$j_x = \frac{E_x}{\rho} - \frac{2}{3}\frac{\nabla_x u}{ne\rho}. \qquad (10.42)$$

Utilizing the expression (10.33) for the local energy density u of the relative electrons and the equation (10.34) relating the local number density n with the local chemical potential μ and the local electron temperature T_e, we can write equation (10.42) in the form

$$j_x = L^{11}\left(E_x - \frac{\nabla_x \mu}{e}\right) + L^{12}\left(-\nabla_x T_e\right), \qquad (10.43)$$

in which the coefficient

$$L^{11} = \frac{1}{\rho}, \tag{10.44}$$

$$L^{12} = \frac{1}{e\rho}\left[\frac{5F_{\frac{3}{2}}(\zeta)}{3F_{\frac{1}{2}}(\zeta)} - \zeta\right], \tag{10.45}$$

$\zeta \equiv \mu/T_e$, and the function $F_\nu(y)$ is defined as

$$F_\nu(y) = \int_0^\infty \frac{x^\nu dx}{\exp(x-y)+1}. \tag{10.46}$$

In the case without a current along the y direction, we have $T_e = T$ and $\nabla_x T_e = \nabla_x T$, and Eq. (10.43) becomes the well-known (weak) current transport equation in the presence of temperature and chemical gradients (Aschcroft and Mermin, 1976). The thermoelectric power S is then

$$S = \frac{L^{12}}{L^{11}} = \frac{1}{e}\left[\frac{5F_{\frac{3}{2}}(\zeta)}{3F_{\frac{1}{2}}(\zeta)} - \zeta\right], \tag{10.47}$$

which is independent of scatterings. This result is physically understandable. In addition to contribute to the possible energy dissipation the only role the scatterings play in the balance-equation theory is to induce a frictional force. Since thermoelectric power is defined under the open circuit (zero current) condition (Aschcroft and Mermin, 1976), and the frictional force in the balance-equation theory always vanishes when the drift velocity is zero, we should expect a thermoelectric power independent of scattering but only depending on the intrinsic band structure of the system. This result has later been confirmed by a similar analysis (Xing et al, 1995). The result (10.47) is also in agreement with the thermoelectric power obtained from a relaxation time approximation calculation of the Boltzmann equation for a parabolic energy band system (Hicks and Dresselhaus, 1993).

In the presence of a strong current density in the y direction electrons are heated. The electron temperature T_e is generally higher than the lattice temperature T. The gradient of the electron temperature $\nabla_x T_e$, however, depends not only on the gradient of the lattice temperature $\nabla_x T$, but also on the electron temperature itself and the way to control the electron heating. The thermoelectric power can be expressed as

$$S = \frac{1}{e}\left[\frac{5F_{\frac{3}{2}}(\zeta)}{3F_{\frac{1}{2}}(\zeta)} - \zeta\right]\frac{\delta T_e}{\delta T}, \tag{10.48}$$

where $\delta T_\mathrm{e}/\delta T$ stands for $\nabla_x T_\mathrm{e}/\nabla_x T$, i.e. the gradient of the electron temperature induced by the existence of the lattice temperature gradient. It should be calculated from the momentum and energy balance equations (10.39) and (10.40) under the condition pertinent to the experiment. The result is of course dependent on the scatterings and on the way to heat electrons.

The above expressions are given for a 3D parabolic band system. The hydrodynamic balance equations (10.20), (10.25) and (10.29) for weakly nonuniform cases are also valid in a two-dimensional system, if the prefactor 2/3 in Eq. (10.25) is replaced by unity, and the prefactor 5/3 in Eq. (10.29) is replaced by 2. The thermoelectric power S for a 2D system (such as a quantum well structure) is easily derived to be

$$S = \frac{1}{e}\left[\frac{2F_1(\zeta)}{3F_0(\zeta)} - \zeta\right]\frac{\delta T_\mathrm{e}}{\delta T}, \qquad (10.49)$$

in agreement with that given by Hicks and Dresselhaus (1993).

To have an idea how the thermoelectric power depends on the carrier density, lattice temperature and the way of heating electrons, we examine an n-type GaAs-based thin (10 nm thickness along the z-direction) quantum well structure (Lei, 1994a). Electrons move in the x-y plane with a parabolic energy spectrum and are assumed to occupy only the lowest subband. Scatterings by the remote (locating at a distance 18 nm from the well center) and background impurities, acoustic phonons (deformation potential and piezoelectric couplings with electrons) and polar optical phonons (Fröhlich coupling with electrons) are included, with typical material and the electron-phonon coupling parameters and full dynamic RPA screening (Lei, Birman and Ting, 1985). The impurity scattering strength is such that the low-temperature linear mobility equals $1.0 \, \mathrm{m^2/Vs}$.

Unlike in the zero-bias case where the thermoelectric power S is determined only by the lattice temperature and the intrinsic band structure of the material, under hot carrier condition S not only depends on the electron temperature T_e itself, but also depends on the ratio factor $\delta T_\mathrm{e}/\delta T$, which varies dramatically with the bias condition. Two kinds of bias conditions are examined: (a) applying a fixed current density along the y direction (fixed-velocity bias), or (b) applying a fixed electric field along the y direction (fixed-field bias). The electron temperature T_e and $\delta T_\mathrm{e}/\delta T$ are calculated from Eqs. (10.39) and (10.40) under the appropriate bias condition. The carrier sheet density is assumed to be $N_\mathrm{s} = 2.5 \times 10^{15}/\mathrm{m^2}$. In the case of fixed-velocity bias, $(\delta T_e/\delta T)_{v_y}$ is always positive, so is the thermoelectric power S. In the case of fixed-field case, however, $(\delta T_e/\delta T)_{E_y}$, thus S, can be negative under high bias fields ($E_y > 1\,\mathrm{kV/cm}$) within certain low-temperature range. For instance, at bias $E_y = 2.0\,\mathrm{kV/cm}$ thermoelectric power S become negative at temperature $T < 85\,\mathrm{K}$, as shown in Fig. 10.1, where S is plotted as a function of lattice temperature T under different bias fields $E_y = 0, 0.1, 0.4, 1.0$ and $2.0\,\mathrm{kV/cm}$. It should be noted that all these results including the prediction of possible negative S under high bias field, which are in agreement with previous finding from solving the coupling kinetic equations for electrons and phonons (Babaev and Gassymov, 1977), are obtained under the assumption that phonons are in local equilibrium with distribution function $n(\Omega_{q\lambda}/T)$ at lattice temperature T, as indicated in the expressions (10.35) and (10.36) for $\boldsymbol{f}$ and w.

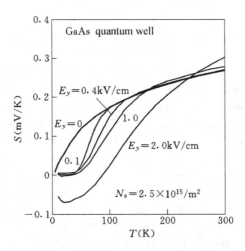

Fig. 10.1 Thermoelectric power S of a n-type GaAs-based quantum well structure (10 nm thickness, effective mass $m = 0.07\,m_e$) having electron sheet density $N_s = 2.5 \times 10^{15}/\mathrm{m}^2$ is shown as a function of lattice temperature T under fixed-field bias condition at several bias fields $E_y = 0, 0.1, 0.4, 1.0$ and $2.0\,\mathrm{kV/cm}$ (b). From Lei(1994).

10.5 Effect of Phonon Drag

In the discussion of thermoelectric power in the preceding section we only considered the effect of the lattice-temperature gradient on the distribution of electrons (i.e. the electron diffusion), appearing as the spatial gradient term $\nabla_x u$ of the electron average energy in Eq. (10.38), while the effect of the lattice-temperature gradient on the distribution of phonons is neglected. The occurrence of lattice temperature gradient may result in a spatial variation of phonon distribution, making the phonon number flowing in a small space volume unequal to that flowing out (phonon diffusion). To take account these effects we need also to deal with phonons in nonequilibrium states (see Sec. 5.2) and consider the rate of change of phonon number of mode $q\lambda$ in a small volume, which is proportional to local phonon occupation or distribution function, $n_{q\lambda}$, in the space. The rate of change of $n_{q\lambda}$ includes that induced by the electron–phonon interaction (5.10),

$$\left[\frac{\partial n_{q\lambda}}{\partial t}\right]_{\mathrm{ep}} = 2|M(\boldsymbol{q},\lambda)|^2 \Pi_2(\boldsymbol{q},\Omega_{q\lambda}-\boldsymbol{q}\cdot\boldsymbol{v})\left[n_{q\lambda} - n\left(\frac{\Omega_{q\lambda}-\boldsymbol{q}\cdot\boldsymbol{v}}{T_{\mathrm{e}}}\right)\right], \tag{10.50}$$

and that induced by the phonon–phonon interaction, $(\partial n_{q\lambda}/\partial t)_{\mathrm{pp}}$. The latter is generally described using an effective phonon relaxation time τ_{p} as (5.15),

$$\left[\frac{\partial n_{q\lambda}}{\partial t}\right]_{\mathrm{pp}} = -\frac{1}{\tau_{\mathrm{p}}}\left[n_{q\lambda} - n\left(\frac{\Omega_{q\lambda}}{T}\right)\right]. \tag{10.51}$$

In the presence of temperature gradient, there is also a diffusion term from the nonvanishing net number of phonons flowing in the volume:

$$\left[\frac{\partial n_{q\lambda}}{\partial t}\right]_{\mathrm{d}} = -\boldsymbol{v}_{\mathrm{p}}(\boldsymbol{q},\lambda)\cdot\nabla n_{q\lambda} \approx -\boldsymbol{v}_{\mathrm{p}}(\boldsymbol{q},\lambda)\cdot\nabla n\left(\frac{\Omega_{q\lambda}}{T}\right), \tag{10.52}$$

where $\boldsymbol{v}_{\mathrm{p}}(\boldsymbol{q},\lambda) = \nabla_{\boldsymbol{q}}\Omega_{q\lambda}$ is the group velocity of mode-$q\lambda$ phonon. The total rate of change of phonon occupation number $n_{q\lambda}$ in a space position can be written as

$$\frac{d}{dt}n_{q\lambda} = \left[\frac{\partial n_{q\lambda}}{\partial t}\right]_{\mathrm{d}} + \left[\frac{\partial n_{q\lambda}}{\partial t}\right]_{\mathrm{ep}} + \left[\frac{\partial n_{q\lambda}}{\partial t}\right]_{\mathrm{pp}}. \tag{10.53}$$

In the steady state $dn_{q\lambda}/dt = 0$, Eq. (10.53) gives

$$n_{q\lambda} = n\left(\frac{\Omega_{q\lambda}-\boldsymbol{q}\cdot\boldsymbol{v}}{T_{\mathrm{e}}}\right) + \frac{n(\Omega_{q\lambda}/T) - n\left((\Omega_{q\lambda}-\boldsymbol{q}\cdot\boldsymbol{v})/T_{\mathrm{e}}\right)}{1+\tau_{\mathrm{p}}\Gamma(\boldsymbol{q},\lambda,\Omega_{q\lambda}-\boldsymbol{q}\cdot\boldsymbol{v})}$$
$$+ \frac{\tau_{\mathrm{p}}}{1+\tau_{\mathrm{p}}\Gamma(\boldsymbol{q},\lambda,\Omega_{q\lambda}-\boldsymbol{q}\cdot\boldsymbol{v})}\left[\frac{\partial n_{q\lambda}}{\partial t}\right]_{\mathrm{d}}, \tag{10.54}$$

with the expression of $\Gamma(\boldsymbol{q},\lambda,\Omega_{q\lambda}-\boldsymbol{q}\cdot\boldsymbol{v})$ given by (5.20). Comparing with the phonon occupation number (5.19) in the spatially uniform case

(Chapter 5) the last term in the above expression is the contribution of phonon diffusion due to lattice temperature gradient.

We note that if $\tau_p \sim 0$, the diffusion gives no contribution to phonon distribution and phonons are in local equilibrium with $n_{q\lambda} = n(\Omega_{q\lambda}/T)$. The effect of temperature-gradient induced phonon diffusion is always associated with a finite τ_p, i.e. with the existence of nonequilibrium phonons. Since in the case of finite τ_p, the phonon distribution is affected by the distribution and the drift motion of electrons, as if phonons are dragged to certain degree by the electrons when they are drifting (both in spatially uniform and nonuniform cases), this phonon nonequilibrium-population and diffusion related phenomenon is thus referred as phonon drag in the literature.

Now we should use $n_{q\lambda}$ given by (10.54) in (10.50) for $[(\partial n_{q\lambda}/\partial t]_{ep}$ to calculate the frictional force $\boldsymbol{f}$ and energy loss rate w from (5.13) and (5.14). Then the phonon scattering induced frictional force can be written as

$$\boldsymbol{f}_{ep} = \boldsymbol{f}_p + \boldsymbol{f}_g. \tag{10.55}$$

$\boldsymbol{f}_p$ is the phonon-induced frictional force (5.21) in the spatially uniform case, and $\boldsymbol{f}_g$ is an additional frictional force induced by the phonon diffusion due to lattice temperature gradient,

$$\boldsymbol{f}_g = \sum_{\boldsymbol{q},\lambda} \boldsymbol{q} \left[\frac{\partial n_{q\lambda}}{\partial t}\right]_d \frac{\tau_p \Gamma(\boldsymbol{q}, \lambda, \Omega_{q\lambda} - \boldsymbol{q} \cdot \boldsymbol{v})}{1 + \tau_p \Gamma(\boldsymbol{q}, \lambda, \Omega_{q\lambda} - \boldsymbol{q} \cdot \boldsymbol{v})}. \tag{10.56}$$

We consider the transport when there is a small lattice temperature gradient along the x direction. The analysis in Sec. 10.4 is still applicable, but the force f_x in the x-direction force-balance equation (10.38) should be replaced by $f_x + f_{gx}$:

$$0 = -\frac{2}{3}\nabla_x u + neE_x + f_x + f_{gx}. \tag{10.57}$$

The resistivity ρ is still given by (10.41), therefore Eq. (10.57) can be written as

$$j_x = \frac{E_x}{\rho} - \frac{2}{3}\frac{\nabla_x u}{ne\rho} + \frac{f_{gx}}{ne\rho}$$

$$= L^{11}\left(E_x - \frac{\nabla_x \mu}{e}\right) + L^{12}(-\nabla_x T_e) + L_g^{12}(-\nabla_x T), \tag{10.58}$$

where L^{11} and L^{12} are just the coefficients introduced in the preceding section, and L_g^{12} is easily obtained from (10.52) and (10.56). The thermoelectric power contains, in addition to S given by (10.48) for 3D or by (10.49) for 2D systems, another part contributed from phonon drag: $S_{\rm g} = L_{\rm g}^{12}/L^{11}$, which can be expressed as (for 3D systems)

$$S_{\rm g} = -\frac{1}{ne}\sum_{\bm{q},\lambda}\frac{q_x^2}{q^2}\frac{\Omega_{\bm{q}\lambda}^2}{T^2}\frac{\tau_{\rm p}\Gamma(\bm{q},\lambda,\Omega_{\bm{q}\lambda}-\bm{q}\cdot\bm{v})}{1+\tau_{\rm p}\Gamma(\bm{q},\lambda,\Omega_{\bm{q}\lambda}-\bm{q}\cdot\bm{v})}n'\!\left(\frac{\Omega_{\bm{q}\lambda}}{T}\right). \tag{10.59}$$

The above $S_{\rm g}$ expression may also be applicable for quasi-2D electron systems in the case of acoustic phonons when the confinement effect is not important (see Chapter 5).

The above formulae are derived by Wu, Horing and Cui (1996). In comparison with the experiment they make use of an effective phonon relaxation time $\tau_{\rm p}$ dependent on the phonon mode $\bm{q}\lambda$ and temperature T, including contributions from boundary scattering,

$$1/\tau_{\rm B\lambda} = v_{\rm s\lambda}/L \tag{10.60}$$

($v_{\rm s\lambda}$ is the sound speed of mode λ, L is a length of order of sample size), and from phonon–phonon scattering (anharmonic process),

$$1/\tau_{\rm Ah} = A_\lambda T^3 \Omega_{\bm{q}\lambda}^2, \tag{10.61}$$

whence

$$1/\tau_{\rm p} = 1/\tau_{\rm B\lambda} + 1/\tau_{\rm Ah}. \tag{10.62}$$

Thermoelectric power in the linear case (zero bias $v_y = 0$ in the y-direction, thus $T_{\rm e} = T$ and $\delta T_{\rm e}/\delta T = 1$) is calculated for several GaAs-based samples, taken into account contributions to $S_{\rm g}$ from longitudinal acoustic-phonon deformation-potential scattering and piezoelectric scattering, and from transverse acoustic-phonon piezoelectric scattering with typical material and coupling parameters (see Sec. 2.6). The coefficient $A_{\rm l}$ and $A_{\rm t}$ are decided by fitting the theoretical result with the experiment for sample 1 of Fletcher et al (1986). The calculated thermoelectric power for a GaAs quantum well of width $a = 10$ nm and electron density $N_{\rm s} = 3.6 \times 10^{16}\,{\rm m}^{-2}$ having sample length $L = 2.5$ mm, is shown in Fig. 10.2, together with the experimental data of Fletcher et al (1994). The theoretical predictions compare well with the experimental trend and in general correspondence with earlier analyses from the Boltzmann equation (Cantrell and Butcher, 1986, 1987; Fletcher et al, 1994).

It should be noted that the balance-equation theory treating the correlated electron and phonon systems, is essentially a linear response theory in the (impurity and electron–phonon) scattering potentials. In the

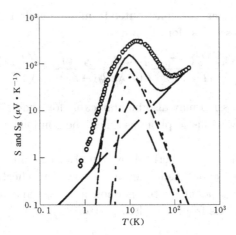

Fig. 10.2 Thermoelectric power of a GaAs-based quantum well having width $a = 10\,\text{nm}$ and electron density $N_s = 3.6 \times 10^{15}/\text{m}^2$ under zero dc bias, is shown as a function of lattice temperature T. The chain line is the thermoelectric power S derived in Sec. 10.4 without phonon drag; the short-dashed, dotted, and long-dashed curves are respectively contributions to S_g from transverse piezoelectric, longitudinal deformation-potential and longitudinal piezoelectric scatterings, and the solid curve is the sum of them all. From Wu, Horing and Cui (1996).

zero-order phonons are assumed in an equilibrium state at the lattice temperature T with $n(\Omega_{q\lambda}/T)$ taken as the unperturbed phonon distribution function, just as relative electrons are assumed in an equilibrium state at the electron temperature T_e with $f(\varepsilon, T_e)$ taken as their zero-order distribution function. Considering nonequilibrium phonon occupation due to electron–phonon interaction and electron drifting in Chapter 5 and due to temperature gradient in this section, corresponds to go an order higher in the scattering coupling in calculating the phonon distribution function in the theory. These calculations are valid only when the electron system and phonon system are weakly coupled, i.e. H_{ep} is weak in comparison with the effective internal couplings H_{ee} and H_{pp} in both systems, that the identification of separate electron system and the phonon system is possible. Otherwise, they become an indistinguishable coupled entity, and the balance equation treatment is no longer meaningful (see Sec. 9.2.3). Therefore, in the balance equation scheme one can never take the limit of $\tau_{\text{p}} \to \infty$ in Eq. (10.54), or enter the strong mutual drag regime, and thus in no way can the phonon distribution function become the form of a drifted Plank distribution at the electron temperature T_e.

10.6 Onsager Relations and Hydrodynamic Balance Equations

In a system weakly deviating from thermal equilibrium, the particle current density $\boldsymbol{j}_\mathrm{n}$ and heat flux $\boldsymbol{j}_\mathrm{Q}$ can be considered as the linear response to the small "forces", the potential gradient $\boldsymbol{X}_1 = -\left(1/T\right)\boldsymbol{\nabla}\left(e\phi+\mu\right)$ (μ is the chemical potential, ϕ is the electric potential) and the temperature gradient $\boldsymbol{X}_2 = -\boldsymbol{\nabla}\left(1/T\right)$, acting on the system:

$$\boldsymbol{j}_\mathrm{n} = \mathcal{L}^{11}\boldsymbol{X}_1 + \mathcal{L}^{12}\boldsymbol{X}_2, \qquad (10.63)$$

$$\boldsymbol{j}_\mathrm{Q} = \mathcal{L}^{21}\boldsymbol{X}_1 + \mathcal{L}^{22}\boldsymbol{X}_2. \qquad (10.64)$$

The Onsager reciprocity principle (Onsager, 1931), or simply the Onsager relation, states that

$$\mathcal{L}^{12} = \mathcal{L}^{21}. \qquad (10.65)$$

The Onsager relation is a manifestation of the microscopic irreversibility for any statistical system near thermal equilibrium (Mahan, 1981). Therefore any properly formulated statistical model should satisfy this relation. It is easy to verify this relation in the Kubo linear-response theory for transport. Moreover, if one can determine the distribution function from the Boltzmann equation, it is also straightforward to verify the Onsager relation by calculating the pertinent moments of the distribution function. However, for the traditional widely used hydrodynamic models, most of which are derived from the Boltzmann equation, verification is elusive. In fact, Anile and Muscato (1995) show that the Onsager relation breaks down in these models. Although an improvement has been made to overcome this difficulty, the existence of this relation has not yet been established within the model itself. This is not unexpected since the hydrodynamic model, which employs a finite number of parameters to simulate the nonequilibrium transport properties and uses the finite number of differential equations to replace the Boltzmann equation, is only an approximate description for transport. As for those hydrodynamic models which use the particle density n, current density $\boldsymbol{j}_\mathrm{n}$ and energy density u as five transport parameters and five equations (one continuity equation, three current or momentum equations and one energy equation derived from the first, second and the third moment of the Boltzmann equation) to close, the requirement to satisfy the Onsager relation in the set of six linear response equations (10.63) and (10.64), is obviously an excessive demand.

Wu, Cui and Horing (1996) carry out a careful examination to see whether or not the hydrodynamic balance equation scheme developed in Sec. 10.3 for weakly nonuniform system can satisfy the Onsager relation. This hydrodynamic scheme also uses five basic field variables n, $\boldsymbol{v}$ and u (T_e and μ are related to n and u through direct algebraic relations) and five nonlinear hydrodynamic balance equations (10.20), (10.25) and (10.29). To discuss the case weakly deviating from thermal equilibrium, they assume that there are small potential gradient and temperature gradient in the x direction. The steady state solution of balance equations (10.25) and (10.29) requires the electron temperature $T_e = T$ and the velocity v_x being a small quantity at all spatial positions. The transport equation for the current density j_x has been derived in Sec. 10.4 to the linear order of the small quantities, i.e. Eq. (10.43). This is the first Onsager equation (10.63) expressing the particle current density as the linear response to the potential and temperature gradients. In fact, it is just the linearization of the momentum balance equation (10.25). Therefore, for a 3D system

$$\mathcal{L}^{11} = \frac{T}{\rho e^2}, \qquad (10.66)$$

$$\mathcal{L}^{12} = \frac{T^2}{\rho e^2}\left[\frac{5F_{\frac{3}{2}}(\zeta)}{3F_{\frac{1}{2}}(\zeta)} - \zeta\right], \qquad (10.67)$$

with the resistivity ρ given by (10.41), the $F_\nu(y)$ function defined by (10.46) and $\zeta \equiv \mu/T$, μ being the chemical potential.

The heat flux $\boldsymbol{j}_Q$ relates the energy flux $\boldsymbol{j}_H$ and the particle current density $\boldsymbol{j}_n$ by (to the linear order of j_n)

$$\boldsymbol{j}_Q = \boldsymbol{j}_H - \mu \boldsymbol{j}_n. \qquad (10.68)$$

The second Onsager equation (10.64) expressing the heat flux as the linear response to the potential and temperature gradients, is beyond the scope of the complete 5-parameter hydrodynamic set of balance equations (10.20), (10.25) and (10.29), but can be obtained in the balance-equation scheme by calculating the rate of change of the energy flux operator (10.18),

$$\frac{d}{dt}\boldsymbol{J}_H(\boldsymbol{R}) = -\mathrm{i}\,[\boldsymbol{J}_H(\boldsymbol{R}), H], \qquad (10.69)$$

and then taking the statistical average based on the density matrix in Sec. 10.3 to the lowest nonzero order of the scattering interaction, to give an energy-flux balance equation

$$\frac{\partial \boldsymbol{j}_\mathrm{H}}{\partial t} + \nabla \cdot \mathcal{A} = \boldsymbol{B} + \frac{5}{3}\frac{eu}{m}\boldsymbol{E} + en\boldsymbol{E}\cdot\boldsymbol{vv} + \frac{1}{2}env^2\boldsymbol{E} + \frac{1}{2}v^2\boldsymbol{f} - w\boldsymbol{v}, \quad (10.70)$$

where $\mathcal{A}$ is a tensor defined as

$$\mathcal{A} = \left\langle \sum_i \frac{p_i^2}{2m}\frac{\boldsymbol{p}_i}{m}\frac{\boldsymbol{p}_i}{m}\delta(\boldsymbol{r}_i - \boldsymbol{R}) \right\rangle$$

$$= \frac{1}{3}[D(\boldsymbol{R}) + uv^2]\mathcal{I} + \boldsymbol{j}_\mathrm{H}\boldsymbol{v} + \boldsymbol{v}\boldsymbol{j}_\mathrm{H} - u\boldsymbol{vv} - \frac{1}{2}mnv^2\boldsymbol{vv}, \quad (10.71)$$

$$D(\boldsymbol{R}) = \left\langle \sum_i \frac{p_i'^4}{2m^3}\delta(r_i') \right\rangle = 2\sum_k \frac{k^4}{2m^3}f\left(\frac{\varepsilon_k - \mu}{T_\mathrm{e}}\right), \quad (10.72)$$

and the vector $\boldsymbol{B} = \boldsymbol{B}_\mathrm{i} + \boldsymbol{B}_\mathrm{p}$ includes contributions from impurity and phonon scatterings, having explicit expressions in the appendix of Wu, Cui and Horing (1996). (10.70) amounts to establish another three equations independent of (10.20), (10.25) and (10.29). In principle, it is impossible to require a transport model with 5 independent parameters to satisfy 8 independent equations. What we can do here is treating (10.20), (10.25) and (10.29) as fundamental equations to determine all the transport parameters, and then to see whether these transport quantities satisfy (10.70) or not. In the case of small potential and temperature gradients, Eq. (10.70) is reduced to

$$\frac{5}{3}\frac{eu(\boldsymbol{R})}{m}E_x - \frac{1}{3}\nabla_x D(\boldsymbol{R}) + B_x = 0, \quad (10.73)$$

and B_x is proportional to $j_{\mathrm{H}x}$ (to linear order of the small quantity $j_{\mathrm{H}x} = \frac{5}{3}uv_x$). Therefore we can define an energy-flux related resistivity

$$\rho_\mathrm{H} = -\frac{m}{ne^2}\frac{B_x}{j_{\mathrm{H}x}}. \quad (10.74)$$

Eq. (10.73) is easily rewritten in the form of the second Onsager equation (10.64), with the coefficient

$$\mathcal{L}^{21} = \frac{T^2}{\rho e^2}\left[\frac{\rho}{\rho_\mathrm{H}}\frac{5F_{\frac{3}{2}}(\zeta)}{3F_{\frac{1}{2}}(\zeta)} - \zeta\right], \quad (10.75)$$

$$\mathcal{L}^{22} = \frac{T^3}{\rho_\mathrm{H} e^2}\left[\frac{7F_{\frac{5}{2}}(\zeta)}{3F_{\frac{1}{2}}(\zeta)}F_{\frac{3}{2}}(\zeta) - \frac{5F_{\frac{3}{2}}(\zeta)}{3F_{\frac{1}{2}}(\zeta)}\right] - \frac{\zeta T^3}{\rho e^2}\left[\frac{5F_{\frac{3}{2}}(\zeta)}{3F_{\frac{1}{2}}(\zeta)} - \zeta\right].$$

$$(10.76)$$

Comparing (10.75) with (10.67), we can see that if

$$\frac{\rho}{\rho_\text{H}} = 1, \qquad (10.77)$$

the Onsager relation holds.

Both ρ and ρ_H include contributions from impurity, longitudinal optic (LO) phonon, longitudinal acoustic (LA) phonon and transverse acoustic (TA) phonon scatterings. Wu, Cui and Horing (1996) examine the ratio ρ/ρ_H in an n-type bulk GaAs for each type of scattering mechanisms (impurity scattering, LO-phonon Fröhlich scattering, TA-phonon piezoelectric scattering and LA-phonon deformation potential and piezoelectric scattering). They found that for any type of scattering mechanism at any temperature, the relation (10.77) always holds in the high electron density limit:

$$\lim_{n \to \infty} \frac{\rho}{\rho_\text{H}} = 1. \qquad (10.78)$$

This shows that when the electron density is high enough the hydrodynamic model of balance equations (10.20), (10.25) and (10.33) satisfy the Onsager relation. In other words, under the condition of small potential and temperature gradients, the energy-flux balance equation (10.70) are in consistent with the first five hydrodynamic balance equations (10.20), (10.25), and (10.29). This is the first set of 5-parameter hydrodynamic balance equations which automatically obeys the Onsager relation without introducing extra terms from outside the model. When the electron density is not high enough ρ/ρ_H deviates from 1. The degree of deviation depends on the temperature and the scattering mechanism and generally is not large. Fig. 10.3 shows ρ/ρ_H as a function of the electron density for impurity and LO-phonon Fröhlich scatterings.

As far as the Onsager relation is concerned the above examination is sufficient since the Onsager relation itself is a statement in the case of weak deviation from thermal equilibrium. Nevertheless, it should be noted that the energy flux balance equation (10.70) can be embodied in the first five balance equations (10.20), (10.25), and (10.29) only in the weakly nonuniform and linear response limit. When the temperature gradient or the electric potential gradient becomes large or there is a large energy flow in the system the energy-flux equation (10.70) is no longer consonant with the first five (the particle density, the momentum and the energy) balance equations. This equation is very important in describing phenomena with large heat flow and should be served as an independent equation in an extended hydrodynamic model including more transport parameters in the

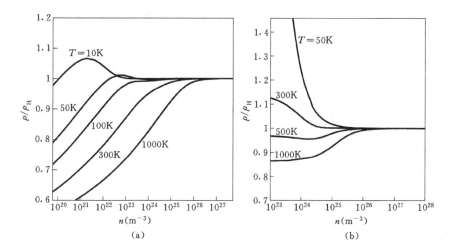

Fig. 10.3 ρ/ρ_H versus electron density n in bulk GaAs. (a) Impurity scattering, (b) LO-phonon Fröhlich scattering. From Wu, Cui and Horing (1996).

initial density matrix to match all eight independent hydrodynamic equations. Of course, this is a complete different scheme.

10.7 Carrier Transport in Semiconductor Devices

10.7.1 *Hydrodynamic equation and Poisson equation*

The traditional simulation of semiconductor devices is based on drift-diffusion equation. This phenomenological model for carrier transport in spatially nonuniform systems can only be used in the low electric field and slowly varying situation. With the size of semiconductor device shrinking to submicrometer range, the drift-diffusion model is generally no longer suitable for describing the transport process of carriers in devices.

In the regime where the classical theory applies, the Boltzmann equation provides reliable description for transport in nonuniform systems. The Monte-Carlo method, which is equivalent to numerically solve the complete Boltzmann equation and, in principle, is able to deal with arbitrarily complicated energy structure and scattering potential, has now become the most important tool to investigate the physical process of carrier transport in semiconductor devices. However, the excessive computational burden limits its direct application in many realistic device applications.

Hydrodynamic models can be used to describe hot-electron, nonstationary and nonlocal transport in submicron semiconductor devices. These macroscopic equations, in which each terms have distinct physical meanings, promise an acceptable accuracy with a much reduced computational burden, thus represent a useful engineering compromise between the simpler, static type drift-diffusion model and the exact, purely numerical Monte-Carlo method.

Most hydrodynamic models are derived by taking the first three, sometimes even higher, moments of the Boltzmann equation, appearing in the form of conservation laws or balance equations, such as particle number (charge) conservation, momentum and energy balance equations. As each moment equation thus obtained contains higher order moment, the set of moment equations form an infinite chains. The hydrodynamic approach is based on the truncation of this hierarchy of equations after certain finite number of the moments. Since the first finite-number moment equations by themselves do not form a closed system, some kind of simplifications is necessary for closure, together with the proper calculation of the collision terms without knowing the exact distribution functions. Different orders of truncation and different ways of closure and simplification lead to different hydrodynamic models. They apply in different situations (Blotekjaer, 1970; Ruden and Odeh, 1986; Woolard $et\ al$, 1991; Tang, Ramaswamy and Nam, 1993).

The hydrodynamic model presented in this Chapter is based on the Heisenberg equations of motion for the rates of changes of the particle number, momentum and energy in a macroscopically small fluid element, and the statistical average of them with respect to the parametrized unperturbed state. The hydrodynamic balance equations thus derived are the particle-number balance equation or the continuity equation (10.20), momentum balance equation (10.25), and energy balance equation (10.29):

$$\frac{\partial n}{\partial t} + \boldsymbol{\nabla} \cdot (n\boldsymbol{v}) = 0, \tag{10.79}$$

$$\frac{\partial \boldsymbol{v}}{\partial t} + \boldsymbol{v} \cdot \boldsymbol{\nabla} \boldsymbol{v} = -\frac{2}{3}\frac{\boldsymbol{\nabla} u}{mn} + \frac{e}{m}\boldsymbol{E} + \frac{\boldsymbol{f}}{mn}, \tag{10.80}$$

$$\frac{\partial u}{\partial t} + \boldsymbol{v} \cdot \boldsymbol{\nabla} u = -\frac{5}{3}u\boldsymbol{\nabla} \cdot \boldsymbol{v} - w - \boldsymbol{v} \cdot \boldsymbol{f}. \tag{10.81}$$

They are formally similar to the hydrodynamic equations derived from the first three moments of the Boltzmann equation. But the production or the collision terms (related to the impurity and phonon scatterings) have been included in the frictional force function f and the energy-loss rate function w. They, as functions of electron density n, drift velocity v and electron temperature T_e, $f = f(n, v, T_e)$ and $w = w(n, v, T_e)$, are explicitly given in (10.35) and (10.36). These equations, supplemented by two relations (10.33) and (10.34) plus the Poisson equation connecting the electric charge density and the electric potential ($E = -\nabla\phi$, κ is the dielectric constant of the material, n_D is the density of the positive charge background),

$$\nabla^2\phi = -\frac{e}{\varepsilon_0\kappa}(n - n_D), \qquad (10.82)$$

form a close set of equations. Usually one can choose n, v, T_e and ϕ as the basic field variables.

For systems with parabolic bands, $\varepsilon_k = k^2/2m$, Eq. (10.34) for the electron number density and Eq. (10.33) for the energy density can be written as ($\zeta \equiv \mu/T_e$, μ is the chemical potential)

$$n = N_c(T_e)\mathcal{F}_{\frac{1}{2}}(\zeta), \qquad (10.83)$$

$$u = \frac{3}{2}N_c(T_e)T_e\mathcal{F}_{\frac{3}{2}}(\zeta), \qquad (10.84)$$

in which $N_c(T_e) \equiv 2(mT_e/2\pi)^{3/2}$ and $\mathcal{F}_\nu(y)$ is the Fermi integral of order ν, defined by

$$\mathcal{F}_\nu(y) = \frac{1}{\Gamma(\nu+1)}\int_0^\infty \frac{x^\nu\,dx}{\exp(x-y)+1}. \qquad (10.85)$$

Eq. (10.83) determines ξ as a function of n and T_e, then (10.84) indicates that u is also a function of n and T_e. In solving these equations, it is often using the particle current density $j_n = nv$ or the current density $j = ej_n = env$ to replace the variable v, and writing the continuity equation and the momentum balance equation in the form

$$\frac{\partial n}{\partial t} + \nabla \cdot j_n = 0, \qquad (10.86)$$

$$\frac{\partial j_n}{\partial t} + \nabla \cdot \left(\frac{j_n j_n}{n}\right) = -\frac{2}{3}\frac{\nabla u}{m} + \frac{neE}{m} + \frac{f}{m}. \qquad (10.87)$$

For the energy balance equation one can take Eq. (10.81) and write it as

$$\frac{\partial u}{\partial t} + \frac{1}{n}j_n \cdot \nabla u = -\frac{5u}{3}\nabla\cdot\left(\frac{j_n}{n}\right) - w - \frac{1}{n}j_n \cdot f, \qquad (10.88)$$

or take the equivalent Eq. (10.31):

$$\frac{\partial h_e}{\partial t} + \nabla \cdot \boldsymbol{j}_H = e\boldsymbol{j}_n \cdot \boldsymbol{E} - w, \qquad (10.89)$$

where

$$h_e = u + \frac{1}{2}nmv^2 \qquad (10.90)$$

is the average kinetic energy density of electrons (including the center of mass and relative electrons) given by (10.14), and

$$\boldsymbol{j}_H = \left(\frac{5u}{3n} + \frac{1}{2}mv^2\right)\boldsymbol{j}_n \qquad (10.91)$$

is the energy-flow density (energy flux) vector defined by (10.19). The energy density u, treated as a function of n and T_e, gives the relation

$$\nabla u = \left(\frac{\partial u}{\partial t}\right)_{T_e}\nabla n + \left(\frac{\partial u}{\partial t}\right)_n \nabla T_e = \frac{3}{2}\Lambda_1 T_e \nabla n + \frac{3}{2}\Lambda_2 n \nabla T_e, \qquad (10.92)$$

in which

$$\Lambda_1 \equiv \frac{\mathcal{F}_{\frac{1}{2}}(\zeta)}{\mathcal{F}_{-\frac{1}{2}}(\zeta)}, \qquad (10.93)$$

$$\Lambda_2 \equiv \frac{2.5\mathcal{F}_{\frac{3}{2}}(\zeta)}{\mathcal{F}_{\frac{1}{2}}(\zeta)} - \frac{1.5\mathcal{F}_{\frac{1}{2}}(\zeta)}{\mathcal{F}_{-\frac{1}{2}}(\zeta)}. \qquad (10.94)$$

It should be noted that in the derivation of above hydrodynamic balance equations, we neglect the interactions between different fluid elements and treat different elements as dynamically independent systems. In writing the particle number, momentum and energy balance equations we keep only the first orders of the spatial differential coefficients of the main parameters n, $\boldsymbol{v}$ and T_e and neglect all the second and higher orders of them. Thus, the heat conduction, diffusion and viscosity (they are second order spatial differentiation of the basic variables) are not included in the model and the hydrodynamic equations derived are those for an ideal fluid. We explain this further using the energy balance equation (10.89) as an example. According to (10.91), the energy flow density $\boldsymbol{j}_H$ is a function of n, $\boldsymbol{v}$, and T_e only, not dependent on ∇n, $\nabla \boldsymbol{v}$, or ∇T_e. Apparently this is the convective part of the energy flow. In the presence of temperature gradient, there may be a diffusion part of the energy flow, i.e. the heat flow $\boldsymbol{j}_Q$, which, according to the Fourier law, is proportional to ∇T_e. But any part proportional to ∇T_e in the energy flow vector $\boldsymbol{j}_H$ appears in the energy balance equation as the second order in the spatial differentiation of T_e, and thus neglected in our weakly nonuniform model. Therefore, the hydrodynamic model truncated at energy balance equation is unable to cover the second Onsager equation (the linear response equation of heat flux). However, as pointed out in Sec. 10.6, if we further consider the balance equation for the energy flux in the framework of balance equation theory, the second Onsager equation is recovered. This should be a natural way for the extension of hydrodynamic model to the case of larger temperature gradient. Unfortunately, this makes things much more complicated. There were some explorations along this direction (Anile and Muscato, 1995), but it has not yet been applied to realistic device simulation.

An often used simple and intuitive treatment within the hydrodynamic model up to the energy balance equation, is to artificially add a heat flux

$$\boldsymbol{j}_Q = -\kappa \nabla T_e \qquad (10.95)$$

into the expression of energy flux

$$\boldsymbol{j}_H = \left(\frac{5u}{3n} + \frac{1}{2}mv^2\right)\boldsymbol{j}_n + \boldsymbol{j}_Q \qquad (10.96)$$

in the $\nabla \cdot \boldsymbol{j}_H$ term of the energy balance equation. Here the "thermal conductivity" κ is taken as an adjustable parameter, expressed by

$$\kappa = \gamma \frac{3}{2} \frac{k_B^2}{e} \mu_0 n T_e, \qquad (10.97)$$

where μ_0 is the linear mobility at lattice temperature in the uniform system and γ is a dimensionless fitting parameter. The addition of this $\nabla^2 T_e$ term into the energy balance equation previously derived by neglecting all the second order differential terms, can be thought as partially taking account of the interaction between neighboring fluid elements through boundaries.

10.7.2 Simulation of transport process in 1D semiconductor devices

We consider a one-dimensional device made of n-type semiconductor having isotropic parabolic energy band under the influence of a constant dc potential field, assuming that the electric field, the current as well as the spatial inhomogeneity, are all along the x direction. In the steady state all the field variables are constant in time. The Poisson equation, particle-number, momentum and energy balance equations can be written as

$$\frac{d^2\phi}{dx^2} = \frac{e}{\epsilon_0 \kappa}(n_D - n), \tag{10.98}$$

$$\frac{dj_n}{dx} = 0, \tag{10.99}$$

$$\left(\Lambda_1 T_e - \frac{m}{n^2}\right)\frac{dn}{dx} + \Lambda_2 n \frac{dT_e}{dx} + en\frac{d\phi}{dx} = f, \tag{10.100}$$

$$\kappa \frac{d^2 T_e}{dx^2} + \frac{3}{2}\Lambda_2 n j_n \frac{dT_e}{dx} - T_e \frac{dn}{dx} = w + \frac{j_n}{n} f. \tag{10.101}$$

They form the complete set of coupled nonlinear differential equations. The doping profile $n_D = n_D(x)$ is the known input, ϕ, n, T_e and j_n are field variables to solve for, and the boundary conditions can be determined by the contact property and bias setup.

Employing these equations Cai and Cui (1995) and Cao and Lei (1996b) analyze the steady state transport of carriers in n^+nn^+ structure diodes of single crystal Si at room temperature. This is a simple unipolar device with only one type of carriers, while exhibiting most of important transport phenomena that occur in semiconductor devices, therefore is traditionally used as the example for testing the devise modeling approach. The simulated Si n^+nn^+ diode has an inner n-type region of $0.4\,\mu$m long with a constant doping density $n_D = 2 \times 10^{15}\,\text{cm}^{-3}$ in between two outer n^+ regions a little longer than $0.1\,\mu$m with constant doping density $n_D = 5 \times 10^{17}\,\text{cm}^{-3}$. The Ohmic contacts are formed at the ends of two outer n^+ regions with metal electrodes, bewteen which a fixed dc voltage is applied. Going from the electrode into the n^+ semiconductor, there may be narrow space charge and band bending region in the immediate vicinity of the interface, then space charge and band bending disappear and electrical neutrality restores. We treat these positions as the left and right ends L and R, assuming a distance $0.1\,\mu$m apart from the n region. The carriers in the narrow regions from the metal plate to the device end are assumed in a local thermal equilibrium state throughout the whole transport process with the electron temperature T_e equal to the lattice temperature $T = 300\,\text{K}$. Therefore, at the two ends of the device,

$$T_e(L) = T_e(R) = T, \tag{10.102}$$

$$n(L) = n_D(L), \quad n(R) = n_D(R), \tag{10.103}$$

$$\phi(R) - \phi(L) = \frac{T}{e}\ln\left(\frac{n(R)}{n(L)}\right) + V. \tag{10.104}$$

These boundary conditions completely determine the solution of the steady state equation (10.98)–(10.101), i.e. the field variables ϕ, n, T_e and j_n.

In executing numerical analysis, the device from the left end to the right end is divided into many small segments and the partial differential equations are transformed into difference equations. Since the steady state continuity equation (10.99) requires j_n to be constant, only the other three equations are needed to handle. The discretization form of the Poison equation (10.98) is derived in a standard manner from the continuous one after linearizing it around a given electric potential. The discretization of momentum and energy balance equations is obtained by using the generalized Scharfetter-Gummel method (Ruden and Odeh, 1986). The resulting algebraic equations of ϕ, n, and T_e can be solved by iteration. For details please refer to the paper of Cai and Cui (1995). For instance, starting from a given particle current density j_n and a given electron temperature distribution T_e^k, one can solve for (ϕ^k, n^k) by the Poisson equation and the momentum balance equation, then from (ϕ^k, n^k) one can solve for T_e^{k+1} by the energy balance equation. This process is repeated until the desired accuracies are reached.

Fig. 10.4 shows the calculated steady state distributions of drift velocity v, electric field E, electron density n and electron temperature T_e/T in this Si n^+nn^+ structure under different values of the applied voltage V by Cao and Lei (1996b). Here the frictional force and energy loss rate functions $f = f(n, T_e, v)$ and $w = w(n, T_e, v)$ are calculated by (10.35) and (10.36) for single crystal Si with the electric field applied in $\langle 111 \rangle$ direction that six valleys contribute equivalently to transport. The scattering mechanisms considered in calculating f and w are those due to charged impurities (donors), acoustic phonon deformation potential and nonpolar optical phonon deformation potential.

Lee et al (1996) carry out a simulation analysis of transient transport for this system. To follow the evolution process the time derivatives in the original particle conservation equation (10.79), momentum balance equation (10.80) and energy balance equation (10.81) should be retained in the equations (10.99), (10.100) and (10.101), and the discretization of these equations should be done with respect to both spatial and temporal variables. When a step or a time-dependent voltage $V(t)$ is applied between the two ends of the device at time $t = 0$, one can take the steady state solution for $V = 0$ as the initial condition. The boundary condition is still

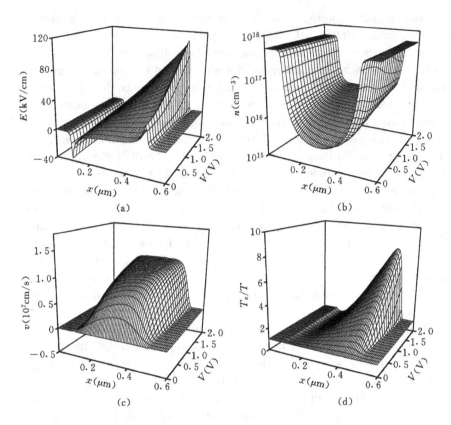

Fig. 10.4 Simulation of the steady state transport in a Si n^+nn^+ diode. (a) Electric field distribution, (b) Electron density distribution, (c) Velocity distribution, and (d) Electron temperature distribution. From Cao and Lei, 1996b.

that given by (10.102) and (10.103) with $V = V(t)$. For details please refer to their original paper.

Chapter 11

Balance Equations for Hot Electron Transport in an Arbitrary Energy Band

So far in most parts of this book we have assumed that the carrier in each energy band (or valley) possesses a constant effective mass m, i.e. the state of the carrier is described by a plane wave with wavevector $\bm{k}$ within an infinite wavevector space and has an energy relating to the wavevector by a quadric or parabolic form (the simplest case is $\varepsilon_{\bm{k}} = k^2/2m$). In semiconductors, especially in narrow-gap semiconductors, the effective mass approximation is valid only in the vicinity of the energy band bottom or top in the $\bm{k}$ space. When the wavevector $\bm{k}$ locates somewhat further from the band bottom or top, the energy dispersion $\varepsilon_{\bm{k}}$ may markedly deviate from a parabolic form. This indicates that the parabolic band model is applicable only for carrier transport driven by a relatively weak electric field. When the electric field becomes sufficiently large that carriers would move further from the band bottom or top or even across the whole Brillouin zone, the parabolic description is certainly invalid. Fortunately in a naturally existing crystal semiconductor, for which the Brillouin zone size π/a (a is the lattice constant) is of order of 10^8 cm^{-1} and the energy band width of order of eV, this could not happen until the electric field reaches the order of 10^5 V/cm. For the description of electron transport in ordinary semiconductor devices, which are working at electric fields much smaller than such a strength and thus the carrier motion is restricted within small ranges in the Brillouin zone, a modified parabolic band model or nonparabolic band model, i.e. a model based on $\bm{k}$ states in the infinite space with a spectrum dispersion modified to include some nonparabolic features, is good enough. However, for minibands in semiconductor superlattices, where the Brillouin zone size ($\pi/d \sim 10^6$ cm^{-1}, d is the period of the superlattice) and the band width ($\sim$ tenths of meV) are only around one percent of those for an ordinary crystal, an electric field of 10^3 V/cm strength is able to drive the carrier moving across the whole Brillouin zone. Therefore, it is of paramount importance to establish a transport model capable of describing carrier motion within the whole Brillouin zone.

Esaki and Tsu (1970) were the first to deal with such extremely nonparabolic transport and consider electrons moving across the whole Brillouin zone accelerated by the electric field. They proposed the idea of superlattice and predicted the occurrence of negative differential mobility in the electron conduction through a superlattice miniband. Büttiker and Thomas (1977) set forth a momentum balance equation to describe the electron moving in an arbitrary energy band. The main feature of this equation is that the effective force exerting on the electron of charge e by the electric field $\boldsymbol{E}$ equals $e\boldsymbol{E}$ multiplied by a reduction factor, rather than $e\boldsymbol{E}$ itself. This reflects the fact that an electron moving to the boundary of the Brillouin zone due to the acceleration of the electric field, will suddenly lose a reciprocal lattice vector, i.e. the Bragg scattering.

The success of balance equation transport theory for parabolic band systems has stimulated further efforts to extend the basic physical ideas of this theoretical approach to general energy band structure. There have been several attempts in the literature, using different models and equations (Magnus, Sala and De Meyer, 1990; Lei, 1992; Huang and Wu, 1994, 1995; Lei, 1995; Wu and Wu, 1995). This chapter will focus on the basic physics ideas in establishing balance equations for Bloch electron transport in an arbitrary energy band.

11.1 Heisenberg Equation of Motion for Electrons in a Brillouin Zone

11.1.1 *Electrons in a periodical potential*

We consider a general system of N interacting electrons subject to a periodical potential $U(\boldsymbol{r})$ and scattered by impurities and phonons. Under the influence of a uniform electric field $\boldsymbol{E}$ the total Hamiltonian of this electron and phonon system, is written as

$$H = H_\mathrm{e} + H_\mathrm{E} + H_\mathrm{I} + H_\mathrm{ph}. \tag{11.1}$$

Here the electron part ($j = 1, 2, \cdots, N$) is

$$H_\mathrm{e} = \sum_j h_j + H_\mathrm{ee}, \quad h_j = \frac{p_j^2}{2m_\mathrm{e}} + U(\boldsymbol{r}_j), \tag{11.2}$$

where r_j and p_j are the position and momentum operators of the jth electron, m_e is the free electron mass,

$$H_{ee} = \frac{e^2}{8\pi\epsilon_0} \sum_{i \neq j} \frac{1}{|r_i - r_j|} \tag{11.3}$$

is the electron–electron Coulomb interaction, and e is the electron charge. In Eq. (11.1)

$$H_E = -NeE \cdot R \tag{11.4}$$

represents the potential of the uniform electric filed E, in which

$$R = \frac{1}{N} \sum_j r_j \tag{11.5}$$

is the position operator of the center of mass of the N electron system. $H_I = H_{ei} + H_{ep}$ represents the electron–impurity and electron–phonon scatterings, which can be written respectively as

$$H_{ei} = \sum_{q,a} u(q) e^{-i q \cdot r_a} \rho_q, \tag{11.6}$$

$$H_{ep} = \sum_{q,\lambda} M(q,\lambda) \phi_{q\lambda} \rho_q, \tag{11.7}$$

where r_a stands for the impurity position, $u(q)$ and $M(q,\lambda)$ are the Fourier representations of the impurity potential and electron–phonon matrix element, $\phi_{q\lambda} \equiv b_{q\lambda} + b^\dagger_{-q\lambda}$ is the phonon field operator, with $b^\dagger_{q\lambda}$ and $b_{q\lambda}$ the creation and annihilation operators of a phonon having wavevector q in branch λ.

$$H_{ph} = \sum_{q,\lambda} \Omega_{q\lambda} b^\dagger_{q\lambda} b_{q\lambda} \tag{11.8}$$

is the phonon Hamiltonian. In Eqs. (11.6) and (11.7)

$$\rho_q = \sum_j e^{i q \cdot r_j} \tag{11.9}$$

is the density operator of the electron system.

Different from the parabolic band case discussed in Chapter 1, due to the existence of the periodical potential $U(r)$, the Hamiltonian of the system is no longer translationally invariant even without an external electric field. However, the electron–electron Coulomb interaction H_{ee}, which depends only on the relative positions of two electrons, is still translational invariant. Therefore, the total momentum operator of the system,

$$P = \sum_j p_j, \qquad (11.10)$$

i.e. the center-of-mass momentum operator defined in Sec. 1.1, which represents the collective translation of the electron system, still commutes with the electron–electron (e–e) interaction H_{ee}. The existence of strong e–e interactions is a fundamental requirement for establishing the balance equation transport theory. Its major role is to promote rapid thermalization of electrons in the reference frame without the integrated motion. This effect is always included in the basic assumption of the balance equation theory, such as the selection of the initial condition. For metals and semiconductors which can be described by the Fermi liquid theory, the e–e Coulomb coupling may lead to a renormalization of the electron energy and a screening of the scattering potential, while keeping the basic picture of single particle excitation essentially unchanged. The former effect is assumed already embedded in the effective Hamiltonian and the latter can be approximately treated using a screening function. As to some other e–e interaction effects specific to electrons moving in a periodical potential, they are generally not important in transport while very difficult to handle, and thus neglected in our treatment. To concentrate on major issues resulting from the electron motion in a finite Brillouin zone and from the band nonparabolicity, we will exclude H_{ee} from H_e in the analyses in this and next sections.

The eigen state of the single particle Hamiltonian h in (11.2) is a Bloch wave characterized by a band index n and a lattice wavevector $\bm{k}$ within the Brillouin zone (BZ),

$$|n\bm{k}\rangle \equiv \psi_{n\bm{k}}(\bm{r}) = e^{i\bm{k}\cdot\bm{r}} u_n(\bm{k},\bm{r}) \qquad (11.11)$$

[$u_n(\bm{k},\bm{r})$ is the cell periodic function], with the eigen energy $\varepsilon_n(\bm{k})$:

$$h|n\bm{k}\rangle = \varepsilon_n(\bm{k})|n\bm{k}\rangle. \qquad (11.12)$$

For convenience, people often employ a periodic Brillouin zone picture, which allows the lattice wavevector $\bm{k}$ to take values outside the Brillouin zone in such a way that for a given band index n, any one point in the $\bm{k}$ space represents an electron state, and for any reciprocal lattice vector $\bm{G}$, $\bm{k}$ and $\bm{k}+\bm{G}$ represent the same state. Therefore, for a given energy band under the description of periodic Brillouin zone picture, any physical quantity determined by the single particle state must be a periodic function of $\boldsymbol{k}$. The energy function, for instance, is a periodic function: $\varepsilon_n(\boldsymbol{k} + \boldsymbol{G}) = \varepsilon_n(\boldsymbol{k})$. Further, when the electron state has only a small change, any directly observable physical quantity determined by the single particle state can only have a small change. Therefore, when the lattice vector $\boldsymbol{k}$ varies in the whole $\boldsymbol{k}$ space, a directly observable physical quantity must be a continuous function of $\boldsymbol{k}$. The energy function is such a continuous function.

The whole Bloch functions $\psi_{n\boldsymbol{k}}$ (plus the spin states) form a complete set of the wave function space for an electron moving in the periodical potential. All physical quantities of a many electron system can be written in the second quantization representation on the basis of single electron Bloch states. For instance, the Hamiltonian $H_{\rm e}$, the center-of-mass position operator $\boldsymbol{R}$ and the total momentum operator $\boldsymbol{P}$ can be expressed as (Callaway, 1986; Kittle, 1986)

$$H_{\rm e} = \sum_{n,\boldsymbol{k},\sigma} \varepsilon_n(\boldsymbol{k})\, c^\dagger_{n\boldsymbol{k}\sigma} c_{n\boldsymbol{k}\sigma}, \tag{11.13}$$

$$\boldsymbol{R} = \frac{1}{N} \sum_{n,n',\boldsymbol{k},\boldsymbol{k}',\sigma} \left[{\rm i}\,\delta_{nn'} \boldsymbol{\nabla}\delta(\boldsymbol{k}-\boldsymbol{k}') + \boldsymbol{\chi}_{nn'}(\boldsymbol{k})\,\delta(\boldsymbol{k}-\boldsymbol{k}') \right] c^\dagger_{n\boldsymbol{k}\sigma} c_{n'\boldsymbol{k}'\sigma}, \tag{11.14}$$

$$\boldsymbol{P} = \sum_{n,n',\boldsymbol{k},\sigma} \left[\boldsymbol{k}\,\delta_{nn'} + \boldsymbol{P}_{nn'}(\boldsymbol{k}) \right] c^\dagger_{n\boldsymbol{k}\sigma} c_{n\boldsymbol{k}\sigma}. \tag{11.15}$$

Here $c^\dagger_{n\boldsymbol{k}\sigma}$ and $c_{n\boldsymbol{k}\sigma}$ are creation and annihilation operators of an electron at the Bloch state $|n\boldsymbol{k}\rangle$ and spin state σ,

$$\boldsymbol{P}_{nn'}(\boldsymbol{k}) = -{\rm i} \int d\boldsymbol{r}\, u^*_n(\boldsymbol{k},\boldsymbol{r}) \boldsymbol{\nabla}_{\boldsymbol{r}} u_{n'}(\boldsymbol{k},\boldsymbol{r}) \tag{11.16}$$

and

$$\boldsymbol{\chi}_{nn'}(\boldsymbol{k}) = {\rm i} \int d\boldsymbol{r}\, u^*_n(\boldsymbol{k},\boldsymbol{r}) \boldsymbol{\nabla}_{\boldsymbol{k}} u_{n'}(\boldsymbol{k},\boldsymbol{r}) \tag{11.17}$$

are integrals associated with the cell periodic functions $u_n(\boldsymbol{k},\boldsymbol{r})$.

11.1.2 Single band subspace, rates of changes of momentum and energy

In many cases when the applied electric field is not too strong that the tunneling between different energy bands is negligible, it is a good approximation to assume that the electric-field induced change of the carrier state is within a single energy band. When electrons are restricted to move within a single energy band of index n with energy $\varepsilon_n(\bm{k}) = \varepsilon(\bm{k})$ and state vector $|n\bm{k}\rangle \equiv |\bm{k}\rangle$ (for simplicity we neglect the band index n), what we deal with is no longer the complete space for electrons moving in the periodical potential, rather than a subspace of it (Lei, 1996a). The Hamiltonian H_e in the single band subspace is written as

$$H_\mathrm{e} = \sum_{\bm{k},\sigma} \varepsilon(\bm{k})\, c^\dagger_{\bm{k}\sigma} c_{\bm{k}\sigma}. \tag{11.18}$$

$c^\dagger_{\bm{k}\sigma}$ and $c_{\bm{k}\sigma}$ represent the electron creation and annihilation operators at the corresponding single electron Bloch state $\bm{k}$ and spin σ. The center-of-mass position operator is given by

$$\bm{R} = \frac{\mathrm{i}}{N} \sum_{\bm{k},\bm{k}',\sigma} \left[\bm{\nabla}\delta(\bm{k}-\bm{k}')\right] c^\dagger_{\bm{k}\sigma} c_{\bm{k}'\sigma} + \frac{1}{N} \sum_{\bm{k},\sigma} \bm{\chi}(\bm{k})\, c^\dagger_{\bm{k}\sigma} c_{\bm{k}\sigma}, \tag{11.19}$$

in which $\bm{\chi}(\bm{k}) = \bm{\chi}_{nn}(\bm{k})$ [see (11.17)]. For an arbitrary $\bm{q}$, the electron density operator $\rho_{\bm{q}}$ can be written as (Lei, 1992a)

$$\rho_{\bm{q}} = \sum_{\bm{k},\bm{k}',\sigma} \langle \bm{k}'|\mathrm{e}^{\mathrm{i}\bm{q}\cdot\bm{r}}|\bm{k}\rangle\, c^\dagger_{\bm{k}'\sigma} c_{\bm{k}\sigma}, \tag{11.20}$$

where both $\bm{k}$ and $\bm{k}'$ are restricted within the Brillouin zone (BZ). Expressing $|\bm{k}\rangle = \mathrm{e}^{\mathrm{i}\bm{k}\cdot\bm{r}} u_k(\bm{r})$ through the cell periodic function $u_k(\bm{r})$, we have

$$\langle \bm{k}'|\mathrm{e}^{\mathrm{i}\bm{q}\cdot\bm{r}}|\bm{k}\rangle = N_\mathrm{L} \sum_{\bm{G}} \delta_{\bm{k}'-\bm{k}-\bm{q},\bm{G}} \int_\Omega d\bm{r}\, \mathrm{e}^{\mathrm{i}(\bm{k}'-\bm{k}-\bm{q})\cdot\bm{r}} u^*_{k'}(\bm{r}) u_k(\bm{r}),$$

where the integral is taken over the volume Ω of a unit cell, and $N_\mathrm{L} = 1/\Omega$ is the total number of unit cells in the system (unit volume). The summation is over all possible reciprocal lattice vectors $\bm{G}$. Thus the matrix element $\langle \bm{k}'|\mathrm{e}^{\mathrm{i}\bm{q}\cdot\bm{r}}|\bm{k}\rangle$ can be nonzero only when $\bm{k}'-\bm{k}-\bm{q}$ equals a reciprocal vector $\bm{G}$. Using the periodic-zone picture description, in which $\bm{k}+\bm{q}$ and $\bm{k}'$ represent the same state, we can write the density operator in the form

$$\rho_{\bm{q}} = \sum_{\bm{k}} \rho_{\bm{k}\bm{q}}, \quad \rho_{\bm{k}\bm{q}} = \sum_\sigma g(\bm{k},\bm{q})\, c^\dagger_{\bm{k}+\bm{q}\sigma} c_{\bm{k}\sigma}, \tag{11.21}$$

where

$$g(\bm{k},\bm{q}) = \sum_{\bm{k}'\in\mathrm{BZ}} \langle \bm{k}'|\mathrm{e}^{\mathrm{i}\bm{q}\cdot\bm{r}}|\bm{k}\rangle = \sum_{\bm{G}} \int d\bm{r}\, \mathrm{e}^{\mathrm{i}\bm{G}\cdot\bm{r}} u^*_{k+q+G}(\bm{r}) u_k(\bm{r}) \tag{11.22}$$

is a form factor depending on $\boldsymbol{k}$ and $\boldsymbol{q}$ and related to electron Bloch function. Note that, the integral in (11.22) is taken over the whole volume of the sample, and the sum over $\boldsymbol{G}$ is just a symbol to indicate that for given $\boldsymbol{k}$ and $\boldsymbol{q}$, the reciprocal vector $\boldsymbol{G}$ takes such a value which gives $\boldsymbol{k}' = \boldsymbol{k} + \boldsymbol{q} + \boldsymbol{G}$ inside the Brillouin zone.

When electrons move within a single energy band, the total momentum is given by

$$\boldsymbol{P} = \sum_{\boldsymbol{k},\sigma} \langle \boldsymbol{k}|\boldsymbol{p}|\boldsymbol{k}\rangle \, c^\dagger_{\boldsymbol{k}\sigma} c_{\boldsymbol{k}\sigma} = m_{\rm e} \sum_{\boldsymbol{k},\sigma} \boldsymbol{v}(\boldsymbol{k}) \, c^\dagger_{\boldsymbol{k}\sigma} c_{\boldsymbol{k}\sigma}. \qquad (11.23)$$

In writing the above expression we have made use of the results of the quantum mechanics average for the single-particle momentum operator $\boldsymbol{p}$ in the Bloch $\boldsymbol{k}$ state: $\langle \boldsymbol{k}|\boldsymbol{p}|\boldsymbol{k}\rangle = m_{\rm e}\boldsymbol{v}(\boldsymbol{k})$ (Kittle, 1986). Here $m_{\rm e}$ is the free electron mass, $\boldsymbol{v}(\boldsymbol{k}) = \nabla_{\boldsymbol{k}}\varepsilon(\boldsymbol{k})$ is the velocity function. Thus, $m_{\rm e}\boldsymbol{v}(\boldsymbol{k})$ is the quantum mechanics average momentum of an electron in the Bloch state $|\boldsymbol{k}\rangle$, while $\boldsymbol{P}$ in (11.23) is the total momentum (operator) of the many electron system in the single energy band. Therefore, the quantity

$$\hat{\boldsymbol{v}} = \frac{\boldsymbol{P}}{Nm_{\rm e}} = \frac{1}{N}\sum_{\boldsymbol{k},\sigma} \boldsymbol{v}(\boldsymbol{k}) \, c^\dagger_{\boldsymbol{k}\sigma} c_{\boldsymbol{k}\sigma} \qquad (11.24)$$

is the average velocity (drift velocity) operator of the many electron system in the single energy band.

According to Heisenberg equation of motion, the rate of change of electron total momentum $\boldsymbol{P}$ is

$$\frac{d\boldsymbol{P}}{dt} = -{\rm i}[\boldsymbol{P}, H] = -{\rm i}[\boldsymbol{P}, H_{\rm E}] - {\rm i}[\boldsymbol{P}, H_{\rm I}]. \qquad (11.25)$$

Directly calculating the commutation relation between $\boldsymbol{R}$ and $\boldsymbol{P}$ from (11.19) and (11.23), we obtain

$$[\boldsymbol{R}, \boldsymbol{P}] = {\rm i}\frac{m_{\rm e}}{N} \sum_{\boldsymbol{k},\boldsymbol{k}',\boldsymbol{q},\sigma,\sigma_1} \left[\nabla\delta(\boldsymbol{k}-\boldsymbol{k}')\right] \boldsymbol{v}(\boldsymbol{q}) \left[c^\dagger_{\boldsymbol{k}\sigma} c_{\boldsymbol{k}'\sigma}, c^\dagger_{\boldsymbol{q}\sigma_1} c_{\boldsymbol{q}\sigma_1}\right]$$

$$= {\rm i}\frac{m_{\rm e}}{N} \sum_{\boldsymbol{k},\boldsymbol{k}',\sigma} \left[\nabla\delta(\boldsymbol{k}-\boldsymbol{k}')\right] \left[\boldsymbol{v}(\boldsymbol{k}) - \boldsymbol{v}(\boldsymbol{k}')\right] c^\dagger_{\boldsymbol{k}\sigma} c_{\boldsymbol{k}'\sigma}$$

$$= {\rm i}\, m_{\rm e} \hat{\mathcal{K}}, \qquad (11.26)$$

where

$$\hat{\mathcal{K}} = \frac{1}{N}\sum_{\boldsymbol{k},\sigma} \nabla\nabla\varepsilon(\boldsymbol{k}) \, c^\dagger_{\boldsymbol{k}\sigma} c_{\boldsymbol{k}\sigma} \qquad (11.27)$$

is the inverse effective mass tensor operator.[1] We then have

$$-\mathrm{i}[\boldsymbol{P}, H_{\mathrm{E}}] = m_{\mathrm{e}} N e \boldsymbol{E} \cdot \hat{\mathcal{K}}, \qquad (11.28)$$

and the equation (11.25) for the rate of change of the total momentum of electrons in the single energy band can be written as

$$\frac{d}{dt}\boldsymbol{P} = m_{\mathrm{e}} N e \boldsymbol{E} \cdot \hat{\mathcal{K}} + \boldsymbol{F}^{\mathrm{s}}, \qquad (11.29)$$

in which

$$\boldsymbol{F}^{\mathrm{s}} \equiv -\mathrm{i}[\boldsymbol{P}, H_{\mathrm{I}}] \qquad (11.30)$$

is the frictional force induced by impurity and phonon scatterings. Eq. (11.29) can also be written as the equation for the rate of change of the average carrier drift velocity:

$$\frac{d\hat{\boldsymbol{v}}}{dt} = e\boldsymbol{E} \cdot \hat{\mathcal{K}} + \hat{\boldsymbol{a}}, \qquad (11.31)$$

where $\hat{\boldsymbol{a}} = \boldsymbol{F}^{\mathrm{s}}/Nm_{\mathrm{e}}$ is the frictional force acceleration due to impurity and phonon scatterings, having the expression

$$\begin{aligned}\hat{\boldsymbol{a}} = &-\frac{\mathrm{i}}{N}\sum_{\boldsymbol{k},\boldsymbol{q},a} u(q)\,\mathrm{e}^{\mathrm{i}\boldsymbol{q}\cdot\boldsymbol{r}_a}\left[\boldsymbol{v}(\boldsymbol{k}+\boldsymbol{q}) - \boldsymbol{v}(\boldsymbol{k})\right]\rho_{\boldsymbol{k}\boldsymbol{q}} \\ &- \frac{\mathrm{i}}{N}\sum_{\boldsymbol{k},\boldsymbol{q},\lambda} M(\boldsymbol{q},\lambda)\phi_{\boldsymbol{q}\lambda}\left[\boldsymbol{v}(\boldsymbol{k}+\boldsymbol{q}) - \boldsymbol{v}(\boldsymbol{k})\right]\rho_{\boldsymbol{k}\boldsymbol{q}},\end{aligned} \qquad (11.32)$$

in which the summation over $\boldsymbol{k}$ is restricted within one Brillouin zone. The inverse effective mass tensor $\hat{\mathcal{K}}$ in Eqs. (11.29) and (11.31) reflects the characteristics of the energy band. For parabolic band $\varepsilon(\boldsymbol{k}) = k^2/m$,

$$\hat{\mathcal{K}} = \frac{\mathcal{I}}{m},$$

and Eq. (11.29) reduces to the equation given in Chapter 1:

$$\frac{d}{dt}\boldsymbol{P} = mN\frac{d\hat{\boldsymbol{v}}}{dt} = Ne\boldsymbol{E} + \boldsymbol{F}, \qquad (11.33)$$

where $\boldsymbol{F} \equiv (m/m_{\mathrm{e}})\boldsymbol{F}^{\mathrm{s}}$. For nonparabolic band, $\hat{\mathcal{K}}$ can not be expressed as a simple constant, and Eq. (11.29) may be quite different from the equation of parabolic band.

---

[1] In deriving the last equality of (11.26), we have assumed that $c^{\dagger}_{\boldsymbol{k}\sigma}c_{\boldsymbol{k}\sigma}$ is a continuous function of $\boldsymbol{k}$. In some case (see Sec. 11.2) this condition is not satisfied at the boundary of the BZ, there will be an additional term on the right-hand-side of the last equality.

Balance Equations for Hot Electron Transport in an Arbitrary Energy Band 369

The Heisenberg equation of motion (11.29) or (11.31) for the rate of change of the momentum can be served as a basis to construct the momentum balance equations. Similarly, for establishing the energy balance equation we need to consider the rate of change of the energy H_e of the electron system. Direct calculation from the Heisenberg equation of motion yields

$$\frac{d}{dt}H_e = -\mathrm{i}[H_e, H] = -\mathrm{i}[H_e, H_E] - \mathrm{i}[H_e, H_I]$$
$$= Ne\boldsymbol{E}\cdot\hat{\boldsymbol{v}} - \mathrm{i}\sum_{\boldsymbol{k},\boldsymbol{q},a} u(\boldsymbol{q})\,\mathrm{e}^{\mathrm{i}\boldsymbol{q}\cdot\boldsymbol{r}_a}\left[\varepsilon(\boldsymbol{k}+\boldsymbol{q})-\varepsilon(\boldsymbol{k})\right]\rho_{\boldsymbol{k}\boldsymbol{q}}$$
$$-\mathrm{i}\sum_{\boldsymbol{k},\boldsymbol{q},\lambda} M(\boldsymbol{q},\lambda)\phi_{\boldsymbol{q}\lambda}\left[\varepsilon(\boldsymbol{k}+\boldsymbol{q})-\varepsilon(\boldsymbol{k})\right]\rho_{\boldsymbol{k}\boldsymbol{q}}. \qquad (11.34)$$

where $\hat{\boldsymbol{v}}$ is the drift velocity operator given by (11.24).

11.1.3 The rate of change of the lattice momentum

In analyzing the single particle property of electrons moving in a periodical potential, another state function of momentum dimension, i.e. the lattice momentum function $\boldsymbol{p}(\boldsymbol{k})$ (Lei, 1995c, 1996a), are also frequently used in addition to the momentum function $\langle\boldsymbol{k}|\boldsymbol{p}|\boldsymbol{k}\rangle = m_e\boldsymbol{v}(\boldsymbol{k})$. The lattice momentum function $\boldsymbol{p}(\boldsymbol{k})$ is defined as a function of the Bloch state and $\boldsymbol{p}(\boldsymbol{k}) = \boldsymbol{k}$ when $\boldsymbol{k}$ locates inside the Brillouin zone. Thereby the lattice momentum function $\boldsymbol{p}(\boldsymbol{k})$ obeys the well known acceleration theorem within the BZ: an electric field $\boldsymbol{E}$ induces a lattice momentum change according to

$$\frac{d}{dt}\boldsymbol{p}(\boldsymbol{k}) = \frac{d\boldsymbol{k}}{dt} = e\boldsymbol{E}. \qquad (11.35)$$

On the other hand, as a state function, it must be a periodic function of $\boldsymbol{k}$ under periodic zone description, i.e. for any vector $\boldsymbol{k}$ and reciprocal lattice vector $\boldsymbol{G}$, $\boldsymbol{p}(\boldsymbol{k}+\boldsymbol{G}) = \boldsymbol{p}(\boldsymbol{k})$. Therefore, $\boldsymbol{p}(\boldsymbol{k})$ is not a continuous function under periodic zone description, but has a sudden jump at the boundary of the BZ. Fig. 1.1 shows the difference between the lattice momentum function $\boldsymbol{p}(\boldsymbol{k})$ and the momentum function $m_e\boldsymbol{v}(\boldsymbol{k})$. Despite its discontinuity, we can still define the total lattice momentum (operator) for the many particle system:

$$\boldsymbol{P}_\ell = \sum_{\boldsymbol{k},\sigma} \boldsymbol{p}(\boldsymbol{k})\, c^\dagger_{\boldsymbol{k}\sigma} c_{\boldsymbol{k}\sigma}. \qquad (11.36)$$

Here the summation over k covers the entire interior of the Brillouin zone and a half of its boundary. The commutation relation between P_ℓ and R can be calculated (Lei, 1996a) directly from the expressions (11.19) and (11.36), to give[2]

$$[R, P_\ell] = \frac{i}{N} \sum_{k,k',q,\sigma,\sigma_1} [\nabla \delta(k-k')] p(q) [c^\dagger_{k\sigma} c_{k'\sigma}, c^\dagger_{q\sigma_1} c_{q\sigma_1}]$$
$$= \frac{i}{N} \sum_{k,\sigma} [\nabla p(k)] c^\dagger_{k\sigma} c_{k\sigma}. \qquad (11.37)$$

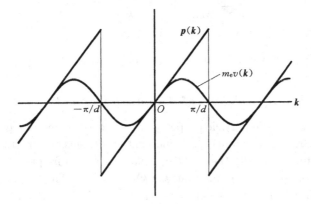

Fig. 11.1 Momentum function $m_e v(k)$ and lattice momentum function $p(k)$.

Inside the BZ, $p(k) = k$ and $\nabla p(k) = \mathcal{I}$, so the summation over the interior of the BZ on the right-hand-side of (11.37) yields

$$\frac{i}{N} \mathcal{I} \sum_{k,\sigma} c^\dagger_{k\sigma} c_{k\sigma} = i\mathcal{I}.$$

On the boundary of the BZ, since the lattice-momentum function $p(k)$ has a jump of a reciprocal lattice vector, $\nabla p(k)$ is a δ function. Let k_b be a wavevector on the BZ boundary and n the unit vector normal to the BZ boundary at k_b. In the immediate vicinity of k_b, $\nabla p(k)$ can be written as $-n G \delta(k_n - k_{bn})$, where $k_n = k \cdot n$, $k_{bn} = k_b \cdot n$, and G is the reciprocal lattice vector connecting the boundary position k_b and the position k'_b on the opposite boundary: $G = k_b - k'_b$. The contribution in (11.37) from the

---

[2] In obtaining the last equality of (11.37), we have assumed that $c^\dagger_{k\sigma} c_{k\sigma}$ is a continuous function of k.

k summation over the half of the BZ boundary can be written in the form of a volume integral in the k space,

$$-\frac{1}{4\pi^3 n}\int_{S/2} d\boldsymbol{s} \int dk_n \boldsymbol{G}\delta(k_n - k_{bn})\, c_{\boldsymbol{k}\sigma}^\dagger c_{\boldsymbol{k}\sigma}$$
$$= -\frac{1}{4\pi^3 n}\int_{S/2} d\boldsymbol{s}\, \boldsymbol{G}\, c_{\boldsymbol{k}\sigma}^\dagger c_{\boldsymbol{k}\sigma} = -\frac{1}{4\pi^3 n}\int_{S/2}(d\boldsymbol{s}\,\boldsymbol{k}_\mathrm{b} + d\boldsymbol{s}'\,\boldsymbol{k}_\mathrm{b}')\, c_{\boldsymbol{k}\sigma}^\dagger c_{\boldsymbol{k}\sigma}$$
$$= -\frac{1}{4\pi^3 n}\oint_{S_\mathrm{BZ}} d\boldsymbol{s}\,\boldsymbol{k}_\mathrm{b}\, c_{\boldsymbol{k}\sigma}^\dagger c_{\boldsymbol{k}\sigma}.$$

Here $S/2$ stands for the half of the BZ boundary that contributes to the summation and S_BZ for the whole BZ boundary, $d\boldsymbol{s} = \boldsymbol{n}ds$ is the area element on the BZ boundary around $\boldsymbol{k}_\mathrm{b}$, $d\boldsymbol{s}' = -d\boldsymbol{s}$ is the area element on the opposite side of the BZ boundary around $\boldsymbol{k}_\mathrm{b}'$, and n is the electron density. Eq. (11.37) then can be written as

$$[\boldsymbol{R}, \boldsymbol{P}_\ell] = \mathrm{i}\hat{\mathcal{R}}, \tag{11.38}$$

and

$$\hat{\mathcal{R}} = \mathcal{I} - \frac{1}{4\pi^3 n}\oint_{S_\mathrm{BZ}} d\boldsymbol{s}\,\boldsymbol{k}\, c_{\boldsymbol{k}\sigma}^\dagger c_{\boldsymbol{k}\sigma} \tag{11.39}$$

is another tensor operator different from $\hat{\mathcal{K}}$ in (11.27).

According to Heisenberg equation of motion, the rate of change of the total lattice momentum of electrons within a single energy band is

$$\frac{d\boldsymbol{P}_\ell}{dt} = -\mathrm{i}[\boldsymbol{P}_\ell, H_\mathrm{E}] - \mathrm{i}[\boldsymbol{P}_\ell, H_\mathrm{I}]. \tag{11.40}$$

The first term on the right-hand-side gives

$$-\mathrm{i}[\boldsymbol{P}_\ell, H_\mathrm{E}] = Ne\boldsymbol{E}\cdot\hat{\mathcal{R}}, \tag{11.41}$$

and the second term is the frictional force related to lattice momentum:

$$\boldsymbol{F}^\ell = -\mathrm{i}[\boldsymbol{P}_\ell, H_\mathrm{I}]. \tag{11.42}$$

Eq. (11.40) is then written as

$$\frac{d}{dt}\boldsymbol{P}_\ell = Ne\boldsymbol{E}\cdot\hat{\mathcal{R}} + \boldsymbol{F}^\ell, \tag{11.43}$$

or

$$\frac{d}{dt}\hat{\boldsymbol{p}}_\ell = e\boldsymbol{E}\cdot\hat{\mathcal{R}} + \hat{\boldsymbol{f}}^\ell, \tag{11.44}$$

where $\hat{\boldsymbol{p}}_\ell \equiv \boldsymbol{P}_\ell/N$ is an equivalent lattice momentum brought by each particle, and $\hat{\boldsymbol{f}}^\ell \equiv \boldsymbol{F}^\ell/N$ corresponds to the lattice-momentum related frictional force exerting on each particle, which is given by

$$\hat{\boldsymbol{f}}^\ell = -\frac{\mathrm{i}}{N} \sum_{\boldsymbol{k},\boldsymbol{q},a} u(\boldsymbol{q})\, \mathrm{e}^{\mathrm{i}\boldsymbol{q}\cdot\boldsymbol{r}_a} \left[\boldsymbol{p}(\boldsymbol{k}+\boldsymbol{q}) - \boldsymbol{p}(\boldsymbol{k})\right] \rho_{\boldsymbol{k}\boldsymbol{q}}$$
$$- \frac{\mathrm{i}}{N} \sum_{\boldsymbol{k},\boldsymbol{q},\lambda} M(\boldsymbol{q},\lambda) \phi_{\boldsymbol{q}\lambda} \left[\boldsymbol{p}(\boldsymbol{k}+\boldsymbol{q}) - \boldsymbol{p}(\boldsymbol{k})\right] \rho_{\boldsymbol{k}\boldsymbol{q}}. \qquad (11.45)$$

11.1.4 Derivation using effective Hamiltonian

The equations derived in the preceding two subsections can also be obtained from the method of single band effective Hamiltonian (Lei, 1992a, 1995c). It is known from the basic energy band theory (Blount, 1962; Anderson, 1963) that in analyzing electron motion in a single energy band with energy function $\varepsilon(\boldsymbol{k})$ and wave function $\psi_{\boldsymbol{k}}(\boldsymbol{r})$ under a spatially slowly varying potential field, it is possible to use the effective Hamiltonian

$$\varepsilon(-\mathrm{i}\boldsymbol{\nabla}_j) \equiv \varepsilon(\hat{\boldsymbol{k}}_j) \qquad (11.46)$$

to replace the single particle Hamiltonian h_j in (11.2), thus writing H_e as

$$H_\mathrm{e} = \sum_j \varepsilon(\hat{\boldsymbol{k}}_j). \qquad (11.47)$$

Here $\hat{\boldsymbol{k}}_j$ can be treated as the lattice momentum operator of the jth electron, having the following commutation relation with the slowly varying space coordinate $\boldsymbol{r}_i$, as if $\hat{\boldsymbol{k}}_j$ is simply the operator $-\mathrm{i}\boldsymbol{\nabla}_j$:

$$[r_{i\alpha}, \hat{k}_{j\beta}] = \mathrm{i}\,\delta_{\alpha\beta}\delta_{ij}. \qquad (11.48)$$

The Bloch function $\psi_{\boldsymbol{k}}(\boldsymbol{r})$ can be considered as the eigenfunction of operator $\hat{\boldsymbol{k}}_j$ with eigenvalue $\boldsymbol{k}$. The total Hamiltonian $H = H_\mathrm{e} + H_\mathrm{E} + H_\mathrm{I} + H_\mathrm{ph}$ is still that given in Sec. 11.1 except H_e is replaced by (11.47).

According to the definition (11.5) of the center-of-mass position operator $\boldsymbol{R}$, the center-of-mass velocity is (Lei, 1992a)

$$\hat{\boldsymbol{v}} = -\mathrm{i}[\boldsymbol{R}, H] = -\mathrm{i}\Big[\boldsymbol{R}, \sum_j \varepsilon(\hat{\boldsymbol{k}}_j)\Big] = \frac{1}{N} \sum_j \boldsymbol{v}(\hat{\boldsymbol{k}}_j), \qquad (11.49)$$

$\boldsymbol{v}(\boldsymbol{k}) \equiv \nabla\varepsilon(\boldsymbol{k})$ being the velocity function. The equation for the rate of change of the velocity is

$$\frac{d\hat{\boldsymbol{v}}}{dt} = -\mathrm{i}[\hat{\boldsymbol{v}}, H] = -\mathrm{i}[\hat{\boldsymbol{v}}, H_\mathrm{e} + H_\mathrm{E} + H_\mathrm{I}]. \tag{11.50}$$

In the above equation, $[\hat{\boldsymbol{v}}, H_\mathrm{e}] = 0$, the contribution from H_E gives

$$-\mathrm{i}[\hat{\boldsymbol{v}}, H_\mathrm{E}] = \frac{\mathrm{i}e}{N}\sum_{i,j}[\boldsymbol{v}(\hat{\boldsymbol{k}}_j), \boldsymbol{r}_i \cdot \boldsymbol{E}] = e\boldsymbol{E} \cdot \hat{\mathcal{K}}, \tag{11.51}$$

$$\hat{\mathcal{K}} = \frac{1}{N}\sum_j \nabla\nabla\varepsilon(\hat{\boldsymbol{k}}_j), \tag{11.52}$$

in agreement with (11.27), and the contribution from H_I,

$$-\mathrm{i}[\hat{\boldsymbol{v}}, H_\mathrm{I}] = \hat{\boldsymbol{a}}, \tag{11.53}$$

is the same as the frictional acceleration operator (11.32). Therefore the equation derived from (11.50) is exactly the same as Eq. (11.31).

In the same vein, we can deal with the lattice momentum operator

$$\boldsymbol{P}_\ell = \sum_j \boldsymbol{p}(\hat{\boldsymbol{k}}_j) = \sum_{\boldsymbol{k},\sigma} \boldsymbol{p}(\boldsymbol{k}) c^\dagger_{\boldsymbol{k}\sigma} c_{\boldsymbol{k}\sigma}. \tag{11.54}$$

Calculating its rate of change using the effective Hamiltonian (Lei, 1995c),

$$\frac{d\boldsymbol{P}_\ell}{dt} = -\mathrm{i}[\boldsymbol{P}_\ell, H] = -\mathrm{i}[\boldsymbol{P}_\ell, H_\mathrm{e} + H_\mathrm{E} + H_\mathrm{I}], \tag{11.55}$$

we see that $[\boldsymbol{P}_\ell, H_\mathrm{e}] = 0$, the H_I contribution yields $\boldsymbol{F}^\ell$ as given in (11.42), and the H_E contribution

$$-\mathrm{i}[\boldsymbol{P}_\ell, H_\mathrm{E}] = -\mathrm{i}e\boldsymbol{E} \cdot \sum_{i,j}[\boldsymbol{r}_i, \boldsymbol{p}(\hat{\boldsymbol{k}}_j)] = e\boldsymbol{E} \cdot \sum_j \nabla\boldsymbol{p}(\hat{\boldsymbol{k}}_j)$$

$$= e\boldsymbol{E} \cdot \sum_{\boldsymbol{k},\sigma}[\nabla\boldsymbol{p}(\boldsymbol{k})]c^\dagger_{\boldsymbol{k}\sigma}c_{\boldsymbol{k}\sigma} = Ne\boldsymbol{E} \cdot \hat{\mathcal{R}}, \tag{11.56}$$

being the same as (11.41). Therefore, exactly the same results as equation (11.43) is obtained when calculating the rate of change of the lattice momentum with single band effective Hamiltonian (11.47).

The electron motion within a single energy band driven by a spatially slowly varying potential can be described by the effective Hamiltonian is an important result in the energy band theory of solids. For more detailed analysis about its physical bases, we refer the readers to the monographs of Blount (1962) and Anderson (1963).

11.2 Initial Density Matrix and Distribution Function

Starting from the Heisenberg equation of motion, we have obtained operator equations for the rates of changes of the momentum, lattice momentum, and energy for systems of many electrons moving in an arbitrary energy band. To derive transport equations, one has to take statistical average over these operator equations and thus needs the density matrix or the distribution function of the system in transport. The exact solution of the nonequilibrium density matrix or distribution function is a very difficult task. The balance equation theory allowing sufficiently strong electric field, deals with weak scattering potentials H_{ei} and H_{ep}, and calculates the density matrix and the statistical average of dynamical variables to the lowest nonzero order of scatterings. As a parameterized theory, it selects the initial density matrix $\hat{\rho}_0$ in a way that it is as simple as possible for easy handling (it is generally an equilibrium or quasi-equilibrium state) while as close to the final transport state as possible, that the system can quickly evolve to the latter after turning on the perturbation.

The Hamiltonian of the system consisting of many electrons (moving in a single energy band) and phonons under the influence of an electric field,

$$H = H_e + H_E + H_I + H_{ph},$$

has been discussed in the last section. In the spirit of the balance equation theory there are two different schemes in the literature to select the initial density matrix and to proceed the statistical average of a dynamical variable.

One scheme (Lei, 1992a) is to treat impurity and phonon scattering potentials as perturbation, which is turned on at time $t = 0$, when the electric field has already been there, i.e. H_E is included in the unperturbed Hamiltonian. Therefore, for any operator $\mathcal{O}$, the interaction picture $\mathcal{O}(t)$ is defined as

$$\mathcal{O}(t) = e^{i(H_e + H_E + H_{ph})t} \mathcal{O} e^{-i(H_e + H_E + H_{ph})t}. \qquad (11.57)$$

The statistical average of a dynamical variable can be calculated in the interaction picture as

$$\langle \mathcal{O} \rangle = \mathrm{tr}\{\hat{\rho}(t)\mathcal{O}(t)\}, \tag{11.58}$$

where $\hat{\rho}(t)$ is the density matrix in the interaction picture, which satisfies the Liouville equation of the form

$$\hat{\rho}(t) = \hat{\rho}_0 - \mathrm{i}\int_0^t dt_1 [H_\mathrm{I}(t_1), \hat{\rho}(t_1)], \tag{11.59}$$

$\hat{\rho}_0$ being the initial density matrix, i.e. the density matrix at $t = 0$. To the lowest order of H_I, the solution of the density matrix is

$$\hat{\rho}(t) = \hat{\rho}_0 - \mathrm{i}\int_0^t dt_1 [H_\mathrm{I}(t_1), \hat{\rho}_0]. \tag{11.60}$$

Therefore the average of the dynamic variable $\mathcal{O}$ can be written as

$$\langle \mathcal{O} \rangle = \langle \mathcal{O} \rangle_0 - \mathrm{i}\int_0^t dt_1 \langle [H_\mathrm{I}(t_1), \mathcal{O}(t)] \rangle_0, \tag{11.61}$$

where $\langle \cdots \rangle_0 = \mathrm{tr}\{\hat{\rho}_0 \cdots\}$ stands for the average with respect to the initial density matrix $\hat{\rho}_0$.

Now the initial state is a state having strong carrier–carrier interaction with the presence of the electric field. In the absence of the electric field, the inter-carrier interaction promotes rapid thermalization of the many electron system towards a Fermi distribution on the energy spectrum $\varepsilon(\boldsymbol{k})$. Strong carrier–carrier interaction implies a strong ability to keep this distribution. On the other hand, for electron in a periodical potential the role of an electric field $\boldsymbol{E}$ is to change its lattice vector according to $d\boldsymbol{k}/dt = e\boldsymbol{E}$ within a time interval τ before being scattered out: an electron original (before turning on the electric field) in state $\boldsymbol{k}$ will change, after turning on the electric field, to the state $\boldsymbol{k} + e\boldsymbol{E}\tau$ (τ is the effective scattering time). The ability of keeping the original distribution unchanged indicates that the electron distribution in the presence of an electric field is approximately a displacement in the $\boldsymbol{k}$ space from the distribution without the electric field (Lei, 1995b). This displacement represents the integrative motion of the electron system in an arbitrary band induced by the electric field. Therefore, Lei (1992a) proposed to choose the initial density matrix as

$$\hat{\rho}_0 = \frac{1}{Z}\mathrm{e}^{-(H_\mathrm{er} - \mu \hat{N})/T_\mathrm{e}}\mathrm{e}^{-H_\mathrm{ph}/T}, \tag{11.62}$$

in which μ is the chemical potential,

$$H_{\text{er}} = \sum_{\bm{k},\sigma} \varepsilon(\bm{k}-\bm{p}_d) c^\dagger_{\bm{k}\sigma} c_{\bm{k}\sigma} \qquad (11.63)$$

can be considered as the "relative electron" Hamiltonian, and

$$\hat{N} = \sum_{\bm{k},\sigma} c^\dagger_{\bm{k}\sigma} c_{\bm{k}\sigma} \qquad (11.64)$$

is the particle number operator. The distribution function $\text{tr}\{c^\dagger_{\bm{k}\sigma} c_{\bm{k}\sigma} \hat{\rho}_0\}$ corresponding to this initial density matrix is

$$f_d(\bm{k}) = \frac{1}{\mathrm{e}^{[\varepsilon(\bm{k}-\bm{p}_d)-\mu]/T_\mathrm{e}}+1} = f(\varepsilon(\bm{k}-\bm{p}_d),T_\mathrm{e}). \qquad (11.65)$$

Both the initial density matrix $\hat{\rho}_0$ and the distribution function $f_d(\bm{k})$ contain the lattice momentum displacement $\bm{p}_d$ and electron temperature T_e as parameters.

In the scheme putting H_E in the unperturbed Hamiltonian [i.e. using (11.57) as the interaction picture], the effect of the electric field acceleration during the correlation time interval may lead to the appearance of an additional term (intracollisional field effect). We have discussed this effect in Sec. 9.1 for parabolic band systems and concluded: it is not important in the electric field range usually dealt with. We consider this result still valid for systems of arbitrary energy band and simply neglect it in the treatment.

Another scheme (Wu and Wu, 1995) is to assume that both the electric field H_E and the scattering potential H_I are turned on from time $t=0$, and the unperturbed state is the one without the electric field. The interaction picture of an operator $\mathcal{O}$ is then

$$\mathcal{O}(t) = \mathrm{e}^{\mathrm{i}(H_\mathrm{e}+H_\mathrm{ph})t} \mathcal{O} \mathrm{e}^{-\mathrm{i}(H_\mathrm{e}+H_\mathrm{ph})t}. \qquad (11.66)$$

The statistical average of a dynamic variable is still calculated by (11.57), but the density matrix in the interaction picture satisfies the following equation:

$$\hat{\rho}(t) = \hat{\rho}_0 - \mathrm{i}\int_0^t dt_1 [H_\mathrm{I}(t_1) + H_\mathrm{E}(t_1), \hat{\rho}(t_1)]. \qquad (11.67)$$

In view of the absence of the electric field, the total momentum of the unperturbed state

$$\bm{P} = \sum_{\bm{k},\sigma} \langle \bm{k}|\bm{p}|\bm{k}\rangle c^\dagger_{\bm{k}\sigma} c_{\bm{k}\sigma} = m_\mathrm{e} \sum_{\bm{k},\sigma} \bm{v}(\bm{k}) c^\dagger_{\bm{k}\sigma} c_{\bm{k}\sigma} \qquad (11.68)$$

is a conserved quantity. Wu and Wu (1995) then suggest to choose

$$\hat{\rho}_0 = \frac{1}{Z} e^{-\left(H_e - \bm{v}_{\mathrm{p}} \cdot \bm{P} - \mu \hat{N}\right)/T_e} e^{-H_{\mathrm{ph}}/T}. \tag{11.69}$$

as the initial density matrix, with the corresponding distribution function

$$f_v(\bm{k}) = \frac{1}{e^{[\varepsilon(\bm{k}) - \bm{p}_v \cdot \bm{v}(\bm{k}) - \mu]/T_e} + 1}, \tag{11.70}$$

containing $\bm{p}_v$ (or $\bm{v}_{\mathrm{p}} \equiv \bm{p}_v/m_{\mathrm{e}}$) and T_{e} as parameters. This distribution function was first proposed by Huang and Wu (1994, 1995). They derived this distribution function by maximizing the entropy for the electron system described by the Hamiltonian H_{e}, subject to a given total number of electrons and a given total energy, and subject to the restriction condition that the average drift velocity is also prescribed.

In addition to selecting (11.69) as the initial density matrix and (11.70) as the distribution function, there is another possible choice (Lei, 1995b) in the case of arbitrary energy band for this scheme (the electric filed is turned on from $t = 0$ together with scattering H_{I}). Since in the unperturbed state without electric field, the total lattice momentum

$$\bm{P}_\ell - \sum_{\bm{k},\sigma} \bm{p}(\bm{k}) c^\dagger_{\bm{k}\sigma} c_{\bm{k}\sigma} \tag{11.71}$$

is also a conserved quantity. Starting with it, the initial density matrix should be chosen as

$$\hat{\rho}_0 = \frac{1}{Z} e^{-\left(H_e - \bm{v}_\ell \cdot \bm{P}_\ell - \mu \hat{N}\right)/T_e} e^{-H_{\mathrm{ph}}/T}, \tag{11.72}$$

and the corresponding distribution function is

$$f_\ell(\bm{k}) = \frac{1}{e^{[\varepsilon(\bm{k}) - \bm{v}_\ell \cdot \bm{p}(\bm{k}) - \mu]/T_e} + 1}, \tag{11.73}$$

containing $\bm{v}_\ell$ and T_{e} as parameters. This distribution function can also be obtained by maximizing the entropy for the electron system described by the Hamiltonian H_{e}, subject to the restriction conditions of a given total number of electrons, a given total energy and a given total lattice momentum. Huang and Wu (1995) later suggested to use $f_\ell(\bm{k})$ as the distribution function instead. It should be noticed that, different from functions $f_d(\bm{k})$ and $f_v(\bm{k})$, the function $f_\ell(\bm{k})$ is not a continuous function at the Brillouin zone boundary. One has to take care of this when performing calculation with it or with the initial density matrix (11.72).

The problems that may possibly occur with the scheme treating both H_E and H_I in the perturbation, have been discussed in Sec. 7.2 for the case of a parabolic band. Because of lack of better way to solve the equation (11.67), one can but expand it perturbatively in $H_E + H_I$, and thus the statistical average of any dynamic variable is also expanded in powers of H_E and H_I. For this perturbative series to be able to describe the transport of the system, the condition (7.25) should be satisfied. Furthermore, to achieve a compact result one can only keep the expansion to the linear order in H_I and to the zeroth order in H_E, i.e. still use the expression (11.61)

$$\langle \mathcal{O} \rangle = \langle \mathcal{O} \rangle_0 - i \int_0^t dt_1 \langle [H_I(t_1), \mathcal{O}(t)] \rangle_0 \qquad (11.74)$$

to calculate the average of a dynamic variable. The condition (7.25) and the approximation neglecting all terms with explicit H_E factors, will add extra restriction for the maximum electric field strength applicable.

The multiple possibilities of choosing the initial density matrix and distribution function is a reflection of the complexity of nonparabolic band system. In the case of parabolic band, the motion of relative electrons is completely separated from that of the center of mass and the density matrix of the whole system is a direct multiplication of the relative electron density matrix and a single particle (center of mass) pure quantum mechanic state, without an electric field exerting on the relative electrons. The relative electron density matrix or distribution function is essentially not affected no matter whether H_E is retained in the unperturbed Hamiltonian or it is treated as a perturbation together with H_I. Furthermore, since the total momentum is a conserved quantity and there is no difference between the momentum and the lattice momentum in a parabolic band system, the three initial density matrices (11.62), (11.69) and (11.72) or three distribution functions (11.65), (11.70) and (11.73) are equivalent and in agreement with that in Chapter 1. For systems of nonparabolic energy band, however, a convincing justification is still lacking as to which one of these three is more appropriate. Therefore, each one of them may be regarded as one possible approximation, having its applicable range, and the effectiveness should be judged by the experiments. For systems of nonparabolic Kane band and of superlattice miniband, Lei (1994c) compared the difference of the nonlinear transport properties calculated from the distribution function (11.65) and from the distribution function (11.70).

11.3 Momentum Balance Equation and Lattice Momentum Balance Equation

As pointed out in the preceding section that, in the balance equation theory, the statistical average of a dynamic variable is always performed according to the formula (11.61) or (11.74). All the three schemes [(11.62), (11.69) and (11.72)] of the initial density matrix ρ_0 are diagonal in the $\boldsymbol{k}$ space, thus the statistical averages are easily carried out.

11.3.1 *Momentum and energy balance equations*

The statistical averages of Eq. (11.31) and (11.34) yield the following momentum and energy balance equations derived by Lei (1992a):

$$\frac{d\boldsymbol{v}_{\rm d}}{dt} = e\boldsymbol{E}\cdot\mathcal{K} + \boldsymbol{a}_{\rm i} + \boldsymbol{a}_{\rm p}, \tag{11.75}$$

$$\frac{d\varepsilon_{\rm e}}{dt} = e\boldsymbol{E}\cdot\boldsymbol{v}_{\rm d} - w. \tag{11.76}$$

In this,

$$\boldsymbol{v}_{\rm d} = \langle\hat{\boldsymbol{v}}\rangle = \frac{2}{N}\sum_{\boldsymbol{k}} \boldsymbol{v}(\boldsymbol{k}) f(\bar{\varepsilon}(\boldsymbol{k}), T_{\rm e}) \tag{11.77}$$

is the average drift velocity,

$$\mathcal{K} = \langle\hat{\mathcal{K}}\rangle = \frac{2}{N}\sum_{\boldsymbol{k}} \nabla\nabla\varepsilon(\boldsymbol{k}) f(\bar{\varepsilon}(\boldsymbol{k}), T_{\rm e}) \tag{11.78}$$

is the average inverse effective mass tensor,

$$\varepsilon_{\rm e} = \frac{\langle H_{\rm e}\rangle}{N} = \frac{2}{N}\sum_{\boldsymbol{k}} \varepsilon(\boldsymbol{k}) f(\bar{\varepsilon}(\boldsymbol{k}), T_{\rm e}) \tag{11.79}$$

is average energy per carrier. $\boldsymbol{a}_{\rm i}$ and $\boldsymbol{a}_{\rm p}$ are the frictional accelerations due to impurity and phonon scatterings,

$$\boldsymbol{a}_{\rm i} = \frac{2\pi n_{\rm i}}{N}\sum_{\boldsymbol{k},\boldsymbol{q}} |u(q)|^2 |g(\boldsymbol{k},\boldsymbol{q})|^2 \left[\boldsymbol{v}(\boldsymbol{k}+\boldsymbol{q}) - \boldsymbol{v}(\boldsymbol{k})\right] \delta(\varepsilon(\boldsymbol{k}+\boldsymbol{q}) - \varepsilon(\boldsymbol{k}))$$
$$\times \left[f(\bar{\varepsilon}(\boldsymbol{k}), T_{\rm e}) - f(\bar{\varepsilon}(\boldsymbol{k}+\boldsymbol{q}), T_{\rm e})\right], \tag{11.80}$$

$$a_{\mathrm{p}} = \frac{4\pi}{N} \sum_{\bm{k},\bm{q},\lambda} |M(\bm{q},\lambda)|^2 |g(\bm{k},\bm{q})|^2 [v(\bm{k}+\bm{q}) - v(\bm{k})]$$
$$\times \delta(\varepsilon(\bm{k}+\bm{q}) - \varepsilon(\bm{k}) + \Omega_{\bm{q}\lambda}) [f(\bar{\varepsilon}(\bm{k}),T_{\mathrm{e}}) - f(\bar{\varepsilon}(\bm{k}+\bm{q}),T_{\mathrm{e}})]$$
$$\times \left[n\left(\frac{\Omega_{\bm{q}\lambda}}{T}\right) - n\left(\frac{\bar{\varepsilon}(\bm{k}) - \bar{\varepsilon}(\bm{k}+\bm{q})}{T_{\mathrm{e}}}\right) \right], \qquad (11.81)$$

and w is the average electron energy loss rate (per carrier) due to electron–phonon interaction,

$$w = \frac{4\pi}{N} \sum_{\bm{k},\bm{q},\lambda} |M(\bm{q},\lambda)|^2 |g(\bm{k},\bm{q})|^2 \Omega_{\bm{q}\lambda}\, \delta(\varepsilon(\bm{k}+\bm{q}) - \varepsilon(\bm{k}) + \Omega_{\bm{q}\lambda})$$
$$\times [f(\bar{\varepsilon}(\bm{k}),T_{\mathrm{e}}) - f(\bar{\varepsilon}(\bm{k}+\bm{q}),T_{\mathrm{e}})] \left[n\left(\frac{\Omega_{\bm{q}\lambda}}{T}\right) - n\left(\frac{\bar{\varepsilon}(\bm{k}) - \bar{\varepsilon}(\bm{k}+\bm{q})}{T_{\mathrm{e}}}\right) \right].$$
$$(11.82)$$

Here $g(\bm{k},\bm{q})$ is the form factor given by (11.22) related to the wave function of electrons, $n(x) \equiv (\mathrm{e}^x - 1)^{-1}$ is the Bose function,

$$f(\bar{\varepsilon}(\bm{k}),T_{\mathrm{e}}) = \frac{1}{\exp\left[(\bar{\varepsilon}(\bm{k}) - \mu)/T_{\mathrm{e}}\right] + 1} \qquad (11.83)$$

is the distribution function in the initial state, containing the electron temperature T_{e} and the chemical potential μ as parameters. The total number of electrons, N, relates to T_{e} and μ through the equation

$$N = 2 \sum_{\bm{k}} f(\bar{\varepsilon}(\bm{k}),T_{\mathrm{e}}). \qquad (11.84)$$

The energy function $\bar{\varepsilon}(\bm{k})$ in the above equations represents either

$$\bar{\varepsilon}(\bm{k}) = \varepsilon(\bm{k} - \bm{p}_d) \qquad (11.85)$$

if $f_d(\bm{k})$ of (11.65) is used as the initial distribution function, or

$$\bar{\varepsilon}(\bm{k}) = \varepsilon(\bm{k}) - \bm{p}_v \cdot \bm{v}(\bm{k}) \qquad (11.86)$$

if $f_v(\bm{k})$ of (11.70) is used as the initial distribution function. However, in view of the discontinuity of function $f_\ell(\bm{k})$ (11.73) at the Brillouin zone boundary, if it is used as the distribution function additional terms will occur in the averages of Eqs. (11.29) and (11.31). Therefore, taking (11.72) as ρ_0 or $f_\ell(\bm{k})$ of (11.73) as the distribution function is inconvenient, unless the energy is much higher at the Brillouin boundary that carriers rarely go there.

The effect of dynamic Coulomb screening of electrons can be considered under random phase approximation (RPA). This is achieved by multiplying the following factor (Lei, 1992a) inside the summation signs of the expressions of a_i, a_p and w,

$$|\varepsilon(\boldsymbol{q}, \bar{\varepsilon}(\boldsymbol{k}) - \bar{\varepsilon}(\boldsymbol{k}+\boldsymbol{q}))|^{-2}. \tag{11.87}$$

Here

$$|\varepsilon(\boldsymbol{q}, \omega)|^2 = \left[1 - \nu_\text{c}(q) \sum_{\boldsymbol{k}} \Pi_1^0(\boldsymbol{k}, \boldsymbol{q}, \omega)\right]^2 + \left[\nu_\text{c}(q) \sum_{\boldsymbol{k}} \Pi_2^0(\boldsymbol{k}, \boldsymbol{q}, \omega)\right]^2, \tag{11.88}$$

$\Pi_1^0(\boldsymbol{k}, \boldsymbol{q}, \omega)$ and $\Pi_2^0(\boldsymbol{k}, \boldsymbol{q}, \omega)$ are the real and imaginary parts of the electron density correlation function

$$\Pi^0(\boldsymbol{k}, \boldsymbol{q}, \omega) = 2|g(\boldsymbol{k}, \boldsymbol{q})|^2 \frac{f(\bar{\varepsilon}(\boldsymbol{k}+\boldsymbol{q}), T_\text{e}) - f(\bar{\varepsilon}(\boldsymbol{k}), T_\text{e})}{\bar{\varepsilon}(\boldsymbol{k}+\boldsymbol{q}) - \bar{\varepsilon}(\boldsymbol{k}) + \omega + \text{i}\delta}, \tag{11.89}$$

and $\nu_\text{c}(q) = e^2/(\varepsilon_0 \kappa q^2)$ is the Coulomb potential.

11.3.2 Lattice momentum and energy balance equations

The equation (11.32) for the momentum change rate can be replaced by the equation (11.44) for the lattice momentum change rate. The statistical averages of Eqs.(11.44) and (11.34) give the lattice momentum balance equation and energy balance equation:

$$\frac{d\boldsymbol{p}_\ell}{dt} = e\boldsymbol{E} \cdot \mathcal{R} + \boldsymbol{f}^\ell, \tag{11.90}$$

$$\frac{d\varepsilon_\text{e}}{dt} = e\boldsymbol{E} \cdot \boldsymbol{v}_\text{d} - w. \tag{11.91}$$

In this,

$$\boldsymbol{p}_\ell = \frac{\langle \boldsymbol{P}_\ell \rangle}{N} = \frac{2}{N} \sum_{\boldsymbol{k}} \boldsymbol{p}(\boldsymbol{k}) f(\bar{\varepsilon}(\boldsymbol{k}), T_\text{e}) \tag{11.92}$$

is the average lattice momentum per carrier,

$$\mathcal{R} = \langle \hat{\mathcal{R}} \rangle = \mathcal{I} - \frac{1}{4\pi^3 n} \oint_{S_\text{BZ}} d\boldsymbol{s}\, \boldsymbol{k} f(\bar{\varepsilon}(\boldsymbol{k}), T_\text{e}) \tag{11.93}$$

is a dimensionless tensor factor, $\boldsymbol{f}^\ell = \boldsymbol{f}_\text{i}^\ell + \boldsymbol{f}_\text{p}^\ell$ is the frictional force related to lattice momentum change due to impurity and phonon scatterings, given by

$$f_i^\ell = \frac{2\pi n_i}{N} \sum_{k,q} |u(q)|^2 |g(k,q)|^2 [p(k+q) - p(k)] \delta(\varepsilon(k+q) - \varepsilon(k))$$
$$\times [f(\bar\varepsilon(k), T_e) - f(\bar\varepsilon(k+q), T_e)], \quad (11.94)$$

$$f_p^\ell = \frac{4\pi}{N} \sum_{k,q,\lambda} |M(q,\lambda)|^2 |g(k,q)|^2 [p(k+q) - p(k)]$$
$$\times \delta(\varepsilon(k+q) - \varepsilon(k) + \Omega_{q\lambda}) [f(\bar\varepsilon(k), T_e) - f(\bar\varepsilon(k+q), T_e)]$$
$$\times \left[n\left(\frac{\Omega_{q\lambda}}{T}\right) - n\left(\frac{\bar\varepsilon(k) - \bar\varepsilon(k+q)}{T_e}\right) \right]. \quad (11.95)$$

Here the energy function $\bar\varepsilon(k)$ is still

$$\bar\varepsilon(k) = \varepsilon(k - p_d) \quad (11.96)$$

for the initial distribution function $f_d(k)$ of (11.65), or

$$\bar\varepsilon(k) = \varepsilon(k) - p_v \cdot v(k) \quad (11.97)$$

for the initial distribution function $f_v(k)$ of (11.70). The lattice-momentum balance equation is the modification and extension of the equation proposed by Büttiker and Thomas (1977). The tensor factor $\mathcal{R}$, which we call Büttiker–Thomas reduction tensor, reflects the effect of Bragg scattering: when an electron in the k space is accelerated by the electric field to the Brillouin zone boundary, its lattice momentum will lose a reciprocal lattice vector, resulting in a reduction of the electric-field acceleration effect on the lattice momentum. The second term on the right hand side of Eq. (11.93), i.e. the closed area integral over the whole boundary of the Brillouin, just reflects this effect. In addition to this, the Bragg scattering also occurs in the expressions of frictional forces f_i^ℓ and f_p^ℓ. Since when $k+q$ is outside the BZ, $p(k+q) - p(k) = q - G$ (G is a reciprocal lattice vector). That is to say, when an electron in state k is scattered to state $k+q$ outside the BZ, its lattice momentum will lose a magnitude of a reciprocal vector.

In the original derivation Büttiker and Thomas (1977) employ a frictional force of constant relaxation time form, $f^\ell = -\gamma p_\ell$ (γ is a constant), and disregard the difference between the average lattice momentum p_ℓ and average momentum (or the average velocity v_d). They assume that v_d is proportional to p_ℓ:

$$v_d = p_\ell / m \quad (11.98)$$

(m is a constant). This relation, though correct in parabolic band system, can not be justified in an arbitrary energy band. An example will be given in Sec. 12.5.

The equation (11.75) for the rate of change of average momentum (or average velocity) contains the inverse effective mass tensor $\mathcal{K}$, which, given by (11.78) as a volume integral over the whole Brillouin zone, is also relevant with Bragg scattering. In fact, if, due to the acceleration of an electric field or some other reason, more electrons occupy the $\bm{k}$ states above the inflection point of the energy spectrum, $\mathcal{K}$ can become negative. This is the result of Bragg scattering.

The hot electron transport in an arbitrary energy band can be studied either from the momentum balance equation (11.75) and energy balance equation (11.76), or from the lattice momentum balance equation (11.90) and energy balance equation (11.91) [(11.76)]. Mathematically, these two sets of equations are not equivalent. But numerical analyses for electron transport in superlattice minibands indicate that the predictions of major transport quantities derived from these two sets of equations are quite close (see Sec. 12.5).

11.4 Boltzmann Equation and Balance Equation, Approximate Distribution Functions

11.4.1 *Derivation of momentum, lattice momentum and energy balance equations from Boltzmann equation*

The momentum balance equation, lattice momentum balance equation and energy balance equation obtained in the last section can also be derived from the moment equations of the Boltzmann equation.

In the energy band theory the state of an electron moving in a single energy band is characterized by a lattice wavevector $\bm{k}$ in the Brillouin zone. We will employ the periodic zone scheme, i.e. allowing the lattice wavevector $\bm{k}$ to extend to the whole $\bm{k}$ space outside the Brillouin zone in such a way that for any lattice wavevector $\bm{k}$ and any reciprocal lattice vector $\bm{G}$, $\bm{k}$ and $\bm{k}+\bm{G}$ represent the same electron state. In this periodic zone description, any function of the electron state, such as the energy function $\varepsilon(\bm{k})$ and the velocity function $\bm{v}(\bm{k})$, must be a periodic function of $\bm{k}$ in the whole extended $\bm{k}$ space. The energy function and the velocity function, which are related to directly observable physical quantities, are also continuous functions of k in the whole extended k space.

For a spatially homogeneous system under the influence of a uniform electric field E, the Boltzmann equation is

$$\frac{d}{dt} f_{\bm k} = -e\bm E \cdot \nabla_{\bm k} f_{\bm k} + \left[\frac{\partial f_{\bm k}}{\partial t}\right]_{\rm c}. \tag{11.99}$$

Here the distribution function $f_{\bm k}$ is assumed to be a periodic and continuous function of $\bm k$ with $f_{\bm k}\, d^3r d^3k/(4\pi^3)$ being the particle number in the phase space $d^3r d^3k$, and $[\partial f_{\bm k}/\partial t]_{\rm c}$ represents the collision term, i.e. the rate of change of particle number induced by the scatterings. We assume that the scatterings are due to randomly distributed impurities and phonons. Multiplying Eq. (11.99) with a scalar or vector function $g(\bm k)$ of the electron state and summing over all the states (i.e. integrating over $\bm k$ within the Brillouin zone and summing over the spin index), we obtain, after a simple transformation,

$$\frac{d}{dt} \int_{\rm BZ} g(\bm k) f_{\bm k} \frac{d^3k}{4\pi} = -\frac{e\bm E}{4\pi^3} \cdot \oint_{S_{\rm BZ}} d\bm s\, g(\bm k) f_{\bm k} + e\bm E \cdot \int_{\rm BZ} \nabla g(\bm k) f_{\bm k} \frac{d^3k}{4\pi^3}$$
$$+ \int_{\rm BZ} \left[\frac{\partial f_{\bm k}}{\partial t}\right]_{\rm c} g(\bm k) \frac{d^3k}{4\pi^3}. \tag{11.100}$$

This is the general moment equation derived from the Boltzmann equation. The first term on the right hand side of Eq. (11.100) is an area integral over the whole (closed) BZ boundary $S_{\rm BZ}$. Note that, as a function of the electron state, $g(\bm k)$ must be a periodic function of $\bm k$: $g(\bm k + \bm G) = g(\bm k)$, but not necessary a continuous function of $\bm k$.

Momentum balance equation. Taking $g(\bm k)$ to be the velocity function $\bm v(\bm k) \equiv \nabla \varepsilon(\bm k)$ [$m_e \bm v(\bm k)$ is the momentum function $\langle \bm k|p|\bm k\rangle$], we can see that the closed area integration on the right hand side of Eq. (11.100) vanishes,

$$\oint_{S_{\rm BZ}} d\bm s\, \bm v(\bm k) f_{\bm k} = 0. \tag{11.101}$$

This is because that $\bm v(\bm k)$ and $f_{\bm k}$ are periodic and continuous function of $\bm k$, that $\bm v(\bm k) f_{\bm k}$ takes the identical values at the two points separated by a reciprocal lattice vector on the Brillouin zone boundary. The Eq. (11.100) then becomes

$$\frac{d\bm v_{\rm d}}{dt} = e\bm E \cdot \mathcal{K} + \bm a, \tag{11.102}$$

in which
$$v_d = \frac{1}{4\pi^3 n} \int_{BZ} d^3k\, \boldsymbol{v}(\boldsymbol{k}) f_{\boldsymbol{k}} \qquad (11.103)$$
is the electron average velocity,
$$\mathcal{K} = \frac{1}{4\pi^3 n} \int_{BZ} d^3k\, [\boldsymbol{\nabla}\boldsymbol{v}(\boldsymbol{k})]\, f_{\boldsymbol{k}} \qquad (11.104)$$
is the inverse effective mass tensor,
$$n = \frac{1}{4\pi^3} \int_{BZ} d^3k\, f_{\boldsymbol{k}} \qquad (11.105)$$
is the electron density. The frictional acceleration $\boldsymbol{a}$ can be written as
$$\boldsymbol{a} = \frac{n_i}{4\pi^2 n} \int_{BZ} d^3k \sum_{\boldsymbol{q}} |u(\boldsymbol{q})|^2 |g(\boldsymbol{k},\boldsymbol{q})|^2 \, [\boldsymbol{v}(\boldsymbol{k}+\boldsymbol{q}) - \boldsymbol{v}(\boldsymbol{k})]$$
$$\times \delta(\varepsilon(\boldsymbol{k}+\boldsymbol{q}) - \varepsilon(\boldsymbol{k}))\,(f_{\boldsymbol{k}} - f_{\boldsymbol{k}+\boldsymbol{q}}) + \frac{1}{2\pi^2 n} \int_{BZ} d^3k$$
$$\times \sum_{\boldsymbol{q},\lambda} |M(\boldsymbol{q},\lambda)|^2 |g(\boldsymbol{k},\boldsymbol{q})|^2 \, [\boldsymbol{v}(\boldsymbol{k}+\boldsymbol{q}) - \boldsymbol{v}(\boldsymbol{k})] \,\delta(\varepsilon(\boldsymbol{k}+\boldsymbol{q}) - \varepsilon(\boldsymbol{k}) + \Omega_{\boldsymbol{q}\lambda})$$
$$\times [n(\Omega_{\boldsymbol{q}\lambda}/T)(f_{\boldsymbol{k}} - f_{\boldsymbol{k}+\boldsymbol{q}}) - f_{\boldsymbol{k}}(1 - f_{\boldsymbol{k}+\boldsymbol{q}})]. \qquad (11.106)$$

Lattice momentum balance equation. We can take $g(\boldsymbol{k})$ to be the lattice momentum function $\boldsymbol{p}(\boldsymbol{k})$ defined in Sec. 11.1.3, which has the property: $\boldsymbol{p}(\boldsymbol{k}) = \boldsymbol{k}$ for $\boldsymbol{k}$ inside the BZ, and for arbitrary $\boldsymbol{k}$ and any reciprocal lattice vector $\boldsymbol{G}$, $\boldsymbol{p}(\boldsymbol{k}+\boldsymbol{G}) = \boldsymbol{p}(\boldsymbol{k})$. Combining together the first and the second terms on the right hand side of Eq. (11.100), we can write
$$\frac{d\boldsymbol{p}_\ell}{dt} = e\boldsymbol{E}\cdot\mathcal{R} + \boldsymbol{f}^\ell, \qquad (11.107)$$
where
$$\boldsymbol{p}_\ell = \frac{1}{4\pi^3 n}\int_{BZ} d^3k\, \boldsymbol{k}\, f_{\boldsymbol{k}} \qquad (11.108)$$
is the average lattice momentum per carrier, and the tensor
$$\mathcal{R} = \mathcal{I} - \frac{1}{4\pi^3 n} \oint_{S_{BZ}} d\boldsymbol{s}\, \boldsymbol{k}\, f_{\boldsymbol{k}} \qquad (11.109)$$
is just the Büttiker–Thomas reduction tensor derived in Sec. 11.3 ($\mathcal{I}$ stands for the unit tensor).

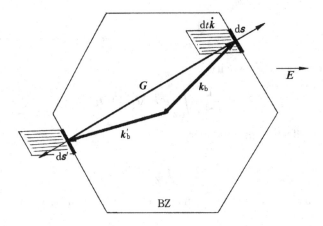

Fig. 11.2 Bragg scattering and reduction factor

The second term on the right hand side of (11.109) comes from the Bragg scattering related to the electric field acceleration. To have an intuitive idea, we consider an area element $d\bm{s}$ on the BZ boundary at position $\bm{k}_\mathrm{b}$ (see Fig. 11.2). Due to the electric field acceleration, $\dot{\bm{k}} = e\bm{E}$, during the time period dt, electrons of number $dt\,\dot{\bm{k}}\cdot d\bm{s}\,f_{\bm{k}}/(4\pi^3)$ reach the boundary $d\bm{s}$ from inside the BZ. As soon as an electron reaches the boundary $d\bm{s}$, its lattice momentum immediately loses a magnitude of the reciprocal lattice vector $\bm{G}$ and becomes $\bm{k}'_\mathrm{b} = \bm{k}_\mathrm{b} - \bm{G}$. Therefore, the rate of lattice momentum loss on the area element $d\bm{s}$ is

$$\dot{\bm{k}}\cdot d\bm{s}\,\bm{G}\,\frac{f_{\bm{k}}}{4\pi^3} = e\bm{E}\cdot\left(d\bm{s}\,\bm{k}_\mathrm{b} + d\bm{s}'\bm{k}'_\mathrm{b}\right)\frac{f_{\bm{k}}}{4\pi^3}.$$

Here $d\bm{s}' = -d\bm{s}$ is the area element at position $\bm{k}'_\mathrm{b}$ of the BZ boundary which is opposite to that of $\bm{k}_\mathrm{b}$. Since under the influence of a fixed-direction electric field acceleration the electrons can only reach a half of BZ boundary from the interior of the BZ, to obtain the total lattice-momentum loss rate it is needed only to integral the above expression over the half of the entire BZ boudary. The result is exactly what the second term of expression (11.109) states. Therefore, Eq. (11.107) is just the lattice momentum balance equation, formally the same as Eq. (11.90) derived in the preceding section, with the frictional force related to lattice momentum $\boldsymbol{f}^\ell$ given by

$$\boldsymbol{f}^\ell = \frac{n_i}{4\pi^2 n} \int_{BZ} d^3k \sum_{\boldsymbol{q}} |u(\boldsymbol{q})|^2 |g(\boldsymbol{k},\boldsymbol{q})|^2 \left[\boldsymbol{p}(\boldsymbol{k}+\boldsymbol{q}) - \boldsymbol{p}(\boldsymbol{k})\right]$$

$$\times \delta(\varepsilon(\boldsymbol{k}+\boldsymbol{q}) - \varepsilon(\boldsymbol{k}))(f_{\boldsymbol{k}} - f_{\boldsymbol{k}+\boldsymbol{q}}) + \frac{1}{2\pi^2 n} \int_{BZ} d^3k$$

$$\times \sum_{\boldsymbol{q},\lambda} |M(\boldsymbol{q},\lambda)|^2 |g(\boldsymbol{k},\boldsymbol{q})|^2 \left[\boldsymbol{p}(\boldsymbol{k}+\boldsymbol{q}) - \boldsymbol{p}(\boldsymbol{k})\right] \delta(\varepsilon(\boldsymbol{k}+\boldsymbol{q}) - \varepsilon(\boldsymbol{k}) + \Omega_{\boldsymbol{q}\lambda})$$

$$\times \left[n(\Omega_{\boldsymbol{q}\lambda}/T)(f_{\boldsymbol{k}} - f_{\boldsymbol{k}+\boldsymbol{q}}) - f_{\boldsymbol{k}}(1 - f_{\boldsymbol{k}+\boldsymbol{q}})\right]. \tag{11.110}$$

Energy balance equation. If we take $g(\boldsymbol{k})$ to be the energy function $\varepsilon(\boldsymbol{k})$, Eq. (11.100) becomes

$$\frac{d\varepsilon_e}{dt} = e\boldsymbol{E} \cdot \boldsymbol{v}_d - w, \tag{11.111}$$

in which

$$\varepsilon_e = \frac{1}{4\pi^3 n} \int_{BZ} d^3k \, \varepsilon(\boldsymbol{k}) f_{\boldsymbol{k}} \tag{11.112}$$

is the average electron energy, and

$$w = \frac{1}{2\pi^2 n} \int_{BZ} d^3k \sum_{\boldsymbol{q},\lambda} |M(\boldsymbol{q},\lambda)|^2 |g(\boldsymbol{k},\boldsymbol{q})|^2 \left[\varepsilon(\boldsymbol{k}+\boldsymbol{q}) - \varepsilon(\boldsymbol{k})\right]$$

$$\times \delta(\varepsilon(\boldsymbol{k}+\boldsymbol{q}) - \varepsilon(\boldsymbol{k}) + \Omega_{\boldsymbol{q}\lambda}) \left[n(\Omega_{\boldsymbol{q}\lambda}/T)(f_{\boldsymbol{k}} - f_{\boldsymbol{k}+\boldsymbol{q}}) - f_{\boldsymbol{k}}(1 - f_{\boldsymbol{k}+\boldsymbol{q}})\right]. \tag{11.113}$$

is the electron energy-loss rate.

11.4.2 Approximate distribution functions

The function $f_{\boldsymbol{k}}$ appearing in the formulas of the preceding subsection is the distribution function. From the point of view of Boltzmann equation theory, it is the solution of the Boltzmann equation (11.99) in the presence of the electric field $\boldsymbol{E}$. However, if the exact distribution function satisfying the Boltzmann equation is obtained, any transport quantity can be calculated directly from it and there is no need for the momentum and energy balance equations. The point is that the exact solution of the Boltzmann equation is very difficult, and the purpose of using momentum and energy balance equations is to avoid solving the Boltzmann equation. In fact, since a part of the main transport information has already been included in the structure of momentum and energy balance equations, it is possible to employ a structurally simpler parameterized approximate distribution function, rather than the exact distribution function, to serve as the starting point in the balance equation transport theory. The values of the involved parameters are determined by the balance equations.

The physical consideration for the selection of one kind of approximation distribution functions was discussed in Sec. 11.2. The system investigated consists of a large number of electrons subject to a periodic potential having the energy spectrum $\varepsilon(\mathbf{k})$. In the absence of the electric field, the electron–electron interaction promotes the system to rapidly thermalize, forming a Fermi distribution with respect to the energy spectrum $\varepsilon(\mathbf{k})$ at temperature T_e: $f(\varepsilon(\mathbf{k}), T_\mathrm{e}) = 1/\{\exp\left[(\varepsilon(\mathbf{k}) - \mu)/T_\mathrm{e}\right] + 1\}$. Strong electron–electron interaction implies that the system possesses a strong ability to keep this distribution unchanged. On the other hand, the presence of the electric field $\mathbf{E}$ forces electron to change its wavevector according to the acceleration theorem: $\dot{\mathbf{k}} = e\mathbf{E}$. If the time interval each electron can be accelerated freely by the electric field is τ (effective scattering time), then an electron originally in state $\mathbf{k}$ in the absence of the electric field will transit to state $\mathbf{k}+e\mathbf{E}\tau$ after turning on the electric field. The ability of keeping the original distribution unchanged means that the distribution function of electrons in the presence of the electric field can be obtained from the distribution function in the absence of the electric field by shifting it in $\mathbf{k}$ space an amount $e\mathbf{E}\tau$:

$$f_{\mathbf{k}} = f_d(\mathbf{k}) = f(\varepsilon(\mathbf{k} - \mathbf{p}_d), T_\mathrm{e}), \qquad (11.114)$$

in which $\mathbf{p}_d \equiv -e\tau\mathbf{E}$ and T_e are parameters to be determined. This is just the initial state selected in Sec. 11.2 when putting the electric field part H_E in the unperturbed Hamiltonian. Replacing $f_{\mathbf{k}}$ in all the expressions in the preceding section by $f_d(\mathbf{k})$, we get exactly the same equations as those derived in Sec. 11.3.

One may introduce an effective scattering time or relaxation time τ from the probability point of view: the probability for an electron to be accelerated freely up to time t is proportional to $\mathrm{e}^{-t/\tau}$. The distribution of these accelerated (time t) electrons is that of the original (without field) distribution $f(\varepsilon(\mathbf{k}), T_\mathrm{e})$ displaced an amount $e\mathbf{E}t = -\mathbf{p}_d t/\tau$ in the $\mathbf{k}$ space. By considering all possible t, the distribution function in the presence of the electric field should be taken as

$$f_{\bm{k}} = f_w(\bm{k}) = \frac{1}{\tau}\int_0^\infty e^{-t/\tau} f(\varepsilon(\bm{k}-\bm{p}_d t/\tau), T_e)\, dt$$
$$= \int_0^\infty e^{-\xi} f(\varepsilon(\bm{k}-\bm{p}_d\xi), T_e)\, d\xi. \tag{11.115}$$

This distribution function is proposed by Wu, Huang and Weng (1997). They point out that $f_w(\bm{k})$ is the solution of the steady state Boltzmann equation of relaxation time approximation,

$$e\bm{E}\cdot\nabla_{\bm{k}} f_{\bm{k}} = -\frac{f_{\bm{k}} - f(\varepsilon(\bm{k}), T_e)}{\tau}. \tag{11.116}$$

It is easy to verify that $f_w(\bm{k})$ satisfies Eq. (11.116). If one replaces all the $f_{\bm{k}}$ functions in the equations of the preceding section by $f_w(\bm{k})$, we obtain another set of balance equations somewhat different from those derived in Sec. 11.3, still using $\bm{p}_d$, T_e and μ as basic parameters.

Chapter 12

Miniband Transport in Semiconductor Superlattices

12.1 Superlattice Miniband

Semiconductor superlattices, which was first proposed by Esaki and Tsu (1970), is a system having one-dimensional periodical structure formed by the spatial variation of the alloy component or impurity density. As a model for theoretical investigation, we treat it as a periodical arrangement of two-dimensional potential wells and barriers, in which carriers move freely in the x–y plane while subject to a periodical potential along the z direction. The single electron states of a superlattice can be described by a miniband index α and a wavevector $\bm{k} = (k_x, k_y, k_z) = (\bm{k}_\parallel, k_z)$, where $-\infty < k_x, k_y < \infty$, $-\pi/d < k_z \leqslant \pi/d$, and d is the period of the superlattice. The Bloch wave function is given by (S is the system area in the x–y plane)

$$\psi_{\alpha\bm{k}}(\bm{r}) \equiv \psi_{\alpha\bm{k}}(\bm{r}_\parallel, z) = \frac{1}{S^{1/2}} e^{i\bm{k}_\parallel \cdot \bm{r}_\parallel} \varphi_{\alpha k_z}(z), \qquad (12.1)$$

and the eigen energy is

$$\varepsilon_\alpha(\bm{k}) = \varepsilon_{\bm{k}_\parallel} + \varepsilon_\alpha(k_z), \qquad (12.2)$$

with $\varepsilon_{\bm{k}_\parallel} = k_\parallel^2/m$, m being the effective mass of the carrier in the bulk material forming the superlattice. The envelope function $\varphi_{\alpha k_z}(z)$ is determined by the periodical potential of the superlattice along the z direction. In this chapter we mainly concentrate on the case where the carriers are mostly bound within a single quantum well, that the envelope function can be represented by the following tight-binding sum:

$$\varphi_{\alpha k_z}(z) = \frac{A_{\alpha k_z}}{L^{1/2}} \sum_l e^{i k_z l d} \phi_\alpha(z - ld), \qquad (12.3)$$

in which $A_{\alpha k_z}$ is a normalized coefficient and $\phi_\alpha(z)$ is the wave function of the single well locating at $z = 0$, having energy ε_α and normalized to $\int |\phi_\alpha(z)|^2 dz = d$. Under tight-binding and nearest neighbor overlapping approximation, the miniband energy spectrum can be written as

$$\varepsilon_\alpha(k_z) = \varepsilon_{\alpha 0} + \frac{\Delta_\alpha}{2}\left[1 + (-1)^\alpha \cos(k_z d)\right]. \tag{12.4}$$

Here Δ_α is the band width and $\varepsilon_{\alpha 0}$ is the energy bottom of the αth miniband. The normalized coefficient is

$$A_{\alpha k_z} = [1 + 2\lambda_\alpha \cos(k_z d)]^{-1/2}, \tag{12.5}$$

in which the overlapping integral of the neighboring well wave functions, $\lambda_\alpha = d^{-1}\int \phi_\alpha^*(z)\phi_\alpha(z-d)dz$, is a small quantity.

In most of the discussions in this chapter we consider only the lowest miniband ($\alpha = 1$). For simplicity we will omit the miniband index and write the electron energy spectrum as

$$\varepsilon(\mathbf{k}) = \varepsilon_{\mathbf{k}_\parallel} + \varepsilon(k_z), \tag{12.6}$$

$$\varepsilon(k_z) = \frac{\Delta}{2}\left[1 - \cos(k_z d)\right]. \tag{12.7}$$

Here we have chosen the energy zero at the bottom of the miniband (at $\mathbf{k}_\parallel = 0$ and $k_z = 0$), and use $\Delta \equiv \Delta_1$ to denote the miniband width.

Substituting the tight-binding wave function (12.3) into the expression

$$g(\mathbf{k}, \mathbf{q}) = \sum_{\mathbf{k'}} \int d\mathbf{r} e^{i\mathbf{q}\cdot\mathbf{r}} \psi_{\mathbf{k'}}^*(\mathbf{r})\psi_{\mathbf{k}}(\mathbf{r}), \tag{12.8}$$

we obtain the form factor for the superlattice miniband, which is approximately (neglecting the small contribution from the wavefunction overlap)

$$g(\mathbf{k}, \mathbf{q}) \approx \frac{1}{d}\int dz e^{i q_z z}|\phi(z)|^2 \equiv g(q_z), \tag{12.9}$$

depending only on q_z. Here $\phi(z)$ is the ground state wave function of the single well.

In Chapter 6 we investigated the in-plane (x and y directions) transport (transverse transport) of semiconductor superlattices. The present chapter will study properties of superlattice longitudinal transport, i.e. the transport along the growth axis (z direction) of a superlattice with wave function overlap between neighboring wells (tunneling superlattice), considering it to form minibands in the z direction. The transport is that of electrons in extended Bloch wave states of energy bands, so called Bloch miniband transport or miniband transport.

12.2 Esaki–Tsu Model of Miniband Transport

12.2.1 *The original analysis of Esaki and Tsu*

The earliest model investigating superlattice longitudinal transport was proposed by Esaki and Tsu in 1970. They consider a single electron moving in a one-dimensional energy band described by (12.7), assuming that at time $t = 0$ the electron is at the band bottom, with wavevector $\bm{k}_\parallel = 0$, $k_z = 0$ and velocity $v(k_z) = 0$. Under the influence of an electric field E along the z direction, according to the acceleration theorem, $\dot{k}_z = eE$ while $\bm{k}_\parallel$ keeps unchanged. If there is no scattering, at time t the electron will transit to the state of $\bm{k}_\parallel = 0$ and $k_z = eEt$, having velocity $v(k_z) = (\Delta d/2)\sin(eEdt)$ along the z direction. The presence of scattering indicates that electron can not be accelerated continuously by the electric field. If the scattering is characterized by a relaxation time τ, with the meaning that the probability an electron can be accelerated freely by the electric field up to the time t is proportional to $e^{-t/\tau}$, then the average electron velocity v_d is given by

$$\left(\frac{\Delta d}{2}\right)\frac{1}{\tau}\int_0^\infty dt\, e^{-t/\tau}\sin(eEdt),$$

or, the electron drift velocity equals

$$v_d = v_m \frac{E/E_\tau}{1 + (E/E_\tau)^2}. \qquad (12.10)$$

Here $v_m = \Delta d/2$ and $E_\tau = 1/ed\tau$. Note that, along the parallel direction the electron is not accelerated by the electric field and the scattering does not affect its in-plane motion. The relation (12.10) shows that when the electric field E equals E_τ, the average drift velocity v_d reaches a maximum value $v_m/2$, and then decreases with increasing electric field when $E > E_\tau$. The velocity v_d versus the electric field E as given by (12.10) is shown in Fig. 12.1, exhibiting negative differential velocity or negative differential mobility. According to (12.10), the longitudinal weak field mobility of a superlattice is $\mu_0 = e\tau\Delta d^2/2$. Despite that the specific form (12.10) of the drift velocity as a function of E is relevant with the miniband energy dispersion (12.7) of the tight-binding type, the essential fact that superlattice miniband transport exhibits negative differential mobility, even the qualitatively behavior of v_d–E similar to that shown in Fig. 12.1, remains true in the case of more general miniband energy dispersion (Esaki and Tsu, 1970, Lebwohl and Tsu, 1970).

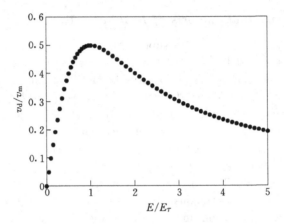

Fig. 12.1 $v_\mathrm{d}/v_\mathrm{m}$ versus E/E_τ as given by Esaki–Tsu formula (12.10).

12.2.2 *Calculation with carrier statistical distribution*

The original model of Esaki and Tsu (1970) deals with only one electron at the band bottom, without any statistical effect. This subsection will extend the Esaki-Tsu model to the case where the carrier occupation in the miniband exhibits a statistical distribution. We will follow the derivation of Huang and Zhu (1992).

We consider that at time $t = 0$ an electron is in the state $(\boldsymbol{k}_\parallel, k_z)$, having $v(k_z) = \partial \varepsilon / \partial k_z$ in the z direction. Under the influence of an electric field E in the z direction, at time t the axial (z direction) wavevector of the electron becomes $k_z + eEt$ (the transverse wavevector $\boldsymbol{k}_\parallel$ keeps unchanged), and the velocity becomes $v(k_z + eEt)$. In the presence of scatterings the electron has a finite lifetime during which it can be accelerated by the electric filed. If the scattering effect is described by a relaxation time τ that, under the scattering, the probability for an electron to survive up to the time t is $(1/\tau)\,\mathrm{e}^{-t/\tau}$, then an electron originally in state $(\boldsymbol{k}_\parallel, k_z)$ will get, due to the electric field acceleration, a velocity increment

$$\Delta \bar{v}(k_z) = \frac{1}{\tau} \int_0^\infty dt\, e^{-t/\tau} [v(k_z + eEt) - v(k_z)]$$

$$= \frac{v_m}{\tau} \int_0^\infty dt\, e^{-t/\tau} [\sin(k_z d + eEdt) - \sin(k_z d)]$$

$$= \frac{E/E_\tau}{1 + (E/E_\tau)^2} v_m [\cos(k_z d) - (E/E_\tau) \sin(k_z d)], \quad (12.11)$$

in which $v_m = \Delta d/2$ and $E_\tau \equiv (ed\tau)^{-1}$. In the absence of the electric field the distribution function of electrons is $f_0(\varepsilon)$ [$\varepsilon = \varepsilon_\| + \varepsilon(k_z)$ is the electron energy given by (12.6), $\varepsilon_\| = \varepsilon_{\mathbf{k}_\|} = k_\|^2/2m$], and the average drift velocity is zero. After turning on the electric field E in the z direction, the electron average drift velocity will be

$$v_d = \frac{m}{2\pi^2 n} \int_0^\infty d\varepsilon_\| \int_{-\pi/d}^{\pi/d} dk_z\, f_0(\varepsilon) \Delta \bar{v}(k_z) = v_p \frac{2E/E_\tau}{1 + (E/E_\tau)^2}, \quad (12.12)$$

$$v_p = \left(\frac{v_m}{2}\right) \frac{m}{2\pi^2 n} \int_0^\infty d\varepsilon_\| \int_{-\pi/d}^{\pi/d} dk_z\, f_0(\varepsilon) \cos(k_z d), \quad (12.13)$$

where the electron volume density n is given by

$$n = \frac{m}{2\pi^2} \int_0^\infty d\varepsilon_\| \int_{-\pi/d}^{\pi/d} dk_z\, f_0(\varepsilon). \quad (12.14)$$

The integral over $d\varepsilon_\| dk_z$ in (12.13) and (12.14) can be rewritten as the integration first over k_z along the iso-energy surface of ε, and then over the energy ε (Huang and Zhu, 1992). For instance,

$$v_p = \left(\frac{v_m}{2}\right) \frac{m}{\pi^2 n} \int_0^\infty f_0(\varepsilon)\, d\varepsilon \int_0^{k_{\max}(\varepsilon)} \cos(k_z d)\, dk_z$$

$$= \left(\frac{v_m}{2}\right) \frac{m}{\pi^2 n d} \int_0^\infty f_0(\varepsilon) \sin(k_{\max}(\varepsilon) d)\, d\varepsilon, \quad (12.15)$$

in which $k_{\max}(\varepsilon)$ denotes the maximum value of k_z in the iso-energy surface of ε. Apparently, when $0 < \varepsilon < \Delta$, $k_{\max}(\varepsilon)$ is determined by the equation

$$\varepsilon = \frac{\Delta}{2} [1 - \cos(k_{\max} d)]; \quad (12.16)$$

when $\varepsilon > \Delta$, $k_{\max}(\varepsilon) = \pi/d$. Therefore, the upper limit of the integral ε in (12.15) is in fact $\varepsilon = \Delta$. This shows that electrons with energy above the miniband width ($\varepsilon > \Delta$) do not contribute to the average velocity, or the current.

If the zero-temperature Fermi energy ε_F of the system locates above the band bottom ($\varepsilon_F > 0$), then at low temperature ($T \ll \varepsilon_F$)

$$\begin{aligned}v_p &= \left(\frac{v_m}{2}\right)\frac{m}{\pi^2 nd}\int_0^{\varepsilon_F}\sin(k_{\max}(\varepsilon)d)\,d\varepsilon \\ &= \left(\frac{v_m}{2}\right)\frac{m\Delta}{\pi^2 n}\int_0^{k_{\max}(\varepsilon_F)}\sin^2(kd)\,dk \\ &= \left(\frac{v_m}{2}\right)\frac{m\Delta}{4\pi^2 nd}\left[k_{\max}(\varepsilon_F)d - \frac{1}{2}\sin(2k_{\max}(\varepsilon_F)d)\right].\end{aligned} \quad (12.17)$$

ε_F relates to n by Eq. (12.14), which in the degenerate case can be written as

$$n = \frac{m\Delta}{2\pi^2 d}\left[k_{\max}(\varepsilon_F)d\,\cos(k_{\max}(\varepsilon_F)d) - \sin(k_{\max}(\varepsilon_F)d)\right]. \quad (12.18)$$

The current density $j = nev_d$, and the weak field ($E \approx 0$) conductivity σ is

$$\sigma = \frac{me^2\Delta^2 d\tau}{8\pi^2}\left[k_{\max}(\varepsilon_F)d - \frac{1}{2}\sin(k_{\max}(\varepsilon_F)d)\right]. \quad (12.19)$$

With increasing electron density the Fermi energy ε_F rises, and so does j at the beginning until the Fermi level increases to $\varepsilon_F = \Delta$ when $k_{\max}(\varepsilon_F)$ reaches the maximum value π/d. There the current j reaches the maximum $j_{\max} = me\Delta^2/(16\pi)$, and the weak field ($E \approx 0$) conductivity also have its maximum value (Huang and Zhu, 1992)

$$\sigma_{\max} = \frac{me^2\Delta^2 d\tau}{8\pi}. \quad (12.20)$$

On the other hand, if the electron distribution is nondegenerate, $f_0(\varepsilon) = C\exp(-\varepsilon/T)$, the electron density can be directly obtained form (12.14) as

$$n = C\frac{mT}{\pi d}e^{-\Delta/2T}I_0\left(\frac{\Delta}{2T}\right). \quad (12.21)$$

At the same time, according to (12.15)

$$\begin{aligned}v_p &= \left(\frac{v_m}{2}\right)\frac{Cm\Delta}{2\pi^2 dn}e^{-\Delta/2T}\int_0^\pi dz\,\exp\left(\frac{\Delta}{2T}\cos z\right)\sin^2 z \\ &= \left(\frac{v_m}{2}\right)\frac{I_1(\Delta/2T)}{I_0(\Delta/2T)},\end{aligned} \quad (12.22)$$

where $I_n(x)$ is the modified Bessel function of the nth order. The linear mobility in the nondegenerate case is obtained from (12.12) and (12.22) to be

$$\mu = \frac{e\tau \Delta d^2}{2} \frac{I_1(\Delta/2T)}{I_0(\Delta/2T)}, \qquad (12.23)$$

which approaches zero like Δ/T. This is the "thermal saturation" of the linear resistivity in superlattice miniband transport. Huang and Zhu (1992) provide a physically very clear explanation: only those electrons with energy ε locating within the range of miniband $(0 < \varepsilon < \Delta)$, can contribute to the current. As temperature increases, more carriers go to the states having energy higher than Δ, and the number of carriers within the minibnad decreases, leading to the mobility decreasing. Note that when $T \to 0$, (12.22) gives $v_\mathrm{p} = v_\mathrm{m}/2$, returning to the results of Esaki and Tsu (12.10). Therefore, the preliminary Esaki-Tsu model corresponds to the nondegenerate and zero-temperature limit of Huang and Zhu (1992).

12.3 Boltzmann Equation with Relaxation Time Approximation

12.3.1 *One-dimensional theory*

A similar analysis can be carried out from the Boltzmann equation with relaxation time approximation. The simplest way is to use a single relaxation time representing the collision term and write the Boltzmann equation in the form

$$\frac{\partial f_{\boldsymbol{k}}}{\partial t} + eE\frac{\partial f_{\boldsymbol{k}}}{\partial k_z} = -\frac{f_{\boldsymbol{k}} - f_0(\varepsilon)}{\tau}, \qquad (12.24)$$

where $f_0(\varepsilon)$ is the distribution function in thermal equilibrium, $f_{\boldsymbol{k}}$ is the distribution function in the presence of an electric field E in the z direction. The relaxation time τ is assumed to be a constant.

At low electric field, to its linear order the distribution function for the steady transport state is

$$f_{\boldsymbol{k}} = -eE\tau \left(\frac{df_0(\varepsilon)}{d\varepsilon}\right)\left(\frac{\partial \varepsilon}{\partial k_z}\right), \qquad (12.25)$$

giving the electron average drift velocity

$$v_\mathrm{d} = \frac{2eE\tau}{N} \sum_{\mathbf{k}} \left(-\frac{df_0(\varepsilon)}{d\varepsilon}\right)\left(\frac{\partial\varepsilon}{\partial k_z}\right)^2, \qquad (12.26)$$

and the linear conductivity

$$\sigma = \frac{me^2\Delta^2 d^2\tau}{8\pi^2} \int_0^\infty d\varepsilon_\| \int_{-\pi/d}^{\pi/d} dk_z \left(-\frac{df_0(\varepsilon)}{d\varepsilon}\right)\sin^2(k_z d). \qquad (12.27)$$

Eq. (12.27) reduces to (12.19) in the degenerate limit.

If the thermal equilibrium distribution function of electrons is nondegenerate, $f_0(\varepsilon) \sim \exp(-\varepsilon/T)$, Eq. (12.24) can be solved analytically (Bass and Rubinstein, 1977; Xia and Zhu, 1995), yielding an expression for the drift velocity exactly the same as given in the preceding section:

$$v_\mathrm{d} = v_\mathrm{p} \frac{2E/E_\tau}{1 + (E/E_\tau)^2}, \qquad (12.28)$$

where

$$E_\tau = (ed\tau)^{-1}, \qquad (12.29)$$

$$v_\mathrm{p} = \frac{\Delta d}{4}\frac{I_1(\Delta/2T)}{I_0(\Delta/2T)}. \qquad (12.30)$$

Further analysis (Ktitorov et al, 1972; Ignatov and Shashkin, 1983; Suris and Shchamkhalova, 1984) introduces two relaxation times in the collision term of the Boltzmann equation: inelastic scattering time (or energy relaxation time) τ_ε and elastic scattering time τ_el. The Boltzmann equation for superlattice miniband transport under the influence of an electric field E along the z direction is written as

$$\frac{\partial f_\mathbf{k}}{\partial t} + eE\frac{\partial f_\mathbf{k}}{\partial k_z} = -\frac{f_\mathbf{k} - f_0(\varepsilon)}{\tau_\varepsilon} - \frac{f_{\mathbf{k}_\|,k_z} - f_{\mathbf{k}_\|,-k_z}}{2\tau_\mathrm{el}}. \qquad (12.31)$$

For systems with minibands described by (12.6) and (12.7), the steady state solution of the equation still gives the following drift velocity as a function of the electric field:

$$v_\mathrm{d} = v_\mathrm{p} \frac{2E/E_\tau}{1 + (E/E_\tau)^2}, \qquad (12.32)$$

but with

$$E_\tau = 1/[ed(\tau_\varepsilon \tau)^{1/2}], \tag{12.33}$$

$$v_p = \frac{\Delta d}{4}\frac{\mathrm{I}_1(\Delta/2T)}{\mathrm{I}_0(\Delta/2T)}\left(\frac{\tau}{\tau_\varepsilon}\right)^{1/2}, \tag{12.34}$$

and

$$\frac{1}{\tau} = \frac{1}{\tau_\varepsilon} + \frac{1}{\tau_{\mathrm{el}}}. \tag{12.35}$$

12.3.2 Three-dimensional theory

All the above theories yield exactly the same velocity–field relation as that of the Esaki-Tsu original model [(12.10), (12.12), (12.28) and (12.32)], and essentially the same critical field $E_c = E_\tau$ [(12.29) and (12.33)], independent of the miniband width Δ. The direct calculation using Monte-Carlo method (Price, 1973; Anderson and Aas, 1973) without relaxation time approximation also gave similar result for the drift velocity.

It was almost 20 years after the proposal of the superlattice idea by Esaki and Tsu, when people really saw the negative differential velocity in the longitudinal conduction related to miniband transport in semiconductor superlattices (Sibille *et al*, 1990; Grahn *et al*, 1991). The experimentally observed drift velocity v_d versus the electric field, though essentially in agreement with the prediction of the Esaki-Tsu model, exhibits much more complicated behavior than the relation (12.6). For instance, the electric field at which the velocity reaches the maximum, i.e. the critical field E_c, is found to change with miniband width Δ.

These experimental findings stimulated further investigations (Lei, Horing and Cui, 1991; Ignatov, Dodin and Shashkin, 1991; Huang and Zhu, 1992) on the theory of miniband transport. Starting from the balance equations introduced in Chapter 11, Lei, Horing and Cui (1991) systematically analyzed v_d–E behavior of miniband transport in GaAs-based superlattices, predicting the peak velocity v_p and the critical field E_c as functions of the miniband width in agreement with experiments. Their balance-equation treatment will be discussed in more detail in the next section. Here, we would like to point out that the main physics for reaching such a conclusion is to take into account the three-dimensional characteristics of realistic superlattices.

As stated before, the superlattices introduced in Sec. 12.1 is a three-dimensional system. So far in all the treatments in this Chapter, however, the 3D feature of the system shows up only in the carrier occupation $f_0(\varepsilon)$ of the equilibrium state, which distributes according to the 3D energy dispersion $\varepsilon(\boldsymbol{k}) = \varepsilon_\parallel + \varepsilon(k_z)$. Under the influence of an axial electric field, the electrons are accelerated along the z direction (k_z change only), and the collisions induce $f_{\boldsymbol{k}}$ to relax only towards $f_0(\varepsilon(\boldsymbol{k}))$, without coupling between the axial and the transverse degrees of freedom. Even in the Boltzmann equation (12.31) of two relaxation times, the correlation happens only between the z direction and the $-z$ direction. Therefore, all these theories are in fact one-dimensional in nature. The universal v_d–E behavior of (12.10) is just the feature of one-dimensional models.

Gerhardts (1993) presents a three-dimensional Boltzmann theory for nonlinear miniband transport. He writes the Boltzmann equation of relaxation time approximation in the form

$$eE\frac{\partial f_{\boldsymbol{k}}}{\partial k_z} = -\frac{f_{\boldsymbol{k}} - f_0(\varepsilon(\boldsymbol{k}))}{\tau_\varepsilon} - \frac{f_{\boldsymbol{k}} - \Phi_f(\varepsilon)}{\tau_\mathrm{el}}, \qquad (12.36)$$

in which

$$\Phi_f(\varepsilon) = \frac{1}{4\pi^3}\int d^3k'\,\delta(\varepsilon - \varepsilon(\boldsymbol{k}'))f_{\boldsymbol{k}'}/D(\varepsilon) \qquad (12.37)$$

is the average of the distribution function taken over the surface of constant energy, $\varepsilon(\boldsymbol{k}') = \varepsilon$, $D(\varepsilon) = (2m/\pi^2 d)z(\varepsilon)$ is the density of states, and function

$$z(\varepsilon) = \begin{cases} \arcsin\sqrt{\varepsilon/\Delta}, & 0 < x < \Delta, \\ \pi/2, & x \geqslant \Delta. \end{cases} \qquad (12.38)$$

This ansatz of the collision term describes back and forth scatterings with equal weights between the state $\boldsymbol{k}$ and all states $\boldsymbol{k}'$ of the same energy at different directions. It effectively couples the motion along the superlattice axis to the lateral degrees of freedom. Denoting $\xi = E/E_\tau$ ($E_\tau = ed\tau$, $\tau^{-1} = \tau_\varepsilon^{-1} + \tau_\mathrm{el}^{-1}$), Eq. (12.36) has the formal solution

$$f_{\boldsymbol{k}} = \left(\frac{d}{\xi}\right)e^{k_z d/\xi}\int_{k_z}^\infty dk'_z\, e^{-k'_z d/\xi}g(\varepsilon(\boldsymbol{k})), \qquad (12.39)$$

$$g(\varepsilon) = (1 - r_\mathrm{e})f_0(\varepsilon) + r_\mathrm{e}\Phi_f(\varepsilon), \qquad (12.40)$$

where $r_\mathrm{e} = \tau/\tau_\mathrm{el}$. Inserting (12.39) into the definition (12.37), we obtain an integral equation for $g(\varepsilon)$ which can be written in the form

$$g(\varepsilon) - (1 - r_\mathrm{e})f_0(\varepsilon) = \frac{r_\mathrm{e}}{z(\varepsilon)}\int_{-z(\varepsilon)}^{z(\varepsilon)} d\chi \int_0^\infty du\, e^{-2u}g(\varepsilon - \Delta b), \qquad (12.41)$$

with $b \equiv \sin^2 \chi + \sin^2(\chi + \xi u)$. The drift velocity is calculated according to

$$v_{\mathrm{d}} = \frac{1}{4\pi^2 n} \int d^3k \, v(k_z) f_{\boldsymbol{k}}, \qquad (12.42)$$

where

$$n = \frac{1}{4\pi^2} \int d^3k f_{\boldsymbol{k}} = \int_0^\infty d\varepsilon \, D(\varepsilon) \, \Phi_f(\varepsilon) \qquad (12.43)$$

is the electron density, which is the same as that in the absence of the electric field. Inserting the formal solution (12.39) into (12.42), we have

$$v_{\mathrm{d}} = \left(\frac{\Delta d}{2}\right) \frac{E/E_\tau}{1 + (E/E_\tau)^2} Q, \qquad (12.44)$$

in which

$$Q \equiv Q(r_{\mathrm{e}}, \xi, t) = \frac{\displaystyle\int_0^\Delta d\varepsilon \, g(\varepsilon) \left[(\varepsilon/\Delta)(1 - (\varepsilon/\Delta))\right]^{1/2}}{\displaystyle\int_0^\infty d\varepsilon \, z(\varepsilon) f_0(\varepsilon)}, \qquad (12.45)$$

depending on r_{e}, ξ and $t = \Delta/T$. In the absence of elastic scattering, $r_{\mathrm{e}} = 0$, one has $g(\varepsilon) = f_0(\varepsilon)$, and Q is independent of ξ. For nondegenerate distribution $f_0(\varepsilon) = C \exp(-\varepsilon/T)$, $Q = \mathrm{I}_1(1/2t)/\mathrm{I}_0(1/2t) = Q(0, 0, t)$, exactly recovering (12.28). In the presence of an elastic scattering $r_{\mathrm{e}} > 0$, $g(\varepsilon)$ and Q depend on $\xi \equiv E/E_\tau$. Then the shape of the v_{d}–E curves is different from the Esaki-Tsu expression and changes with changing temperature. Two limits can be examined analytically and return to one-dimensional result. In the linear response regime, $\xi \ll 1$, one obtains from Eq. (12.41) $g = f_0 + o(\xi^2)$. Thus, up to the first order in ξ, the distribution function and the drift velocity depend only on the total scattering rate, and it is enough to take the approximation $Q = Q(0, 0, t)$. At high temperatures when $f_0(\varepsilon)$ becomes a constant independent of ε, the solution of Eq. (12.41) is the constant $g_f = f_0$ and again $Q = Q(0, 0, t)$ is sufficient.

In the case having elastic scattering, Gerhardts (1993) calculates the nonlinear conduction for nondegenerate statistics by numerically solved the integral equation (12.41). Fig. 12.2 shows the calculated drift velocity $v_{\mathrm{d}}/v_{\mathrm{m}}$ ($v_{\mathrm{m}} = \Delta d/2$) versus the electric field E/E_τ ($E_\tau = 1/ed\tau$) for three different values of $T/\Delta = 0.1, 0.5$ and 1.0, and three different elastic scattering ratios $r_{\mathrm{e}} = 0$ (dashed lines), 0.5 (solid lines) and 0.9 (dash-dotted lines) while keeping the total scattering rate $1/\tau$ fixed. The $v_{\mathrm{d}}/v_{\mathrm{m}}$–$E/E_\tau$ curves are apparently different at different ratios of the elastic scattering. Only in the case of $r_e = 0$, E_τ equals the critical electric field E_c at which the drift velocity reaches maximum. In the presence of elastic scattering ($r_e > 0$) E_c deviates from E_τ and depends on T/Δ, as shown in Fig. 12.3.

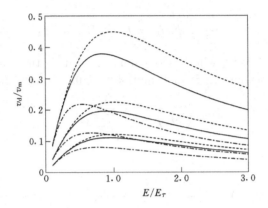

Fig. 12.2 v_d/v_m versus E/E_τ obtained from 3D Boltzmann theory, for $r_e \equiv \tau_{el}/\tau = 0$ (dashed lines), 0.5 (solid lines) and 0.9 (dash-dotted lines) while keeping τ fixed, and in each case, for $T/\Delta = 0.1$, 0.5 and 1.0 (from top to bottom). Here $v_m = \Delta d/2$ and $E_\tau = 1/ed\tau$. From Gerhardts (1993).

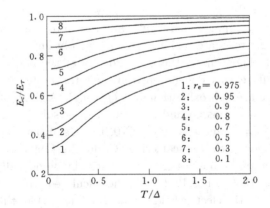

Fig. 12.3 The critical electric field E_c normalized by E_τ versus the reduced temperature T/Δ for several values of r_e. From Gerhardts (1993).

12.4 Superlattice Miniband Transport Treated with Momentum and Energy Balance Equations

The superlattice miniband transport can be investigated by different sets of balance equations for arbitrary energy band developed in Chapter 11. In comparison with methods discussed in the last two sections the balance equation approach has following features: (i) It goes beyond the relaxation time approximation by directly treating the microscopic scattering mechanisms in realistic systems; (ii) It explicitly considers the heating of the electron system due to the application of the electric field; (iii) It includes the coupling between the longitudinal and transverse motions induced by the scattering, therefore is truly three-dimensional in nature. In view of these advantages, when the Esaki-Tsu miniband conduction was experimentally observed in the early 1990s, the balance equation approach was immediately employed in a systematic theoretical investigation of miniband transport in GaAs-based superlattices, obtaining results in agreement with experimental observations (Lei, Horing and Cui, 1991, 1992).

We will first treat miniband transport using the momentum and Energy Balance Equations (11.75) and (11.76).

For a superlattice miniband characterized by the energy spectrum (12.6) and (12.7), when an electric field $\boldsymbol{E} = (0, 0, E)$ is applied along the growth axis (z direction) the electron average drift velocity and the scattering-induced frictional accelerations are in the z direction: $\boldsymbol{v} = (0, 0, v)$, $\boldsymbol{a}_\mathrm{i} = (0, 0, a_\mathrm{i})$, $\boldsymbol{a}_\mathrm{p} = (0, 0, a_\mathrm{p})$. The momentum and energy balance equations can be written as

$$\frac{dv_\mathrm{d}}{dt} = \frac{eE}{m_z^*} + a_\mathrm{i} + a_\mathrm{p}, \tag{12.46}$$

$$\frac{d\varepsilon_\mathrm{e}}{dt} = eEv_\mathrm{d} - w, \tag{12.47}$$

where

$$v_\mathrm{d} = \frac{2}{N} \sum_{\boldsymbol{k}} v(k_z) f(\bar{\varepsilon}(\boldsymbol{k}), T_\mathrm{e}) \tag{12.48}$$

is the average drift velocity [$v(k_z) = d\varepsilon(k_z)/dk_z$ is the velocity function of electrons in the miniband],

$$1/m_z^* = \mathcal{K}_{zz} = \frac{2}{N} \sum_{\boldsymbol{k}} \frac{d^2\varepsilon(k_z)}{dk_z^2} f(\bar{\varepsilon}(\boldsymbol{k}), T_\mathrm{e}) \tag{12.49}$$

is the zz component of the inverse effective mass tensor $\mathcal{K}$, and $\varepsilon_e = \varepsilon_\| + \varepsilon_z$ is the average electron energy per carrier with

$$\varepsilon_\| = \frac{2}{N} \sum_{\bm{k}} \frac{k_\|^2}{2m} f(\bar{\varepsilon}(\bm{k}), T_e), \tag{12.50}$$

$$\varepsilon_z = \frac{2}{N} \sum_{\bm{k}} \varepsilon(k_z) f(\bar{\varepsilon}(\bm{k}), T_e). \tag{12.51}$$

The total number of carriers expressed by the equation

$$N = 2 \sum_{\bm{k}} f(\bar{\varepsilon}(\bm{k}), T_e) \tag{12.52}$$

determines the chemical potential μ. The frictional accelerations a_i and a_p, and electron energy loss rate w per carrier are given by

$$a_i = \frac{2\pi n_i}{N} \sum_{\bm{k},\bm{q}} |\bar{u}(\bm{q})|^2 g(q_z) [v(k_z + q_z) - v(k_z)] \delta(\varepsilon(\bm{k} + \bm{q}) - \varepsilon(\bm{k}))$$
$$\times [f(\bar{\varepsilon}(\bm{k}), T_e) - f(\bar{\varepsilon}(\bm{k} + \bm{q}), T_e)], \tag{12.53}$$

$$a_p = \frac{4\pi}{N} \sum_{\bm{k},\bm{q},\lambda} |\bar{M}(\bm{q},\lambda)|^2 g(q_z) [v(k_z + q_z) - v(k_z)] \delta(\varepsilon(\bm{k} + \bm{q}) - \varepsilon(\bm{k}) + \Omega_{\bm{q}\lambda})$$
$$\times [f(\bar{\varepsilon}(\bm{k}), T_e) - f(\bar{\varepsilon}(\bm{k} + \bm{q}), T_e)] \left[n\left(\frac{\Omega_{\bm{q}\lambda}}{T}\right) - n\left(\frac{\bar{\varepsilon}(\bm{k}) - \bar{\varepsilon}(\bm{k} + \bm{q})}{T_e}\right) \right], \tag{12.54}$$

$$w = \frac{4\pi}{N} \sum_{\bm{k},\bm{q},\lambda} |\bar{M}(\bm{q},\lambda)|^2 g(q_z) \Omega_{\bm{q}\lambda} \delta(\varepsilon(\bm{k} + \bm{q}) - \varepsilon(\bm{k}) + \Omega_{\bm{q}\lambda})$$
$$\times [f(\bar{\varepsilon}(\bm{k}), T_e) - f(\bar{\varepsilon}(\bm{k} + \bm{q}), T_e)] \left[n\left(\frac{\Omega_{\bm{q}\lambda}}{T}\right) - n\left(\frac{\bar{\varepsilon}(\bm{k}) - \bar{\varepsilon}(\bm{k} + \bm{q})}{T_e}\right) \right]. \tag{12.55}$$

Here, we have assumed the tight-binding wave function and the approximate form factor (12.9) depending only on q_z for the miniband. Note that, since electrons are limited to move within a single miniband the summation over k_z is limited to the range of $(-\pi/d, \pi/d)$. The summation over q_z, however, is not so restricted. The upper limit of q_z is determined mainly by the rapid decay of the form factor $g(q_z)$ at large $|q_z|$. In addition to these, in the treatment of this chapter, the effect of dynamical screening induced by the electron–electron Coulomb interaction is neglected, and the static screening is considered to be included in the electron–impurity and electron–phonon scattering matrix elements: $|\bar{u}(q)|^2 = |u(q)|^2/|\varepsilon(q,0)|^2$, $|\bar{M}(\boldsymbol{q},\lambda)|^2 = |M(\boldsymbol{q},\lambda)|^2/|\varepsilon(q,0)|^2$. The initial distribution function

$$f(\bar{\varepsilon}(\boldsymbol{k}), T_e) = 1/\left\{\exp\left[(\bar{\varepsilon}(\boldsymbol{k}) - \mu)/T_e\right] + 1\right\} \tag{12.56}$$

in the above equations can have two choices: (11.85) or (11.86). We first use $f_d(\boldsymbol{k})$ function, i.e. with a lattice-momentum shifted $\bar{\varepsilon}(\boldsymbol{k})$ in (12.56):

$$\bar{\varepsilon}(\boldsymbol{k}) = \varepsilon(\boldsymbol{k} - \boldsymbol{p}_d). \tag{12.57}$$

For the superlattice longitudinal transport induced by an axial electric field, $\boldsymbol{p}_d = (0, 0, p_d)$ is along the z direction. Denoting $z_d \equiv p_d d$, we can write the drift velocity, the zz component of the inverse effective mass tensor and the average electron energy in the form:

$$v_d = v_m \alpha(T_e) \sin(z_d), \tag{12.58}$$

$$1/m_z^* = (1/M^*)\alpha(T_e)\cos(z_d), \tag{12.59}$$

$$\varepsilon_z = \frac{\Delta}{2}\left[1 - \alpha(T_e)\cos(z_d)\right], \tag{12.60}$$

$$\varepsilon_\| = \frac{\Delta}{2}\beta(T_e), \tag{12.61}$$

in which $v_m \equiv \Delta d/2$, $1/M^* \equiv \Delta d^2/2$,

$$\alpha(T_e) = \frac{2}{N}\sum_{\boldsymbol{k}}\cos(k_z d)f(\varepsilon(\boldsymbol{k}), T_e), \tag{12.62}$$

$$\beta(T_e) = \frac{2}{N}\sum_{\boldsymbol{k}}\frac{k_\|^2}{m\Delta}f(\varepsilon(\boldsymbol{k}), T_e). \tag{12.63}$$

The equation (12.52) for the total particle number and the expressions (12.62) and (12.63) for $\alpha(T_e)$ and $\beta(T_e)$ can be written in numerically more convenient forms:

$$1 = \left(\frac{T_e}{\varepsilon_{F0}^{(2)}}\right)\frac{1}{\pi}\int_0^\pi d\theta \ln\left\{1 + \exp\left[\frac{\mu}{T_e} - \frac{\Delta}{2T_e}(1 - \cos\theta)\right]\right\}, \tag{12.64}$$

$$\alpha(T_e) = \left(\frac{T_e}{\varepsilon_{F0}^{(2)}}\right)\frac{1}{\pi}\int_0^\pi d\theta \ln\left\{1 + \exp\left[\frac{\mu}{T_e} - \frac{\Delta}{2T_e}(1 - \cos\theta)\right]\right\}\cos\theta, \tag{12.65}$$

$$\beta(T_e) = \left(\frac{T_e}{\varepsilon_{F0}^{(2)}}\right)\left(\frac{2T_e}{\Delta}\right)\frac{1}{\pi}\int_0^\pi d\theta\, F_1\left(\frac{\mu}{T_e} - \frac{\Delta}{2T_e}(1 - \cos\theta)\right), \tag{12.66}$$

where

$$F_1(y) \equiv \int_0^\infty \frac{x\,dx}{\exp(x-y)+1}, \qquad (12.67)$$

and

$$\varepsilon_{F0}^{(2)} = \pi N_s/m \qquad (12.68)$$

is the zero temperature Fermi level (with band bottom as the energy zero) of a 2D electron gas having mass m and sheet density equal to the carrier density per period of the superlattice, $N_s = nd$.

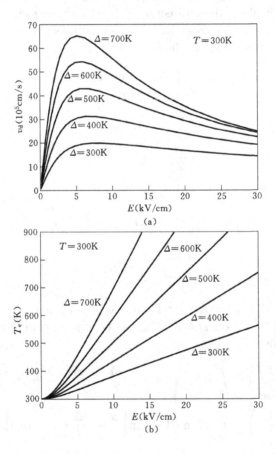

Fig. 12.4 The calculated drift velocity v_d and electron temperature T_e are shown as functions of the electric field for GaAs-based superlattices of different miniband widths Δ having the same spatial period $d = 5.7$ nm and electron sheet density $N_s = 2 \times 10^{10}$ cm^{-2} at lattice temperature $T = 300$ K.

Given the electron density N_s, the electric field E and lattice temperature T, the time dependent momentum and energy balance equations

$$eE/m_z^* + a_i + a_p = 0, \qquad (12.69)$$

$$eEv_d - w = 0, \qquad (12.70)$$

plus equation (12.52) for the electron density, determine parameters z_d, T_e and μ, and thus all the transport quantities in the steady state.

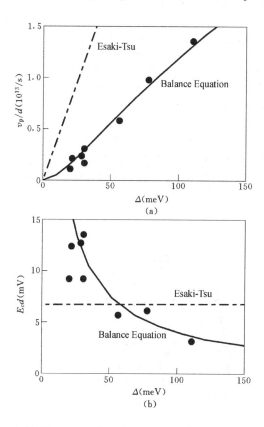

Fig. 12.5 The peak drift velocity v_p and the critical field E_c are shown as functions of miniband widths Δ for GaAs-based superlattices. The solid and dashed curves are calculated results of the balance equation theory and Esaki-Tsu model respectively. The Black dots are experimental value of Sibille et al (1990). After Lei, Horing and Cui (1991).

This set of momentum and energy balance equations has been applied to the investigation of miniband transport in many semiconductor superlattices with different structures under different conditions. The earliest example is the calculation of miniband transport properties at room temperature ($T = 300\,\mathrm{K}$) for a series of GaAs-based superlattices having different miniband widths but the same spatial period $d = 5.7\,nm$. The scatterings from impurities, longitudinal and transverse acoustic phonons and longitudinal optic phonons are taken into account. The scattering matrix elements and the material parameters are taken typical expressions and values as used in Sec. 2.6. All the systems of different miniband widths are assumed to have the same electron sheet density $N_s = 2 \times 10^{10}\,\mathrm{cm}^{-2}$ and the same impurity scattering rate yielding a linear mobility $\mu_0 = 2.0\,\mathrm{m^2 V^{-1} s^{-1}}$ at $4.2\,\mathrm{K}$ in the $\Delta = 500\,\mathrm{K}$ system. Fig. 12.4 illustrates the calculated drift velocity v_d and electron temperature T_e as functions of the electric field. The larger miniband width of the system, the quicker the electron temperature rises, the higher the peak drift velocity and the stronger the negative differential mobility exhibit. The peak drift velocity v_p and the critical field E_c (the field at which the drift velocity reaches its maximum) extracted from this set of v_d–E curves are plotted in Fig. 12.5 as functions of miniband width Δ. The theoretical predictions are in good agreement with the experimental results of Sibille *et al* (1990).

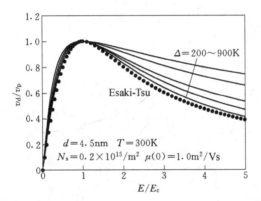

Fig. 12.6 The reduced drift velocity $v_\mathrm{d}/v_\mathrm{p}$ versus the reduced electric field E/E_c. The dots are the Esaki–Tsu relation (12.71).

Note that, the v_d–E curves of Fig. 12.4 derived from the momentum and energy balance equations deviate from the Esaki–Tsu relation

$$v_\mathrm{d} = v_\mathrm{p} \frac{2E/E_\mathrm{c}}{1+(E/E_\mathrm{c})^2}. \tag{12.71}$$

This can be seen clearly if one redraws the v_d–E curves of Fig. 12.4 as that of v_d/v_p versus E/E_c in Fig. 12.6. The form of v_d/v_p as a function of E/E_c is different for different miniband widths, in disagreement with (12.71).

12.5 Superlattice Miniband Transport Treated with Lattice Momentum and Energy Balance Equations

Under the influence of an electric field along the growth axis of the superlattice, the average lattice momentum and the frictional acceleration of the system are in the z direction: $\boldsymbol{p}_\ell = (0, 0, p_\ell)$, and $\boldsymbol{f}^\ell = (0, 0, f^\ell)$. The lattice-momentum balance equation (11.90) and energy balance equation (11.91) for this transport state are

$$\frac{dp_\ell}{dt} = eE\mathcal{R}_{zz} + f^\ell, \tag{12.72}$$

$$\frac{d\varepsilon_\text{e}}{dt} = eEv_\text{d} - w. \tag{12.73}$$

The energy balance equation (12.73) and the expressions for v_d, ε_e and w are the same as (12.47) and related expressions (12.48)–(12.51) in Sec. 12.4. In Eq. (12.72)

$$p_\ell = \frac{2}{N} \sum_{\boldsymbol{k}} p(k_z) f(\bar{\varepsilon}(\boldsymbol{k}), T_\text{e}) \tag{12.74}$$

is the average lattice momentum, and $p(k_z)$ is the z component of the lattice momentum function $\boldsymbol{p}(\boldsymbol{k})$ defined in Sec. 11.1.3. It depends only on k_z in such a way that $p(k_z) = k_z$ when $-\pi/d < k_z < \pi/d$, and, for arbitrary k_z, $p(k_z + 2\pi/d) = p(k_z)$. $\mathcal{R}_{zz}$ is the zz component of the Büttiker–Thomas reduction tensor

$$\mathcal{R} = \mathcal{I} - \frac{1}{4\pi^3 n} \oint_{S_\text{BZ}} d\boldsymbol{s}\,\boldsymbol{k} f(\bar{\varepsilon}(\boldsymbol{k}), T_\text{e}). \tag{12.75}$$

The lattice-momentum related frictional force $f^\ell = f_\text{i}^\ell + f_\text{p}^\ell$ is given by

$$f_\text{i}^\ell = \frac{2\pi n_\text{i}}{N} \sum_{\boldsymbol{k},\boldsymbol{q}} |\bar{u}(\boldsymbol{q})|^2 g(q_z) \left[p(k_z + q_z) - p(k_z)\right] \delta(\varepsilon(\boldsymbol{k}+\boldsymbol{q}) - \varepsilon(\boldsymbol{k}))$$

$$\times \left[f(\bar{\varepsilon}(\boldsymbol{k}), T_\text{e}) - f(\bar{\varepsilon}(\boldsymbol{k}+\boldsymbol{q}), T_\text{e})\right], \tag{12.76}$$

$$f_p^\ell = \frac{4\pi}{N} \sum_{\boldsymbol{k},\boldsymbol{q},\lambda} |\bar{M}(\boldsymbol{k},\lambda)|^2 g(q_z) [p(k_z + q_z) - p(k_z)]$$
$$\times \delta(\varepsilon(\boldsymbol{k}+\boldsymbol{q}) - \varepsilon(\boldsymbol{k}) + \Omega_{\boldsymbol{q}\lambda})[f(\bar{\varepsilon}(\boldsymbol{k}),T_e) - f(\bar{\varepsilon}(\boldsymbol{k}+\boldsymbol{q}),T_e)]$$
$$\times \left[n\left(\frac{\Omega_{\boldsymbol{q}\lambda}}{T}\right) - n\left(\frac{\bar{\varepsilon}(\boldsymbol{k}) - \bar{\varepsilon}(\boldsymbol{k}+\boldsymbol{q})}{T_e}\right) \right]. \quad (12.77)$$

For comparison with results obtained from the momentum and energy balance equations in Sec. 12.4, we still use the lattice-momentum shifted $f_d(\boldsymbol{k})$ (11.65) as the distribution function, i.e. describing the transport state with $\boldsymbol{p}_d$ and T_e, then $\bar{\varepsilon}(\boldsymbol{k}) = \varepsilon(\boldsymbol{k} - \boldsymbol{p}_d)$. For superlattices having energy spectrum (12.6) and (12.7) subject to an axial electric field, $\boldsymbol{p}_d = (0, 0, p_d)$ is in the z direction, and the average lattice momentum can be written as

$$p_\ell = p_d \left\{ 1 - \frac{T_e}{\varepsilon_{F0}^{(2)}} \frac{1}{z_d} \int_0^{z_d} dz \, \ln\left[1 + \exp\left(\frac{\mu}{T_e} - \frac{\Delta}{2T_e}(1+\cos z)\right)\right] \right\}. \quad (12.78)$$

Noticing that, only the two planar surfaces passing through $k_z = \pi/d$ and $k_z = -\pi/d$ and perpendicular to the k_z axis contribute to the closed area integral over the Brillouin zone boundary $S_{\rm BZ}$, we have

$$\mathcal{R}_{zz} = 1 - \frac{T_e}{\varepsilon_{F0}^{(2)}} \ln\left[1 + \exp\left(\frac{\mu}{T_e} - \frac{\Delta}{2T_e}(1+\cos z_d)\right)\right]. \quad (12.79)$$

in which μ is determined by the particle number equation (12.64).

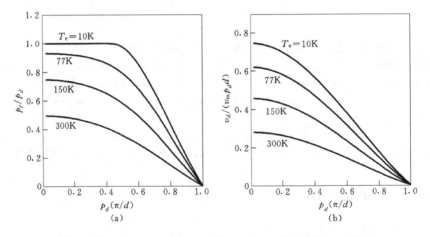

Fig. 12.7 The average lattice momentum p_ℓ and average drift velocity v_d are plotted as functions of p_d, for a GaAs-based superlattice with $\Delta = 400$ K and $N_s = 2 \times 10^{15}$ m^{-2} at several electron temperatures T_e.

For a GaAs-based superlattice with $\Delta = 400$ K and $N_\text{s} = 2 \times 10^{15}\,\text{m}^{-2}$, we plot the average lattice momentum p_ℓ/p_d versus p_d in Fig. 12.7(a) at different electron temperatures. Only in the case of very low electron temperature ($T_\text{e} = 10$ K) and small p_d ($p_d d < \pi/2$), p_ℓ agrees with p_d. Otherwise, p_ℓ deviates from p_d quite significantly. For comparison, in Fig. 12.7(b) we show the average drift velocity $v_\text{d}/(v_\text{m} p_d d)$, which is proportional to the average momentum given by (12.58), versus p_d. v_d/p_d can be nearly a constant only at very small p_d ($p_d d < 0.1\pi$). These indicate that the assumption (11.98), i.e. v_d proportional to p_ℓ, can be valid only within very small ranges of p_d and T_e.

As an example, we examine the steady state transport at lattice temperature $T = 300$ for a GaAs-based superlattice having spatial period $d = 4.5\,\text{nm}$, miniband width $\Delta = 900$ K and low-temperature (4.2 K) linear mobility $\mu(0) = 1.0\,\text{m}^2\text{V}^{-1}\text{s}^{-1}$, using the time independent lattice-momentum and energy balance equations discussed in this section:

$$eE\mathcal{R}_{zz} + f^\ell = 0, \tag{12.80}$$

$$eEv_\text{d} - w = 0. \tag{12.81}$$

The calculated drift velocity v_d as a function of the electric field E is shown in Fig. 12.8 for two different values of the carrier density: $N_\text{s} = 0.2 \times 10^{15}\,\text{m}^{-2}$ and $4.0 \times 10^{15}\,\text{m}^{-2}$.

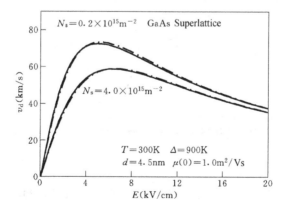

Fig. 12.8 The drift velocity versus the electric field in miniband transport at $T = 300$ K for a superlattice described in the text with two different carrier densities. The solid lines are calculated from the momentum and energy balance equations (12.69) and (12.70); and the dot-dashed lines are calculated from the lattice-momentum and energy balance equations (12.80) and (12.81). After Lei (1996b).

For comparison, calculation for the same system has also been carried out with the momentum and energy balance equations discussed in Sec. 12.4. The results are also presented in Fig. 12.8. We can see that the predictions from these two methods are quite close throughout the whole electric field range. As pointed out in Sec. 11.3, both the momentum balance equation (12.75) and the lattice-momentum balance equation (12.90) contain the effect of Bragg scattering of the nonparabolic band. In the momentum balance equation (12.75) this effect is included in the average inverse effective mass $\mathcal{K}_{zz} \equiv 1/m_z^*$. In the lattice-momentum balance equation (12.90), this effect is mainly reflected in the Büttiker–Thomas reduction factor $\mathcal{R}_{zz}$. The v_d-vs-E curve exhibiting feature of Fig. 12.8 is intimately related with these two factors. To have an intuitive idea, we plot in Fig. 12.9 the effective mass ratio m/m_z^* and the reduction factor $\mathcal{R}_{zz}$ in the transport process described by Fig. 12.8 as functions of the electric field. When the electric field increases from zero to $20\,\mathrm{kV/cm}$, both m/m_z^* and $\mathcal{R}_{zz}$ shrink almost an order of magnitude.

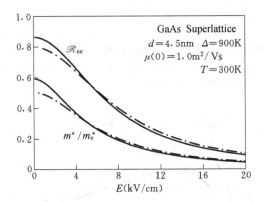

Fig. 12.9 The variation of m/m_z^* and $\mathcal{R}_{zz}$ with the electric field. The solid lines are for $N_\mathrm{s} = 0.2 \times 10^{15}\,\mathrm{m}^{-2}$ and the dot-dashed lines are for $N_\mathrm{s} = 4.0 \times 10^{15}\,\mathrm{m}^{-2}$.

12.6 Comparison of Using Two Different Initial Distribution Functions

In the last two sections we analyzed superlattice miniband transport with two different sets of balance equations: momentum and energy balance equations and lattice-momentum and energy balance equations, but using the same kind of initial distribution function

$$f_d(\bm{k}) = \frac{1}{\mathrm{e}^{[\varepsilon(\bm{k}-\bm{p}_d)-\mu]/T_\mathrm{e}}+1}, \tag{12.82}$$

that is to employ $\bm{p}_d$ and T_e to describe the transport state $[\bar{\varepsilon}(\bm{k}) = \varepsilon(\bm{k}-\bm{p}_d)]$. As discussed in Sec. 11.2, one may instead take

$$f_v(\bm{k}) = \frac{1}{\mathrm{e}^{[\varepsilon(\bm{k})-\bm{p}_v\cdot\bm{v}(\bm{k})-\mu]/T_\mathrm{e}}+1} \tag{12.83}$$

as the initial distribution function, that is to employ $\bm{p}_v$ and T_e to describe the transport state $[\bar{\varepsilon}(\bm{k}) = \varepsilon(\bm{k}) - \bm{p}_v \cdot \bm{v}(\bm{k})]$. In this section we will start

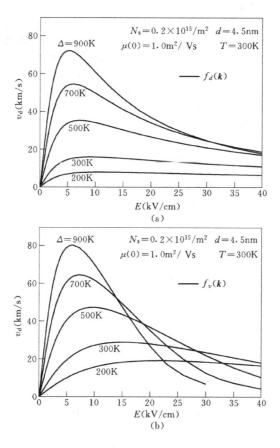

Fig. 12.10 The v_d–E behavior of superlattice miniband transport obtained from momentum and energy balance equations, with distribution function $f_d(\bm{k})$ (a) and with distribution function $f_v(\bm{k})$ (b). From Lei (1994c).

from the momentum and energy balance equation set

$$eE/m_z^* + a_i + a_p = 0, \tag{12.84}$$

$$eEv_d - w = 0, \tag{12.85}$$

respectively associated with $f_d(\mathbf{k})$ (12.82) and $f_v(\mathbf{k})$ (12.83) as the initial distribution function, to calculate the steady state miniband transport at $T = 300\,\mathrm{K}$ in a series of GaAs-based superlattices having period $d = 4.5$ nm, carrier sheet density $N_s = 0.2 \times 10^{15}\,\mathrm{m}^{-2}$, and low-temperature (4.2 K) linear mobility $\mu(0) = 1.0\,\mathrm{m^2/V\,s}$, but different miniband widths Δ. Fig. 12.10(a) plots the calculated v_d–E curves using $f_d(\mathbf{k})$ as the initial distribution function. Fig. 12.10(b) shows the results obtained using $f_v(\mathbf{k})$ as the initial distribution function. One can see that though qualitatively the v_d–E behavior predicted in two cases is similar, the peak velocity v_p is higher and the drift velocity v_d descent with increasing E in the negative differential mobility range is steeper in the $f_v(\mathbf{k})$ case than in the $f_d(\mathbf{k})$ case. The predicted v_p and E_c from both cases are shown in Fig. 12.11 as functions of the miniband width Δ.

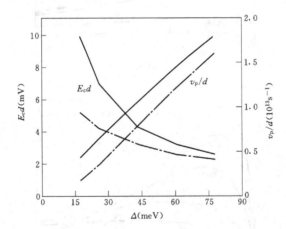

Fig. 12.11 Peak drift velocity v_p and critical field E_c versus miniband width Δ. The solid lines are results using $f_v(\mathbf{k})$ as the initial distribution function, and dot-dashed lines are results using $f_d(\mathbf{k})$ as the initial distribution function.

To see these features more clearly, we redraw the data of Fig. 12.10 in Fig. 12.12 by plotting v_d/v_p as a function of E/E_c. There, the dot-dashed lines are results with $f_d(\mathbf{k})$ function, and the solid curves are those with

$f_v(\boldsymbol{k})$ function. The small circles represent the Esaki-Tsu relation

$$\frac{v_\mathrm{d}}{v_\mathrm{p}} = \frac{2E/E_\mathrm{c}}{1+(E/E_\mathrm{c})^2}. \tag{12.86}$$

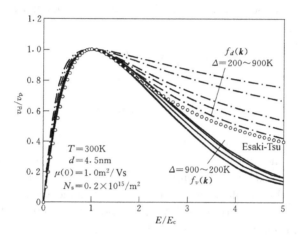

Fig. 12.12 The normalized drift velocity $v_\mathrm{d}/v_\mathrm{p}$ is plotted as a function of the normalized electric field E/E_c. The dot-dashed lines are results with distribution function $f_d(\boldsymbol{k})$, and solid lines are results with distribution function $f_v(\boldsymbol{k})$. The small circles represent the Esaki-Tsu relation (12.86). From Lei (1994c).

12.7 Laterally Confined Superlattices

Superlattices discussed so far in this Chapter have the structure in which carriers are free to move in the x–y plane without confinement. They may be called the planar superlattice or laterally unconfined superlattice. In this section we are going to deal with such a system in which carriers can move in the z direction within the miniband formed by the periodical potential wells and barriers of finite heights, while in the x–y plane they are spatially confined inside a small cylindrical region of diameter d_r by infinitely high potential walls (a cylindrically 2D-confined superlattice). Of course we can also consider a 1D-confined superlattice in which carriers move freely in the y direction, while in the x direction they are confined within a narrow range of width d_x. Apparently, if d_r is large enough or d_x is large enough, the system returns to the case of a laterally unconfined superlattice.

12.7.1 The momentum and energy balance equations of a laterally confined superlattice

Assuming that the energy separation between the longitudinal lowest and second minibands are large enough that electrons occupy only the lowest miniband, we can describe the energy spectrum of a single electron as the sum of the transverse energy ε_n and a tight-binding type longitudinal miniband energy $\varepsilon(k_z)$:

$$\varepsilon_n(k_z) = \varepsilon_n + \varepsilon(k_z), \qquad (12.87)$$

in which

$$\varepsilon(k_z) = \frac{\Delta}{2}\left[1 - \cos(k_z d)\right], \qquad (12.88)$$

$-\pi/d < k_z \leqslant \pi/d$, d is the period in the z direction of the superlattice, and Δ is the miniband width. The transverse state of the electron is characterized by the transverse quantum number n (transverse subband index), determined by the specific form of the lateral confinement. In the case that electrons are confined within a cylindrical region of diameter d_r the transverse quantum number $n = (l, m)$ ($l = 1, 2, \cdots$; $m = 0, \pm 1, \cdots, \pm l$), and the transverse wave function is

$$\varphi_n(\mathbf{r}_\parallel) = c_l^m \mathrm{J}_m\left(\frac{2 x_l^m}{d_r} r_\parallel\right) \mathrm{e}^{\mathrm{i}\, m\phi}, \qquad (12.89)$$

where $c_l^m = 2/(\sqrt{\pi} d_r y_l^m)$ is the normalization factor, $r_\parallel \equiv (x, y)$ and ϕ are respectively the transverse coordinate vector and azimuth, and x_l^m represents the lth zero of the mth order Bessel function, i.e. $\mathrm{J}_m(x_l^m) = 0$, and $y_l^m = \mathrm{J}_{m+1}(x_l^m)$. The corresponding eigenenergy is

$$\varepsilon_n = \frac{2(x_l^m)^2}{m^* d_r^2}, \qquad (12.90)$$

m^* being the effective mass of the background bulk material. From the point of view of the band structure, the difference between the laterally confined superlattice and the quantum wire discussed in Sec. 2.9 is that in the latter electrons are free to move along the z direction (parabolic band), while in the former electrons are subject to periodical potential (extremely nonparabolic miniband).

When a uniform electric field E is applied parallel to the superlattice axis, the carriers are accelerated by the field and scattered by impurities and phonons, resulting in an overall drift motion and a possible heating or cooling of the carrier system. The transport state of the system can be described by the momentum and energy balance equations

$$\frac{dv_{\rm d}}{dt} = \frac{eE}{m_z^*} + a_{\rm i} + a_{\rm p}, \qquad (12.91)$$

$$\frac{d\varepsilon_{\rm e}}{dt} = eEv_{\rm d} - w. \qquad (12.92)$$

The expressions of relevant quantities in the equations, such as the drift velocity $v_{\rm d}$, the average inverse effective mass $1/m_z^*$, the average energy per carrier $\varepsilon_{\rm e} = \varepsilon_\| + \varepsilon_z$ and the total electron number N, are easily written out directly from the corresponding expressions in Sec. 12.4, by replacing the transverse electron state wavevector $\boldsymbol{k}_\| \equiv (k_x, k_y)$ there with the transverse quantum number n:

$$v_{\rm d} = \frac{2}{N} \sum_{n,k_z} v(k_z) f(\bar{\varepsilon}_n(k_z), T_{\rm e}), \qquad (12.93)$$

$$\frac{1}{m_z^*} = \frac{2}{N} \sum_{n,k_z} \frac{d^2 \varepsilon(k_z)}{dk_z^2} f(\bar{\varepsilon}_n(k_z), T_{\rm e}), \qquad (12.94)$$

$$\varepsilon_\| = \frac{2}{N} \sum_{n,k_z} \varepsilon_n f(\bar{\varepsilon}_n(k_z), T_{\rm e}), \qquad (12.95)$$

$$\varepsilon_z = \frac{2}{N} \sum_{n,k_z} \varepsilon(k_z) f(\bar{\varepsilon}_n(k_z), T_{\rm e}), \qquad (12.96)$$

$$N = 2 \sum_{n,k_z} f(\bar{\varepsilon}_n(k_z), T_{\rm e}). \qquad (12.97)$$

The impurity and phonon scattering induced frictional accelerations $a_{\rm i}$ and $a_{\rm p}$, and the electron energy loss rate to the lattice, w, can also be obtained from the corresponding expressions in Sec. 12.4 without difficulty. The only thing needed to take care is that, in view of loss of translational invariance of the transverse wave function in the laterally confined systems, an intersubband form factor $J_{nn'}(q_\|)$ should appear in all the expressions and the previous sum over the transverse wavevector should be replaced by the sum over the transverse quantum numbers n, n' to give

$$a_{\rm i} = \frac{2\pi n_{\rm i}}{N} \sum_{n,n',k_z,\boldsymbol{q}} |\bar{u}(\boldsymbol{q})|^2 |J_{nn'}(q_\|)|^2 g(q_z) [v(k_z + q_z) - v(k_z)]$$
$$\times \delta(\varepsilon_n(k_z + q_z) - \varepsilon_{n'}(k_z)) [f(\bar{\varepsilon}_n(k_z), T_{\rm e}) - f(\bar{\varepsilon}_{n'}(k_z + q_z), T_{\rm e})],$$
$$(12.98)$$

$$a_{\mathrm{p}} = \frac{4\pi}{N} \sum_{n,n',k_z,\boldsymbol{q},\lambda} |\bar{M}(\boldsymbol{q},\lambda)|^2 |J_{nn'}(\boldsymbol{q}_{\|})|^2 g(q_z) [v(k_z+q_z) - v(k_z)]$$
$$\times \delta(\varepsilon_n(k_z+q_z) - \varepsilon_{n'}(k_z) + \Omega_{\boldsymbol{q}\lambda}) [f(\bar{\varepsilon}_n(k_z), T_{\mathrm{e}}) - f(\bar{\varepsilon}_{n'}(k_z+q_z), T_{\mathrm{e}})]$$
$$\times \left[n\left(\frac{\Omega_{\boldsymbol{q}\lambda}}{T}\right) - n\left(\frac{\bar{\varepsilon}_n(k_z) - \bar{\varepsilon}_{n'}(k_z+q_z)}{T_{\mathrm{e}}}\right) \right], \tag{12.99}$$

$$w = \frac{4\pi}{N} \sum_{n,n',k_z,\boldsymbol{q},\lambda} |\bar{M}(\boldsymbol{q},\lambda)|^2 |J_{nn'}(\boldsymbol{q}_{\|})|^2 g(q_z) \Omega_{\boldsymbol{q}\lambda}$$
$$\times \delta(\varepsilon_n(k_z+q_z) - \varepsilon_{n'}(k_z) + \Omega_{\boldsymbol{q}\lambda}) [f(\bar{\varepsilon}_n(k_z), T_{\mathrm{e}}) - f(\bar{\varepsilon}_{n'}(k_z+q_z), T_{\mathrm{e}})]$$
$$\times \left[n\left(\frac{\Omega_{\boldsymbol{q}\lambda}}{T}\right) - n\left(\frac{\bar{\varepsilon}_n(k_z) - \bar{\varepsilon}_{n'}(k_z+q_z)}{T_{\mathrm{e}}}\right) \right]. \tag{12.100}$$

Here, the wave vector $\boldsymbol{q} \equiv (\boldsymbol{q}_{\|}, q_z)$ and $\boldsymbol{q}_{\|} \equiv (q_x, q_y)$, the form factor $J_{nn'}(\boldsymbol{q}_{\|})$ is related to the electron transverse subband wave functions $\varphi_n(\boldsymbol{r}_{\|})$ and $\varphi_{n'}(\boldsymbol{r}_{\|})$,

$$J_{nn'}(\boldsymbol{q}_{\|}) = \int d\boldsymbol{r}_{\|} e^{i\boldsymbol{q}_{\|} \cdot \boldsymbol{r}_{\|}} \varphi_{n'}^*(\boldsymbol{r}_{\|}) \varphi_n(\boldsymbol{r}_{\|})$$
$$= 2 \int_0^1 d\xi\, \xi J_m(x_l^m \xi) J_{m'}(x_{l'}^{m'} \xi) J_{m-m'}(q_{\|} d_r \xi) \left(y_l^m y_{l'}^{m'}\right)^{-1}, \tag{12.101}$$

and $g(q_z)$ is the form factor related to the longitudinal miniband wave functions of the superlattice given by (12.9).

When using the lattice-momentum shifted distribution $f(\bar{\varepsilon}_n(k_z), T_{\mathrm{e}})$ of (12.57) for the superlattice with longitudinal energy spectrum $\varepsilon(k_z)$ (12.88), the particle number equation (12.97) can be written in terms of electron line density N_1 along the z direction as

$$N_1 d = \frac{2}{\pi} \sum_n \int_0^\pi \frac{d\theta}{\exp\{[\varepsilon_n + (\Delta/2)(1 - \cos\theta) - \mu]/T_{\mathrm{e}}\} + 1}. \tag{12.102}$$

The average drift velocity, inverse effective mass and electron average energy in the z direction can be written as

$$v_d = v_{\mathrm{m}} \alpha(T_{\mathrm{e}}) \sin(z_d), \tag{12.103}$$
$$1/m_z^* = (1/M^*) \alpha(T_{\mathrm{e}}) \cos(z_d), \tag{12.104}$$
$$\varepsilon_z = \frac{\Delta}{2}[1 - \alpha(T_{\mathrm{e}}) \cos(z_d)], \tag{12.105}$$

in which $v_{\rm m} \equiv \Delta d/2$, $1/M^* \equiv \Delta d^2/2$, and

$$\alpha(T_{\rm e}) = \frac{2}{\pi N_1 d} \sum_n \int_0^\pi \frac{\cos\theta\, d\theta}{\exp\left\{[\varepsilon_n + (\Delta/2)(1-\cos\theta) - \mu]/T_{\rm e}\right\} + 1}. \tag{12.106}$$

Note that $1/m_z^*$ is related to ε_z by

$$M^*/m_z^* = 1 - 2\varepsilon_z/\Delta. \tag{12.107}$$

All the quantities listed above as well as $\varepsilon_\parallel$, are functions of z_d and $T_{\rm e}$. And all the expressions go to those of laterally unconfined superlattice in Sec. 12.4 if $\varepsilon_n \to k_\parallel^2/2m$ and $\sum_n \to \sum_{\boldsymbol{k}_\parallel}$.

The above equations can be conveniently used to investigate the electron transport in laterally confined superlattices, especially the variation of transport properties with the confined diameter.

12.7.2 DC steady state transport

When a constant electric field E is applied along the superlattice growth axis, the steady state transport is determined by the time-independent balance equations

$$eE/m_z^* + a_{\rm i} + a_{\rm p} = 0, \tag{12.108}$$

$$eEv_{\rm d} - w = 0. \tag{12.109}$$

Lei and Wang (1993) calculate linear and nonlinear transport properties of cylindrically 2D-confined GaAs superllatices with longitudinal spatial period d and miniband width Δ but having different degrees of lateral confinement (transverse diameters d_r ranging from 10 nm to 50 nm), as well as those of an unconfined superlattice (referred as 3D superlattice) at lattice temperatures from 45 to 300 K. To facilitate comparison the carrier sheet density per period is kept unchanged for all the 2D-confined systems and the 3D system, $N_{\rm s} = 4.0 \times 10^{15}\,{\rm m}^{-2}$. Thus, with increasing laterally diameters the carrier line density $N_1 = \pi d_r^2 N_{\rm s}/(4d)$ and the total number of carriers accommodated in the miniband, $N_1 d$, increase like d_r^2, and, more and more transverse subbands are occupied. On the other hand, the position of the subband energy bottom ε_n decreases like d_r^{-2} as lateral diameter increases. Since linear and nonlinear conductions are strongly affected by the relative position of the Fermi energy in a partially occupied band, a strongly lateral-diameter-dependent transport behavior is anticipated. They include 21 transverse subbands in the calculation. The

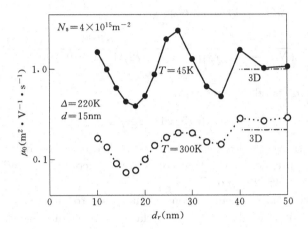

Fig. 12.13 Miniband linear mobility μ_0 of 15 GaAs superlattices having the same period $d = 15$ nm, miniband width $\Delta = 220$ K, carrier sheet density $N_s = 4.0 \times 10^{15}$ m^{-2} and low temperature (4.2 K) mobility $\mu_0 = 1.0$ m^{2}V^{-1}s^{-1}, but different transverse diameter d_r. From Lei and Wang (1993).

form factors $|J_{nn'}(q_\parallel)|^2$ ($n, n' = 1, 2, \cdots, 21$) are numerically obtained from (12.101). Scatterings due to charged impurities, acoustic phonons (through the deformation potential and piezoelectric couplings with electrons) and polar optic phonons (through the Fröhlich coupling with electrons) are considered, with 3D Thomas–Fermi static screening. All the material and coupling parameters are taken typical values of GaAs. The strength of the impurity scattering is so chosen that the low temperature (4.2 K) mobility equals $\mu_0 = 1.0$ m^{2}V^{-1}s^{-1} for all the systems.

The calculated miniband linear mobilities at temperatures $T = 45$ K and $T = 300$ K for a family of fifteen 2D-confined superlattices having spatial period $d = 15$ nm, miniband width $\Delta = 200$ K and sheet density $N_s = 4.0 \times 10^{15}$ m^{-2} but with different lateral diameters d_r, are plotted in Fig. 12.13. Also shown is the linear mobility of the 3D superlattice with the same d, Δ and N_s, calculated from balance equations in Sec. 12.4 (dot-dashed lines labeled 3D in the figure). We see that, even at lattice temperature $T = 300$ K, the linear mobility still strongly oscillates with changing d_r. This is the feature of miniband conductions. In the case of quantum wires discussed in Sec. 2.9, the linear mobility oscillation with changing transverse size almost completely disappears when temperature rises to 100 K.

An example of the calculated drift velocity v_d and electron temperature T_e in the nonlinear steady state transport for several 2D-confined super-

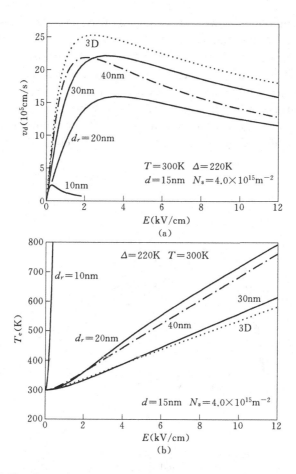

Fig. 12.14 Drift velocity v_d and electron temperature T_e are shown as functions of the electric field for the GaAs superlattices described in Fig. 12.13 having transverse diameters $d_r = 10, 20, 30$ and 40 nm. The dashed curves (3d) are for laterally unconfined superlattice. From Lei and Wang (1993).

lattices and 3D superlattices are shown in Fig. 12.14 as functions of the electric field at $T = 300$ K. All the velocity–field curves exhibit negative differential mobility. But the shape of the curve may be quite different for systems of different d_r. Especially in the $d_r = 10$ nm system in which the energy gap between two neighboring subbands is larger and the miniband width $\Delta = 220$ K is smaller than the energy of an optic phonon, neither the interband nor the intraband optical phonon scattering can happen. Since the acoustic phonons are not very efficient in dissipating the system energy out, the electron temperature will rise rapidly with increasing electric field. This greatly limits the maximum value the drift velocity can attain, and the peak drift velocity v_p is then reached at a very low critical field E_c. For system of $d_r \geqslant 20$ nm, the intersubband energy gaps are greatly reduced or even neighboring subbands overlap with each other allowing polar optic-phonon scatterings, which provide a very efficient energy dissipation for the electrons. The rise of the electron temperature then becomes much slower and the peak drift velocity is reached at a much larger electric field.

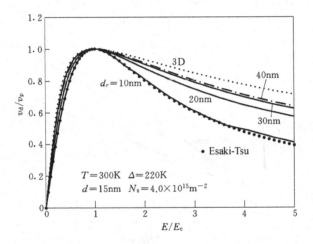

Fig. 12.15 Reduced drift velocity $v_\mathrm{d}/v_\mathrm{p}$ versus the reduced electric field E/E_c for systems described in Fig. 12.14. From Lei and Wang (1993).

The curves of Fig. 12.14(a) is redrawn in Fig. 12.15 showing the reduced drift velocity $v_\mathrm{d}/v_\mathrm{p}$ as a function of the reduced electric field E/E_c, together with the Esaki–Tsu formula (12.71). We notice that, with increasing transverse diameters from $d_r = 10, 20, 30$ to 40 nm, none of v_d–E and $v_\mathrm{d}/v_\mathrm{p}$–$E/E_\mathrm{c}$ behavior goes monotonically towards that of 3D. These curves oscillate around the corresponding 3D curves. But the $v_\mathrm{d}/v_\mathrm{p}$–$E/E_\mathrm{c}$ behavior of $d_r = 10$ nm system is quite close to that of the Esaki and Tsu. This indicates that the strict one-dimensional superlattice (laterally limited to only one subband) is essentially described by the Esaki-Tsu formula (12.71).

12.8 Transient and High-Frequency Transport in Superlattices

The time-dependent momentum and energy balance equations

$$\frac{dv_{\rm d}}{dt} = \frac{eE(t)}{m^*} + a, \qquad (12.110)$$

$$\frac{d\varepsilon_{\rm e}}{dt} = eE(t)v_{\rm d} - w, \qquad (12.111)$$

can be directly used to investigate the transient response of a semiconductor superlattice to a time-dependent electric field. Since all the quantities in the equations ($v_{\rm d}$, $\varepsilon_{\rm e} = \varepsilon_{\|} + \varepsilon_z$, $1/m^*$, $a = a_{\rm i} + a_{\rm p}$ and w) can be directly expressed as functions of $z_d \equiv p_d d$ and $T_{\rm e}$ ($\varepsilon_{\|}$ depends only on $T_{\rm e}$), it is generally convenient taking z_d, $T_{\rm e}$ and their derivatives as the basic parameters. In terms of these variables Eqs. (12.110) and (12.111) are written as

$$\alpha(T_{\rm e})\cos(z_d)\frac{dz_d}{dt} + \left(\frac{d\alpha(T_{\rm e})}{dT_{\rm e}}\right)\sin(z_d)\frac{dT_{\rm e}}{dt} = $$
$$edE(t)\alpha(T_{\rm e})\cos(z_d) + \frac{2a}{\Delta d}, \qquad (12.112)$$

$$\alpha(T_{\rm e})\sin(z_d)\frac{dz_d}{dt} - \left[\left(\frac{d\alpha(T_{\rm e})}{dT_{\rm e}}\right)\cos(z_d) - \frac{2}{\Delta}\left(\frac{d\varepsilon_{\|}}{dT_{\rm e}}\right)\right]\frac{dT_{\rm e}}{dt} = $$
$$edE(t)\alpha(T_{\rm e})\sin(z_d) - \frac{2w}{\Delta}. \qquad (12.113)$$

12.8.1 *The Transient response of a superlattice to a step electric field*

Starting from Eqs. (12.112) and (12.113), Lei (1994b) performs a numerical calculation for the transient response at lattice temperature $T = 77\,{\rm K}$ to an electric field E suddenly turned on at time $t = 0$ in a laterally confined superlattice with transverse diameter $d_r = 10$ nm, period $d = 15$ nm, miniband width $\Delta = 220$ K, carrier sheet density $N_{\rm s} = 4.0 \times 10^{11}$ cm^{-2} (corresponding to linear density $N_1 d = 0.314$) and low-temperature (4.2 K) linear mobility $\mu(0) = 1.0$ m^{2}V^{-1}s^{-1}, assuming that the system is in an initial equilibrium state of $z_d = 0$ and $T_{\rm e} = T$ at $t = 0$.

If only the lowest transverse subband is taken into account, for this narrow miniband width the polar optic-phonon scattering is prohibited by the requirement of energy conservation. This greatly reduces the critical

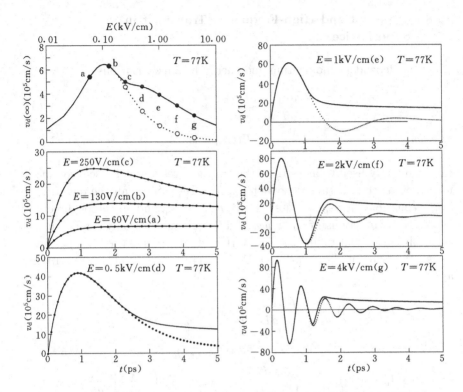

Fig. 12.16 The steady state drift velocity as a function of applied electric field (upper left part), and the transient velocity response to step electric fields of different strengths turned at $t = 0$. The system is a laterally confined GaAs superlattice with $d_r = 10$ nm, $d = 15$ nm, $\Delta = 220$ K, $\mu(0) = 1.0$ m^{2}V^{-1}s^{-1} and $N_s = 4.0 \times 10^{11}$ cm^{-2}. The lattice temperature is $T = 77$ K. The solid curves are results calculated by including 21 transverse subbands, and the dotted curves are those with the lowest subband only. From Lei (1994b).

field E_c at which the steady-state drift velocity peaks. Considering scatterings from impurity, longitudinal and transverse acoustic phonons, the calculated steady state drift velocity $v_d(\infty)$ as a function of the electric field E is shown as the dotted curve in the left-upper part of Fig. 12.16, with a critical field E_c about 105 V/cm. The transient velocity response to step electric fields turned on at time $t = 0$ having different strengths are plotted as dotted curves in the other parts of Fig. 12.16. In the case of $E < E_c$ (a), there is no velocity overshoot in the transient response. Once the field strength $E > E_c$ (b, c and d) the velocity overshoot shows up, and becomes more pronounced at larger E. The oscillatory transient velocity begins to appear at a field strength about 8 times the critical field E_c. At $E = 1\,\text{kV/cm}$ (e) clear oscillations of the drift velocity are already seen, with an oscillatory period about 3.0 ps, a little larger than the free Bloch oscillation period $T_B = 2\pi/|eEd| = 2.76\,\text{ps}$ at this field. When the electric field further increases the transient velocity oscillation becomes increasingly stronger (f and g). For the cases of $E = 2\,\text{kV/cm}$ (f) and $E = 4\,\text{kV/cm}$ (g), the oscillation periods are 1.41 ps and 0.695 ps respectively, and the corresponding free Bloch periods are $T_B = 1.38\,\text{ps}$ and $T_B = 0.69\,\text{ps}$.

When higher transverse subbands are taken into consideration, both the steady-state velocity–field characteristic and the transient response, though remain the same at low electric fields, change remarkably at high electric fields. The solid curves shown in Fig. 12.16 are calculated results of steady state v_d–E curve and transient v_d response to step fields of different strengths by including 21 transverse subbands for the same system as described above. Since the electron temperature increases with increasing electric field, high-lying subbands are expected to affect transport behavior greatly at high electric fields. We find that by taking many subbands into account the steady-state electron temperature T_e (not shown) at high field becomes much lower than that in the one-subband-only case and the steady-state drift velocity becomes higher. The oscillatory response of the drift velocity weakens markedly: oscillatory behavior begins to appear at higher electric field strength than the single-subband case and at the same field strength fewer oscillatory periods survive before approaching a final steady state. The main difference between single-subband and multi-subband cases is that in the latter the average transverse energy $\varepsilon_\parallel$ increases with increasing electron temperature ($d\varepsilon_\parallel/dT_e > 0$). This indicates that the transverse movement of carriers in the superlattice gives rise to an extra damping and thus suppresses the oscillation of its longitudinal transient velocity.

With increasing transverse diameter of the confined superlattice, the energies of higher subbands are reduced quickly, and the role of the carrier transverse movement is expected to become more important. To see how the transient transport behavior is affected by the transverse diameter d_r, we calculate the steady-state velocity–field curve and the transient drift velocity response at lattice temperature $T = 77\,\text{K}$ for two laterally confined superlattices having transverse diameters $d_r = 20\,\text{nm}$ and $d_r = 40\,\text{nm}$, with same other parameters as the $d_r = 10\,\text{nm}$ system described in Fig. 12.16 ($d = 15\,\text{nm}$, $\Delta = 220\,\text{K}$, $\mu(0) = 1.0\,\text{m}^2\text{V}^{-1}\text{s}^{-1}$ and $N_s = 4.0 \times 10^{11}\,\text{cm}^{-2}$) to step electric fields (turned-on at time $t = 0$) of strengths $E = 2, 5, 10$

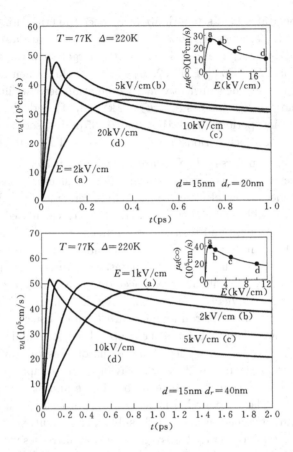

Fig. 12.17 Transient drift velocity response to step electric fields of different strengths turned on at $t = 0$ in two confined superlattices having $d_r = 20$ nm (upper figure) and $d_r = 40$ nm (lower figure), with same parameters $d = 15$ nm, $\Delta = 220$ K, $\mu(0) = 1.0$ m^{2}V^{-1}s^{-1} and $N_s = 4.0 \times 10^{11}$ cm^{-2}. The insets show the corresponding locations of these fields on the steady-state velocity–field curve. The lattice temperature is $T = 77$ K and 21 transverse subbands are included in the calculation. From Lei (1994b).

and 20 kV/cm respectively, including 21 transverse subbands. The results are plotted in Fig. 12.17. Note that in the system of $d_r = 20$ nm the energy gaps between neighboring transverse subbands are small enough to allow polar optical phonon scatterings to occur. This drastically changes both the steady-state and transient transport behavior. In the steady-state case the rise of the electron temperature with increasing electric field in the $d_r = 20$ nm system becomes much slower than in the $d_r = 10$ nm system, and the peak drift velocity is reached at a much larger field strength ($E_c =$

2.6 kV/cm). The transient drift velocity response curve of the $d_r = 20$ nm system looks completely different from that of the $d_r = 10$ nm system: drift velocity exhibits only overshoot, without oscillation. In the system of $d_r = 40$ nm, the neighboring subbands overlap with each other, and the steady-state transport behavior is close to a corresponding unconfined superlattice. The transient drift velocity exhibits only overshoot, without oscillatory behavior up to field strength $E = 20$ kV/cm (the curve for $E = 20$ kV/cm is similar but not shown in the figure).

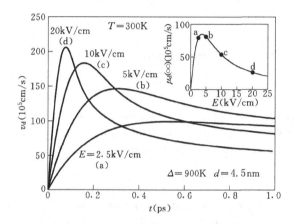

Fig. 12.18 Transient drift velocity response of a 3D superlattice with period $d = 4.5$ nm, miniband width $\Delta = 900$ K, carrier sheet density $N_s = 2.0 \times 10^{11}$ cm^{-2} and low-temperature linear mobility $\mu(0) = 1.0$ m^{2}V^{-1}s^{-1}, to step electric fields turned on at time $t = 0$ having various strengths. The inset shows the corresponding locations of these fields on the steady-state velocity–field curve. The lattice temperature is $T = 300$ K. From Lei (1994b).

Fig. 12.18 shows the calculated results for a 3D superlattice of period $d = 4.5$ nm, miniband width $\Delta = 900$ K, carrier sheet density $N_s = 2.0 \times 10^{11}$ cm^{-2} and low-temperature linear mobility $\mu(0) = 1.0$ m^{2}V^{-1}s^{-1} at $T = 300$ K, based on the unconfined superlattice formulation ($\varepsilon_n \to k_\parallel^2/2m$, $\sum_n \to \sum_{\bm{k}_\parallel}$). Apparently, there is stronger damping showing up and the Bloch oscillation of the drift velocity does not appear in such a superlattice of high carrier density.

The discussion of this section shows that because of the damping induced by the thermal excitation of the carrier transverse movement, the Bloch oscillation of the longitudinal velocity in GaAs-based superlattices can appear only in systems with strong lateral confinement ($d_r \leqslant 20$ nm).

In unconfined or weakly confined superlattices the Bloch oscillation of the carrier average drift velocity may not appear. Of course, this conclusion is reached in the balance-equation transport theory, which uses a unique electron temperature to describe the occupations of electrons in both the longitudinal and transverse states. Such a requirement of the rapid thermalization along all the directions can be satisfied only for a system with strong carrier–carrier interaction, i.e. having high electron density. It is at variance with what is anticipated from purely one-dimensional theories, which essentially describe transport of electrons having dilute density in a superlattice miniband.

12.8.2 *High-frequency small-signal transport*

The time-dependent momentum and energy balance equations (12.112) and (12.113) can be conveniently used to analyze the high-frequency small-signal transport in an arbitrary dc biased superlattice (Lei, 1994d). We consider that a dc bias electric field E_0 and a small signal ac electric field δE of single frequency ω are applied along the superlattice axis,

$$E(t) = E_0 + \delta E. \tag{12.114}$$

After a transient process, the system reaches a steady state in which z_d and T_e can be written in terms of a dc bias part and a small ac response: $z_d = z_0 + \delta z$ and $T_e = T_0 + \delta T$, and all other quantities in the balance equations can be expanded to linear order in the small ac quantities about the bias point z_0 and T_0. For the zero-th order we have just the dc steady-state equations:

$$edE_0\alpha_0 \cos(z_0) + \frac{2a_0}{\Delta d} = 0, \tag{12.115}$$

$$edE_0\alpha_0 \sin(z_0) - \frac{2w_0}{\Delta} = 0, \tag{12.116}$$

where $\alpha_0 \equiv \alpha(T_0)$, $a_0 \equiv a(z_0, T_0)$ and $w_0 = w(z_0, T_0)$. These two equations determine z_0 and T_0, and thus the dc drift velocity at the bias point $v_0 = v_m\alpha_0 \sin(z_0)$. The equations for linear-order ac quantities are as follows:

$$\alpha_0 \cos(z_0)\frac{d}{dt}(\delta z) + \alpha_0' \sin(z_0)\frac{d}{dt}(\delta T) = ed\alpha_0 \cos(z_0)\delta E$$
$$- \left[edE_0\alpha_0 \sin(z_0) - \frac{2}{\Delta d}\left(\frac{\partial a_0}{\partial z_d}\right)\right]\delta z + \left[edE_0\alpha_0' \cos(z_0) + \frac{2}{\Delta d}\left(\frac{\partial a_0}{\partial T_e}\right)\right]\delta T,$$
$$\tag{12.117}$$

$$\alpha_0 \sin(z_0)\frac{d}{dt}(\delta z) - [\alpha'_0 \cos(z_0) - 2\epsilon'_{\|0}/\Delta]\frac{d}{dt}(\delta T) = ed\alpha_0 \sin(z_0)\delta E$$
$$+ \left[edE_0\alpha_0 \cos(z_0) - \frac{2}{\Delta}\left(\frac{\partial w_0}{\partial z_d}\right)\right]\delta z + \left[edE_0\alpha'_0 \sin(z_0) - \frac{2}{\Delta}\left(\frac{\partial w_0}{\partial T_e}\right)\right]\delta T.$$
(12.118)

Here we have used the symbols $\alpha'_0 \equiv d\alpha(T_0)/dT_e$ and $\epsilon'_{\|0} \equiv d\epsilon_\|(z_0, T_0)/dT_e$. This set of equations linear in small ac quantities is conveniently solved for a monochromatic driving field by using a complex form: $\delta E = E_1 e^{-i\omega t}$. For the ac steady state, δz and δT oscillate at the same single frequency: $\delta z = z_1 e^{-i\omega t}$ and $\delta T = T_1 e^{-i\omega t}$, leading to an oscillating ac drift velocity given by $\delta v = v_1 e^{-i\omega t}$, superposed on the dc drift velocity, with

$$v_1 = v_m\bigl[\alpha_0 \cos(z_0) z_1 + \alpha'_0 \sin(z_0) T_1\bigr]. \tag{12.119}$$

The complex frequency-dependent mobility (small-signal dynamic mobility) is defined as

$$\mu_\omega = \frac{v_1}{E_1}. \tag{12.120}$$

Fig. 12.19 shows the real part $\mathrm{Re}\,\mu_\omega$ of the high-frequency complex mobility μ_ω as a function of frequency $\nu = \omega/2\pi$, calculated based on these equations for a laterally 2D-confined (transverse diameter $d_r = 20\,\mathrm{nm}$) and an unconfined ($d_r = \infty$) superlattice, having parameters $d = 7\,\mathrm{nm}$, $\Delta = 900\,\mathrm{K}$, effective bulk 3D carrier density $n = 2.14 \times 10^{23}\,\mathrm{m}^{-3}$, and low-temperature mobility $\mu(0) = 1.0\,\mathrm{m}^2\mathrm{V}^{-1}\mathrm{s}^{-1}$, under various bias electric fields E_0.

The high-frequency small-signal mobility in superlattice longitudinal transport exhibits peculiar behavior. We will discuss it using Fig. 12.19(b) of the unconfined system as an example. Under zero and small dc bias ($E_0 \leqslant 0.1\,\mathrm{kV/cm}$) the behavior of μ_ω is similar to that of Drude formula: $\mathrm{Re}\,\mu_\omega$ is essentially a constant up to the frequency $\nu \approx 100\,\mathrm{GHz}$ and then monotonically decreases towards zero with continuing increase of the frequency. $\mathrm{Re}\,\mu_\omega$ is about a half of its low-frequency value at $\nu \approx 0.9\,\mathrm{THz}$. With enhancing bias dc field, $\mathrm{Re}\,\mu_\omega$ drops from the maximum low-frequency value $0.48\,\mathrm{m}^2\mathrm{V}^{-1}\mathrm{s}^{-1}$ and reaches a minimum at around $E_0 \approx 7.0\,\mathrm{kV/cm}$ before it slowly increases with continuing rise of the bias field. The most interesting feature of μ_ω behavior is that, at finite dc biases, $\mathrm{Re}\,\mu_\omega$ increases with increasing frequency starting from several tens of GHz and exhibits a broad hump at around 400 GHz then approaching zero at about the same frequency as in the zero bias case.

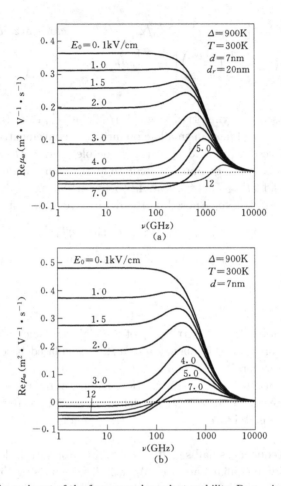

Fig. 12.19 The real part of the frequency-dependent mobility, $\text{Re}\,\mu_\omega$, is plotted against the signal frequency $\nu = \omega/2\pi$ at several different dc bias fields E_0 for two GaAs-based superlattices with period $d = 7\,\text{nm}$, miniband width $\Delta = 900\,\text{K}$, low-temperature linear dc mobility $\mu(0) = 1.0\,\text{m}^2\text{V}^{-1}\text{s}^{-1}$, and effective carrier bulk density $n = 2.14 \times 10^{23}/\text{m}^3$: (a) laterally confined system $d_r = 20\,\text{nm}$; (b) unconfined system $d_r = \infty$. The lattice temperature is $T = 300\,\text{K}$.

This kind of frequency dependence of the differential mobility is characteristic of superlattice miniband conduction, in which the average inverse effective mass $1/m_z^*$ decreases with increasing average electron energy ε_z in the z direction [see (12.107)], and the "energy relaxation time" τ_ε implicitly contained in the energy-balance equation to reflect the response speed of the average electron energy to the signal is much longer than the "momentum relaxation time" τ implicitly contained in the momentum-balance equation to reflect the response speed of the average electron velocity to the signal. It can be roughly estimated from the curves in Fig. 12.19(b) that $\tau_\varepsilon \approx 1.0 \times 10^{-12}$ s and $\tau \approx 1.5 \times 10^{-13}$ s. For laterally confined superlattice (Fig. 12.19(a) $\tau_\varepsilon \approx 0.4 \times 10^{-12}$ s.

For all the superlattices discussed above, even biased in the region where negative differential mobility show up in the dc steady state transport, the high-frequency small-signal mobility can be negative only at frequency range lower than a upper limit (Lei et al, 1994). For laterally unconfined superlattices, this frequency limit is around $100 \sim 200$ GHz. For laterally confined superlattices, the upper frequency for the existence of negative differential mobility can be much higher. This is because that the existence of transverse degrees of freedom makes it difficult for the system average energy (including the energy of longitudinal and transverse motions) to follow the high-frequency signal, i.e. prolonging the energy relaxation time of the system. Of course, this result obtained from the balance-equation theory can applicable only to system with high carrier density. In superlattices of low carrier density, where the coupling between the transverse and longitudinal motions will not be so strong, the miniband conduction behavior may be closer to laterally confined systems.

12.8.3 *High-frequency large-signal response*

To simulate the carrier transport in a superlattice under the irradiation of an electromagnetic wave of arbitrary strength, we consider a dc electric field E_0 and a sinusoidal ac field of frequency ω and amplitude E_ω,

$$E(t) = E_0 + E_\omega \sin(\omega t), \tag{12.121}$$

are applied along the superlattice growth axis. With this $E(t)$ in the momentum and energy balance equations (12.110) and (12.111) or (12.112) and (12.113), we can solve for the time-dependent response of the superlattice starting from the initial values of v_d and T_e. Since this electric field is a periodic function of time, after the transient dies out, the system will approach a time-dependent steady state, in which $v_d(t)$ and $T_e(t)$ and all other transport quantities are periodic time functions with period $T_\omega = 2\pi/\omega$ and can be written in the form of Fourier series consisting of dc and harmonic components. For the drift velocity of the steady state we write

$$v_\text{d}(t) = v_0 + \sum_{n=1}^{\infty} v_{n1}\sin(n\omega t) + v_{n2}\cos(n\omega t), \quad (12.122)$$

where the dc component v_0 and the harmonic coefficients v_{n1} and v_{n2} are calculated from $v_\text{d}(t)$ within a period after reaching the steady state:

$$v_0 = \frac{1}{T_\omega}\int_{t_1}^{t_1+T_\omega} v_\text{d}(t)\,dt, \quad (12.123)$$

$$v_{n1} = \frac{2}{T_\omega}\int_{t_1}^{t_1+T_\omega} v_\text{d}(t)\sin(n\omega t)\,dt, \quad (12.124)$$

$$v_{n2} = \frac{1}{T_\omega}\int_{t_1}^{t_1+T_\omega} v_\text{d}(t)\cos(n\omega t)\,dt. \quad (12.125)$$

We have calculated the time-dependent response of a GaAs-based n-type superlattice with period $d = 4.8$ nm, miniband width $\Delta = 50$ meV and electron bulk density $n = 8.2 \times 10^{16}$ cm^{-3}, to an electric field $E(t)$ of form (12.121) turned on at $t = 0$ with the initial condition $v_\text{d} = 0$ ($p_\text{d} = 0$) and $T_\text{e} = T$. At lattice temperature $T = 300$ K, the drift velocity $v_\text{d}(t)$ and electron temperature $T_\text{e}(t)$ essentially approach a time-dependent steady state within a time delay of a few picosecond. This time delay exhibits no apparent dependence on the frequency of the ac field. However, the steady-state behavior of $v_\text{d}(t)$ and $T_\text{e}(t)$ varies rather strongly with changing frequency. Fig. 12.20 shows one cycle of the steady time-dependent response of the drift velocity $v_\text{d}(t)$ and electron temperature $T_\text{e}(t)$ to an electric field of (12.121) with the dc-field strength $E_0 = 8$ kV/cm, ac-field amplitude $E_\omega = 8$ kV/cm and period $T_\omega = 1.0$ ps. We can see that at this frequency the oscillation of the electron temperature is quite out of phase of the ac electric field and the amplitude of the electron temperature oscillation is much reduced (less than 0.3 of that in the low-frequency ac response). For fixed $E_0 = 8$ kV/cm and $E_\omega = 8$ kV/cm, the maximum amplitude of the temperature oscillation, $\Delta T_\text{M} \equiv T_\text{e}|_\text{max} - T_\text{e}|_\text{min}$, and the dc component v_0 and the harmonic coefficients v_{11}, v_{12}, v_{21}, v_{22}, v_{31} and v_{32} of the drift velocity in the steady state, are presented in Fig. 12.21, as functions of the temporal period T_ω of the high-frequency field from $T_\omega = 0.1$ to 10 ps (frequency 100 GHz to 10 THz). The rapid decrease of ΔT_M with increasing frequency shows that the electron temperature does not follow the rapid oscillation of the applied field, such that T_e remains to be a relatively slowly varying quantity. When the frequency of the driven field becomes high, the electron temperature essentially keeps unchanged at a value determined by the strength and frequency of the field. On the other hand, even at $T_\omega = 10\,\text{ps}$ (100 GHz) the steady-state velocity response shown in Fig. 12.21 is still quite different from the static (zero frequency) case.

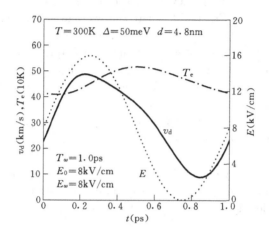

Fig. 12.20 Steady time-dependent response (one cycle) of drift velocity v_d (solid line) and electron temperature T_e (chain line) to an electric field $E(t) = E_0 + E_\omega \sin(2\pi t/T_\omega)$ with $E_0 = 8\,\text{kV/cm}$, $E_\omega = 8\,\text{kV/cm}$ and $T_\omega = 1.0\,\text{ps}$ (dashed line) in a GaAs-based superlattice of $d = 4.8\,\text{nm}$, $\Delta = 50\,\text{meV}$, and $n = 8.2 \times 10^{16}/\text{cm}^3$. From Lei (1997).

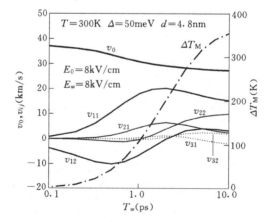

Fig. 12.21 ΔT_M, v_0 and the harmonic coefficients v_{11}, v_{12}, v_{21}, v_{22}, v_{31} and v_{32} in the steady time-dependent response of the superlattice described in Fig. 12.20, are plotted as functions of the period of the high-frequency field, T_ω. From Lei (1997).

One of the remarkable features of the steady-state response to large-signal high-frequency signal is the suppression of the average current by the strong high-frequency field. We can see from Fig. 12.21 that for fixed E_ω, high-frequency field of larger T_ω (lower frequency) produces stronger effect on v_0. For fixed frequency, ac field of larger amplitude has stronger effect as illustrated in Fig. 12.22, where the average velocity change $-\delta v_0$, defined by $\delta v_0 \equiv [v_0(E_0, E_\omega) - v_0(E_0, 0)]$, is shown as a function of the bias dc field E_0 for different values of the high-frequency field amplitude $E_\omega = 12, 20, 30$ and $50\,\mathrm{kV/cm}$ (solid curves) of fixed frequency (3 THz), together with the drift-velocity $v_0(E_0, 0)$ without high-frequency field (chain line) at lattice temperature $T = 300\,\mathrm{K}$. In the presence of an intense high-frequency field, the dc component v_0 of the steady-state velocity response is strongly reduced in comparison with that at $E_\omega = 0$. These are in qualitative agreement with the experimental results of Schomburg et al (1996) and Winnerl et al (1997), which are indicated in the inset as the measured current change $-\delta I$ versus the applied voltage U across a similar superlattice under the influence of a 3.9 THz field of three different power levels.

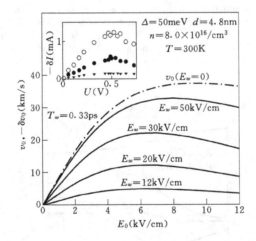

Fig. 12.22 The calculated average velocity change $-\delta v_0 \equiv [v_0(E_0, 0) - v_0(E_0, E_\omega)]$, is plotted as a function of the bias dc field E_0 at different values of the high-frequency field amplitude $E_\omega = 12, 20, 30$ and $50\,\mathrm{kV/cm}$ (solid curves), for the same superlattice as described in Fig. 12.20. The chain line represents the drift velocity in the case without high-frequency field, i.e. $v_0(E_0, 0)$. The inset shows the experimental data of Winnerl et al (1997) on the current change $-\delta I$ versus the applied voltage U across a similar superlattice under the influence of a 3.9 THz field of three different power levels. From Lei (1997).

12.9 Miniband Transport of Superlattices in a Quantized Magnetic Field, Magneto-Phonon Resonance

We consider a system which is composed of N electrons moving in a model superlattice with periodical potential wells of period d along the z direction under the influence of a longitudinal magnetic field $\bm{B} = (0,0,B)$. The electron energy spectrum still forms minibands in the longitudinal direction due to the periodical potential, while in the x–y plane it is quantized into Landau levels due to the magnetic field. Taking into account only the lowest miniband the electron state can be described, in the Landau representation, by the quantum number n of the Landau level, the wave vectors k_x and k_z ($-\pi/d < k_z \leqslant \pi/d$) and a spin index σ. The spatial wave function can be written as

$$\varphi_{nk_xk_z}(x,y,z) = \frac{1}{\sqrt{L_x}} e^{i k_x x} \chi_{nk_x}(y) \xi_{k_z}(z), \qquad (12.126)$$

L_x is the sample length along the x direction, $\xi_{k_z}(z)$ stands for the Bloch function of the superlattice miniband along the z direction, and

$$\chi_{nk_x}(y) = \left(\frac{1}{\sqrt{\pi} l_\mathrm{B} 2^n n!}\right)^{1/2} \exp\left[-\frac{(y-y_0)^2}{2 l_\mathrm{B}^2}\right] \mathrm{H}_n\!\left(\frac{y-y_0}{l_\mathrm{B}}\right), \qquad (12.127)$$

in which $y_0 = -k_x/(eB)$, $l_\mathrm{B} = |eB|^{-1/2}$ is the magnetic length, and $\mathrm{H}_n(x)$ are the Hermite polynomials. The eigen energy is

$$\varepsilon_n(k_z) = \left(n + \frac{1}{2}\right) \omega_\mathrm{c} + \varepsilon(k_z) \quad (n = 0, 1, 2, \cdots), \qquad (12.128)$$

where $\omega_\mathrm{c} = |eB|/m$ is the cyclotron frequency, m is the effective mass of the (parabolic band) electrons of the background material, and $\varepsilon(k_z)$ is the energy spectrum of the lowest superlattice miniband, assumed to be of the tight-binding form:

$$\varepsilon(k_z) = \frac{\Delta}{2}\big[1 - \cos(k_z d)\big], \qquad (12.129)$$

Δ being the miniband width. The eigen energy is degenerate with respect to k_x, and for simplicity, we neglect the energy difference between different spin states. The degeneracy of each Landau level is proportional to the strength of the magnetic field and is given by

$$\sum_{k_x,\sigma} \sim \frac{L_x}{\pi} \int dk_x \sim \frac{L_x L_y}{\pi l_\mathrm{B}^2} = \frac{S}{\pi l_\mathrm{B}^2}, \qquad (12.130)$$

where L_y is the sample length along the y direction, and $S = L_x L_y$ is the transverse area of the sample. In obtaining the above expression we have made use of the condition that y_0 is within L_y to determine the integral range of dk_x.

We can see that the effect of applying a quantized magnetic field along the axial direction in an unconfined superlattice is equivalent to adding a laterally confined potential to the system: changing the original continuous energy spectrum into a quantized separate spectrum. Therefore, all the formulas from (12.91) to (12.100) in Sec. 12.7 for laterally confined superlattices are applicable in the present case as long as the indices n and n' there are understood as the Landau-level numbers and the form factor $|J_{nn'}(q_\parallel)|^2$ is replaced by the function $C_{nn'}(l_B^2 q_\parallel^2/2)$ defined as

$$C_{nn'}(x) = \frac{n_2!}{n_1!} x^{n_1-n_2} e^{-x} \left[L_{n_2}^{n_1-n_2}(x) \right]^2 \qquad (12.131)$$

[$L_n^l(x)$ is the associated Laguerre polynomials], and the total electron number N in all the formulas is changed to

$$N \to 2\pi l_B^2 N_s Z_n, \qquad (12.132)$$

where N_s is the electron area density per period, Z_n is the total number of periods in the z direction ($L_z = Z_n d$, L_z is the sample length in the z direction).

These formulae are applied by Shu and Lei (1994) to the investigation of linear and nonlinear transport properties of three GaAs-based quantum well superlattices with the same period $d = 9.0\,\text{nm}$, well width $a = 6.0\,\text{nm}$ and electron sheet density $N_s = 2.0 \times 10^{14}\,\text{m}^{-2}$ per period, but having different miniband widths $\Delta = 8.5, 21.3$ and $60\,\text{meV}$, subject to strong longitudinal magnetic fields ranging from 6 to 30 T. The elastic scatterings in the systems are assumed due to randomly distributed background charge impurities with density $n_i = 8.6 \times 10^{21}\,\text{m}^{-3}$, which produces low temperature (4.2 K) linear mobilities $\mu_0 = 0.17, 0.78$ and $2.8\,\text{m}^2\text{V}^{-1}\text{s}^{-1}$ for systems of $\Delta = 8.5$, 21.3 and 60 meV respectively. Both longitudinal and transverse acoustic phonons (through the deformation potential and piezoelectric couplings with electrons) and longitudinal optic (LO) phonons (through the Fröhlich coupling with electrons) are taken into account. The maximum number of Landau subbands n_M in the calculation is determined by the condition that more than 99% of total electrons is accommodated inside the levels $n = 0, 1, \cdots, n_M$ at the highest temperature. Up to 25 Landau levels are included in the real calculation for magnetic field as low as $B = 6\,\text{T}$.

The calculated linear mobility as a function of the longitudinal magnetic field B for the superlattice of miniband width $\Delta = 8.5\,\text{meV}$ is shown in Fig. 12.23 at three lattice temperatures $T = 50, 150$ and $300\,\text{K}$. Since the LO-phonon energy $\Omega_{\text{LO}} = 35.6\,\text{meV}$, the intra-miniband LO-phonon scatterings are always excluded in this narrow miniband system. At low temperature ($T = 50$ K), impurities and acoustic phonons (spread energy spectrum) are main scatterers and the mobility decreases monotonically with increasing magnetic field. At higher lattice temperatures when the LO-phonon scattering is greatly enhanced ($T = 150$ K) or becomes dominant ($T = 300$ K), it appears a strong mobility oscillation superposed on the decreasing background component. There clearly exhibit three minima at $B \approx 7, 11$ and $22\,\text{T}$. They can be easily explained by the maximum in the density of states at the bottom and top of each separated Landau miniband under a quantizing magnetic field as pointed out by Noguchi et al (1992). When the magnetophonon resonance condition

$$n\omega_c = \Omega_{\text{LO}} \quad (n = \text{integer}) \tag{12.133}$$

is satisfied, the final states exist for all electrons in the miniband to scatter

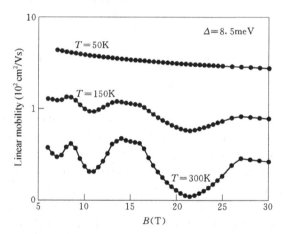

Fig. 12.23 Linear mobility is shown as a function of the longitudinal magnetic field B for a superlattice of miniband width $\Delta = 8.5\,\text{meV}$ at three lattice temperatures $T = 50, 150$ and $300\,\text{K}$. From Shu and Lei (1994).

to if they absorb an optic phonon, resulting in a larger magnetoresistance. On the contrary, deviation from the above resonance condition greatly reduces the available final states for electron scattering by optic phonons, leading to a smaller resistance. The three minima just corresponding to the resonance condition of $n = 1, 2$ and 3. The $n = 1$ valley, which corresponds to the highest magnetic field to satisfy the resonance condition (12.133), is the deepest one since the density of states in each Landau level is proportional to B. The oscillation amplitudes are larger at $T = 300\,\text{K}$ than at $T = 150\,\text{K}$, simply because that LO-phonon scatterings are much stronger at higher temperature.

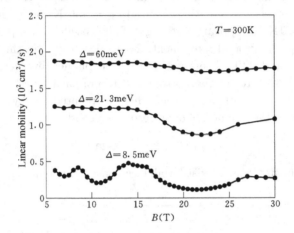

Fig. 12.24 Linear mobility is shown as a function of the longitudinal magnetic field B for three superlattices having miniband widths $\Delta = 8.5, 21.3$ and $60\,\text{meV}$ at lattice temperature $T = 300\,\text{K}$. From Shu and Lei (1994).

For fixed spatial period d of the superlattice, the total number of electron states (unit volume) in each miniband is identical. In wider miniband these electron states spread over a wider energy range and the singularity of their density of states reduces. The magnetophonon resonance oscillation is anticipated to weaken, as is clearly seen from Fig. 12.24, in which we plot the calculated linear mobility as a function of the longitudinal magnetic field for superlattices having $\Delta = 8.5, 21.3$ and $60\,\text{meV}$ at lattice temperature $T = 300\,\text{K}$. Though all $n = 1, 2$ and 3 minima appear clearly for system of $\Delta = 8.5\,\text{meV}$, in the case of $\Delta = 21.3\,\text{meV}$ the $n = 2$ and $n = 3$ minima almost disappear and only $n = 1$ minimum retains. It further weakens in the $\Delta = 60\,\text{meV}$ system. At lower lattice temperatures, these structures are further smoothed out, exhibiting a monotonic decrease of the mobility with increasing magnetic field. All these features are in good qualitatively agreement with the experimental results of Sibille *et al* (1990b), Noguchi,

Leburton and Sakaki (1993), and Hutchinson *et al* (1994).

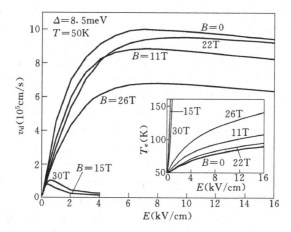

Fig. 12.25 Nonlinear drift velocity v_d and electron temperature T_e are plotted as functions of the applied electric field E for a superlattice of $\Delta = 8.5\,\text{meV}$ at lattice temperature $T = 50\,\text{K}$, under the influence of longitudinal magnetic fields having strengths $B = 0, 11, 15, 22, 26$ and $30\,\text{T}$, respectively. From Shu and Lei (1994).

Strong oscillation also appears in the nonlinear transport in narrow miniband superlattices. In Fig. 12.25 we plot the calculated the drift velocity v_d as a function of the electric field E for the superlattice of $\Delta = 8.5\,\text{meV}$ at lattice temperature $T = 50\,\text{K}$, under the influence of a longitudinal magnetic field having strengths $B = 0, 11, 15, 22, 26$ and $30\,\text{T}$, respectively. Two curves, with $B = 30\,\text{T}$ and $B = 15\,\text{T}$, are distinctively different from the other four. This is caused by the prohibition of the LO-phonon scattering. There are two cases in which the optic-phonon scattering is forbidden. One is when the magnetic field is large enough that electron can never jump to the higher Landau miniband even if they absorb an optic phonon, i.e. $\omega_c > \Omega_{\text{LO}} + \Delta$. The other is when the magnetic field satisfies the condition $n\omega_c + \Delta < \Omega_{\text{LO}} < (n+1)\omega_c - \Delta$, such that the whole range of the electron energy in a miniband, after absorbing an optic phonon, will completely fall inside the gap between two neighboring Landau minibands. Since acoustic phonons are far less efficient in dissipating energy from the carrier system, the electron temperature increases rapidly with increasing electric field. This quick rise of the electron temperature drastically limits the maximum average drift velocity of the carriers in the miniband and thus the peak drift velocity v_p is reached at a small critical electric field E_c. This case is very similar to that of a laterally confined superlattice discussed in Sec. 12.7. $B = 30\,\text{T}$ and $B = 15\,\text{T}$ correspond to the first and the second cases of the prohibition of the LO-phonon scattering. The electron temperature in these two cases grow very quickly, as can be seen in the inset of Fig. 12.25, where the electron temperature T_e is shown as a function of the applied electric field for all five magnetic fields. In Fig. 12.25 the $B = 22\,\text{T}$ curve and $B = 11\,\text{T}$ curve correspond to cases when the magnetophonon resonance condition (12.133) is satisfied and the optic-phonon scatterings are strong. Since the LO-phonon scattering is very efficient in transferring energy from the carrier system, the increase of the electron temperature with increasing electric field slows down greatly and the peak drift velocity reaches at a much higher electric field. The $B = 26\,\text{T}$ curve is an intermediate case when only a small portion of electrons in the miniband can absorb an optic phonon and jump to a higher Landau miniband. Though this part of electrons is only about 20% of all electrons in this special case, the allowed LO-phonon scattering has already shown the ability to dissipate the electron energy and to keep the carrier system much cooler than at $B = 15\,\text{T}$ and $B = 30\,\text{T}$.

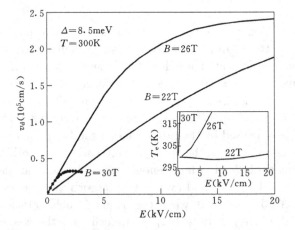

Fig. 12.26 Nonlinear drift velocity v_d and electron temperature T_e are shown as functions of the applied electric field E at lattice temperature $T = 300\,\text{K}$, for the same $\Delta = 8.5\,\text{meV}$ superlattice as described in Fig. 12.25, subjected to longitudinal magnetic fields having strengths $B = 22, 26$ and $30\,\text{T}$, respectively. From Shu and Lei (1994).

Figure 12.26 shows the calculated v_d and T_e as functions of the applied electric field E for the same $\Delta = 8.5\,\text{meV}$ system at lattice temperature $T =$

300 K. At this temperature a sufficient number of LO phonons is excited and available to be absorbed by the carriers if the electron–LO-phonon scattering is energetically allowed. In the case of resonance, $B = 22\,\text{T}$, LO-phonon scattering is so strong that the electron temperature experiences almost no change with increasing the electric field, even becomes lower than the lattice temperature within a wide range of the field strength between $E = 0$ and $E = 15\,\text{kV/cm}$, as is seen in the inset of the figure. In the absence of a magnetic field this type of electron cooling may appear for systems where the impurity scattering is very weak (Lei and Ting, 1984, 1985c; Lei and Horing, 1987b). The present result indicates that a magnetic field may promote the appearance of the electron cooling in a system with less weak impurity scattering due to the enhancement of the LO-phonon scattering by the magnetophonon resonance.

12.10 Balance Equations for One-Dimensional Superlattices, Relaxation Time Approximation

12.10.1 *Balance equations for one-dimensional superlattices*

In Secs. 12.7 and 12.8 we have discussed miniband transport in laterally confined and unconfined superlattices from the momentum and energy balance equations ($a \equiv a_\text{i} + a_\text{p}$),

$$\frac{dv_\text{d}}{dt} = \frac{eE(t)}{m_z^*} + a, \tag{12.134}$$

$$\frac{d\varepsilon_\text{e}}{dt} = eE(t)v_\text{d} - w. \tag{12.135}$$

These superlattices are essentially three-dimensional systems. Their single electron state is described by a quantum number n describing the transverse motion and a longitudinal wavevector k_z (one miniband only, $-\pi/d < k_z \leqslant \pi/d$), and the energy $\varepsilon_n(k_z)$ is written as the sum of a transverse motion energy ε_n and a longitudinal miniband energy $\varepsilon(k_z)$:

$$\varepsilon_n(k_z) = \varepsilon_n + \varepsilon(k_z), \tag{12.136}$$

$$\varepsilon(k_z) = \frac{\Delta}{2}\big[1 - \cos(k_z d)\big]. \tag{12.137}$$

For systems in which carriers move freely in the lateral directions (x–y plane) the transverse quantum number and energy are $\boldsymbol{k}_\parallel = (k_x, k_y)$ and

$\varepsilon_{\boldsymbol{k}_\parallel} = k_\parallel^2/2m$. If carriers are laterally confined within a cylindrical region of diameter d_r, the transverse quantum number is $n = (l, m)$ and the transverse motion energy is given by (12.90). The three-dimensional nature of balance equations (12.134) and (12.135) resides in the fact that the average electron energy ε_e contains, in addition to the longitudinal motion energy ε_z, a transverse motion energy $\varepsilon_\parallel$:

$$\varepsilon_e = \varepsilon_\parallel + \varepsilon_z. \tag{12.138}$$

Note that, with the miniband energy dispersion (12.137), the average inverse effective mass component $1/m_z^*$ in the momentum balance equation (12.134) depends only on ε_z as given by (12.107), or

$$m_z^*(\varepsilon_z) = \frac{M^*}{1 - 2\varepsilon_z/\Delta}, \tag{12.139}$$

where $M^* \equiv \Delta d^2/2$.

If carriers are confined to very small dimensions in both directions in the lateral plane (one-dimensional superlattice) that they are frozen in the lowest transverse subband during the transport process, $\varepsilon_\parallel$ will be a constant. In this case, if the frictional acceleration a and energy-transfer rate w also depend only on v_d and ε_z, the momentum and energy balance equations (12.134) and (12.135) may reduce to pure one-dimensional forms. The simplest way for this purpose is to take the constant relaxation time approximation, i.e. assuming that the frictional acceleration a is proportional to the drift velocity

$$a = -v_d/\tau, \tag{12.140}$$

(τ is called the momentum relaxation time) and the energy transfer rate w is proportional to the energy deviation from the equilibrium value ϵ_T at the lattice temperature T in the absence of the electric field:

$$w = (\varepsilon_z - \varepsilon_T)/\tau_\varepsilon, \tag{12.141}$$

(τ_ε is call the energy relaxation time). Here

$$\varepsilon_T = \frac{\Delta}{2}[1 - \alpha(T)], \tag{12.142}$$

$$\alpha(T) = \frac{2}{\pi N_1 d} \int_0^\pi \frac{\cos\theta \, d\theta}{\exp\{[\Delta/2(1 - \cos\theta) - \mu]/T\} + 1}, \tag{12.143}$$

N_1 is the number of electrons in the superlattice of unit length and μ is the chemical potential, satisfying the equation of the particle number

$$N_1 d = \frac{2}{\pi} \int_0^\pi \frac{d\theta}{\exp\{[\Delta/2(1-\cos\theta) - \mu]/T\} + 1}. \tag{12.144}$$

Under the relaxation-time approximation the momentum and energy balance equations (12.134) and (12.135) become pure one-dimensional equations containing two variables v_d and ε_z only:

$$\frac{dv_d}{dt} = \frac{eE(t)}{m^*(\varepsilon_z)} - v_d/\tau, \tag{12.145}$$

$$\frac{d\varepsilon_z}{dt} = eE(t)v_d - (\varepsilon_z - \varepsilon_T)/\tau_\varepsilon. \tag{12.146}$$

This set of equations is similar to the equations used by Ignatov, Dodin and Shashkin (1991) in discussing superlattice miniband transport, with the difference that the equilibrium energy ε_T used by these authors is the nondegenerate limit of ε_T given by (12.142). In this nodegenerate limit $(-\mu/T \gg 1)$, $f(\varepsilon, T) \sim e^{(\mu-\varepsilon)/T}$, $\alpha(T) \sim I_1(\Delta/2T)/I_0(\Delta/2T)$, thus

$$\varepsilon_T = \frac{\Delta}{2}\left[1 - \frac{I_1(\Delta/2T)}{I_0(\Delta/2T)}\right]. \tag{12.147}$$

The nondegenerate limit always exists in three-dimensional superlattices, as long as the temperature is high enough. However, for these three-dimensional systems the carrier transverse energy $\varepsilon_\parallel$ has to be included in ε_e and the energy balance equation can not be simply written in the form of (12.146), unless the carrier transverse motion has no correlation with its longitudinal motion that $\varepsilon_\parallel$ keeps unchanged during the process of longitudinal miniband transport.

In a pure one-dimensional superlattice, in which carriers laterally are frozen in one subband, the zero-temperature Fermi energy ε_F obtained from the particle number equation (12.144) is

$$\frac{2\varepsilon_F}{\Delta} = 1 - \cos\left(\frac{\pi}{2}N_1 d\right). \tag{12.148}$$

At enough high temperature ($\Delta/2T \ll 1$) the chemical potential is

$$\mu/T \simeq -\ln\left(\frac{2}{N_1 d} - 1\right) + \frac{\Delta}{2T}. \tag{12.149}$$

Since the maximum particle number which can be accommodated in a miniband is $N_1 d = 2$, Eq. (12.149) shows that in the high temperature limit μ/T

approaches a constant:

$$\lim_{T \to \infty} \frac{\mu}{T} = -\ln\left(\frac{2}{N_1 d} - 1\right), \quad (12.150)$$

which is negative when $N_1 d < 1$ and positive when $N_1 d > 1$. Therefore, the nondegenerate limit generally does not exist in a pure one dimensional system unless the carrier density is very low, $N_1 d \approx 0$, that $\ln[2/(N_1 d) - 1] \gg 1$ (Lei, 1992b). By the way, when $N_1 d = 1$ (half filled), the Fermi energy does not change with temperature: $\mu = \Delta/2$. For pure one-dimensional superlattices the behavior of $\alpha(T)$ (12.143) is easily seen: in the low temperature limit

$$\alpha(T) = \sin\left(\frac{\pi}{2} N_1 d\right)\left(\frac{\pi}{2} N_1 d\right)^{-1}, \quad (12.151)$$

while in high temperature limit ($\Delta/2T \ll 1$)

$$\alpha(T) \simeq \left(1 - \frac{N_1 d}{2}\right)\frac{\Delta}{4T}. \quad (12.152)$$

Only when $N_1 d \ll 1$, this behavior agrees with the nondegenerate limit of $\alpha(T)$,

$$\alpha(T)|_{\text{nondeg}} = \frac{I_1(\Delta/2T)}{I_0(\Delta/2T)} \sim \frac{\Delta}{4T}. \quad (12.153)$$

12.10.2 DC and small-signal solutions of one-dimensional balance equations for superlattices

The analytical solution of one-dimensional momentum and energy balance equations (12.145) and (12.146) can be obtained in most cases when the applied electric field $E(t)$ is given.

We first consider the case of applying a dc electric field $E(t) = E_0$. By denoting the steady state drift velocity $v_d = v_0$ and average energy $\varepsilon_z = \varepsilon_0$, we can write the one-dimensioanl momentum and energy balance equations (12.145) and (12.146) in the form

$$\frac{eE_0}{M^*}\left(1 - \frac{2\varepsilon_0}{\Delta}\right) - \frac{v_0}{\tau} = 0, \quad (12.154)$$

$$eE_0 v_0 - \frac{(\varepsilon_0 - \varepsilon_T)}{\tau_\varepsilon} = 0. \quad (12.155)$$

This set of equations yields

$$v_0 = v_\mathrm{p} \frac{2b}{1+b^2}. \tag{12.156}$$

Here $b = E_0/E_\mathrm{c}$,

$$E_\mathrm{c} = 1/[ed(\tau\tau_\varepsilon)^{1/2}], \tag{12.157}$$

$$v_\mathrm{p} = \frac{\Delta d}{4}\left(\frac{\tau}{\tau_\varepsilon}\right)^{1/2}\alpha(T). \tag{12.158}$$

Therefore, the reduced drift velocity v_0/v_p versus the reduced electric field E_0/E_c derived from pure one-dimensional balance equation still follows the Esaki-Tsu formula. Here the factor $\alpha(T)$ is given by (12.151). If $\alpha(T)$ in (12.158) is replaced by its nondegenerate limit, $\mathrm{I}_1(\Delta/2T)/\mathrm{I}_0(\Delta/2T)$, the above results are equivalent to the solutions (12.32)–(12.34) of the Boltzmann equation with relaxation time approximation. However, because of the possible absence of the nondegenerate limit in a pure one-dimensional system, there may be some difference between the two results when the carrier density is not small.

Next, we consider a dc bias electric field E_0 and a small signal ac electric field of angular frequency ω,

$$\delta E = E_1 \mathrm{e}^{-\mathrm{i}\omega t} + E_1^* \mathrm{e}^{\mathrm{i}\omega t}, \tag{12.159}$$

apply simultaneously to the superlattice that the total electric field is $E(t) = E_0 + \delta E$. After the transient dies out the system arrives a time-dependent steady state, in which all the transport quantities, such as the drift velocity and the average energy, can be written as the sum of a dc part and a small ac part, $v_\mathrm{d} = v_0 + \delta v$ and $\varepsilon_z = \varepsilon_0 + \delta\varepsilon$. Substituting these expressions into the momentum and energy balance equations (12.145) and (12.146), we immediately get Eqs. (12.154) and (12.155) for the zero-th order quantities, and thus the relation (12.156) between the dc bias velocity v_0 and dc electric field E_0. The equations of the first order in the small-signal quantities are (Lei, 1994d)

$$\frac{d}{dt}\delta v + \frac{\delta v}{\tau} = \frac{e\,\delta E}{m_z^*(\varepsilon_0)} - eE_0 d^2 \delta\varepsilon, \tag{12.160}$$

$$\frac{d}{dt}\delta\varepsilon + \frac{\delta\varepsilon}{\tau_\varepsilon} = ev_0\,\delta E + eE_0\,\delta v. \tag{12.161}$$

Under the influence of the small-signal ac electric field (12.159) of frequency ω, the steady-state responses of drift velocity and average energy are also small signals of the same frequency ω:

$$\delta v = v_1 e^{-i\omega t} + v_1^* e^{i\omega t}, \quad (12.162)$$

$$\delta\varepsilon = \varepsilon_1 e^{-i\omega t} + \varepsilon_1^* e^{i\omega t}. \quad (12.163)$$

Their amplitudes v_1 and ε_1 can be easily solved from Eqs. (12.160) and (12.161), and then all other relevant transport quantities are obtained. For instance, the complex frequency-dependent mobility is

$$\mu_\omega \equiv \frac{v_1}{E_1}. \quad (12.164)$$

The main difference of Eq. (12.160) from the ordinary equation lies in the second term on the right-hand side of it. This term comes from the fact that the increase in the system internal energy, $\delta\epsilon$, due to a small increment of the signal field, results in a decrease in the inverse effective mass. This effect has a profound influence on the differential mobility. In fact, Eqs. (12.160) and (12.161) yield a complex differential mobility at frequency ω of the form (Lei, Horing and Cui, 1993; Lei, 1994d)

$$\mu_\omega = \mu_c \frac{1 - b^2(1 - i\omega\tau_\varepsilon)^{-1}}{1 - i\omega\tau + b^2(1 - i\omega\tau_\varepsilon)^{-1}}, \quad (12.165)$$

where

$$\mu_c = \frac{v_0}{E_0} = \frac{\Delta d^2 e\tau\alpha(T)}{2(1+b^2)} \quad (12.166)$$

is the dc mobility at the bias electric field E_0. According to (12.165), the zero-frequency differential mobility is

$$\mu_0 = \mu_c \frac{1-b^2}{1+b^2} \quad (12.167)$$

$$= \mu_0(0) \frac{1-b^2}{(1+b^2)^2}, \quad (12.168)$$

in which

$$\mu_0(0) = \frac{1}{2}\Delta d^2 e\tau\alpha(T) \quad (12.169)$$

is the zero-frequency differential mobility at zero dc bias ($b = 0$).

The terms containing b^2 in the expressions (12.165) and (12.167) are related to the second term on the right-hand side of Eq. (12.160). We can see that when a small ac field δE is applied together with a finite dc bias $E_0 \neq 0$, there are two influential effects. First, at finite bias field E_0 the inverse effective mass $1/m_z^*(\varepsilon_0)$ is smaller than at zero bias, and it monotonically decreases with increasing E_0 because the average longitudinal energy ε_0 increases. However, since $1/m_z^*(\varepsilon_0)$, determined by the steady-state dc balance equations, is always positive, this effect, though tending to decrease the differential mobility, can not bring it into negative regime. The occurrence of negative differential mobility devolves upon the fact that a positive δE induces a positive $\delta\varepsilon$, as required by the energy balance equation, and this $\delta\varepsilon$ in turn results in an additional negative change of the inverse effective mass, which contributes a negative term [the term containing b^2 in the numerator of Eq. (12.160)] to the differential mobility at a non-zero dc bias. It is just this contribution that gives rise to negative differential mobility at low frequency in superlattice miniband transport. This effect, which stems from the change of average longitudinal energy with a small ac electric-field signal, however, is frequency-dependent. At higher frequency, when $\delta\varepsilon$ can not synchronously follow δE as in the low-frequency limit, this effect weakens and consequently $\mathrm{Re}\,\mu_\omega$ may exhibit an increase with increasing frequency at finite dc bias. Fig. 12.27 shows the real part $\mathrm{Re}\,\mu_\omega$

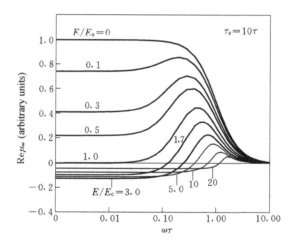

Fig. 12.27 The real part of frequency-dependent mobility μ_ω, as predicted by Eq. (12.165) in the case of $\tau_\varepsilon = 10\tau$, is shown as a function of $\omega\tau$ for different bias dc fields.

of μ_ω given by (12.165) as a function of frequency $\omega\tau$ at different bias dc fields $b \equiv E_0/E_c = 0$, 0.1, 0.3, 0.5, 1, 1.7, 3, 5, 10 and 20, for the case of $\tau_\varepsilon = 10\tau$. It is just this effect that gives rise to the humps in $\mathrm{Re}\mu_\omega$ as a function of $\omega\tau$ before it finally vanishes at large $\omega\tau$. Without this effect (i.e. $b=0$ and thus $\epsilon_0 = \epsilon_T$) μ_ω will be a Drude form:

$$\mu_\omega = \frac{\mu_0(0)}{1 - i\omega\tau}. \tag{12.170}$$

Though a series of major approximations are contained in the one-dimensional balance equations (12.154) and (12.155) of constant relaxation times, they have been widely used in the analyses of many superlattice transport problems for their simplicity and possibility of obtaining analytical results (Ignatov, Renk and Dodin, 1993; Lei *et al*, 1997).

Chapter 13

Nonparabolic Systems with Magnetic field, Impact Ionization or under Nonuniform Condition

In this chapter we will address several other issues of electronic transport under different conditions in nonparabolic band semiconductors in the framework of balance equation theory, including the following issues: (i) The hot-electron transport in an arbitrary energy band in crossed magnetic and electric fields (the classical magnetotransport in 3D tight-binding energy bands, quantum-wire arrays and superlattice minibands); (ii) The carrier transport in a nonparabolic Kane band; (iii) The high electric field transport in nonparabolic multiband (multivalley) systems (hot-electron transport in Si, GaAs and the longitudinal transport of a superlattice with high-lying minibands); (iv) Transport in semiconductors with impact ionization and Auger recombination effect; (v) The hydrodynamic balance equation description of spatially nonuniform nonparabolic band systems and the simulation of transport process in semiconductor devices made from materials with nonparabolic Kane bands and superlattice minibands.

13.1 Balance Equations for Electron Transport in an Arbitrary Energy Band in Crossed Magnetic and Electric Fields

We consider N electrons moving in a single general energy band with the energy dispersion $\varepsilon(\bm{k}) = \varepsilon(k_x, k_y, k_z)$. There is a spatially uniform electric field $\bm{E} = (E_x, E_y, 0)$ in the x–y plane and a uniform magnetic field $\bm{B} = (0, 0, B)$ applies in the z direction. The magnetic field is assumed to be weak that the Landau quantization is not important and its role can be considered as a classical external field acting on electrons in an energy band. Taking the vector potential to be $\bm{A} = (0, Bx, 0)$ and using the single-band effective Hamiltonian description (Blount, 1962; Anderson, 1963), we can write the electron part of the Hamiltonian as

$$H_e = \sum_j h_j, \quad h_j = \varepsilon(\hat{k}_{xj}, \hat{k}_{yj} - eBx_j, \hat{k}_{zj}), \tag{13.1}$$

in which $\hat{\boldsymbol{k}}_j = (\hat{k}_{xj}, \hat{k}_{yj}, \hat{k}_{zj}) = -i\boldsymbol{\nabla}_j$, is the lattice momentum operator of the jth electron (Sec. 11.1.4). The total Hamiltonian of the electron–phonon system is $H = H_e + H_E + H_I + H_{ph}$, with H_E, $H_I = H_{ei} + H_{ep}$ and H_{ph} given by (11.4), (11.6), (11.7) and (11.8).

The electron average velocity operator $\hat{\boldsymbol{v}}$ is the rate of change of the average position operator $\boldsymbol{R} = N^{-1}\sum_j \boldsymbol{r}_j$ (i.e. the center-of-mass position operator), $\hat{\boldsymbol{v}} = -i[\boldsymbol{R}, H]$, and thus the rate of change of the velocity is

$$\frac{d\hat{\boldsymbol{v}}}{dt} = -[[\boldsymbol{R}, H], H]. \tag{13.2}$$

Calculating the above commutation relations and taking its statistical average, we obtain the following equations in the case of weak magnetic fields (Lei, 1995a)

$$\frac{dv_x}{dt} = \frac{eE_x}{m^*_{xx}} + \frac{eE_y}{m^*_{xy}} + \frac{eBv_y}{m^*_{xx}}\gamma_{y,xx} - \frac{eBv_x}{m^*_{xy}}\gamma_{x,xy} + a_x, \tag{13.3}$$

$$\frac{dv_y}{dt} = \frac{eE_x}{m^*_{yx}} + \frac{eE_y}{m^*_{yy}} + \frac{eBv_y}{m^*_{yx}}\gamma_{y,xy} - \frac{eBv_x}{m^*_{yy}}\gamma_{x,yy} + a_y, \tag{13.4}$$

$$\frac{dv_z}{dt} = \frac{eE_x}{m^*_{zx}} + \frac{eE_y}{m^*_{zy}} + \frac{eBv_y}{m^*_{zx}}\gamma_{y,zx} - \frac{eBv_x}{m^*_{zy}}\gamma_{x,zy} + a_z, \tag{13.5}$$

where $\boldsymbol{a} = (a_x, a_y, a_z) = \boldsymbol{a}_i + \boldsymbol{a}_p$ is the damping acceleration induced by the impurity and phonon scatterings, expressed by (11.80) and (11.81). The equation set (13.3)–(13.5) is the equation of motion of a single particle which carries a charge e, possesses an inverse effective mass tensor $\mathcal{K}_{ab} \equiv 1/m_{ab}$ ($a, b = x, y, z$) and moves at a velocity $\boldsymbol{v} \equiv (v_x, v_y, v_z)$ in crossed magnetic and electric fields. Here the velocity $\boldsymbol{v}$ and the inverse effective mass tensor $\mathcal{K}$,

$$\boldsymbol{v} = \langle \boldsymbol{v}(\boldsymbol{k}) \rangle, \tag{13.6}$$

$$\mathcal{K} = \langle \boldsymbol{\nabla}\boldsymbol{\nabla}\varepsilon(\boldsymbol{k}) \rangle, \tag{13.7}$$

are the same as given by (11.77) and (11.78). In addition, six dimensionless coefficients $\gamma_{y,xx}$, $\gamma_{x,yy}$, $\gamma_{x,xy}$, $\gamma_{y,xy}$, $\gamma_{x,yz}$ and $\gamma_{y,zx}$, are introduced to describe the modification of a nonparabolic band on the role of the magnetic field. They are defined as $(a, b, c = x, y, z)$

$$\gamma_{a,bc} = \frac{\langle \varepsilon'_a \varepsilon''_{bc} \rangle}{\langle \varepsilon'_a \rangle \langle \varepsilon''_{bc} \rangle}, \qquad (13.8)$$

with $\varepsilon'_a \equiv \partial \varepsilon(\mathbf{k})/\partial k_a$, $\varepsilon''_{ab} \equiv \partial^2 \varepsilon(\mathbf{k})/\partial k_a \partial k_b$. Here $\langle \cdots \rangle$ still represents the average with respect to the distribution function $f(\bar{\varepsilon}(\mathbf{k}), T_e)$ as in (13.6) and (13.7):

$$\langle \cdots \rangle = \frac{2}{N} \sum_{\mathbf{k}} \cdots f(\bar{\varepsilon}(\mathbf{k}), T_e). \qquad (13.9)$$

On the other hand, calculating the rate of change of the electron energy,

$$\frac{d}{dt} H_e = -i [H_e, H], \qquad (13.10)$$

we obtain the energy balance equation:

$$\frac{d\varepsilon_e}{dt} = eE_x v_x + eE_y v_y - w, \qquad (13.11)$$

which is exactly the same as equation (11.76) in the absence of the magnetic field and the expression of the electron energy dissipation rate w is given by (11.82).

For an isotropic parabolic band, $\mathcal{K} = \mathcal{I}/m$ and $\gamma_{a,bc} = 1$, the equation of motion of the electron, (13.3)–(13.5), reduces to

$$\frac{d\mathbf{v}}{dt} = \frac{e\mathbf{E}}{m} + \frac{e}{m} \mathbf{v} \times \mathbf{B} + \mathbf{a}. \qquad (13.12)$$

The statistical averages in (13.6)–(13.8) can be carried out using the distribution function (11.65),

$$f(\bar{\varepsilon}(\mathbf{k}), T_e) = f_d(\mathbf{k}) = \frac{1}{e^{[\varepsilon(\mathbf{k}-\mathbf{p}_d)-\mu]/T_e} + 1}, \qquad (13.13)$$

or using the distribution function (11.70),

$$f(\bar{\varepsilon}(\mathbf{k}), T_e) = f_v(\mathbf{k}) = \frac{1}{e^{[\varepsilon(\mathbf{k})-\mathbf{p}_v \cdot \mathbf{v}(\mathbf{k})-\mu]/T_e} + 1}, \qquad (13.14)$$

or the distribution function (11.115): $f(\bar{\varepsilon}(\mathbf{k}), T_e) \to f_w(\mathbf{k})$. In this chapter we generally use the distribution function $f_d(\mathbf{k})$, i.e. taking the lattice momentum shift $\mathbf{p}_d$, electron temperature T_e and chemical potential μ as the basic parameters.

Equations (13.3)–(13.5) and equation (13.11), together with the equation for the total electron number,

$$N = 2\sum_{k} f(\bar{\varepsilon}(\bm{k}), T_e), \tag{13.15}$$

forms a complete set of equations to determine the transport properties of electrons in an arbitrary energy band in crossed magnetic and electric fields.

Balance equations (13.3)–(13.5) are greatly simplified if the energy dispersion $\varepsilon(\bm{k})$ can be written as a sum of three separate terms:

$$\varepsilon(\bm{k}) = \varepsilon_1(k_x) + \varepsilon_2(k_y) + \varepsilon_3(k_z). \tag{13.16}$$

For this spectrum $\partial^2 \varepsilon / \partial k_a \partial k_b = 0$ ($a \neq b$) and thus the nondiagonal elements of the inverse effective-mass tensor vanish: $(m^*_{xy})^{-1} = (m^*_{yz})^{-1} = (m^*_{xz})^{-1} = 0$. Equation (13.5) then yields $\bm{p}_d = (p_x, p_y, 0)$ and $\bm{v} = (v_x, v_y, 0)$, and we are left with equations

$$\frac{dv_x}{dt} = \frac{eE_x}{m^*_{xx}} + \frac{eBv_y}{m^*_{xx}} \gamma_{y,xx} + a_x, \tag{13.17}$$

$$\frac{dv_y}{dt} = \frac{eE_y}{m^*_{yy}} - \frac{eBv_x}{m^*_{yy}} \gamma_{x,yy} + a_y, \tag{13.18}$$

together with the energy-balance equation (13.11). Therefore, only two inverse effective-mass tensor components, $(m^*_{xx})^{-1}$ and $(m^*_{yy})^{-1}$, and two γ coefficients, $\gamma_{y,xx}$ and $\gamma_{x,yy}$, are needed to describe the center-of-mass motion in crossed magnetic and electric fields.

Balance equations (13.3)–(13.5) can also be greatly simplified if $\hat{\bm{x}}$, $\hat{\bm{y}}$ and $\hat{\bm{z}}$ are the principal axes of the iso-energy surface [though $\varepsilon(\bm{k})$ is not necessarily written as the separable form (13.16)] and the momentum shift $\bm{p}_d$, thus the drift velocity $\bm{v}$, is along the x direction (or along the y direction). In this case we also have vanishing nondiagonal elements of the inverse effective-mass tensor $\mathcal{K}$, and the steady-state balance equations (13.17) and (13.18) are reduced to

$$\frac{eE_x}{m^*_{xx}} + a_x = 0, \tag{13.19}$$

$$\frac{eE_y}{m^*_{yy}} - \frac{eBv}{m^*_{yy}} \gamma_{x,yy} = 0. \tag{13.20}$$

These two equations facilitate determination of the steady-state longitudinal and transverse resistivities under hot-electron condition. Since the current density (in the x direction) is given by

$$j_x = nev, \qquad (13.21)$$

where n is the genuine electron number density and e is the electron charge, the (nonlinear) longitudinal resistivity, defined as $\rho_{xx} \equiv E_x/j_x$, is obtained from (13.19) as

$$\rho_{xx} = -\frac{m^*_{xx} a_x}{ne^2 v}. \qquad (13.22)$$

On the other hand, the electric field along the y direction is determined from (13.20) to be

$$E_y = Bv\gamma_{x,yy}. \qquad (13.23)$$

This gives the nonlinear transverse (Hall) resistivity

$$\rho_{yx} \equiv \frac{E_y}{j_x} = \frac{\gamma_{x,yy}}{ne} B, \qquad (13.24)$$

and the Hall coefficient

$$R_{Hx} = \frac{\gamma_{x,yy}}{ne} \qquad (13.25)$$

in this configuration.

The longitudinal and transverse resistivities ρ_{yy} and ρ_{xy}, as well as the Hall coefficient R_{Hy} in the configuration that current follows along the y direction, are obtained if we make an index exchange $x \leftrightarrow y$. Note that in the case of nonlinear conduction, not only the longitudinal resistivity but also the transverse resistivity and the Hall coefficient may depend on the current direction in an anisotropic energy band.

It is worth noticing that, a drastic simplification may appear if the energy dispersion is separable [$\varepsilon(\mathbf{k})$ of form (13.16)] and if there exists a nondegenerate limit (Maxwell-Boltzmann distribution) at, for instance, high electron temperature. In the nondegenerate limit,

$$f(\bar{\varepsilon}(\mathbf{k}), T_e) \approx \exp\left[-\left(\bar{\varepsilon}(\mathbf{k}) - \mu\right)/T_e\right],$$

all the γ coefficients $\gamma_{y,xx} = \gamma_{x,yy} = 1$ irrespective of the form of $\varepsilon(\mathbf{k})$, and the Hall coefficient in both linear (weak electric field) and nonlinear (strong electric field) transport is always given by the simple expression

$$R_{Hx} = R_{Hy} = R_H = \frac{1}{ne}. \qquad (13.26)$$

Eq. (13.26) is also a well-known result valid for an isotropic parabolic band in both degenerate and nondegenerate statistics.

The formulation presented in this section provides a convenient tool to investigate hot-electron quasi-classical mangetotransport in an arbitrary energy band. Important properties in linear and nonlinear magnetotransport, such as the Hall resistivity and the type of conduction carriers and its change with electron density and electron temperature, are immediately obtained as long as the γ coefficients are calculated.

13.2 Examples of Quasi-Classical Magnetotransport in Arbitrary Energy Bands

In this section we will discuss the properties of magnetotransport directly derived from balance equations (13.3)–(13.5) in several specific nonparabolic bands.

13.2.1 3D tight-binding systems

We consider the tight-binding s band of a simple cubic lattice having lattice constant d. The energy dispersion can be written as

$$\varepsilon(\bm{k}) = \frac{3}{2}\Delta - \frac{\Delta}{2}\bigl[\cos(k_x d) + \cos(k_y d) + \cos(k_z d)\bigr], \qquad (13.27)$$

with the band width 3Δ, the energy bottom at $\bm{k} = 0$ and energy top at $\bm{k} = (\pm\pi/2d, \pm\pi/2d, \pm\pi/2d)$. Maximally two electrons per unit cell can be accommodated in this band, i.e., the density of electrons that fills up the whole band is $nd^3 = 2$. Other energy bands are assumed energetically far away and thus negligible.

In the scheme with lattice-momentum shift distribution function $f(\bar{\varepsilon}(\bm{k}), T_e) = \{\exp[(\varepsilon(\bm{k} - \bm{p}_d) - \mu)/T_e] + 1\}^{-1}$, the transport state of the electron system in crossed electric and magnetic fields is described by a lattice momentum shift $\bm{p}_d = (p_x, p_y, p_z)$ and an electron temperature T_e, while the chemical potential μ which relates to the total number of particles N by the equation

$$N = 2\sum_{\bm{k}} f(\varepsilon(\bm{k}), T_e), \qquad (13.28)$$

is independent of $\bm{p}_d$. The drift velocity $\bm{v} = (v_x, v_y, v_z)$ is given by ($a =$

x, y, z)

$$v_a = v_m \alpha(T_e) \sin(p_a d), \qquad (13.29)$$

where $v_m \equiv \Delta d/2$, and

$$\alpha(T_e) \equiv \frac{2}{N} \sum_{\mathbf{k}} \cos(k_x d) f(\varepsilon(\mathbf{k}), T_e) \qquad (13.30)$$

is a function of T_e only (independent of $\mathbf{p}_d$). The inverse effective-mass tensor $\mathcal{K}_{ab} = 1/m^*_{ab}$ is diagonal with nonzero components ($1/M^* \equiv \Delta d^2/2$)

$$\frac{1}{m^*_{aa}} = \frac{1}{M^*} \alpha(T_e) \cos(p_a d) \qquad (13.31)$$

depending on T_e and p_a. From the band symmetry it is apparent that $\gamma_{y,xx} = \gamma_{x,yy} = \gamma$, and

$$\gamma = \frac{2}{N\alpha(T_e)^2} \sum_{\mathbf{k}} \cos(k_x d) \cos(k_y d) f(\varepsilon(\mathbf{k}), T_e) \qquad (13.32)$$

depends only on T_e, independent of $\mathbf{p}_d$.

For the 3D tight-binding band both the magnitude and sign of this γ coefficient can change with changing electron density. We write the Hall coefficient of the system in the form

$$R_H = \frac{\gamma}{ne} = \mathrm{sgn}(\gamma) \frac{|\gamma|}{ne} = \frac{1}{n^* \mathrm{sgn}(\gamma) e}, \qquad (13.33)$$

where

$$n^* = n/|\gamma|. \qquad (13.34)$$

Equation (13.33) can be thought as the Hall coefficient of a kind of carriers having effective density n^* and effective charge $\mathrm{sgn}(\gamma)e$.

It is interesting to see how γ varies with changing electron density at zero temperature. In the vicinity of the band bottom, i.e. $nd^3 \ll 1$, the zero-temperature Fermi level ε_F is

$$\frac{\varepsilon_F}{\Delta} \simeq \frac{1}{4}(3\pi^2 nd^3)^{\frac{2}{3}}. \qquad (13.35)$$

and the parameters

$$\alpha \simeq 1, \quad \gamma \simeq 1. \qquad (13.36)$$

The Hall coefficient

$$R_H \simeq 1/ne. \tag{13.37}$$

This is a natural result: near the bottom the energy band is essentially parabolic and is occupied by electrons of density n and charge e. On the other hand, in the vicinity of the band top, i.e. $1 - nd^3/2 \ll 1$, the zero-temperature Fermi level is

$$\frac{\varepsilon_F}{\Delta} \simeq \frac{1}{4}\left[\frac{3\pi^2}{2}\left(1 - \frac{nd^3}{2}\right)\right]^{2/3}, \tag{13.38}$$

and

$$\alpha \simeq \frac{2}{nd^3} - 1, \tag{13.39}$$

$$\gamma \simeq -\frac{nd^3}{2}\left(1 - \frac{nd^3}{2}\right)^{-1}. \tag{13.40}$$

The Hall coefficient is

$$R_H = \frac{\gamma}{ne} = -\frac{1}{n^*e}, \tag{13.41}$$

with the effective carrier density

$$n^* = (2/d^3 - n). \tag{13.42}$$

This is equivalent to the case of a parabolic band occupied by carriers having charge $-e$ and density n^* given by Eq. (13.42). This n^* is just the number density of the states unoccupied by electrons in this 3D tight-binding band, i.e. the density of holes.

The zero-temperature α and γ, calculated as functions of the electron density nd^3, are plotted in Fig. 13.1, together with γ/nd^3, which is proportional to the Hall coefficient: $R_H = (\gamma/nd^3)(d^3/e)$. The transition from the electron type conduction to the hole type conduction appears naturally. Carriers are electronlike when the electron occupation is less than the half filled ($nd^3 < 1$), and are holelike when electron occupation exceeds the half filled ($nd^3 > 1$). The effective carrier number density $n/|\gamma|$ diverges at both sides of $nd^3 = 1$, but the Hall coefficient R_H is zero at half filled electron occupation, as indicated by the dotted line in the figure.

To see the temperature dependence of the Hall resistivity we show in Fig. 13.2 the calculated γ against Δ/T_e for several different electron densities $nd^3 = 0.2, 0.5, 0.8, 1.0, 1.2, 1.4$ and 1.6. Except for the half filled case

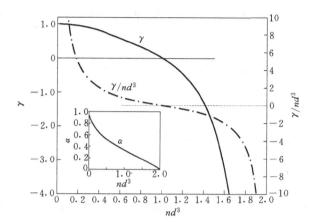

Fig. 13.1 Zero-temperature coefficient γ [(Eq. (13.32)], γ/nd^3, and α [Eq. (13.30)] are shown as functions of the electron density nd^3 for the 3D tight-binding system. From Lei (1995a).

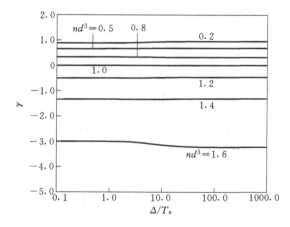

Fig. 13.2 Temperature dependence of the coefficient γ of the 3D tight-binding system is shown at different values of the electron population: $nd^3 = 0.2, 0.5, 0.8, 1.0, 1.2, 1.4$ and 1.6. From Lei (1995a).

of $nd^3 = 1$ where γ keeps constant over the whole temperature range, the coefficient γ depends weakly on the temperature. This weak temperature dependence of γ indicates that in a single 3D tight-binding energy band, the influence of a strong electric field may slightly show up in the Hall resistivity through its effect on the electron temperature, but the effect is generally quite small. Note that the nondegenerate limit does not exist in this system except for very small electron density $nd^3 \sim 0$. At finite electron density γ approaches a value less than 1 at high electron temperature.

Two-dimensional tight-binding energy band

$$\varepsilon(\boldsymbol{k}) = \Delta - \frac{\Delta}{2}\bigl[\cos(k_x d) + \cos(k_y d)\bigr] \qquad (13.43)$$

is a model for lateral semiconductor superlattices (Lei, 1993). Its behavior in classical magnetotransport is quite similar to that of 3D tight-binding systems. The momentum and energy balance equations for this system are used in the investigation of magneto-miniband transport in lateral semiconductor superlattices. We refer the reader to the original paper (Lei, 1993) for detail.

13.2.2 Quantum-wire arrays

In this section we are going to discuss another 3D energy band which differs from the above 3D tight-binding band by allowing electrons to move freely along the z direction:

$$\varepsilon(\boldsymbol{k}) = \frac{\Delta}{2}\bigl[2 - \cos(k_x d) - \cos(k_y d)\bigr] + \frac{k_z^2}{2m}. \qquad (13.44)$$

Here $-\pi/d < k_x \leqslant \pi/d$, $-\pi/d < k_y \leqslant \pi/d$ and $-\infty < k_z < \infty$. This dispersion represents the lowest energy band of a model system which is composed of a square arrays of parallel long quantum wires along the z direction. d is the lattice constant of the square array in the x–y plane. This band has an infinite width with an energy minimum at $\boldsymbol{k} = (0,0,0)$ and saddle points at $\boldsymbol{k} = (\pm\pi/d, \pm\pi/d, 0)$.

Near the band bottom $(0,0,0)$, the energy dispersion is an anisotropic parabolic structure:

$$\varepsilon(\boldsymbol{k}) \simeq \frac{k_x^2 + k_y^2}{2M^*} + \frac{k_z^2}{2m}. \qquad (13.45)$$

with $1/M^* = \Delta d^2/2$. In the vicinity of a saddle point, e.g. $(\pi/d, \pi/d, 0)$, the energy dispersion has the form

$$\varepsilon(\boldsymbol{k}) \simeq 2\Delta - \frac{\kappa_x^2 + \kappa_y^2}{2M^*} + \frac{k_z^2}{2m}, \qquad (13.46)$$

with $\kappa_a \equiv k_a - \pi/d$ $(a = x, y)$.

We assume that the electric field $\boldsymbol{E}$ is in the x–y plane and the magnetic field $\boldsymbol{B}$ is along the z direction, and that the transport state is described by $\boldsymbol{p}_d = (p_x, p_y, 0)$ and T_e. The expressions (13.29) and (13.31) for v_x, v_y, $1/m^*_{xx}$ and $1/m^*_{yy}$, and Eqs. (13.28), (13.30) and (13.32) for N, $\alpha(T_e)$ and γ are still valid for this system except that the sum over k_z now goes from $-\infty$ to ∞. They are written as

$$0 = 1 - \frac{\sqrt{\pi}}{2\pi^2}\left(\frac{T_e}{\varepsilon_{F1}}\right)^{\frac{1}{2}} \int_0^\pi dx \int_0^\pi dy\, F_{-\frac{1}{2}}\left(\zeta - \frac{\Delta}{2T_e}(2 - \cos x - \cos y)\right), \tag{13.47}$$

$$\alpha = \frac{\sqrt{\pi}}{2\pi^2}\left(\frac{T_e}{\varepsilon_{F1}}\right)^{\frac{1}{2}} \int_0^\pi dx \int_0^\pi dy\, \cos x\, F_{-\frac{1}{2}}\left(\zeta - \frac{\Delta}{2T_e}(2 - \cos x - \cos y)\right), \tag{13.48}$$

$$\gamma = \frac{1}{\alpha^2}\frac{\sqrt{\pi}}{2\pi^2}\left(\frac{T_e}{\varepsilon_{F1}}\right)^{\frac{1}{2}} \int_0^\pi dx \int_0^\pi dy\, \cos x\, \cos y$$
$$\times F_{-\frac{1}{2}}\left(\zeta - \frac{\Delta}{2T_e}(2 - \cos x - \cos y)\right). \tag{13.49}$$

Here we have introduced an energy dimension quantity ε_{F1} defined by

$$nd^2 = \frac{2}{\pi}\sqrt{2m\varepsilon_{F1}}, \tag{13.50}$$

which is the zero-temperature Fermi energy of a one-dimensional electron gas with line density $N_1 = nd^2$ and effective mass m, and n is the electron density of the present 3D system. The $F_{-\frac{1}{2}}(y)$ function is defined by

$$F_{-\frac{1}{2}}(y) = \frac{1}{\sqrt{\pi}}\int_0^\infty \frac{x^{-\frac{1}{2}}dx}{\exp(x-y)+1}. \tag{13.51}$$

At zero temperature $T_e \to 0$, Eqs. (13.47)–(13.49) become

$$0 = 1 - \frac{1}{\pi^2}\sqrt{\frac{\Delta}{\varepsilon_{F1}}} \int_0^\pi dx \int_0^\pi dy \left[\frac{\varepsilon_F}{\Delta} - 1 + \frac{1}{2}(\cos x + \cos y)\right]^{\frac{1}{2}}, \tag{13.52}$$

$$\alpha = \frac{1}{\pi^2}\sqrt{\frac{\Delta}{\varepsilon_{F1}}} \int_0^\pi dx \int_0^\pi dy\, \cos x \left[\frac{\varepsilon_F}{\Delta} - 1 + \frac{1}{2}(\cos x + \cos y)\right]^{\frac{1}{2}}, \tag{13.53}$$

$$\gamma = \frac{1}{\alpha^2 \pi^2}\sqrt{\frac{\Delta}{\varepsilon_{F1}}} \int_0^\pi dx \int_0^\pi dy\, \cos x \cos y \left[\frac{\varepsilon_F}{\Delta} - 1 + \frac{1}{2}(\cos x + \cos y)\right]^{\frac{1}{2}}. \tag{13.54}$$

Here ε_F is the Fermi energy of this quantum-wire array at zero temperature. Fig. 13.3 shows ε_F/Δ (chain line) determined from Eq. (13.52) and zero-temperature γ obtained from Eqs. (13.53) and (13.54), as functions of ε_{F1}/Δ. We see that γ is positive when $\varepsilon_{F1}/\Delta < 0.245$ and becomes negative when $\varepsilon_{F1}/\Delta > 0.245$, indicating the conduction change from the electron type to the hole type at the electron density corresponding to $\varepsilon_{F1}/\Delta = 0.245$.

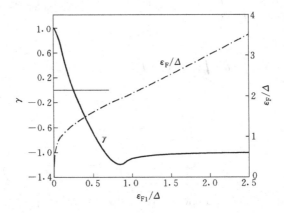

Fig. 13.3 Zero-temperature coefficient γ [Eq. (13.54)] and Fermi energy ε_F/Δ are shown as functions of the electron occupation represented by ε_{F1}/Δ for the quantum-wire array. From Lei (1995a).

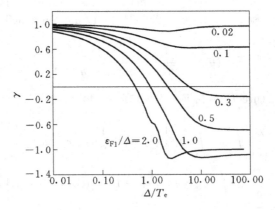

Fig. 13.4 Temperature dependence of the coefficient γ in the quantum-wire array is shown at several values of the electron occupations represented by $\varepsilon_{F1}/\Delta = 0.02, 0.1, 0.3, 0.5, 1.0$ and 2.0. From Lei (1995a).

To see the temperature variation of the γ coefficient we plot in Fig. 13.4 the calculated γ as a function of Δ/T_e for several different electron densities in terms of $\varepsilon_{F1}/\Delta = 0.02, 0.1, 0.3, 0.5, 1.0$ and 2.0. In this system the γ coefficient exhibits a much stronger temperature dependence than in a 3D tight-binding band. Furthermore, there exists a nondegenerate limit for any value of the electron density in this system, such that γ always approaches unity at high enough temperature. Therefore, if the electron density is high that $\varepsilon_{F1}/\Delta \geqslant 0.3$ and the conduction is hole type at low temperature, the system will, with increasing electron temperature, experience a conduction transition from the hole type to the electron type.

13.2.3 Superlattice vertical transport in the presence of a transverse magnetic field

We consider electrons moving freely along the y and z directions, but subject to a strong periodic potential in the x direction such that the energy dispersion takes the form

$$\varepsilon(\boldsymbol{k}) = \frac{\Delta}{2}\left[1 - \cos(k_x d)\right] + \frac{k_y^2 + k_z^2}{2m}. \tag{13.55}$$

Here $-\pi/d < k_x \leqslant \pi/d$, and $-\infty < k_y, k_z < \infty$. This dispersion represents the lowest miniband band of a planar superlattice with period d in the x direction.

When a magnetic field is applied along the z direction, $\boldsymbol{B} = (0, 0, B)$, and an electric field is in the x–y plane, $\boldsymbol{E} = (E_x, E_y, 0)$, the transport state of the system can be described by the lattice momentum shift $\boldsymbol{p}_d = (p_x, p_y, 0)$ and an electron temperature T_e. The average drift velocity is given by

$$v_x = v_m \alpha(T_e) \sin(p_x d), \tag{13.56}$$
$$v_y = p_y/m, \tag{13.57}$$

and $v_z = 0$, and the nonzero components of the inverse effective-mass tensor are

$$1/m_{xx}^* = (1/M^*)\alpha(T_e)\cos(p_x d) \tag{13.58}$$

and

$$1/m_{yy}^* = 1/m_{zz}^* = 1/m. \tag{13.59}$$

Here $v_m \equiv \Delta d/2$, $1/M^* \equiv \Delta d^2/2$, and $\alpha(T_e)$ is given by (13.30).

It is easily seen that for this system we always have

$$\gamma_{y,xx} = \gamma_{x,yy} = 1, \tag{13.60}$$

for arbitrary value of $\boldsymbol{p}_d$ and even in the case of degenerate statistics. The effective momentum-balance equations (13.17) and (13.18) thus reduce to

$$\frac{dv_x}{dt} = \frac{eE_x}{m_{xx}^*} + \frac{eBv_y}{m_{xx}^*} + a_x, \tag{13.61}$$

$$\frac{dv_y}{dt} = \frac{eE_y}{m} - \frac{eBv_x}{m} + a_y. \tag{13.62}$$

These equations have been applied to the discussion of the effect of a transverse magnetic field on superlattice vertical transport (Lei, 1992c). Despite the neglect of Landau quantization in these quasi-classical equations, the predictions from them are still in qualitative agreement with relevant experiments.

13.3 Electronic Transport in Nonparabolic Kane Bands

13.3.1 The Kane band model

Different from the case of a superlattice miniband described by a tight-binding energy band model, in a conventional bulk semiconductor, even in the presence of a relatively high electric field, carriers usually move within a small region in the $\boldsymbol{k}$ space in the vicinity of one (or several) energy extremum position $\boldsymbol{k}_0$, and thus the problem resulting from the carrier motion across the Brillouin zone boundary can be disregarded. When the wavevector $\boldsymbol{k}$ is sufficiently close to an energy extremum point $\boldsymbol{k}_0$ the energy dispersion is always close to a parabolic form. For wide-gap semiconductors, the parabolic spectrum applies in relatively wide range and the nonparabolicity is generally negligible. For narrow-gap semiconductors, especially direct narrow-gap semiconductors, such as GaAs, InAs, InSb, etc, in which the separation between the valence band top and conduction band bottom is quite small, a parabolic spectrum can apply only within a very small range in the $\boldsymbol{k}$ space and nonparabolicity has to be taken into account in many realistic transport calculations.

Based on the $\mathbf{k} \cdot \mathbf{p}$-method, Kane (1957) treats a direct-gap semiconductor with the conduction-band bottom and valence-band top at the zone center and expresses the energy dispersions of both conduction and valence bands analytically in terms of the Γ-point band parameters. Using the wavefunction set consisting of energetically closest conduction band (c with Γ_6 symmetry), heavy hole valence band (v_1 with Γ_8 symmetry), light hole valence band (v_2 with Γ_8 symmetry) and spin-orbit split band (v_3 with Γ_7 symmetry) as the basis, one can write the electron wave function around $\boldsymbol{k} = 0$ in the form

$$\psi_{\boldsymbol{k}}(\boldsymbol{r}) = \left[a_{\boldsymbol{k}} u_{\mathrm{c}}(\boldsymbol{r}) + b_{\boldsymbol{k}} u_{\mathrm{v}_1}(\boldsymbol{r}) + c_{\boldsymbol{k}} u_{\mathrm{v}_2}(\boldsymbol{r}) + d_{\boldsymbol{k}} u_{\mathrm{v}_3}(\boldsymbol{r})\right] e^{i\boldsymbol{k}\cdot\boldsymbol{r}}, \qquad (13.63)$$

where $u_{\mathrm{c}}(\boldsymbol{r})$, $u_{\mathrm{v}_1}(\boldsymbol{r})$, $u_{\mathrm{v}_2}(\boldsymbol{r})$ and $u_{\mathrm{v}_3}(\boldsymbol{r})$ are respectively the cell periodic functions (with implicit spin indices) of the conduction and three valence bands at the state $\boldsymbol{k} = 0$. Solution of the secular equation including the $\mathbf{k} \cdot \mathbf{p}$ interaction gives rise to the following expression for the conduction band spectrum $\varepsilon(\boldsymbol{k})$ around $\boldsymbol{k} = 0$ in most of the III-V compound semiconductors:

$$\varepsilon(\boldsymbol{k})\left[1 + \alpha \varepsilon(\boldsymbol{k})\right] = k^2/2m, \qquad (13.64)$$

or

$$\varepsilon(\boldsymbol{k}) = \frac{1}{2\alpha} \left[\left(1 + 2\alpha \frac{k^2}{m}\right)^{1/2} - 1 \right], \qquad (13.65)$$

in which m is the effective mass at the conduction band edge ($\boldsymbol{k} = 0$),

$$\alpha = 1/\varepsilon_{\mathrm{g}} \qquad (13.66)$$

is the nonparabolic coefficient, with ε_{g} the energy gap between the conduction and valence bands. The energy spectrum (13.65) is derived from the $\mathbf{k} \cdot \mathbf{p}$ method by considering couplings between the closest conduction and valence bands and of course can apply only within a certain range around $\boldsymbol{k} = 0$. But, people often treat it as a model of the nonparabolic energy spectrum for narrow-gap semiconductors having conduction band bottom at the Brillouin zone center, called the Kane model (Kane, 1957).

For those semiconductors having conduction band bottom $\boldsymbol{k}_0$ outside the the Brillouin zone center ($\boldsymbol{k}_0 \neq 0$), such as GaP and Si, the nonparabolic effect can be described approximately using the extended Kane model:

$$\varepsilon(\boldsymbol{k})\left[1 + \alpha \varepsilon(\boldsymbol{k})\right] = \frac{1}{2} \left(\frac{k_\perp^2}{m_\perp} + \frac{k_\parallel^2}{m_\parallel} \right). \qquad (13.67)$$

Here $k_\parallel$ and $k_\perp$ respectively represent the wavevector components parallel and perpendicular to $\boldsymbol{k}_0$.

The Kane spectrum (13.64) represents an isotropic energy band. For small $\mathbf{k}$, $\varepsilon(\mathbf{k})$ approaches $k^2/2m$, i.e. an isotropic parabolic band; for large wavevector, $\varepsilon(\mathbf{k}) \simeq k/\sqrt{2m\alpha}$, exhibiting a linear dependence on k. In a Kane band system the velocity function is

$$\mathbf{v}(\mathbf{k}) \equiv \nabla_{\mathbf{k}}\varepsilon(\mathbf{k}) = \frac{\mathbf{k}}{m(1+2\alpha k^2/m)^{1/2}}, \qquad (13.68)$$

and the components of the inverse effective mass tensor are given by ($a, b = x, y, z$)

$$\frac{\partial^2 \varepsilon(\mathbf{k})}{\partial k_a \partial k_b} = \frac{1}{m(1+2\alpha k^2/m)^{1/2}}\left(\delta_{ab} - \frac{2\alpha k_a k_b/m}{1+2\alpha k^2/m}\right). \qquad (13.69)$$

The wave function of nonparabolic Kane band [i.e. the coefficients $a_{\mathbf{k}}$, $b_{\mathbf{k}}$, $c_{\mathbf{k}}$ and $d_{\mathbf{k}}$ in (13.63)] can be obtained from solving the $\mathbf{k}\cdot\mathbf{p}$ eigen equation. Since the Brillouin zone in the Kane model is considered as infinite, the form factor $g(\mathbf{k},\mathbf{q})$ (11.22) appearing in the electron density function (11.21) and in the expressions of the frictional force and energy-loss rate, can be written as the overlap integral of the cell periodic functions $u_{\mathbf{k}}$ and $u_{\mathbf{k}+\mathbf{q}}$ in the form

$$g(\mathbf{k},\mathbf{q}) = \int d\mathbf{r}\, u^*_{\mathbf{k}+\mathbf{q}}(\mathbf{r}) u_{\mathbf{k}}(\mathbf{r}), \qquad (13.70)$$

thus easily expressed in terms of the coefficients $a_{\mathbf{k}}$, $b_{\mathbf{k}}$, $c_{\mathbf{k}}$ and $d_{\mathbf{k}}$.

13.3.2 Hot-electron transport in Kane band systems

For Kane band systems, in view of infinite Brillouin zone, the transport properties can be conveniently studied by starting from the momentum and energy balance equation (11.75) and (11.76). We will use the shifted lattice-momentum function $f(\bar{\varepsilon}(\mathbf{k}), T_e) = f(\varepsilon(\mathbf{k}-\mathbf{p}_d), T_e)$ as the initial distribution, i.e. take the lattice momentum shift $\mathbf{p}_d$, electron temperature T_e and chemical potential μ as basic transport parameters. The equation for the total carrier number

$$1 = \frac{2}{N}\sum_{\mathbf{k}} f(\varepsilon(\mathbf{k}-\mathbf{p}_d), T_e) = \frac{2}{N}\sum_{\mathbf{k}} f(\varepsilon(\mathbf{k}), T_e) \qquad (13.71)$$

can be written as

$$n = \frac{2}{\sqrt{\pi}} N_c(T_e) \int_0^\infty dy\, g(\lambda, \zeta, y), \qquad (13.72)$$

in which n is the carrier volume density, $N_c(T_e) \equiv 2(mT_e/2\pi)^{3/2}$, $\zeta \equiv \mu/T_e$, $\lambda \equiv \alpha T_e$, and

$$g(\lambda,\zeta,y) \equiv \frac{(1+2\lambda y)[y(1+\lambda y)]^{1/2}}{\exp(y-\zeta)+1}. \qquad (13.73)$$

For nonzero T_e, Eq. (13.72) determines ζ as a function of (n/N_c) and λ, independent of $\boldsymbol{p}_d$. This fucntion can be approximately expressed as (Altschul and Finkman, 1991)

$$\frac{\mu}{T_e} = \ln\left(\frac{n}{B_0 N_c}\right) + B_1 \frac{n}{N_c}, \qquad (13.74)$$
$$B_0 = 1 + 3.75\lambda + 3.291\lambda^2 - 2.416\lambda^3,$$
$$B_1 = (0.325 + 0.5115\lambda + 3.5166\lambda^2 - 2.3234\lambda^3)/B_0^3.$$

The Fermi energy ε_F at $T=0\,\mathrm{K}$ is given by the degenerate limit of Eq. (13.72):

$$n = \frac{\sqrt{2}}{\pi^2} m^{3/2} \int_0^{\varepsilon_F} d\varepsilon \, (1+2\alpha\varepsilon)[\varepsilon(1+\alpha\varepsilon)]^{1/2}. \qquad (13.75)$$

In view of isotropy of the system, when an electric field $\boldsymbol{E}=(E,0,0)$ is applied along the x direction, the lattice-momentum shift $\boldsymbol{p}_d=(p_d,0,0)$, the drift velocity $\boldsymbol{v}=(v,0,0)$ and the frictional accelerations $\boldsymbol{a}_i=(a_i,0,0)$ and $\boldsymbol{a}_p=(a_p,0,0)$ are all along the x direction. The steady state momentum and energy balance equations are

$$eE\mathcal{K}_{xx} + a_i + a_p = 0, \qquad (13.76)$$
$$eEv - w = 0. \qquad (13.77)$$

For a given carrier density n, the quantities v, $\mathcal{K}_{xx}$, a_i, a_p and w are all functions of p_d and T_e. The expressions (11.77) and (11.78) for v and $\mathcal{K}_{xx}$ can be written as

$$v = \frac{2}{\sqrt{\pi}} \frac{N_c(T_e)}{n} \int_0^\infty dy\, J(\eta,p_d) g(\lambda,\zeta,y), \qquad (13.78)$$
$$\mathcal{K}_{xx} = \frac{2}{\sqrt{\pi}} \frac{N_c(T_e)}{n} \int_0^\infty dy\, K(\eta,p_d) g(\lambda,\zeta,y), \qquad (13.79)$$

where $\eta \equiv [2mT_e y(1+\lambda y)]^{1/2}$, and functions

$$J(k,p_d) \equiv \frac{1}{2}\int_{-1}^{1} dx \frac{kx+p_d}{m\left[1+\frac{2\alpha}{m}(k^2+p_d^2+2kp_dx)\right]^{1/2}}, \quad (13.80)$$

$$K(k,p_d) \equiv \frac{1}{2}\int_{-1}^{1} dx \frac{1+\frac{2\alpha}{m}k^2(1-x^2)}{m\left[1+\frac{2\alpha}{m}(k^2+p_d^2+2kp_dx)\right]^{1/2}}. \quad (13.81)$$

a_i, a_p and w, which are dependent on the scattering potential and the form factor, can be derived from (11.80)–(11.82) for expressions directly usable in the numerical calculation.

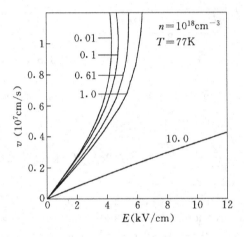

Fig. 13.5 The drift velocity is shown as a function of the electric field in Kane band systems at lattice temperature $T = 77\,\text{K}$. The electron density $n = 1.0 \times 10^{18}\,\text{cm}^{-3}$, being equal to the impurity density. The number pointing to the curve is the value of the nonparabolic coefficient (in units of eV^{-1}). From Weng and Lei (1994).

Starting from equations (13.76) and (13.77), Weng and Lei (1994) calculate the hot-electron transport in a model Kane band system at lattice temperature $T = 77\,\text{K}$, assuming the polar optical phonons and impurities to be the dominant scattering mechanisms. The material parameters are taken as typical values of GaAs: conduction band bottom effective mass $m = 0.068 m_e$, longitudinal optical phonon frequency $\Omega_{\text{LO}} = 35.4\,\text{meV}$, low-frequency dielectric constant $\kappa = 12.9$, high-frequency dielectric constant $\kappa_\infty = 10.8$. The electron density $n = 1.0 \times 10^{18}\,\text{cm}^{-3}$, being equal to the impurity density. Fig. 13.5 shows the calculated drift velocity v versus the electric field E at different values of $\alpha = 0.01, 0.1, 0.61, 1.0$ and $10\,\text{eV}^{-1}$.

In weakly nonparabolic bands ($\alpha = 0.01$ and 0.1 eV^{-1}), there exist positive, zero and negative differential resistivity regions in the v–E curve. In medially nonparabolic systems ($\alpha = 0.61$ and 1.0 eV^{-1}), the negative differential resistivity region disappears, but the v–E curve is still upturned. At very strong nonparabolicity ($\alpha = 10$ eV^{-1}) v–E curve becomes somewhat downward at high field.

13.3.3 Hall resistivity of hot-electron transport in nonparabolic Kane bands

Consider electrons moving within an isotropic, nonparabolic Kane band with the energy dispersion (13.65). Under the influence of an electric field in the x–y plane, $\boldsymbol{E} = (E_x, E_y, 0)$, and a magnetic field in the z direction, $\boldsymbol{B} = (0, 0, B)$, the transport state of the system can be described with the momentum shift $\boldsymbol{p}_d = (p_x, p_y, 0)$ and electron temperature T_e, and the drift velocity and damping acceleration are all in the plane: $\boldsymbol{v} = (v_x, v_y, 0)$, and $\boldsymbol{a} = (a_x, a_y, 0)$. The momentum-balance equations (13.3) and (13.4), together with the energy-balance equation (13.12), form a complete set of equations for the determination of parameters p_x, p_y and T_e. We will not write the detailed expressions of $1/m^*_{ab}$ and $\gamma_{a,bc}$ here, but only point out that since the Kane energy spectrum $\varepsilon(\boldsymbol{k})$ (13.65) can not be written as a sum of three separate terms, the nondiagonal elements of the inverse effective mass tensor, $1/m^*_{xy}$ and $1/m^*_{yx}$ are generally finite when both momentum shift components p_x and p_y are not zero. Nevertheless, for this isotropic band we can always choose the x-axis to be along the velocity direction, $\boldsymbol{v} = (v, 0, 0)$ and $\boldsymbol{p}_d = (p_d, 0, 0)$, to let $1/m^*_{xy} = 1/m^*_{yx} = 0$. The steady-state momentum balance equations are then reduced to the form of (13.19) and (13.20), and expressions (13.22) for the longitudinal resistivity ρ_{xx}, (13.24) for the Hall resistivity ρ_{yx}, and (13.25) for the Hall coefficient R_H are all valid. Of course, these results are independent of the direction of the current flow for this isotropic Kane band. Thus we write the Hall coefficient as

$$R_H = \frac{\gamma}{ne}, \qquad (13.82)$$

where the Hall factor

$$\gamma = \gamma_{x,yy}. \qquad (13.83)$$

The Hall factors γ of Kane band systems with parameters corresponding to GaAs, InAs and InSb are investigated by Lei and Weng (1994). In Fig. 13.6 we plot the weak electric field ($p_d \approx 0$) Hall factor γ as a function of electron temperature T_e for a nonparabolic Kane band pertinent to n-InSb ($\alpha = 4.32\,\text{eV}^{-1}$, $m = 0.0138 m_e$, m_e is the free electron mass) at different values of the electron density. For comparison we also show the corresponding γ for a very weakly nonparabolic Kane band ($\alpha = 0.01\,\text{eV}^{-1}$, $m = 0.138 m_e$, and electron density $n = 2 \times 10^{14}\,\text{cm}^{-3}$). Such a weak nonparabolicity implies that it represents almost a parabolic system, and thus $\gamma \simeq 1$ except at the highest-temperature region. Since the deviation of a Kane band from the corresponding parabolic band is proportional to the square of the wave vector k, the nonparabolic effect should appear stronger when higher wave vector states are occupied, i.e. at higher electron density or at higher electron temperature. For sufficiently low electron density and at sufficiently low electron temperature a Kane band behaves parabolically. This can be seen clearly in Fig. 13.6, where for the lowest electron density shown, $n = 2 \times 10^{14}\,\text{cm}^{-3}$, the value of γ approaches a limit very close to unity at low electron temperature, but monotonically decreases with rising electron temperature. In the case of high electron density, γ is always far from unity even at low temperature.

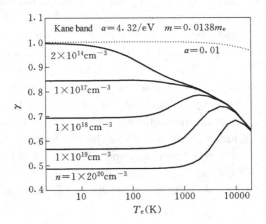

Fig. 13.6 Temperature dependence of the Hall factor γ of an n-InSb Kane band ($\alpha = 4.32\,\text{eV}^{-1}$, $m = 0.0138 m_e$) is shown at different values of the electron density n. The dashed line indicates the corresponding γ for a Kane band with $\alpha = 0.01\,\text{eV}^{-1}$ and $m = 0.138 m_e$ at $n = 2 \times 10^{14}\,\text{cm}^{-3}$. From Lei and Weng (1994).

Unlike the systems discussed in Sec. 13.2, the Hall factor for a nonparabolic Kane band depends also on the momentum shift p_d, thus γ

changes with changing strength of the current flowing in the system. In Fig. 13.7 we plot the calculated γ as a function of electron temperature T_e for an InAs Kane band ($\alpha = 2.73$ eV^{-1}, $m = 0.022 m_e$) at electron density $n = 1 \times 10^{19}$ cm^{-3} for several values of p_d represented by a dimensionless parameter $z_d \equiv (2\alpha/m)^{1/2} p_d$.

These results compare favorably with the earlier unexplained galvanomagnetic measurement of Alberga et al (1982), who reported a Hall factor decrease about 10% when increasing the applied electric field from 10 V/cm to 400 V/cm in n-type InSb. The temperature dependence of γ exhibited in the low density curve in Fig. 13.6 is in agreement with this finding. Numerical analysis of the steady-state balance equations determines that at an applied field of 400 V/cm the momentum shift z_d is less than 0.1, for an n-InSb system of electron density $n = 1 \times 10^{16}$ having polar optical phonon and impurity scatterings with impurity density $n_i = n$. Therefore, up to this strength of the electric field, the Hall factor γ is essentially the same as that of $z_d = 0$ and its field variation results almost solely from the change of the electron temperature, which increases from 77 K to about 400 K with increasing field strength from 10 to 400 V/cm, leading to a decrease $\sim 10\%$ in γ.

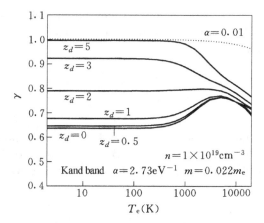

Fig. 13.7 Hall factor γ is plotted as a function of electron temperature T_e at different values of the dimensionless momentum shift $z_d = 0, 0.5, 1.0, 2.0, 3.0$ and 5.0, for an n-InAs Kane band ($\alpha = 2.73$ eV^{-1}, $m = 0.022 m_e$) at electron density $n = 1 \times 10^{19}$ cm^{-3}. From Lei and Weng (1994).

13.3.4 Transport of a Kane band system under a strong longitudinal magnetic field

In the last subsection we discussed transport of narrow-gap semiconductors in a classical transverse magnetic field using nonparabolic Kane model, assuming that the energy dispersion and the scattering probability are not influenced by the magnetic field. This assumption will no longer be valid when the magnetic field become strong, since in a strong magnetic field the electron transverse energy, i.e. the energy related to the wave vector perpendicular to the magnetic field, is quantized and the conduction band of the semiconductor splits into Landau levels. Such a quantization effect has remarkable influence on electron transport. Though the issue of Landau quantization for nonparabolic energy bands has not yet a general solution, it seems reasonable to take the following treatment (Wu and Spector, 1971) for a Kane band.

In the absence of a magnetic field the relation between the energy ε and wave vector $\boldsymbol{k}$ of a nonparabolic Kane band is

$$\varepsilon(\boldsymbol{k}) = \frac{1}{2\alpha}\left[\left(1 + \frac{2\alpha}{m}k^2\right)^{1/2} - 1\right] \equiv K_\alpha\left(\frac{k^2}{2m}\right). \tag{13.84}$$

According to the single band effective Hamiltonian method (Sec. 11.1.4), the effective Hamiltonian of this system in the presence of the magnetic field can be written as

$$h = K_\alpha\left(\frac{(\boldsymbol{p}-e\boldsymbol{A})^2}{2m}\right). \tag{13.85}$$

Here $\boldsymbol{p} = -\mathrm{i}\nabla$ is the momentum operator, $\boldsymbol{A}$ is the vector potential of the magnetic field. With the Landau gauge $\boldsymbol{A} = (-By, 0, 0)$ for the magnetic field along the z direction, $\boldsymbol{B} = (0, 0, B)$, the Landau wave function (for system of unit volume)

$$\psi_{nk_xk_z}(\boldsymbol{r}) = \mathrm{e}^{\mathrm{i}(k_x x + k_z z)}\chi_{nk_x}(y) \tag{13.86}$$

can also be considered as the eigenfunction of the effective Hamiltonian (13.85), with the eigenenergy

$$\varepsilon_n(k_z) = K_\alpha\big(\varepsilon_n^{(0)}(k_z)\big), \tag{13.87}$$

where
$$\varepsilon_n^{(0)}(k_z) = \left(n + \frac{1}{2}\right)\omega_c + \frac{k_z^2}{2m} \tag{13.88}$$

is the Landau level of an isotropic, parabolic energy band.

When applying a dc electric field of arbitrary strength along the z direction, $\boldsymbol{E} = (0,0,E)$, the transport state of this system can be described by a lattice-momentum shift along the z direction, $\boldsymbol{p}_d = (0,0,p_d)$, and an electron temperature T_e, which satisfy the following momentum and energy balance equations in the steady state:

$$eE/m_z^* + a_\mathrm{i} + a_\mathrm{p} = 0, \tag{13.89}$$
$$eEv - w = 0. \tag{13.90}$$

Here the drift velocity v and the inverse effective mass $1/m_z^*$ are respectively given by

$$v = \frac{1}{\pi l_B^2 N} \sum_{n,k_z} \frac{d\varepsilon_n(k_z)}{dk_z} f(\varepsilon_n(k_z - p_d), T_e), \tag{13.91}$$

$$\frac{1}{m_z^*} = \frac{1}{\pi l_B^2 N} \sum_{n,k_z} \frac{d^2\varepsilon_n(k_z)}{dk_z^2} f(\varepsilon_n(k_z - p_d), T_e), \tag{13.92}$$

in which the chemical potential μ in the Fermi distribution function $f(\varepsilon, T_e) = \{\exp[(\varepsilon - \mu)/T_e] + 1\}^{-1}$ is determined by the particle number equation

$$N = \frac{1}{\pi l_B^2} \sum_{n,k_z} f(\varepsilon_n(k_z), T_e). \tag{13.93}$$

The expressions for damping accelerations a_i and a_p, and for the energy-dissipation rate w can be obtained from (12.98)–(12.100) by understanding indices n and n' as the Landau quantum numbers and $\varepsilon_n(k_z)$ as the energy defined by (13.87) $[\bar{\varepsilon}_n(k_z) \equiv \varepsilon_n(k_z - p_d)]$, changing $J_{nn'}(q_\parallel)$ to the $C_{nn'}(l_B^2 q_\parallel^2/2)$ function defined by (8.25) and replacing the total electron number N (the electron density for unit volume system) by $2\pi l_B^2 N$ ($2\pi l_B^2$ comes from the summation over degenerate k_x with transverse unit area).

The main role of a longitudinal magnetic field is to confine the carriers in the range of radius l_B and to quantize the energy to form Landau subbands. Therefore, the longitudinal transport of a Kane band system corresponds to that of a laterally confined nonparabolic quantum wire.

Using the momentum and energy balance equations (13.89) and (13.90), Weng and Lei (1995) investigate the longitudinal hot-electron magnetotransport of narrow-gap semiconductor n-InSb with electron density $n = 1 \times 10^{16}$ cm^{-3} at $T = 42$ K. They consider acoustic phonon deformation potential and longitudinal phonon Fröhlich scatterings and use typical parameters for InSb: lattice constant $a_0 = 0.648$ nm, mass density $\rho_d = 5.8$ g/cm^3, conduction band bottom effective mass $m = 0.014 m_e$ (m_e is the free electron mass), nonparabolic coefficient $\alpha = 5.56$ eV^{-1}, acoustic phonon deformation potential $\Xi = 30.0$ eV, sound speed $v_s = 3.75 \times 10^5$ cm/s, optic phonon energy $\Omega_{\text{LO}} = 24.2$ meV, low-frequency dielectric constant $\kappa = 18.7$, and optic dielectric constant $\kappa_\infty = 16.8$. For magnetic fields from $B = 2$ T to $B = 25$ T, the calculated drift velocity v and electron temperature T_e are shown in Fig. 13.8. The behavior of drift velocity and electron temperature versus the electric field is qualitatively similar to that without a magnetic field: T_e increases quickly with increasing E, and v increases with rising E at the beginning, but saturates at high electric field. For a fixed electric field both the electron temperature and the drift velocity decrease with increasing magnetic field strength.

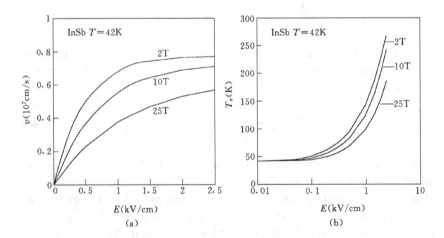

Fig. 13.8 The drift velocity v (a) and electron temperature T_e (b) versus the electric field for an n-InSb Kane band system under longitudinal strong magnetic fields. The lattice temperature $T = 42$ K and the band parameters are $\alpha = 5.56$ eV^{-1}, $m = 0.014 m_e$. The numbers near the curves are the strength of the magnetic fields. From Weng and Lei (1995).

13.4 Transport Balance Equations for Nonparabolic Multi-band Systems

Multivalley semiconductors, or systems having multiple energy bands or multiple subbands, can always be treated as systems consisting of several species of carriers with or without particle exchange. The transport balance equations for these systems were discussed in some detail in Chapter 6, but there, each species of carriers was assumed to have a parabolic energy dispersion. In this section we will extend these transport equations to allow nonparabolic or arbitrary band structure for all the subsystems (Lei, Cao and Dong, 1996).

The analyses in Sec. 11.1 for single-band system suggest the necessity of distinguishing the momentum (or physics momentum) from the lattice momentum. There are two different methods to establish a momentum balance equation for a nonparabolic or arbitrary energy band system: either by considering the rate of change of the total lattice momentum, or by considering the rate of change of the total (physics) momentum. In either case, there appears an additional factor in associated with the electric field acceleration (stream) term $e\bm{E}$ in the equation for a general energy band defined within a Brillouin zone in the lattice momentum $\bm{k}$ space. This factor reflects the most important nonparabolic effect or Bragg scattering effect. In the case of lattice-momentum related equations, this factor is the Büttiker–Thomas reduction tensor $\mathcal{R}$ (11.93), expressed as a closed-area integral over the surface of the Brillouin zone, thus its calculation requires the knowledge of the accurate electron population and depopulation there. This is usually not convenient for studying the effects of band nonparabolicity on transport, in which most carriers rarely move to the zone boundary under the acceleration of the electric field. In the case of physics momentum related equations, on the other hand, this factor is the ensemble-averaged inverse effective mass tensor $\mathcal{K}$ (11.78), expressed as a volume integral over the Brillouin zone and in most cases the dominant contribution comes from a small range of the most occupied states, thus is much more convenient to obtain in a realistic calculation. Therefore, we will develop the transport balance equations for nonparabolic multiband systems on the basis of the (physics) momentum balance.

We consider a semiconductor consisting of n energy bands (or valleys) with the energy dispersion

$$\varepsilon_a(\bm{k}) \quad (a = 1, 2, ...n), \tag{13.94}$$

the wave function $\psi_{a\bm{k}}$, and the particle creation (annihilation) operators $c^{\dagger}_{a\bm{k}\sigma}$ ($c_{a\bm{k}\sigma}$) for the ath band in the $\bm{k}$-space, having intraband and interband impurity and phonon scatterings as well as intercarrier Coulomb interactions. Treating electrons in each band as a species of carriers, we can describe the transport state of this system under the influence of a uniform electric field $\bm{E}$ by a lattice momentum shift $\bm{p}_a$, an electron temperature T_a and a chemical potential μ_a for each energy band ($a = 1, 2, ...n$). Following the same procedure as deriving the balance equations for single energy band in Secs. 11.1–11.3, i.e. starting from the Heisenberg equations of motion for the rates of changes of the particle number $\hat{N}_a = \sum_{\bm{k},\sigma} c^{\dagger}_{a\bm{k}\sigma}c_{a\bm{k}\sigma}$, the particle energy $H_a = \sum_{a,\bm{k},\sigma} \varepsilon_a(\bm{k}) c^{\dagger}_{a\bm{k}\sigma}c_{a\bm{k}\sigma}$, and the particle momentum $\bm{P}_a = m_e \sum_{\bm{k},\sigma} \bm{v}_a(\bm{k}) c^{\dagger}_{a\bm{k}\sigma}c_{a\bm{k}\sigma}$ for each band [$\bm{v}_a(\bm{k}) \equiv \bm{\nabla}_{\bm{k}}\varepsilon_a(\bm{k})$ is the velocity function of the ath band and m_e is the free electron mass] and then taking statistical averages to the lowest nonzero order in scattering interactions, we can obtain the following equations for the momentum, energy and carrier-population balance in the ath energy band:

$$\frac{d}{dt}(N_a \bm{v}_a) = N_a e \bm{E} \cdot \mathcal{K}_a + \bm{A}^a_{\text{ei}} + \bm{A}^a_{\text{ep}} + \sum_{b(\neq a)} \bm{A}^{ab}_{\text{ei}} + \sum_{b(\neq a)} \bm{A}^{ab}_{\text{ep}} + \sum_{b(\neq a)} \bm{A}^{ab}_{\text{ee}}, \tag{13.95}$$

$$\frac{d}{dt}(N_a \varepsilon_a) = N_a e \bm{E} \cdot \bm{v}_a - W^a_{\text{ep}} - \sum_{b(\neq a)} W^{ab}_{\text{ep}} - \sum_{b(\neq a)} W^{ab}_{\text{ee}}, \tag{13.96}$$

$$\frac{d}{dt} N_a = \sum_{b(\neq a)} X^{ab}_{\text{ei}} + \sum_{b(\neq a)} X^{ab}_{\text{ep}}. \tag{13.97}$$

Here N_a is the average number, $\bm{v}_a$ is the average velocity, ε_a is the average energy and $\mathcal{K}_a$ is the ensemble-averaged inverse effective mass tensor of carriers populating the ath energy band, given by

$$N_a = 2 \sum_{\bm{k}} f((\bar{\varepsilon}_a(\bm{k}) - \mu_a)/T_a), \tag{13.98}$$

$$\bm{v}_a = \frac{2}{N_a} \sum_{\bm{k}} \bm{v}_a(\bm{k}) f((\bar{\varepsilon}_a(\bm{k}) - \mu_a)/T_a), \tag{13.99}$$

$$\varepsilon_a = \frac{2}{N_a} \sum_{\bm{k}} \varepsilon_a(\bm{k}) f((\bar{\varepsilon}_a(\bm{k}) - \mu_a)/T_a), \tag{13.100}$$

$$\mathcal{K}_a = \frac{2}{N_a} \sum_{\bm{k}} [\bm{\nabla}\bm{\nabla}\varepsilon_a(\bm{k})] f((\bar{\varepsilon}_a(\bm{k}) - \mu_a)/T_a), \tag{13.101}$$

in which $f(x) \equiv 1/[1+\exp(x)]$ stands for the Fermi function, and

$$\bar{\varepsilon}_a(\boldsymbol{k}) \equiv \varepsilon_a(\boldsymbol{k}-\boldsymbol{p}_a). \tag{13.102}$$

These equations state that the frictional acceleration experienced by the carriers in the ath band consists of contributions from the intraband electron–impurity scattering, $\boldsymbol{A}_{\text{ei}}^a$, the intraband electron-phonon scattering, $\boldsymbol{A}_{\text{ep}}^a$, the interband electron–impurity scattering, $\boldsymbol{A}_{\text{ei}}^{ab}$, the interband electron–phonon scattering, $\boldsymbol{A}_{\text{ep}}^{ab}$, and the electron intraband–intraband Coulomb scattering, $\boldsymbol{A}_{\text{ee}}^{ab}$. The energy-loss rate of the carriers in the ath band consists of contributions from the intraband electron–phonon scattering, W_{ep}^a, the interband electron–phonon scattering, W_{ep}^{ab}, and the electron intraband–intraband Coulomb scattering, W_{ee}^{ab}. The rate of change of the carrier population in the ath band is due to the interband electron–impurity scattering, X_{ei}^{ab}, and the interband electron–phonon scattering, X_{ep}^{ab}. The contributions from carrier interband–interband Coulomb coupling to the frictional acceleration, the energy-loss rate and the rate of change of carrier population have been neglected, which are expected to be much smaller than those from intraband–intraband couplings or much smaller than those from interband electron–impurity and electron–phonon scattering.

The intraband frictional acceleration and energy-loss rate have the form

$$\boldsymbol{A}_{\text{ei}}^a = -n_{\text{i}} \sum_{\boldsymbol{k},\boldsymbol{q}} |u_a(\boldsymbol{k},\boldsymbol{q})|^2 [\boldsymbol{v}_a(\boldsymbol{k}) - \boldsymbol{v}_a(\boldsymbol{k}+\boldsymbol{q})] \, \Pi^{(a)}(\boldsymbol{k},\boldsymbol{q},0), \tag{13.103}$$

$$\boldsymbol{A}_{\text{ep}}^a = -2 \sum_{\boldsymbol{k},\boldsymbol{q},\lambda} |M_a(\boldsymbol{k},\boldsymbol{q},\lambda)|^2 [\boldsymbol{v}_a(\boldsymbol{k}) - \boldsymbol{v}_a(\boldsymbol{k}+\boldsymbol{q})] \, \Lambda^{(a)}(\boldsymbol{k},\boldsymbol{q},\Omega_{\boldsymbol{q}\lambda}), \tag{13.104}$$

$$W_{\text{ep}}^a = 2 \sum_{\boldsymbol{k},\boldsymbol{q},\lambda} |M_a(\boldsymbol{k},\boldsymbol{q},\lambda)|^2 \, \Omega_{\boldsymbol{q}\lambda} \, \Lambda^{(a)}(\boldsymbol{k},\boldsymbol{q},\Omega_{\boldsymbol{q}\lambda}), \tag{13.105}$$

with the intraband carrier screening considered in the random phase approximation. The relevant interband quantities are expressed as follows:

$$\boldsymbol{A}_{\text{ei}}^{ab} = -2n_{\text{i}} \sum_{\boldsymbol{k},\boldsymbol{q}} |u_{ab}(\boldsymbol{k},\boldsymbol{q})|^2 \, \boldsymbol{v}_a(\boldsymbol{k}) \, \Pi^{ab}(\boldsymbol{k},\boldsymbol{q},0), \tag{13.106}$$

$$\boldsymbol{A}_{\text{ep}}^{ab} = -2 \sum_{\boldsymbol{k},\boldsymbol{q},\lambda} |M_{ab}(\boldsymbol{k},\boldsymbol{q},\lambda)|^2 \, \boldsymbol{v}_a(\boldsymbol{k}) \, \Lambda_{+}^{ab}(\boldsymbol{k},\boldsymbol{q},\Omega_{\boldsymbol{q}\lambda}), \tag{13.107}$$

$$W_{\text{ep}}^{ab} = 2 \sum_{\boldsymbol{k},\boldsymbol{q},\lambda} |M_{ab}(\boldsymbol{k},\boldsymbol{q},\lambda)|^2 \, \varepsilon_a(\boldsymbol{k}) \, \Lambda_{+}^{ab}(\boldsymbol{k},\boldsymbol{q},\Omega_{\boldsymbol{q}\lambda}), \tag{13.108}$$

$$X_{\text{ei}}^{ab} = -2n_{\text{i}} \sum_{\boldsymbol{k},\boldsymbol{q}} |u_{ab}(\boldsymbol{k},\boldsymbol{q})|^2 \, \Pi^{ab}(\boldsymbol{k},\boldsymbol{q},0), \tag{13.109}$$

$$X_{\text{ep}}^{ab} = -2 \sum_{\bm{k},\bm{q},\lambda} |M_{ab}(\bm{k},\bm{q},\lambda)|^2 \, \Lambda_+^{ab}(\bm{k},\bm{q},\Omega_{\bm{q}\lambda}), \tag{13.110}$$

$$\bm{A}_{\text{ee}}^{ab} = -2 \sum_{\bm{k},\bm{k}',\bm{q}} |\nu_{ab}(\bm{k},\bm{k}',\bm{q})|^2 \, [\bm{v}_a(\bm{k}) - \bm{v}_a(\bm{k}+\bm{q})] \, \Upsilon^{ab}(\bm{k},\bm{k}',\bm{q}), \tag{13.111}$$

$$W_{\text{ee}}^{ab} = 2 \sum_{\bm{k},\bm{k}',\bm{q}} |\nu_{ab}(\bm{k},\bm{k}',\bm{q})|^2 \, [\varepsilon_a(\bm{k}) - \varepsilon_a(\bm{k}+\bm{q})] \, \Upsilon^{ab}(\bm{k},\bm{k}',\bm{q}), \tag{13.112}$$

having symmetrical properties:

$$W_{\text{ep}}^{ab} + W_{\text{ep}}^{ba} = 2 \sum_{\bm{k},\bm{q},\lambda} |M_{ab}(\bm{k},\bm{q},\lambda)|^2 \, \Omega_{\bm{q}\lambda} \, \Lambda_-^{ab}(\bm{k},\bm{q},\Omega_{\bm{q}\lambda}), \tag{13.113}$$

$$X_{\text{ei}}^{ab} = -X_{\text{ei}}^{ba}, \qquad X_{\text{ep}}^{ab} = -X_{\text{ep}}^{ba}. \tag{13.114}$$

Here the relevant functions are defined as

$$\Pi^{ab}(\bm{k},\bm{q},\Omega) = 2\pi \left[f\!\left(\frac{\bar{\xi}_{a\bm{k}}}{T_a}\right) - f\!\left(\frac{\bar{\xi}_{b\bm{k}+\bm{q}}}{T_b}\right) \right] \Delta(\varepsilon_a(\bm{k}) - \varepsilon_b(\bm{k}+\bm{q}) - \Omega), \tag{13.115}$$

$$\Pi^{(a)}(\bm{k},\bm{q},\Omega) = \Pi^{aa}(\bm{k},\bm{q},\Omega) \big| \epsilon_a\!\left(\bm{q}, \bar{\varepsilon}_a(\bm{k}) - \bar{\varepsilon}_a(\bm{k}+\bm{q})\right) \big|^{-2}, \tag{13.116}$$

$$\Lambda^{(a)}(\bm{k},\bm{q},\Omega) = \Pi^{(a)}(\bm{k},\bm{q},\Omega) \left[n\!\left(\frac{\Omega}{T}\right) - n\!\left(\frac{\bar{\varepsilon}_a(\bm{k}) - \bar{\varepsilon}_a(\bm{k}+\bm{q})}{T_a}\right) \right], \tag{13.117}$$

$$\Lambda_\pm^{ab}(\bm{k},\bm{q},\Omega) = \Pi^{ab}(\bm{k},\bm{q},\Omega) \left[n\!\left(\frac{\Omega}{T}\right) - n\!\left(\frac{\bar{\xi}_{a\bm{k}}}{T_a} - \frac{\bar{\xi}_{b\bm{k}+\bm{q}}}{T_b}\right) \right]$$
$$\pm \Pi^{ab}(\bm{k},\bm{q},-\Omega) \left[n\!\left(\frac{\Omega}{T}\right) - n\!\left(\frac{\bar{\xi}_{b\bm{k}+\bm{q}}}{T_b} - \frac{\bar{\xi}_{a\bm{k}}}{T_a}\right) \right], \tag{13.118}$$

$$\Upsilon^{ab}(\bm{k},\bm{k}',\bm{q}) = 2\pi \, \Delta\!\left(\varepsilon_a(\bm{k}) - \varepsilon_a(\bm{k}+\bm{q}) + \varepsilon_b(\bm{k}'+\bm{q}) - \varepsilon_b(\bm{k}')\right)$$
$$\times \big| \epsilon_a\!\left(\bm{q},\bar{\varepsilon}_a(\bm{k}) - \bar{\varepsilon}_a(\bm{k}+\bm{q})\right) \big|^{-2} \big| \epsilon_b\!\left(\bm{q},\bar{\varepsilon}_b(\bm{k}') - \bar{\varepsilon}_b(\bm{k}'+\bm{q})\right) \big|^{-2}$$
$$\times \left[f\!\left(\frac{\bar{\xi}_{a\bm{k}}}{T_a}\right) - f\!\left(\frac{\bar{\xi}_{a\bm{k}+\bm{q}}}{T_a}\right) \right] \left[f\!\left(\frac{\bar{\xi}_{b\bm{k}'}}{T_b}\right) - f\!\left(\frac{\bar{\xi}_{b\bm{k}'+\bm{q}}}{T_b}\right) \right]$$
$$\times \left[n\!\left(\frac{\bar{\varepsilon}_a(\bm{k}) - \bar{\varepsilon}_a(\bm{k}+\bm{q})}{T_a}\right) - n\!\left(\frac{\bar{\varepsilon}_b(\bm{k}') - \bar{\varepsilon}_b(\bm{k}'+\bm{q})}{T_b}\right) \right]. \tag{13.119}$$

In above equations

$$\bar{\xi}_{a\bm{k}} \equiv \bar{\varepsilon}_a(\bm{k}) - \mu_a, \qquad (13.120)$$

$n(x) \equiv 1/[\exp(x) - 1]$ stands for the Bose distribution function, n_i is the impurity density, $\Omega_{\bm{q}\lambda}$ is the energy of a phonon with wavevector $\bm{q}$ in the λth branch,

$$u_a(\bm{k}, \bm{q}) = u(\bm{q})\, g_{aa}(\bm{k}, \bm{q}),$$

and

$$u_{ab}(\bm{k}, \bm{q}) = u(\bm{q})\, g_{ab}(\bm{k}, \bm{q})$$

are respectively the intraband and interband impurity potentials,

$$M_a(\bm{k}, \bm{q}, \lambda) = M(\bm{q}, \lambda)\, g_{aa}(\bm{k}, \bm{q}),$$

and

$$M_{ab}(\bm{k}, \bm{q}, \lambda) = M(\bm{q}, \lambda)\, g_{ab}(\bm{k}, \bm{q})$$

are respectively the intraband and interband electron–phonon matrix elements, and

$$\nu_{ab}(\bm{k}, \bm{k}', \bm{q}) = \nu_\mathrm{c}(\bm{q})\, g_{aa}(\bm{k}, \bm{q})\, g_{bb}(\bm{k}', \bm{q})$$

is the electron intraband–intraband Coulomb potential. Here $g_{aa}(\bm{k}, \bm{q}) = \sum_{\bm{k}'} \langle a\bm{k}' | e^{\mathrm{i}\bm{q}\cdot\bm{r}} | a\bm{k}\rangle$ and $g_{ab}(\bm{k}, \bm{q}) = \sum_{\bm{k}'} \langle a\bm{k}' | e^{\mathrm{i}\bm{q}\cdot\bm{r}} | b\bm{k}\rangle$ are respectively the intraband and interband form factors related to the wave functions of the ath and bth energy bands, $u(\bm{q})$, $M(\bm{q}, \lambda)$ and $\nu_\mathrm{c}(\bm{q})$ are respectively the electron–impurity potential, the electron–phonon matrix element, and the electron–electron Coulomb potential in the plane-wave representation. In Eqs. (13.116) and (13.119) $\epsilon_a(\bm{q}, \omega)$ is the dielectric function of the ath band. The Coulomb couplings between different bands are treated perturbatively in the present model, therefore they do not provide additional screening.

The total number of carriers in this system is

$$N = \sum_a N_a, \qquad (13.121)$$

and the average drift velocity of the whole system is given by

$$\bm{v} = \frac{1}{N} \sum_a N_a \bm{v}_a. \qquad (13.122)$$

In the case of multiple parabolic valleys ($\bm{k}_a$ is position of the ath valley),

$$\varepsilon_a(\bm{k}) = \sum_{i,j} \frac{(k_i - k_{ai})(k_j - k_{aj})}{2m_{aij}} \quad (i,j = x, y, z), \tag{13.123}$$

$\mathcal{K}_a$ defined in Eq. (13.101) is just the inverse band-mass tensor ($\mathcal{K}_{aij} = 1/m_{aij}$), and balance equations (13.95)–(13.97) reduce to the equations equivalent to those proposed in Chapter 6 for multiple parabolic bands.

13.5 Applications of Nonparabolic Multiband Balance Equations

As examples of the application of these nonparabolic multiband balance equations, this section will discuss high electric field transport in Si (Lei, Cao and Dong, 1996) and GaAs (Cao and Lei, 1997), and in superlattice longitudinal transport with high-lying minibands (Lei, da Cunha Lima and Troper, 1997).

13.5.1 High electric field transport in Si

We describe the bulk Si as a system consisting of six valleys locating at $\bm{k}_1 = (2\pi/a_0)(0.85, 0, 0)$, $\bm{k}_2 = (2\pi/a_0)(-0.85, 0, 0)$, $\cdots$, and $\bm{k}_6 = (2\pi/a_0)(0, 0, -0.85)$, having nonparabolic energy dispersions of Kane-type for each valley,

$$\varepsilon_a(\bm{k}) = \frac{1}{2\alpha}\left[\sqrt{1 + 4\alpha \varepsilon_a^{(\mathrm{p})}(\bm{k})} - 1\right] \quad (a = 1, 2, \cdots, 6), \tag{13.124}$$

where a_0 is the lattice constant, α is the nonparabolic parameter, $\varepsilon_a^{(\mathrm{p})}(\bm{k})$ is the parabolic energy dispersion pertinent to the ath valley ($a = 1, 2, \cdots, 6$):

$$\varepsilon_a^{(\mathrm{p})}(\bm{k}) = \frac{(k_x - k_{ax})^2}{2m_{ax}} + \frac{(k_y - k_{ay})^2}{2m_{ay}} + \frac{(k_z - k_{az})^2}{2m_{az}}, \tag{13.125}$$

where $1/m_{ax}$, $1/m_{ay}$ and $1/m_{az}$ are respectively the xx, yy and zz components of the inverse-mass tensor at the bottom of the ath valley. For instance, taking the first valley to be the one centered at the positive x-axis, we have $m_{1x} = m_{1y} = m_t = 0.1905\, m_\mathrm{e}$ and $m_{1z} = m_l = 0.9163\, m_\mathrm{e}$ (m_e is the free electron mass), m_t and m_l being the transverse and longitudinal mass at the band bottom.

If an electric field is applied along the $\langle 111 \rangle$ direction, $\boldsymbol{E} = (E/\sqrt{3})(1,1,1)$, the overall spatial symmetry of the transport configuration indicates that all six valleys perform equivalently, such that they have identical electron temperature, identical chemical potential and identical lattice momentum shift: $T_a = T_e$, $\mu_a = \mu$, and $\boldsymbol{p}_a = \boldsymbol{p}_d = (p_d/\sqrt{3})(1,1,1)$ ($a = 1, 2, \cdots, 6$). The drift velocity $\boldsymbol{v}_d$ is of course also along the $\langle 111 \rangle$ direction. The equations for determining these parameters are obtained by summing Eqs. (13.95), (13.96) and (13.97) over all six valleys. The x, y and z components of the overall electric-field acceleration term of the resultant equation are identical. Its x component can be written in the form ($\hat{\boldsymbol{x}}$ denotes the unit vector in the x direction)

$$\hat{\boldsymbol{x}} \cdot \sum_{a=1}^{6} N_a e\boldsymbol{E} \cdot \mathcal{K}_a = \frac{4}{\sqrt{3}} \left(1 + \frac{2m_l}{m_t}\right) \left(\frac{m_t}{m_l}\right)^{1/2} \left(\frac{m_t}{m_s}\right)^{1/2} \frac{eE}{m_s}$$

$$\times \sum_{\boldsymbol{k}} \frac{f((\varepsilon^{(0)}(\boldsymbol{k} - \boldsymbol{p}_1^*) - \mu)/T_e)}{\sqrt{1 + 2\alpha\, k^2/m_s}} \left[1 - \frac{4\alpha}{m_s}\left(\frac{\boldsymbol{k} \cdot \boldsymbol{p}_1^*}{|\boldsymbol{p}_1^*|}\right)^2 \left(1 + \frac{2\alpha\, k^2}{m_s}\right)^{-1}\right],$$

(13.126)

in which

$$\varepsilon^{(0)}(\boldsymbol{k}) = \frac{1}{2\alpha}\left(\sqrt{1 + \frac{2\alpha\, k^2}{m_s}} - 1\right) \tag{13.127}$$

is the Kane dispersion of an isotropic energy band with a bottom mass m_s arbitrarily chosen, and $\boldsymbol{p}_1^*$ is the first-valley Herring–Vogt (1956) transformed lattice-momentum shift:

$$\boldsymbol{p}_1^* = \frac{p_d}{\sqrt{3}}\left(\sqrt{\frac{m_s}{m_l}}, \sqrt{\frac{m_s}{m_t}}, \sqrt{\frac{m_s}{m_t}}\right). \tag{13.128}$$

Identical results will be obtained if one uses any other $\boldsymbol{p}_a^*$ to replace $\boldsymbol{p}_1^*$ in the equation.

We consider intravalley scatterings from acoustic phonons through the deformation potential and intervalley scatterings from both acoustic and optic phonons. Each term of the overall frictional acceleration and energy-loss rate can also be simplified by performing a Herring-Vogt transformation respectively for the summation variables of each valley. For instance, the x component of the frictional acceleration term due to intervalley phonon scattering can be written in the form

$$\hat{x} \cdot \sum_{a=1}^{6}\sum_{b\neq a} A_{\text{ep}}^{ab} = -2 \sum_{a=1}^{6}\sum_{b\neq a}\sum_{\mathbf{k},\mathbf{k}'} |M_{ab}|^2 v_{ax}(\mathbf{k})$$
$$\times L\left[\varepsilon_a^{(\text{p})}(\mathbf{k}), \varepsilon_a^{(\text{p})}(\mathbf{k}-\mathbf{p}_d), \varepsilon_b^{(\text{p})}(\mathbf{k}'), \varepsilon_b^{(\text{p})}(\mathbf{k}'-\mathbf{p}_d), \Omega_{ab}\right]$$
$$= -\frac{8\pi}{m_s}\left(1 + \frac{2m_l}{m_t}\right)^{1/2}\left(\frac{m_t}{m_s}\right)^2\left(\frac{m_l}{m_s}\right)^{1/2}$$
$$\times \sum_{b=2}^{6}\sum_{\mathbf{k},\mathbf{k}'} |M_{1b}|^2 \left(\frac{\mathbf{k}\cdot\mathbf{p}_1^*}{|\mathbf{p}_1^*|}\right)\left(1 + \frac{2\alpha k^2}{m_s}\right)^{1/2}$$
$$\times L\left[\frac{k^2}{2m_s}, \frac{(\mathbf{k}-\mathbf{p}_1^*)^2}{2m_s}, \frac{k'^2}{2m_s}, \frac{(\mathbf{k}'-\mathbf{p}_b^*)^2}{2m_s}, \Omega_{1b}\right], \qquad (13.129)$$

where we have denoted the function $\Lambda_+^{ab}(\mathbf{k}, \mathbf{q}, \Omega_{ab})$ [see (13.118)] as ($\mathbf{k}' = \mathbf{k} + \mathbf{q}$)

$$L\left[\varepsilon_a^{(\text{p})}(\mathbf{k}), \varepsilon_a^{(\text{p})}(\mathbf{k}-\mathbf{p}_d), \varepsilon_b^{(\text{p})}(\mathbf{k}'), \varepsilon_b^{(\text{p})}(\mathbf{k}'-\mathbf{p}_d), \Omega_{ab}\right], \qquad (13.130)$$

explicitly indicating its functional dependence on the relevant electron and phonon energies and wavevectors. The summation over b in the above equation consists of contributions from the "parallel" valley ($b = 2$) or f scattering, and from perpendicular valleys ($b = 3, 4, 5, 6$) or g scattering. We consider three f and three g scatterings. Their squared matrix elements are respectively ($\lambda = 1, 2, 3$)

$$|M_{12}|^2 = D_{f\lambda}^2/(2d_c\Omega_{f\lambda}),$$
$$|M_{1b}|^2 = D_{g\lambda}^2/(2d_c\Omega_{g\lambda}) \quad (b = 3, 4, 5, 6).$$

The squared matrix element of the intravalley acoustic phonon scattering (with intraband carrier screening) is approximated by

$$|M_{aa}|^2 = \Xi^2 q/(2d_c v_s).$$

To facilitate comparison we use the same set of material and electron-phonon coupling parameters as in the previous Monte-Carlo calculation (Canali et al, 1975): the dielectric constant $\kappa = 11.7$, the mass density $d_c = 2.329\,\text{g/cm}^3$, the intravalley deformation potential $\Xi = 9.2\,\text{eV}$, the longitudinal sound speed $v_{sl} = 9.04 \times 10^3\,\text{m/s}$, the equivalent phonon frequencies (in units of absolute temperature) and coupling constants for three f scatterings are taken as $\Omega_{f\lambda} = 210, 500, 630\,\text{K}$ and $D_{f\lambda} = 0.15, 3.4, 4.0 \times$

10^8 eV/cm ($\lambda = 1, 2, 3$), and the equivalent phonon frequencies and coupling constants for three g scatterings are taken as $\Omega_{g\lambda} = 140, 210, 700$ K and $D_{g\lambda} = 0.5, 0.8, 3.0 \times 10^8$ eV/cm ($\lambda = 1, 2, 3$).

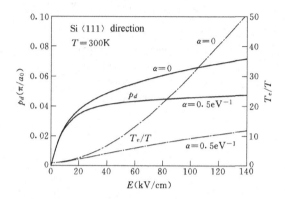

Fig. 13.9 The calculated lattice momentum shift p_d (solid curves, in units of π/a_0) and electron temperature T_e (chain curves, in units of lattice temperature T) for dc steady state transport along the $\langle 111 \rangle$ direction in a six-valley band model for bulk Si described in the text, are shown as functions of the applied field strength E at lattice temperature $T = 300$ K for a nonparabolic ($\alpha = 0.5$ eV^{-1}) case and the parabolic ($\alpha = 0$) case. From Lei, Cao and Dong (1996).

The lattice momentum shift p_d and the electron temperature T_e obtained from the dc steady-state solution to the above-mentioned equations (derived by summing Eqs. (13.95), (13.96) and (13.97) over all six valleys), are shown in Fig. 13.9 for a nonparabolic case with coefficient $\alpha = 0.5$ eV^{-1} and the parabolic case ($\alpha = 0$). At low fields ($E < 10$ kV/cm) such a moderate nonparabolicity ($\alpha = 0.5$ eV^{-1}) yields almost the same result as in the parabolic case. At higher strength of the electric field the nonparabolic band has a smaller lattice momentum shift p_d and much lower electron temperature T_e. At highest field strength ($E = 140$ kV/cm), the electron temperature T_e in the $\alpha = 0.5$ eV^{-1} case is only one-fifth that required in the parabolic case ($\alpha = 0$).

The average drift velocities v for the nonparabolic (solid curve) and the parabolic (dotted curve) cases, calculated from Eqs. (13.99) and (13.122), are plotted in Fig. 13.10 as functions of the electric field strength E. For comparison, the theoretical result of Monte-Carlo simulation (chain curve) reported by Canali, Nava and Reggiani (1985) and the experimental data obtained by a time of flight (Tof) technique (crosses) and by a microwave time of flight technique (dots and circles, obtained with microwaves of wave-

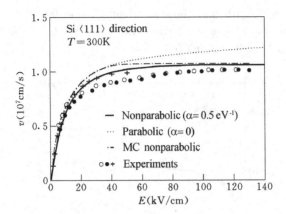

Fig. 13.10 Drift velocity v versus the strength of the electric field E in the dc steady state transport along the $\langle 111 \rangle$ direction of bulk Si at lattice temperature $T = 300$ K. The solid and dotted lines are respectively the nonparabolic ($\alpha = 0.5\,\text{eV}^{-1}$) and parabolic ($\alpha = 0$) results, the chain line is the Monte-Carlo prediction with similar nonparabolic energy dispersion, the crosses, dots and circles are respectively the experimental data obtained by a time of flight (Tof) technique (crosses) and by a microwave time of flight (MTof) technique (dots and circles, obtained with microwave of wavelengths 3.13 and 5.39 μm respectively). From Lei, Cao and Dong (1996).

length 3.13 and 5.39 μm respectively) of Smith, Inoue and Frey (1980) are also shown in the figure. The theoretical predictions of the present balance-equation approach with no adjustable parameter, are in good agreement with the experimental data over the whole range of the electric field. Note that the balance-equation curve, which are obtained using the same set of material and electron–phonon coupling parameters as in the Monte-Carlo calculation and at a computational CPU cost of only a few percent that needed in a Monte-Carlo simulation, seems better than the Monte-Carlo curve. And a slight adjustment of the parameters can further improve the agreement between the present nonparabolic predictions and the measured data. On the other hand, the theoretical predictions of parabolic band are far from the experimental results except in the low field regime. This shows that the nonparabolicity of the energy band is crucial for interpreting the experiments in the range of mediate to high strength of the electric field.

13.5.2 High electric field transport in GaAs

The conduction band of GaAs is considered here to consist of three non-parabolic isotropic valleys (see Sec. 6.8): a center valley at Γ point ($d_\Gamma = 1$),

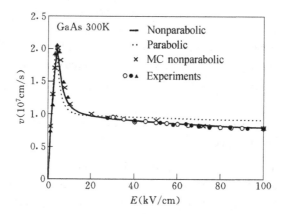

Fig. 13.11 The average drift velocity v versus the electric field E in the dc steady-state transport of n-type GaAs at lattice temperature $T = 300$ K. The solid and dashed curves are the balance equation calculations for nonparabolic and parabolic energy dispersions, respectively. The discrete symbols represent the Monte-Carlo results (Fischetti and Laux, 1988) and the experimental data. From Cao and Lei (1997).

four equivalent valleys at L points ($d_\mathrm{L} = 4$) and three equivalent valleys at X points ($d_\mathrm{X} = 3$). All of them have nonparabolic Kane energy dispersion ($a = \Gamma, \mathrm{L}, \mathrm{X}$),

$$\varepsilon_a(\boldsymbol{k}) = \frac{1}{2\alpha_a}\left\{\left[1 + \frac{2\alpha_a}{m_a}(\boldsymbol{k}-\boldsymbol{k}_a)^2\right]^{1/2} - 1\right\}, \qquad (13.131)$$

in which $\boldsymbol{k}_a$ is the position and m_a is the effective mass of the band bottom of the ath valley, and α_a is the band nonparabolic coefficient of the ath valley. We take $\alpha_\Gamma = 0.62\,\mathrm{eV}^{-1}$, $\alpha_\mathrm{L} = 0.5\,\mathrm{eV}^{-1}$, and $\alpha_\mathrm{X} = 0.3\,\mathrm{eV}^{-1}$. When $\alpha_a \to 0$, the energy dispersions (13.131) reduce to the parabolic spectra discussed in Sec. 6.8. The calculated average drift velocity $\boldsymbol{v} = N^{-1}\sum_a N_a \boldsymbol{v}_a$ ($N = N_\Gamma + N_\mathrm{L} + N_\mathrm{X}$ is the total electron number, and N_a is the electron number in the ath valley) using the nonparabolic multiband balance equations at $T = 300$ K with the scattering mechanisms described in Sec. 6.8, is shown in Fig. 13.11 (solid curve) as a function of the electric field E. For comparison the results obtained from the balance equations by using the parabolic energy dispersion $\alpha_a = 0$ (dashed line) and that from nonparabolic Monte-Carlo simulation (crosses), are also shown, together with the experimental data (small circles, black dots and triangles). The prediction of nonparabolic balance equation approach agrees well with the Monte-Carlo simulation and with the experimental measurements through the whole electric field range up to 100 KV/cm. The parabolic result, though agrees with nonparabolic calculation at low fields ($E < 5\,\text{kV/cm}$), is lower at mediate electric fields ($5 \sim 20\,\text{kV/cm}$) and higher at high electric fields ($E > 20\,\text{kV/cm}$), indicating that nonparabolicity has to be taken into account at high electric fields in GaAs.

13.5.3 Effect of high-lying minibands on superlattice longitudinal transport

The analyses of superlattice longitudinal transport in Chapter 12 were based on the assumption that carriers are moving within a single miniband and effects of high-lying miniband are neglected completely. Although the basic physical feature of Bragg-diffraction-related phenomenon is contained in the single-miniband model, the carrier population of high-lying minibands is not negligible for steady-state transport when the electric field is close to or falls in the negative differential mobility regime, where the electron temperature T_e (in equivalent to the energy) can be as high as, or even higher than, the energy distance between the bottoms of the first and second minibands of the superlattice. Therefore the effect of high-lying minibands on superlattice transport can be appreciable and a single-miniband model is not sufficient for a quantitative examination of the miniband conduction. The nonparabolic multiband balance equations developed in Sec. 13.4 provide a way to include high-lying minibands in the study of superlattice longitudinal transport.

The superlattice considered here is the same as that discussed in Sec. 12.1, having its single electron state characterized by a miniband index α ($\alpha = 1, 2, \cdots, s$) and a three-dimensional wave vector $\boldsymbol{k} = (k_x, k_y, k_z) = (\boldsymbol{k}_\parallel, k_z)$, where $-\infty < k_x, k_y < \infty$, $-\pi/d < k_z \leqslant \pi/d$, and d is the period of the superlattice. The wave function $\psi_{\alpha\boldsymbol{k}}(\boldsymbol{r})$ of this state is described by (12.1) with the eigen energy given by $\varepsilon_\alpha(\boldsymbol{k}) = k_\parallel^2/m + \varepsilon_\alpha(k_z)$, m being the effective mass of the carrier motion in the plane. For the miniband energy dispersion $\varepsilon_\alpha(k_z)$ we take the nearest neighbor tight-binding overlap approximation:

$$\varepsilon_\alpha(k_z) = \varepsilon_{\alpha 0} + \frac{\Delta_\alpha}{2}\left[1 + (-1)^\alpha \cos(k_z d)\right], \qquad (13.132)$$

where Δ_α is the energy width, and $\varepsilon_{\alpha 0}$ is the energy bottom of the αth miniband. Choosing the lowest miniband bottom as the energy zero ($\varepsilon_{10} = 0$) and denoting $\Delta \equiv \Delta_1$ for simplicity, we write the energy spectrum of the first miniband as

$$\varepsilon_1(k_z) = \frac{\Delta}{2}\left[1 - \cos(k_z d)\right] \tag{13.133}$$

(band bottom at $k_z = 0$), the energy spectrum of the second miniband as

$$\varepsilon_2(k_z) = \varepsilon_{20} + \frac{\Delta_2}{2}\left[1 + \cos(k_z d)\right] \tag{13.134}$$

(band bottom at $k_z = \pi/d$), and so on. For square well superlattices, these miniband parameters are easily calculated based on the Kronig-Penny model (Kittel, 1976) once the superlattice period d, well width a and barrier height V_b are given. As an example, we consider a GaAs/Al$_x$Ga$_{1-x}$As-based superlattice with $d = 10$ nm and $a = 9$ nm. Based on the potential barrier height $V_b = 0.258$ eV determined from the band offset between two materials, we have $\Delta = 400$ K, $\Delta_2 = 1400$ K and $\varepsilon_{20} = 1100$ K. On the other hand, we approximately take the following function of an infinitely deep potential well of width d,

$$\phi_\alpha(z) = \begin{cases} \sin(\alpha\pi z/d) & \text{when } 0 \leqslant z < d, \\ 0 & \text{when } z < 0 \text{ or } z > d, \end{cases} \tag{13.135}$$

for the single well functions $\phi_\alpha(z)$ ($\alpha = 1, 2, \cdots, s$) in the tight-binding envelope function (12.3).

Assume that there are N electrons in the above-mentioned multi-miniband system. They may populate over several minibands and we treat electrons in each miniband as one species of carriers. When a uniform electric field $\boldsymbol{E} = (0, 0, E)$ is applied along the superlattice growth axis (z direction), these electrons residing in different minibands, are accelerated by the electric field and scattered by impurities, phonons and among themselves. Generally the intra-miniband Coulomb interactions are considered strong, that the transport state of each species of carriers can be described by an electron temperature T_α, a chemical potential μ_α and a momentum shift $\boldsymbol{p}_\alpha = (0, 0, p_\alpha)$ (by symmetry, the average drift velocity and the frictional acceleration, thus the momentum shift for each miniband, are in the z direction). Among the inter-miniband collision terms, the type with subscript $(\alpha, \alpha; \alpha', \alpha')$ ($\alpha \neq \alpha'$), which represents collisions of two electrons belong to different minibands α and α' and which are scattered respectively within the same miniband, is the strongest and tends to give a common electron temperature for both minibands without inducing the particle exchange between them. The remaining nonzero inter-miniband terms $(\alpha, \alpha'; \alpha', \alpha)$ and $(\alpha, \alpha'; \alpha, \alpha')$ ($\alpha \neq \alpha'$), which involve the exchange of particles between two minibands, are generally small as compared with those of $(\alpha,\alpha;\alpha',\alpha')$ type. Therefore, we assume that intraband and interband Coulomb couplings are strong enough to give a unique electron temperature within the whole electron system, and that the frictional mutual drag and particle exchange between different minibands due to interband electron–electron scattering are less important in comparison with those due to electron–phonon and electron–impurity scatterings and are neglected. In this way, we describe the transport state of electrons in a multi-miniband system under the influence of an uniform electric field $\boldsymbol{E} = (0,0,E)$ along the superlattice axis (z direction), by a lattice momentum shift $p_\alpha = (0,0,p_d)$ and a chemical potential μ_α for each miniband, together with a unique electron temperature T_e for the whole system.

For the determination of these transport parameters, one needs the equations for carrier population conservation and effective force balance in each miniband,[1]

$$\frac{d}{dt}N_\alpha = \sum_{\beta(\neq\alpha)} X_{\text{ei}}^{\alpha\beta} + \sum_{\beta(\neq\alpha)} X_{\text{ep}}^{\alpha\beta}, \qquad (13.136)$$

$$\frac{d}{dt}(N_\alpha v_\alpha) = N_\alpha \frac{eE}{m_{\alpha z}^*} + A_{\text{ei}}^\alpha + A_{\text{ep}}^\alpha + \sum_{\beta(\neq\alpha)} A_{\text{ei}}^{\alpha\beta} + \sum_{\beta(\neq\alpha)} A_{\text{ep}}^{\alpha\beta}, \qquad (13.137)$$

as well as the equation of energy balance for the whole system (summing Eq. (13.96) over all s minibands),

$$\frac{d}{dt}\mathcal{E} = NeEv_d - W_{\text{ep}}^{\text{intra}} - W_{\text{ep}}^{\text{inter}}. \qquad (13.138)$$

Here N, $\mathcal{E}$ and v_d are respectively the total particle number, the average total energy and the average drift velocity of the system:

$$N = \sum_\alpha N_\alpha, \qquad (13.139)$$

$$\mathcal{E} = \sum_\alpha N_\alpha \varepsilon_\alpha, \qquad (13.140)$$

$$v_d = \sum_\alpha rn_\alpha v_\alpha, \qquad (13.141)$$

with N_α the number and $rn_\alpha \equiv N_\alpha/N$ the fraction of carriers populating the αth miniband. In the above equations, the drift velocity v_α, the average

---
[1] The contribution of interband Coulomb scattering on the frictional acceleration [corresponding to $A_{\text{ee}}^{\alpha\beta}$ in (13.95)] is neglected in (13.137).

electron energy (per carrier) ε_α and the zz component $1/m^*_{\alpha z}$ of the inverse effective mass tensor $\mathcal{K}_\alpha$ of the αth miniband, have the same expressions as (13.98)–(13.101). The intraband and interband frictional accelerations and population change rates due to impurity and phonon scatterings, A^α_{ei}, A^α_{ep}, $A^{\alpha\beta}_{\mathrm{ei}}$, $A^{\alpha\beta}_{\mathrm{ep}}$, $X^{\alpha\beta}_{\mathrm{ei}}$ and $X^{\alpha\beta}_{\mathrm{ep}}$, are given respectively by (13.103), (13.104), (13.106), (13.107), (13.109) and (13.110). The intraband and interband electron–phonon scattering induced energy-loss rate of the whole system, $W^{\mathrm{intra}}_{\mathrm{ep}}$ and $W^{\mathrm{inter}}_{\mathrm{ep}}$, can be written as

$$W^{\mathrm{intra}}_{\mathrm{ep}} = 2 \sum_{\alpha,\bm{k},\bm{q},\lambda} |M(\bm{q},\lambda)|^2 |g_{\alpha\alpha}(q_z)|^2 \Omega_{\bm{q}\lambda} \Lambda^{(\alpha)}(\bm{k},\bm{q},\Omega_{\bm{q}\lambda}), \qquad (13.142)$$

$$W^{\mathrm{inter}}_{\mathrm{ep}} = 2 \sum_{\alpha,\beta(\neq\alpha),\bm{k},\bm{q},\lambda} |M(\bm{q},\lambda)|^2 |g_{\alpha\beta}(q_z)|^2 \Omega_{\bm{q}\lambda} \Lambda^{\alpha\beta}_-(\bm{k},\bm{q},\Omega_{\bm{q}\lambda}). \qquad (13.143)$$

Note that we always have relations $X^{\alpha\beta}_{\mathrm{ei}} = -X^{\beta\alpha}_{\mathrm{ei}}$ and $X^{\alpha\beta}_{\mathrm{ep}} = -X^{\beta\alpha}_{\mathrm{ep}}$. For a system having s minibands, there are $2s+1$ variables: p_α, μ_α ($\alpha = 1, 2, ...s$) and T_e. In the steady transport state $[dN_\alpha/dt = 0,\ d(N_\alpha v_\alpha)/dt = 0$ and $d\mathcal{E}/dt = 0]$, (13.136) contains $s-1$ independent equations for carrier number conservation, (13.137) contains s independent equations for the momentum balance and (13.138) gives one independent equation for the energy balance. These $2s$ independent equations, together with the constrain (13.139), form a complete set of equations for the determination of the above $2s + 1$ variables if the total number of carrier N, the electric field E and the lattice temperature T are given, allowing us to calculate the drift velocity, mobility and other transport quantities.

Based on the above equations, Lei, da Cunha Lima and Troper (1997) investigate the effect of high-lying minibands on superlattice vertical transport from a two-miniband model of above-mentioned GaAs superlattice ($d = 10\,\mathrm{nm}$, $\Delta = 400\,\mathrm{K}$, $\Delta_2 = 1400$ K, $\varepsilon_{20} = 1100$ K) with carrier density $N_\mathrm{s} = 2.0 \times 10^{15}\,\mathrm{m^{-2}}$ and low-temperature linear mobility $\mu(0) = 1.0$ $\mathrm{m^2/Vs}$, considering the first and the second minibands. The partial results are presented here in Figs. 13.12 and 13.13.

Fig. 13.12 plots the lattice momentum shifts p_1 and p_2 and drift velocities v_1 and v_2 for the first and second minibands as functions of the electric field E in the steady state transport. The chemical potential difference between these two minibands, $\Delta\mu \equiv \mu_1 - \mu_2$, and the fraction of the number of carriers accommodating the second miniband, rn_2, are shown in the inset of this figure. As expected, both p_1 and p_2 grow with increasing electric field, but the growth rate of p_2 is much slower than that of p_1 and the magnitude

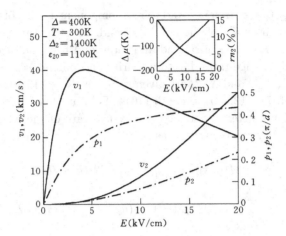

Fig. 13.12 The transport parameters of a GaAs/AlGaAs two-miniband model superlattice. From Lei, da Cunha Lima and Troper (1997).

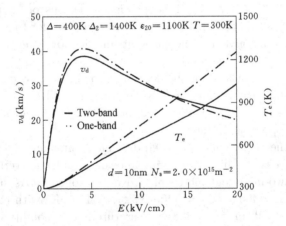

Fig. 13.13 The average drift velocity v_d and electron temperature T_e, obtained from the two-miniband model (solid curves) for the system described in Fig. 13.12. The chain curves are results obtained from a one-miniband model with only the first miniband. From Lei, da Cunha Lima and Troper (1997).

of p_1 is much larger than p_2 at small to medium field strength. the velocity of the upper miniband increases monotonically with increasing E, while the velocity of the lower miniband exhibits a peak around $E \sim 4.5\,\text{kV/cm}$ before it decreases with increasing field. Note that the chemical potential of the upper miniband is lower than that of the lower miniband, and this difference is enhanced with increasing field strength.

The total average velocity v_d obtained from (13.141) and the electron temperature T_e of the two-miniband model are shown in Fig. 13.13 (solid curves) as functions of the electric field E. For comparison, in the same figure we also plot the theoretical prediction of v_d and T_e obtained by assuming that only the lowest miniband is relevant (one-miniband model). It turns out that the existence of a high-lying miniband not only greatly reduces the electron temperature T_e, but also results in an appreciable decrease of the peak drift velocity and a slow-down of the descending rate of v_d in the negative differential mobility regime.

13.6 Balance Equations with Impact Ionization and Auger Recombination Processes

Impact ionization and Auger recombination processes are basic mechanisms for the generation and annihilation of free charge carriers in semiconductors. They are transition processes between the conduction band and valence band induced by carrier–carrier Coulomb interactions in semiconductors (Ridley, 1988). Due to direct or indirect collision, an electron in a higher energy state $c\bm{k}_1\sigma_1$ of the conduction band transits to a lower energy state $c\bm{k}'_1\sigma_1$, and the released energy is absorbed by an electron which is initially in the state $v\bm{k}_2\sigma_2$ of the valence band and then transits to the state $c\bm{k}'_1\sigma_1$ of the conduction band, resulting in the generation of a hole in the valence band. This is an impact ionization process. The Auger recombination is the inverse process of the impact ionization: two electrons in the conduction band collide with the result that one transits to the valence band and the other one, absorbed the released energy, goes to the higher energy state of the conduction band. The impact ionization and Auger recombination processes have a great influence on the behavior of high electric-field carrier transport and the property of semiconductor devices.

13.6.1 *Impact ionization and Auger recombination processes in semiconductors*

So far we have not yet considered the impact ionization and Auger recombination in this book, therefore the applicable electric field ranges of the derived balance equations are restricted. To deal with carrier transport under higher strengths of the electric fields, the role of impact ionization and Auger recombination processes has to be included in the formulation of the balance equation theory.

We consider a simplest semiconductor composed of a conduction band with the energy dispersion $\varepsilon_c(\mathbf{k}) \equiv \varepsilon(\mathbf{k})$ and wave function $\psi_{c\mathbf{k}\sigma}(\mathbf{r}) = e^{i\mathbf{k}\cdot\mathbf{r}} u_{c\mathbf{k}}(\mathbf{r})\chi_\sigma$, and a valence band with the energy dispersion $\varepsilon_v(\mathbf{k})$ and wave function $\psi_{v\mathbf{k}\sigma}(\mathbf{r}) = e^{i\mathbf{k}\cdot\mathbf{r}} u_{v\mathbf{k}}(\mathbf{r})\chi_\sigma$. Here $u_{c\mathbf{k}}(\mathbf{r})$ and $u_{v\mathbf{k}}(\mathbf{r})$ are cell periodic functions of the conduction band and valence band, and χ_σ is the spin function ($\sigma = \pm 1$). In terms of creation (annihilation) operators of electrons in the conduction and valence bands, $c_{\mathbf{k}\sigma}^\dagger$ ($c_{\mathbf{k}\sigma}$) and $c_{v\mathbf{k}\sigma}^\dagger$ ($c_{v\mathbf{k}\sigma}$), we write the part of direct electron–electron Coulomb interaction contribution to the impact ionization and Auger recombination in the form

$$H_{\rm ii} = \sum_{\substack{\mathbf{k}_1,\mathbf{k}_2,\mathbf{k}'_1,\mathbf{k}'_2 \\ \sigma,\sigma'}} \langle c\mathbf{k}'_2, c\mathbf{k}'_1|V_c|v\mathbf{k}_1, c\mathbf{k}_2\rangle\, c_{\mathbf{k}'_2\sigma'}^\dagger c_{\mathbf{k}'_1\sigma}^\dagger c_{v\mathbf{k}_1\sigma} c_{\mathbf{k}_2\sigma'} + {\rm c.c.}, \quad (13.144)$$

where V_c is the screened Coulomb potential, and c.c. represents the complex conjugate of the preceding part in the equation. When neglecting the U processes, the above matrix elements can be written as

$$\langle c\mathbf{k}'_2, c\mathbf{k}'_1|V_c|v\mathbf{k}_1, c\mathbf{k}_2\rangle = \frac{e^2}{\varepsilon_0 \kappa} \frac{I_{\rm cv}(\mathbf{k}'_1,\mathbf{k}_1) I_{\rm cc}(\mathbf{k}'_2,\mathbf{k}_2)}{|\mathbf{k}'_1 - \mathbf{k}_1|^2 + q_{\rm s}^2} \delta(\mathbf{k}_1 + \mathbf{k}_2 - \mathbf{k}'_1 - \mathbf{k}'_2), \tag{13.145}$$

in which κ is the dielectric constant of the background material, $q_{\rm s}$ is the screening wave vector, and $I_{\rm cv}$ and $I_{\rm cc}$ are, respectively, the overlap integrals between the conduction and valence bands and intra-conduction band:

$$I_{\rm cv}(\mathbf{k}'_1,\mathbf{k}_1) = \int u_{c\mathbf{k}'_1}^*(\mathbf{r}_1)\, u_{v\mathbf{k}_1}(\mathbf{r}_1)\, d\mathbf{r}_1, \tag{13.146}$$

$$I_{\rm cc}(\mathbf{k}'_2,\mathbf{k}_2) = \int u_{c\mathbf{k}'_2}^*(\mathbf{r}_2)\, u_{c\mathbf{k}_2}(\mathbf{r}_2)\, d\mathbf{r}_2. \tag{13.147}$$

Note that, though the spin index does not show up in the matrix elements (13.145), the importance of the scattering processes in which the two electrons in the initial state (final state) having parallel spins or anti-parallel spins are quite different. For the scattering events with parallel spins in the initial (final) state, the direct process is not distinguishable from the exchange processes and the interference between these two processes occurs, resulting in the cancellation of their contribution to the scattering rate to a large degree (Anderson, 1963). This can be seen from the expression of $H_{\rm ii}$ in (13.144). By first exchanging the summation indices $\mathbf{k}'_1 \leftrightarrow \mathbf{k}'_2$ and then exchanging the order of two creation operators, we can write the $\sigma' = \sigma$ part of H_{ii} as

$$H_{\text{ii}}(\sigma' = \sigma) = \frac{1}{2} \sum_{\bm{k}_1,\bm{k}_2,\bm{k}'_1,\bm{k}'_2,\sigma} [\langle c\bm{k}'_2, c\bm{k}'_1 | V_c | v\bm{k}_1, c\bm{k}_2 \rangle -$$
$$\langle c\bm{k}'_1, c\bm{k}'_2 | V_c | v\bm{k}_1, c\bm{k}_2 \rangle] c^\dagger_{\bm{k}'_2\sigma} c^\dagger_{\bm{k}'_1\sigma} c_{v\bm{k}_1\sigma} c_{\bm{k}_2\sigma} + \text{c.c.} \quad (13.148)$$

Because in the dominant scattering processes the energy difference between two initial (final) states $c\bm{k}'_1$ and $c\bm{k}'_2$ is small, two wavevectors $\bm{k}'_1$ and $\bm{k}'_2$ are not far away and the overlap integrals (13.146) and (13.147) change slowly with changing wavevector. The two matrix elements in the square bracket on the right-hand-side of (13.148) are quite close that they are almost canceled with each other. Therefore, people generally consider only the spin anti-parallel ($\sigma' = \bar{\sigma}$) part in H_{ii} for simplicity, and treat the overlap integrals I_{cv} and I_{cc} as constants.

In an n-type semiconductor many states in the conduction band are populated by electrons, but generally only a few unoccupied states in the valence band. At thermodynamic equilibrium there are a unique temperature T and a unique chemical potential μ for the whole system. It is usually more convenient to use the concept of hole to describe the behavior of the valence band. A hole is a quasiparticle to characterize an unoccupied state (such as $v\bm{k}\sigma$) in such a way that the creation of a hole in the $\bm{k}\sigma$ state is equivalent to the annihilation of an electron in the $v\bm{k}\sigma$ state. If $d^\dagger_{\bm{k}\sigma}$ ($d_{\bm{k}\sigma}$) denotes the creation (annihilation) operator of a hole, then $d^\dagger_{\bm{k}\sigma} = c_{v\bm{k}\sigma}$ and $d_{\bm{k}\sigma} = c^\dagger_{v\bm{k}\sigma}$. The single electron Hamiltonian of the semiconductor composed of conduction and valence bands can be written as

$$H_{\text{s}} = \sum_{\bm{k},\sigma} \varepsilon(\bm{k}) c^\dagger_{\bm{k}\sigma} c_{\bm{k}\sigma} + \sum_{\bm{k},\sigma} \varepsilon_{\text{v}}(\bm{k}) c^\dagger_{v\bm{k}\sigma} c_{v\bm{k}\sigma}$$
$$= \sum_{\bm{k},\sigma} \varepsilon(\bm{k}) c^\dagger_{\bm{k}\sigma} c_{\bm{k}\sigma} - \sum_{\bm{k},\sigma} \varepsilon_{\text{v}}(\bm{k}) (c_{v\bm{k}\sigma} c^\dagger_{v\bm{k}\sigma} - 1)$$
$$= H_{\text{e}} + H_{\text{h}} + \sum_{\bm{k},\sigma} \varepsilon_{\text{v}}(\bm{k}).$$

In the above Hamiltonian (the last constant term is not important),

$$H_{\text{e}} = \sum_{\bm{k},\sigma} \varepsilon(\bm{k}) c^\dagger_{\bm{k}\sigma} c_{\bm{k}\sigma} \quad (13.149)$$

is the conduction band Hamiltonian with $c^\dagger_{\bm{k}\sigma}$ ($c_{\bm{k}\sigma}$) the creation (annihilation) operator of the conduction band electron, and

$$H_{\rm h} = \sum_{\bm{k},\sigma} \varepsilon^{\rm h}(\bm{k})\, d^\dagger_{\bm{k}\sigma} d_{\bm{k}\sigma} \qquad (13.150)$$

is the equivalent valence band Hamiltonian with $\varepsilon^{\rm h}(\bm{k}) = -\varepsilon_{\rm v}(\bm{k})$, and $d^\dagger_{\bm{k}\sigma}$ ($d_{\bm{k}\sigma}$) the creation (annihilation) operator of the valence band hole. In other words, a hole in the state $\bm{k}\sigma$ of the valence band is a free particle having charge $-e$ (e is the charge of an electron), spin σ, and energy $\varepsilon^{\rm h}(\bm{k})$.[2]

Since the effective mass of the hole in the valence band is usually much larger than that of the electron in the conduction band, the drift motion and the temperature rise of the hole system under the influence of an electric field will be much smaller than those of the conduction band electron system. People usually treat the hole system as the one without participating transport, i.e. with zero drift velocity and the same temperature T as the lattice. Therefore, in addition to the Hamiltonian of a single-band electron system subject to a uniform electric field, which we have discussed in detail in the preceding Chapters, what we need to add into the total Hamiltonian for including the impact ionization and Auger recombination process are the part of $H_{\rm ii}$ and that of the hole system $H_{\rm h}$ without being influenced by the electric field:

$$H = H_{\rm E} + H_{\rm e} + H_{\rm h} + H_{\rm ph} + H_{\rm ei} + H_{\rm ep} + H_{\rm ii}, \qquad (13.151)$$

in which $H_{\rm E} = -e\bm{E} \cdot \sum_j \bm{r}_j$ relates only to the total sum of the electron coordinates in the conduction band, and the impact ionization and Auger recombination part can be written as

$$H_{\rm ii} = \sum_{\substack{\bm{k},\bm{k}',\bm{q} \\ \sigma}} M_{\rm ii}(\bm{q})\, c^\dagger_{\bm{k}'+\bm{q}\sigma} c^\dagger_{\bm{k}-\bm{q}\sigma} d^\dagger_{\bm{k}\sigma} c_{\bm{k}'\bar{\sigma}} + M^*_{\rm ii}(\bm{q})\, c^\dagger_{\bm{k}'\bar{\sigma}} d_{\bm{k}\sigma} c_{\bm{k}-\bm{q}\sigma} c_{\bm{k}'+\bm{q}\bar{\sigma}},$$

$$\qquad (13.152)$$

in which $\bar{\sigma} = -\sigma$, and

$$M_{\rm ii}(\bm{q}) = \frac{e^2}{\varepsilon_0 \kappa} \frac{I_{\rm cv} I_{\rm cc}}{(q^2 + q_s^2)} = \frac{e^2}{\varepsilon_0 (q^2 + q_s^2)} C_{\rm ii}, \qquad (13.153)$$

where the parameter $C_{\rm ii} \equiv I_{\rm cv} I_{\rm cc}/\kappa$ represents the overlap integral part.

---

[2] $\varepsilon(\bm{k})$, $\varepsilon_{\rm v}(\bm{k})$ and $\varepsilon^{\rm h}(\bm{k})$ should always be written with respect to a common energy zero. For a direct gap semiconductor near the band edge as an example, when taking the energy zero at the conduction band bottom, we have $\varepsilon(\bm{k}) = k^2/2m$, $\varepsilon_{\rm v}(\bm{k}) = -\varepsilon_{\rm g} - k^2/2m_{\rm h}$, and $\varepsilon^{\rm h}(\bm{k}) = \varepsilon_{\rm g} + k^2/2m_{\rm h}$. Here $\varepsilon_{\rm g}$ is the gap, m and $m_{\rm h}$ are electron and hole effective mass.

13.6.2 Balance equations with ionization and recombination processes

On the basis of Hamiltonian (13.151)–(13.153), Wang and Lei (1995) derive the balance equations of high electric field transport with impact ionization and Auger recombination processes. They calculate the rates of changes of the total momentum $\boldsymbol{P} = m_e \sum_{\boldsymbol{k},\sigma} \boldsymbol{v}(\boldsymbol{k}) c^\dagger_{\boldsymbol{k}\sigma} c_{\boldsymbol{k}\sigma}$ [$\boldsymbol{v}(\boldsymbol{k}) \equiv \nabla \varepsilon(\boldsymbol{k})$, and m_e is the free electron mass], the total energy H_e given by (13.149), and the total particle number operator $\hat{N}_e = \sum_{\boldsymbol{k},\sigma} c^\dagger_{\boldsymbol{k}\sigma} c_{\boldsymbol{k}\sigma}$, of the the electron system with the help of the Heisenberg equation of motion (all the symbols without being explained in this section have the same meaning as those in Sec. 11.1):

$$\frac{\dot{\boldsymbol{P}}}{m_e} = -\frac{\mathrm{i}}{m_e}[\boldsymbol{P}, H]$$

$$= e\boldsymbol{E} \cdot \sum_{\boldsymbol{k},\sigma} \nabla \nabla \varepsilon(\boldsymbol{k}) c^\dagger_{\boldsymbol{k}\sigma} c_{\boldsymbol{k}\sigma} - \mathrm{i} \sum_{\boldsymbol{q},a} u(\boldsymbol{q}) \mathrm{e}^{\mathrm{i}\boldsymbol{q}\cdot\boldsymbol{r}_a} \sum_{\boldsymbol{k},\sigma} [\boldsymbol{v}(\boldsymbol{k}+\boldsymbol{q}) - \boldsymbol{v}(\boldsymbol{k})] c^\dagger_{\boldsymbol{k}+\boldsymbol{q}\sigma} c_{\boldsymbol{k}\sigma}$$

$$- \mathrm{i} \sum_{\boldsymbol{q},\lambda} M(\boldsymbol{q},\lambda)(b_{\boldsymbol{q}\lambda} + b^\dagger_{-\boldsymbol{q}\lambda}) \sum_{\boldsymbol{k},\sigma} [\boldsymbol{v}(\boldsymbol{k}+\boldsymbol{q}) - \boldsymbol{v}(\boldsymbol{k})] c^\dagger_{\boldsymbol{k}+\boldsymbol{q}\sigma} c_{\boldsymbol{k}\sigma}$$

$$- \mathrm{i} \sum_{\substack{\boldsymbol{k},\boldsymbol{k}' \\ \boldsymbol{q},\sigma}} [\boldsymbol{v}(\boldsymbol{k}-\boldsymbol{q}) + \boldsymbol{v}(\boldsymbol{k}'+\boldsymbol{q}) - \boldsymbol{v}(\boldsymbol{k}')] M_{\mathrm{ii}}(\boldsymbol{q}) \, c^\dagger_{\boldsymbol{k}-\boldsymbol{q}\bar{\sigma}} d^\dagger_{\boldsymbol{k}\bar{\sigma}} c^\dagger_{\boldsymbol{k}'+\boldsymbol{q}\sigma} c_{\boldsymbol{k}'\sigma}$$

$$- \mathrm{i} \sum_{\substack{\boldsymbol{k},\boldsymbol{k}' \\ \boldsymbol{q},\sigma}} [\boldsymbol{v}(\boldsymbol{k}') - \boldsymbol{v}(\boldsymbol{k}-\boldsymbol{q}) - \boldsymbol{v}(\boldsymbol{k}'+\boldsymbol{q})] M_{\mathrm{ii}}(-\boldsymbol{q}) \, d_{\boldsymbol{k}\bar{\sigma}} c_{\boldsymbol{k}-\boldsymbol{q}\bar{\sigma}} c^\dagger_{\boldsymbol{k}'\sigma} c_{\boldsymbol{k}'+\boldsymbol{q}\sigma},$$

$$(13.154)$$

$$\dot{H}_e = -\mathrm{i}[H_e, H] = -\mathrm{i} \sum_{\boldsymbol{k},\sigma} \varepsilon(\boldsymbol{k}) [c^\dagger_{\boldsymbol{k}\sigma} c_{\boldsymbol{k}\sigma}, H]$$

$$= e\boldsymbol{E} \cdot \sum_{\boldsymbol{k},\sigma} \nabla \varepsilon(\boldsymbol{k}) c^\dagger_{\boldsymbol{k}\sigma} c_{\boldsymbol{k}\sigma} - \mathrm{i} \sum_{\boldsymbol{q},a} u(\boldsymbol{q}) \mathrm{e}^{\mathrm{i}\boldsymbol{q}\cdot\boldsymbol{r}_a} \sum_{\boldsymbol{k},\sigma} [\varepsilon(\boldsymbol{k}+\boldsymbol{q}) - \varepsilon(\boldsymbol{k})] c^\dagger_{\boldsymbol{k}+\boldsymbol{q}\sigma} c_{\boldsymbol{k}\sigma}$$

$$- \mathrm{i} \sum_{\boldsymbol{q},\lambda} M(\boldsymbol{q},\lambda)(b_{\boldsymbol{q}\lambda} + b^\dagger_{-\boldsymbol{q}\lambda}) \sum_{\boldsymbol{k},\sigma} [\varepsilon(\boldsymbol{k}+\boldsymbol{q}) - \varepsilon(\boldsymbol{k})] c^\dagger_{\boldsymbol{k}+\boldsymbol{q}\sigma} c_{\boldsymbol{k}\sigma}$$

$$- \mathrm{i} \sum_{\substack{\boldsymbol{k},\boldsymbol{k}' \\ \boldsymbol{q},\sigma}} [\varepsilon(\boldsymbol{k}-\boldsymbol{q}) + \varepsilon(\boldsymbol{k}'+\boldsymbol{q}) - \varepsilon(\boldsymbol{k}')] M_{\mathrm{ii}}(\boldsymbol{q}) \, c^\dagger_{\boldsymbol{k}-\boldsymbol{q}\bar{\sigma}} d^\dagger_{\boldsymbol{k}\bar{\sigma}} c^\dagger_{\boldsymbol{k}'+\boldsymbol{q}\sigma} c_{\boldsymbol{k}'\sigma}$$

$$- \mathrm{i} \sum_{\substack{\boldsymbol{k},\boldsymbol{k}' \\ \boldsymbol{q},\sigma}} [\varepsilon(\boldsymbol{k}') - \varepsilon(\boldsymbol{k}-\boldsymbol{q}) - \varepsilon(\boldsymbol{k}'+\boldsymbol{q})] M_{\mathrm{ii}}(-\boldsymbol{q}) \, d_{\boldsymbol{k}\bar{\sigma}} c_{\boldsymbol{k}-\boldsymbol{q}\bar{\sigma}} c^\dagger_{\boldsymbol{k}'\sigma} c_{\boldsymbol{k}'+\boldsymbol{q}\sigma},$$

$$(13.155)$$

$$\dot{\hat{N}}_e = -i\,[\hat{N}_e, H]$$
$$= i\sum_{\substack{k,k'\\q,\sigma}} \left[M_{ii}(-q)\, d_{k\bar{\sigma}} c_{k-q\bar{\sigma}} c^\dagger_{k'\sigma} c_{k'+q\sigma} - M_{ii}(q)\, c^\dagger_{k-q\bar{\sigma}} d^\dagger_{k\bar{\sigma}} c^\dagger_{k'+q\sigma} c_{k'\sigma} \right].$$

(13.156)

Performing the statistical average with respect to the initial density matrix discussed in Sec. 11.2 to the lowest order in the scattering interaction $H_I = H_{ei} + H_{ep} + H_{ii}$, and identifying

$$\left\langle \frac{\dot{\boldsymbol{P}}}{m_e} \right\rangle = N_e \frac{d\boldsymbol{v}_d}{dt} + \boldsymbol{v}_d \frac{dN_e}{dt},$$

$$\langle \dot{H}_e \rangle = N_e \frac{d\varepsilon_e}{dt} + \varepsilon_e \frac{dN_e}{dt},$$

$$\langle \dot{\hat{N}}_e \rangle = \frac{dN_e}{dt},$$

they obtain the following balance equations:

$$\frac{d\boldsymbol{v}_d}{dt} = e\boldsymbol{E}\cdot\mathcal{K} + \boldsymbol{a}_i + \boldsymbol{a}_p + \boldsymbol{a}_{ii} - \boldsymbol{v}_d g_{ii}, \qquad (13.157)$$

$$\frac{d\varepsilon_e}{dt} = e\boldsymbol{E}\cdot\boldsymbol{v}_d - w - w_{ii} - \varepsilon_e g_{ii}, \qquad (13.158)$$

$$\frac{dN_e}{dt} = N_e g_{ii}. \qquad (13.159)$$

Here N_e is the average number of electrons, $\boldsymbol{v}_d$ is the average drift velocity of the electron system, ε_e is the average energy per electron, $\mathcal{K}$ is the average inverse effective mass tensor, $\boldsymbol{a}_i$ and $\boldsymbol{a}_p$ are respectively the impurity and phonon induced frictional accelerations, and w is the phonon-induced electron energy loss rate. Their expressions are exactly the same as those in Sec. 11.3. The newly appeared quantities are: the impact ionization and Auger recombination induced net generation rate of the electron–hole pairs,

$$g_{ii} = \frac{4\pi}{N_e} \sum_{k,k',q} |M_{ii}(q)|^2\, I(k,k',q), \qquad (13.160)$$

the impact ionization and Auger recombination induced electron damping acceleration

$$\boldsymbol{a}_{ii} = \frac{4\pi}{N_e} \sum_{k,k',q} |M_{ii}(q)|^2 \left[\boldsymbol{v}(k'+q) + \boldsymbol{v}(k-q) - \boldsymbol{v}(k')\right] I(k,k',q), \qquad (13.161)$$

and the impact ionization and Auger recombination induced electron energy loss rate

$$w_{ii} = \frac{4\pi}{N_e} \sum_{k,k',q} |M_{ii}(q)|^2 \, \varepsilon^h(k) \, I(k,k',q). \qquad (13.162)$$

In these expressions

$$I(k,k',q) \equiv \delta\big(\varepsilon(k-q) + \varepsilon^h(k) + \varepsilon(k'+q) - \varepsilon(k')\big) f\left(\frac{\bar{\varepsilon}(k') - \mu}{T_e}\right)$$

$$\times \left[1 - f\left(\frac{\bar{\varepsilon}(k'+q) - \mu}{T_e}\right)\right] \left[1 - f\left(\frac{\bar{\varepsilon}(k-q) - \mu}{T_e}\right)\right]$$

$$\times \left[1 - \frac{f[(\varepsilon^h(k) - \mu_h)/T]}{f[(\bar{\varepsilon}(k') - \bar{\varepsilon}(k-q) - \bar{\varepsilon}(k'+q) + \mu)/T_e]}\right], \quad (13.163)$$

$\bar{\varepsilon}(k) \equiv \varepsilon(k - p_d)$, and p_d is the lattice momentum shift. The quantities a_{ii} and w_{ii} account for the direct momentum and energy exchange between the electron system and the hole system due to impact ionization and Auger recombination processes. Besides, the change of the total momentum and energy of the electron (hole) system due to ionization and recombination induced change of the electron (hole) number, would also yield extra effective frictional force and energy loss rate on electrons.

The transport state is described by the lattice momentum shift p_d, the electron temperature T_e and the chemical potential μ of the electron system. The chemical potential of the hole system, μ_h, also shows up in the equations, though holes do not provide any drift motion. In the absence of the electric field, the semiconductor composed of conduction and valence bands is in a thermodynamic equilibrium state having a unique temperature T and chemical potential μ^0 without drift motion. The electron number in the conduction band is

$$N_e^0 = \sum_{k,\sigma} f\left(\frac{\varepsilon(k) - \mu^0}{T}\right). \qquad (13.164)$$

According to the definition of hole system, its chemical potential should be $\mu_h^0 = -\mu^0$, and the number of hole, or the number of the unoccupied states of the valence band is

$$N_h^0 = \sum_{k,\sigma} f\left(\frac{\varepsilon^h(k) - \mu_h^0}{T}\right) = \sum_{k,\sigma} f\left(\frac{\mu^0 - \varepsilon_v(k)}{T}\right) = \sum_{k,\sigma} \left[1 - f\left(\frac{\varepsilon_v(k) - \mu^0}{T}\right)\right].$$

$$(13.165)$$

When an electric field E is applied to the semiconductor, the electron system, besides undertaking a drift motion ($\boldsymbol{p}_d \neq 0$), is characterized by an electron temperature T_e, a chemical potential μ, and the particle number N_e; while the hole system is described by the temperature T, a chemical potential μ_h, and the hole number N_h. We have

$$N_e = \sum_{\boldsymbol{k},\sigma} f\left(\frac{\varepsilon(\boldsymbol{k}) - \mu}{T_e}\right), \qquad (13.166)$$

$$N_h = \sum_{\boldsymbol{k},\sigma} f\left(\frac{\varepsilon^h(\boldsymbol{k}) - \mu_h}{T}\right). \qquad (13.167)$$

Since the generation and annihilation of electrons and holes are always in pair, the increment of the electron number must be equal to that of the hole number, i.e.

$$N_e - N_e^0 = N_h - N_h^0. \qquad (13.168)$$

This equation give a relation between μ_h and μ in the transport state. Equations (13.157), (13.158) and (13.159), together with this relation, constitute a complete set of equations to determine all the parameters in the transport state under the influence of an electric field.

13.6.3 *Quasi-steady and steady transport under a constant electric field*

The time-dependent balance equations (13.157)–(13.159) involve several time scales of different orders of magnitude. The intraband carrier thermalization time τ_{th}, reflecting the speed of the carriers in an energy band to approach equilibrium in the relative reference frame without integrative motion and determined by the intraband Coulombic and other coupling mediated two-body interactions, is assumed to be sufficiently short. The momentum or the energy relaxation time, τ_m or τ_ε, which represents the time needed for the system to approach a steady transport state after a step electric field is suddenly impressed and will be denoted by τ_s, depends mainly on the impurity and phonon scatterings. The balance-equation approach applies for the case of $\tau_{\text{th}} \ll \tau_s$. The impact ionization and Auger recombination processes, due to its dependence on the wavefunction overlap integral I_{cv} between the conduction and valence bands, have a much smaller effective scattering rate. Therefore, the relaxation time τ_{ii} related to the particle generation process is generally much longer to satisfy $\tau_s \ll \tau_{ii}$ (McGroddy and Nathan, 1966).

In consequence, if a step electric field of strength E is applied to the system at time $t = 0$, after a time elapse somewhat longer than τ_s the electron drift velocity v_d and average energy ε_e (per carrier) will arrive at an essentially steady, or, more accurately, very slowly changing state, which is called a quasi-steady state. From the steady version ($d\boldsymbol{v}_d/dt = 0$ and $d\varepsilon_e/dt = 0$) of Eqs. (13.157) and (13.158),

$$0 = e\boldsymbol{E} \cdot \mathcal{K} + \boldsymbol{a}_i + \boldsymbol{a}_p + \boldsymbol{a}_{ii} - \boldsymbol{v}_d g_{ii}, \qquad (13.169)$$

$$0 = e\boldsymbol{E} \cdot \boldsymbol{v}_d - w - w_{ii} - \varepsilon_e g_{ii}, \qquad (13.170)$$

one can immediately solve for the transport parameters p_d (or v_d) and T_e. Though $dN_e/dt = g_{ii} \neq 0$ at this time, g_{ii} is small that the numbers of electrons and holes are essentially the same as those before applying the electric field, N_e^0 and N_h^0. The electron–hole pair generation rate

$$g_{ii} = g(E) \qquad (13.171)$$

and the impact ionization coefficient

$$\alpha = g_{ii}/v_d \qquad (13.172)$$

are of course functions of the electric field strength E. Many experimental measurements for the electron–hole generation rate and impact ionization coefficient are performed under the quasi-steady condition.

Now if the electric field E continues to apply, even v_d and T_e may no longer change, the numbers of electrons and holes, and thus the current flowing through the sample, will increase continually. There will be two possibilities with a given electric field strength. One possibility is that the increase of the electron and hole densities gradually saturates due to cancellation of the ionization generation by the Auger recombination, that the net generation rate of electron–hole pair becomes zero ($g_{ii} = 0$), and the system reaches a steady transport state. This steady state is determined by the following balance equations:

$$0 = e\boldsymbol{E} \cdot \mathcal{K} + \boldsymbol{a}_i + \boldsymbol{a}_p + \boldsymbol{a}_{ii}, \qquad (13.173)$$

$$0 = e\boldsymbol{E} \cdot \boldsymbol{v}_d - w - w_{ii}. \qquad (13.174)$$

$$0 = g_{ii}. \qquad (13.175)$$

At this steady state, the numbers of electrons and holes, N_e and N_h, may be much higher than the values of N_e^0 and N_h^0 in the absence of the electric field. The equation

$$N_e - N_e^0 = N_h - N_h^0 \qquad (13.176)$$

is a supplemental condition to equations (13.173)–(13.175). The solution to these equations exists only when the electric field strength E is lower than a certain value. For higher field strength the condition $g_{ii} = 0$ can not be realized and the number of the electron–hole pairs will increase without saturation. This is what is called the avalanche breakdown. From the above analysis one can see that even at an electric field lower than the threshold field for avalanche, the detected behavior of the drift-velocity versus electric-field may still depend on the way to measure it. Measurement at short time (of order of τ_s) elapse reflects the quasi-steady state behavior of v_d–E before appreciable change of the carrier number; measurement after long time ($> \tau_{ii}$) elapse gives the steady state transport behavior with the impact ionization generation of electrons; the results in between the above two limit cases may be obtained if a measurement is carried out in the intermediate range of time elapse.

On the basis of above equations, Wang et al (1998) calculate the electron–hole generation rate g in the quasi-steady condition and the current–field curve in the steady state for an n-type Hg$_{0.8}$Cd$_{0.2}$Te semiconductor at $T = 77$ K. They describe the conduction band with a nonparabolic energy dispersion of Kane type:

$$\varepsilon(\boldsymbol{k}) = \frac{\varepsilon_g}{2}\left[\left(1 + \frac{2}{\varepsilon_g}\frac{k^2}{m_c}\right)^{1/2} - 1\right], \qquad (13.177)$$

with the conduction band edge effective mass $m_c = 0.011 m_0$ (m_0 is the free electron mass), and band gap $\varepsilon_g = 0.1$ eV. The hole energy dispersion is taken to be the parabolic form: $\varepsilon^h(\boldsymbol{k}) = k^2/2m_h + \varepsilon_g$ with $m_h = 0.46 m_0$. Before applying the electric field the system is assumed to have an electron density $n_e^0 = 5.4 \times 10^{14}$ cm^{-3} and a hole density $n_h^0 \approx 0$. The electrons are considered to suffer mainly from longitudinal optic phonon Fröhlich scattering with LO phonon energy $\Omega_{LO} = 18$ meV, static dielectric constant $\kappa = 17.4$ and optic dielectric constant $\kappa_\infty = 12.5$. The Thomas-Fermi wavevector $q_s = 6\pi n_e^0 e^2/\mu^0$ (μ^0 is the conduction band Fermi level at zero field) is used in the impact ionization matrix element $M_{ii}(\boldsymbol{q})$ (13.153), together with the parameter C_{ii} taken as an adjustable constant to best fit the experimental data (Nimtz et al, 1974) of generation rate as a function of the electric field in the quasi-steady condition (see the inset of Fig. 13.14). With this C_{ii} in the quasi-steady equations (13.169) and (13.170), the effective electron drift velocity $v_d N_e/N_e^0$ (v_d is the electron drift velocity) obtained is plotted in Fig. 13.14 as a function of the electric field E (solid curve). The same quantity $v_d N_e/N_e^0$ is also calculated from the steady state equations (13.173)–(13.175) and shown as the dot-dashed curve in the figure. The experimental data of Nimtz et al (1974) shown as small black squares, are in between the quasi-steady and steady curves. It is worthy noticing that for the system in discussion, the quasi-steady velocity–field curve is almost the same as that derived from the steady-state balance equations without impact ionization processes [i.e. Eqs. (13.169) and (13.170) with $a_{ii} = 0$, $w_{ii} = 0$ and $g_{ii} = 0$] within the electric field range $E < 500\,\text{V/cm}$.

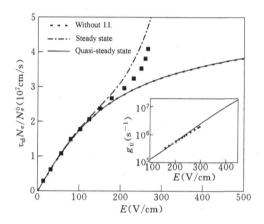

Fig. 13.14 The effective drift velocity $v_d N_e/N_e^0$ versus the electric field E in an n-type $\text{Hg}_{0.8}\text{Cd}_{0.2}\text{Te}$ at lattice temperature $T = 77\,\text{K}$. The solid curve is obtained from the quasi-steady equations (13.169) and (13.170). The dot-dashed curve is from the steady state equations (13.173)–(13.175). The small black dots are results without impact ionization (II). The black squares are the experimental data of Nimtz et al (1974). The inset shows the calculated electron–hole generation rate g_{ii} under quasi-steady condition (solid line) together with experimental data. From Wang et al (1998).

13.6.4 Multivalley balance equations with impact ionization

The balance equation treatment with impact ionization and Auger recombination developed in the preceding subsections is easily extended to the semiconductors having multiple energy valleys (Cao and Lei, 1998a; 1999a), as long as the impact ionization and Auger recombination processes between the conduction and valence bands in each valley are taken into account separately. The corresponding equations can be obtained directly from the balance equations for nonparabolic multivalley (band) systems, (13.95)–(13.97), by adding an impact-ionization related electron–hole pair generation rate g_{ii}^a, a frictional acceleration $\boldsymbol{A}_{ii}^a$ and an energy-loss rate W_{ii}^a into the equation of each valley,

$$\frac{d}{dt}(N_a\boldsymbol{v}_a) = N_a e \boldsymbol{E} \cdot \mathcal{K}_a + \boldsymbol{A}_{ei}^a + \boldsymbol{A}_{ep}^a + \boldsymbol{A}_{ii}^a$$
$$+ \sum_{b(\neq a)} \boldsymbol{A}_{ei}^{ab} + \sum_{b(\neq a)} \boldsymbol{A}_{ep}^{ab} + \sum_{b(\neq a)} \boldsymbol{A}_{ee}^{ab}, \qquad (13.178)$$

$$\frac{d}{dt}(N_a\varepsilon_a) = N_a e \boldsymbol{E} \cdot \boldsymbol{v}_a - W_{ep}^a - W_{ii}^a - \sum_{b(\neq a)} W_{ep}^{ab} - \sum_{b(\neq a)} W_{ee}^{ab}, \qquad (13.179)$$

$$\frac{dN_a}{dt} = \sum_{b(\neq a)} X_{ei}^{ab} + \sum_{b(\neq a)} X_{ep}^{ab} + N_a g_{ii}^a, \qquad (13.180)$$

together with the equations for the hole number balance in each valley,

$$\frac{dN_h^a}{dt} = N_a g_{ii}^a. \qquad (13.181)$$

Here N_a, $\boldsymbol{v}_a$ and ε_a are the number (density), the average velocity and average energy of electrons, and N_h^a is the number (density) of holes in the ath valley. The expressions of g_{ii}^a, $\boldsymbol{A}_{ii}^a$ and W_{ii}^a for each valley can be directly written down from those of g_{ii} (13.160), $N_e \boldsymbol{a}_{ii}$ (13.161) and $N_e w_{ii}$ (13.162), with the impact-ionization relevant matrix element given by

$$M_{ii}^a(\boldsymbol{q}) = \frac{e^2}{\varepsilon_0 \kappa} \frac{I_{cv} I_{cc}}{(q^2 + q_{sa}^2)} = \frac{e^2}{\varepsilon_0(q^2 + q_{sa}^2)} C_{ii}, \qquad (13.182)$$

in which q_{sa} is the screening wave vector of the ath valley related to the carrier density N_a and electron temperature T_a of the valley. The impact ionization rate of the whole system is expressed as

$$g_{ii} = \frac{1}{N} \sum_a N_a g_{ii}^a, \qquad (13.183)$$

and the impact ionization coefficient

$$\alpha_{ii} = g_{ii}/v_d, \qquad (13.184)$$

where $N = \sum_a N_a$ is the total number (density) of electrons and $v_d = \sum N_a v_a / N$ is total average velocity of the system.

In a multivalley system, in addition to the impact ionization processes, the intervalley impurity and phonon scatterings can also contribute to the increase of the electron number in each valley, as stated in (13.180). Equation (13.181) assumes that the increase of the hole number in each valley is purely due to the impact ionization within the valley (without other mechanisms for intervalley hole exchange), therefore it is equal to the impact-ionization induced increase of electrons in the same valley.

The quasi-steady-state transport can be calculated by setting $d\bm{v}_a/dt = 0$ and $d\varepsilon_a/dt = 0$ on the left-hand sides of Eqs. (13.178) and (13.179) and dN_a/dt is given by (13.180).

These equations are first applied to the investigation of high-field transport in $\langle 111 \rangle$ direction in bulk Si (Cao and Lei, 1998a). Since for the impact ionization effect to show up the electric fields experienced by the electrons must be quite large, we therefore take into account, in addition to the X valleys (as in Sec. 13.5.1), also the role of L valleys which are energetically around 1.1 eV higher above the former. The phonon scatterings considered in the calculation are intravalley acoustic deformation scattering, including both three f and three g types of intervalley scatterings (Sec. 13.5.1), and four X–L nonequivalent intervalley scatterings. The interband Coulomb scatterings among nonequivalent conduction valleys are neglected. We use the Kane-type ellipsoidal energy dispersion (13.124) for both X and L valleys in calculating $\bm{A}_{\rm ep}^{a}$, $W_{\rm ep}^{a}$, $\bm{A}_{\rm ep}^{ab}$, $W_{\rm ep}^{a}$, and $X_{\rm ep}^{ab}$, but for the sake of simplicity, a parabolic dispersion is used in the evaluation of $\bm{A}_{\rm ii}^{a}$, $W_{\rm ii}^{a}$ and $g_{\rm ii}^{a}$, and take all the form factors in the electron–phonon matrix elements to be unity. A screening wavevector of Debye-type, $q_{sa}^2 = N_a e^2/\kappa T_a$, is used for $M_{\rm ii}^{a}(q)$, and the value of the overlap integral $I_{\rm cc}I_{\rm cv}$, or the parameter $C_{\rm ii}$, which is relevant to the threshold energy for impact ionization, is assumed to be a constant and treated as the only adjustable parameter in the quasi-steady state calculation to best fit the experimental data of impact ionization coefficient $\alpha_{\rm ii}$. It turns out that such a single parameter fitting is able to give a good agreement between the calculated $\alpha_{\rm ii}$ versus the electric field and the experimental data.

The calculated average electron velocities at lattice temperature $T = 77, 160$ and $300\,{\rm K}$ are shown in Fig. 13.15 as functions of the electric field. It is seen that when the applied electric field is larger than $200\,{\rm kV/cm}$, the impact ionization process begins to contribute to electron transport in Si. The average velocity with impact ionization is higher than that without impact ionization. It is also seen from this figure that a slight negative differential mobility occurs around $200\,{\rm kV/cm}$ due to the impact ionization process and the X–L nonequivalent intervalley scattering. These results are in agreement with the Monte-Carlo simulations (Fischetti and Laux, 1988; Sano *et al*, 1990; Martin *at el*, 1993).

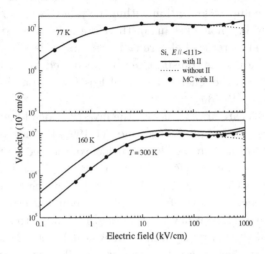

Fig. 13.15 The calculated average velocities with (solid lines) and without (dashed lines) impact ionization (II) process are shown as functions of the electric field applied along the ⟨111⟩ direction in bulk Si at lattice temperature $T = 77, 160$ and 300 K. The black circles are the Monte-Carlo results taken from Sano *et al* (1990). From Cao and Lei (1998a).

The above equations are also applied to investigate the high-field electron transport in compound semiconductors GaN (Cao and Lei, 1999a) and ZnS (Cao, 2004) with impact ionization. For GaN, a model consisting of nonequivalent Γ_1, L–M, and Γ_2 valleys is used for the band structure and the Kane-type nonparabolic energy-wavevector relations are taken for all the conduction band dispersions, while the parabolic approximation is assumed for the valence bands. The intra- and intervalley scatterings from ionized impurities, acoustic and polar optical phonons are considered as the main scattering mechanisms, together with impact ionization process. The calculated average electron velocity v_d and the electron temperature T_1 of the Γ_1 valley in the quasi-steady state condition, are shown in Fig. 13.16, for a bulk GaN system having an ionized impurity density equal to the zero-field electron density of $1 \times 10^{17}\,\mathrm{cm}^{-3}$ at $T = 300$ K. The velocity–field curve, as well as the average total electron energy and the fractional electron density for each valley versus the electric field, exhibits a behavior typical for compound semiconductors in that all these transport quantities changes abruptly around the threshold field ($\simeq 180\,\text{kV/cm}$) for the onset of negative differential mobility. The impact ionization process contributes to transport when the electric field is higher than a value $E_\text{T} = 530\,\text{kV/cm}$. It reduces the electron temperature and enhances the average electron velocity at a given electric field.

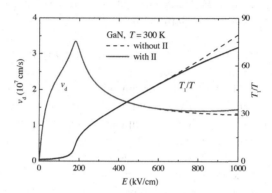

Fig. 13.16 The average electron drift velocity v_d and the electron temperature T_1 of the Γ_1 valley are shown as functions of the applied electric field in an n-type GaN at lattice temperature $T = 300\,\text{K}$ with (solid lines) and without (dashed lines) impact ionization (II) process. The system has an ionized impurity density of $1 \times 10^{17}\,\text{cm}^{-3}$ equal to the zero-field electron density. From Cao and Lei (1999a).

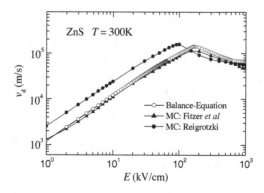

Fig. 13.17 The calculated electron drift velocity v_d (open circles) is shown as a function of the electric field in ZnS at lattice temperature $T = 300\,\text{K}$. Also shown are results of Monte-Carlo simulations from Fitzer et al (2003) (solid triangles) and from Reigrotzki (1998) (solid circles). After Cao (2004).

Similar results are obtained with a three-valley model (Γ, X and L) calculation of high-field transport for ZnS at $T = 300\,\mathrm{K}$ (Cao, 2004), as shown in Fig. 13.17.

13.7 Hydrodynamic Balance Equations for Systems with Arbitrary Energy Dispersion

To deal with electron transport in systems of arbitrary energy dispersion under weakly nonuniform conditions, we need to extend the balance equations derived in Secs. 11.3 and 13.4 for spatially homogeneous configuration to spatially inhomogeneous cases, as did in Chapter 10 for parabolic bands. We first consider a system which is composed of many electrons moving in an energy band having an arbitrary energy-wavevector relation. The single electron state is characterized by a lattice wavevector $\boldsymbol{k}$ in the Brillouin zone and a spin index σ, with the energy dispersion $\varepsilon(\boldsymbol{k})$. In the spatially homogeneous case, the transport state of the system under the influence of a uniform electric field is completely determined by the momentum balance equation and energy balance equation. As far as the momentum balance equation is concerned, one can choose either the equation for the total momentum balance or the equation for the total lattice-momentum balance (Sec. 11.3). Since in most real cases the calculation of the inverse effective mass tensor $\mathcal{K}$ is easier than the calculation of Büttiker–Thomas reduction factor $\mathcal{R}$, we will use the equation for the total momentum balance in the following discussion. Furthermore, we will take the initial density matrix or the distribution function to be the lattice-momentum shift type, i.e., describe the spatially uniform transport state of electrons in an arbitrary energy band by a lattice momentum shift $\boldsymbol{p}_d$ and an electron temperature T_e. For the description of spatially nonuniform transport, we need, in addition to $\boldsymbol{p}_d$ and T_e, also a chemical potential μ or the ratio of chemical potential to electron temperature, $\zeta \equiv \mu/T_\mathrm{e}$, as one of the basic parameters. To deal with spatially inhomogeneous problems, we allow $\boldsymbol{p}_d$, T_e and μ, thus all the transport quantities, to be slowly varying function of space and time. Once $\boldsymbol{p}_d$, T_e and μ are determined, all other transport quantities, such as the carrier density n, drift velocity v_d, electric potential or electric field, are immediately obtained. Considering the electron number, momentum and energy balances (conservations) in a small volume element around a spatial position, we can derive the the following hydrodynamic balance equations (Lei, 1995):

$$\frac{\partial n}{\partial t} = -\boldsymbol{\nabla} \cdot (n\boldsymbol{v}), \tag{13.185}$$

$$\frac{\partial}{\partial t}(n\boldsymbol{v}) = -\boldsymbol{\nabla} \cdot (n\mathcal{B}) + ne\boldsymbol{E} \cdot \mathcal{K} + n\boldsymbol{a}, \tag{13.186}$$

$$\frac{\partial}{\partial t}(n\varepsilon) = -\boldsymbol{\nabla} \cdot (n\boldsymbol{S}) + ne\boldsymbol{E} \cdot \boldsymbol{v} - nw. \tag{13.187}$$

Here the electron density

$$n = \frac{1}{4\pi^3} \int_{\mathrm{BZ}} d^3k\, f(\varepsilon(\boldsymbol{k})/T_\mathrm{e} - \zeta), \tag{13.188}$$

the average electron velocity

$$\boldsymbol{v} = \langle \boldsymbol{\nabla}\varepsilon(\boldsymbol{k})\rangle, \tag{13.189}$$

the average electron energy (per carrier)

$$\varepsilon = \langle \varepsilon(\boldsymbol{k})\rangle, \tag{13.190}$$

the average inverse effective mass tensor

$$\mathcal{K} = \langle \boldsymbol{\nabla}\boldsymbol{\nabla}\varepsilon(\boldsymbol{k})\rangle, \tag{13.191}$$

the average velocity dyadic tensor

$$\mathcal{B} = \langle (\boldsymbol{\nabla}\varepsilon(\boldsymbol{k}))(\boldsymbol{\nabla}\varepsilon(\boldsymbol{k}))\rangle, \tag{13.192}$$

and the average energy-flow vector

$$\boldsymbol{S} = \langle \varepsilon(\boldsymbol{k})(\boldsymbol{\nabla}\varepsilon(\boldsymbol{k}))\rangle. \tag{13.193}$$

In these expressions $\langle \cdots \rangle$ stands for the following weighted integral over the Brillouin zone of the $\boldsymbol{k}$ space:

$$\langle \cdots \rangle = \frac{1}{4\pi^3 n} \int_{\mathrm{BZ}} d^3k\, f(\varepsilon(\boldsymbol{k}-\boldsymbol{p}_d)/T_\mathrm{e} - \zeta)\cdots, \tag{13.194}$$

with $f(x) = 1/[\exp(x)+1]$ the Fermi function. The frictional acceleration $\boldsymbol{a} = \boldsymbol{a}_\mathrm{i} + \boldsymbol{a}_\mathrm{p}$ (due to impurity and phonon scatterings) in Eq. (13.186), and the per-carrier energy-loss rate w (due to phonon scattering) in Eq. (13.187), are given by (11.80), (11.81) and (11.82). In addition to Eqs. (13.185)–(13.187), the spatial divergence of the electric field relates to the local charge density by the Gauss theorem (Poisson equation),

$$\nabla \cdot \boldsymbol{E} = \frac{e}{\epsilon_0 \kappa}(n - n_\mathrm{D}), \tag{13.195}$$

where e is the electron charge, κ is the dielectric constant of the semiconductor, and n_D is the background density distribution of the positive charge which is determined by the design and fabrication of the device and assumed to be known.

As such, for a weakly nonuniform system with electrons moving in an arbitrary energy band, we have particle-number conservation equation (continuity equation) (13.185), momentum-balance equation (13.186), energy-balance equation (13.187) and Poisson equation (13.195). They constitute a complete set of first order partial differential equations, and can be solved to determine all the space-time field quantities of transport for given initial and boundary conditions.

In the case of parabolic energy dispersion, $\varepsilon(\boldsymbol{k}) = k^2/2m$ and $\boldsymbol{v} - \boldsymbol{p}_d/m$, these equations become the hydrodynamic balance equations obtained in Chapter 10.

It is straightforward to extend the homogeneous transport balance equations of nonparabolic multiband systems introduced in Sec. 13.4 to the weakly nonuniform case. For system consisting of s energy bands, the transport state can be described by a lattice-momentum shift $\boldsymbol{p}_a$, an electron temperature T_a and a chemical potential μ_a for each energy band. The hydrodynamic balance equations for spatially weakly nonuniform transport are obtained from spatially uniform balance equations (13.95)–(13.97) by adding the divergence terms of the particle flow vector $n_a \boldsymbol{v}_a$, the velocity dyadic tensor (or the momentum flow tensor) $n_a \mathcal{B}_a$, and the energy-flow vector $n_a \boldsymbol{S}_a$, respectively into them to give ($a = 1, 2, \cdots, s$)

$$\frac{\partial n_a}{\partial t} = -\nabla \cdot (n_a \boldsymbol{v}_a) + \sum_{b(\neq a)} X_\mathrm{ei}^{ab} + \sum_{b(\neq a)} X_\mathrm{ep}^{ab}, \tag{13.196}$$

$$\frac{\partial}{\partial t}(n_a \boldsymbol{v}_a) = -\nabla \cdot (n_a \mathcal{B}_a) + n_a e \boldsymbol{E} \cdot \mathcal{K}_a + \boldsymbol{A}_\mathrm{ei}^a + \boldsymbol{A}_\mathrm{ep}^a$$
$$+ \sum_{b(\neq a)} \boldsymbol{A}_\mathrm{ei}^{ab} + \sum_{b(\neq a)} \boldsymbol{A}_\mathrm{ep}^{ab} + \sum_{b(\neq a)} \boldsymbol{A}_\mathrm{ee}^{ab}, \tag{13.197}$$

$$\frac{\partial}{\partial t}(n_a \varepsilon_a) = -\nabla \cdot (n_a \boldsymbol{S}_a) + n_a e \boldsymbol{E} \cdot \boldsymbol{v}_a - W_\mathrm{ep}^a - \sum_{b(\neq a)} W_\mathrm{ep}^{ab} - \sum_{b(\neq a)} W_\mathrm{ee}^{ab}.$$
$$\tag{13.198}$$

13.8 Modeling of Transport in Semiconductor Devices from Nonparabolic Band Material

The parabolic band description must be modified in semiconductor devices, if an appreciable portion of carriers can move to the states not in the vicinity of the band bottom in the $\boldsymbol{k}$ space, due to increasing electric field, rising temperature or some other reasons. The nonparabolic hydrodynamic balance equations introduced in the last section provide a convenient tool to deal with carrier transport in semiconductor devices, taking into account the nonparabolicity of the energy band structure of the constituent materials.

As an example, we assume that the energy dispersion $\varepsilon(\boldsymbol{k})$ of the material can be described by an isotropic Kane model,

$$\varepsilon(1+\alpha\varepsilon) = k^2/2m, \tag{13.199}$$

and examine the transport in one-dimensional devices as that analyzed in Sec 10.7 for parabolic material. For this isotropic material, when the spatial variations of the doping density $n_D = n_D(x)$ and electric potential $\phi = \phi(x)$ are along the x direction, the lattice-momentum shift $\boldsymbol{p}_d = (p_d, 0, 0)$, particle flow density $\boldsymbol{j}_n = n\boldsymbol{v} = (j_n, 0, 0)$, and frictional acceleration $\boldsymbol{a} = (a, 0, 0)$ have only nonzero x components and all the transport quantities can have spatial variations only along the x direction. The Poisson equation and hydrodynamic balance equations are written as

$$\frac{d^2\phi}{dx^2} = \frac{e}{\varepsilon_0\kappa}(n_D - n), \tag{13.200}$$

$$\frac{dn}{dt} + \frac{dj_n}{dx} = 0, \tag{13.201}$$

$$\frac{dj_n}{dt} + \frac{d}{dx}(n\mathcal{B}_{xx}) = -en\mathcal{K}_{xx}\frac{d\phi}{dx} + na, \tag{13.202}$$

$$\frac{d}{dt}(n\varepsilon) + \frac{d}{dx}(nS_x) = -ej_n\frac{d\phi}{dx} - nw. \tag{13.203}$$

All the quantities in the equations, n, j_n, ε, $\mathcal{K}_{xx}$, $\mathcal{B}_{xx}$, S_x, a and w are functions of chemical potential μ, electron temperature T_e and lattice-momentum shift p_d. For instance,

$$n = n(\mu, T_e) = \frac{2}{\sqrt{\pi}} N_c(T_e) \int_0^\infty dx\, f_\alpha(x, \mu, T_e), \tag{13.204}$$

$$j_n = j_n(\mu, T_e, p_d) = \frac{2}{\sqrt{\pi}} N_c(T_e) \int_0^\infty dx\, J(\eta, p_d) f_\alpha(x, \mu, T_e), \tag{13.205}$$

$$n\varepsilon = \frac{2}{\sqrt{\pi}} N_{\rm c}(T_{\rm e}) \int_0^\infty dx\, \varepsilon(\eta, p_d) f_\alpha(x, \mu, T_{\rm e}), \qquad (13.206)$$

$$n\mathcal{K}_{xx} = \frac{2}{\sqrt{\pi}} N_{\rm c}(T_{\rm e}) \int_0^\infty dx\, K(\eta, p_d) f_\alpha(x, \mu, T_{\rm e}), \qquad (13.207)$$

$$n\mathcal{B}_{xx} = \frac{2}{\sqrt{\pi}} N_{\rm c}(T_{\rm e}) \int_0^\infty dx\, B(\eta, p_d) f_\alpha(x, \mu, T_{\rm e}), \qquad (13.208)$$

$$nS_x = \frac{2}{\sqrt{\pi}} N_{\rm c}(T_{\rm e}) \int_0^\infty dx\, S(\eta, p_d) f_\alpha(x, \mu, T_{\rm e}). \qquad (13.209)$$

Here $N_{\rm c}(T_{\rm e}) \equiv 2(mT_{\rm e}/2\pi)^{3/2}$, $\eta \equiv [2mT_{\rm e} x(1+\alpha T_{\rm e} x)]^{1/2}$,

$$f_\alpha(x, \mu, T_{\rm e}) = \frac{(1+2\alpha T_{\rm e} x)[x(1+\alpha T_{\rm e} x)]^{1/2}}{1 + \exp(x - \mu/T_{\rm e})}, \qquad (13.210)$$

$$J(k, p_d) = \frac{1}{2} \int_{-1}^1 dy\, \frac{ky + p_d}{m\left[1 + \frac{2\alpha}{m}(k^2 + p_d^2 + 2kp_d y)\right]^{1/2}}, \qquad (13.211)$$

$$\varepsilon(k, p_d) = \frac{1}{2} \int_{-1}^1 dy\, \frac{k^2 + p_d^2 + 2kp_d y}{m\left[1 + \sqrt{1 + \frac{2\alpha}{m}(k^2 + p_d^2 + 2kp_d y)}\right]}, \qquad (13.212)$$

$$K(k, p_d) = \frac{1}{2} \int_{-1}^1 dy\, \frac{1 + \frac{2\alpha}{m} k^2(1 - y^2)}{m\left[1 + \frac{2\alpha}{m}(k^2 + p_d^2 + 2kp_d y)\right]^{3/2}}, \qquad (13.213)$$

$$B(k, p_d) = \frac{1}{2} \int_{-1}^1 dy\, \frac{(ky + p_d)^2}{m^2\left[1 + \frac{2\alpha}{m}(k^2 + p_d^2 + 2kp_d y)\right]}, \qquad (13.214)$$

$$S(k, p_d) = \frac{1}{2} \int_{-1}^1 dy\, \frac{(ky + p_d)}{m^2\left[1 + \frac{2\alpha}{m}(k^2 + p_d^2 + 2kp_d y)\right]^{1/2}}$$
$$\times \frac{k^2 + p_d^2 + 2kp_d y}{1 + \sqrt{1 + \frac{2\alpha}{m}(k^2 + p_d^2 + 2kp_d y)}}. \qquad (13.215)$$

In a real calculation, $n(x,t)$ is given by the spatial derivative of the potential $\phi(x,t)$ from the Poisson equation, $T_{\rm e}(x,t)$ and $j_n(x,t)$ are often used as the basic variables. Then $\mu = \mu(n, T_{\rm e})$ can be solved from Eq. (13.204) and $p_d = p_d(n, T_{\rm e}, j_n)$ is then obtained from (13.205). With these two relations, $n\varepsilon$, $n\mathcal{K}_{xx}$, $n\mathcal{B}_{xx}$ and nS_x, as well as a and w, all become functions of the basic variables $(n, T_{\rm e}, j_n)$. Eqs. (13.200)–(13.203) are the starting point of the transport process simulation in one-dimensional devices for given doping profile $n_{\rm D}(x)$.

For steady state transport, Eqs. (13.201)–(13.203) are reduced to

$$\frac{dj_n}{dx} = 0, \tag{13.216}$$

$$j_n = \left(\frac{v}{a}\right)\left[\left.\frac{\partial(n\mathcal{B}_{xx})}{\partial n}\right|_{T_e,j}\frac{dn}{dx} + \left.\frac{\partial(n\mathcal{B}_{xx})}{\partial T_e}\right|_{n,j}\frac{dT_e}{dx} + en\mathcal{K}_{xx}\frac{d\phi}{dx}\right], \tag{13.217}$$

$$\left.\frac{\partial(nS_x)}{\partial n}\right|_{T_e,j}\frac{dn}{dx} + \left.\frac{\partial(nS_x)}{\partial T_e}\right|_{n,j}\frac{dT_e}{dx} + ej_n\frac{d\phi}{dx} + nw = 0. \tag{13.218}$$

Taking the boundary conditions expressed by (10.102)–(10.104), Cao and Lei (1998b) perform a simulation of steady state transport process in a Si n^+nn^+ system. The input quantities are the doping profile $n_D(x)$, the lattice temperature T, the voltage applied on the two ends of the device, as well as the band structure and scattering parameters of the material. The output quantities are ϕ, n, T_e and v. We refer the readers to their original paper for detail. Fig. 13-18 shows the calculated profile of the drift velocity under the applied voltage $V = 0.5, 1.0$ and 2.0 V. The nonparabolic coefficient is taken to be $\alpha = 0.5\,\text{eV}^{-1}$. For comparison, the simulation results with parabolic energy dispersion ($\alpha = 0$) at $V = 2.0$ V are also plotted in the figure. The nonparabolic effect appears quite clear in this device even for a rather weak nonparabolicity $\alpha = 0.5\,\text{eV}^{-1}$.

Note that the present simulation from the nonparabolic hydrodynamic balance equation is already quite close to the result from a Monte-Carlo simulation using the same nonparabolic model (Tang and Ramaswamy, 1993), except that v is slight higher in the $0.1 \sim 0.2\,\mu\text{m}$ range and slight lower in the $0.4 \sim 0.5\,\mu\text{m}$ range, as shown in the figure. This is not unexpected because the present hydrodynamic equations are essentially a model for weakly nonuniform systems. All the second and higher orders of spatial derivatives of the fundamental quantities are neglected in the equations. One of possible remedies is to add into the expression (13.207) of the energy flow density nS_x a small "heat flow" term

$$j_Q = -\kappa\frac{\partial T_e}{\partial x}, \tag{13.219}$$

where κ takes the form of (10.97). By taking a γ around 0.1 the resulting velocity profile of the hydrodynamic simulation will approach that of the Monte-Carlo simulation quite accurately, as shown in Fig. 13.19.

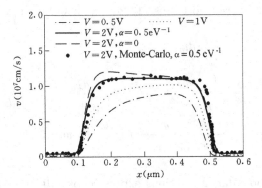

Fig. 13.18 The drift velocity v profile in a Si n^+nn^+ diode. The solid, dotted and dash-dotted curves are obtained from the hydrodynamic balance equation simulation with nonparabolic Kane dispersion ($\alpha = 0.5$) at bias voltage $V = 2, 1$ and $0.5\,\text{V}$ respectively. The dashed curve are parabolic band ($\alpha = 0$) result at $V = 2\,\text{V}$. The small black circles are Monte-Carlo simulation at $V = 2\,\text{V}$ (Tang and Ramaswamy, 1993). The lattice temperature is $T = 300\,\text{K}$. From Cao and Lei (1998b).

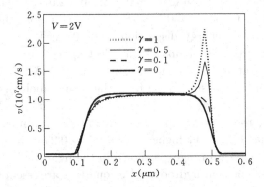

Fig. 13.19 The drift velocity v profile at $V = 2\,\text{V}$, obtained from nonparabolic hydrodynamic simulation when adding a heat-flow term (13.219) in the energy balance equation. From Cao and Lei (1998b).

13.9 Instabilities in Superlattice Miniband Transport and Spatiotemporal Domains

In Chapter 12 we have discussed the spatially uniform transport of electrons in superlattice minibands driven by dc and high-frequency electric fields. The most remarkable feature of superlattice miniband transport is the negative differential mobility showing up under strong dc biases. The occurrence of the negative differential mobility, which is able to supply a compensation for the energy dissipation, may induce current instabilities and self-sustained oscillations in the system. Most theoretical analyses on these phenomena rely on the drift-diffusion model. The hydrodynamic balance equations introduced in Sec. 13.7 provide a more accurate alternative.

13.9.1 Hydrodynamic balance equations for superlattice miniband transport

Consider a laterally unconfined planar superlattice in which electrons move freely in the transverse (x–y) plane and can travel along the growth axis (z direction) through the lowest miniband formed by the periodical potential with the tight-binding-type energy dispersion

$$\varepsilon(\boldsymbol{k}) = \frac{k_\parallel^2}{2m} + \varepsilon(k_z), \tag{13.220}$$

$$\varepsilon(k_z) = \frac{\Delta}{2}\left[1 - \cos(k_z d)\right], \tag{13.221}$$

where m is the band effective mass of the bulk semiconductor, d is the superlattice period along the z direction, Δ is the miniband width, and $\boldsymbol{k} \equiv (\boldsymbol{k}_\parallel, k_z)$, with $\boldsymbol{k}_\parallel \equiv (k_x, k_y)$ and k_z as the in-plane and longitudinal wavevectors: $-\infty < k_x, k_y < \infty$ and $-\pi/d < k_z \leqslant \pi/d$.

According to the nonparabolic hydrodynamic model, the transport state of the system is described by the lattice-momentum shift $\boldsymbol{p}_d$, electron temperature T_e and chemical potential μ (or the ratio of chemical potential to electron temperature $\zeta \equiv \mu/T_e$), which, in spatially inhomogeneous cases, are treated as time- and space-dependent. If the electric field is applied along the superlattice growth axis (z direction), $\boldsymbol{E} = (0, 0, E)$, the lattice momentum shift, the drift velocity and the frictional acceleration are all in the z direction: $\boldsymbol{p}_d = (0, 0, p_d)$, $\boldsymbol{v} = (0, 0, v)$ and $\boldsymbol{a} = (0, 0, a)$, and the spatial inhomogeneity of all the transport quantities can only be in the z direction (independent of x and y). The hydrodynamic balance equations are then written as

$$\frac{\partial n}{\partial t} = -\frac{\partial}{\partial z}(nv), \tag{13.222}$$

$$\frac{\partial}{\partial t}(nv) = -\frac{\partial}{\partial z}(n\mathcal{B}_z) + \frac{neE}{m_z^*} + na, \tag{13.223}$$

$$\frac{\partial}{\partial t}(n\varepsilon) = -\frac{\partial}{\partial z}(nS_z) + neEv - nw. \tag{13.224}$$

Here, n is the average carrier density, v the average carrier drift velocity, ε the average carrier energy, $1/m_z^*$ the zz component of the inverse effective-mass tensor, $\mathcal{B}_z$ the zz component of the average velocity dyadic, and S_z the z component of the energy flux vector. The Poisson equation is

$$\frac{\partial E}{\partial z} = \frac{e}{\varepsilon_0 \kappa}(n - n_0), \qquad (13.225)$$

where e is the electron charge, κ is the dielectric constant of the semiconductor, and n_0 is the background density of positive charge which is assumed to be uniformly distributed over the whole superlattice. This equation shows that if the electric field is uniformly distributed, $\partial E/\partial z = 0$, the electron charge density is $n = n_0$. The other transport quantities, such as n, v, ε, $1/m_z^*$, $\mathcal{B}_z$ and S_z, are all functions of $z_d \equiv p_d d$, T_e and ζ. For the superlattice miniband described by the energy dispersion (13.220) and (13.221), they are explicitly given by

$$n = n_0 \alpha_0(T_e, \zeta), \qquad (13.226)$$

$$nv = n_0 v_m \alpha_1(T_e, \zeta) \sin z_d, \qquad (13.227)$$

$$n/m_z^* = \frac{n_0}{M^*} \alpha_1(T_e, \zeta) \cos z_d, \qquad (13.228)$$

$$n\mathcal{B}_z = n_0 (v_m^2/2) [\alpha_0(T_e, \zeta) - \alpha_2(T_e, \zeta) \cos 2z_d], \qquad (13.229)$$

$$n\varepsilon = n\varepsilon_z + n\varepsilon_\parallel,$$

$$n\varepsilon_z = n_0(\Delta/2)[\alpha_0(T_e, \zeta) - \alpha_1(T_e, \zeta) \cos z_d], \qquad (13.230)$$

$$n\varepsilon_\parallel = n_0(\Delta/2)\beta_0(T_e, \zeta), \qquad (13.231)$$

$$nS_z = n_0 \frac{v_m \Delta}{2}\left[\alpha_1(T_e, \zeta) \sin z_d - \frac{1}{2}\alpha_2(T_e, \zeta) \sin 2z_d + \beta_1(T_e, \zeta) \sin z_d\right]. \qquad (13.232)$$

In the above equations the relevant quantities are defined as: $v_m \equiv \Delta d/2$, $1/M^* \equiv \Delta d^2/2$, $\alpha_0(T_e, \zeta) \equiv \langle 1 \rangle_0$, $\alpha_1(T_e, \zeta) \equiv \langle \cos(k_z d) \rangle_0$, $\alpha_2(T_e, \zeta) \equiv \langle \cos(2k_z d) \rangle_0$, $\beta_0(T_e, \zeta) \equiv \langle (k_\parallel^2/m\Delta) \rangle_0$, $\beta_1(T_e, \zeta) \equiv \langle (k_\parallel^2/m\Delta) \cos(k_z d) \rangle_0$, and the symbol $\langle \cdots \rangle_0$ denotes the statistical average

$$\langle \cdots \rangle_0 = \frac{1}{4\pi^3 n_0} \int\int dk_x dk_y \int_{-\pi/d}^{\pi/d} dk_z \, f(\varepsilon(\mathbf{k})/T_e - \zeta) \cdots,$$

where $f(x) = 1/[\exp(x) + 1]$ is the Fermi function.

In addition, the frictional acceleration $a = a_\mathrm{i} + a_\mathrm{p}$ in (13.223) and the electron energy-loss rate w in (13.224) are given by the expressions (12.53)–(12.55), as functions of z_d, T_e and ζ: $a = a(z_d, T_e, \zeta)$, $w = w(z_d, T_e, \zeta)$.

13.9.2 Analyses with small wave-like perturbations, convective and absolute instabilities

To investigate the dynamic mobility in the bias condition and possible spatial amplification and temporal growth of a high-frequency fluctuation in superlattice miniband transport, we consider a small perturbation superimposed on a dc bias. We write z_d, T_e, ζ, v and any other transport quantity Q involved in the balance equations (13.222)–(13.224) as the sum of the dc bias part and a small fluctuation: $z_d = z_0 + \delta z_d$, $T_e = T_0 + \delta T_e$, $\zeta = \zeta_0 + \delta\zeta$, $v = v_0 + \delta v$, and $Q = Q_0 + \delta Q$ (Q stands for E, n, $1/m_z^*$, B_z, S_z, ε, ε_z, or $\varepsilon_\parallel$).

In the zeroth order, the continuity equation (13.222) and the Poisson equation (13.225) are identities, and the other two equations are just the dc steady-state equations for the effective-force and energy balance in the spatially homogeneous case:

$$0 = eE_0/m_{z0}^* + a_0, \qquad (13.233)$$

$$0 = eE_0 v_0 - w_0, \qquad (13.234)$$

where

$$v_0 = v_m \alpha_{10} \sin z_0, \qquad (13.235)$$

$$1/m_{z0}^* = (1 - 2\varepsilon_{z0}/\Delta)/M^*, \qquad (13.236)$$

$$\varepsilon_{z0} = (\Delta/2)(1 - \alpha_{10} \cos z_0), \qquad (13.237)$$

and the equation

$$1 = \alpha_0(T_0, \zeta_0). \qquad (13.238)$$

Here we have used the simplified notations: $\alpha_{10} \equiv \alpha_1(T_0, \zeta_0)$, $a_0 \equiv a(z_0, T_0, \zeta_0)$, and $w_0 \equiv w(z_0, T_0, \zeta_0)$. Eq. (13.238), which is just the equation (12.64) for the determination of chemical potential in spatially uniform transport, gives ζ_0 as a function of T_0 in the spatially homogeneous condition. Eqs. (13.233) and (13.234) are the same as Eqs. (12.69) and (12.70). Their solution and its dependence on the miniband width, superlattice period, carrier density, lattice temperature and the strength of impurity scattering have been discussed in Chapter 12. The main feature of the velocity v_0 versus the electric field E_0 is the appearance of negative differential conductance when E_0 is large than a critical value, typical of those shown in Sec. 12.4.

The balance equations for the linear order fluctuations are as follows:

$$\frac{\partial}{\partial t}\delta n = -\frac{\partial}{\partial z}\delta(nv), \qquad (13.239)$$

$$\frac{\partial}{\partial t}\delta(nv) = -\frac{\partial}{\partial z}\delta(n\mathcal{B}_z) + \frac{en_0}{m_{z0}^*}\delta E + eE_0\,\delta(n/m_z^*) + \delta(na), \qquad (13.240)$$

$$\frac{\partial}{\partial t}\delta(n\varepsilon) = -\frac{\partial}{\partial z}\delta(nS_z) + en_0v_0\,\delta E + eE_0\,\delta(nv) - \delta(nw), \qquad (13.241)$$

and the Poisson equation reads

$$\frac{\partial}{\partial z}\delta E = \frac{e}{\epsilon_0\kappa}\delta n. \qquad (13.242)$$

To investigate the instability related to the space-charge waves we consider wave-like fluctuations (Lei, Horing and Cui, 1995; 1998), i.e. δz_d, δT_e, $\delta\zeta$, δE, and δQ, having the spatiotemporal dependence of the form

$$e^{i(kz-\omega t)}. \qquad (13.243)$$

The Poisson equation (13.242) and the continuity equation (13.239) are then

$$\frac{en_0}{\epsilon_0\kappa}(\alpha_{0t}\delta T_e + \alpha_{0\zeta}\delta\zeta) - i\,k\,\delta E = 0, \qquad (13.244)$$

$$(kv_m\alpha_{10}\cos z_0)\,\delta z_d + (kv_m\alpha_{1t}\sin z_0 - \omega\alpha_{0t})\,\delta T_e$$
$$+ (kv_m\alpha_{1\zeta}\sin z_0 - \omega\alpha_{0\zeta})\,\delta\zeta = 0. \qquad (13.245)$$

Equations (13.240) and (13.241) can be written as

$$c_{31}\,\delta z_d + c_{32}\,\delta T_e + c_{33}\,\delta\zeta + c_{34}\,\delta E/E_0 = 0, \qquad (13.246)$$

$$c_{41}\,\delta z_d + c_{42}\,\delta T_e + c_{43}\,\delta\zeta + c_{44}\,\delta E/E_0 = 0. \qquad (13.247)$$

In these equations, $\alpha_{it} \equiv \partial\alpha_i(T_0,\zeta_0)/\partial T_e$ ($i = 0, 1$), $\alpha_{i\zeta} \equiv \partial\alpha_i(T_0,\zeta_0)/\partial\zeta$ ($i = 0, 1$); and c_{3j} and c_{4j} ($j = 1, 2, 3$), which are functions of k and ω and explicitly given in the original paper (Lei and Cui, 1997), depends on the bias dc point and scattering.

Equations (13.244)–(13.247) constitute a set of four linear algebraic equations for four variables: δz_d, δT_e, $\delta\zeta$ and δE. The condition for it to have nonzero solution requires that the determinant of its coefficients vanishes:

$$\mathcal{D}(k,\omega) = 0. \qquad (13.248)$$

The equation (13.248) is the dispersion equation. It determines all the possible space-charge-wave modes and their dispersion relation between k and ω in the system.

In the spatially uniform case, $k = 0$, both Eq. (13.244) and Eq. (13.245) become $\alpha_{0t}\delta T_e + \alpha_{0\zeta}\delta\zeta = 0$, hence $\mathcal{D}(0,\omega) = 0$ is satisfied for arbitrary ω, i.e. the wave (13.243) of $k = 0$ is always the eigenmode. Therefore, we can, in principle, apply a small uniform ac field δE of (real) frequency ω superimposed on a bias dc field E_0, and consider the linear velocity response δv of the same frequency in the system. The small-signal complex mobility (dynamic mobility) is defined as [see (12.120)]

$$\mu_\omega = \frac{\delta v}{\delta E}. \tag{13.249}$$

The behavior of this small-signal mobility has been examined in Sec. 12.8.2, where Fig. 12.19(b) plots its real part $\operatorname{Re}\mu_\omega$ for a laterally unconfined GaAs-based superlattices with $\Delta = 900\,\text{K}$, $d = 7\,\text{nm}$, $N_s = 1.5 \times 10^{15}\,\text{m}^{-2}$ and $\mu(0) = 1.0\,\text{m}^2/\text{Vs}$.

In the spatially inhomogeneous case of $k \neq 0$, Eq. (13.248) can be satisfied only for complex $k = k_1 + ik_2$ and/or complex $\omega = \omega_1 + i\omega_2$. We first examine the case of a real wavevector: $k = k_1$ and $k_2 = 0$ (Lei, Horing and Cui, 1995). The real wavevector k_1 and the imaginary part of frequency, ω_2, obtained from Eq. (13.248) as functions of real frequency $\nu = \omega_1/2\pi$ for the superlattice mentioned above, are shown in Figs. 13.20(a) and (b) for several dc bias fields E_0 from 2.5 to 7.0 kV/cm. When biased in the negative differential mobility regime, ω_2 is positive for low ω_1 (low k_1), and continues to be positive with increasing frequency until ω_1 reaches a maximum value dependent on the bias field E_0. A positive ω_2 implies a traveling wave

$$\exp[\omega_2 t + i(k_1 z - \omega_1 t)] \tag{13.250}$$

propagating along the positive z direction with its amplitude growing exponentially as a function of time. The phase velocity of this space-charge wave, $v_{\text{ph}} = \omega_1/k_1$, is shown in the inset of Fig. 13.20(a). Depending on the bias field and wavevector k_1 (frequency), v_{ph} varies from $1.05\,v_0$ to $1.3\,v_0$, v_0 being the bias drift velocity at bias field E_0. The existence of such an amplitude-growth space-charge-wave solution indicates a convective instability in the superlattice biased in the negative differential mobility range.

Next we examine the case of real frequency (Lei, Horing and Cui, 1998).

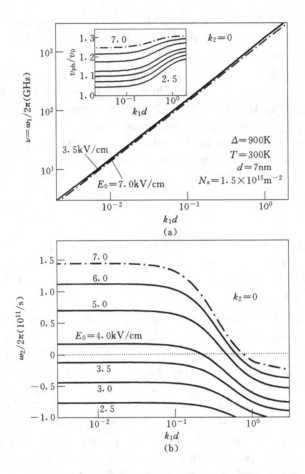

Fig. 13.20 The real wavevector ($k_2 = 0$) space-charge waves propagating along the positive z direction. The real part $\nu = \omega_1/2\pi$ (a) and the imaginary part $\omega_2/2\pi$ (b) of the frequency, and the phase velocity $v_{\rm ph}/v_0$ (inset) are shown as functions of wavevector k_1. The numbers near the curves are the bias field in kV/cm.

A solution of Eq. (13.248) for the same superlattice in the case of real frequency ($\omega_2 = 0$) are shown in Fig. 13.21, plotting the real part k_1 and imaginary part k_2 of the wavevector k as functions of frequency $\nu = \omega_1/2\pi$. When biased in the negative differential mobility range, k_2 can be negative. A negative k_2 implies a wave turbulence of the form

$$\exp[|k_2|z + {\rm i}\,(k_1 z - \omega_1 t)], \tag{13.251}$$

representing a traveling wave along the positive z direction with its amplitude enlarged when propagating forward. When $E_0 \leqslant 6.0\,\mathrm{kV/cm}$ the phase velocity is in between $1.05 \sim 1.3\,v_0$. In the case of $E_0 = 7.0\,\mathrm{kV/cm}$, v_{ph} can be as low as $0.8\,v_0$ at low frequency.

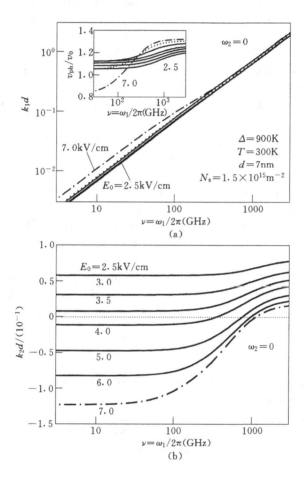

Fig. 13.21 The real frequency ($\omega_2 = 0$) space-charge waves propagating along the positive z direction. The real part k_1 (a) and the imaginary part k_2 (b) of the wavevector, and the phase velocity v_{ph}/v_0 (inset) are shown as functions of frequency $\nu = \omega_1/2\pi$.

It should be noted that when the superlattice is biased in the negative differential mobility regime, the solutions of real wavevector with positive ω_2 and real frequency with negative k_2 exist only when the frequency $\nu = \omega_1/2\pi$ is lower than a certain maximum value ν_{m}. But this ν_{m} is much higher than the maximum frequency for the negative differential mobility to appear in the spatially homogeneous case (see Sec 12.8.2). This is because that both carriers and the space-charge wave move along the same direction, and the carrier drift velocity is close to the phase velocity of the space-charge wave, such that the real oscillating frequency of carriers is much lower than the frequency of the space-charge wave. The occurrence of the convective instability (the traveling space-charge waves of amplitude growing with time or space) in the small-signal analysis for a miniband superlattice biased in the negative differential mobility regime implies the possible realization of propagating field domains and high-frequency self-sustained oscillations in the system (see the next subsection).

In the case of real frequency, in addition to the solution of traveling wave along the positive direction with growing amplitude as described in Fig. 13.21, there is another solution of traveling wave along the negative ($-z$) direction which is damped when propagating. Generally, at a give dc bias the dispersion equation (13.248), which has two roots, maps a point of the complex ω plane onto two points of the complex k plane. The instability of the superlattice system can be analyzed based on this mapping (Gueret, 1971; Lei and Cui, 1997). For the specific point $\omega = \mathrm{i}\omega_\mathrm{b}$ on the imaginary axis of the complex ω plane, the two roots coincide and thus are mapped to a single point $k = \mathrm{i}k_\mathrm{b}$ on the imaginary axis of the complex k plane. In this case, the space-time dependence of the system response is of the form

$$\exp(\omega_\mathrm{b} t - k_\mathrm{b} z), \qquad (13.252)$$

exhibiting, for positive ω_b, temporal growth without spatial propagation. This instability featuring with temporal growth of local domains without traveling, is called the absolute instability. The onset of the absolute instability would switch the superlattice from the propagating field-domain state to a spatially stationary nonuniform field distribution state, leading to the disappearance of the current oscillation.

The range for the absolute instability to occur, i.e. the range in which the double roots ($\mathrm{i}\omega_\mathrm{b}, \mathrm{i}k_\mathrm{b}$) of the dispersion equation (13.249) has a positive ω_b, are numerically examined by Lei and Cui (1997) for a typical laterally unconfined superlattice. They find that the system may become absolutely unstable only when it is biased within a range deep in the negative differential mobility regime, and the enhancement of the elastic scattering suppresses the occurrence of the absolute instability.

The occurrence of convective and absolute instabilities in semiconductors exhibiting negative differential mobility is usually analyzed with a phenomenological drift-diffusion model (Gueret, 1971), which works with a much simpler drift-diffusion equation (Lei et al, 1997)

$$\left[\tau_m \frac{\partial}{\partial t} + 1\right] \delta(nv) = \mu_0 n_0 \,\delta E + v_0 \,\delta n - D\frac{\partial}{\partial z}(\delta n), \qquad (13.253)$$

in place of the effective momentum-balance equation (13.240) and without the energy balance equation (13.241). Here, v_0 is the bias drift velocity, $\mu_0 = \partial v_0/\partial E_0$ is the zero frequency differential mobility, D is a diffusion constant, and τ_m is a momentum relaxation time. In most of drift-diffusion-model analyses of space charge wave instability, a very small τ_m is assumed. This yields a frequency-independent dynamic mobility in the spatially uniform case ($\mu_\omega = \mu_0$), and a dispersion relation of the following form for small wave-like fluctuations (13.243) in the case of real $k = k_1$ ($k_2 = 0$):

$$\omega = v_0 k - \mathrm{i}(\omega_c + Dk^2), \qquad (13.254)$$

where

$$\omega_c = (en_0/\epsilon_0 \kappa)\mu_0. \qquad (13.255)$$

Eq. (13.254) describes a convective space-charge wave which propagates with a phase velocity equal to the carrier drift velocity v_0 and has an imaginary frequency (amplitude growth rate) $\omega_2 = -\omega_c$ at small wavevectors. In addition to the convective instability, the absolute instability may also occur if μ_0 is negative and

$$|\omega_c| > v_0^2/4D. \qquad (13.256)$$

Estimation for a typical miniband superlattice (Lei and Cui, 1997) indicates that the criterion (13.256) can be satisfied (thus the system becomes absolutely unstable) very soon after it enters negative differential regime. The other calculations using hydrodynamic balance equations with a relaxation time approximation (Büttiker and Thomas, 1977; 1979) or using a simplified Boltzmann-equation (Ignatov and Shaskin, 1987) yield essentially the same behavior.

Unlike these basically one-dimensional calculations, the above hydrodynamic investigation based on Eqs. (13.233), (13.234) and (13.239)–(13.242), are fully three-dimensional in nature with strong interactions between the longitudinal and transverse motions of carriers, thus give rise to significantly different predictions for laterally unconfined systems having high carrier densities.

13.9.3 Spatiotemporal domains in voltage-biased superlattices

The small-signal theory in the preceding subsection is useful for examining the condition for the onset of instabilities in superlattice miniband transport. Once a traveling space-charge wave having amplitude growth with time or space, or a temporally growing local fluctuation domain appears, the small-signal theory will soon be inapplicable. To investigate the real situation of the system after the onset of an instability, we have to proceed with a large-signal analysis by starting from the full time-dependent hydrodynamic balance equations (13.222)–(13.224).

As an example, Cao and Lei (1999a,1999b) study a GaAs-based n-type superlattice structure sandwiched in between two heavily doped n^+ GaAs bulk contact layers, and a bias dc or time-dependent voltage V is applied across the two ends. The n-region, which is composed of a quantum-well superlattice with period d and miniband width Δ, has a total length $L = 0.55\,\mu$m and is assumed uniformly doped with a doping concentration $n_\mathrm{D} = 7 \times 10^{16}\,\mathrm{cm}^{-3}$. Outside in the two contacts the doping concentration increases exponentially to $n_\mathrm{D} = 2 \times 10^{18}\,\mathrm{cm}^{-3}$ within the ranges of $0.1\,\mu$. The electric potential $\phi(z,t)$, carrier density $n(z,t)$, particle-flow density $j_n(z,t) \equiv n(z,t)v(z,t)$ and electron temperature $T_\mathrm{e}(z,t)$ are chosen as fundamental field variables in the numerical treatment. In both n and n^+ regions, the hydrodynamic and Poission equations can be written in the form ($A \equiv na$, $W \equiv nw$)

$$\frac{\partial n}{\partial t} = -\frac{\partial j_n}{\partial z}, \tag{13.257}$$

$$\frac{\partial j_n}{\partial t} = -\frac{\partial}{\partial z}(n\mathcal{B}_z) - \frac{en}{m_z^*}\frac{\partial \phi}{\partial z} + A, \tag{13.258}$$

$$\frac{\partial}{\partial t}(n\varepsilon) = -\frac{\partial}{\partial z}(nS_z) - ej_n\frac{\partial \phi}{\partial z} - W, \tag{13.259}$$

$$\frac{\partial^2 \phi}{\partial z^2} = -\frac{e}{\epsilon_0 \kappa}(n - n_\mathrm{D}). \tag{13.260}$$

In the n-region, where a tight-binding type energy dispersion (13.220) and (13.221) for the lowest miniband superlattice is assumed and thus the expressions (13.226)–(13.229) hold at any time t and z throughout the range. We can express ζ and $z_d = p_d d$ as functions of n and T_e by means of Eqs. (13.226) and (13.227), and then all the other quantities in Eqs. (13.257)–(13.260), including $n\varepsilon$, $n\mathcal{B}_z$, nS_z, A and W, can be expressed in terms of n, j_n and T_e. In the n^+ regions of two contacts, where the energy dispersions are approximately isotropic and parabolic, $\varepsilon(\mathbf{k}) = k^2/2m$ (m is the effective mass at the conduction band edge of the the Γ valley), we have (see Sec. 10.7.)

$$n = N_c(T_e)\mathcal{F}_{1/2}(\zeta), \tag{13.261}$$

$$u = \frac{3}{2}N_c(T_e)T_e\mathcal{F}_{3/2}(\zeta), \tag{13.262}$$

$$n\varepsilon = u + \frac{n}{2}mv^2, \tag{13.263}$$

$$n\mathcal{B}_z = \frac{2u}{3m} + nv^2, \tag{13.264}$$

$$nS_z = \left(\frac{5}{3}u + \frac{n}{2}mv^2\right)v, \tag{13.265}$$

where $N_c(T_e) = 2(mT_e/2\pi)^{3/2}$, $\mathcal{F}_\nu(x)$ is the the Fermi integral of order ν defined in (10.85). From these relations and the expressions of A and W, all the quantities in Eqs. (13.257)–(13.260) can be expressed in terms of n, j_n and T_e. As a result, Eqs. (13.257)–(13.260) are expressed as the partial differential equations for the fundamental field variables $\phi(z,t)$, $n(z,t)$, $j_n(z,t) \equiv n(z,t)v(z,t)$ and $T_e(z,t)$ throughout the n^+nn^+ device.

We assume that at the initial time $t = 0$ the above n^+nn^+ system is in an equilibrium state:

$$T_e = T, \tag{13.266}$$

$$n(z,0) = n_D(z), \tag{13.267}$$

$$j_n(z,0) = 0, \tag{13.268}$$

$$\phi(z,0) = \frac{T}{e}\ln\left(\frac{n_D(z)}{n_i}\right) \tag{13.269}$$

(n_i is an arbitrary density constant). If a constant dc electric potential V_{dc} is applied between the two ends (z_L and z_R) of the n^+nn^+ structure from $t = 0$, the boundary conditions are

$$T_e(z_L,t) = T_e(z_R,t) = T, \tag{13.270}$$

$$n(z_L,t) = n_D(z_L), \quad n(z_R,t) = n_D(z_R), \tag{13.271}$$

$$\phi(z_R,t) - \phi(z_L,t) = \frac{T}{e}\ln\left(\frac{n_D(z_R)}{n_D(z_L)}\right) + V. \tag{13.272}$$

The time-dependent transport quantities $\phi(z,t)$, $n(z,t)$, $j_n(z,t)$ and $T_e(z,t)$ can be obtained by numerically solving the discretized equations (13.257)–

(13.260) with the above initial and boundary conditions. It is convenient to define a total current density $J(t)$ as the sum of the conduction current density $j = ej_n$ and the displacement current density $\varepsilon_0 \kappa (\partial E/\partial t)$,

$$J(t) \equiv j(z,t) + \varepsilon_0 \kappa \frac{\partial E(z,t)}{\partial t}. \tag{13.273}$$

From the continuity equation (13.257) and Poisson equation (13.260) we have $\partial J/\partial z = 0$, i.e. $J(t)$ is independent of z, or it takes a constant value throughout the system at any given time. This total current density $J(t)$ will be used as a quantity to characterize the behavior of the system response.

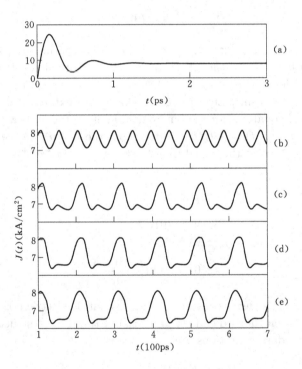

Fig. 13.22 Temporal evolution of the total current density $J(t)$ in the n^+nn^+ structure described in the text at $T = 300$ K, under different dc bias voltages: (a) $V_{\rm dc} = 0.6$ V, (b) $V_{\rm dc} = 0.71$ V, (c) $V_{\rm dc} = 0.75$ V, (d) $V_{\rm dc} = 0.8$ V, and (e) $V_{\rm dc} = 0.9$ V. From Cao and Lei (1999b).

Numerical calculations are performed (Cao and Lei, 1999b) for a GaAs-based superlattice structure with period $d = 5.1$ nm, well width $a = 3.1$ nm and miniband width $\Delta = 17\,\mathrm{meV}$ under a dc bias voltage $V = V_\mathrm{dc}$ at $T = 300\,\mathrm{K}$, taking account of scatterings from impurities, longitudinal and transverse acoustic phonons and polar optic phonons. For comparison with experiments, the impurity scattering strength is chosen to yield the low-temperature linear mobility $\mu_0 = 0.26\,\mathrm{m^2/V\,s}$, so that the threshold voltage V_th for the onset of negative differential conductance is 0.69 V for this superlattice system (Schomburg et al, 1996). To mimic a realistic situation, a slight doping notch is assumed near the cathode end of the structure. The calculated temporal evolution of the total current density $J(t)$ is shown in Fig. 13.22 at several dc bias voltages marked by (a) $V_\mathrm{dc} = 0.6\,\mathrm{V}$, (b) $V_\mathrm{dc} = 0.71\,\mathrm{V}$, (c) $V_\mathrm{dc} = 0.75\,\mathrm{V}$, (d) $V_\mathrm{dc} = 0.8\,\mathrm{V}$, and (e) $V_\mathrm{dc} = 0.9\,\mathrm{V}$. In the case of $V_\mathrm{dc} = 0.6\,\mathrm{V} < V_\mathrm{th}$ (a), $J(t)$ approaches a constant steady value after a short transient overshoot and slight oscillations in picosecond time interval. Once $V_\mathrm{dc} > V_\mathrm{th}$ [(b), (c), (d) and (e)], $J(t)$ exhibits steady self-sustained oscillations. This temporal oscillation of the total current density is related to the formation of traveling high-field domains due to the spatial amplification and temporal growth in the superlattice exhibiting negative differential mobility.

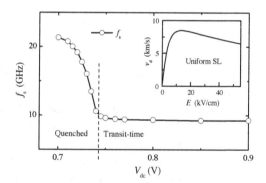

Fig. 13.23 The calculated self-oscillating frequency f_s is shown as a function of the applied dc bias V_dc for a superlattice structure as described in the text. The inset shows the corresponding steady-state velocity–field relation calculated under spatially uniform condition. From Cao and Lei (1999b).

The calculated self-oscillating frequency f_s is shown in Fig. 13.23 as a function of applied dc bias V_dc for the above superlattice system, together with the velocity–field curve under uniform condition (inset). Both the quenched [Fig. 13.22(b)] and transit-time [Fig. 13.22(c), (d) and (e)] dipole modes exhibit, which are similar to the domain modes appearing in

Gunn transferred devices. In the region of quenched mode ($V_{dc} < 0.74\,\text{V}$) the self-sustained oscillating frequency decreases rapidly with increasing dc bias voltage, while in the region of transit-time mode ($V_{dc} > 0.74\,\text{V}$) the high-field domains travel at almost constant frequencies close to the inverse electron transit time. The transit-time effect stems from the time delay between a voltage excitation and the arrival of the electrons at the anode. As a demonstration of this transit-time effect, we show in Fig. 13.24 the spatial- and temporal-dependent electric-field domain $E(z,t)$ and the contour line of the normalized electron temperature T_e/T in the case of $V_{dc} = 0.8\,\text{V}$ [Fig. 13.22(d)]. The growing electric-filed domain periodically travels from the cathode to the anode end with a period $T_s \simeq 103\,\text{ps}$ (corresponding to $f_s = 9.7\,\text{GHz}$). At the same time, electrons are periodically heated by the electric field during the time evolution. At the largest field strength $E \simeq 45\,\text{kV/cm}$, the normalized temperature T_e/T is about 1.8.

To compare the theoretical results directly with experiments, we calculate the total average current flowing through the superlattice, $I = J_{av}S$, where J_{av} is the time-average total current density $J(t)$ over one oscillating period T_s in the time-dependent steady state,

$$J_{av} = \frac{1}{T_s} \int_t^{t+T_s} J(s)\,ds, \qquad (13.274)$$

and S is the effective area of the superlattice. The calculated total average current I for $S = 2.2\,\mu\text{m}^2$ as a function of the bias dc voltage V_{dc}, shown in Fig. 13.24 as the solid line, is in good agreement with the experimental result (small black dots) of Schomburg *et al* (1996) within the whole voltage range of 0.1–0.9 V. The I–V relation obtained by using the time-independent hydrodynamic equations (shown as dashed curve in the figure) deviates from the experiments when $V_{dc} > V_{th}$. This indicates that the observed negative differential conductance results from the self-sustained oscillations of the electric current.

Numerical examinations are also carried out (Cao and Lei, 1999c) for the above GaAs-based n^+nn^+ superlattice structure under both a dc and a sinusoidal ac bias voltage of frequency f_{ac},

$$V = V_{dc} + V_{ac}\sin(2\pi f_{ac}t), \qquad (13.275)$$

which enters through the boundary condition (13.272). In this nonlinear dynamic system having negative differential mobility in the velocity–field relation, the spatiotemporal behavior of the system response depends on

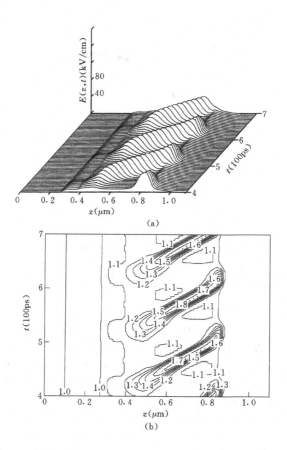

Fig. 13.24 (a) The spatial- and temporal-dependent electric field domains $E(z,t)$, and (b) the contour line of normalized electron temperature $T_e(z,t)/T$ in the n^+nn^+ structure described in the text at bias potential $V_{dc} = 0.8$ V. From Cao and Lei (1999b).

the interplay of V_{dc}, V_{ac} and f_{ac}. At fixed V_{dc} and fixed f_{ac}, large V_{ac} usually leads to a periodic solution of the current density $J(t)$ having its fundamental frequency synchronized with the driving ac frequency f_{ac}. In this case the time-averaged current density J_{av}, which should be expressed as the average over one oscillating period $T_{ac} = 1/f_{ac}$,

$$J_{av} = \frac{1}{T_{ac}} \int_t^{t+T_{ac}} J(s)\, ds, \qquad (13.276)$$

is strongly reduced in comparison with that of $V_{ac} = 0$ case, and the negative differential conductance showing up in the J_{av}-vs-V_{dc} curve at $V_{ac} = 0$

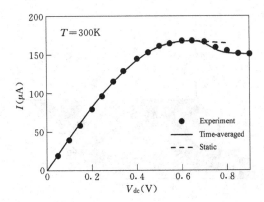

Fig. 13.25 Average current versus bias voltage for a GaAs-based n^+nn^+ superlattice structure as described in the text. The solid line is the theoretical result obtained from the time average of the oscillating solution to the time-dependent hydrodynamic equations (13.257)–(13.260); the dashed line is the solution of the static (time-independent) and spatially uniform equations (13.233) and (13.234); the dots are the experimental data of Schomburg et al (1996). From Cao and Lei (1999b).

disappears when $V_{ac} = 0.74, 0.94$ and $1.27\,\mathrm{V}$ at frequency $f_{ac} = 78\,\mathrm{GHz}$, as shown in Fig. 13.25. Furthermore, at a fixed ac amplitude, lower frequency ac driving has stronger effect in suppressing the average current of the superlattice system. These are in good agreement with the experimental measurement (Schomburg et al, 1996).

When the amplitude of the ac voltage decreases the situation becomes complicated. The examination of the spatiotemporal solution at a fixed $V_{dc} = 0.85\,\mathrm{V}$ and a fixed ac frequency $f_{ac} = \frac{1}{2}(\sqrt{5}+1)f_s = 1.61803 f_s$ (f_s is the self-oscillating frequency at $V_{ac} = 0$ and $V_{dc} = 0.85\,\mathrm{V}$), indicates that the behavior of the current density $J(t)$ sensitively depends on the amplitude ratio $c = V_{ac}/V_{dc}$. When $c > 0.2507$ the solution goes to a period-1 state with the ac frequency f_{ac}. For small ac amplitude $c < 0.23067$ the system response becomes chaotic. In the range between them, most frequency-locked oscillations, such as period-3 and period-6, also some period-1 and period-8, are observed. For more detail we refer the readers to the original paper of Cao and Lei (1999c).

Chapter 14

Carrier Transport in Semiconductors Driven by THz Radiation Fields

In several preceding Sections in this book (such as Secs. 3.3–3.5, 3.6, 8.5, 12.8, 13.9) the time-dependent steady-state transport properties of various semiconductor systems driven by high-frequency electric fields of small or large amplitudes, were studied directly using the time-dependent momentum-balance equation (1.40) and energy-balance equation (1.55), or their extensions to the case in the presence of a magnetic field or to the system with nonparabolic energy band. For small high-frequency fields, analytical expressions for linear high-frequency resistivity and conductivity can be obtained by expanding the relevant quantities in the equations to the linear order in the field strength. For strong high-frequency fields, one has to follow the temporal evolution by integrating these ordinary time-differential equations starting from the initial state, until a steady transport state is reached.

As pointed out in the Chapter 1, the time t in equations (1.40) and (1.55) is introduced as a slowly varying macroscopic parameter in such a way that a full microscopic evolution from the initial to the final state in the Liouville equation can be realized without appreciable change of the time t. Therefore, these equations can only be used in the case in which the time variations of the electric field $\boldsymbol{E}(t)$, the center-of-mass velocity $\boldsymbol{v}(t)$, the electron temperature T_e and all other macroscopic statistical quantities can appear only in a time scale much larger than the microscopic relaxation time or the thermalization time τ_{th}. In other words, these equations can apply to describe a high-frequency field driven transport only when its frequency $\nu = \omega/2\pi$ is much smaller than the inverse thermalization time $1/\tau_{\text{th}}$ of the system.

The microscopic relaxation time τ_{th} varies greatly in different systems.

For conventional 3D or 2D semiconductors with high carrier density, $\tau_{\rm th}$ is roughly of order of a few tenths of a picosecond (see Chapter 7). For such systems, the preceding time-dependent treatments can be useful in transport driven by an ac field of frequency no higher than a few THz's.

A kind of "time-independent" balance equation method (Lei, 1998a; 1998b; 1998c) has been developed with the initial purpose to deal with electron transport in semiconductor driven by a strong ac field of frequency in the range of or higher than a terahertz.

14.1 Balance Equations for Electron Transport in Semiconductors Driven by a THz Electric Field

14.1.1 *Center-of-mass and relative-electron variables in a uniform high-frequency electric field*

We first consider an isotropic three-dimensional (3D) parabolic band system containing N electrons of effective mass m and charge e. They are interacting with each other through the Coulomb potential and also scattered by randomly distributed impurities and by phonons in the lattice. In the absence of the electric field, this coupled electron–phonon system is described by the Hamiltonian (1.1) without the electric potential term.

Consider that a uniform dc (or slowly time-varying) electric field $\boldsymbol{E}_0$ and a uniform sinusoidal radiation field of frequency ω and amplitude $\boldsymbol{E}_\omega$,

$$\boldsymbol{E}(t) = \boldsymbol{E}_0 + \boldsymbol{E}_\omega \sin(\omega t) \tag{14.1}$$

($\boldsymbol{E}_0$ and $\boldsymbol{E}_\omega$ being of arbitrary strengths and directions), are applied in the system. We can represent the electric field $\boldsymbol{E}(t)$ by means of a vector potential $\boldsymbol{A}(t)$ and a scalar potential $\varphi(\boldsymbol{r})$ of the form

$$\boldsymbol{A}(t) = (\boldsymbol{E}_\omega/\omega) \cos(\omega t), \tag{14.2}$$

$$\varphi(\boldsymbol{r}) = -\boldsymbol{r} \cdot \boldsymbol{E}_0. \tag{14.3}$$

The Hamiltonian of the system in the presence of such electric fields can be expressed in the form:

$$H = H_{\rm e}(t) + H_{\rm ei} + H_{\rm ep} + H_{\rm ph}, \tag{14.4}$$

in which

$$H_{\rm e}(t) = \sum_j \left\{ \frac{1}{2m} \left[\boldsymbol{p}_j - e\boldsymbol{A}(t) \right]^2 + \varphi(\boldsymbol{r}_j) \right\} + H_{\rm ee} \tag{14.5}$$

is the Hamiltonian of the electrons under the influence of the electric field $\boldsymbol{E}(t)$ with $H_{\rm ee}$ being the electron–electron Coulomb interaction, $H_{\rm ei}$ and $H_{\rm ep}$ are respectively the electron–impurity and electron–phonon couplings, and $H_{\rm ph}$ stands for the phonon Hamiltonian. Their expressions are the same as those given in Sec. 1.1. In equation (14.5), $\boldsymbol{r}_j$ and $\boldsymbol{p}_j$ are the coordinate and momentum of the jth electron. As in Chapter 1, we introduce, for this N-electron system, the center-of-mass variables $\boldsymbol{R}$ and $\boldsymbol{P}$, and relative-electron variables $\boldsymbol{r}'_j$ and $\boldsymbol{p}'_j$ ($j = 1, 2, \cdots, N$) [see (1.2) and (1.3)]. They satisfy the commutation relations (1.4)–(1.8) and other relevant relations. In terms of the center-of-mass and relative electron variables the Hamiltonian (14.4) can be rewritten in the form

$$H_{\rm e}(t) = H_{\rm c} + H_{\rm er}, \tag{14.6}$$

$$H_{\rm c} = \frac{\boldsymbol{P}^2}{2Nm} - \boldsymbol{P} \cdot \boldsymbol{v}_\omega \cos(\omega t) - Ne\boldsymbol{E}_0 \cdot \boldsymbol{R} + \frac{N}{m} e^2 \boldsymbol{A}^2(t), \tag{14.7}$$

where

$$\boldsymbol{v}_\omega \equiv \frac{e\boldsymbol{E}_\omega}{m\omega}, \tag{14.8}$$

and $H_{\rm er}$ has exactly the same expression as the relative electron Hamiltonian (1.11) in the absence of the high-frequency field:

$$H_{\rm er} = \sum_j \frac{(\boldsymbol{p}'_j)^2}{2m} + \frac{e^2}{4\pi\epsilon_0} \sum_{i<j} \frac{1}{|\boldsymbol{r}'_i - \boldsymbol{r}'_j|}$$

$$= \sum_{\boldsymbol{k},\sigma} \varepsilon_{\boldsymbol{k}} c^\dagger_{\boldsymbol{k}\sigma} c_{\boldsymbol{k}\sigma} + \frac{1}{2} \sum_{\boldsymbol{q}} v_{\rm c}(q)(\rho_{\boldsymbol{q}}\rho_{-\boldsymbol{q}} - N), \tag{14.9}$$

in which $v_{\rm c}(q) = e^2/(\epsilon_0 \kappa q^2)$ is the Coulomb potential between electrons, $\varepsilon_{\boldsymbol{k}} = k^2/2m$ is the kinetic energy, $c^\dagger_{\boldsymbol{k}\sigma}$ ($c_{\boldsymbol{k}\sigma}$) is the creation (annihilation) operator, and

$$\rho_{\boldsymbol{q}} = \sum_j e^{i\boldsymbol{q}\cdot\boldsymbol{r}'_j} = \sum_{\boldsymbol{k},\sigma} c^\dagger_{\boldsymbol{k}+\boldsymbol{q}\sigma} c_{\boldsymbol{k}\sigma} \tag{14.10}$$

is the density operator, of relative electrons in the wavevector representation. We have taken the volume of system to be unity for clarity.

The nice thing here is that both the dc electric field $\boldsymbol{E}_0$ and the radiation electric field $\boldsymbol{E}_\omega \sin(\omega t)$ exert only on the center of mass, and relative electrons do not directly experience the existence of either one of them ($\boldsymbol{E}_0$ and $\boldsymbol{E}_\omega$). In addition, we assume that $H_{\rm ph}$ is not affected by the radiation field, and $H_{\rm ei}$ and $H_{\rm ep}$ also have exactly the same expressions as those given by (1.13)–(1.15) in terms of center-of-mass coordinate $\boldsymbol{R}$ and the density operator $\rho_{\boldsymbol{q}}$ of relative electrons.

To derive the momentum- and energy-balance equations, we calculate the rates of changes of the center-of-mass momentum $\boldsymbol{P}$ and the relative electron energy H_{er} according to the Heisenberg equation of motion, yielding the following operator equations:

$$\dot{\boldsymbol{P}} = -\mathrm{i}\,[\boldsymbol{P}, H]$$
$$= Ne\boldsymbol{E}_0 - \mathrm{i} \sum_{\boldsymbol{q},a} \boldsymbol{q}\, u(\boldsymbol{q})\, e^{\mathrm{i}\,\boldsymbol{q}\cdot(\boldsymbol{R}-\boldsymbol{r}_a)} \rho_{\boldsymbol{q}} - \mathrm{i} \sum_{\boldsymbol{q},\lambda} \boldsymbol{q}\, M(\boldsymbol{q},\lambda)\, \phi_{\boldsymbol{q}\lambda}\, e^{\mathrm{i}\,\boldsymbol{q}\cdot\boldsymbol{R}} \rho_{\boldsymbol{q}}, \tag{14.11}$$

$$\dot{H}_{\text{er}} = -\mathrm{i}\,[H_{\text{er}}, H]$$
$$= -\mathrm{i} \sum_{\boldsymbol{q},a} u(\boldsymbol{q})\, e^{\mathrm{i}\,\boldsymbol{q}\cdot(\boldsymbol{R}-\boldsymbol{r}_a)} \sum_{\boldsymbol{k}} (\varepsilon_{\boldsymbol{k}+\boldsymbol{q}} - \varepsilon_{\boldsymbol{k}})\, c^{\dagger}_{\boldsymbol{k}+\boldsymbol{q}\sigma} c_{\boldsymbol{k}\sigma}$$
$$-\mathrm{i} \sum_{\boldsymbol{q},\lambda} M(\boldsymbol{q},\lambda)\, \phi_{\boldsymbol{q}\lambda}\, e^{\mathrm{i}\,\boldsymbol{q}\cdot\boldsymbol{R}} \sum_{\boldsymbol{k}} (\varepsilon_{\boldsymbol{k}+\boldsymbol{q}} - \varepsilon_{\boldsymbol{k}})\, c^{\dagger}_{\boldsymbol{k}+\boldsymbol{q}\sigma} c_{\boldsymbol{k}\sigma}. \tag{14.12}$$

Here $\boldsymbol{r}_a$ and $u(\boldsymbol{q})$ are the impurity position and its potential, $M(\boldsymbol{q},\lambda)$ is the matrix element for the electron coupling with a phonon of wavevector $\boldsymbol{q}$ in branch λ having energy $\Omega_{\boldsymbol{q}\lambda}$, and $\phi_{\boldsymbol{q}\lambda} = b_{\boldsymbol{q}\lambda} + b^{\dagger}_{-\boldsymbol{q}\lambda}$ stands for the phonon field operator. We note that, in addition to impurity, phonon and relative-electron variables, the center-of-mass coordinate $\boldsymbol{R}$ is also involved in these operator equations.

To deal with the center-of-mass coordinate $\boldsymbol{R}$ in these equations, we notice that the center-of-mass velocity $\boldsymbol{V}$ is the time derivative of its coordinate: $\boldsymbol{V} \equiv \dot{\boldsymbol{R}}$. Therefore, from the Heisenberg equation of motion and the total Hamiltonian (14.4),

$$\boldsymbol{V} = -\mathrm{i}\,[\boldsymbol{R}, H] = \frac{\boldsymbol{P}}{Nm} - \boldsymbol{v}_\omega \cos(\omega t) \tag{14.13}$$

containing a rapid oscillatory term directly induced by the high-frequency electric field.

Based on the analyses in Chapter 1, in view of the enormous mass of the center of mass, we can treat its coordinate $\boldsymbol{R}$ and velocity $\boldsymbol{V}$ classically, and, by neglecting their small fluctuations we will, from now on, regard them as the time-dependent expectation (or average) values of the center-of-mass coordinate and velocity, denoting as $\boldsymbol{R}(t)$ and $\boldsymbol{V}(t)$. Obviously

$$\boldsymbol{R}(t) = \int_{t_0}^{t} \boldsymbol{V}(s)\, ds + \boldsymbol{R}(t_0). \tag{14.14}$$

Furthermore, when the frequency of the radiation field is high enough, the center of mass would not be able to follow its rapid oscillation, that in the gauge used, the statistical average of the center-of-mass momentum can be regarded as a slowly-varying quantity

$$\left\langle \frac{\boldsymbol{P}}{Nm} \right\rangle = \boldsymbol{v}_0, \tag{14.15}$$

which is identified as the average velocity of the center of mass, or the average drift velocity of the carrier system. Therefore we write

$$\boldsymbol{V}(t) = \boldsymbol{v}_0 - \boldsymbol{v}_\omega \cos(\omega t). \tag{14.16}$$

By this consideration, the scattering interaction $H_{\mathrm{I}t} \equiv H_{\mathrm{ei}} + H_{\mathrm{ep}}$ and all the operators in equations (14.11) and (14.12) involving the center-of-mass coordinate $\boldsymbol{R}$ are time-dependent operators in the relative-electron–phonon space. In the presence of a uniform high-frequency field, the relative-electron–phonon system turns out to behave exactly the same as that discussed in Chapter 1 for the case without the radiation field ($E_\omega = 0$), and thus we can proceed in exactly the same way as in Chapter 1. For instance, we can solve for the density matrix of the relative-electron–phonon system, $\hat{\rho}$, from the Liouville equation

$$\mathrm{i}\frac{d\hat{\rho}}{dt} = [H_{\mathrm{er}} + H_{\mathrm{ph}} + H_{\mathrm{I}t}, \hat{\rho}] \tag{14.17}$$

by starting from the initial ($t = -\infty$) state

$$\hat{\rho}|_{t=-\infty} = \hat{\rho}_0 = \frac{1}{Z} e^{-H_{\mathrm{er}}/T_{\mathrm{e}}} e^{-H_{\mathrm{ph}}/T}, \tag{14.18}$$

in which the phonon system is in equilibrium at the lattice temperature T and the relative electron system is in equilibrium at an electron temperature T_{e}. To the lowest order in $H_{\mathrm{I}t}$, the statistical average of a dynamical variable $\mathcal{O}$ can be written in the form

$$\langle \mathcal{O} \rangle = \langle \mathcal{O} \rangle_0 - \mathrm{i} \int_{-\infty}^{t} dt' \langle [H_{\mathrm{I}t'}(t'), \mathcal{O}(t)] \rangle_0, \tag{14.19}$$

where $\langle \cdots \rangle_0$ stands for the average with respect to the initial density matrix $\hat{\rho}_0$, and $\mathcal{O}(t)$ is defined as

$$\mathcal{O}(t) \equiv e^{\mathrm{i}(H_{\mathrm{er}}+H_{\mathrm{ph}})t} \mathcal{O} e^{-\mathrm{i}(H_{\mathrm{er}}+H_{\mathrm{ph}})t}. \tag{14.20}$$

14.1.2 Momentum- and energy-balance equations under a high-frequency electric field

The momentum-balance equation is obtained by taking the statistical average of the operator equation (14.11) and identifying $\langle \dot{P} \rangle$ on the left-hand-side of the equation as $Nm(d\bm{v}_0/dt)$. We have[1]

$$Nm\frac{d}{dt}\bm{v}_0 = Ne\bm{E}_0 + \bm{F}. \qquad (14.21)$$

It states that the increase rate of the average center-of-mass momentum equals the sum of the dc-field induced force and the frictional fore due to impurity and phonon scatterings, $\bm{F} = \bm{F}_i + \bm{F}_p$. The frictional forces $\bm{F}_i$ and $\bm{F}_p$, due to impurity and phonon scatterings, are given, respectively by

$$\bm{F}_i = -n_i \sum_{\bm{q}} \bm{q}\, |u(\bm{q})|^2 \int_{-\infty}^{t} dt'\, A(\bm{q},t,t') \langle [\rho_{\bm{q}}(t-t'), \rho_{-\bm{q}}] \rangle_0, \qquad (14.22)$$

$$\bm{F}_p = -\sum_{\bm{q},\lambda} \bm{q}\, |M(\bm{q},\lambda)|^2 \int_{-\infty}^{t} dt'\, A(\bm{q},t,t') \langle [\phi_{\bm{q}\lambda}(t-t')\rho_{\bm{q}}(t-t'), \phi_{-\bm{q}\lambda}\rho_{-\bm{q}}] \rangle_0. \qquad (14.23)$$

Here n_i is the impurity density, $\rho_{\bm{q}}(t)$ and $\phi_{\bm{q}\lambda}(t)$ are defined by (14.20),

$$A(\bm{q},t,t') \equiv \exp\left[i\bm{q}\cdot(\bm{R}(t) - \bm{R}(t'))\right] \qquad (14.24)$$

is the function defined in Sec. 1.5. Now according to (14.14) and (14.16), we have

$$A(\bm{q},t,t') \equiv \exp\left[i\bm{q}\cdot\int_{t'}^{t}\bm{V}(s)ds\right]$$
$$= e^{i\bm{q}\cdot\bm{v}_0(t-t')}e^{-i\bm{q}\cdot\bm{r}_\omega[\sin(\omega t) - \sin(\omega t')]}, \qquad (14.25)$$

in which

$$\bm{r}_\omega \equiv \frac{\bm{v}_\omega}{\omega} = \frac{e\bm{E}_\omega}{m\omega^2} \qquad (14.26)$$

is the electro-optic coupling parameter determined by the strength and frequency of the radiation electric field. Here we have made use of the fact that $\bm{v}_0$ is a slowing-varying quantity during the microscopic evolution

---

[1] In most cases in this chapter when we write the balance equations for a unit volume system the symbol N is naturally understood as the electron volume density to avoid using symbol n for this purpose.

process or the thermalization process. Using the general equality related to Bessel functions,

$$e^{-iz\sin x} = \sum_{n=-\infty}^{\infty} J_n(z) e^{-inx}, \tag{14.27}$$

we can rewrite the $A(\boldsymbol{q},t,t')$ function of (14.25) into a sum of two terms:

$$A(\boldsymbol{q},t,t') = \sum_{n=-\infty}^{\infty} J_n^2(\boldsymbol{q}\cdot\boldsymbol{r}_\omega) e^{i\boldsymbol{q}\cdot\boldsymbol{v}_0(t-t')} e^{-in\omega(t-t')} + \sum_{m\neq 0} e^{im\omega t}$$
$$\times \left[\sum_{n=-\infty}^{\infty} J_n(\boldsymbol{q}\cdot\boldsymbol{r}_\omega) J_{n-m}(\boldsymbol{q}\cdot\boldsymbol{r}_\omega) e^{i\boldsymbol{q}\cdot\boldsymbol{v}_0(t-t')} e^{-in\omega(t-t')} \right].$$
$$\tag{14.28}$$

The first term on the right-hand-side of (14.28) is a function of $(t-t')$ only. Thus, after the integration over t' in Eqs. (14.22) and (14.23), it yields contributions to $\boldsymbol{F}_i$ and $\boldsymbol{F}_p$ no longer dependent on t. The contribution of the second term appears to be rapidly oscillating at the fundamental frequency ω and its harmonics, since the integration over t' renders its inner part (inside the bracket) a finite constant value while leaving the outer oscillatory factor intact. If what one measures is an average over a time interval much longer than the period of the radiation field, the contribution of the second term is not detectable. Therefore we are left with

$$\boldsymbol{F} = n_\mathrm{i} \sum_{\boldsymbol{q}} \boldsymbol{q} |u(\boldsymbol{q})|^2 \sum_{n=-\infty}^{\infty} J_n^2(\boldsymbol{q}\cdot\boldsymbol{r}_\omega) \Pi_2(\boldsymbol{q},\omega_0 - n\omega)$$
$$+ \sum_{\boldsymbol{q},\lambda} \boldsymbol{q} |M(\boldsymbol{q},\lambda)|^2 \sum_{n=-\infty}^{\infty} J_n^2(\boldsymbol{q}\cdot\boldsymbol{r}_\omega) \Lambda_2^+(\boldsymbol{q},\lambda,\omega_0 - n\omega), \tag{14.29}$$

in which

$$\Lambda_2^+(\boldsymbol{q},\lambda,\omega_0 - n\omega) \equiv 2\Pi_2(\boldsymbol{q},\Omega_{\boldsymbol{q}\lambda} + \omega_0 - n\omega)$$
$$\times \left[n\left(\frac{\Omega_{\boldsymbol{q}\lambda}}{T}\right) - n\left(\frac{\Omega_{\boldsymbol{q}\lambda} + \omega_0 - n\omega}{T_\mathrm{e}}\right) \right]. \tag{14.30}$$

Here $\omega_0 \equiv \boldsymbol{q}\cdot\boldsymbol{v}_0$, $J_n(x)$ is the Bessel function of order n, $\Pi_2(\boldsymbol{q},\omega)$ is the imaginary part of the electron density correlation function at electron temperature T_e, and $n(x) \equiv 1/(e^x - 1)$ is the Bose function.

The energy-balance equation is obtained by taking the statistical average of the operator equation (14.12) and identifying $\langle \dot{H}_\mathrm{er} \rangle$ as the rate of change of the relative-electron energy U. By the same consideration as above we arrive at

$$\frac{d}{dt}U = -\boldsymbol{v}_0 \cdot \boldsymbol{F} - W + S_{\mathrm{p}}. \qquad (14.31)$$

Here, $-\boldsymbol{v}_0 \cdot \boldsymbol{F}$ is the work done by the center of mass on the relative-electron system,

$$W = \sum_{\boldsymbol{q},\lambda} \Omega_{\boldsymbol{q}\lambda} |M(\boldsymbol{q},\lambda)|^2 \sum_{n=-\infty}^{\infty} \mathrm{J}_n^2(\boldsymbol{q}\cdot\boldsymbol{r}_\omega) \Lambda_2^+(\boldsymbol{q},\lambda,\omega_0 - n\omega) \qquad (14.32)$$

is the energy-transfer rate from the electron system to the phonon system, and

$$\begin{aligned}S_{\mathrm{p}} = &\; n_{\mathrm{i}} \sum_{\boldsymbol{q}} |u(\boldsymbol{q})|^2 \sum_{n=-\infty}^{\infty} \mathrm{J}_n^2(\boldsymbol{q}\cdot\boldsymbol{r}_\omega)\, n\omega\, \Pi_2(\boldsymbol{q},\omega_0 - n\omega) \\ &+ \sum_{\boldsymbol{q},\lambda} |M(\boldsymbol{q},\lambda)|^2 \sum_{n=-\infty}^{\infty} \mathrm{J}_n^2(\boldsymbol{q}\cdot\boldsymbol{r}_\omega)\, n\omega\, \Lambda_2^+(\boldsymbol{q},\lambda,\omega_0 - n\omega)\end{aligned} \qquad (14.33)$$

is the net rate of the energy the electron system gains from the radiation field through multiphoton (absorption and emission) processes ($n = \pm 1, \pm 2, \cdots$) in associated with intraband transition of electrons. Such processes are possible only with the assistance of the impurity or phonon scattering (indirect photon absorption and emission).

Note that in writing $\boldsymbol{F}$, W and S_{p} in the above form we have somewhat mixed contributions from absorption and emission processes of photons. For instance, in Eq. (14.33) the part with negative (positive) values of the summation index n does not imply a pure emission (absorption) process. Instead, the combination of both n and $-n$ terms represents the total (net) contribution to S_{p} from the n-photon ($n \geqslant 1$) emission and absorption processes. We therefore write S_{p} as a sum of all orders of n-photon contributions:

$$S_{\mathrm{p}} = \sum_{n=1}^{\infty} S_{\mathrm{p}}^{(n)}, \qquad (14.34)$$

where $S_{\mathrm{p}}^{(n)}$ is the total contribution from terms having index n and $-n$ in Eq. (14.33).

The frictional force $\boldsymbol{F}$, energy-loss rate w and energy-gain rate S_{p} of electrons are functions of $\boldsymbol{v}_0$ and T_{e}, and contain Bessel functions $\mathrm{J}_n^2(\boldsymbol{q}\cdot\boldsymbol{r}_\omega)$ having an argument proportional to the electro-optic coupling parameter

$r_\omega = eE_\omega/(m\omega^2)$. The effect of a radiation field on carrier transport emerges, by changing the frictional force $\bm{F}$ and the electron energy-loss rate W, and by supplying an energy S_p to the electron system through multiphoton processes ($n = \pm 1, \pm 2, \cdots$). Even without absorption nor emission of real photons (virtual photon process) represented by $n = 0$ term, the frictional force and the energy-loss rate are still affected by the presence of a finite radiation field $E_\omega \neq 0$. In the case without a radiation field ($E_\omega = 0$), in view of $J_0(0) = 1$ and $J_n(0) = 0$ ($n \neq 0$) we have $S_\mathrm{p} = 0$, and $\bm{F}$ and W reduce to the corresponding expressions (2.4)–(2.6). Likewise, for radiation fields of very high frequency, since $J_n^2(\bm{q} \cdot \bm{r}_\omega) \propto \omega^{-4|n|}$, we also have a vanishing S_p and the same $\bm{F}$ and W as those without radiation. This indicates that a very high frequency radiation field has no influence on the intraband carrier transport. Far-infrared or THz field is the regime of the electromagnetic waves, which would strongly affect the transport behavior of carriers in semiconductors. The momentum- and energy-balance equations (14.21) and (14.31) and the expressions (14.29), (14.32) and (14.33) for $\bm{F}$, W and S_p, provide a convenient tool to analyze this effect.

The basic methodology in developing the above balance equation approach is summarized as follows. (1) The scheme is based on the separation of the center-of-mass motion from the relative degrees of freedom of electrons in the system: spatially uniform dc and high-frequency electric fields exert only on the center of mass, and the relative electrons do not directly experience the existence of either one of them. (2) The average center-of-mass velocity $\bm{v}_0$ and the electron temperature T_e are used as the fundamental parameters to describe the steady-state transport of the system, and the density matrix (14.18) is used as the initial state for the solution to the Liouville equation. Such a description requires the system to have a short thermalization time. Therefore, the balance equations derived here apply to systems with high carrier density. (3) In addition, another approximation (14.15) has been taken in deriving these equations, i.e. the statistical average of the center-of-mass momentum is assumed to be a slowly time-varying quantity and $\langle \bm{P}/Nm \rangle$ is identified as the average velocity $\bm{v}_0$ of the center of mass. In other words, in the chosen gauge (14.2) for the high-frequency electric field $\bm{E}_\omega \sin(\omega t)$, only an oscillating part with $\pi/2$ out of phase of the ac field exists in the center-of-mass velocity function $\bm{V}(t)$. It is just this approximation that enables us to obtain very concise momentum- and energy-balance equations, with multiphoton process contributed frictional force, energy-transfer rate to the lattice and energy absorption rate from the radiation field.

Such an approximation is valid for radiation fields of sufficiently high frequency. When the period of the radiation field is much shorter than the center-of-mass motion related relaxation time τ_m (i.e. the momentum relaxation time), the center of mass will not follow the rapidly oscillation of the driving field. But this is apparently not true if the frequency of the driving ac field becomes low. For a system having momentum relaxation time around several tenths of 10^{-12} s, the balance equation method developed in this section is expected to apply to the radiation field of frequency higher than 0.5 THz.

On the other hand, the time-dependent balance equation approach developed in Chapter 1 can be used to deal with the system response to a time-dependent electric field $\bm{E}(t)$ of the form (14.1), if its frequency $\nu = \omega/2\pi \ll \tau_{\text{th}}^{-1}$. For systems with τ_{th} of order of several tenths of 10^{-12} s, the applicable frequency is lower than $1 \sim 2$ THz.

14.2 Transport in Quasi-Two-Dimensional Semiconductors Irradiated by a THz Field

14.2.1 *Balance equations for THz-driven quasi-2D systems*

It is straightforward to extend the discussion in the last subsection to quasi-two-dimensional systems, for which the single electron state is described by a 2D wavevector $\bm{k}_\| = (k_x, k_y)$ and a subband index s. We consider a unit area system containing N_s electrons, with the eigenenergy and the wave function given by

$$\varepsilon_{s\bm{k}_\|} = \varepsilon_s + k_\|^2/2m, \tag{14.35}$$

$$\psi_{s\bm{k}_\|}(\bm{r}, z) = e^{i\bm{k}_\| \cdot \bm{r}} \zeta_s(z), \tag{14.36}$$

where m is the band effective mass of the material, $\bm{r} \equiv (x, y)$ stands for the 2D position vector, and $\zeta_s(z)$ is the envelope function of the sth subband. When both a uniform dc electric filed $\bm{E}_0$ and a uniform high-frequency electric filed $\bm{E}_\omega \sin(\omega t)$ simultaneously apply in the x–y plane, the carrier transport of the quasi-2D system can be analyzed by means of the 2D center-of-mass and relative electron variables (Sec. 2.5). Taking the average center-of-mass velocity $\bm{v}_0$ and the electron temperature T_e as the fundamental parameters to describe the transport state of the system, we can derive the momentum- and energy-balance equations for quasi-2D semiconductors, formally the same as (14.21) and (14.31). In the steady state, the balance equations are written as

$$0 = N_s e \boldsymbol{E}_0 + \boldsymbol{F}, \qquad (14.37)$$
$$0 = -\boldsymbol{v}_0 \cdot \boldsymbol{F} - W + S_\mathrm{p}. \qquad (14.38)$$

Here, the frictional force due to impurity and phonon scatterings, $\boldsymbol{F}$, the energy-transfer rate from the electron system to phonon system, W, and the rate of the electron energy-gain from the radiation field, S_p, are respectively given by

$$\boldsymbol{F} = \sum_{s',s,\boldsymbol{q}_\|} |U_{s's}(\boldsymbol{q}_\|)|^2 \boldsymbol{q}_\| \sum_{n=-\infty}^{\infty} \mathrm{J}_n^2(\boldsymbol{q}_\| \cdot \boldsymbol{r}_\omega) \Pi_2(s',s,\boldsymbol{q}_\|,\omega_0 - n\omega)$$
$$+ \sum_{s',s,\boldsymbol{q},\lambda} |M_{s's}(\boldsymbol{q},\lambda)|^2 \boldsymbol{q}_\| \sum_{n=-\infty}^{\infty} \mathrm{J}_n^2(\boldsymbol{q}_\| \cdot \boldsymbol{r}_\omega) \Lambda_2^+(s',s,\boldsymbol{q}_\|,\lambda,\omega_0 - n\omega),$$
$$(14.39)$$

$$W = \sum_{s',s,\boldsymbol{q},\lambda} |M_{s's}(\boldsymbol{q},\lambda)|^2 \Omega_{q\lambda} \sum_{n=-\infty}^{\infty} \mathrm{J}_n^2(\boldsymbol{q}_\| \cdot \boldsymbol{r}_\omega) \Lambda_2^+(s',s,\boldsymbol{q}_\|,\lambda,\omega_0 - n\omega),$$
$$(14.40)$$

$$S_\mathrm{p} = \sum_{s',s,\boldsymbol{q}_\|} |U_{s's}(\boldsymbol{q}_\|)|^2 \sum_{n=-\infty}^{\infty} \mathrm{J}_n^2(\boldsymbol{q}_\| \cdot \boldsymbol{r}_\omega) n\omega \, \Pi_2(s',s,\boldsymbol{q}_\|,\omega_0 - n\omega)$$
$$+ \sum_{s',s,\boldsymbol{q},\lambda} |M_{s's}(\boldsymbol{q},\lambda)|^2 \sum_{n=-\infty}^{\infty} \mathrm{J}_n^2(\boldsymbol{q}_\| \cdot \boldsymbol{r}_\omega) n\omega \, \Lambda_2^+(s',s,\boldsymbol{q}_\|,\lambda,\omega_0 - n\omega).$$
$$(14.41)$$

In this, the effective electron–impurity potential $U_{s's}(\boldsymbol{q}_\|)$, the effective electron–phonon matrix element $M_{s's}(\boldsymbol{q},\lambda)$ and the imaginary part of the electron density correlation function, $\Pi_2(s',s,\boldsymbol{q}_\|,\Omega)$, related to subbands s' and s, are those given explicitly in Sec. 2.5, and the function

$$\Lambda_2^+(s',s,\boldsymbol{q}_\|,\lambda,\omega_0 - n\omega) \equiv 2\Pi_2(s',s,\boldsymbol{q}_\|,\Omega_{q\lambda} + \omega_0 - n\omega)$$
$$\times \left[n\left(\frac{\Omega_{q\lambda}}{T}\right) - n\left(\frac{\Omega_{q\lambda} + \omega_0 - n\omega}{T_\mathrm{e}}\right)\right]. \quad (14.42)$$

Here a 3D phonon model is assumed, with $\Omega_{q\lambda}$ the energy of a phonon of mode $q\lambda$, and the electro-optic coupling parameter

$$r_\omega \equiv \frac{eE_\omega}{m\omega^2} \tag{14.43}$$

is related to the strength and frequency of the radiation field.

14.2.2 Transport of a GaAs-based quantum well irradiated by THz fields

Balance equations (14.37) and (14.38) can be directly used to study the transport of THz-driven quasi-2D semiconductors. Given E_0, E_ω and ω, the steady-state average velocity v_0, electron temperature T_e and other transport quantities are completely determined by these two equations.

Numerical calculations (Lei, 1998a; 1998b) are performed for a GaAs-based quantum well having well width $a = 12.5\,\text{nm}$, electron sheet density $N_s = 5.5 \times 10^{15}\,\text{m}^{-2}$ and low-temperature (4.2 K) linear mobility $\mu_0 = 31\,\text{m}^2\text{V}^{-1}\text{s}^{-1}$ at lattice temperature $T = 10\,\text{K}$. The elastic scattering is assumed due to remote charged impurities located at a distance 40 nm from the center plane of the well. Both the electron–polar-optic-phonon scattering (via the Fröhlich coupling) and the electron–acoustic-phonon scattering (via the deformation potential and the piezoelectric couplings) are included, and phonons are assumed the same as those of a bulk GaAs and remain in equilibrium at lattice temperature T during the transport process. We consider the role of two lowest subbands ($s = 0, 1$). The energy separation between the bottoms of the zeroth and the first subbands, $\varepsilon_{10} \equiv \varepsilon_1 - \varepsilon_0$, is taken to be 69 meV. This ε_{10} value is estimated from a well depth of 280 meV according to the band offset of $GaAs/Al_{0.3}Ga_{0.7}As$. On the other hand, since the transport properties are not sensitive to the exact form of the wave functions, we use the corresponding subband functions of an infinitely deep potential well having the same well width to calculate the related matrix elements for simplicity. The other material parameters are taken to be typical values of bulk GaAs as those listed in Sec. 2.6.

The effect of a radiation field shows up through the electro-optic coupling parameter $r_\omega \equiv eE_\omega/(m\omega^2)$ entering in the argument of Bessel function factors $J_n^2(q_\parallel \cdot r_\omega)$ inside the summation over all multiphoton processes ($n = \pm 1, \pm 2, \cdots$). Apparently, larger $|n|$ processes have to be included in the calculation for a radiation field of larger amplitude E_ω and/or lower frequency ω.

To determine the ohmic mobility of a quasi-2D system subjected to an intense high-frequency radiation, we only need to consider the small v_0 limit of Eqs. (14.37) and (14.38). In the limit of $v_0 \to 0$, the energy balance equation (14.38) reduces to (E_ω is assumed along the x direction)

$$0 = S_{\mathrm{p}} - W$$
$$= \sum_{s',s,\mathbf{q}_\parallel} |U_{s's}(\mathbf{q}_\parallel)|^2 \sum_{n=-\infty}^{\infty} J_n^2(q_x r_\omega)\, n\omega\, \Pi_2(s',s,\mathbf{q}_\parallel,-n\omega)$$
$$+ \sum_{s',s,\mathbf{q},\lambda} |M_{s's}(\mathbf{q},\lambda)|^2 \sum_{n=-\infty}^{\infty} J_n^2(q_x r_\omega)\, (n\omega - \Omega_{\mathbf{q}\lambda})\, \Lambda_2^+(s',s,\mathbf{q}_\parallel,\lambda,-n\omega).$$
(14.44)

This equation determines the electron temperature T_{e}, which may be greatly different from the lattice temperature T when the system is irradiated by an intense THz field even in the vanishing dc field limit. Then, with this T_{e} the force balance equation (14.37) directly gives the resistivity, which depends on the relative directions of $\mathbf{E}_0$ and $\mathbf{E}_\omega$. In the case of $\mathbf{E}_0 \| \mathbf{E}_\omega$ or $\mathbf{E}_0 \perp \mathbf{E}_\omega$, v_0 lies in the same direction as $\mathbf{E}_0$, and the inverse mobility, defined by $1/\mu \equiv E_0/v_0$, is obtained as

$$\frac{1}{\mu} = \frac{1}{\mu_{\mathrm{i}}} + \frac{1}{\mu_{\mathrm{p}}}, \tag{14.45}$$

$$\frac{1}{\mu_{\mathrm{i}}} = \frac{1}{N_{\mathrm{s}} e} \sum_{s',s,\mathbf{q}_\parallel} q_\alpha^2 |U_{s's}(\mathbf{q}_\parallel)|^2 \sum_{n=-\infty}^{\infty} J_n^2(q_x r_\omega) \left[\frac{\partial}{\partial \Omega} \Pi_2(s',s,\mathbf{q}_\parallel,\Omega)\right]_{\Omega=n\omega}, \tag{14.46}$$

$$\frac{1}{\mu_{\mathrm{p}}} = \frac{2}{N_{\mathrm{s}} e} \sum_{s',s,\mathbf{q},\lambda} q_\alpha^2 |M_{s's}(\mathbf{q},\lambda)|^2 \sum_{n=-\infty}^{\infty} J_n^2(q_x r_\omega)$$
$$\times \left\{ \Pi_2(s',s,\mathbf{q}_\parallel,\Omega_{\mathbf{q}\lambda}-n\omega) \left[\frac{1}{T_{\mathrm{e}}} n'\left(\frac{\Omega_{\mathbf{q}\lambda}}{T_{\mathrm{e}}}\right)\right] \right.$$
$$\left. + \left[n\left(\frac{\Omega_{\mathbf{q}\lambda}-n\omega}{T_{\mathrm{e}}}\right) - n\left(\frac{\Omega_{\mathbf{q}\lambda}}{T}\right)\right] \left[\frac{\partial}{\partial \Omega} \Pi_2(s',s,\mathbf{q}_\parallel,\Omega)\right]_{\Omega=\Omega_{\mathbf{q}\lambda}-n\omega} \right\}, \tag{14.47}$$

Here $\alpha = x$ or y for the parallel ($\mathbf{E}_0 \| \mathbf{E}_\omega$) or perpendicular ($\mathbf{E}_0 \perp \mathbf{E}_\omega$) configuration.

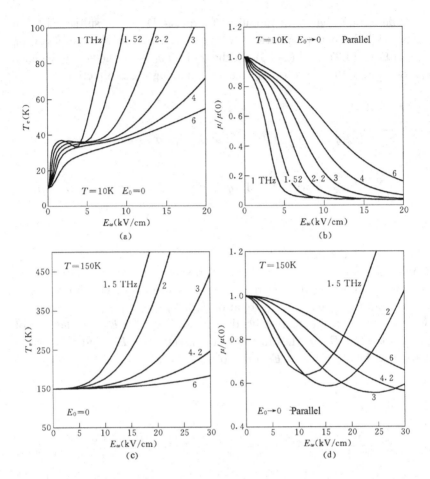

Fig. 14.1 The electron temperature T_e and the normalized mobility $\mu/\mu(0)$ [$\mu(0)$ is the mobility without radiation] in the small dc field limit $E_0 \sim 0$ are shown as functions of the amplitude E_ω of radiation fields having different frequencies respectively at lattice temperature $T = 10$ and 150 K. After Lei (1998b).

The calculated electron temperature T_e from Eq. (14.44) and the normalized mobility $\mu/\mu(0)$ [$\mu(0)$ is the mobility in the absence of radiation] from (14.45)–(14.47) in the small $\boldsymbol{v}_0$ limit in the parallel configuration ($\boldsymbol{E}_0 \parallel \boldsymbol{E}_\omega$) are shown as functions of the amplitude E_ω of radiation fields of different frequencies in Fig. 14.1(a) and (b) at lattice temperature $T = 10$ K and in Fig. 14.1(c) and (d) at lattice temperature $T = 150$ K.

We see that, at $T = 10$ K and $T = 150$ K, the electron temperature is always higher than the lattice temperature, and, except in the case of

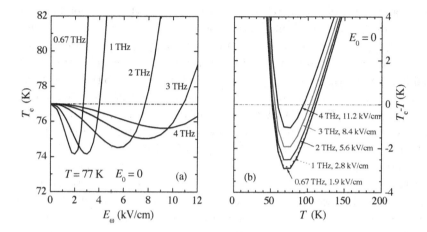

Fig. 14.2 The electron temperature T_e versus the amplitude E_ω of radiation fields having frequency $\omega/2\pi = 0.67, 1, 2, 3$ and 4 THz at lattice temperature $T = 77$ K (a), and the lattice-temperature dependence of $\Delta T \equiv T_e - T$ induced by five THz fields of different frequencies and amplitudes (b). After Lei and Liu (2003b).

$T = 10$ K and lower frequencies ($1 \sim 3$ THz) where the electron temperature exhibits a slight minimum around $E_\omega \sim 5$ kV/cm, T_e generally increases with increasing E_ω. However, within a lattice temperature range around 72 K the electron temperature T_e may descend with increasing radiation intensity from the beginning and reaches a minimum which is about a few degrees lower than the lattice temperature T, then rises with further increase of the radiation field, as seen in Fig. 14.2. This is the electron cooling induced by the irradiation of a high-frequency field (Lei, 1998b; Lei and Liu, 2003b).

The effect of the radiation field on the mobility looks different. At low lattice temperature $T = 10$ K, $\mu/\mu(0)$ monotonically decreases with increasing E_ω at a fixed frequency. Lower frequency generally has stronger effect in suppressing μ than higher frequency. μ can be suppressed down to about $0.05\,\mu(0)$ by the radiation field. At $T = 150$ K the $\mu/\mu(0)$-vs-E_ω curve for fixed lower frequencies (1.5, 2 and 3 THz) first descends down with E_ω increasing from zero, then reaches a minimum before rising with further increasing E_ω. For fixed radiation field amplitude E_ω, the effects are enhanced with decreasing frequency. The above features can be understood from the role of a radiation field embedded in the argument of Bessel function factors $J_n^2(\boldsymbol{q}_\parallel \cdot \boldsymbol{r}_\omega)$ in the expressions of $\boldsymbol{F}$, W and S_p. With the increase of the electro-optic coupling parameter $\boldsymbol{r}_\omega \equiv e\boldsymbol{E}_\omega/(m\omega^2)$, i.e. the increase of the amplitude or the decrease of the frequency of the radiation field, the contributions from the virtual-photon process ($n = 0$) are weakened, while the multiphoton processes ($|n| \geqslant 1$) of increasing order $|n|$ become more important in determining the carrier transport. Since contributions to $\boldsymbol{F}$ and W from multiphoton process are generally small due to the cancellation between the n and $-n$ terms, an increase in E_ω (or a decrease in ω) leads to a decrease in $\boldsymbol{F}$ and W if the electron temperature keeps unchanged. On the other hand, the increasing order of multiphoton processes greatly enhances S_p, the rate of the energy supplied from the radiation field. This, together with the decreased W, results in the rise of the electron temperature with growing E_ω, which, in turn, enhances the frictional force, i.e. reduces the mobility. This electron-temperature-rising effect is more remarkable at lower lattice temperature than at higher lattice temperature because the main electron energy-loss mechanism (polar-optic-phonon scattering) is much less effective at low temperature. It is the interplay of these two effects that gives rise to the major numerical features: dc mobility decreases at low lattice temperature ($T = 10\,\mathrm{K}$), increases at room temperature (not shown, see Lei, 1998b), while exhibits minima at lattice temperatures in between, with increasing strength of the radiation field.

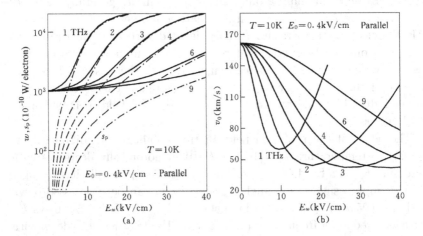

Fig. 14.3 The energy-loss rate w, energy-gain rate s_p and drift velocity v_0 versus the amplitude E_ω of radiation fields of different frequencies at lattice temperature $T = 10\,\mathrm{K}$ under a finite bias field $E_0 = 0.4\,\mathrm{kV/cm}$ in the parallel configuration. After Lei (1998a).

In the case of a finite dc bias electric field $\boldsymbol{E}_0$ the effect of a strong THz radiation field on the transport properties of a quasi-2D system can be calculated directly from Eqs. (14.37) and (14.38). Fig. 14.3(a) and (b) show the calculated (per carrier) electron energy-loss rate to the lattice, $w = W/N_s$, the electron energy-gain rate from the radiation field, $s_p = S_p/N_s$, and the average drift velocity v_0 as functions of the amplitude E_ω of the radiation field of several different frequencies $\nu = \omega/2\pi = 1, 2, 3, 4, 6$ and $9\,\text{THz}$, in the case of a finite bias dc field $E_0 = 0.4\,\text{kV/cm}$ in the parallel configuration ($\bm{E}_0 \parallel \bm{E}_\omega$) at lattice temperature $T = 10\,\text{K}$. We see that, for a given frequency, at this strength of dc bias field the average drift velocity v_0 first descends with rising E_ω and reaches a minimum at certain strength of the radiation field, then rebounds with further increase in E_ω.

14.3 Miniband Transport in Semiconductor Superlattices Irradiated by THz Fields

14.3.1 *Balance equations for THz-driven electron transport in an arbitrary energy band*

We consider a system consisting of N electrons moving in a single energy band, with the electron state described by a lattice wave vector $\bm{k}$ in the Brillouin zone having wave function $\psi_{\bm{k}}(\bm{r})$ and energy $\varepsilon(\bm{k})$. These electrons are coupled with phonons and scattered by randomly distributed impurities. Under the influence of a uniform dc electric field $\bm{E}_0$ and a high-frequency electric field $\bm{E}_\omega \sin(\omega t)$, the total Hamiltonian of the electron–phonon system is written in the effective Hamiltonian form as

$$H = H_\text{e}(t) + H_\text{E} + H_\text{I} + H_\text{ph}, \tag{14.48}$$

where H_ph stands for the phonon Hamiltonian,

$$H_\text{E} = -e\bm{E}_0 \cdot \sum_j \bm{r}_j, \tag{14.49}$$

$$H_\text{e}(t) = \sum_j \varepsilon\left(\bm{p}_j - e\bm{A}(t)\right) \tag{14.50}$$

are respectively the potential energy and kinetic energy parts of the electron Hamiltonian, $\bm{p}_j$ and $\bm{r}_j$ represent the momentum and coordinate of the jth electron, and

$$\bm{A}(t) = (\bm{E}_\omega/\omega)\cos(\omega t) \tag{14.51}$$

is the vector potential of the high-frequency electric field. The expressions for the electron–impurity and electron–phonon scattering, $H_{\rm I} = H_{\rm ei} + H_{\rm ep}$, are given by (11.6) and (11.7), involving the electron density operator

$$\rho_q = \sum_j e^{i q \cdot r_j}, \qquad (14.52)$$

which can be written in the second quantization representation of $\psi_k(r)$ basis as

$$\rho_q = \sum_k \rho_{kq} = \sum_{k,\sigma} g(k,q) c^\dagger_{k+q\sigma} c_{k\sigma}, \qquad (14.53)$$

$g(k,q)$ being the form factor (11.22) related to the wave function $\psi_k(r)$. The center-of-mass position of the electron system is defined by

$$R = \frac{1}{N} \sum_j r_j, \qquad (14.54)$$

and the center-of-mass velocity is thus

$$V \equiv \dot{R} = -i\,[R,H] = \frac{1}{N} \sum_j v(p_j - eA(t)) = \frac{1}{N} \sum_{k,\sigma} v(k - eA(t)) c^\dagger_{k\sigma} c_{k\sigma}. \qquad (14.55)$$

Here $v(k) \equiv \nabla \varepsilon(k)$ is the velocity function of the energy band. Writing the center-of-mass velocity V as the sum of a slowly varying part V_0 and a rapid oscillating part V_A due to the high-frequency field,

$$V = V_0 + V_A, \qquad (14.56)$$

with V_0 defined as

$$V_0 = \frac{1}{N} \sum_{k,\sigma} v(k)\, c^\dagger_{k\sigma} c_{k\sigma}, \qquad (14.57)$$

we consider the rate of change of V_0,

$$\frac{dV_0}{dt} = -i\,[V_0, H]$$

$$= \frac{eE_0}{N} \cdot \sum_{k,\sigma} \nabla v(k)\, c^\dagger_{k\sigma} c_{k\sigma} - \frac{i}{N} \sum_{k,q,a} u(q)\, e^{i q \cdot r_a} [v(k+q) - v(k)]\, \rho_{kq}$$

$$\quad - \frac{i}{N} \sum_{k,q,\lambda} M(q,\lambda) \phi_{q\lambda}[v(k+q) - v(k)]\, \rho_{kq}, \qquad (14.58)$$

and the rate of change of an electron energy defined by

$$H_{e0} \equiv \sum_{k,\sigma} \varepsilon(k)\, c^\dagger_{k\sigma} c_{k\sigma}, \tag{14.59}$$

$$\begin{aligned}\frac{dH_{e0}}{dt} &= -\mathrm{i}\,[H_{e0}, H] \\ &= NeE_0 \cdot V_0 - \mathrm{i} \sum_{k,q,a} u(q)\, \mathrm{e}^{\mathrm{i}q\cdot r_a}[\varepsilon(k+q) - \varepsilon(k)]\, \rho_{kq} \\ &\quad - \mathrm{i} \sum_{k,q,\lambda} M(q,\lambda)\phi_{q\lambda}[\varepsilon(k+q) - \varepsilon(k)]\, \rho_{kq}. \end{aligned} \tag{14.60}$$

For the calculation of the statistical average of the above operator equations, we need the density of matrix of the transport state, which can be solved from the Liouville equation by starting from a parameterized initial state at time $t = -\infty$, in which the phonon system is in equilibrium at the lattice temperature T and the relative electron system is in equilibrium at an electron temperature T_e with a shifted lattice wavevector p_d:

$$\hat{\rho}|_{t=-\infty} = \hat{\rho}_0 = \frac{1}{Z} \mathrm{e}^{-H_{\mathrm{er}}/T_e} \mathrm{e}^{H_{\mathrm{ph}}/T}, \tag{14.61}$$

in which

$$H_{\mathrm{er}} = \sum_{k,\sigma} \bar{\varepsilon}(k)\, c^\dagger_{k\sigma} c_{k\sigma} = \sum_{k,\sigma} \varepsilon(k - p_d)\, c^\dagger_{k\sigma} c_{k\sigma}. \tag{14.62}$$

The balance equation method treats the coupling part $H_{\mathrm{I}} = H_{\mathrm{ei}} + H_{\mathrm{ep}}$ as a perturbation. In the presence of the radiation field, the zero-order Hamiltonian of the electron-phonon system $H_0(t) = H_{\mathrm{e}}(t) + H_{\mathrm{ph}}$ is time-dependent. Nevertheless, the density matrix can still be obtained to the linear order in H_{I}, and the statistical average of a dynamical variable $\mathcal{O}$ can be written in the form

$$\langle \mathcal{O} \rangle = \langle \mathcal{O} \rangle_0 - \mathrm{i} \int_{-\infty}^{t} dt'\, \langle [H_{\mathrm{I}}(t'), \mathcal{O}(t)] \rangle_0, \tag{14.63}$$

where $\langle \cdots \rangle_0$ stands for the average with respect to the initial density matrix $\hat{\rho}_0$, and the interaction picture of an operator $\mathcal{O}$ (in the schröndinger picture) is defined as

$$\mathcal{O}(t) \equiv U_0^\dagger(t)\mathcal{O} U_0(t). \tag{14.64}$$

Here $U_0(t)$ is the evolution operator, satisfying the equation

$$i\frac{d}{dt}U_0(t) = H_0(t)U_0(t) \tag{14.65}$$

and the initial condition $U_0(0) = 1$. It can be written as

$$U_0(t) = \exp\left[-i\sum_{k,\sigma} S_k(t)\, c_{k\sigma}^\dagger c_{k\sigma} - i(H_E + H_{ph})\,t\right], \tag{14.66}$$

where

$$S_k(t) = \int_0^t dt'\, \varepsilon(k - eA(t')). \tag{14.67}$$

Note that the interaction picture of electron density operator ρ_q is

$$\rho_q(t) = \sum_{k,\sigma} g(k,q)\, e^{i[S_{k+q}(t) - S_k(t)]} c_{k+q\sigma}^\dagger c_{k\sigma}, \tag{14.68}$$

and, by neglecting the intracollisional field effect as analyzed in Sec. 11.2, we can take the interaction picture of operator ρ_{kq} as

$$\rho_{kq}(t) = g(k,q)\, e^{i[S_{k+q}(t) - S_k(t)]} c_{k+q\sigma}^\dagger c_{k\sigma}. \tag{14.69}$$

Furthermore, we make the following approximation in the exponential factor in the above expression,

$$S_{k+q}(t) - S_k(t) \approx [\varepsilon(k+q) - \varepsilon(k)]t - [v(k+q) - v(k)] \cdot e_\omega \sin(\omega t), \tag{14.70}$$

in which

$$e_\omega \equiv eE_\omega/\omega^2. \tag{14.71}$$

This approximation is valid if one of the following three cases applies: (i) high frequency but not too strong radiation field; or (ii) narrow energy band (such as superlattice miniband) that the maximum value of the velocity function $v(k)$ is not high; or (iii) weak nonparabolicity of the energy band. With the approximation (14.70) we have the following exponential factors in the expressions for the friction force and energy transfer rate in carrying out the statistical averages of operator equations (14.58) and (14.60):

$$\int_{-\infty}^t dt'\, e^{i[\varepsilon(k+q)-\varepsilon(k)](t-t')} e^{-i[v(k+q)-v(k)]\cdot e_\omega[\sin(\omega t) - \sin(\omega t')]}. \tag{14.72}$$

Further treatments proceed in a way similar to that in Sec. 14.1.2: using the equality of the Bessel functions

$$e^{-iz\sin x} = \sum_{n=-\infty}^{\infty} J_n(z) e^{-inx},$$

and taking the average over a time interval much longer than the period of the high-frequency field, we arrive at the force and energy balance equations of the form:

$$e\,\boldsymbol{E}_0 \cdot \mathcal{K} + \boldsymbol{a} = 0, \tag{14.73}$$

$$e\,\boldsymbol{E}_0 \cdot \boldsymbol{v}_0 - w + s_\mathrm{p} = 0. \tag{14.74}$$

In this, we have identified

$$\boldsymbol{v}_0 = \langle \boldsymbol{V}_0 \rangle = \frac{2}{N} \sum_{\boldsymbol{k}} \boldsymbol{v}(\boldsymbol{k}) f(\bar{\varepsilon}(\boldsymbol{k}), T_\mathrm{e}) \tag{14.75}$$

as the average velocity of the system, and

$$\mathcal{K} = \langle \hat{\mathcal{K}} \rangle = \frac{2}{N} \sum_{\boldsymbol{k}} \nabla\nabla \varepsilon(\boldsymbol{k}) f(\bar{\varepsilon}(\boldsymbol{k}), T_\mathrm{e}) \tag{14.76}$$

is the average inverse effective mass tensor, $\boldsymbol{a}$, w and s_p are respectively the frictional acceleration, the average per-electron energy-loss rate and the average energy-gain rate per electron from the radiation field:

$$\boldsymbol{a} = \frac{2\pi n_\mathrm{i}}{N} \sum_{\boldsymbol{k},\boldsymbol{q}} |u(\boldsymbol{q})|^2 \sum_{n=-\infty}^{\infty} J_n^2(\xi) \left[\boldsymbol{v}(\boldsymbol{k}+\boldsymbol{q}) - \boldsymbol{v}(\boldsymbol{k})\right] \Pi_2(\boldsymbol{k},\boldsymbol{q},-n\omega)$$

$$+ \frac{2\pi}{N} \sum_{\boldsymbol{k},\boldsymbol{q},\lambda} |M(\boldsymbol{q},\lambda)|^2 \sum_{n=-\infty}^{\infty} J_n^2(\xi) \left[\boldsymbol{v}(\boldsymbol{k}+\boldsymbol{q}) - \boldsymbol{v}(\boldsymbol{k})\right] \Lambda_2^+(\boldsymbol{k},\boldsymbol{q},\lambda,-n\omega), \tag{14.77}$$

$$w = \frac{2\pi}{N} \sum_{\boldsymbol{k},\boldsymbol{q},\lambda} |M(\boldsymbol{q},\lambda)|^2 \sum_{n=-\infty}^{\infty} J_n^2(\xi) \Omega_{\boldsymbol{q}\lambda} \Lambda_2^+(\boldsymbol{k},\boldsymbol{q},\lambda,-n\omega), \tag{14.78}$$

$$s_\mathrm{p} = \frac{2\pi n_\mathrm{i}}{N} \sum_{\boldsymbol{k},\boldsymbol{q}} |u(\boldsymbol{q})|^2 \sum_{n=-\infty}^{\infty} J_n^2(\xi)\, n\omega\, \Pi_2(\boldsymbol{k},\boldsymbol{q},-n\omega)$$

$$+ \frac{2\pi}{N} \sum_{\boldsymbol{k},\boldsymbol{q},\lambda} |M(\boldsymbol{q},\lambda)|^2 \sum_{n=-\infty}^{\infty} J_n^2(\xi)\, n\omega\, \Lambda_2^+(\boldsymbol{k},\boldsymbol{q},\lambda,-n\omega). \tag{14.79}$$

In the above expressions,

$$\xi \equiv [v(k+q) - v(k)] \cdot e_\omega, \qquad (14.80)$$

$$\Pi_2(k, q, -n\omega) \equiv |g(k,q)|^2 \left[f(\bar{\varepsilon}(k), T_\mathrm{e}) - f(\bar{\varepsilon}(k+q), T_\mathrm{e}) \right]$$
$$\times \delta(\varepsilon(k+q) - \varepsilon(k) - n\omega), \qquad (14.81)$$

$$\Lambda_2^+(k, q, \lambda, -n\omega) \equiv 2\Pi_2(k, q, \Omega_{q\lambda} - n\omega)$$
$$\times \left[n\left(\frac{\Omega_{q\lambda}}{T}\right) - n\left(\frac{\bar{\varepsilon}(k) - \bar{\varepsilon}(k+q)}{T_\mathrm{e}}\right) \right]. \qquad (14.82)$$

Here,

$$\bar{\varepsilon}(k) \equiv \varepsilon(k - p_d),$$

$$f(\varepsilon(k), T_\mathrm{e}) = \{\exp[(\varepsilon(k) - \mu)/T_\mathrm{e}] + 1\}^{-1}$$

is the Fermi distribution function at electron temperature T_e, and μ is the chemical potential determined by the total particle number (density):

$$N = 2 \sum_k f(\varepsilon(k), T_\mathrm{e}). \qquad (14.83)$$

14.3.2 Superlattice miniband transport subject to a strong THz field

Miniband transport in semiconductor superlattices driven by a THz electric field can be conveniently studied directly using the time-independent balance equations (14.73) and (14.74). Consider a GaAs-based planar superlattice in which electrons travel along its growth axis (the z-direction) through the (lowest) miniband formed by periodically spaced potential wells and barriers of finite heights. The electron energy dispersion can be written as the sum of the transverse energy $\varepsilon_{k_\parallel} = k_\parallel^2/2m$ [$k_\parallel \equiv (k_x, k_y)$] and a tight-binding-type miniband energy $\varepsilon(k_z)$ related to the longitudinal motion:

$$\varepsilon(k_\parallel, k_z) = k_\parallel^2/2m + \varepsilon(k_z), \qquad (14.84)$$

$$\varepsilon(k_z) = \frac{\Delta}{2} \left[1 - \cos(k_z d) \right], \qquad (14.85)$$

where d is the superlattice period, $-\pi/d < k_z \leqslant \pi/d$, and Δ is the miniband width. In view of the axial symmetry of the system, when the dc electric field E_0 and the sinusoidal high-frequency field E_ω both are polarized along the superlattice growth axis, the carrier drift motion, i.e. $\bm{p}_d$ and $\bm{v}_0$, is in the z direction.

The numerical method for solving balance equations (14.73) and (14.74) in the presence of a THz radiation is similar to that for solving Eqs. (12.69) and (12.70), except that an ac-field related Bessel-function factor shows up and an additional summation over multiphoton processes is needed for calculating $\bm{a}$, w and s_p.

An important effect of the irradiation of a high-frequency field on superlattice miniband transport is the reduction of the carrier average drift velocity. Fig. 14.4 shows this effect of a radiation field of 1 THz at lattice temperature $T = 77\,\mathrm{K}$ on a GaAs-based superlattice having period $d = 4.8\,\mathrm{nm}$, miniband width $\Delta = 56.8\,\mathrm{meV}$, carrier sheet density $N_\mathrm{s} = 2.0 \times 10^{15}\,\mathrm{m}^{-2}$ per layer ($N = N_\mathrm{s}/d$), and low-temperature linear dc mobility (in the absence of the radiation field) $\mu_0 = 10\,\mathrm{m^2V^{-1}s^{-1}}$. For a fixed strength of the dc field E_0, the average drift velocity $v_0(E_\omega)$ normalized by its value in the absence of the THz field, $v_0(0)$, decreases with increasing strength of the radiation field. This dc current suppression is stronger at lower dc field than at higher dc field. In the limit of $E_0 \to 0$, the dc current at $E_\omega = 20\,\mathrm{kV/cm}$ appears to be less than 0.1 of its zero E_ω value.

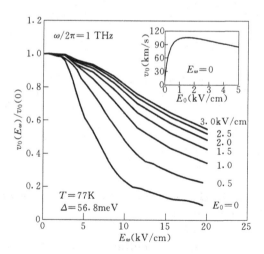

Fig. 14.4 Average drift velocity $v_0(E_\omega)$ normalized by its value at $E_\omega = 0$, $v_0(0)$, is plotted as a function of the amplitude E_ω of 1 THz radiation field subject to different bias dc fields $E_0 = 0, 0.5, 1.0, 1.5, 2.0, 2.5$ and $3.0\,\mathrm{kV/cm}$ at lattice temperature $T = 77\,\mathrm{K}$ for a GaAs-based superlattice described in the text. The inset show the v_0–E_0 curve of the system in the absence of the radiation field. From Lei and Cui (1998).

Fig. 14.5 plots the average drift velocity v_0 versus the dc field E_0 for another GaAs superlattice at lattice temperature $T = 300\,\text{K}$ under the influence of a 3 THz radiation field of different amplitudes $E_\omega = 0, 20, 30, 50$ and $80\,\text{kV/cm}$. Under the influence of an enhanced radiation field the v_0–E_0 curve, which exhibits negative differential mobility, shifts downward, in agreement with the experimental findings (Schomburg et al, 1996; Winnerl et al, 1997).

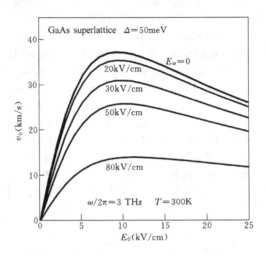

Fig. 14.5 Average drift velocity v_0 versus dc field E_0 under the influence of 3 THz radiation fields of amplitudes $E_\omega = 0, 20, 30, 50$ and $80\,\text{kV/cm}$ in a GaAs-based superlattice having period $d = 4.8\,\text{nm}$, miniband width $\Delta = 50\,\text{meV}$, carrier sheet density $N_\text{s} = 0.39 \times 10^{15}\,\text{m}^{-2}$, and low-temperature linear dc mobility (in the absence of the radiation field) $\mu_0 = 0.15\,\text{m}^2\text{V}^{-1}\text{s}^{-1}$. The lattice temperature $T = 300\,\text{K}$. From Lei and Cui (1998).

14.4 Nonlinear Free-Carrier Absorption of Intense THz Radiation in Semiconductors

14.4.1 *Nonlinear absorption coefficient of a THz radiation in a bulk semiconductor*

The quantity S_p in the energy-balance equation (14.33) represents a part of the average energy of the radiation field $\boldsymbol{E}_\omega \sin(\omega t)$ which is absorbed in unit time by the carriers in a 3D bulk system of unit volume in the presence of a bias dc electric field E_0. We consider, in the semiconductor, a small cylindrical element having its axis parallel to the propagating direction of the radiation field and volume $\Delta\tau = \Delta S \Delta z$ (ΔS and Δz are respectively the area of the upper/lower bottom and the height of the cylinder). The (one-period) average rate of the electromagnetic radiation energy flowing into the small volume is $\frac{1}{2}v\kappa\epsilon_0 E_\omega^2 \Delta S$, in which $v = c/\sqrt{\kappa}$ is the propagating velocity of the electromagnetic wave in the semiconductor, c is the light speed in vacuum and κ is the dielectric constant of the material. The energy of the radiation field, which is absorbed in unit time by the carriers inside the small volume, $S_p \Delta S \Delta z$, is just the energy-loss of the electromagnetic wave after propagating a distance Δz. Therefore, the absorption coefficient is given by

$$\alpha = \frac{2}{\sqrt{\kappa}\epsilon_0 c} \frac{S_p}{E_\omega^2}. \tag{14.86}$$

In the limit of weak high-frequency field (small E_ω), the argument $\mathbf{q} \cdot \mathbf{r}_\omega$ of the Bessel functions on the right-hand-side of the S_p expression (14.33) is small. It suffices, for the calculation of S_p, to retain the single-photon process ($n = 1$ and -1) and keep just the leading terms of the expansion of the Bessel function: $J_{\pm1}^2(\mathbf{q} \cdot \mathbf{r}_\omega) \approx \frac{1}{4}(\mathbf{q} \cdot \mathbf{r}_\omega)^2$. We have the linear absorption coefficient for weak high-frequency fields,

$$\alpha = \frac{1}{\sqrt{\kappa}\epsilon_0 c} \frac{Ne^2}{m\omega^2} M_2(\omega, \mathbf{v}_0), \tag{14.87}$$

where $M_2(\omega, \mathbf{v}_0)$ is the imaginary part of the memory function $M(\omega, \mathbf{v}_0)$ under a dc bias velocity $\mathbf{v}_0$, which consists of contributions from impurity and phonon scatterings,

$$M_2(\omega, \mathbf{v}_0) = M_2^{(i)}(\omega, \mathbf{v}_0) + M_2^{(p)}(\omega, \mathbf{v}_0), \tag{14.88}$$

with the imaginary parts of the impurity and phonon contributed memory functions given by (3.54) and (3.55):

$$M_2^{(i)}(\omega, \mathbf{v}_0) = \frac{n_i}{Nm\omega} \sum_{\mathbf{q}} |u(\mathbf{q})|^2 q_\omega^2 \Pi_2(\mathbf{q}, \mathbf{q} \cdot \mathbf{v}_0 - \omega), \tag{14.89}$$

$$M_2^{(p)}(\omega, \mathbf{v}_0) = \frac{1}{Nm\omega} \sum_{\mathbf{q},\lambda} |M(\mathbf{q}, \lambda)|^2 q_\omega^2 \Big\{ \Pi_2(\mathbf{q}, \Omega_{\mathbf{q}\lambda} + \mathbf{q} \cdot \mathbf{v}_0 - \omega)$$
$$\times \left[n\left(\frac{\Omega_{\mathbf{q}\lambda}}{T}\right) - n\left(\frac{\Omega_{\mathbf{q}\lambda} + \mathbf{q} \cdot \mathbf{v}_0 - \omega}{T_e}\right) \right] - (\omega \to -\omega) \Big\}, \tag{14.90}$$

where q_ω stands for the component of $\boldsymbol{q}$ along the $\boldsymbol{E}_\omega$-direction. $M_2^{(i)}(\omega, \boldsymbol{v}_0)$ and $M_2^{(p)}(\omega, \boldsymbol{v}_0)$ depend on the scattering matrix element, the frequency of the radiation field, the magnitude of $\boldsymbol{v}_0$, and the relative direction of $\boldsymbol{v}_0$ and $\boldsymbol{E}_\omega$.

The expression (14.87) for the linear absorption coefficient of a weak radiation field is in agreement with the formula of the linear response theory for the free carrier absorption. It is equivalent to the high frequency ($\omega\tau \gg 1$) result of the extended Drude-type formula (3.58) and (3.59) [$D(\omega, \boldsymbol{v}_0)$ vanishes at high frequency] for the dynamic conductivity under a dc bias: $\alpha \sim \text{Re}[\sigma(\omega, \boldsymbol{v}_0)]$,

$$\sigma(\omega, \boldsymbol{v}_0) \simeq \frac{Ne^2}{m} \frac{i}{\omega + M(\omega, \boldsymbol{v}_0)} = \frac{iNe^2}{m^*(\omega + i/\tau)}, \qquad (14.91)$$

where $m^* = m[1 + M_1(\omega, \boldsymbol{v}_0)/\omega]$ and $1/\tau = M_2(\omega, \boldsymbol{v}_0)/[1 + M_1(\omega, \boldsymbol{v}_0)/\omega]$. (14.87) is apparently invalid in the low-frequency ($\omega \to 0$) limit, reflecting the fact that the present approach works only at high frequencies.

In the case of an intense high-frequency field, the absorption coefficient given by Eq. (14.86) depends on the amplitude of the high-frequency field. To calculate this nonlinear absorption coefficient, one needs to solve the steady-state momentum and energy balance equations

$$Ne\boldsymbol{E}_0 + \boldsymbol{F} = 0, \qquad (14.92)$$

$$\boldsymbol{v}_0 \cdot \boldsymbol{F} + W - S_p = 0, \qquad (14.93)$$

for $\boldsymbol{v}_0$ and T_e, and to obtain the pertinent energy absorption rate S_p for using in the general expression (14.86). In the absence of a bias field $\boldsymbol{E}_0 = 0$, however, it suffices to solve the energy balance equation to get S_p and α:

$$W - S_p = 0. \qquad (14.94)$$

Fig. 14.6 shows the calculated absorption coefficient α as a function of wavelength λ (or the frequency $\nu = \omega/2\pi$) at $E_\omega \sim 0$ (linear case) and several finite amplitudes $E_\omega = 5, 7, 10, 12$, and $14\,\text{kV/cm}$ at lattice temperature $T = 150\,\text{K}$ and zero dc bias, for a bulk model n-type GaAs system with electrons of density $N = 10^{23}\,\text{m}^{-3}$ moving in a parabolic Γ valley. The elastic scattering is assumed due to randomly distributed charged impurities with density n_i equal to the electron density (This results in a 4.2 K linear mobility $\mu_0 = 8\,\text{m}^2\text{V}^{-1}\text{s}^{-1}$). The inelastic scatterings are due to polar optic phonons (via Fröhlich coupling with electrons), longitudinal acoustic phonons (via deformation potential coupling with electrons),

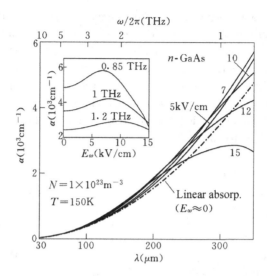

Fig. 14.6 The absorption coefficient α of a bulk GaAs subject to a THz radiation in the absence of a dc bias, is plotted as a function of the wavelength λ (or frequency $\omega/2\pi$) of the radiation field for vanishing (chain line) and several finite amplitudes $E_\omega = 5, 7, 10, 12,$ and $14\,\text{kV/cm}$ (solid lines). The lattice temperature is $T = 150\,\text{K}$. The inset shows α-vs-E_ω at several fixed frequencies. From Lei and Liu (2000a).

and transverse acoustic phonons (via piezoelectric coupling with electrons). The material and electron-phonon coupling parameters are taken as typical values of GaAs. The absorption coefficient α for these intense fields can be higher or lower than the linear absorption coefficient (as shown in the Fig. 14.6 by the chain line). Its wavelength dependence deviates markedly from that in the linear case. At fixed frequency, the absorption coefficient increases with increasing amplitude of the radiation field from $E_\omega = 0$, then reaches a maximum at around 7–14 kV/cm before decreasing quickly with further increase of E_ω, as shown in the inset of Fig. 14.6.

To see the role of individual multiphoton processes, we define

$$\alpha_n \equiv \frac{2}{\sqrt{\kappa}\epsilon_0 c} \frac{S_\text{p}^{(n)}}{E_\omega^2} \tag{14.95}$$

as the contribution of the n-photon (absorption and emission) process to the absorption coefficient: $\alpha = \sum_{n=1}^{\infty} \alpha_n$. Here $S_\text{p}^{(n)}$ represents the combination of both n- and $-n$-terms within the summation over n in the expression (14.33), that $S_\text{p} = \sum_{n=1}^{\infty} S_\text{p}^{(n)}$. At a given amplitude $E_\omega = 10\,\text{kV/cm}$, the n-photon contributions α_n ($n = 1, 2, 3, 5,$ and 10) are plotted in Fig. 14.7(a) as functions of the wavelength λ (or frequency $\omega/2\pi$) of the radiation field, together with the total absorption coefficient α ($\alpha_4, \alpha_6, \ldots$ are not shown in the figure for clarity). At given frequency $\omega/2\pi = 1.2\,\text{THz}$, the n-photon contributions α_n ($n = 1, 2, 3, 5,$ and 10) and the total absorption coefficient α are shown in Fig. 14.7(b) as functions of the amplitude E_ω of the radiation field. The single-photon process dominates only when $E_\omega \leqslant 1\,\text{kV/cm}$. At higher strength of the radiation field, multiphoton ($n \geqslant 2$) channels provide major contributions to the absorption.

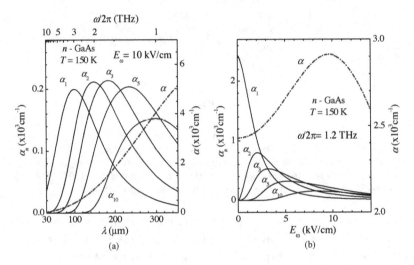

Fig. 14.7 The n-photon contributions α_n ($n = 1, 2, 3, 5,$ and 10) and the total absorption coefficient α are plotted against the wavelength λ (or frequency $\omega/2\pi$) of the radiation field at a fixed amplitude $E_\omega = 10\,\text{kV/cm}$ (a) and against the amplitude E_ω of the radiation field at a fixed frequency $\omega/2\pi = 1.2\,\text{THz}$ (b), for the n-GaAs system described in the text. After Lei and Liu (2000a).

14.4.2 Absorption rate of a THz radiation passing through a 2D sheet

In a quasi-2D system with single electron state described by (14.36), the transport balance equations under the influence of a dc and high-frequency radiation field parallel to the 2D plane, are given by (14.37) and (14.38):

$$N_s e \boldsymbol{E}_0 + \boldsymbol{F} = 0, \qquad (14.96)$$

$$\boldsymbol{v}_0 \cdot \boldsymbol{F} + W - S_\text{p} = 0. \qquad (14.97)$$

In this, the quantities, such as $\boldsymbol{F}$, W and S_p given by (14.39)–(14.41), are written for carriers in a unit area of the quasi-2D sheet. For such a thin planar structure, we can use the ratio of the energy loss of the electromagnetic radiation after passing through the 2D sheet to the energy of the electromagnetic wave impinging into the system as a measure of the absorption of the radiation. We call it the absorption rate or percentage and also denote it by α. In the case of perpendicular incidence ($\boldsymbol{E}_\omega$ parallel to the 2D plane), the absorption rate is given by

$$\alpha = \frac{2}{\sqrt{\kappa}\epsilon_0 c} \frac{S_\mathrm{p}}{E_\omega^2}, \qquad (14.98)$$

formally the same as (14.86) for the 3D case. But now S_p is the energy-gain by the carriers in a unit area of the 2D sheet from the radiation field, the absorption rate defined by (14.98) is dimensionless.

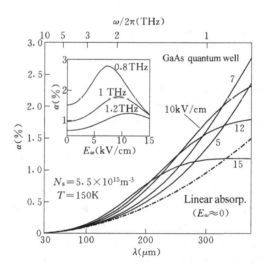

Fig. 14.8 The absorption rate (percentage) α of a THz radiation passing through the 2D sheet of a GaAs-based quantum well described in the text in the absence of dc bias, is plotted as a function of the wavelength λ (or frequency $\omega/2\pi$) of the radiation field for vanishing (chain line) and several finite amplitudes $E_\omega = 5, 7, 10, 12$, and $14\,\mathrm{kV/cm}$ (solid lines). The lattice temperature is $T = 150\,\mathrm{K}$. The inset shows α versus E_ω at several fixed frequencies. From Lei and Liu (2000a).

Numerical calculations are carried out for the GaAs-based quantum well described in Sec. 14.2.2, considering the role of two lowest subbands. The calculated absorption rate α at lattice temperature $T = 150\,\mathrm{K}$ in the absence of dc bias, is shown in Fig. 14.8 as a function of wavelength λ or frequency $\omega/2\pi$ of the radiation field for several different field strengths, $E_\omega = 5, 7, 10, 12$ and $14\,\text{kV/cm}$, as well as for vanishing E_ω (chain line). The absorption rate of the GaAs quantum well exhibits a stronger nonlinearity than in bulk GaAs. The maxima showing up in the α-versus-E_ω curves for fixed frequencies, are higher than in the 3D case (see the inset of Fig. 14.8).

14.5 Transport in Quasi-2D Semiconductors with Impact Ionization Irradiated by a Strong THz Field

We consider a quasi-2D electron–hole system, with the electron state characterized by a 2D wavevector $\bm{k}_\parallel = (k_x, k_y)$ and a subband index s having the energy dispersion

$$\varepsilon_{s\bm{k}_\parallel} = \varepsilon_s + k_\parallel^2/2m, \tag{14.99}$$

and the hole state having the energy dispersion

$$\varepsilon_{\bm{k}_\parallel}^{\text{h}} = \varepsilon_{\text{g}} + k_\parallel^2/2m_{\text{h}}. \tag{14.100}$$

In this, m and m_{h} are respectively the electron and hole effective mass, ε_s is the energy bottom of the sth electron subband with the ground subband bottom as the energy zero, and ε_{g} is the band gap between the conduction and valence bands. The energy separations between different hole subbands are neglected and the hole subband indices are not explicitly written for simplicity. Despite that a large number of holes may be created in the impact ionization process, their contribution to transport is still negligible due to large effective mass of holes in the valence subbands. Therefore, what we discuss is still the transport of a quasi-two-dimensional electron system of multiple subbands having a Hamiltonian given by Sec. 2.5 with the impact-ionization process as an additional scattering mechanism H_{ii} to it:

$$\begin{aligned}H_{\text{ii}} = \sum_{\substack{\bm{k}_\parallel, \bm{k}'_\parallel, \bm{q}_\parallel \\ s', s, \sigma}} & \Big[M_{s's}^{\text{ii}}(\bm{q}_\parallel)\, c_{s'\bm{k}'_\parallel+\bm{q}_\parallel\bar{\sigma}}^\dagger c_{s'\bm{k}_\parallel-\bm{q}_\parallel\sigma}^\dagger d_{\bm{k}_\parallel\sigma}^\dagger c_{s\bm{k}'_\parallel\bar{\sigma}} \\ & + M_{s's}^{\text{ii}*}(\bm{q}_\parallel)\, c_{s\bm{k}'_\parallel\bar{\sigma}}^\dagger d_{\bm{k}\sigma} c_{s'\bm{k}_\parallel-\bm{q}_\parallel\sigma} c_{s'\bm{k}'_\parallel+\bm{q}_\parallel\bar{\sigma}} \Big], \end{aligned} \tag{14.101}$$

where $\bar{\sigma} \equiv -\sigma$, $c_{s\bm{k}_\parallel\sigma}^\dagger$ ($c_{s\bm{k}_\parallel\sigma}$) are creation (annihilation) operators of electrons in the sth subband, $d_{\bm{k}_\parallel\sigma}^\dagger$ ($d_{\bm{k}_\parallel\sigma}$) are creation (annihilation) operators of holes in one of the hole subbands, and

$$M^{ii}_{s's}(\boldsymbol{q}_{\parallel}) = I^{s'}_{cc} I^{s}_{cv} \frac{e^2}{2\epsilon_0 \kappa q_{\parallel}} = C_{ii} \frac{e^2}{2\epsilon_0 \kappa q_{\parallel}} \quad (14.102)$$

are the matrix elements for the impact ionization process that creates two electrons in the s'th subband while annihilates an electron in the sth subband and creates a hole in one of the hole subband (see Sec. 2.5). The parameter $C_{ii} \equiv I^{s'}_{cc} I^{s}_{cv}$ denotes the overlap integral.

When a uniform dc (or slowly time-varying) electric filed $\boldsymbol{E}_0$ and a uniform high-frequency electric filed $\boldsymbol{E}_\omega \sin(\omega t)$ of frequency ω simultaneously apply in the x–y plane of the 2D system from time $t=0$, the carrier transport state can be described by means of the 2D center-of-mass and relative electron variables (Sec. 2.5). We take the average electron drift velocity $\boldsymbol{v}$ and the electron temperature T_e as the fundamental parameters to characterize the transport of the system and, following the derivation as presented in Secs. 14.1, 14.2 and 13.62, to obtain the balance equations as follows (Cao and Lei, 2002; 2003)

$$\frac{d\boldsymbol{v}}{dt} = \frac{e}{m}\boldsymbol{E} + \boldsymbol{a} + \boldsymbol{a}_{ii} - \boldsymbol{v} g_{ii}, \quad (14.103)$$

$$\frac{d\varepsilon_e}{dt} = e\boldsymbol{E} \cdot \boldsymbol{v} - w - w_{ii} - \varepsilon_e g_{ii} + s_p + s^{ii}_p, \quad (14.104)$$

$$\frac{dN_e}{dt} = N_e g_{ii}. \quad (14.105)$$

Here N_e is the average number of electrons in the 2D sheet of unit area; the others, including the impurity and phonon induced frictional acceleration $\boldsymbol{a} = \boldsymbol{a}_i + \boldsymbol{a}_p$, the average electron energy ε_e, the phonon-induced electron energy-loss rate w and the impurity and phonon induced electron energy-gain rate (from the high-frequency field) s_p, are quantities per electron. Their expressions are exactly the same as those in Sec. 14.2. The newly appearing per-electron quantities

$$\boldsymbol{a}_{ii} = \frac{2}{N_e} \sum_{s',s,\boldsymbol{k}_{\parallel},\boldsymbol{q}_{\parallel}} |M^{ii}_{s's}(\boldsymbol{q}_{\parallel})|^2 \frac{\boldsymbol{k}}{m} \sum_{n=-\infty}^{\infty} J^2_n(\boldsymbol{q}_{\parallel} \cdot \boldsymbol{r}_\omega) I_n(s',s,\boldsymbol{k}_{\parallel},\boldsymbol{q}_{\parallel}), \quad (14.106)$$

$$w_{ii} = \frac{2}{N_e} \sum_{s',s,\boldsymbol{k}_{\parallel},\boldsymbol{q}_{\parallel}} |M^{ii}_{s's}(\boldsymbol{q}_{\parallel})|^2 \varepsilon^h_{\boldsymbol{k}_{\parallel}} \sum_{n=-\infty}^{\infty} J^2_n(\boldsymbol{q}_{\parallel} \cdot \boldsymbol{r}_\omega) I_n(s',s,\boldsymbol{k}_{\parallel},\boldsymbol{q}_{\parallel}), \quad (14.107)$$

$$s^{ii}_p = \frac{2}{N_e} \sum_{s',s,\boldsymbol{k}_{\parallel},\boldsymbol{q}_{\parallel}} |M^{ii}_{s's}(\boldsymbol{q}_{\parallel})|^2 \sum_{n=-\infty}^{\infty} J^2_n(\boldsymbol{q}_{\parallel} \cdot \boldsymbol{r}_\omega) n\omega I_n(s',s,\boldsymbol{k}_{\parallel},\boldsymbol{q}_{\parallel}), \quad (14.108)$$

and

$$g_{\text{p}}^{\text{ii}} = \frac{2}{N_{\text{e}}} \sum_{s',s,\bm{k}_\|,\bm{q}_\|} |M_{s's}^{\text{ii}}(\bm{q}_\|)|^2 \sum_{n=-\infty}^{\infty} \text{J}_n^2(\bm{q}_\| \cdot \bm{r}_w) I_n(s',s,\bm{k}_\|,\bm{q}_\|), \qquad (14.109)$$

are the impact ionization and Auger recombination induced net generation rate of the electron–hole pairs, electron damping acceleration, electron energy-loss rate and electron energy-gain rate from the high-frequency field, and the function involved is given by

$$I_n(s',s,\bm{k}_\|,\bm{q}_\|) = \Pi_2(s',s,\bm{q}_\|,\omega_2-n\omega)\left[f\left(\frac{\varepsilon_{s\bm{k}_\|-\bm{q}_\|}-\mu}{T_{\text{e}}}\right)+n\left(\frac{\omega_2-n\omega}{T_{\text{e}}}\right)\right]$$
$$\times \left[f\left(\frac{\varepsilon_{s\bm{k}_\|}-\mu}{T}\right)-f\left(\frac{\omega_1+\varepsilon_{\bm{k}_\|}^{\text{h}}+\mu}{T_{\text{e}}}\right)\right]. \qquad (14.110)$$

with $\omega_1 \equiv \bm{k}_\| \cdot \bm{v} + \frac{1}{2}mv^2$ and $\omega_2 \equiv \omega_1 + \varepsilon_{\bm{k}_\|}^{\text{h}} + \varepsilon_{s\bm{k}_\|-\bm{q}_\|}$.

As discussed in Sec. 13.62 without a radiation field, in view of the relaxation time τ_{ii} related to the particle generation process is much longer than the relaxation time τ_s related to the impurity and phonon scattering, in the case of high-frequency field driving, one can still follow the slow part of the transport time evolution from the turn-on of the fields at $t = 0$ and distinguish the quasi-steady state and the steady state. The former is reached at a time elapse $t \geqslant \tau_s$, during which the carrier drift velocity $\bm{v}$ and electron temperature T_{e} become almost constant, while the total number of carriers is still increasing due to the impact ionization process with a net electron–hole generation rate $g_{\text{ii}} \neq 0$. The latter may be reached when the time elapse $t \geqslant \tau_{\text{ii}}$, in which all the transport quantities are constant and can be solved by setting the right-hand-sides of Eqs. (14.103), (14.104) and (14.105) to be zero ($d\bm{v}/dt = 0$, $d\varepsilon_{\text{e}}/dt = 0$ and $dN_{\text{e}}/dt = 0$):

$$0 = \frac{e}{m}\bm{E} + \bm{a} + \bm{a}_{\text{ii}} - \bm{v}g_{\text{ii}}, \qquad (14.111)$$
$$0 = e\bm{E} \cdot \bm{v} - w - w_{\text{ii}} - \varepsilon_{\text{e}} g_{\text{ii}} + s_{\text{p}} + s_{\text{p}}^{\text{ii}}, \qquad (14.112)$$
$$0 = g_{\text{ii}}. \qquad (14.113)$$

Numerical calculations are performed for transport properties in THz-driven InAs/AlSb heterojunctions (Cao and Lei, 2003; Cao, 2003), taking account three lowest electron subbands with $\varepsilon_0 = 0$, $\varepsilon_1 = 35\,\text{meV}$ and $\varepsilon_2 = 200\,\text{meV}$ and up to 10 hole subbands. They set the carrier sheet density of the InAs well to be $N_{\text{e}}^0 = 5 \times 10^{12}\,\text{cm}^{-2}$, and depletion layer charge density $N_{\text{dep}} = 5 \times 10^{10}\,\text{cm}^{-2}$. The electrons are assumed to experience, in addition to the elastic scatterings from background impurities of density $n_i = 6.86 \times 10^{15}\,\text{cm}^{-3}$ and from remote impurities located in the AlSb barrier at a distance of $s = 10\,\text{nm}$ from the interface of the heterojunction having density $N_I = 1.53 \times 10^{11}\,\text{cm}^{-2}$, scatterings from acoustic phonons (via the deformation potential and the piezoelectric couplings) and polar optic phonons (via the Fröhlich coupling). The material and coupling parameters are taken to be typical values of InAs/AlSb systems listed in Cao and Lei (2002), To contact with the experiment of THz absorption (Asmar et al, 1995) the lattice temperature is chosen as $T = 300\,\text{K}$, and the dc field is $E_0 = 1.5\,\text{V/cm}$. The only adjustable parameter C_{ii} in (14.102) in the model is taken to be $C_{ii} = 0.114$ throughout the calculation.

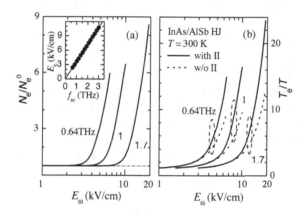

Fig. 14.9 (a) Electron–hole pair generation rate N_e/N_e^0 and (b) electron temperature T_e/T are shown as functions of the amplitude E_ω of radiation fields of different frequencies $f_{ac} = 0.64, 1.0$ and $1.7\,\text{THz}$ in an InAs/AlSb heterojunction at lattice temperature $T = 300\,\text{K}$. The inset indicates the critical field strength E_c versus radiation frequency f_{ac}. After Cao (2003).

Under the influence of a perpendicularly incident THz radiation of frequency $f_{ac} = \omega/2\pi$ and amplitude E_ω, the solution of the steady state balance equations gives rise to all the transport quantities, such as N_e, v, T_e, s_p and s_p^{ii}. At a given frequency f_{ac} the onset of the impact ionization process, which can be determined when the ratio N_e/N_e^0 begins to increase appreciably from 1, occurs when the radiation field strength E_ω becomes larger than a threshold strength $E_c(f_{ac})$, as seen in Fig. 14.9(a). The calculation indicates that $E_c(f_{ac})/f_{ac}$ is nearly a constant. At high radiation field the impact ionization process plays an important role and N_e grows exponentially with increasing E_ω. This exponentially growing electron–hole pair generation rate under a radiation field plays a crucial role in the rapid increase in the photon absorption and leads to a higher T_e than in the case without impact ionization [Fig. 14.9(b)].

With impact ionization the absorption rate (percentage) α of a THz radiation passing through the 2D sheet is given by

$$\alpha = \frac{2}{\sqrt{\kappa\epsilon_0}c}\frac{N_e(s_p + s_p^{ii})}{E_\omega^2}. \tag{14.114}$$

Fig. 14.10 demonstrates the calculated absorption rate α of the InAs/AlSb heterojunction described above versus the amplitude E_ω of the THz radiation at different frequencies $f_{ac} = 0.64, 1.0$ and 1.7 THz. The solid circles are the experimental results for $f_{ac} = 0.64$ THz derived from the transmission T and the reflection rate R of Asmar et al (1995): $\alpha = 1 - T - R$. Good agreement is achieved between the calculated results and the experimental data. The significance of the impact ionization process can be clearly seen: impurity and phonon scattering induced absorption increases slowly at low E_ω and reaches a maximum then decreases with continuing increase of E_ω. The occurrence of the impact ionization process provides an additional absorption which enhances rapidly with increasing field strength, resulting in a total absorption rate evenly increasing at high radiation fields.

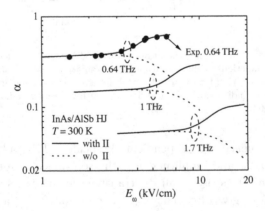

Fig. 14.10 Absorption rate (percentage) α of a THz radiation passing through an InAs/AlSb heterojunction versus the amplitude E_ω of radiation field of different frequencies $f_{ac} = 0.64, 1.0$ and 1.7 THz at lattice temperature $T = 300$ K. The solid circles are experimental results of Asmar et al (1995). After Cao (2003).

14.6 THz-Field-Driven Miniband Transport of a Superlattice in a Longitudinal Magnetic Field

We consider a planar superlattice which is formed by periodical potential wells and barriers of finite heights along the z direction with period d and is subjected to a longitudinal magnetic field $\boldsymbol{B} = (0,0,B)$. The electron energy spectrum of the system forms minibands in the z direction and is quantized into Landau levels in the x–y plane. Assuming that the energy separation between the longitudinal lowest and second minibands is large enough that only the lowest miniband needs to be taken into account, the electron state is characterized by a quantum number s of the Landau level ($s = 0, 1, 2, ...$), a transverse wavevector k_x ($-\infty < k_x < \infty$), a longitudinal wavevector k_z ($-\pi/d < k_z \leqslant \pi/d$), and a spin index σ, with the energy given by (neglecting the spin-related energy for simplicity)

$$\varepsilon_s(k_z) = \left(s + \frac{1}{2}\right)\omega_c + \varepsilon(k_z), \tag{14.115}$$

$$\varepsilon(k_z) = \frac{\Delta}{2}\left[1 - \cos(k_z d)\right]. \tag{14.116}$$

They were discussed in Sec. 12.9 in detail. Here $\omega_c = |eB|/m$ is the cyclotron frequency and Δ is the miniband width.

When a uniform dc electric field $\boldsymbol{E}_0$ and a sinusoidal terahertz ac field $\boldsymbol{E}_\omega \sin(\omega t)$ are applied along the superlattice growth axis, $\boldsymbol{E}_0 = (0, 0, E_0)$ and $\boldsymbol{E}_\omega = (0, 0, E_\omega)$, the carrier drift motion, is in the z direction and the transport state can be described by a lattice-momentum shift p_d (in the z-direction) and an electron temperature T_e, which are to be determined by the following momentum- and energy-balance equations formally the same as (14.73) and (14.74) without magnetic field:

$$eE_0/m_z^* + a_i + a_p = 0, \tag{14.117}$$

$$eE_0 v_0 - w + s_p = 0. \tag{14.118}$$

The effect of a longitudinal magnetic field is to confine the electrons transversely inside a circle with radius of order of a magnetic length l_B and to quantize the electron energy spectrum related to its transverse motion, such that the corresponding quantities v_0, $1/m_z^*$, a_i and a_p in Eqs. (14.117) and (14.118) are given by

$$v_0 = \frac{1}{\pi l_B^2 N_s n_z} \sum_{s,k_z} v(k_z) f(\varepsilon_s(k_z - p_d), T_e), \tag{14.119}$$

$$\frac{1}{m_z^*} = \frac{1}{\pi l_B^2 N_s n_z} \sum_{s,k_z} \frac{d^2\varepsilon(k_z)}{dk_z^2} f(\varepsilon_s(k_z - p_d), T_e), \qquad (14.120)$$

$$\begin{aligned}
a_i &= \frac{n_i}{2\pi l_B^2 N_s Z_n} \sum_{s,s',\boldsymbol{q},k_z} |u(\boldsymbol{q})|^2 c_{ss'}(l_B^2 q_\parallel^2/2) |g(q_z)|^2 \sum_{n=-\infty}^{\infty} J_n^2(\xi) \\
&\quad \times [v(k_z + q_z) - v(k_z)] \, \delta(\varepsilon_{s'}(k_z + q_z) - \varepsilon_s(k_z) - n\omega) \\
&\quad \times [f(\varepsilon_s(k_z - p_d), T_e) - f(\varepsilon_{s'}(k_z + q_z - p_d), T_e)], \qquad (14.121)
\end{aligned}$$

$$\begin{aligned}
a_p &= \frac{1}{\pi l_B^2 N_s Z_n} \sum_{s,s',\boldsymbol{q},k_z} |M(\boldsymbol{q},\lambda)|^2 c_{ss'}(l_B^2 q_\parallel^2/2) |g(q_z)|^2 \sum_{n=-\infty}^{\infty} J_n^2(\xi) \\
&\quad \times [v(k_z + q_z) - v(k_z)] \, \delta(\varepsilon_{s'}(k_z + q_z) - \varepsilon_s(k_z) + \Omega_{\boldsymbol{q}\lambda} - n\omega) \\
&\quad \times [f(\varepsilon_s(k_z - p_d), T_e) - f(\varepsilon_{s'}(k_z + q_z - p_d), T_e)] \\
&\quad \times \left[n\left(\frac{\Omega_{\boldsymbol{q}\lambda}}{T}\right) - n\left(\frac{\varepsilon_s(k_z) - \varepsilon_{s'}(k_z + q_z)}{T_e}\right) \right]. \qquad (14.122)
\end{aligned}$$

In these equations, $v(k_z) \equiv d\varepsilon(k_z)/dk_z$, $\xi \equiv [v(k_z + q_z) - v(k_z)]eE_\omega/\omega^2$, $N_s \equiv N/(Z_n S)$ represents the electron sheet density per period (S denotes the lateral area of the superlattice, and Z_n is the total number of its longitudinal periods), $u(\boldsymbol{q})$ is the 3D fourier transform of the impurity potential, $M(\boldsymbol{q}, \lambda)$ is the 3D plane-wave representation of the electron–phonon matrix element for the phonon with wavevector $\boldsymbol{q} = (q_x, q_y, q_z)$ in branch λ, having frequency $\Omega_{\boldsymbol{q}\lambda}$. In addition, $g(q_z)$ and $c_{ss'}(l_B^2 q_\parallel^2/2)$ [$q_\parallel^2 \equiv q_x^2 + q_y^2$] are form factors due to the longitudinal miniband wave function and the transverse Landau quantized wave function respectively [see (12.9) in Sec. 12.1 and (12.131) in Sec. 12.9]. Furthermore, $n(x) = (e^x - 1)^{-1}$ is the Bose function, and $f(\varepsilon, T_e) = \{\exp[(\varepsilon - \mu)/T_e] + 1\}^{-1}$ is the Fermi distribution function at electron temperature T_e, with μ the chemical potential determined by the sheet density of electrons from the equation

$$1 = \frac{1}{\pi l_B^2 N_s Z_n} \sum_{s,k_z} f(\varepsilon_s(k_z), T_e). \qquad (14.123)$$

The expression for w can be obtained from that of a_p by replacing the factor $[v(k_z + q_z) - v(k_z)]$ on the right hand of Eq. (14.122) by $\Omega_{\boldsymbol{q}\lambda}$, and the expression for s_p is obtained from that of $a_i + a_p$ by replacing the factor $[v(k_z + q_z) - v(k_z)]$ on the right hand sides of both Eqs. (14.121) and (14.122) by $n\omega$.

Equations (14.117) and (14.118) facilitate the evaluation of linear and nonlinear miniband transport properties of semiconductor superlattices under the influence of a strong longitudinal magnetic field and an intense THz radiation field. As a numerical example, we consider a GaAs-based quantum-well superlattice having period $d = 9.0$ nm, well width $a = 6.0$ nm, miniband width $\Delta = 8.5$ meV, and electron sheet density $N_s = 2.0 \times 10^{14}$ m^{-2} per period. We use randomly distributed background impurities to mimic the overall elastic scattering and assume an impurity density $n_i = 1 \times 10^{21}$ m^{-3}. For phonon scatterings we also take into account both longitudinal and transverse acoustic phonons, in addition to the longitudinal optic (LO) phonons.

In the absence of the THz field, the calculated linear mobility at lattice temperature $T = 300$ K as a function of the strength of the magnetic field is shown as the solid line in Fig. 14.11, which is essentially the same as that plotted in Fig. 12.23 in Sec. 12.9, exhibiting strong mobility oscillation. The mobility minima at $B \simeq 7, 10.5$, and 21 T, correspond to the positions satisfying the magnetophonon resonance condition

$$M\omega_c = \Omega_{\rm LO} \qquad (14.124)$$

of $M = 1, 2$, and 3 (Sec 12.9).

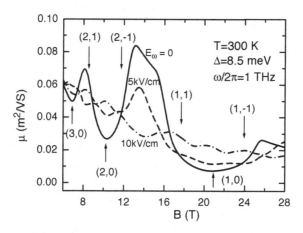

Fig. 14.11 Linear (low electric field) mobility is shown as a function of the longitudinal magnetic field B for a GaAs-based superlattice having miniband width $\Delta = 8.5$ meV at lattice temperature $T = 300$ K, in the absence (solid curve) of THz field and in the presence of a 1 THz high-frequency field of strengths $E_\omega = 5$ and 10 kV/cm, respectively. From Yan and Lei (2001).

Such an oscillation behavior is drastically affected when the superlattice is exposed to a radiation field of THz frequency and some kind of resonance may appear around where the condition ($M = 1, 2, \cdots$ and $n = 0, \pm 1, \pm 2, \cdots$)

$$M\omega_c + n\omega = \Omega_{\text{LO}} \tag{14.125}$$

is satisfied. The dashed line and chain line in Fig. 14.11 are the linear mobility of the same superlattice subjected to a radiation field of frequency $\omega/2\pi = 1\,\text{THz}$ having strength $E_\omega = 5$ and $10\,\text{kV/cm}$ respectively. We can see that, in the $E_\omega = 5\,\text{kV/cm}$ curve, in addition to the minima around the resonant magnetic fields in the absence of the THz radiation, which are identified by an integer $(M, 0)$ ($M = 1, 2, \cdots$), there appear two new minima around $B = 18$ and $24\,\text{T}$, identified by two integers (M, n) as $(1, 1)$ and $(1, -1)$, and another two new minima around $B = 8.6$ and $11.6\,\text{T}$, identified as $(2, 1)$ and $(2, -1)$. The THz irradiation enhances the mobility around the magnetic field, at which the mobility minima locate in the absence of the radiation field. This trend remains true in the case of $E_\omega = 10\,\text{kV/cm}$ irradiation: the mobility around the resonant magnetic field continues to increase. The multiphoton peaks, however, are further smeared and barely seen.

In Fig. 14.12(a) we plot the calculated nonlinear drift velocity v_0 as a function of the dc electric field E_0 for the same superlattice as discussed in Fig. 14.11 at lattice temperature $T = 150\,\text{K}$, subjected to a longitudinal magnetic field having strength $B = 11, 15, 22, 25$ and $30\,\text{T}$ in the absence of the radiation. These curves are similar to those calculated in Sec. 12.9. An intense radiation field at THz frequency drastically affects the v_0–E_0 behavior. As shown in Fig. 14.12(b), when an ac field of frequency $\omega/2\pi = 1\,\text{THz}$ and amplitude $E_\omega = 5\,\text{kV/cm}$ applies to the system, each v_0–E_0 curve for nonzero magnetic field differs remarkably from the corresponding one without THz field. The most dramatic changes are for $B = 15\,\text{T}$ and $B = 30\,\text{T}$ curves: the magnetic-field quenching of the drift velocity showing up previously in the absence of the high-frequency field disappears almost completely. On the other hand, in the case of zero magnetic field, where LO-phonon scattering can always take place through intra-miniband transverse transition, the effect of a THz radiation on the conduction of this narrow miniband is small. Strong magnetic field quantizes the transverse energy of the electron, leading to the alternating appearance of permitted or forbidden transition between Landau minibands by LO phonons. When an intense THz field applies, the forbidden LO scatterings become available due to multiphoton-assisted processes.

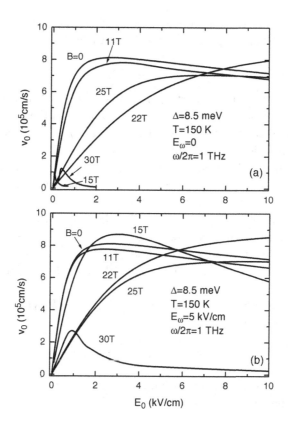

Fig. 14.12 Nonlinear drift velocity v_0 is plotted as a function of the dc electric field E_0 for the same superlattice as discussed in Fig. 14.14 at lattice temperature $T = 150\,\text{K}$ subject to a longitudinal magnetic field of strength $B = 0, 11, 15, 22, 25$ and $30\,\text{T}$, respectively in the absence of the THz field (a), and in the presence (b) of an intense radiation field of frequency $\omega/2\pi = 1\,\text{THz}$ and amplitude $E_\omega = 5\,\text{kV/cm}$.

14.7 Extension to Radiation Fields of Lower Frequency

The balance equations developed so far in this chapter for electron transport driven by a uniform dc (or slowly time-varying) electric field $\bm{E}_0$ and a uniform sinusoidal radiation field of frequency ω and amplitude $\bm{E}_\omega$,

$$\bm{E}(t) = \bm{E}_0 + \bm{E}_\omega \sin(\omega t), \qquad (14.126)$$

are valid for enough high frequency of the ac field. When the period of the radiation field is much shorter than the center-of-mass-motion related relaxation time τ_m (momentum relaxation time), the center of mass will not be able to follow the rapidly oscillation of the driving ac field, that the center-of-mass velocity takes the form of (14.16) ($\boldsymbol{v}_\omega \equiv e\boldsymbol{E}_\omega/m\omega$),

$$\boldsymbol{V}(t) = \boldsymbol{v}_0 - \boldsymbol{v}_\omega \cos(\omega t), \qquad (14.127)$$

consisting of a slowly-varying part v_0 and a frequency-ω oscillating part $\pi/2$ out of phase of the driving ac field, which results from the selected gauge of the vector and scalar potentials (14.2) and (14.3). For a system having momentum relaxation time around several tenths of 10^{-12} s, the method developed so far in this chapter is expected to apply to the radiation field of frequency higher than 0.5 THz. Such an approximation is apparently not valid if the frequency of the driving ac field becomes lower and the derived expressions, such as (14.86) and (14.87) for the free-carrier absorption coefficient, are generally not allowed to go to the limit of $\omega \to 0$. To deal with transport driven by ac-fields of lower frequency, one should consider that the center-of-mass may somehow move following the oscillatory driving field.

We notice that the center-of-mass velocity operator $\boldsymbol{V}$ is the time derivative of its coordinate:

$$\boldsymbol{V} = -\mathrm{i}\,[\boldsymbol{R}, H] = \frac{\boldsymbol{P}}{Nm} - \boldsymbol{v}_\omega \cos(\omega t), \qquad (14.128)$$

which is explicitly time dependent, and the equation of motion for the time derivative of the velocity,

$$\dot{\boldsymbol{V}} = -\mathrm{i}\,[\boldsymbol{V}, H] + \frac{\partial \boldsymbol{V}}{\partial t} = \frac{\dot{\boldsymbol{P}}}{Nm} + \boldsymbol{v}_\omega \omega \sin(\omega t) \qquad (14.129)$$

[$\dot{\boldsymbol{P}}$ is given by (14.11)] can serve as the operator equation for the momentum balance. Now we treat the center-of-mass coordinate $\boldsymbol{R}$ and velocity $\boldsymbol{V}$ classically, i.e. to regard them as the time-dependent expectation (or average) values of the center-of-mass coordinate and velocity, $\boldsymbol{R}(t)$ and $\boldsymbol{V}(t)$, and assume that $\boldsymbol{V}(t)$ may contain, in addition to a part $\pi/2$ out of phase of the driving ac field, also a part in phase of the driving ac field:

$$\boldsymbol{V}(t) = \boldsymbol{v}_0 + \boldsymbol{v}_c \cos(\omega t) + \boldsymbol{v}_s \sin(\omega t), \qquad (14.130)$$

in which $\boldsymbol{v}_c$ and $\boldsymbol{v}_s$ are parameters. The center-of-mass coordinate is then given by

$$\boldsymbol{R}(t) = \int_{t_0}^{t} \boldsymbol{V}(s)\, ds + \boldsymbol{R}(t_0), \qquad (14.131)$$

and the function $A(\boldsymbol{q}, t, t')$ in the expressions of the frictional force, energy-loss and energy-gain rates takes the form

$$A(\boldsymbol{q}, t, t') = \exp\left[\mathrm{i}\boldsymbol{q} \cdot \int_{t'}^{t} \boldsymbol{V}(s)ds\right] = \sum_{n=-\infty}^{\infty} \mathrm{J}_n^2(\xi)\, \mathrm{e}^{\mathrm{i}(\boldsymbol{q}\cdot\boldsymbol{v}_0 - n\omega)(t-t')}$$

$$+ \sum_{m \neq 0} \mathrm{e}^{\mathrm{i}\, m(\omega t - \varphi)} \left[\sum_{n=-\infty}^{\infty} \mathrm{J}_n(\xi)\mathrm{J}_{n-m}(\xi)\, \mathrm{e}^{\mathrm{i}(\boldsymbol{q}\cdot\boldsymbol{v}_0 - n\omega)(t-t')}\right]. \quad (14.132)$$

Here the argument in the Bessel functions

$$\xi \equiv \frac{1}{\omega}\left[(\boldsymbol{q} \cdot \boldsymbol{v}_c)^2 + (\boldsymbol{q} \cdot \boldsymbol{v}_s)^2\right]^{\frac{1}{2}}, \quad (14.133)$$

and $\tan\varphi = (\boldsymbol{q} \cdot \boldsymbol{v}_s)/(\boldsymbol{q} \cdot \boldsymbol{v}_c)$.

Using the form (14.130) for the center-of-mass velocity implies that we are considering the oscillatory velocity response of the electron system only to the first harmonics of the high-frequency field. In accordance with this, since the lowest nonzero response of the system energy (i.e. the electron temperature) to an ac field beyond the average is of the second harmonics in the zero dc bias, the rapidly oscillatory part of the electron temperature can be ignored. In the case of nonzero dc bias, the base frequency response of the electron temperature, though possible, is much weaker because the energy relaxation time is generally much longer than the momentum relaxation time. Therefore, we describe the transport state of the electron system driven by the dc and high-frequency fields (14.126) with four parameters $\boldsymbol{v}_0$, $\boldsymbol{v}_c$, $\boldsymbol{v}_s$, and T_e.

Taking the statistical average of the operator equation (14.129) to the lowest order in scattering interactions as in Sec. 14.1, and keeping the zero and the first harmonics of the high-frequency field in the frictional force by retaining the $m = 0, \pm 1$ terms on the right-hand side of $A(\boldsymbol{q}, t, t')$ expression (14.132), we obtain the force balance equation as follows:

$$\frac{d}{dt}\boldsymbol{v}_0 - \boldsymbol{v}_c\,\omega\sin(\omega t) + \boldsymbol{v}_s\,\omega\cos(\omega t) = \frac{e}{m}\boldsymbol{E}_0 + \frac{1}{Nm}\boldsymbol{F} + \boldsymbol{v}_\omega\omega\sin(\omega t), \quad (14.134)$$

with the frictional force $\boldsymbol{F}$ in the form

$$\boldsymbol{F} = \boldsymbol{F}_0 + \boldsymbol{F}_s \sin(\omega t) + \boldsymbol{F}_c \cos(\omega t). \quad (14.135)$$

Or, writing the time average part and the first harmonic oscillating parts respectively as

$$\frac{d}{dt}\bm{v}_0 = \frac{e}{m}\bm{E}_0 + \frac{\bm{F}_0}{Nm}, \qquad (14.136)$$

$$\bm{v}_c = -\bm{v}_\omega - \frac{\bm{F}_s}{Nm\omega}, \qquad (14.137)$$

$$\bm{v}_s = \frac{\bm{F}_c}{Nm\omega}. \qquad (14.138)$$

Here, the frictional forces $\bm{F}_0$, $\bm{F}_s = \bm{F}_{2s} - \bm{F}_{1c}$, and $\bm{F}_c = \bm{F}_{1s} + \bm{F}_{2c}$, are given by

$$\bm{F}_0 = n_\mathrm{i} \sum_{\bm{q}} \bm{q}\,|u(\bm{q})|^2 \sum_{n=-\infty}^{\infty} \mathrm{J}_n^2(\xi)\,\Pi_2(\bm{q}, \omega_0 - n\omega)$$

$$+ \sum_{\bm{q},\lambda} \bm{q}\,|M(\bm{q},\lambda)|^2 \sum_{n=-\infty}^{\infty} \mathrm{J}_n^2(\xi)\,\Lambda_2(\bm{q},\lambda,\,\omega_0 - n\omega), \qquad (14.139)$$

$$\bm{F}_{1j} = n_\mathrm{i} \sum_{\bm{q}} \bm{q}\,\eta_j\,|u(\bm{q})|^2 \sum_{n=-\infty}^{\infty} [\mathrm{J}_n^2(\xi)]'\,\Pi_1(\bm{q}, \omega_0 - n\omega)$$

$$+ \sum_{\bm{q},\lambda} \bm{q}\,\eta_j\,|M(\bm{q},\lambda)|^2 \sum_{n=-\infty}^{\infty} [\mathrm{J}_n^2(\xi)]'\,\Lambda_1(\bm{q},\lambda,\,\omega_0 - n\omega), \qquad (14.140)$$

$$\bm{F}_{2j} = -n_\mathrm{i} \sum_{\bm{q}} \bm{q}\,\frac{\eta_j}{\xi}\,|u(\bm{q})|^2 \sum_{n=-\infty}^{\infty} 2n\,\mathrm{J}_n^2(\xi)\,\Pi_2(\bm{q}, \omega_0 - n\omega)$$

$$- \sum_{\bm{q},\lambda} \bm{q}\,\frac{\eta_j}{\xi}\,|M(\bm{q},\lambda)|^2 \sum_{n=-\infty}^{\infty} 2n\,\mathrm{J}_n^2(\xi)\,\Lambda_2(\bm{q},\lambda,\,\omega_0 - n\omega), \qquad (14.141)$$

in which $\omega_0 \equiv \bm{q}\cdot\bm{v}_0$, $\eta_j \equiv \bm{q}\cdot\bm{v}_j/\omega\xi$ and $j = c$ and s. In these equations, $\Pi_2(\bm{q},\Omega)$ and $\Pi_1(\bm{q},\Omega)$ are the imaginary and real parts of the electron density correlation function $\Pi(\bm{q},\Omega)$; $\Lambda_2(\bm{q},\lambda,\Omega)$ and $\Lambda_1(\bm{q},\lambda,\Omega)$ are the imaginary and real parts of the electron–phonon correlation function $\Lambda(\bm{q},\lambda,\Omega)$, which can be expressed in terms of $\Pi_2(\bm{q},\Omega)$ and $\Pi_1(\bm{q},\Omega)$ functions according to (3.9) and (3.10):

$$\Lambda_2(\bm{q},\lambda,\Omega) = \Pi_2(\bm{q},\Omega_{q\lambda} + \Omega)\left[n\!\left(\frac{\Omega_{q\lambda}}{T}\right) - n\!\left(\frac{\Omega_{q\lambda} + \Omega}{T_\mathrm{e}}\right)\right]$$

$$- \Pi_2(\bm{q},\Omega_{q\lambda} - \Omega)\left[n\!\left(\frac{\Omega_{q\lambda}}{T}\right) - n\!\left(\frac{\Omega_{q\lambda} - \Omega}{T_\mathrm{e}}\right)\right], \qquad (14.142)$$

$$\Lambda_1(\boldsymbol{q},\lambda,\Omega) = [\Pi_1(\boldsymbol{q},\Omega_{q\lambda}+\Omega) + \Pi_1(\boldsymbol{q},\Omega_{q\lambda}-\Omega)]\, n\!\left(\frac{\Omega_{q\lambda}}{T}\right)$$
$$+\frac{1}{\pi}\int_{-\infty}^{\infty} d\omega_1\, \Pi_2(\boldsymbol{q},\omega_1)\, n\!\left(\frac{\omega_1}{T_{\mathrm{e}}}\right)$$
$$\times\left(\frac{1}{\Omega_{q\lambda}+\Omega-\omega_1} + \frac{1}{\Omega_{q\lambda}-\Omega-\omega_1}\right). \qquad (14.143)$$

The energy balance equation can be obtained by taking the statistical average of the operator equation (14.12) and identifying $\langle \dot{H}_{\mathrm{er}}\rangle$ as the rate of change of the relative-electron energy U. Retaining only slowly time-varying part we have exactly the same equation as (14.31):

$$\frac{d}{dt}U = -\boldsymbol{v}_0\cdot\boldsymbol{F}_0 - W + S_{\mathrm{p}}. \qquad (14.144)$$

Here, $-\boldsymbol{v}_0\cdot\boldsymbol{F}_0$ is the work done by the center of mass on the relative-electron system,

$$W = \sum_{\boldsymbol{q},\lambda} \Omega_{q\lambda}\, |M(\boldsymbol{q},\lambda)|^2 \sum_{n=-\infty}^{\infty} \mathrm{J}_n^2(\xi)\, \Lambda_2^{+}(\boldsymbol{q},\lambda,\omega_0-n\omega) \qquad (14.145)$$

is the energy-transfer rate from the electron system to the phonon system, and

$$S_{\mathrm{p}} = n_{\mathrm{i}} \sum_{\boldsymbol{q}} |u(\boldsymbol{q})|^2 \sum_{n=-\infty}^{\infty} n\omega\, \mathrm{J}_n^2(\xi)\, \Pi_2(\boldsymbol{q},\omega_0-n\omega)$$
$$+ \sum_{\boldsymbol{q},\lambda} |M(\boldsymbol{q},\lambda)|^2 \sum_{n=-\infty}^{\infty} n\omega\, \mathrm{J}_n^2(\xi)\, \Lambda_2(\boldsymbol{q},\lambda,\omega_0-n\omega). \qquad (14.146)$$

is the net rate of the energy the electron system gains from the radiation field through multiphoton (absorption and emission) processes ($n = \pm 1, \pm 2, ...$) in associated with the intraband transition of electrons. The $\Lambda_2^{+}(\boldsymbol{q},\lambda,\Omega)$ function in the expression (14.145) of W is given by

$$\Lambda_2^{+}(\boldsymbol{q},\lambda,\Omega) \equiv 2\Pi_2(\boldsymbol{q},\Omega_{q\lambda}+\Omega)\left[n\!\left(\frac{\Omega_{q\lambda}}{T}\right) - n\!\left(\frac{\Omega_{q\lambda}+\Omega}{T_{\mathrm{e}}}\right)\right]. \qquad (14.147)$$

Note that the $\Lambda_2(\boldsymbol{q},\lambda,\Omega)$ function in the expressions (14.139) and (14.146) of $\boldsymbol{F}_0$ and S_{p} can also be replaced by the $\Lambda_2^{+}(\boldsymbol{q},\lambda,\Omega)$ function.

In the present scheme the average work done by the ac field $\boldsymbol{E}_\omega \sin(\omega t)$ on the electron system having the drift velocity (14.130) can be directly calculated to be

$$\langle [NeE_\omega \sin(\omega t)] \cdot [v_c \cos(\omega t) + v_s \sin(\omega t)] \rangle = \frac{1}{2} NeE_\omega \cdot v_s.$$

It can be seen, from the force-balance equations (14.137) and (14.138) and expressions (14.140) and (14.141), that this quantity exactly equals S_p:

$$\frac{1}{2} NeE_\omega \cdot v_s = -\frac{1}{2}(F_c \cdot v_c + F_s \cdot v_s) = -\frac{1}{2}(F_{2c} \cdot v_c + F_{2s} \cdot v_s) = S_p, \quad (14.148)$$

as given by (14.146).

In the limit of weak high-frequency field (small E_ω), v_c and v_s and thus the argument ξ of the Bessel functions in the expressions of F_{1j}, F_{2j}, W and S_p, are small. Retaining only the zero- and single-photon processes ($n = 0$ and ± 1) in the summation and keeping the leading terms of the expansions of the Bessel functions, $J_0(\xi) \approx 1$ and $J_{\pm 1}(\xi) \approx \pm \xi/2$, we have ($j = c, s$)

$$F_{1j} = -Nm\,v_j\, M_1(\omega, v_0), \quad (14.149)$$
$$F_{2j} = -Nm\,v_j\, M_2(\omega, v_0). \quad (14.150)$$

Here $M_1(\omega, v_0)$ and $M_2(\omega, v_0)$ are respectively the real part and the imaginary part of the memory function $M(\omega, v_0)$ under a dc bias velocity v_0 due to impurity and phonon scatterings:[2]

$$M_1(\omega, v_0) = \frac{n_i}{Nm\omega} \sum_q |u(q)|^2 q_\omega^2 \left[\Pi_1(q, q \cdot v_0) - \Pi_1(q, q \cdot v_0 + \omega) \right]$$
$$+ \frac{1}{2Nm\omega} \sum_{q,\lambda} |M(q,\lambda)|^2 q_\omega^2 \left[2\Lambda_1(q,\lambda, q \cdot v_0) \right.$$
$$\left. - \Lambda_1(q,\lambda, q \cdot v_0 - \omega) - \Lambda_1(q,\lambda, q \cdot v_0 + \omega) \right], \quad (14.151)$$

$$M_2(\omega, v_0) = \frac{n_i}{Nm\omega} \sum_q |u(q)|^2 q_\omega^2 \Pi_2(q, q \cdot v_0 - \omega),$$
$$+ \frac{1}{2Nm\omega} \sum_{q,\lambda} |M(q,\lambda)|^2 q_\omega^2 \left[\Lambda_2(q,\lambda, q \cdot v_0 - \omega) - \Lambda_2(q,\lambda, q \cdot v_0 + \omega) \right],$$

(14.152)

where q_ω stands for the component of q along the E_ω-direction (see also (14.88) in Sec. 14.4.1).

---

[2] In the case of E_ω parallel or perpendicular to bias dc field, v_c, v_s and E_ω are in the same direction.

With (14.149) and (14.150) for $\boldsymbol{F}_{1j}$ and $\boldsymbol{F}_{2j}$ in Eqs. (14.137) and (14.138), $\boldsymbol{v}_c$ and $\boldsymbol{v}_s$ are solved to be (the arguments of the memory functions are neglected for simplicity)

$$\boldsymbol{v}_c = \boldsymbol{v}_\omega \frac{1 + M_1/\omega}{(1 + M_1/\omega)^2 + (M_2/\omega)^2}, \qquad (14.153)$$

$$\boldsymbol{v}_s = -\boldsymbol{v}_\omega \frac{M_2/\omega}{(1 + M_1/\omega)^2 + (M_2/\omega)^2}. \qquad (14.154)$$

The energy absorbed per unit time by the electron system from the high-frequency field is then

$$S_\mathrm{p} = \frac{Ne^2}{2m} \frac{M_2(\omega, \boldsymbol{v}_0)}{(\omega + M_1)^2 + M_2^2} E_\omega^2, \qquad (14.155)$$

and the absorption coefficient

$$\alpha = \frac{1}{\sqrt{\kappa \epsilon_0} c} \frac{Ne^2}{m} \frac{M_2(\omega, \boldsymbol{v}_0)}{(\omega + M_1)^2 + M_2^2}. \qquad (14.156)$$

This expression for the linear absorption coefficient of a weak radiation field is in agreement with the result of perturbation theory for the free carrier absorption. It is equivalent to the result of the extended Drude-type formula (3.58) and (3.59) for the dynamic conductivity under a dc bias with a negligible $D(\omega, \boldsymbol{v}_0)$: $\alpha \sim \mathrm{Re}[\sigma(\omega, \boldsymbol{v}_0)]$ with

$$\sigma(\omega, \boldsymbol{v}_0) \simeq \frac{Ne^2}{m} \frac{\mathrm{i}}{\omega + M(\omega, \boldsymbol{v}_0)}. \qquad (14.157)$$

Formally, the expressions (14.155) and (14.156) are finite even in the zero-frequency limit. In reality, in view of the neglect of the electron temperature oscillation in the treatment, a growing inaccuracy may appear when the frequency of the radiation field descends down much lower than the sub-terahertz range.

Chapter 15

Radiation Driven Magnetotransport in Two-Dimensional Systems in Faraday Geometry

The balance-equation approach developed in the last chapter (Secs. 14.1–14.7) for hot-electron transport in semiconductors driven by a high-frequency electric field of single frequency, $\bm{E}_\omega \sin(\omega t)$, makes use of the fact that, when the harmonic generation is small and the frequency gets into THz regime or higher, the electron drift velocity in the steady transport state oscillates almost $\pi/2$ out of phase of the electric field, i.e., the oscillating drift velocity is essentially of the form $-\bm{v}_\omega \cos(\omega t)$. At the same time, all orders of this (frequency ω) photon assisted impurity and phonon scatterings are included in the relaxation processes. This method has been successfully applied to discuss THz photoabsorption and THz-induced dc conductivity response in three-dimensional (3D) and two-dimensional (2D) semiconductors in the case without a magnetic field or with a magnetic field in the Voigt configuration. As analyzed in Sec. 14.6, this approximation may no longer hold if the radiation-field frequency gets lower. Here we would like to point out that this assumption for the time-dependent form of the electron drift velocity will also be invalid when there is a strong magnetic field $\bm{B}$ not parallel to the high-frequency electric field $\bm{E}_\omega$ or the electron velocity $-\bm{v}_\omega \cos(\omega t)$ induced by it. Since the Lorentz force $-e\bm{v}_\omega \times \bm{B} \cos(\omega t)$ acting on the moving electron will drive it to move in the direction perpendicular to its velocity $\bm{v}_\omega$, the steady-state drift velocity of the electron will contain a term proportional to $(\omega_c/\omega)v_\omega \sin(\omega t)$, where $\omega_c = |eB|/m$ is the cyclotron frequency. This velocity, which is perpendicular to $\bm{v}_\omega$ and $\bm{B}$ and oscillates $\pi/2$ out of phase of $\bm{v}_\omega \cos(\omega t)$, may be of the same order of magnitude as, or even greater than, v_ω, if ω_c is of the same order of ω or $\omega_c > \omega$. Because of this, the balance-equation method presented in the last chapter is not able to deal with radiation-induced magnetotransport in

Faraday configuration if the condition $\omega_c/\omega \ll 1$ is not satisfied.

This limitation can be cured as long as we take a more generally form

$$\bm{V}(t) = \bm{v}_0 + \bm{v}_c \cos(\omega t) + \bm{v}_s \sin(\omega t)$$

to replace (14.16) as the center-of-mass velocity. As discussed in Sec. 14.6, by doing so, we are able not only to establish balance equations capable of dealing with radiation-induced magnetotransport in Faraday configuration for arbitrarily strong magnetic field but also capable of extending its applicable frequency to a much lower range.

In this chapter we will focus on quasi-two-dimensional electron systems outside the quantum Hall regime, that the ordinary quasi-particle description of extended Landau states is applicable. The balance-equation formulations (the equations and the expressions of the main physical quantities) in radiation-induced magnetotransport in Faraday configuration are similar for both 3D and 2D systems. Since the expressions of most of fundamental quantities for 3D semiconductors are the same as those presented in Sec. 14.6, one can easily pick them up to construct the framework for 3D magnetotransport in Faraday configuration in the same spirit.

15.1 Balance Equations for Radiation-Induced Magnetotransport in Two-Dimensional Electron Systems in Faraday Configuration

15.1.1 *Hamiltonian in terms of center-of-mass and relative electron variables*

We consider N_s electrons in a unit area of an infinite quasi-2D system, such as a semiconductor heterojunction or quantum well, in which electrons are free to move in the x–y plane but subjected to a confining potential $V(z)$ in the z direction. These electrons, besides interacting with each other, are coupled with phonons and scattered by randomly distributed impurities in the lattice.

To include possible elliptically polarized electromagnetic radiation we assume that a uniform dc electric field $\bm{E}_0$ and a high-frequency ac field of angular frequency ω,

$$\bm{E}(t) \equiv \bm{E}_s \sin(\omega t) + \bm{E}_c \cos(\omega t), \tag{15.1}$$

are applied inside the 2D system in the x–y plane, together with a magnetic field $\boldsymbol{B} = (0, 0, B)$ along the z direction. These magnetic and electric fields can be described by a vector potential $\boldsymbol{A}(\boldsymbol{r})$ and a scalar potential $\varphi(\boldsymbol{r}, t)$ of the form

$$\nabla \times \boldsymbol{A}(\boldsymbol{r}) = \boldsymbol{B}, \tag{15.2}$$

$$\varphi(\boldsymbol{r}, t) = -\boldsymbol{r} \cdot \boldsymbol{E}_0 - \boldsymbol{r} \cdot \boldsymbol{E}(t). \tag{15.3}$$

In the presence of these electric and magnetic fields the Hamiltonian of the system can be written as

$$H = H_\mathrm{e}(t) + H_\mathrm{ei} + H_\mathrm{ep} + H_\mathrm{ph}. \tag{15.4}$$

Here

$$H_\mathrm{e}(t) = \sum_j \left[\frac{1}{2m} (\boldsymbol{p}_{j\|} - e\boldsymbol{A}(\boldsymbol{r}_{j\|}))^2 + \varphi(\boldsymbol{r}_{j\|}, t) + \frac{p_{jz}^2}{2m_z} + V(z_j) \right]$$

$$+ \sum_{i<j} V_\mathrm{c}(\boldsymbol{r}_{i\|} - \boldsymbol{r}_{j\|}, z_i, z_j) \tag{15.5}$$

is the Hamiltonian of electrons subjected to the electric and magnetic fields with V_c standing for the electron–electron Coulomb interaction, H_ph is the Hamiltonian of phonons, H_ei and H_ep are, respectively, the electron–impurity and electron–phonon couplings. In equation (15.5) $\boldsymbol{r}_{j\|} \equiv (x_j, y_j)$ and $\boldsymbol{p}_{j\|} \equiv (p_{jx}, p_{jy})$ are the coordinate and momentum of the jth electron in the 2D plane, and z_j and p_{jz} are those perpendicular to the plane; m and m_z are, respectively, the effective mass parallel and perpendicular to the plane. The Zeeman splitting related to the electron spin is neglected in the following discussions in this chapter.

The spatial homogeneity of the fields and the parabolic band structure in the plane allows to describe the transport state of this system in terms of its center-of-mass and relative electron motions. The center-of-mass momentum $\boldsymbol{P}$ and coordinate $\boldsymbol{R}$ of a 2D electron system are defined as (Lei, Birman and Ting, 1985)

$$\boldsymbol{P} = \sum_j \boldsymbol{p}_{j\|}, \qquad \boldsymbol{R} = \frac{1}{N_\mathrm{s}} \sum_j \boldsymbol{r}_{j\|} \tag{15.6}$$

with $\boldsymbol{p}_{j\|} \equiv (p_{jx}, p_{jy})$ and $\boldsymbol{r}_{j\|} \equiv (x_j, y_j)$ being the momentum and coordinate of the jth electron in the 2D plane, and the relative electron momentum and coordinate are defined as

$$\boldsymbol{p}'_{j\|} = \boldsymbol{p}_{j\|} - \frac{1}{N_\mathrm{s}} \boldsymbol{P}, \qquad \boldsymbol{r}'_{j\|} = \boldsymbol{r}_{j\|} - \boldsymbol{R}. \tag{15.7}$$

In terms of the center-of-mass variables and relative electron variables, the electron Hamiltonian of the system, $H_e(t)$, can be written as the sum of a center-of-mass part H_c and a relative electron part H_{er}:

$$H_e(t) = H_c + H_{er} \tag{15.8}$$

with

$$H_c = \frac{1}{2N_s m}(\boldsymbol{P} - N_s e \boldsymbol{A}(\boldsymbol{R}))^2 - N_s e(\boldsymbol{E}_0 + \boldsymbol{E}(t)) \cdot \boldsymbol{R}, \tag{15.9}$$

$$H_{er} = \sum_j \left[\frac{1}{2m}(\boldsymbol{p}'_{j\|} - e\boldsymbol{A}(\boldsymbol{r}'_{j\|}))^2 + \frac{p_{jz}^2}{2m_z} + V(z_j) \right]$$

$$+ \sum_{i<j} V_c(\boldsymbol{r}'_{i\|} - \boldsymbol{r}'_{j\|}, z_i, z_j). \tag{15.10}$$

This relative-electron Hamiltonian H_{er} is just that of a quasi-2D system subjected to a uniform magnetic field in the z direction without dc nor high-frequency electric field. The finite confining potential in the z direction leads to the formation of electron subbands. In this chapter we will limit to the case that 2D electrons occupy only the lowest subband and thus ignore its subband index.

The phonon system is assumed to consist of bulk modes described by the 3D wavevector $\boldsymbol{q} = (\boldsymbol{q}_\|, q_z)$ and branch index λ having energy $\Omega_{q\lambda}$ and creation (annihilation) operator $b_{q\lambda}^\dagger (b_{q\lambda})$:

$$H_{ph} = \sum_{\boldsymbol{q},\lambda} \Omega_{q\lambda} b_{q\lambda}^\dagger b_{q\lambda}. \tag{15.11}$$

In terms of center-of-mass variables and relative electron variables, the electron–impurity and electron–phonon interactions are given by

$$H_{ei} = \sum_{j,a,\boldsymbol{q}_\|} U(\boldsymbol{q}_\|, z_a) \, e^{i\boldsymbol{q}_\| \cdot (\boldsymbol{R} - \boldsymbol{r}_{a\|})} \rho_{\boldsymbol{q}_\|}, \tag{15.12}$$

$$H_{ep} = \sum_{j,\boldsymbol{q}} M(\boldsymbol{q},\lambda) \phi_{\boldsymbol{q}\lambda} \, e^{i\boldsymbol{q}_\| \cdot \boldsymbol{R}} \rho_{\boldsymbol{q}_\|}, \tag{15.13}$$

expressed with the phonon field operator $\phi_{\boldsymbol{q}\lambda} = b_{\boldsymbol{q}\lambda} + b_{-\boldsymbol{q}\lambda}^\dagger$ and the density operator of the relative electrons in the 2D plane-wave representation,

$$\rho_{\boldsymbol{q}_\|} = \sum_j e^{i\boldsymbol{q}_\| \cdot \boldsymbol{r}'_{j\|}}. \tag{15.14}$$

Here $U(\boldsymbol{q}_\|, z_a)$ is the potential of the ath impurity locating at $(\boldsymbol{r}_{a\|}, z_a)$ and $M(\boldsymbol{q}_\|, q_z)$ is the matrix element of the electron–phonon interaction in the 3D plane-wave representation. Note that the uniform electric fields (dc and ac) appear only in H_c, and that H_{er} is just the Hamiltonian of the quasi-2D system subjected to the magnetic field without electric field. The coupling between the center-of-mass and the relative electrons enters only through the exponential factor $\exp(i\boldsymbol{q}_\| \cdot \boldsymbol{R})$ inside the 2D momemtum $\boldsymbol{q}_\|$ summation in H_{ei} and H_{ep}.

The derivation of balance equations starts with the Heisenberg equation for the rate of change of the center-of-mass velocity (operator) $\boldsymbol{V}$, which is the rate of change of the center-of-mass coordinate $\boldsymbol{R}$,

$$\boldsymbol{V} \equiv \dot{\boldsymbol{R}} = -i[\boldsymbol{R}, H] = \frac{1}{N_s m}(\boldsymbol{P} - N_s e \boldsymbol{A}(\boldsymbol{R})), \qquad (15.15)$$

to get the operator equation for the force balance:

$$\begin{aligned}\dot{\boldsymbol{V}} &= -i[\boldsymbol{V}, H] + \frac{\partial \boldsymbol{V}}{\partial t} \\ &= \frac{e}{m}\left[\boldsymbol{E}_0 + \boldsymbol{E}(t) + \boldsymbol{V} \times \boldsymbol{B}\right] + \frac{\hat{\boldsymbol{F}}}{N_s m},\end{aligned} \qquad (15.16)$$

in which the frictional force operator is given by

$$\hat{\boldsymbol{F}} = -i \sum_{\boldsymbol{q}_\|, a} U(\boldsymbol{q}_\|, z_a)\, \boldsymbol{q}_\| \, e^{i\boldsymbol{q}_\| \cdot (\boldsymbol{R}-\boldsymbol{r}_{a\|})} \rho_{\boldsymbol{q}_\|} - i \sum_{\boldsymbol{q}, \lambda} M(\boldsymbol{q}, \lambda) \phi_{\boldsymbol{q}\lambda}\, e^{i\boldsymbol{q}_\| \cdot \boldsymbol{R}} \rho_{\boldsymbol{q}_\|}. \qquad (15.17)$$

The equation for the rate of change of the relative electron energy,

$$\begin{aligned}\dot{H}_{er} &= -i[H_{er}, H] \\ &= -i\sum_{\boldsymbol{q}_\|, a} U(\boldsymbol{q}_\|, z_a)\, e^{i\boldsymbol{q}_\| \cdot (\boldsymbol{R}-\boldsymbol{r}_{a\|})} \dot\rho_{\boldsymbol{q}_\|} - i\sum_{\boldsymbol{q}, \lambda} M(\boldsymbol{q}, \lambda) \phi_{\boldsymbol{q}\lambda}\, e^{i\boldsymbol{q}_\| \cdot \boldsymbol{R}} \dot\rho_{\boldsymbol{q}_\|},\end{aligned} \qquad (15.18)$$

will serve as the operator equation for the energy balance, in which $\dot\rho_{\boldsymbol{q}_\|} \equiv -i[\rho_{\boldsymbol{q}_\|}, H_{er}]$.

15.1.2 Force- and energy-balance equations

The center-of-mass coordinate $\boldsymbol{R}$ and velocity $\boldsymbol{V}$ will be treated classically and their fluctuations neglected. Thus the variables $\boldsymbol{R}$ and $\boldsymbol{V}$ in the above equations are the time-dependent expectation values of the center-of-mass coordinate and velocity, $\boldsymbol{R}(t)$ and $\boldsymbol{V}(t)$. The determination of the statistical average of the operator equation (15.16) subject to an arbitrary time-dependent electric field $\boldsymbol{E} = \boldsymbol{E}_0 + \boldsymbol{E}(t)$ follows the same procedure as in Chapter 1 to the lowest order in the scattering interaction, $H_\mathrm{I} = H_\mathrm{ei} + H_\mathrm{ep}$, using the initial density matrix $\hat{\rho}_0$ given by (1.30). By identifying $\langle \dot{\boldsymbol{V}} \rangle = d\boldsymbol{V}/dt$ as the rate of change of the center-of-mass velocity in the presence of electric and magnetic fields, we obtain the force balance equation of the form

$$N_\mathrm{s} m \frac{d\boldsymbol{V}}{dt} = N_\mathrm{s} e \boldsymbol{E} + N e \boldsymbol{V} \times \boldsymbol{B} + \boldsymbol{F}_\mathrm{i} + \boldsymbol{F}_\mathrm{p}. \qquad (15.19)$$

Here the frictional forces $\boldsymbol{F}_\mathrm{i}$ and $\boldsymbol{F}_\mathrm{p}$ are of the form

$$\boldsymbol{F}_\mathrm{i} = -\mathrm{i} \sum_{\boldsymbol{q}} \boldsymbol{q}_\| |U(\boldsymbol{q}_\|)|^2 \int_{-\infty}^{\infty} dt'\, A(\boldsymbol{q}_\|, t, t')\, \Pi(\boldsymbol{q}_\|, t - t'), \qquad (15.20)$$

$$\boldsymbol{F}_\mathrm{p} = -\mathrm{i} \sum_{\boldsymbol{q},\lambda} \boldsymbol{q}_\| |M(\boldsymbol{q}, \lambda)|^2 \int_{-\infty}^{\infty} dt'\, A(\boldsymbol{q}_\|, t, t')\, \Lambda(\boldsymbol{q}, \lambda, t - t'), \qquad (15.21)$$

in which the correlation functions are given by

$$\Pi(\boldsymbol{q}_\|, t - t') = -\mathrm{i}\, \theta(t - t') \langle [\rho_{\boldsymbol{q}_\|}(t), \rho_{-\boldsymbol{q}_\|}(t')] \rangle_0, \qquad (15.22)$$

$$\Lambda(\boldsymbol{q}, \lambda, t - t') = -\mathrm{i}\, \theta(t - t') \langle [\phi_{\boldsymbol{q}\lambda}(t) \rho_{\boldsymbol{q}_\|}(t), \phi_{-\boldsymbol{q}\lambda}(t') \rho_{-\boldsymbol{q}_\|}(t')] \rangle_0, \qquad (15.23)$$

with the step function $\theta(t) = 1$ for $t \geqslant 0$ and $\theta(t) = 0$ for $t < 0$, and operators in the interaction picture:

$$\rho_{\boldsymbol{q}_\|}(t) = \mathrm{e}^{\mathrm{i} H_\mathrm{er} t} \rho_{\boldsymbol{q}_\|} \mathrm{e}^{-\mathrm{i} H_\mathrm{er} t}, \qquad (15.24)$$

$$\phi_{\boldsymbol{q}\lambda}(t) = \mathrm{e}^{\mathrm{i} H_\mathrm{ph} t} \phi_{\boldsymbol{q}\lambda} \mathrm{e}^{-\mathrm{i} H_\mathrm{ph} t} = b_{\boldsymbol{q}\lambda} \mathrm{e}^{-\mathrm{i}\Omega_{\boldsymbol{q}\lambda} t} + b^\dagger_{-\boldsymbol{q}\lambda} \mathrm{e}^{\mathrm{i}\Omega_{\boldsymbol{q}\lambda} t}. \qquad (15.25)$$

The $A(\boldsymbol{q}_\|, t, t')$ function in (15.20) and (15.21) is solely determined by the moving center of mass,

$$A(\boldsymbol{q}_\|, t, t') \equiv \exp\left[\mathrm{i} \boldsymbol{q}_\| \cdot (\boldsymbol{R}(t) - \boldsymbol{R}(t')) \right] \simeq \exp\left[\mathrm{i} \boldsymbol{q}_\| \cdot \int_{t'}^{t} \boldsymbol{V}(s) ds \right]. \qquad (15.26)$$

The center-of-mass velocity $\boldsymbol{V}(t)$ and electron temperature $T_\mathrm{e}(t)$ are therefore functionally involved in $\boldsymbol{F}_\mathrm{i}$ and $\boldsymbol{F}_\mathrm{p}$.

For slowly time-varying $\boldsymbol{V}(t)$ with the approximate $A(\boldsymbol{q}_\|, t, t')$ expansion (3.2), a simplified 2D time-dependent force-balance equation resembling Eq. (8.12) can be obtained with the damping forces $\boldsymbol{f}_\mathrm{i}$ and $\boldsymbol{f}_\mathrm{p}$ of 2D forms corresponding to (8.14) and (8.15). These equations, however, are suitable for relatively slowly time-varying transport.

Now we are dealing with electron transport under a high-frequency irradiation for which the small $\dot{V}(t)$ approximation does not apply. However, since we focus mainly on the photoresistance, i.e. the effect of a monochromatic electromagnetic field on the dc resistivity, which is a physical quantity directly relating to the time-averaged part of the center-of-mass velocity, the steady time-dependent transport under a modest radiation can be treated while disregarding the higher harmonic current. As a mater of fact, in an ordinary semiconductor the generated power of even the lowest harmonic current is rather weak as compared to the fundamental. For the radiation field intensity concerned in most of studies in this chapter, which is at least an order of magnitude smaller than that needed for the harmonic generation ($\sim 10\,\mathrm{kV/cm}$, see Sec. 12.8.3), the effect of harmonic current is safely negligible. Therefore, we assume that the center-of-mass velocity consists of a dc part $\boldsymbol{v}_0$ and a stationary time-dependent part $\boldsymbol{v}(t)$ of the form

$$\boldsymbol{V}(t) = \boldsymbol{v}_0 + \boldsymbol{v}_c \cos(\omega t) + \boldsymbol{v}_s \sin(\omega t), \qquad (15.27)$$

i.e. the drift velocity is truncated up to the first harmonics. This approximation, though neglecting higher harmonic components of the ac current, retains the most important effect induced by the multiphoton-assisted scattering processes due to photons of the base frequency ω. With the center-of-mass velocity given by (15.27) we can expand the function

$$A(\boldsymbol{q}_{\parallel}, t, t') = \exp\left[i\boldsymbol{q}_{\parallel} \cdot \int_{t'}^{t} \boldsymbol{V}(s)ds\right] = \sum_{n=-\infty}^{\infty} \mathrm{J}_n^2(\xi)\, e^{i(\boldsymbol{q}_{\parallel} \cdot \boldsymbol{v}_0 - n\omega)(t-t')}$$

$$+ \sum_{m \neq 0} e^{i m(\omega t - \varphi)}\left[\sum_{n=-\infty}^{\infty} \mathrm{J}_n(\xi)\mathrm{J}_{n-m}(\xi)\, e^{i(\boldsymbol{q}_{\parallel} \cdot \boldsymbol{v}_0 - n\omega)(t-t')}\right], \quad (15.28)$$

in terms of Bessel functions $\mathrm{J}_n(\xi)$. Here $\tan\varphi = (\boldsymbol{q}_{\parallel} \cdot \boldsymbol{v}_s)/(\boldsymbol{q}_{\parallel} \cdot \boldsymbol{v}_c)$, and

$$\xi \equiv \frac{1}{\omega}\left[(\boldsymbol{q}_{\parallel} \cdot \boldsymbol{v}_c)^2 + (\boldsymbol{q}_{\parallel} \cdot \boldsymbol{v}_s)^2\right]^{\frac{1}{2}}. \qquad (15.29)$$

On the other hand, we treat the electron temperature as a constant in the steady transport state driven by radiation field of high frequency. The frictional force $\boldsymbol{F} = \boldsymbol{F}_\mathrm{i} + \boldsymbol{F}_\mathrm{p}$ derived from (15.21) and (15.22), then can be written, up to the terms oscillating with the base frequency ω, as

$$\boldsymbol{F} = \boldsymbol{F}_0 + \boldsymbol{F}_s \sin(\omega t) + \boldsymbol{F}_c \cos(\omega t) = \boldsymbol{F}(t). \qquad (15.30)$$

In this, the frictional forces $\bm{F}_0$, $\bm{F}_s = \bm{F}_{2s} - \bm{F}_{1c}$, and $\bm{F}_c = \bm{F}_{1s} + \bm{F}_{2c}$, are given by

$$\bm{F}_0 = \sum_{\bm{q}_\|} \bm{q}_\| \, |U(\bm{q}_\|)|^2 \sum_{n=-\infty}^{\infty} \mathrm{J}_n^2(\xi)\, \Pi_2(\bm{q}_\|, \omega_0 - n\omega)$$

$$+ \sum_{\bm{q},\lambda} \bm{q}_\| \, |M(\bm{q},\lambda)|^2 \sum_{n=-\infty}^{\infty} \mathrm{J}_n^2(\xi)\, \Lambda_2(\bm{q},\lambda, \omega_0 - n\omega), \qquad (15.31)$$

$$\bm{F}_{1j} = \sum_{\bm{q}_\|} \bm{q}_\| \eta_j \, |U(\bm{q}_\|)|^2 \sum_{n=-\infty}^{\infty} [\mathrm{J}_n^2(\xi)]'\, \Pi_1(\bm{q}_\|, \omega_0 - n\omega)$$

$$+ \sum_{\bm{q},\lambda} \bm{q}_\| \eta_j \, |M(\bm{q},\lambda)|^2 \sum_{n=-\infty}^{\infty} [\mathrm{J}_n^2(\xi)]'\, \Lambda_1(\bm{q},\lambda, \omega_0 - n\omega), \qquad (15.32)$$

$$\bm{F}_{2j} = -\sum_{\bm{q}_\|} \bm{q}_\| \frac{\eta_j}{\xi} |U(\bm{q}_\|)|^2 \sum_{n=-\infty}^{\infty} 2n\, \mathrm{J}_n^2(\xi)\, \Pi_2(\bm{q}_\|, \omega_0 - n\omega)$$

$$- \sum_{\bm{q},\lambda} \bm{q}_\| \frac{\eta_j}{\xi} |M(\bm{q},\lambda)|^2 \sum_{n=-\infty}^{\infty} 2n\, \mathrm{J}_n^2(\xi)\, \Lambda_2(\bm{q},\lambda, \omega_0 - n\omega), \qquad (15.33)$$

where $\omega_0 \equiv \bm{q}_\| \cdot \bm{v}_0$, $\eta_j \equiv \bm{q}_\| \cdot \bm{v}_j/\omega\xi$ and $j = c$ and s. In these equations, $\Pi_2(\bm{q}_\|, \Omega)$ and $\Pi_1(\bm{q}_\|, \Omega)$ are the imaginary and real parts of the electron density correlation function $\Pi(\bm{q}_\|, \Omega)$ of the 2D system in the presence of the magnetic field; $\Lambda_2(\bm{q},\lambda, \Omega)$ and $\Lambda_1(\bm{q},\lambda, \Omega)$ are the imaginary and real parts of the electron–phonon correlation function $\Lambda(\bm{q},\lambda, \Omega)$, which can be expressed in terms of $\Pi(\bm{q}_\|, \Omega)$ function as

$$\Lambda_2(\bm{q},\lambda, \Omega) = \Pi_2(\bm{q}_\|, \Omega_{q\lambda} + \Omega) \left[n\!\left(\frac{\Omega_{q\lambda}}{T}\right) - n\!\left(\frac{\Omega_{q\lambda} + \Omega}{T_\mathrm{e}}\right) \right]$$

$$- \Pi_2(\bm{q}_\|, \Omega_{q\lambda} - \Omega) \left[n\!\left(\frac{\Omega_{q\lambda}}{T}\right) - n\!\left(\frac{\Omega_{q\lambda} - \Omega}{T_\mathrm{e}}\right) \right], \qquad (15.34)$$

$$\Lambda_1(\bm{q},\lambda, \Omega) = \left[\Pi_1(\bm{q}_\|, \Omega_{q\lambda} + \Omega) + \Pi_1(\bm{q}_\|, \Omega_{q\lambda} - \Omega)\right] n\!\left(\frac{\Omega_{q\lambda}}{T}\right)$$

$$+ \frac{1}{\pi}\int_{-\infty}^{\infty} d\omega_1\, \Pi_2(\bm{q}, \omega_1)\, n\!\left(\frac{\omega_1}{T_\mathrm{e}}\right)$$

$$\times \left(\frac{1}{\Omega_{q\lambda} + \Omega - \omega_1} + \frac{1}{\Omega_{q\lambda} - \Omega - \omega_1} \right). \qquad (15.35)$$

Note that the $\Lambda_2(\bm{q}, \lambda, \Omega)$ function in the frictional force expressions (15.31) and (15.33) can be replaced by the function

$$\Lambda_2^+(q, \lambda, \Omega) \equiv \Pi_2(q_\parallel, \Omega_{q\lambda} + \Omega) \left[n\left(\frac{\Omega_{q\lambda}}{T}\right) - n\left(\frac{\Omega_{q\lambda} + \Omega}{T_e}\right) \right]. \quad (15.36)$$

Up to the base frequency oscillating terms the momentum-balance equation (15.19) then reduces to

$$v_s\omega\cos(\omega t) - v_c\omega\sin(\omega t) = \frac{1}{N_s m} F(t) + \frac{e}{m}\{E_0 + E(t) + [v_0 + v(t)] \times B\}. \quad (15.37)$$

Or, written separately in terms of the average and base frequency oscillating components,

$$N_s e E_0 + N_s e (v_0 \times B) + F_0 = 0, \quad (15.38)$$

$$\omega v_c = -\frac{eE_s}{m} - \frac{F_s}{N_s m} - \frac{e}{m}(v_s \times B), \quad (15.39)$$

$$\omega v_s = \frac{eE_c}{m} + \frac{F_c}{N_s m} + \frac{e}{m}(v_c \times B). \quad (15.40)$$

The energy-balance equation is obtained by taking the long-time average of statistically averaged operator equation (15.18) for the steady state:

$$N_s e E_0 \cdot v_0 + S_p - W = 0, \quad (15.41)$$

in which

$$W = \sum_{q,\lambda} \Omega_{q\lambda} |M(q,\lambda)|^2 \sum_{n=-\infty}^{\infty} J_n^2(\xi) \Lambda_2^+(q,\lambda, \omega_0 - n\omega) \quad (15.42)$$

is the energy-transfer rate from the electron system to the phonon system,

$$S_p = \sum_{q_\parallel} |U(q_\parallel)|^2 \sum_{n=-\infty}^{\infty} n\omega\, J_n^2(\xi)\, \Pi_2(q_\parallel, \omega_0 - n\omega)$$

$$+ \sum_{q,\lambda} |M(q,\lambda)|^2 \sum_{n=-\infty}^{\infty} n\omega\, J_n^2(\xi)\, \Lambda_2^+(q,\lambda, \omega_0 - n\omega) \quad (15.43)$$

is the time-averaged rate of the energy the electron system gains from the high-frequency field. As a matter of fact, the average work done by the high-frequency field $E_s \sin(\omega t) + E_c \cos(\omega t)$ on the electron system having a drift velocity (15.27) can be directly calculated to be

$$\langle N_s e [E_s \sin(\omega t) + E_c \cos(\omega t)] \cdot [v_c \cos(\omega t) + v_s \sin(\omega t)] \rangle = \frac{N_s e}{2}(E_c \cdot v_c + E_s \cdot v_s).$$

According to force-balance equations (15.39) and (15.40), this quantity exactly equals S_p:

$$\frac{N_s e}{2}(\boldsymbol{E}_c \cdot \boldsymbol{v}_c + \boldsymbol{E}_s \cdot \boldsymbol{v}_s) = -\frac{1}{2}(\boldsymbol{F}_c \cdot \boldsymbol{v}_c + \boldsymbol{F}_s \cdot \boldsymbol{v}_s) = S_\mathrm{p}, \qquad (15.44)$$

as expressed in (15.43).

The momentum and energy balance equations (15.38) to (15.41) constitute a close set of equations to determine the parameters $\boldsymbol{v}_0$, $\boldsymbol{v}_c$, $\boldsymbol{v}_s$, and T_e when $\boldsymbol{E}_0$, $\boldsymbol{E}_c$ and $\boldsymbol{E}_s$ are given.

The summation over n in the expressions for $\boldsymbol{F}_0$, $\boldsymbol{F}_{ij}$, W and S_p represents contribution of all orders of multiphoton processes related to the photons of frequency ω. In the present formulation, the role of the single-frequency radiation field is two folds. (1) It induces photon-assisted impurity and phonon scatterings associated with single ($|n| = 1$) and multiple ($|n| \geqslant 1$) photon processes, which are superposed on the direct impurity and phonon scattering ($n = 0$) term. (2) It transfers energy S_p to the electron system through single and multiple photon-assisted processes.

15.1.3 Density correlation function of 2D electrons in a magnetic field, Landau level broadening

The $\Pi_2(\boldsymbol{q}_\|, \Omega)$ and $\Pi_1(\boldsymbol{q}_\|, \Omega)$ in the expressions of $\boldsymbol{F}_0$, $\boldsymbol{F}_{ij}$, W and S_p are respectively the imaginary and real parts of the electron density correlation function $\Pi(\boldsymbol{q}_\|, \Omega)$ of the investigated 2D system in the presence of the magnetic field. In the present approach, in addition to the contribution of a Lorentz force in the momentum balance equations, the only way a magnetic field enters in the formulation is through this electron density correlation function. The density correlation function of a 2D electron system in a magnetic field has been discussed in Chapter 8.

In the random phase approximation the quasi-2D electron density correlation function is given by (lowest subband only)

$$\Pi(\boldsymbol{q}_\|, \omega) = \frac{\Pi_0(\boldsymbol{q}_\|, \omega)}{1 - V(q_\|) \Pi_0(\boldsymbol{q}_\|, \omega)} = \frac{\Pi_0(\boldsymbol{q}_\|, \omega)}{\epsilon(\boldsymbol{q}_\|, \omega)}, \qquad (15.45)$$

where $[H(q_\|) = H_{0000}(q_\|)$ see (2.79), or (2.136)]

$$V(q_\|) = \frac{e^2}{2\epsilon_0 \kappa q_\|} H(q_\|), \qquad (15.46)$$

and $\Pi_0(\boldsymbol{q}_\parallel, \omega)$ is the noninteracting electron density correlation function of a pure 2D system in the presence of the magnetic field.

The density correlation function of noninteracting electrons of a pure 2D system in a magnetic field, $\Pi_0(\boldsymbol{q}_\parallel, \omega)$, can be expressed as weighted sums of the Landau representation correlation function $\Pi_0(n, n', \omega)$ over all the Landau levels (Ting, Ying and Quinn, 1977):

$$\Pi_0(\boldsymbol{q}_\parallel, \omega) = \frac{1}{2\pi l_B^2} \sum_{n,n'} C_{nn'}(l_B^2 q_\parallel^2/2) \Pi_0(n, n', \omega), \qquad (15.47)$$

where $l_B \equiv \sqrt{1/|eB|}$ is the magnetic length and $C_{nn'}(x)$ are functions given by (8.25):

$$C_{nn'}(x) \equiv \frac{n_2!}{n_1!} x^{n_1-n_2} e^{-x} \left[L_{n_2}^{n_1-n_2}(x)\right]^2, \qquad (15.48)$$

in which $n_1 = \max(n, n')$, $n_2 = \min(n, n')$, and $L_m^l(x)$ are associated Laguerre polynomials:

$$L_m^l(x) = \sum_{s=0}^{m} (-1)^s \frac{(m+l)! x^s}{(l+s)!(m-s)!s!}. \qquad (15.49)$$

The correlation function $\Pi_0(n, n', \omega)$ can be constructed using the retarded Green's function $G_n(\omega)$. Its real part and imaginary part are given by (Ting, Ying and Quinn, 1977)

$$\Pi_{01}(n, n', \omega) = -\frac{2}{\pi} \int_{-\infty}^{\infty} d\varepsilon\, f(\varepsilon) \left[\operatorname{Re} G_n(\varepsilon + \omega) \operatorname{Im} G_{n'}(\varepsilon) \right.$$
$$\left. + \operatorname{Re} G_{n'}(\varepsilon - \omega) \operatorname{Im} G_n(\varepsilon)\right], \qquad (15.50)$$

$$\Pi_{02}(n, n', \omega) = -\frac{2}{\pi} \int_{-\infty}^{\infty} d\varepsilon \left[f(\varepsilon) - f(\varepsilon + \omega)\right] \operatorname{Im} G_n(\varepsilon + \omega) \operatorname{Im} G_{n'}(\varepsilon). \qquad (15.51)$$

Here $f(\varepsilon) = f(\varepsilon, T_e)$ is the Fermi distribution function at electron temperature T_e, $\operatorname{Re} G_n(\omega)$ and $\operatorname{Im} G_n(\omega)$ are respectively the real part and the imaginary part of the single-particle retarded Green's function of the nth Landau level. The latter is proportional to the density-of-states $D_n(\varepsilon)$ of the nth Landau level:

$$D_n(\varepsilon) = -\frac{1}{\pi^2 l_B^2} \operatorname{Im} G_n(\varepsilon). \qquad (15.52)$$

The electron density N_s of the 2D system equals the number of all the occupied states of the Landau levels, i.e.

$$N_s = -\frac{1}{\pi^2 l_B^2} \sum_{n=0}^{\infty} \int_{-\infty}^{\infty} d\varepsilon\, f(\varepsilon) \mathrm{Im} G_n(\varepsilon). \tag{15.53}$$

This equation determines the chemical potential.

Different forms of the $\mathrm{Im} G_n(\varepsilon)$ function or the density-of-states function $D_n(\varepsilon)$ have been discussed in Sec. 8.6.2.

15.2 Nonlinear Cyclotron Resonance in Quasi-2D Systems

15.2.1 Cyclotron resonance of drift velocity and dynamic conductivity

From force-balance equations (15.39) and (15.40), we can write

$$\begin{aligned}
\boldsymbol{v}_c &= \frac{-1}{(1-\omega_c^2/\omega^2)} \Big\{ \frac{e}{m\omega}\left[\boldsymbol{E}_s + \frac{e}{m\omega}(\boldsymbol{E}_c \times \boldsymbol{B})\right] \\
&\quad + \frac{1}{N_e m\omega}\left[\boldsymbol{F}_s + \frac{e}{m\omega}(\boldsymbol{F}_c \times \boldsymbol{B})\right] \Big\},
\end{aligned} \tag{15.54}$$

$$\begin{aligned}
\boldsymbol{v}_s &= \frac{1}{(1-\omega_c^2/\omega^2)} \Big\{ \frac{e}{m\omega}\left[\boldsymbol{E}_c - \frac{e}{m\omega}(\boldsymbol{E}_s \times \boldsymbol{B})\right] \\
&\quad + \frac{1}{N_e m\omega}\left[\boldsymbol{F}_c - \frac{e}{m\omega}(\boldsymbol{F}_s \times \boldsymbol{B})\right] \Big\}.
\end{aligned} \tag{15.55}$$

Cyclotron resonance is directly seen in the case of weak scatterings when the terms with $\boldsymbol{F}_s$ or $\boldsymbol{F}_c$ in the above equations are small: both $\boldsymbol{v}_c$ and $\boldsymbol{v}_s$ exhibit peaks around $\omega_c \sim \omega$. Since all the transport quantities can be expressed as functions of drift velocities $\boldsymbol{v}_c$ and $\boldsymbol{v}_s$ at given $\boldsymbol{v}_0$, the cyclotron resonance of $\boldsymbol{v}_c$ and $\boldsymbol{v}_s$ may result in resonance of other transport quantities at $\omega_c \sim \omega$. For instance, the energy absorption of the electron system from the radiation field, S_p, as direct expressed by $\boldsymbol{v}_c$ and $\boldsymbol{v}_s$ [see (15.44)], peaks at $\omega_c \sim \omega$.

Equations (15.54) and (15.55) can be further simplified when the radiation field is weak and the dc field is absent. In this case $\boldsymbol{v}_c$ and $\boldsymbol{v}_s$ can be treated as small quantities. To the first order of these small parameters, the force functions $\boldsymbol{F}_{1j}$ and $\boldsymbol{F}_{2j}$ can be written as ($j = c, s$)

$$\boldsymbol{F}_{1j} = -N_s m\, \boldsymbol{v}_j\, M_1(\omega), \tag{15.56}$$

$$\boldsymbol{F}_{2j} = -N_s m\, \boldsymbol{v}_j\, M_2(\omega), \tag{15.57}$$

where $M_1(\omega)$ and $M_2(\omega)$ are the real and imaginary parts of the memory functions (Sec. 8.5). It is more convenient to write out the relevant formulas for the complex velocities $v_+ \equiv v_{cx} + iv_{sx}$ and $v_- \equiv v_{sy} - iv_{cy}$ rather than for $\boldsymbol{v}_c$ and $\boldsymbol{v}_s$:

$$v_+ = \frac{e\tau}{2m^*}\left[\frac{E_+}{(\omega - \omega_c^*)\tau + i} + \frac{E_-}{(\omega + \omega_c^*)\tau + i}\right], \tag{15.58}$$

$$v_- = -\frac{e\tau}{2m^*}\left[\frac{E_+}{(\omega - \omega_c^*)\tau + i} - \frac{E_-}{(\omega + \omega_c^*)\tau + i}\right]. \tag{15.59}$$

Here, we have defined

$$m^* = m[1 + M_1(\omega)/\omega],$$
$$1/\tau = M_2(\omega)/[1 + M_1(\omega)/\omega],$$
$$\omega_c^* = eB/m^*,$$
$$E_+ = E_{sx} + E_{cy} + i(E_{sy} - E_{cx}),$$
$$E_- = E_{sx} - E_{cy} - i(E_{sy} + E_{cx}),$$

with $\boldsymbol{E}_s \equiv (E_{sx}, E_{sy})$ and $\boldsymbol{E}_c \equiv (E_{cx}, E_{cy})$.

It can been seen that for circularly polarized ac field ($E_{sx} = E_\omega, E_{sy} = 0, E_{cx} = 0, E_{cy} = \pm E_\omega$) the dynamic conductivity [as defined by (8.73), (8.82) and (8.93)] given by (15.58) and (15.59) is the same as (8.105):

$$\sigma_\pm(\omega) = \frac{\sigma^*}{1 - i(\omega \pm \omega_c^*)\tau}. \tag{15.60}$$

And, by defining $\sigma_\pm(\omega) \equiv \sigma_{xx}(\omega) \pm i\sigma_{xy}(\omega)$, one can write

$$\sigma_{xx}(\omega) = \sigma^* \frac{1 - i\omega\tau}{(1 - i\omega\tau)^2 + (\omega_c^*\tau)^2}, \tag{15.61}$$

$$\sigma_{xy}(\omega) = \sigma^* \frac{\omega_c^*\tau}{(1 - i\omega\tau)^2 + (\omega_c^*\tau)^2}, \tag{15.62}$$

in which $\sigma^* \equiv ne^2\tau/m^*$.

15.2.2 Incident electromagnetic field, selfconsistent field and radiative decay

The high-frequency electric field

$$\boldsymbol{E}(t) = \boldsymbol{E}_s \sin(\omega t) + \boldsymbol{E}_c \cos(\omega t) \qquad (15.63)$$

appearing in the force-balance equations (15.39) and (15.40) is the total (external and induced) field really acting on the 2D electrons. Experiments are always performed under the condition of giving external incident radiation. We assume that the electromagnetic wave propagates perpendicularly (along z-axis) towards the 2D electrons from the air with an incident electric field

$$\boldsymbol{E}_{\rm i}(t) = \boldsymbol{E}_{{\rm i}s} \sin(\omega t) + \boldsymbol{E}_{{\rm i}c} \cos(\omega t) \qquad (15.64)$$

at plane $z = 0$. The relation between $\boldsymbol{E}(t)$ and $\boldsymbol{E}_{\rm i}(t)$ is easily obtained by solving the Maxwell equations connecting both sides of the 2D electron gas which is carrying a sheet current density $N_s e \boldsymbol{v}(t)$. If the 2D electron gas locates within a thin layer under the surface plane at $z = 0$ of a thick (treated as semi-infinite) semiconductor substrate having a refraction index $n_{\rm s}$, the field $\boldsymbol{E}(t)$ driving the 2D electrons, which equals the sum of the incident and the reflected fields at $z = 0$ and equals the transmitted field (the field just passes through the 2D layer), can be expressed as

$$\boldsymbol{E}(t) = \frac{N_s e \boldsymbol{v}(t)}{(n_0 + n_{\rm s})\epsilon_0 c} + \frac{2n_0}{n_0 + n_{\rm s}} \boldsymbol{E}_{\rm i}(t). \qquad (15.65)$$

Here n_0 is the refractive index of the air and c and ϵ_0 are respectively the light speed and the dielectric constant in vacuum (Chiu, Lee and Quinn, 1976; Liu and Lei, 2003). If the 2D electron gas is contained in a thin layer suspended in vacuum at the plane $z = 0$, then

$$\boldsymbol{E}(t) = \frac{N_s e \boldsymbol{v}(t)}{2\epsilon_0 c} + \boldsymbol{E}_{\rm i}(t). \qquad (15.66)$$

Thus, the transmitted field $\boldsymbol{E}(t)$ depends on the drift velocity $\boldsymbol{v}(t)$ or the oscillating current of the 2D electrons. For a given incident field $\boldsymbol{E}_{\rm i}(t)$, it should be selfconsistently determined by combining the fore-balance equations (15.39) and (15.40) with Eq. (15.65) or (15.66). This electrodynamic effect, i.e. the high-frequency field inside the 2D electron gas is not solely determined by the incident field but also depends on the oscillating current of 2D electrons themselves, yields an additional damping force to the motion of 2D electrons. The effect of this damping is equivalent to a decay resulting from the radiation produced by the oscillating 2D electrons, thus called the radiative decay (Mikhailov, 2004).

15.2.3 Transmittance and Faraday effect

A detailed numerical study (Liu and Lei, 2003) on the magnetotransport at lattice temperature $T = 4.2\,\text{K}$ in a GaAs/AlGaAs heterojunction driven by a linearly polarized sinusoidal incident radiation field along the x-axis,

$$\boldsymbol{E}_\text{i}(t) = (E_{\text{i}s}\sin(\omega t), 0), \tag{15.67}$$

is carried out in the vicinity of the cyclotron resonance $\omega_\text{c}/\omega \sim 1$. The system has an electron sheet density $N_\text{s} = 2.5 \times 10^{15}\,\text{m}^{-2}$ and a low-temperature linear mobility $50\,\text{m}^2/\text{Vs}$ in the absence of the magnetic field. The elastic scattering due to randomly distributed charged impurity and the nonelastic scattering due to polar optic phonons (via Fröhlich coupling with electrons), longitudinal acoustic phonons (via deformation potential and piezoelectric coupling), and transverse acoustic phonons (via piezoelectric coupling with electrons) are taken into account. The material and electron–phonon coupling parameters are taken as typical values for GaAs. For a given strength of the incident field $E_{\text{i}s}$ of frequency ω, the electron temperature T_e and the oscillating electron velocities $\boldsymbol{v}_c$ and $\boldsymbol{v}_s$ (thus the transmitted field $\boldsymbol{E}(t)$) are determined from the coupled force- and energy-balance equations (15.39), (15.40) and (15.41) in the case of $\boldsymbol{v}_0 = 0$, together with equation (15.65).

We choose a Gaussian-type density-of-states function as given by (8.133), $\text{Im}\,G_n(\varepsilon) = -(\sqrt{2\pi}/\Gamma_n)\exp[-2(\varepsilon - \varepsilon_n)^2/\Gamma_n^2]$, for simulating the Landau-level broadening with a unified broadening parameter $\Gamma_n = \Gamma = (8e\omega_\text{c}/\pi m\mu_0)^{1/2}$ for all the Landau levels. To approximately include the effect of electron heating on the Landau level broadening, we treat μ_0 as the linear mobility of the 2D system in the absence of the magnetic field at temperature T_e. This treatment, though very crude, can still obtain a qualitative agreement between the calculated and experimental results within the magnetic field range considered. In the numerical calculation the maximum Landau level is taken to be 20, and the summations over the multiphoton index n are carried out up to a given accuracy of 10^{-3} for each quantity.

The transmittance $\mathcal{T}$ can be defined as the intensity ratio of the transmitted electromagnetic field to the incident electromagnetic field (Chiu, Lee and Quinn, 1976):

$$\mathcal{T} = \frac{\langle |\boldsymbol{E}(t)|^2 \rangle_t}{\langle |\boldsymbol{E}_\mathrm{i}(t)|^2 \rangle_t}, \tag{15.68}$$

where $\langle \cdots \rangle_t$ denotes the time average. The calculated transmittance and corresponding electron temperature T_e are plotted in Fig. 15.1 as functions of the intensity of the incident radiation field at two frequencies $\omega/2\pi = 0.83$ and 1.6 THz in the center position of the cyclotron resonance, $\omega_\mathrm{c} = \omega$. It can be seen that, the transmittance first decreases gently with increasing intensity of the THz radiation from zero and reaches a minimum at a critical intensity around $10\,\mathrm{W/cm^2}$, then increases rapidly with further increasing field strength. This feature appears more pronounced at lower frequency, in consistent with the experimental observation (Rodriguez et al, 1986) as shown in the inset of Fig. 15.1, where the measured transmittance for 1.6 and 0.24 THz exhibits a similar trend.

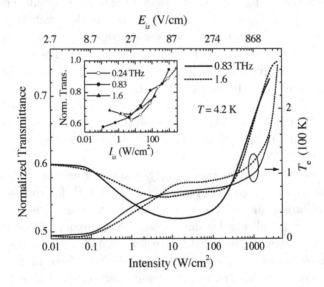

Fig. 15.1 The intensity-dependence of the 2D semiconductor transmittance (normalized to its value at zero magnetic field) and electron temperature is plotted at the cyclotron resonance position, $\omega_\mathrm{c} = \omega$, for incident radiations of two different frequencies $\omega/2\pi = 0.83$ and $\omega/2\pi = 1.6\,\mathrm{THz}$ at lattice temperature $T = 4.2\,\mathrm{K}$. Experimental results of normalized transmittance versus the radiation intensity $I_{\mathrm{i}s}$ (Rodriguez et al, 1986) are reproduced in the inset. From Liu and Lei (2003).

Figure 15.2 displays the line shape of the transmittance cyclotron resonance for incident electromagnetic fields of different intensities at frequency 0.83 THz. The line width exhibits no significant change below the critical intensity but increases rapidly when the intensity of the incident THz field grows above the critical value.

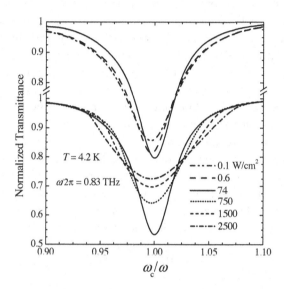

Fig. 15.2 The cyclotron resonance of the 2D semiconductor transmittance for several incident electromagnetic waves with same frequency $\omega/2\pi = 0.83$ THz but different intensities $I_{is} = 0.1, 0.6, 74, 750, 1500, 2500 \, \text{W/cm}^2$. From Liu and Lei (2003).

This kind of intensity-dependent behavior of the 2D transmittance can be compared with the intensity dependence of the absorption rate $\alpha \sim S_\text{p}/E_\omega^2$ examined in Sec. 14.4.2 in the absence of the magnetic field. There the absorption rate of a similar GaAs-based 2D semiconductor increases with increasing strength of the radiation field from the low-field value, then reaches a maximum (of the order of 2 %) at a field amplitude around several kilovolts per centimeter before decreasing with further increase of the radiation field strength, and the maximum is stronger for lower frequency (Fig. 14.8). This behavior of the absorption rate comes mainly from the drift-velocity dependence of multiphoton assisted scattering matrix elements, as described by the Bessel functions $\text{J}_n^2(\xi)$ in the expression of S_p. In fact, except for the single photon process ($n = 1$), all multiphoton ($n \geqslant 2$) contributions to the absorption rate are zero at vanishing velocity and reach maxima at finite (increasing with n) velocities, resulting in the initial increase of the total absorption rate with increasing velocity. When the drift velocity becomes sufficiently large, the reduction of absorption rates, induced by the large argument of the lower order Bessel functions, will exceed the increased contributions from other multiphoton processes, leading to a drop of the total absorption rate. In the present case having a strong magnetic field, the cyclotron resonance greatly enhances the drift velocity $\boldsymbol{v}_c$ and $\boldsymbol{v}_s$ at $\omega_c \sim \omega$ for a given incident strength in comparison with the case without magnetic field. Therefore, the maximum absorption rate should appear at a much smaller strength of the radiation field and has a much larger value in the vicinity of the cyclotron resonance than in the absence of a magnetic field or far away from cyclotron resonance.

The line shape of the transmittance cyclotron resonance is related to the radiative decay and the frictional-force damping. For fields below the critical intensity the electron temperature is less than 60 K (see Fig. 15.1) and the frictional force, which is mainly due to impurities, is relatively small. The line shape is then determined mostly by the radiative decay and exhibits little dependence on the incident field strength. When the radiation field goes above the critical intensity, the electron temperature grows rapidly and the line shape is determined by the growing frictional force which depends strongly on the electron temperature and thus varies with incident field strength.

In the Faraday configuration the electromagnetic wave may experience a change in its polarization after passing through the 2D sheet: for a linearly polarized incident electromagnetic field the transmitted electromagnetic field may become elliptically polarized. This phenomenon, well known as the Faraday rotation or the Faraday effect in dielectric media, has been investigated in the conducting system only under linear transport conditions. The present approach provides a convenient way to examine this effect when the incident field is strong and the nonlinear absorption occurs. For the case that the incident radiation is an x-direction linearly polarized sinusoidal field (15.67), the relevant quantities characterizing the Faraday effect, i.e. the ellipticity η and Faraday rotation angle θ_F, are conveniently determined through the amplitudes of the x and y components of the transmitted field $\mathbf{E}(t)$ in the form (15.63):

$$\tan\eta = \frac{a^+ - a^-}{a^+ + a^-}, \tag{15.69}$$

$$\theta_F = \frac{1}{2}(\phi^+ - \phi^-), \tag{15.70}$$

$$\tan\phi^+ = \frac{E_{sx} + E_{cy}}{E_{cx} - E_{sy}}, \tag{15.71}$$

$$\tan\phi^- = \frac{-E_{sx} + E_{cy}}{E_{cx} + E_{sy}}, \tag{15.72}$$

$$a^+ = \frac{1}{2}\sqrt{(E_{sx} + E_{cy})^2 + (E_{cx} - E_{sy})^2}, \tag{15.73}$$

$$a^- = \frac{1}{2}\sqrt{(E_{sx} - E_{cy})^2 + (E_{cx} + E_{sy})^2}, \tag{15.74}$$

with $(E_{sx}, E_{sy}) \equiv \mathbf{E}_s$ and $(E_{cx}, E_{cy}) \equiv \mathbf{E}_c$.

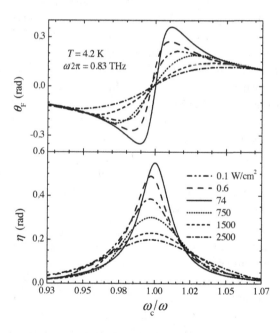

Fig. 15.3 The Faraday angle θ_F and ellipticity η of a 0.83 THz electromagnetic wave passing through the 2D system described in the text, are plotted as functions of the magnetic field strength B in terms of ω_c/ω for different incident radiation intensities $I_{is} = 0.1, 0.6, 74, 750, 1500, 2500\,\text{W/cm}^2$. From Liu and Lei (2003).

The calculated η and θ_F of an electromagnetic wave of frequency $\omega/2\pi = 0.83\,\text{THz}$ passing through the above mentioned two-dimensional system are shown in Fig. 15.3 as functions of the magnetic field strength B for different incident intensities. The resonance in ellipticity and antiresonance in Faraday rotation angle can be seen clearly. Their line shapes also manifest different behaviors when the intensity of the incident field lies below or above the critical value. These intensity-dependent features of the ellipticity and Faraday rotation angle can also be understood on the basis of multiphoton-assisted scattering and hot-electron-effect induced damping-force variation.

15.2.4 *Cyclotron resonance of electron heating and cooling*

As discussed in Sec. 14.2.2, under the influence of an intense THz irradiation, the electron temperature of a GaAs-based 2D system in the absence of the magnetic field, is usually higher than the lattice temperature T and T_e increases with increasing radiation intensity. That is the electron heating. However, within a lattice temperature range around 72 K the electron temperature T_e may descend with increasing radiation intensity from the beginning and reaches a minimum which is about a few degrees lower than the lattice temperature T before rises with further increase of the radiation field, as seen in Fig. 14.2. This is the electron temperature cooling induced by the irradiation of a high-frequency field (Lei, 1998b; Lei and Liu, 2003b). Effects of a Faraday-geometry magnetic field on this radiation-induced electron temperature change (heating or cooling) can be remarkable. This is not only because a strong magnetic field significantly changes the electron density correlation function due to Landau quantization and level broadening, but particularly because the cyclotron resonance strongly shows up in the electron heating and cooling.

Figure 15.4 is an example of the cyclotron resonance of the electron cooling at lattice temperature $T = 77\,\text{K}$ in a GaAs quantum well having well width $a = 12.5\,\text{nm}$, electron sheet density $N_s = 5.5 \times 10^{15}\,\text{m}^{-2}$ and low-temperature linear mobility $\mu_0 = 31\,\text{m}^2\,\text{V}^{-1}\,\text{s}^{-1}$, subject to 4 THz radiation fields and magnetic fields B in Faraday configuration. Under a magnetic field $B = 9.57\,\text{T}$ satisfying the cyclotron resonance condition $(\omega = \omega_c)$, or under a magnetic field $B = 9\,\text{T}$ somewhat deviated from the cyclotron resonance $(\omega_c/\omega = 0.94)$, the electron temperature T_e of the system first goes down with increasing strength of the incident field ampli-

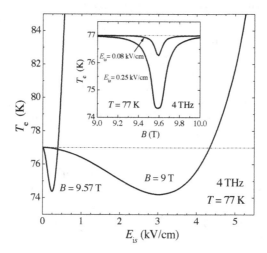

Fig. 15.4 The electron temperature T_e of a GaAs quantum well subject to a 4 THz radiation field and a magnetic field B in Faraday configuration at lattice temperature $T = 77$ K, is shown versus the incident field amplitude E_{is} ($E_{ic} = 0$) at $B = 9.0$ and 9.75 T. The inset shows T_e versus the magnetic field B at $E_{is} = 0.08$ and 0.25 kV/cm. From Lei and Liu (2003b).

tude E_{is} (E_{ic}=0) before reaching a minimum, and then rises with further increase of the radiation strength. The striking feature in the presence of a Faraday-configuration magnetic field is that the sensitivity of the electron temperature change is greatly enhanced by the cyclotron resonance. One can see that although the lowest electron temperature attained are essentially the same in both cases, the critical field at which T_e reaches its minimum is an order of magnitude smaller for $B = 9.57$ T than for $B = 9$ T. This reflects the cyclotron resonance in the electron cooling, as seen in the inset of Fig. 15.4, where at each fixed THz field intensity, $E_{is} = 0.08$ kV/cm or $E_{is} = 0.25$ kV/cm, T_e exhibits deep valley around $B = 9.57$ T.

15.2.5 Cyclotron resonance of THz photoconductivity at high temperatures

The transverse and longitudinal resistivities of a 2D semiconductor in the presence of a high-frequency field is easily obtained from the force-balance equation (15.38). Taking $\boldsymbol{v}_0$ to be in the x direction, $\boldsymbol{v}_0 = (v_0, 0, 0)$, we immediately get the transverse and longitudinal resistivities:

$$R_{yx} \equiv \frac{E_{0y}}{N_s e v_0} = \frac{B}{N_s e}, \tag{15.75}$$

$$R_{xx} \equiv \frac{E_{0x}}{N_s e v_0} = -\frac{F_0}{N_s^2 e^2 v_0}, \tag{15.76}$$

where $F_0 \equiv \boldsymbol{F}_0 \cdot \boldsymbol{v}_0/v_0$. The linear magnetoresistivity is the weak dc current limit ($v_0 \to 0$) of (15.76):

$$\begin{aligned}R_{xx} = &-\frac{1}{N_s^2 e^2} \sum_{\boldsymbol{q}_\|} q_x^2 |U(\boldsymbol{q}_\|)|^2 \sum_{n=-\infty}^{\infty} \mathrm{J}_n^2(\xi) \left[\frac{\partial}{\partial \Omega} \Pi_2(\boldsymbol{q}_\|, \Omega)\right]_{\Omega = -n\omega} \\ &- \frac{1}{N_s^2 e^2} \sum_{\boldsymbol{q},\lambda} q_x^2 |M(\boldsymbol{q},\lambda)|^2 \sum_{n=-\infty}^{\infty} \mathrm{J}_n^2(\xi) \left[\frac{\partial}{\partial \Omega} \Lambda_2(\boldsymbol{q},\lambda,\Omega)\right]_{\Omega = \Omega_{q\lambda} - n\omega}.\end{aligned} \tag{15.77}$$

The parameters $\boldsymbol{v}_c$, $\boldsymbol{v}_s$ and T_e in these expressions should be determined by solving equations (15.39), (15.40) and (15.41) with vanishing $\boldsymbol{v}_0$. The longitudinal photoresistivity is defined as

$$\Delta R_{xx} \equiv R_{xx} - R_{xx}^0(T), \tag{15.78}$$

with $R_{xx}^0(T)$ being the longitudinal magnetoresistivity in the absence of the radiation field at lattice temperature T.

Photoresistivity in semiconductors and the cyclotron resonance of far-infrared or THz photoresistivity of two-dimensional electron systems under high magnetic fields, has long been known at low temperatures and understood to result from the electron heating due to the absorption of the radiation field energy (Maan et al, 1982; Stein, Ebert, and von Klitzing, 1984; Hirakawa et al, 2001; Kawaguchi et al, 2002). But further magneto-photoresistivity measurement of a GaAs/AlGaAs system performing at lattice temperature $T = 150\,\mathrm{K}$ subjected to irradiations of 4 THz frequency, surprisingly shows remarkable cyclotron resonance peaks (Koenraad et al, 1998). At such a high temperature with polar optic phonon scattering providing an efficient energy dissipation channel, the radiation-induced electron-temperature rise is far smaller to account for such a strong cyclotron resonance in photoresistivity.

In the present theory the photoresistivity given by (15.77) and (15.78) arises not only from the effect of electron heating (electron temperature change), but also from photon-assisted electron–impurity and electron–phonon scatterings. It can be seen that, in addition to being evaluated at different temperatures (T_e and T), the main difference between R_{xx} and R_{xx}^0 lies in the multiphoton-related parts: $n = \pm 1, \pm 2, \cdots$. These multiphoton-assisted scattering terms are superposed on the direct impurity and phonon scattering terms already existent without irradiation, therefore give rise to additional contributions to the frictional damping of the moving electrons and lead to an additional dc resistivity upon irradiation even without heating of electrons. In the presence of a magnetic field, this portion of photoresistivity is also resonantly enhanced when the frequency of the electron cyclotron motion matches the photon frequency and dominates the photoresponse cyclotron resonance in two-dimensional polar semiconductors at high temperatures.

To have an idea of how strong the cyclotron resonance of photoresistivity can be at high temperature and how large contribution from the nonthermal mechanism to the photoresponse can appear at cyclotron resonance, we consider a model GaAs/AlGaAs heterostructure having a parabolic band of effective mass $m = 0.068\,m_e$ (m_e is the free electron mass), electron sheet density $N_s = 2.0 \times 10^{15}\,\mathrm{m}^{-2}$, and low-temperature (4.2 K) zero-field linear dc mobility $\mu(0) = 200\,\mathrm{m}^2/\mathrm{Vs}$, subjected to a radiation field of frequency $\omega/2\pi = 4\,\mathrm{THz}$ and a magnetic field B in the vicinity of cyclotron resonance $\omega_c \equiv eB/m = \omega$. We take account of scatterings due to randomly distributed impurities, longitudinal and transverse acoustic phonons and polar optic phonons, with typical coupling parameters for GaAs. Effects of Landau level broadening are included by assuming a Gaussian-type density-of-states function, $\mathrm{Im}\,G_n(\varepsilon) = -(\sqrt{2\pi}/\Gamma)\exp[-2(\varepsilon - \varepsilon_n)^2/\Gamma^2]$, with a unified broadening parameter $\Gamma = (8\alpha e\omega_c/\pi m\mu_0)^{1/2}$ and μ_0 taken to be the linear dc mobility at $T = 150\,\mathrm{K}$ in the absence of magnetic field and $\alpha = 13$. The incident field is linearly polarized $\boldsymbol{E}_i(t) = \boldsymbol{E}_{is}\sin(\omega t)$ and the field inside the 2D sheet is deduced with other transport quantities by selfconsistently solving balance equations (15.39), (15.40) and (15.41) together with the boundary condition (15.65). With these, we calculate the photoresistivity ΔR_{xx} at lattice temperature $T = 150\,\mathrm{K}$ induced by radiation fields of frequency 4 THz having different strengths $E_{is} = 0.1, 0.13, 0.15, 0.27, 0.61, 0.87$ and $1.1\,\mathrm{kV/cm}$. The calculated ΔR_{xx} is shown in Fig. 15.5(a) and Fig. 15.5(b) as a function of ω_c/ω. Cyclotron resonance peaks appear at $\omega_c \sim \omega$. Note that under strong THz illuminations, the electron temperature T_e also exhibits resonance as seen in Fig. 15.5(c). This electron heating, however, contributes only a small fraction of the total photoresistivity. For smaller incident field strengths, e.g. $E_{\text{is}} \leqslant 0.27\,\text{kV/cm}$, the electron temperature has no appreciable difference from T at cyclotron resonance, while photoresistivity still exhibits clear resonant peak.

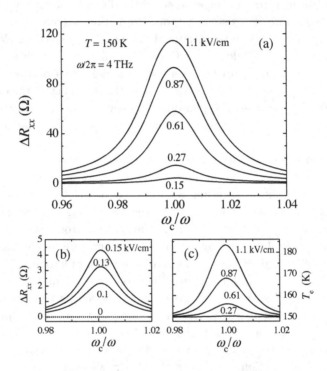

Fig. 15.5 Cyclotron resonance in the longitudinal photoresistivity ΔR_{xx} (a) and (b), and electron temperature T_{e} (c), induced by radiation fields of different incident amplitudes E_{is} but same frequency $\omega/2\pi = 4\,\text{THz}$. The system is described in the text and the lattice temperature is $T = 150\,\text{K}$. From Lei and Liu (2003a)

Although it is difficult to quantitatively distinguish contributions to photoconductivity from different mechanisms when the applied terahertz field is strong, we can formally write the longitudinal photoresistivity as the sum of two terms:

$$\Delta R_{xx} = \Delta R_{xx}^{(\text{h})} + \Delta R_{xx}^{(\text{op})}. \tag{15.79}$$

The first term $\Delta R_{xx}^{(\text{h})}$, defined by $\Delta R_{xx}^{(\text{h})} = R_{xx}^0(T_{\text{e}}) - R_{xx}^0(T)$, can be considered as the part arising from the electron heating $\Delta T_{\text{e}} = T_{\text{e}} - T \neq 0$. The electron temperature T_{e} is determined by the energy-balance equation

(15.41) at $\boldsymbol{v}_0 = 0$, and is related to the rates of electron energy absorption from the radiation field S_p and the electron energy dissipation to the lattice W. The second term $R_{xx}^{(\text{op})}$ is thus the part arising from the nonthermal mechanism, which comes from photon-assisted impurity and phonon scatterings, as well as from the electron distribution change directly induced by the THz field.

As can be seen from Fig. 15.6, although the ratio of the electron-heating induced photoresistivity $\Delta R_{xx}^{(\text{h})}$ to the total photoresistivity ΔR_{xx}, increases with increasing THz field strength from small $E_{\text{i}s}$, the highest percentage is less than 0.12 at $\omega_\text{c} = \omega$ for $T = 150\,\text{K}$.

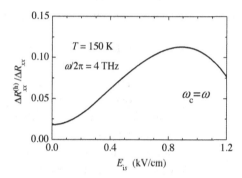

Fig. 15.6 The ratio of the electron-heating induced photoresistivity $\Delta R_{xx}^{(\text{h})}$ to the total photoresistivity ΔR_{xx} is shown as a function of radiation field strength $E_{\text{i}s}$. From Lei and Liu (2003a).

15.3 Radiation-Induced Magnetoresistance Oscillations in High-Mobility Two-Dimensional Electron Systems at Low Temperatures

Interest in magnetotransport in two-dimensional electron system has recently been revived since the experimental discovery of giant magnetoresistance oscillations in high mobility two-dimensional electron gas subjected to a microwave irradiation and moderate magnetic fields corresponding to high filling factors of Landau levels (Zudov et al, 2001; Ye et al, 2001), particularly following the spectacular observation of the zero electric resistance in very clean samples in the Hall configuration (Mani et al, 2002; Zudov et al, 2003), and of the zero electric conductance in the Corbino configuration (Yang et al, 2003). These radiation-induced oscillations of longitudinal magnetoresistivity R_{xx} exhibit a smooth magnetic-field variation and are periodical in the inverse magnetic field $1/B$, with the period determined by the radiation frequency ω, having primary resistivity maxima at $\omega/\omega_c = j - \delta_-$ and minima at $\omega/\omega_c = j + \delta_+$ (ω_c is the cyclotron frequency, $j = 1, 2, 3, \cdots$ and $\delta_\pm$ in the range around $0.1 - 0.25$). With increasing intensity of the radiation the oscillation amplitude increases: the resistivity maximum goes up and the minimum drops downward until a vanishing resistance is reached, while $\omega/\omega_c = j$ are essentially the node points where the magnetoresistivity appears almost fixed with changing radiation intensity. On the other hand the Hall resistivity keeps the classical form $R_{yx} = B/N_s e$ over the whole magnetic field range exhibiting R_{xx} oscillations.

Despite many other possible origins for these "zero-resistance" or "zero-conductance" states which have been explored since the experimental discovery, a majority accepted explanation for this phenomenon is based on the electrodynamic analysis: a homogeneous current-carrying state of a system is unstable if either the absolute dissipative resistivity or the differential resistivity becomes negative. The system will spontaneously develop a non-vanishing local current density which is determined by the condition that the component of electric field parallel to the local current vanishes. Thus the appearance of negative longitudinal resistivity or conductivity in a uniform model suffices to explain the observed vanishing resistance (Andreev, Aleiner and Mills, 2003; Alicea et al, 2005).

The possibility of absolute negative photoconductance associated with impurity scattering in a two-dimensional electron system subject to a perpendicular magnetic field and a microwave radiation was first explored more than 30 years ago by Ryzhii (1970). The balance equation theory developed in Sec. 15.1 for radiation-driven magnetotransport in 2D electron systems in Faraday configuration provides a convenient tool for the investigation of various aspects of this interesting phenomenon (Lei and Liu, 2003c; Lei, 2004).

15.3.1 *Magnetoresistance oscillations under monochromatic irradiation*

We consider a quasi-two-dimensional system consisting of N_s electrons in a unit area in the x–y plane. These electrons, interacting with each other and scattered by randomly distributed impurities and by phonons in the lattice, are subjected to a uniform magnetic field $\boldsymbol{B} = (0, 0, B)$ of modest strength in the z direction and a dc electric field $\boldsymbol{E}_0$ within the 2D plane.

When a monochromatic electromagnetic wave of frequency ω propagates perpendicularly (along z-axis) towards the 2D layer from the air with an incident electric field

$$\boldsymbol{E}_\text{i}(t) = \boldsymbol{E}_\text{is} \sin(\omega t) + \boldsymbol{E}_\text{ic} \cos(\omega t) \tag{15.80}$$

at plane $z = 0$, the high-frequency electric field inside the 2D electron sheet

$$\boldsymbol{E}(t) = \boldsymbol{E}_s \sin(\omega t) + \boldsymbol{E}_c \cos(\omega t) \tag{15.81}$$

relates to the incident field $\boldsymbol{E}_\text{i}(t)$ and the oscillating electron velocity

$$\boldsymbol{v}(t) = \boldsymbol{v}_c \cos(\omega t) + \boldsymbol{v}_s \sin(\omega t) \tag{15.82}$$

through the boundary condition (15.65) or (15.66). The transport state of the 2D electron system can be described by the dc and the high-frequency components of the average electron velocity, $\boldsymbol{v}_0$, $\boldsymbol{v}_c$ and $\boldsymbol{v}_s$, and the electron temperature T_e, which are to be determined by the force and energy balance equations:

$$N_s e \boldsymbol{E}_0 + N_s e (\boldsymbol{v}_0 \times \boldsymbol{B}) + \boldsymbol{F}_0 = 0, \tag{15.83}$$

$$\omega \boldsymbol{v}_c = -\frac{e\boldsymbol{E}_s}{m} - \frac{\boldsymbol{F}_s}{N_s m} - \frac{e}{m}(\boldsymbol{v}_s \times \boldsymbol{B}), \tag{15.84}$$

$$\omega \boldsymbol{v}_s = \frac{e\boldsymbol{E}_c}{m} + \frac{\boldsymbol{F}_c}{N_s m} + \frac{e}{m}(\boldsymbol{v}_c \times \boldsymbol{B}), \tag{15.85}$$

$$N_s e \boldsymbol{E}_0 \cdot \boldsymbol{v}_0 + S_\text{p} - W = 0, \tag{15.86}$$

The expressions for $\boldsymbol{F}_0$, $\boldsymbol{F}_c$, $\boldsymbol{F}_s$, W and S_p are exactly as given in Sec. 15.1.

Experiments of microwave-induced magnetoresistance oscillations are usually performed at quite a low lattice temperature T ($\sim 1\,\text{K}$) in high-mobility 2D semiconductors subject to an irradiation of moderate strength. In this case the electron heating of the system is modest with an electron temperature usually lower than $20\,\text{K}$, whence the phonon contributions to the frictional forces $\boldsymbol{F}_0$, $\boldsymbol{F}_c$ and $\boldsymbol{F}_s$ and to the energy absorption S_p are negligible in comparison with those from impurities. Phonon scatterings, however, have to be taken into account for the evaluation of the electron energy-dissipation rate W to the lattice to keep the energy balance. Furthermore, in view of a large radiative decay always existing in the ordinary experimental setup for which the incident radiation field is the controlled input quantity, the much smaller scattering-induced oscillating damping forces in Eqs. (15.84) and (15.85) are negligible in these high-mobility systems at low temperatures. Therefore, for a given incident field $\boldsymbol{E}_i(t)$ the high-frequency velocity components $\boldsymbol{v}_c$ and $\boldsymbol{v}_s$ can be directly obtained from the force-balance equations

$$-m\omega\boldsymbol{v}_c = e\boldsymbol{E}_s + e(\boldsymbol{v}_s \times \boldsymbol{B}), \tag{15.87}$$

$$m\omega\boldsymbol{v}_s = e\boldsymbol{E}_c + e(\boldsymbol{v}_c \times \boldsymbol{B}), \tag{15.88}$$

together with the boundary relation

$$\boldsymbol{E}(t) = \frac{N_s e \boldsymbol{v}(t)}{(n_0 + n_s)\epsilon_0 c} + \frac{2n_0}{n_0 + n_s}\boldsymbol{E}_i(t), \tag{15.89}$$

(if the 2D electron gas is contained within a thin layer under the surface plane at $z = 0$ of a thick semiconductor substrate having a refraction index n_s), or with the boundary relation

$$\boldsymbol{E}(t) = \frac{N_s e \boldsymbol{v}(t)}{2\epsilon_0 c} + \boldsymbol{E}_i(t) \tag{15.90}$$

(if the 2D electron gas is contained in a thin layer suspended in vacuum at the plane $z = 0$). With these velocity components $\boldsymbol{v}_c$ and $\boldsymbol{v}_s$, thus the electro-optical coupling parameter

$$\xi \equiv \frac{1}{\omega}\left[(\boldsymbol{q}_\| \cdot \boldsymbol{v}_c)^2 + (\boldsymbol{q}_\| \cdot \boldsymbol{v}_s)^2\right]^{\frac{1}{2}} \tag{15.91}$$

in the Bessel functions, the electron temperature T_e and $\boldsymbol{v}_0$ (for given $\boldsymbol{E}_0$) or $\boldsymbol{E}_0$ (for given $\boldsymbol{v}_0$) are determined by the force and energy balance equations:

$$N_s e \boldsymbol{E}_0 + N_s e(\boldsymbol{v}_0 \times \boldsymbol{B}) + \boldsymbol{F}_0 = 0, \tag{15.92}$$

$$N_s e \boldsymbol{E}_0 \cdot \boldsymbol{v}_0 + S_p - W = 0. \tag{15.93}$$

Here the dominant contributions to $\boldsymbol{F}_0$ and S_p come from impurity scatterings,

$$\boldsymbol{F}_0 = \sum_{\boldsymbol{q}_\|} \boldsymbol{q}_\| |U(\boldsymbol{q}_\|)|^2 \sum_{n=-\infty}^{\infty} J_n^2(\xi)\, \Pi_2(\boldsymbol{q}_\|, \omega_0 - n\omega), \tag{15.94}$$

$$S_p = \sum_{\boldsymbol{q}_\|} |U(\boldsymbol{q}_\|)|^2 \sum_{n=-\infty}^{\infty} n\omega\, J_n^2(\xi)\, \Pi_2(\boldsymbol{q}_\|, \omega_0 - n\omega), \tag{15.95}$$

and the phonon scatterings give rise only to the electron energy-loss rate

$$W = \sum_{q,\lambda} \Omega_{q\lambda} |M(q,\lambda)|^2 \sum_{n=-\infty}^{\infty} J_n^2(\xi)\, \Pi_2(q_\parallel, \omega_0 - n\omega)$$
$$\times \left[n\left(\frac{\Omega_{q\lambda}}{T}\right) - n\left(\frac{\Omega_{q\lambda} + \omega_0 - n\omega}{T_e}\right) \right]. \tag{15.96}$$

In this, $\omega_0 \equiv q_\parallel \cdot v_0$, and $\Pi_2(q_\parallel, \Omega)$ is the imaginary part of the electron density correlation function of the 2D system in the presence of the magnetic field. Formally, the magnetoresistivity of a 2D semiconductor in the presence of a high-frequency field can be directly obtained from the force balance equation (15.92). For systems isotropic in the plane, taking v_0 to be in the x direction, $v_0 = (v_0, 0, 0)$, we immediately get the transverse and longitudinal resistivities in the form

$$R_{yx} \equiv \frac{E_{0y}}{N_s e v_0} = \frac{B}{N_s e}, \tag{15.97}$$

$$R_{xx} \equiv \frac{E_{0x}}{N_s e v_0} = -\frac{F_0}{N_s^2 e^2 v_0}, \tag{15.98}$$

in which $F_0 \equiv v_0 \cdot F_0/v_0$. The transverse resistivity R_{yx} has always a classical form independent of dc and radiation fields. The longitudinal resistivity R_{xx}, on the other hand, depends on the average velocity v_0 (i.e. the dc current density $J = N_s e v_0$) via the parameter ω_0 and depends on the incident high-frequency field $E_i(t)$ via the parameter ξ, as well as through the electron temperature T_e implicitly involved in the Π_2 function.

The linear longitudinal resistivity is the weak dc current ($v_0 \to 0$) limit of (15.98), which can be thought as a function of the parameter ξ, directly through the argument of the Bessel functions $J_n(\xi)$ and indirectly through the electron temperature in the Π_2' function:

$$R_{xx}(\xi) = -\frac{1}{N_s^2 e^2} \sum_{q_\parallel} q_x^2 |U(q_\parallel)|^2 \sum_{n=-\infty}^{\infty} J_n^2(\xi)\, \Pi_2'(q_\parallel, n\omega), \tag{15.99}$$

where $\Pi_2'(q_\parallel, \Omega) \equiv \frac{\partial}{\partial \Omega} \Pi_2(q_\parallel, \Omega)$ stands for the drivative of the imaginary part of the electron density correlation function with respect to its frequency variable.

The $\Pi_2(q_\parallel, \Omega)$ function of a 2D system in a magnetic field can be expressed as the sum over all Landau levels l and l' in the Landau representation:

$$\Pi_2(\bm{q}_\|, \Omega) = \frac{1}{2\pi l_B^2} \sum_{l,l'} C_{l,l'}(l_B^2 q_\|^2/2) \Pi_2(l,l',\Omega), \tag{15.100}$$

$$\Pi_2(l,l',\Omega) = -\frac{2}{\pi} \int d\varepsilon \, [f(\varepsilon) - f(\varepsilon+\Omega)] \, \mathrm{Im}G_l(\varepsilon+\Omega) \mathrm{Im}G_{l'}(\varepsilon), \tag{15.101}$$

where $l_B \equiv \sqrt{1/|eB|}$ is the magnetic length, $f(\varepsilon) = \{\exp[(\varepsilon-\mu)/T_e]+1\}^{-1}$ is the Fermi distribution function at electron temperature T_e, and $C_{l,l'}(Y)$ is the function given by (15.48). The density-of-states function $\mathrm{Im}G_l(\varepsilon)$ of the lth Landau level will be modeled with a Gaussian form

$$\mathrm{Im}G_l(\varepsilon) = -(\sqrt{2\pi}/\Gamma) \exp[-2(\varepsilon-\varepsilon_l)^2/\Gamma^2], \tag{15.102}$$

having a half-width

$$\Gamma = \left(\frac{8e\omega_c \alpha}{\pi m \mu_0}\right)^{1/2} \tag{15.103}$$

around the level center $\varepsilon_l = l\omega_c$. Here m is the carrier effective mass, μ_0 is the linear mobility at lattice temperature T in the absence of the magnetic field, and α is a semiempirical parameter to take account of the difference of the transport scattering time determining the mobility μ_0, from the single particle lifetime related to the Landau level broadening.

Some qualitative features of this radiation-driven magnetotransport can be conveniently examined direct from these equations. The effect of a radiation field on transport shows up mainly through the electro-optic coupling parameter $\xi = [(\bm{q}_\| \cdot \bm{v}_c)^2 + (\bm{q}_\| \cdot \bm{v}_s)^2]^{\frac{1}{2}}/\omega$ in the Bessel functions $J_n(\xi)$ within the summation over multiphoton index n in the expressions (15.94), (15.95) and (15.96) of F_0, S_p and W. For give $\bm{q}_\|$ and ω, the magnitude of ξ, which is mainly determined by the amplitude (thus the intensity or power) of the incident radiation field, depends also on its polarization state. On the other hand, since $\xi \sim \bm{q}_\| \cdot \bm{E}_i/\omega^2$, at given incident intensity and polarization, ξ is much larger for the radiation field of lower frequency. The frequency, intensity and polarization dependence of the radiation-driven magnetotransport results mainly from the corresponding behavior of the parameter ξ. Furthermore, since ξ is proportional to $q_\|$, and the behavior of the impurity potential $|U(\bm{q}_\|)|^2$, which plays as a weighted function in F_0 and S_p, can greatly affect the photoresistance and absorption. To induce equivalent photoresistance response, a long-range potential $|U(\bm{q}_\|)|^2$ which weighs heavy in a small wavevector $\bm{q}_\|$ range, requires much stronger incident power of the radiation field than a short-range potential which is evenly weighted throughout the whole wavevector range (Lei and Liu, 2003c, 2005b).

The width of the broadened Landau level has a tremendous influence on the photoresistance response. Experimentally remarkable magnetoresistance oscillations and zero-resistance states are found only in very high-mobility samples. This can be easily understood by the relation (15.103) of the half-width of the Landau levels with the mobility: Γ is proportional to $\mu_0^{-\frac{1}{2}}$. Another important feature of the phenomenon is its sensitive dependence on temperature. The zero-resistance states and radiation-induced magnetoresistance oscillations show up strongly only at low temperatures typically around $T = 1\,\mathrm{K}$ or lower. With increasing temperature the oscillatory structures diminish rapidly and disappear when $T \geqslant 5 \sim 6\,\mathrm{K}$ (Mani et al, 2002; Zudov et al, 2003). The temperature dependence is easily explained by the increase of the Landau level width with increasing temperature: Γ is related to all kinds of elastic and inelastic scatterings, including the electron–impurity and electron–phonon couplings as well as the electron–electron interaction (though it gives no contribution to the mobility). Both electron–phonon and electron–electron scatterings enhance rapidly with rising temperature. This leads to a growth of the Landau level width faster than that given by $\mu_0^{-\frac{1}{2}}$ and a diminishing of magnetoresistance oscillation faster than what is predicted purely from (15.103) of $\Gamma \sim \mu_0^{-\frac{1}{2}}$ (Lei, 2004).

In the following subsections we will demonstrate some detailed and numerical results on this radiation-induced magnetoresistance oscillation phenomenon in GaAs-based high mobility two-dimensional semiconductors at low temperatures.

15.3.2 Magnetoresistivity, electron temperature and energy absorption rate

We consider magnetotransport of 2D electrons in GaAs-based high mobility heterosystems under the influence of a monochromatic microwave radiation at low temperature around $T \simeq 1\,\mathrm{K}$. The basic scheme assumes that the dominant mechanisms for the energy absorption S_p and photoresistivity $R_{xx} - R_{xx}(0)$ come from the impurity-assisted photon absorption and emission processes. At different magnetic field strengths, these processes are associated with electron transitions between either inter-Landau level states or intra-Landau-level states. According to (15.102), the width of each Landau level is about 2Γ. The condition for inter-Landau level transition with impurity-assisted single-photon process[1] is $\omega > \omega_c - 2\Gamma$, or $\omega_c/\omega < a_{\text{inter}} = (\beta + \sqrt{\beta^2 + 4})^2/4$; and that for impurity-assisted intra-Landau level transition is $\omega < 2\Gamma$, or $\omega_c/\omega > a_{\text{intra}} = \beta^{-2}$, where $\beta = (32e\alpha/\pi m\mu_0\omega)^{\frac{1}{2}}$. However, since the density of states of each Landau level is assumed to be Gaussian rather than a clear cutoff function and the multiphoton processes also play roles, the transition boundaries between different regimes may be somewhat smeared. Using a semielliptic form of the density-of-states function for separated Laudau levels may lead to slightly different results (Lei and Liu, 2005a).

As indicated by the experiment (Umansky, Picciotto and Heiblum, 1997), although long range scatterings due to remote donors always exist in semiconductor heterostructures, in ultra-clean GaAs-based 2D samples having mobility of order of $10^3 \, \text{m}^2/\text{Vs}$, the dominant contribution to the momentum scattering rate comes from short-range scatterers such as residual impurities or defects in the background. Furthermore, even with the same momentum scattering rate the long range remote impurity scattering is much less efficient in contributing to microwave-induced magnetoresistance oscillations than short-ranged background impurities or defects (Lei and Liu, 2005b). Therefore, we assume that the elastic scatterings are due to short-range impurities randomly distributed throughout the GaAs region. The impurity densities are determined by the requirement that the electron total linear mobility at zero magnetic field equals the given value at lattice temperature T. Possibly, long-range remote donnor scattering may give rise to important contributions to Landau-level broadening. This effect, together with the role of electron-phonon and electron-electron scatterings, is included in the semiempirical parameter α in the expression of half-width Γ.

In order to obtain the energy dissipation rate from the electron system to the lattice, W, we take into account scatterings from bulk longitudinal acoustic and transverse acoustic phonons (via the deformation potential and piezoelectric couplings with electrons), as well as from longitudinal optical phonons (via the Fröhlich coupling with electrons) in the GaAs-based system. The relevant matrix elements are well known (Lei, Birman and Ting, 1985). The material and coupling parameters for the system are taken to be widely accepted values in bulk GaAs: electron effective mass $m = 0.068 \, m_e$

---

[1] Mulitiphoton ($|n| > 1$) processes of course can take place up to higher magnetic fields.

(m_e is the free electron mass), transverse sound speed $v_{st} = 2.48 \times 10^3$ m/s, longitudinal sound speed $v_{sl} = 5.29 \times 10^3$ m/s, acoustic deformation potential $\Xi = 8.5$ eV, piezoelectric constant $e_{14} = 1.41 \times 10^9$ V/m, dielectric constant $\kappa = 12.9$, and material mass density $d = 5.31$ g/cm^3.

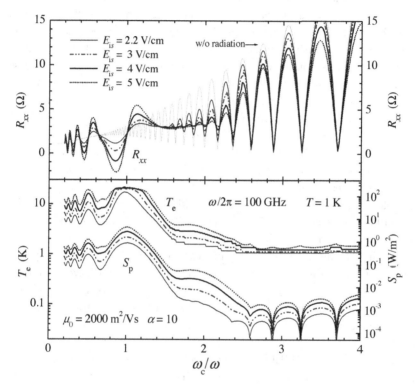

Fig. 15.7 The linear magnetoresistivity R_{xx}, electron temperature T_e and energy absorption rate S_p of a GaAs-based 2D electron system with $N_s = 3.0 \times 10^{15}$ m^{-2}, $\mu_0 = 2000$ m^2/Vs and $\alpha = 10$, subjected to 100 GHz linearly x-polarized incident high-frequency fields $E_{is}\sin(\omega t)$ of four different strengths. The lattice temperature is $T = 1$ K. From Lei and Liu (2005b).

Figure 15.7 shows the calculated energy absorption rate S_p, the electron temperature T_e and the linear magnetoresistivity R_{xx} as functions of ω_c/ω for a GaAs-based 2D system with electron density $N_s = 3.0 \times 10^{15}$ m^{-2}, linear mobility $\mu_0 = 2000$ m^2/Vs and broadening parameter $\alpha = 10$, subject to linearly x-direction polarized incident microwave radiations of frequency $\omega/2\pi = 100$ GHz having four different amplitudes $E_{is} = 2.2, 3, 4$ and 5 V/cm ($E_{ic} = 0$) at lattice temperature $T = 1$ K. The energy absorption rate S_p exhibits a broad main peak at cyclotron resonance $\omega_\text{c}/\omega = 1$ and secondary peaks at harmonics $\omega_\text{c}/\omega = 1/2, 1/3, 1/4$. The electron heating has a similar feature: T_e exhibits peaks around $\omega_\text{c}/\omega = 1, 1/2, 1/3, 1/4$. For this GaAs system, $\beta = 0.65$, $a_\text{inter} = 1.6$, and $a_\text{intra} = 4.7$. We can see that, at lower magnetic fields, especially $\omega_\text{c}/\omega < 1.4$, the system absorbs enough energy from the radiation field via inter-Landau level transitions and T_e is significantly higher than T, with the maximum as high as 21 K around $\omega_\text{c}/\omega = 1$. With increasing strength of the magnetic field the inter-Landau level transition weakens (impurity-assisted single-photon process is mainly allowed when $\omega_\text{c}/\omega < a_\text{inter} = 1.6$) and the absorbed energy decreases rapidly. Within the range $2 < \omega_\text{c}/\omega < 4$ before intra-Landau level transitions can take place, S_p is two orders of magnitude smaller than that in the low magnetic field range. Correspondingly, the electron temperature T_e is only slightly higher than the lattice temperature T. The magnetoresistivity R_{xx} showing in the upper part of Fig. 15.7, exhibits interesting features. Radiation-induced magnetoresistance oscillations (RIMOs) clearly appear at lower magnetic fields, which are insensitive to the electron heating even at T_e of the order of 20 K. Shubnikov–de Haas oscillations (SdHOs) appearing in the higher magnetic field side, however, are damped due to the rise of the electron temperature $T_\text{e} > 1\,\text{K}$ as compared with that without radiation. With an increase in the microwave amplitude from $E_{\text{i}s} = 2.2\,\text{V/cm}$ to $5\,\text{V/cm}$, RIMOs become much stronger and SdHOs are further damped. But the radiation-induced SdHO damping is always relatively smaller within $2.4 < \omega_\text{c}/\omega < 4$ between allowed ranges of inter- and intra-Landau level transitions.

The sensitivity of SdHOs on the electron temperature can result in interesting modulation of magnetoresistivity by a rather weak irradiation in the magnetic field range where SdHOs can show up with or without RIMO (Du et al, 2004; Dorozhkin et al, 2005; Lei and Liu, 2005a,b).

15.3.3 Vanishing linear photoresponse at cyclotron resonance and its harmonics at low temperatures

A salient feature which can be recognized directly from the general expression (15.99) of the linear magnetoresistivity $R_{xx}(\xi)$, is that the photoresponse vanishes at cyclotron resonance and its harmonics. In fact, it can be seen from the expressions (15.100) and (15.101) that, in the case of low electron temperature ($T_\text{e} \ll \epsilon_\text{F}$, the Fermi level) and large Landau-level filling factor ν, for any integer M satisfying $|M| \ll \nu$, $\Pi_2'(\boldsymbol{q}_\parallel, \Omega + M\omega_\text{c}) =$

$\Pi_2'(\boldsymbol{q}_\parallel, \Omega)$. Thus, when $\omega = l\omega_{\rm c}$ ($l = 1, 2, \cdots$), the $\Pi_2'(\boldsymbol{q}_\parallel, n\omega)$ function in (15.99) is independent of n, so that $\sum_n {\rm J}_n^2(\xi)\, \Pi_2'(\boldsymbol{q}_\parallel, n\omega) = \Pi_2'(\boldsymbol{q}_\parallel, 0)$. Therefore we have

$$R_{xx}(\xi)\big|_{\omega = l\omega_{\rm c}} = -\frac{1}{N_{\rm s}^2 e^2} \sum_{\boldsymbol{q}_\parallel} q_x^2 |U(\boldsymbol{q}_\parallel)|^2\, \Pi_2'(\boldsymbol{q}_\parallel, 0). \qquad (15.104)$$

This expression indicates that at cyclotron resonance frequency and its harmonics, $\omega = l\omega_{\rm c}$ ($l = 1, 2, \cdots$), the difference of $R_{xx}(\xi)$ driven by monochromatic frequency-ω fields of different strengths, can appear only through the difference of electron temperature contained in the Π_2' function. Under the condition of $T_{\rm e} \ll \epsilon_{\rm F}$, the effect of electron heating on photoresistance is quite weak; therefore, at any integer-$\omega_{\rm c}$ position of the frequency ω, the magnetoresistivity in the presence of a radiation is essentially the same as that without radiation, $R_{xx}(\xi) \simeq R_{xx}(0)$, i.e. the photoresponse (arising from impurity scattering) vanishes at cyclotron resonance and its harmonics.[2]

This is clearly seen in Fig. 15.8, where we plot the calculated linear magnetoresistivity R_{xx} of a GaAs-based 2D electron system with $N_{\rm s} = 2.4 \times 10^{15}\,{\rm m}^{-2}$, $\mu_0 = 2000\,{\rm m}^2/{\rm Vs}$ and $\alpha = 5$, irradiated by linearly polarized monochromatic waves of frequency $\omega/2\pi = 31\,{\rm GHz}$ having incident amplitudes $E_{\rm is} = 0, 0.5, 0.7, 1.0, 1.3$ and $1.7\,{\rm V/cm}$ ($E_{\rm ic} = 0$) at lattice temperature $T = 1\,{\rm K}$. The scattering is assumed due to short-range impurities and the material parameters used are the same as given in the last subsection. All the curves, with and without radiation, cross at $\omega/\omega_{\rm c} = 1, 2, 3, 4, 5, 6$ and 7. This feature, predicted theoretically (Durst et al, 2003; Lei and Liu, 2003; Ryzhii and Suris, 2003) and observed experimentally (Mani et al, 2002, 2004; Zudov et al, 2003), occurs quite generally, independent of the polarization state of the radiation, the behavior of the elastic scattering potential, and other properties of the two-dimensional electron system. It holds quite precisely up to rather strong radiation field even multiphoton processes play important role, as long as the 2D electron gas remains in degenerate ($T_{\rm e} \ll \epsilon_{\rm F}$), thus provides a convenient and accurate method to determine the effective mass of electrons in two-dimensional systems.

---

[2]This conclusion is valid only for the case of dominant impurity scattering thus only at low temperatures. We have already seen in Sec. 15.2 that, in the case of high temperature (150 K) and low Landau-level filling factor, where the polar-optic phonons are the dominant scatterers, the longitudinal photoresistivity ΔR_{xx} exhibit a large peak in the vicinity of cyclotron resonance $\omega_{\rm c}/\omega = 1$ (Lei and Liu, 2003a).

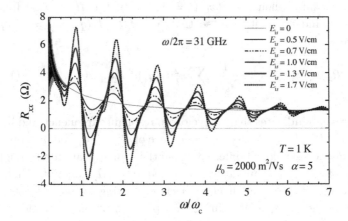

Fig. 15.8 The linear magnetoresistivity R_{xx} induced by monochromatic radiations of frequency $\omega/2\pi = 31$ GHz at lattice temperature $T = 1$ K, in a GaAs-based 2D electron system with $N_s = 2.4 \times 10^{15}$ m^{-2}, $\mu_0 = 2000$ m^2/Vs and $\alpha = 5$. From Lei and Liu (2006b).

Another feature which shows up when the radiation strength is modest that contributions from two and higher-order photon processes are negligible and radiation-induced resistance oscillations result mainly from single-photon-assisted processes:

$$R_{xx}(\xi) \simeq -\frac{1}{N_s^2 e^2} \sum_{\mathbf{q}_\parallel} q_x^2 |U(\mathbf{q}_\parallel)|^2 \left[J_0^2(\xi)\, \Pi_2'(\mathbf{q}_\parallel, 0) \right.$$
$$\left. + J_1^2(\xi)\, \Pi_2'(\mathbf{q}_\parallel, \omega) + J_{-1}^2(\xi)\, \Pi_2'(\mathbf{q}_\parallel, -\omega) \right]. \quad (15.105)$$

Since the density-of-states function decays rapidly when deviating from the center of each Landau level, the magnitude of $\Pi_2'(\mathbf{q}_\parallel, \omega)$ reaches a minimum of almost zero when $\omega/\omega_c = (l+\tfrac{1}{2})$ ($l = 1, 2, 3, \cdots$). Therefore around these half-integer-ω_c positions, the last two terms in (15.105) are negligible and the magnetoresistivity is approximately given by

$$R_{xx}(\xi)\big|_{\omega=(l+\frac{1}{2})\omega_c} \simeq -\frac{1}{N_s^2 e^2} \sum_{\mathbf{q}_\parallel} q_x^2 |U(\mathbf{q}_\parallel)|^2 J_0^2(\xi)\, \Pi_2'(\mathbf{q}_\parallel, 0), \quad (15.106)$$

which is only weakly dependent on the radiation strength through $J_0^2(\xi)$, i.e., half-integer-ω_c positions of frequency ω are approximately node points for modest radiations. This feature can also be seen in Fig. 15.8 where almost all the resistivity curves cross at $\omega/\omega_c = 3.5, 4.5, 5.5$, and 6.5, and the curves of lower radiation strengths cross at $\omega/\omega_c = 1.5$ and 2.5.

15.3.4 Multiple and virtual photon processes

Note that all orders of real ($|n| > 0$) and virtual ($n = 0$) photon processes are included in the summations over n in expressions (15.94), (15.95), (15.96), and (15.99). The results presented in the last subsection are under an irradiation of modest strength, where the multiple photon processes, though show up already, are still not noticeable. When the radiation intensity further increases conspicuous symbols of multiphoton processes appear in the curve of magnetoresistance oscillation. An example is shown in Fig. 15.9, where the calculated magnetoresistivity R_{xx} is plotted as a function of ω/ω_c for the same system as described in Fig. 15.7, illuminated by 50 GHz microwaves of three incident amplitudes $E_{is} = 4, 5$ and $6.5\,\text{V/cm}$ ($E_{ic} = 0$) at lattice temperature $T = 1\,\text{K}$. We see that in addition to the primary maximum-minimum pairs at $\omega/\omega_c = 1, 2, 3$, and 4, prominent peak-valley pairs also appear around $\omega/\omega_c = 1/2, 2/3, 3/5, 2/5$, and $2/7$, and the minima of pairs $\omega/\omega_c = 1/2$ and $\omega/\omega_c = 3/5$ drop down to negative.

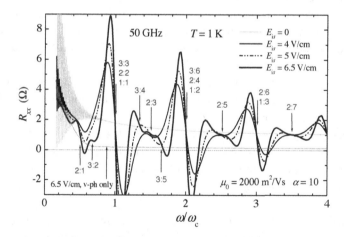

Fig. 15.9 The magnetoresistivity R_{xx} of a GaAs-based 2D electron system with $N_s = 3.0 \times 10^{15}\,\text{m}^{-2}$, $\mu_0 = 2000\,\text{m}^2/\text{Vs}$ and $\alpha = 10$, subjected to 50 GHz radiations of incident amplitudes $E_{is} = 4, 5, 6.5$ and $8.5\,\text{V/cm}$ ($E_{ic} = 0$) at lattice temperature $T = 1\,\text{K}$. From Lei and Liu (2006a).

The appearance of the oscillatory magnetoresistance comes from real photon-assisted electron transitions between different Landau levels as indicated in the summation of the electron density correlation function in Eq. (15.100). We denote a real-photon assisted process in which an electron jumps across l Landau-level spacings with the assistance (emission or absorption) of n photons as $n\omega{:}l\omega_\text{c}$, or simply $n{:}l$. This process contributes, in the R_{xx}–ω_c/ω curve, a pair structure consisting of a minimum and a maximum on both sides of $\omega_\text{c}/\omega = n/l$. The location of its minimum or maximum may change somewhat depending on the strength of the incident microwave, but the node point, which is roughly in the center, essentially keeps at the position $\omega_\text{c}/\omega = n/l$. Therefore we use its node position, rather than that of its minimum or maximum, to identify a pair structure. Thus, the single-phonon process 1:1, the two-phonon process 2:2, and the three-photon process 3:3,$\cdots$, all contribute to the minimum–maximum pair around $\omega_\text{c}/\omega = 1$; the single-phonon process 1:2, the two-phonon process 2:4, and the three-photon process 3:6,$\cdots$, all contribute to the minimum–maximum pair around $\omega_\text{c}/\omega = 1/2$; etc. These are indicated by arrows in Fig. 15.9.

In Fig. 15.10 we plot the evolution of the magnetoresistivity R_{xx} with increasing radiation intensity $E_{\text{i}s} = 1.5, 2, 3$, and $4\,\text{V/cm}$ for a GaAs-based system with $N_\text{s} = 3.0\times 10^{15}\,\text{m}^{-2}$, $\mu_0 = 2000\,\text{m}^2/\text{Vs}$ and $\alpha = 5$, irradiated by $27\,\text{GHz}$ microwaves. Several other valley-peak pairs associated with two-, three-, and four-photon assisted processes are also identified. The predicted two-photon structures 2:1 centered at $\omega_\text{c}/\omega = 2$, 2:3 centered at $\omega_\text{c}/\omega = 2/3$ and 2:5 centered at $\omega_\text{c}/\omega = 2/5$ are in agreement with experimental observations (Mani *et al*, 2002, 2004; Zudov *et al*, 2003; Zudov, 2004).

Note that, with the progressive emergence of new multiphoton-related pairs when increasing radiation power, the peaks (valleys) of the low-order photon related pairs become narrower, as is clearly seen for the single-photon related pairs at $\omega/\omega_\text{c} = 1, 2$, and 3 in Figs. 15.9 and 15.10. This feature and the anticipated positions of peaks and possible zero resistance states, are in agreement with the recent experimental observation (Zudov *et al*, 2006).

Another interesting aspect is that, concomitantly with enhanced R_{xx} oscillation, the average magnetoresistance descends down significantly with increasing radiation power. This resistance drop is due to the effect of virtual photon processes, i.e. intra-Landau-level electron scattering by impurities with intermediate emission and absorption of an arbitrary number of photons. To show this we also demonstrate the resistivity contributed from the virtual photon processes alone, i.e. the J_0 term in (15.99) of

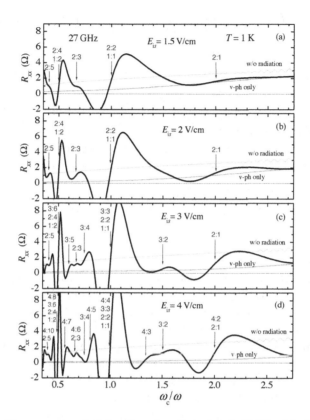

Fig. 15.10 The magnetoresistivity R_{xx} of a GaAs-based 2DEG with $N_e = 3.0 \times 10^{15}\,\text{m}^{-2}$, $\mu_0 = 2000\,\text{m}^2/\text{Vs}$ and $\alpha = 5$, subjected to 27 GHz radiations of incident amplitudes $E_{is} = 1.5, 2, 3$ and $4\,\text{V/cm}$ at lattice temperature $T = 1\,\text{K}$. From Lei and Liu (2006a).

$R_{xx}(\xi)$, in Figs. 15.9 and 15.10. The resistance suppression appears almost throughout the whole magnetic field range. It may not be previously noticed in the region where exhibit strong R_{xx} oscillations and zero resistance states (Dorozhkin, 2003; Mani, 2004; Dorozhkin et al, 2005; Lei and Liu, 2005a,b; Ryzhii, 2005; Zudov et al, 2006).

15.3.5 Nonlinear magnetoresistivity and differential magnetoresistivity

The nonlinear longitudinal magnetoresistivity and differential resistivity of a 2D system in the presence of a radiation field and a finite dc drift velocity

v_0 (i.e. current $\bm{J} = N_s e \bm{v}_0$) are given by [$F_0 \equiv \bm{v}_0 \cdot \bm{F}_0/v_0$, see (15.98)]

$$R_{xx} = -\frac{F_0}{N_s^2 e^2 v_0}, \tag{15.107}$$

$$r_{xx} = -\frac{1}{N_s^2 e^2} \frac{\partial F_0}{\partial v_0}. \tag{15.108}$$

In the absence of microwave radiation these formulas reduce to (8.138) and (8.139) in Sec. 8.6.4, and the results presented there are recovered. The resistance oscillations are controlled by a dimensionless parameter

$$\epsilon_j \equiv \frac{\omega_j}{\omega_c} = \frac{2mk_F v_0}{eB} = \sqrt{\frac{8\pi}{N_s} \frac{m}{e^2} \frac{J}{B}}, \tag{15.109}$$

and the oscillatory minimum-maximum pairs in the magnetoresistivity R_{xx} versus ϵ_j appear around the positions $\epsilon_j = \eta m$ $(m = 1, 2, \cdots)$, with $\eta \sim 1$.

In the case of linear photoresistance, $\bm{v}_0 \to 0$, electron transition (intra- and inter-Landau levels) can take place by absorbing or emitting n photons of frequency ω and jumping from Landau level l to Landau level l' (across m levels). This process gives rise to the terms of $\pm n$ in the expression (15.94) of the frictional force $\bm{F}_0$. The periodicity of the electron density correlation function in the case of low electron temperature and many-Landau-level occupation, $\Pi_2(\bm{q}_\parallel, \Omega + \omega_c) = \Pi_2(\bm{q}_\parallel, \Omega)$, leads to the appearance of the peak-valley pairs in the linear photoresistivity around the positions of $n\omega = m\omega_c$ (see Sec. 15.3.4). The primary peak-valley pairs come from the single-photon process $n = 1$, showing up around the node positions at $\omega = m\omega_c$, or

$$\epsilon_\omega \equiv \frac{\omega}{\omega_c} = m = 1, 2, 3, \cdots \tag{15.110}$$

and the primary period of the photoresistance oscillation is $\Delta \epsilon_\omega = 1$. Two and multiple photon processes $(n \geqslant 2)$ yield secondary peak-valley pairs.

When there is a finite dc current flowing in the 2D system, in view of the extra energy $\omega_0 \equiv \bm{q}_\parallel \cdot \bm{v}_0$ provided by the electron drift motion together with the energy $n\omega$ provided by the absorption or emission of n photons of the radiation field, an electron can be scattered by elastic impurities and jump across m Landau levels of different energies. The resonance condition in this case should be determined by $\eta \omega_j + n\omega = m\omega_c$. When paying attention to the single photon processes $(n = 1)$, the valley-peak pairs of the nonlinear photoresistance oscillation are anticipated to appear around positions

$$\epsilon_\omega + \eta\,\epsilon_j = \epsilon_\omega\left(1 + \eta\frac{\omega_j}{\omega}\right) = m \tag{15.111}$$

or

$$\epsilon_\omega \equiv \frac{\omega}{\omega_c} = \frac{m}{1+\gamma_j}, \tag{15.112}$$

where $m = 1, 2, 3, \cdots$, and

$$\gamma_j \equiv \eta\frac{\omega_j}{\omega}. \tag{15.113}$$

Therefore, the oscillatory behavior of R_{xx} is controlled by the parameter $\epsilon_\omega + \eta\,\epsilon_j$, or the parameter $\epsilon_\omega(1+\gamma_j)$.

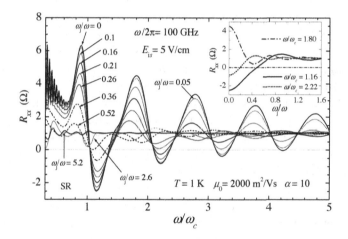

Fig. 15.11 Longitudinal magnetoresistivity R_{xx} versus $\epsilon_\omega = \omega/\omega_c$ for a GaAs-based 2D electron system with $N_s = 3 \times 10^{15}\,\mathrm{m}^{-2}$, $\mu_0 = 2000\,\mathrm{m}^2\mathrm{V}^{-1}\mathrm{s}^{-1}$ and $\alpha = 10$ at lattice temperature $T = 1\,\mathrm{K}$, under linearly polarized irradiation of frequency $\omega/2\pi = 100\,\mathrm{GHz}$ and incident amplitude $E_{\mathrm{is}} = 5\,\mathrm{V/cm}$ and $E_{\mathrm{ic}} = 0$ and subject to different bias current density of $\omega_j/\omega = 0, 0.05, 0.1, 0.16, 0.21, 0.26, 0.36, 0.52, 2.6$ and 5.2. The elastic scatterings are assumed due to short-range impurities. The inset shows R_{xx} as a function of ω_j/ω at three fixed magnetic field strengths $\omega/\omega_c = 1.16, 1.80$ and 2.22. From Lei (2007b).

Figure 15.11 presents the calculated longitudinal magnetoresistivity R_{xx} versus $\epsilon_\omega = \omega/\omega_c$ for a GaAs-based 2D electron system with carrier density $N_s = 3 \times 10^{15}\,\mathrm{m}^{-2}$ and low-temperature linear mobility $\mu_0 = 2000\,\mathrm{m}^2\mathrm{V}^{-1}\mathrm{s}^{-1}$ at lattice temperature $T = 1\,\mathrm{K}$, under the irradiation of a linearly x-direction polarized microwave of frequency $\omega/2\pi = 100\,\mathrm{GHz}$

and incident amplitude $E_{is} = 5\,\text{V/cm}$ and $E_{ic} = 0$, and subject to bias velocities $2v_0/v_F = 0, .001, .002, .003, .004, .005, .007, .01, .05$, and 0.1 ($v_F = k_F/m$ is the Fermi velocity), corresponding to current densities $J = 0.06, 0.11, 0.17, 0.23, 0.29, 0.40, 0.57, 2.9$, and 5.7 A/m or the bias current parameter $\omega_j/\omega = 0, 0.05, 0.1, 0.16, 0.21, 0.26, 0.36, 0.52, 2.6$, and 5.2, respectively. The elastic scatterings are assumed due to short-range impurities and the broadening parameter is taken to be $\alpha = 10$. Linear (zero dc bias $v_0 = 0$) magnetoresistivity exhibits typical feature of radiation-induced magnetoresistance oscillations, with two-photon processes slightly showing up as a shoulder around $\epsilon_\omega \approx 1.5$. In the case of small dc bias, $\gamma_j \ll 1$, the node positions of peak-valley pairs are shifted down from $\epsilon_\omega = m$ at zero dc bias to $\epsilon_\omega = m/(1+\gamma_j) \approx m - m\gamma_j$. The shifted distances relative to the original positions, $m\gamma_j$, are larger for larger m but still locate within the original order range of ϵ_ω as long as $m\gamma_j < 1$. When the shifted distance becomes as large as 0.5, the original peak will move to the range where previously there is a valley. In the case of strong dc bias, $\gamma_j > 1$, the original pairs of higher order (larger m) may shift to lower order range of ϵ_ω. For instance, in the case of short-range impurity scattering $\eta = 0.94$ (Lei, 2007a), for the bias current of $\omega_j/\omega = 0.52$ ($\gamma_j = 4.9$), the original pairs at $\epsilon_\omega = m$ with $m = 1, 2, 3, 4, 5$, and 6 are shifted to the positions $\epsilon_\omega = m/(1+\gamma_j) = 0.17, 0.34, 0.51, 0.68, 0.85$, and 1.02. Except for the first one ($\epsilon_\omega = 0.17$), which is beyond the range plotted, the other five pairs are clearly identified in the figure.

The inset of Fig. 15.11 shows the change of the magnetoresistivity R_{xx} with increasing bias current density J in terms of ω_j/ω at three fixed magnetic field strengths: $\omega/\omega_c = 1.16, 1.80$, and 2.22, respectively, near the first minimum, the second maximum and second minimum of R_{xx}. Here we show the fact that within a certain magnetic field range around a resistivity minimum, the R_{xx} can be negative at small v_0, but increases with increasing v_0 and passes through zero at a finite v_0 (Lei, 2004). This is exactly what is required for the instability of the time-independent small-current solution and for the development of a spatially nonuniform state which exhibits measured zero resistance (Andreev, Aleiner and Mills, 2003). Accordingly, the region where a absolute negative dissipative magnetoresistance develops is identified as that of the measured zero-resistance state.

The calculated longitudinal magnetoresistivity R_{xx} and differential magnetoresistivity r_{xx} for another GaAs-based 2D system with carrier density $N_s = 3.7 \times 10^{15}\,\text{m}^{-2}$ and low-temperature linear mobility $\mu_0 =$

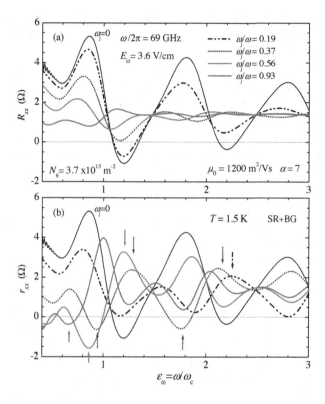

Fig. 15.12 Longitudinal magnetoresistivity R_{xx} (a) and differential magnetoresistivity r_{xx} (b) are shown as functions of $\epsilon_\omega \equiv \omega/\omega_c$ for a GaAs-based 2D system as described in the text at $T = 1.5$ K, under the irradiation of a linearly polarized microwave of frequency $\omega/2\pi = 69$ GHz and incident amplitude $E_{is} = 3.6$ V/cm and $E_{ic} = 0$, and subjected to different bias current density having parameter $\omega_j/\omega = 0, 0.19, 0.37, 0.56$, and 0.93. The elastic scatterings are from a mixture of short-range and background impurities. From Lei (2007b).

$1200\,\text{m}^2\text{V}^{-1}\text{s}^{-1}$ at lattice temperature $T = 1.5$ K are plotted in Fig. 15.12 as functions of $\epsilon_\omega \equiv \omega/\omega_c$, under the irradiation of a linearly x-direction polarized microwave of frequency $\omega/2\pi = 69$ GHz and incident amplitude $E_{is} = 3.6$ V/cm and $E_{ic} = 0$, and subjected to different bias drift velocities $2v_0/v_\text{F} = 0, 0.002, 0.004, 0.006$, and 0.01, corresponding to current densities $J = 0, 0.14, 0.28, 0.42$, and 0.70 A/m or the bias parameter $\omega_j/\omega = 0, 0.19, 0.37, 0.56$, and 0.93, respectively. The elastic scatterings are assumed due to a mixture of short-range and background impurities and the broadening parameter is taken to be $\alpha = 7$. We see that the positions of the peak-valley pairs and their maxima and minima of different ϵ_ω-order m in the zero-bias R_{xx} curves [Fig. 15.12(a)] are shifted downward with increasing current density ω_j/ω following the same rule as stated above in the case of Fig. 15.11. The positions of the peak-valley pairs and their maxima and minima in the differential resistivity r_{xx} [Fig. 15.12(b)] are shifted downward further relative to those of the corresponding R_{xx} at the same order m. Of particular interest is that with increasing bias current density, in the ϵ_ω range where the zero-bias magnetoresistivity ($R_{xx} = r_{xx}$) exhibits positive maximum the differential magnetoresistivity r_{xx} can be driven down to a considerable negative value, as indicated by the up-directed arrows in Fig. 15.12(b); and in the ϵ_ω range where the zero-bias magnetoresistivity ($R_{xx} = r_{xx}$) exhibits a minimum the differential magnetoresistivity r_{xx} can be driven up to exhibit a maximum, as indicated by the down-directed arrows in Fig. 15.12(b).

According to the electrodynamic analyses, a homogeneous state of uniform current J is unstable if either the absolute dissipative resistivity R_{xx} or the differential resistivity r_{xx} becomes negative (Andreev, Aleiner and Mills, 2003; Alicea et al, 2005). Thus, despite the detailed macroscopic structure yet to be explored, it is expected that, under suitable condition, a zero-resistance state (ZRS) or some kind of quasi-ZRS could be measured in the region where R_{xx} or r_{xx} exhibits a negative value in a homogeneous microscopic analysis. The fact that the experiment performed by Zhang et al (2007b) with the setup to measure the differential resistivity did not detect any negative value of r_{xx}, also supports such a consideration. With this in mind, the behavior of R_{xx} and r_{xx} in Fig. 15.12 could predict the following measured results. In the vicinity around $\epsilon_\omega = 2.25$, the ZRS showing up in the zero dc bias will disappear at bias $\omega_j/\omega \geqslant 0.10$ because both R_{xx} and r_{xx} return to positive. Instead, one should observe a bulged r_{xx} in the cases of $\omega_j/\omega = 0.19, 0.37$, and 0.93 and a dented r_{xx} in the case of $\omega_j/\omega = 0.56$, as shown in Fig. 15.12. In the vicinity around $\epsilon_\omega = 1.25$, where the negative R_{xx} (thus the ZRS), maintains up to the dc bias $\omega_j/\omega = 0.35$, one may not detect r_{xx} significantly different from zero at low dc biases but a bulged r_{xx} should be observed at higher dc biases when R_{xx} becomes positive and the ZRS disappears, as seen in Fig. 15.12 in the cases of $\omega_j/\omega = 0.37$ and 0.56. In the vicinity around $\epsilon_\omega = 1.75$ and 0.8, the zero-bias magnetoresistivity R_{xx} ($= r_{xx}$) exhibits main peaks while the differential resistivity r_{xx} may becomes negative at certain strong dc biases, suggesting the possible zero-resistance state induced by the direct current. Fig. 15.12(b) indicates that a ZRS can appear around $\epsilon_\omega = 1.75$ at bias $\omega_j/\omega = 0.56$, but with further increase of the current density it disappears and becomes a bulged differential resistivity. In the range $0.6 \leqslant \epsilon_\omega \leqslant 1$, the differential resistivity r_{xx} at all three strong dc biases ($\omega_j/\omega = 0.37, 0.56$, and 0.93) can become negative, as shown in Fig. 15.12b, implying possible resistance suppression or quasi-ZRSs induced by the bias current. These results indeed showed up in a recent experimental observation (Zhang et al, 2007b).

15.4 Magnetoresistance Oscillations under Bichromatic Irradiation in Two-Dimensional Electron Systems

The balance-equation approach to magnetotransport driven by a monochromatic radiation discussed in the last section is easily extended to the case of simultaneous driving of two radiation fields of different frequencies.

To deal with the case of bichromatic radiation we assume that a dc or slowly time-varying electric field $\boldsymbol{E}_0$ and two high frequency fields,

$$\boldsymbol{E}_1(t) = \boldsymbol{E}_{1s}\sin(\omega_1 t) + \boldsymbol{E}_{1c}\cos(\omega_1 t), \quad (15.114)$$

$$\boldsymbol{E}_2(t) = \boldsymbol{E}_{2s}\sin(\omega_2 t) + \boldsymbol{E}_{2c}\cos(\omega_2 t), \quad (15.115)$$

are applied simultaneously in a quasi-2D system consisting of N_s interacting electrons in a unit area of the x–y plane, subjected to a magnetic field $\boldsymbol{B} = (0, 0, B)$ along the z direction. The frequencies ω_1 and ω_2 are high enough and their difference is large enough that ω_1 and ω_2, as well as $|n_1\omega_1 - n_2\omega_2|$ (for arbitrary integers n_1 and n_2), are all much larger than $1/\tau_0$, where τ_0 stands for the scale of the time variation of the slowly varying field $\boldsymbol{E}_0$, or the time scale within which people carry out the transport measurement, whichever the shorter. The transport state of this high-carrier-density many-electron system under these electric fields can be described in terms of a rapidly time-varying electron drift velocity oscillating at both base radiation frequencies, $\boldsymbol{v}(t) = \boldsymbol{v}_1(t) + \boldsymbol{v}_2(t)$, with

$$\boldsymbol{v}_1(t) = \boldsymbol{v}_{1c}\cos(\omega_1 t) + \boldsymbol{v}_{1s}\sin(\omega_1 t), \quad (15.116)$$

$$\boldsymbol{v}_2(t) = \boldsymbol{v}_{2c}\cos(\omega_2 t) + \boldsymbol{v}_{2s}\sin(\omega_2 t), \quad (15.117)$$

and a dc or slowly varying part $\boldsymbol{v}_0$ of the electron drift motion, together with an electron temperature T_e characterizing the electron heating. In the case of ultra-clean electron gas at low temperatures, the steady-state quantities $\boldsymbol{v}_0$ and T_e satisfy the following force- and energy-balance equations:

$$N_s e\boldsymbol{E}_0 + N_s e(\boldsymbol{v}_0 \times \boldsymbol{B}) + \boldsymbol{F}_0 = 0, \quad (15.118)$$

$$N_s \boldsymbol{E}_0 \cdot \boldsymbol{v}_0 + S_p - W = 0, \tag{15.119}$$

with $\boldsymbol{v}_{1c}$ and $\boldsymbol{v}_{1s}$ determined by

$$-m\omega_1 \boldsymbol{v}_{1c} = e\boldsymbol{E}_{1s} + e(\boldsymbol{v}_{1s} \times \boldsymbol{B}), \tag{15.120}$$

$$m\omega_1 \boldsymbol{v}_{1s} = e\boldsymbol{E}_{1c} + e(\boldsymbol{v}_{1c} \times \boldsymbol{B}); \tag{15.121}$$

and $\boldsymbol{v}_{2c}$ and $\boldsymbol{v}_{2s}$ determined by

$$-m\omega_2 \boldsymbol{v}_{2c} = e\boldsymbol{E}_{2s} + e(\boldsymbol{v}_{2s} \times \boldsymbol{B}), \tag{15.122}$$

$$m\omega_2 \boldsymbol{v}_{2s} = e\boldsymbol{E}_{2c} + e(\boldsymbol{v}_{2c} \times \boldsymbol{B}). \tag{15.123}$$

Here e and m are the electron charge and effective mass,

$$\boldsymbol{F}_0 = \sum_{\boldsymbol{q}_\parallel} |U(\boldsymbol{q}_\parallel)|^2 \sum_{n_1,n_2=-\infty}^{\infty} \mathrm{J}_{n_1}^2(\xi_1)\, \mathrm{J}_{n_2}^2(\xi_2)$$
$$\times \boldsymbol{q}_\parallel\, \Pi_2(\boldsymbol{q}_\parallel, \omega_0 - n_1\omega_1 - n_2\omega_2) \tag{15.124}$$

is the damping force of the moving center of mass of electrons,

$$S_p = \sum_{\boldsymbol{q}_\parallel} |U(\boldsymbol{q}_\parallel)|^2 \sum_{n_1,n_2=-\infty}^{\infty} \mathrm{J}_{n_1}^2(\xi_1)\, \mathrm{J}_{n_2}^2(\xi_2)$$
$$\times (n_1\omega_1 + n_2\omega_2)\, \Pi_2(\boldsymbol{q}_\parallel, \omega_0 - n_1\omega_1 - n_2\omega_2) \tag{15.125}$$

is the averaged rate of the electron energy absorption from the radiation fields, and

$$W = \sum_{\boldsymbol{q},\lambda} |M(\boldsymbol{q},\lambda)|^2 \sum_{n_1,n_2=-\infty}^{\infty} \mathrm{J}_{n_1}^2(\xi_1)\, \mathrm{J}_{n_2}^2(\xi_2)$$
$$\times \Omega_{\boldsymbol{q}\lambda}\, \Lambda_2^+(\boldsymbol{q},\lambda,\omega_0 + \Omega_{\boldsymbol{q}\lambda} - n_1\omega_1 - n_2\omega_2) \tag{15.126}$$

is the average rate of the electron energy dissipation to the lattice. In the above equations, $\omega_0 \equiv \boldsymbol{q}_\parallel \cdot \boldsymbol{v}_0$, $\xi_1 \equiv [(\boldsymbol{q}_\parallel \cdot \boldsymbol{v}_{1c})^2 + (\boldsymbol{q}_\parallel \cdot \boldsymbol{v}_{1s})^2]^{\frac{1}{2}}/\omega_1$, $\xi_2 \equiv [(\boldsymbol{q}_\parallel \cdot \boldsymbol{v}_{2c})^2 + (\boldsymbol{q}_\parallel \cdot \boldsymbol{v}_{2s})^2]^{\frac{1}{2}}/\omega_2$, $\mathrm{J}_{n_1}(\xi_1)$ and $\mathrm{J}_{n_2}(\xi_2)$ are Bessel functions of order n_1 and n_2, $\Pi_2(\boldsymbol{q}_\parallel,\Omega)$ is the imaginary part of the electron density correlation function of the 2D system in the magnetic field, and $\Lambda_2^+(\boldsymbol{q},\lambda,\Omega)$ is the electron–phonon correlation function given by (15.36).

We assume that the 2D electrons are contained in a thin layer suspended in vacuum at plane $z = 0$. When both electromagnetic waves illuminate upon the plane perpendicularly with the incident electric fields $\boldsymbol{E}_{i1}(t) = \boldsymbol{E}_{i1s}\sin(\omega_1 t) + \boldsymbol{E}_{i1c}\cos(\omega_1 t)$ and $\boldsymbol{E}_{i2}(t) = \boldsymbol{E}_{i2s}\sin(\omega_2 t) + \boldsymbol{E}_{i2c}\cos(\omega_2 t)$, the high-frequency fields in the 2D electron gas, determined by the electrodynamic equations, are

$$\boldsymbol{E}_1(t) = \frac{N_s e \boldsymbol{v}_1(t)}{2\epsilon_0 c} + \boldsymbol{E}_{i1}(t), \qquad (15.127)$$

$$\boldsymbol{E}_2(t) = \frac{N_s e \boldsymbol{v}_2(t)}{2\epsilon_0 c} + \boldsymbol{E}_{i2}(t). \qquad (15.128)$$

Using this $\boldsymbol{E}_1(t)$ in Eqs. (15.120) and (15.121), and $\boldsymbol{E}_2(t)$ in Eqs. (15.122) and (15.123), the oscillating velocity $\boldsymbol{v}_{1c}$ and $\boldsymbol{v}_{1s}$ (and thus the argument ξ_1) are explicitly expressed in terms of incident field $\boldsymbol{E}_{i1s}$ and $\boldsymbol{E}_{i1c}$, and the oscillating velocity $\boldsymbol{v}_{2c}$ and $\boldsymbol{v}_{2s}$ (and thus the argument ξ_2) are explicitly expressed in terms of incident field $\boldsymbol{E}_{i2s}$ and $\boldsymbol{E}_{i2c}$. Therefore, for a given $\boldsymbol{v}_0$ we need only to solve the energy balance equation $S_p - W = 0$ to obtain the electron temperature T_e under given incident radiation fields, before directly calculating the transverse and longitudinal magnetoresistivities from

$$R_{yx} = B/N_s e, \qquad (15.129)$$

$$R_{xx} = -\frac{\boldsymbol{F}_0 \cdot \boldsymbol{v}_0}{N_s^2 e^2 v_0^2}. \qquad (15.130)$$

The linear magnetoresistivity is the $\boldsymbol{v}_0 \to 0$ limit of (15.130), and will be denoted as

$$R_{xx}^{\omega_1 \omega_2}(\xi_1, \xi_2) = -\frac{1}{N_s^2 e^2} \sum_{\boldsymbol{q}_\parallel} |U(\boldsymbol{q}_\parallel)|^2 \sum_{n_1, n_2 = -\infty}^{\infty} \mathrm{J}_{n_1}^2(\xi_1) \mathrm{J}_{n_2}^2(\xi_2)$$
$$\times q_x^2\, \Pi_2'(\boldsymbol{q}_\parallel, n_1 \omega_1 + n_2 \omega_2) \quad (15.131)$$

to explicitly indicate that it is the resistivity under simultaneous illumination of ω_1–ξ_1 ($\boldsymbol{E}_{i1}$) and ω_2–ξ_2 ($\boldsymbol{E}_{i2}$) radiations. At the same time we will use $R_{xx}^{\omega}(\xi)$ to denote the linear magnetoresistivity (15.99) under a monochromatic radiation of ω–ξ ($\boldsymbol{E}_i$).

In the degenerate and large Landau-filling-factor case, at positions $\omega_2 = l\omega_c$ ($l = 1, 2, 3, \cdots$), $\Pi_2'(\boldsymbol{q}_\parallel, n_1\omega_1 + n_2\omega_2)$ function does not depend on n_2, thus (15.131) reducing to

$$R_{xx}^{\omega_1 \omega_2}(\xi_1, \xi_2) = -\frac{1}{N_e^2 e^2} \sum_{\boldsymbol{q}_\parallel} q_x^2 |U(\boldsymbol{q}_\parallel)|^2 \sum_{n_1=-\infty}^{\infty} \mathrm{J}_{n_1}^2(\xi_1)\, \Pi_2'(\boldsymbol{q}_\parallel, n_1\omega_1)$$
$$\approx R_{xx}^{\omega_1}(\xi_1), \qquad (15.132)$$

i.e., bichromatic $R_{xx}^{\omega_1\omega_2}(\xi_1,\xi_2)$ equals the monochromatic $R_{xx}^{\omega_1}(\xi_1)$ given by (15.99) under ω_1-$\boldsymbol{E}_{i1}$ radiation (effect of electron temperature difference on photoresistance is negligible). This means that all the bichromatic $R_{xx}^{\omega_1\omega_2}(\xi_1,\xi_2)$ curves with fixed $\boldsymbol{E}_{i1}$ but changing $\boldsymbol{E}_{i2}$, cross with the monochromatic $R_{xx}^{\omega_1}(\xi_1)$ curve of $\boldsymbol{E}_{i1}$ at integer-ω_c points of ω_2. Likewise, at integer-ω_c points of ω_1, $R_{xx}^{\omega_1\omega_2}(\xi_1,\xi_2) \approx R_{xx}^{\omega_2}(\xi_2)$.

In the case of $\omega_2 \sim 1.5\omega_1$, even and odd ω_c points exhibit a somewhat different behavior. At an even-ω_c point of ω_1 ($\omega_1 = 2\omega_c$, $4\omega_c$, or $6\omega_c$), ω_2 is also close to an integer ω_c ($\omega_2 \approx 3\omega_c$, $6\omega_c$, and $9\omega_c$) that $R_{xx}^{\omega_2}(\xi_2)$ is almost identical to the dark resistivity $R_{xx}(0)$, independent of $\boldsymbol{E}_{i2}$. Therefore, all bichro-resistivity $R_{xx}^{\omega_1\omega_2}$, mono-resistivity $R_{xx}^{\omega_1}$ and mono-resistivity $R_{xx}^{\omega_2}$, cross at these points with dark resistivity $R_{xx}(0)$. At an odd-ω_c point of ω_1 ($\omega_1 = \omega_c$, $3\omega_c$, $5\omega_c$, or $7\omega_c$), ω_2 is close to a half-integer ω_c ($\omega_2 \approx 1.5\omega_c$, $4.5\omega_c$, $7.5\omega_c$, or $10.5\omega_c$). It is a node point of $R_{xx}^{\omega_2}(\xi_2)$ when E_{i2} changes modestly. Thus, at these points bichromatic $R_{xx}^{\omega_1\omega_2}$ and monochromatic $R_{xx}^{\omega_1}$ and $R_{xx}^{\omega_2}$ are nearly the same, but may be somewhat different from the dark resistivity.

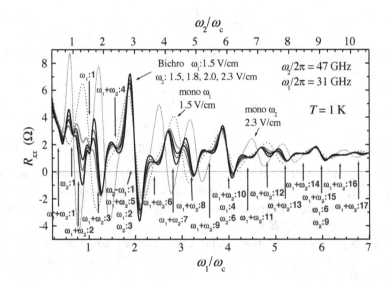

Fig. 15.13 Magnetoresistivity R_{xx} induced by bichromatic or monochromatic radiations ($R_{xx}^{\omega_1\omega_2}$, $R_{xx}^{\omega_1}$ or $R_{xx}^{\omega_2}$) in a GaAs-based 2D electron system with $N_s = 3.0 \times 10^{15}$ m^{-2}, $\mu_0 = 2000$ m^2/Vs and $\alpha = 5$, at lattice temperature $T = 1$ K. From Lei and Liu (2006b).

These results can be seen in Fig. 15.13, where the calculated magnetoresistivity $R_{xx}^{\omega_1\omega_2}$ versus the inverse magnetic field for a GaAs-based 2D system having electron density $N_s = 3.0 \times 10^{15}\,\mathrm{m}^{-2}$, linear mobility $\mu_0 = 2000\,\mathrm{m}^2/\mathrm{Vs}$ and broadening parameter $\alpha = 5$, simultaneously irradiated by two linearly polarized microwaves of frequencies $\omega_1/2\pi = 31\,\mathrm{GHz}$ and $\omega_2/2\pi = 47\,\mathrm{GHz}$ [$\boldsymbol{E}_{\mathrm{i1}s} = (E_{\mathrm{i1}}, 0), \boldsymbol{E}_{\mathrm{i1}c} = 0$ and $\boldsymbol{E}_{\mathrm{i2}s} = (E_{\mathrm{i2}}, 0), \boldsymbol{E}_{\mathrm{i2}c} = 0$] with incident amplitudes $E_{\mathrm{i1}} = 1.5\,\mathrm{V/cm}$ and $E_{\mathrm{i2}} = 1.5, 1.8, 2.0$, and $2.3\,\mathrm{V/cm}$ at $T = 1\,\mathrm{K}$. The monochromatic $R_{xx}^{\omega_1}$ at $E_{\mathrm{i1}} = 1.5\,\mathrm{V/cm}$ and $R_{xx}^{\omega_2}$ at $E_{\mathrm{i2}} = 2.3\,\mathrm{V/cm}$ are also shown.

As in the case of monochromatic illumination, the appearance of magnetoresistance oscillation in the bichromatic irradiation comes from real photon-assisted electron transitions between different Landau levels as indicated in the summation of the electron density correlation function. Apparently, all multiple photon processes related to the ω_1 monochromatic radiation and the ω_2 monochromatic radiation are included. In addition, there are multiple photon processes related to the mixing ω_1 and ω_2 radiation.

With three positive integers n_1, n_2 and l to characterize a multiple photon assisted electron transition, we use the symbol $n_1\omega_1 + n_2\omega_2 : l$ to denote a process during which an electron jumps across l Landau level spacings with simultaneous absorption of n_1 photons of frequency ω_1 and n_2 photons of frequency ω_2, or simultaneous emission of n_1 photons of frequency ω_1 and n_2 photons of frequency ω_2; and use the symbol $n_1\omega_1 - n_2\omega_2 : l$ to denote a process during which an electron jumps across l Landau level spacings with simultaneous absorption of n_1 photons of frequency ω_1 and emission of n_2 photons of frequency ω_2, or simultaneous emission of n_1 photons of frequency ω_1 and absorption of n_2 photons of frequency ω_2 (thus the symbol $n_2\omega_2 - n_1\omega_1 : l$ has the same meaning as $n_1\omega_1 - n_2\omega_2 : l$). The symbol $n_1\omega_1 : l$ indicates a process during which an electron jumps across l Landau level spacings with the assistance of n_1 real (emission or absorption) ω_1-photons and virtual ω_2-photons. So does the symbol $n_2\omega_2 : l$. The processes represented by $n_1\omega_1 + n_2\omega_2 : l$ contribute, in the R_{xx}-versus-ω_c/ω curve, a structure consisting of a minimum and a maximum on both sides of $\omega_c/\omega_1 = n_1(1 + n_2\omega_2/n_1\omega_1)/l$ or $\omega_c/\omega_2 = n_2(1 + n_1\omega_1/n_2\omega_2)/l$. And those by $n_1\omega_1 - n_2\omega_2 : l$ (assume $n_1\omega_1 > n_2\omega_2$) contribute a minimum-maximum pair around $\omega_c/\omega_1 = n_1(1 - n_2\omega_2/n_1\omega_1)/l$ or $\omega_c/\omega_2 = n_2(n_1\omega_1/n_2\omega_2 - 1)/l$.

As shown in the figure the bichromatic curve consists of much more sizable peak-valley pairs in comparison with monochromatic ones. Since there are only minor high-order structures showing up in both monochromatic curves at these radiation strengths, most of the pairs appearing in the bichromatic curves relate to mixing photon processes. We have identified the lowest order ($|n_1| \leq 1$ and $|n_2| \leq 1$) individual (ω_1 or ω_2) and mixing ($\omega_1 - \omega_2$ or $\omega_1 + \omega_2$) photon-assisted electron transition processes to all the pairs in the figure. Except those pairs around $\omega_2/\omega_c = 1$ and $\omega_1/\omega_c = 1$, which relate to single-ω_2 and single-ω_1 processes, and those around $\omega_1/\omega_c = 2, 4$, and 6, which relate to both individual and mixing photon processes, all other pairs are related to mixing processes $\omega_1 + \omega_2 : l$ ($l = 1, 2, \cdots, 17$) with electron transitions jumping l Landau-level spacings.

Bibliography

Abdolsalami F, Khan F S (1990), *Phys. Rev. B*, **41**, 3494.
Abramowitz M, Stegun I A (1965), *Handbook of Mathematical Functions*, US GPU, Washington D C.
Agaeva R G, Askerov B M, Gashimzade D E (1976), *Sov. Phys. Semicond.* **10**, 1268.
Alberga, E G et al (1982), *Appl. Phys. A*, **27**, 107.
Alicea J et al (2005), *Phys. Rev. B*, **71**, 235322.
Altshuler B L, Aronov A G, Khmelmitsky D E (1981), *Solid State Commun.* **39**, 619.
Altschul V, Finkman E (1991), *Appl. Phys. Lett.* **58**, 942.
Anderson P W (1963), *Concepts in Solids*, Benjamin, New York, ch. 2.
Andreev A V, Aleiner I L, Mills A J (2003), *Phys. Rev. Lett.* **91**, 056803.
Anderson D L, Aas E J (1973), *J. Appl. Phys.* **44**, 3721.
Anderson P W, Abrahams E, Ramakrishnan T V (1979), *Phys. Rev. Lett.* **43**, 718.
Ando T (1974), *J. Phys. Soc. Jpn.* **37**, 1233.
Ando T (1976), *Phys. Rev. Lett.* **36**, 1383.
Ando T, Fowler A B, Stern F (1982), *Rev. Mod. Phys.* **54**, 437.
Ando T, Uemura Y (1974), *J. Phys. Soc. Jpn.* **36**, 959.
Anile A M, Muscato O (1995), *Phys. Rev. B*, **51**, 16728.
Arai M R (1983), *Appl. Phys. Lett.* 1983, **42**, 906.
Argyres P N (1960), *Phys. Rev.* **117**, 315.
Argyres P N, Kelley P L (1964), Phys. Rev. **134**, A98
Argyres P N (1963), *Phys. Rev.* **132**, 1527.
Argyres P N (1989), *Phys. Rev. B*, **39**, 2982.
Argyres P N, Adams E N (1956), *Phys. Rev.* **104**, 900.
Argyres P N, Sigel J L (1974), *Phys. Rev. B*, **9**, 3197.
Arora V K (1983), *J. Appl. Phys.* **54**, 824.
Aschcroft N W, Mermin N D (1976), *Solid State Physics*, Holt, Rimehart and Winston, New York, ch. 13.
Asmar N G et al (1995), *Phys. Rev. B*, **51**, 18041.
Babaev M M, Gassymov T M (1977), *Phys. Stat. Sol. (b)*, **84**, 473.

Babaev M M et al (2002), *Phys. Rev. B*, **65**, 165324.
Babiker M (1986), *J. Phys. C*, **19**, 683.
Barker J R (1973), *J. Phys. C*, **6**, 2663.
Barker J R (1978), *Solid State Electron.* **21**, 267.
Barker J R (1979), *Solid State Commun.* **32**, 1013.
Barker J R, Ferry D K (1979), *Phys. Rev. Lett.* **42**, 1779.
Bass F G, Rubinstein E A (1977), *Sov. Phys.: Solid State*, **19**, 800.
Berggren et al (1986), *Phys. Rev. Lett.* **57**, 1769.
Blount E I (1962), *Formalisms of Band Theory*, In: Seitz F, Turnbull D, ed. *Solid State Physics: Advances in Research and Applications*, vol 13, Academic, New York, p. 305.
Bogolyubov N N (1962), *Problems of Dynamic Thorey in Statistical Physics*, North-Holland, Amsterdam.
Büttiker M, Thomas H (1977), *Phys. Rev. Lett.* **38**, 78.
Büttiker M, Thomas H (1979), *Z. Physik B*, **33**, 275.
Cai J, Cui H L (1995), *J. Appl. Phys.* **78**, 6802.
Cai W et al (1989), *Phys. Rev. B*, **40**, 7671.
Cai W, Lax M (1992), *Int. J. Mod. Phys.* **6**, 1007; In: Ting C S ed. *Physics of Hot Electron Transport in Semiconductors*, World Scientific, Singapore, p.203.
Cai W, Lei X L, Ting C S (1985), *Phys. Rev. B*, **31**, 4070.
Cai W, Marchetti M C, Lax M (1986), *Phys. Rev. B*, **34**, 8573.
Cai W, Marchetti M C, Lax M (1988), *Phys. Rev. B*, **37**, 2636.
Cai W, Ting C S (1986), *Phys. Rev. B*, **33**, 3967.
Cai W, Zheng T F, Lax M (1988), *Phys. Rev. B*, **37**, 8205.
Calecki D, Pottier N (1981), *J. Phys. (Paris) Colloq.* **42**, 67.
Callaway J (1986), *Quantum theory of the Solid State, 2nd ed.*, Academic, New York, ch. 6.
Canali C et al (1971), *J. Phys. Chem. Solids*, **32**, 1707.
Canali C et al (1975), *Phys. Rev. B*, **12**, 2265.
Canali C, Nava F and Reggiani L (1985), In: Reggiani L, ed. *Hot-Electron Transport in Semiconductors*, Springer-Verlag, Berlin, p.87.
Cao J C (2003), *Phys. Rev. Lett.* **91**, 237401.
Cao J C (2004), *Phys. Rev. B*, **69**, 165203.
Cao J C et al (2001a), *Phys. Rev. B*, **63**, 115308.
Cao J C et al (2001b), *Appl. Phys. Lett.* **78**, 2524.
Cao J C et al (2001c), *Appl. Phys. Lett.* **79**, 3524.
Cao J C, Lei X L (1996a), *Solid State Electron.* **7**, 971.
Cao J C, Lei X L (1996b), *Int. J. Electron.* **81**, 237.
Cao J C, Lei X L (1997), *Solid State Electron.* **41**, 1181.
Cao J C, Lei X L (1998a), *Solid State Electron.* **42**, 419.
Cao J C, Lei X L (1998b), *Commun. Theor. Phys.* **30**, 381.
Cao J C, Lei X L (1999a), *Eur. Phys. J. B*, **7**, 79.
Cao J C, Lei X L (1999b), *Phys. Rev. B*, **59**, 2199.
Cao J C, Lei X L (1999c), *Phys. Rev. B*, **60**, 1871.
Cao J C, Lei X L (2002), *Eur. Phys. J. B*, **29**, 553.
Cao J C, Lei X L (2002), *Phys. Rev. B*, **67**, 085309.

Cao J C, Liu H C and Lei X L (2000a), *Phys. Rev. B*, **61**, 5546.
Cao J C, Liu H C and Lei X L (2000b), *J. Appl. Rev.* **87**, 2867.
Chen L Y, Ting C S, Horing N J M (1989), *Phys. Rev. B*, **40**, 4756.
Chen L Y, Ting C S, Horing N J M (1990), *Phys. Rev. B*, **42**, 1129.
Chiu K W, Lee T K, Quinn J J (1976), *Surf. Sci.* **58**, 128.
Chou K C, Su Z B, Hao B L, Yu L (1985), *Phys. Rep.* **118**, 1.
Cui H L, Horing N J M (1989), *Phys. Rev. B*, **40**, 2901.
Cui H L, Lei X L, Horing N J M (1988a), *Phys. Rev. B*, **37**, 8223.
Cui H L, Lei X L, Horing N J M (1988b), *Phys. Stat. Sol. (b)*, **146**, 189.
Cui H L, Lei X L, Horing N J M (1993), *Superlatt. and Microstr.* **13**, 221.
da Cunha Lima I C, Wang X F, Lei X L (1997), *Phys. Rev. B*, **55**, 10681.
Das Sarma (1981), *Phys. Rev. B*, **23**, 4592.
Das Sarma S, Stern F (1985), *Phys. Rev. B*, **32**, 8442.
Das Sarma S, Xie X C (1988), *Phys. Rev. Lett.* **61**, 738.
Das Sarma S, Jain J K, Jalabert R (1990), *Phys. Rev. B*, **41**, 3561.
Das Sarma S, Korenman V (1991), *Phys. Rev. Lett.* **67**, 2916.
Davydov B, Pomeranchuk I (1958), *J. Phys. USSR (Moscow)* **2**, 279.
Dharma-wardana M W C (1991), *Phys. Rev. Lett.* **66**, 197.
Dong B, Lei X L (1998a), *Commun. Thoret. Phys.* **29**, 7.
Dong B, Lei X L (1998b), *Comm. Theoret. Phys.* **29**, 195.
Doran N J, Gerlach E (1976), *Phys. Stat. Sol.(b)* **75**, K15.
Dorozhkin S I (2003), *JETP Lett.* **77**, 577.
Dorozhkin S I et al (2005), *Phys. Rev. B*, **71**, 201316(R).
Drummond T J et al (1981), *Electron. Lett.* **17**, 545.
Durst A C et al (2003), *Phys. Rev. Lett.* **91**, 086803.
Ebert G et al (1983), *J. Phys. C*, **16**, 5441.
Edwards S F (1965), *Proc. Phys. Soc.* **86**, 977.
Elesin V F (1969), *Sov. Phys. JETP* **28**, 410.
Esaki L, Tsu R (1970), *IBM J. Res. Develop.* **14**, 61.
Fischetti M V, Laux S E (1988), *Phys. Rev. B*, **38**, 9721.
Fishman R S (1989), *Phys. Rev. B*, **39**, 2994.
Fishman R S, Mahan G D (1989), *Phys. Rev. B*, **39**, 2990.
Fitzer N et al (2003), *Phys. Rev. B*, **67**, 201201.
Fletcher R et al (1986), *Phys. Rev. B*, **33**, 7122.
Fletcher R et al (1994), *Phys. Rev. B*, **50**, 14991.
Fuchs R, Kliewer K L (1965), *Phys. Rev.* **140**, A2076.
Ganguly A K, Ting C S (1977), *Phys. Rev. B*, **16**, 3541.
Gerhardts R R (1975), *Z. Phys. B*, **21**, 275; 283.
Gerhardts R R (1993), *Phys. Rev. B*, **48**, 9178.
Gerlach E (1986), *J. Phys. C*, **19**, 4585.
Gold A, Ghazali A (1990), *Phys. Rev. B*, **41**, 7626.
Gornik E et al (1985), *Phys. Rev. Lett.* **54**, 1820.
Götze W, Wölfle P (1972), *Phys. Rev. B*, **6**, 122.
Grahn H T et al (1991), *Phys. Rev. B*, **43**, 12094.
Gueret P (1971), *Phys. Rev. Lett.* **27**, 256.
Guillemot C et al (1990), *Superlatt. and Microstr.* **8**, 259.

Guillemot C, Clérot F (1993), *Phys. Rev. B*, **47**, 7227.
Guillemot C, Clérot F, Regreny A (1992), *Phys. Rev. B*, **46**, 10152.
Gupta R, Balban R, Ridley B K (1992), *Phys. Rev. B*, **46**, 7745.
Hasbun J E (1993), *J. Phys. Chem. Solids*, **56**, 791.
Hasbun J E (1994), *J. Appl. Phys.* **75**, 270.
Hasbun J E, Nee T W (1991), *Phys. Rev. B*, **44**, 3125.
Herring C, Vogt E (1956), *Phys. Rev.* **101**, 944.
Hicks L D, Dresselhaus M S (1993), *Phys. Rev. B*, **47**, 12727.
Hirakawa K *et al* (1993), *Phys. Rev. B*, **47**, 16651.
Hirakawa K *et al* (1993), *Phys. Rev. B*, **63**, 085320.
Hirakawa K, Sakaki H (1988), *J. Appl. Phys.* **63**, 803.
Höpfel R A *et al* (1986), *Phys. Rev. Lett.* **56**, 2736.
Höpfel R A, Weimann G (1985), *Appl. Phys. Lett.* **46**, 291.
Horing N J M (1965), *Ann. Phys. (NY)*, **31**, 1.
Horing N J M, Argyres P N (1962), In: *Report of the Int. Conf. on the Physics of Semicond.*, Inst. of Phys. and Physical Society, London, p.58.
Horing N J M, Cui H L, Lei X L (1987), *Phys. Rev. B*, **35**, 6438.
Horing N J M, Cui H L, Lei X L (1992), *Int. J. Mod. Phys.* **6**, 937; In: Ting C S ed. *Physics of Hot Electron Transport in Semiconductors*, World Scientific, Singapore, p.133.
Horing N J M, Lei X L, Cui H L (1986), *Phys. Rev. B*, **33**, 6929.
Horing N J M, Yildiz M M (1976), *Ann. Phys. (N Y)*, **97**, 216.
Hu Ben Yu-Kuang, Flensberg K (1996), *Phys. Rev. B*, **53**, 10072.
Hu G Y, O'Connell R F (1987), *Phys. Rev. B*, **36**, 5798.
Hu G Y, O'Connell R F (1988a), *Physica A*, **149**, 1.
Hu G Y, O'Connell R F (1988b), *Physica A*, **151**, 33.
Hu G Y, O'Connell R F (1988c), *Physica A*, **153**, 114.
Hu G Y, O'Connell R F (1988d), *Phys. Rev. B*, **37**, 10391.
Hu G Y, O'Connell R F (1988e), *J. Phys. C*, **21**, 4325.
Hu G Y, O'Connell R F (1989a), *Phys. Rev. B*, **39**, 12717.
Hu G Y, O'Connell R F (1989b), *Phys. Rev. B*, **40**, 3600.
Hu P, Ting C S (1986), *Phys Rev B*, **34**, 7003.
Huang D H *et al* (2004), *Phys. Rev. B*, **69**, 075214.
Huang D H *et al* (2005), *Phys. Rev. B*, **71**, 195205.
Huang K, Wu X G (1994), *Phys. Rev. B*, **49**, 2223.
Huang K, Wu X G (1995), *Phys. Rev. B*, **51**, 5531.
Huang K, Zhu B F (1988), *Phys. Rev. B*, **38**, 2183, 13377.
Huang K, Zhu B F (1992), *Phys. Rev. B*, **45**, 14404.
Huberman M, Chester G V (1975), *Adv. Phys.* **24**, 489.
Hutchinson H J *et al* (1994), *J. Appl. Phys.* **75**, 320.
Ignatov A A *et al* (1991), *Mod. Phys. Lett. B*, **5**, 1087.
Ignatov A A, Renk K F, Dodin E P (1993), *Phys. Rev. Lett.* **70**, 1996.
Ignatov A A, Shaskin, V I (1983), *Phys. Lett. A*, **94**, 169.
Ignatov A A, Shaskin, V I (1987), *Sov. Phys. JETP*, **66**, 526.
Jain J K, Jalabert R, Das Sarma S (1988), *Phys. Rev. Lett.* **60**, 353; *ibid*, **61**, 2005.

Jain J K, Das Sarma S (1989), *Phys. Rev. Lett.* **62**, 2305.
Kalashnikov V P (1970), *Physica*, **48**, 93.
Kane E O (1957), *J. Phys. Chem. Solids*, **1**, 249.
Kawaguchi Y et al (2002), *Appl. Phys. Lett.* **80**, 136.
Keldysh L V, Kirzhritz D A, Maradudin A A (1989), *The Dielectric Function of Condensed Systems*, North-Holland, Amsterdam.
Kenkre V M, Dresden M (1972), *Phys. Rev. A*, **6**, 769.
Kennedy T A, Wagner R J, McCombe B D, Tsui D C (1975), *Phys. Rev. Lett.* **35**, 1109.
Khan F S, Davis J H, Wilkins J W (1987), *Phys. Rev. B*, **36**, 2578.
Kittel C (1976), *Introduction to Solid State Physics*, John Wiley & Sons, New York.
Kittle C (1986), *Quantum Theory of Solids*, Wiley, Chichester, ch. 9.
Klein M V (1986), *IEEE: J. Quantum Electron.* **QE-22**, 1760.
Kleinert P, Asche M (1994), *Phys. Rev. B*, **50**, 11022.
Knox W H et al (1989), *J. Phys.: Condens. Matter*, **1**, 9401.
Koenraad P M et al (1998), *Physica B*, **256–268**, 268.
Kogan S M (1963), *Fiz. Tverd. Tela (Leningrad)*, **4**, 2474; *Sov. Phys.: Solid State*, **4**, 1813.
Kohn W, Luttinger J M (1957), *Phys. Rev.* **108**, 590.
Kocevar P (1985), *Physica B*, **134**, 155.
Kostial H et al (1993), *Phys. Rev. B*, **47**, 4485.
Kotthaus J P (1987), *Phys. Scr.* **T19**, 120.
Krieger J B, Iafrate G J (1987), *Phys. Rev. B*, **36**, 9644.
Ktitorov S A et al (1972), *Sov. Phys.: Solid State*, **13**, 1872.
Kubo R (1959), *J. Phys. Soc. Jpn.* **12**, 570.
Kubo R, Hashitsume N, Miyake S J (1965), In: Seitz F and Turnbull D, ed. *Solid State Physics*, vol. 17, Academic, New York, p.269.
Kubrak V, Kleinert P (1998), *J. Phys.: Condens. Matter*, **10**, 7391.
Kubrak V, Kleinert P, Asche M (1998), *Semicond. Sci. Technol.* **13**, 277.
Lai W Y, Das Sarma S (1986), *Phys. Rev. B*, **33**, 8874.
Landau L D, Lifshitz E M (1977) *Quantum Mechanics*, Pergamon, Oxford, 3rd ed.
Lassnig R, Zawadzki W (1983) *J. Phys. C*, **16**, 5435.
Laux S E, Stern F (1986), *Appl. Phys. Lett.* **49**, 91.
Lax M, Cai W (1992), *Int. J. Mod. Phys.* **6**, 975; In: Ting C S ed. *Physics of Hot Electron Transport in Semiconductors*, World Scientific, Singapore, p.171.
Leadly D R et al (1993), *Phys. Rev. B*, **48**, 5457.
Lebwohl P A, Tsu R (1970), *J. Appl. Phys.* **41**, 2664.
Lee C C et al (1996), *J. Appl. Phys.* **80**, 1891.
Lee P A, Ramakrishnan T V (1985), *Rev. Mod. Phys.* **57**, 287.
Lei X L (1980), *Acta. Phys. Sinica*, **29**, 1935. In Chinese.
Lei X L (1985), *J. Phys. C*, **18**, L593.
Lei X L (1987), *Commun. Theoret. Phys.* **8**, 385.
Lei X L (1990a), *Phys. Lett. A*, **148**, 384.
Lei X L (1990b), *Phys. Rev. B*, **41**, 8085.

Lei X L (1990c), In: Xia J B ed. *Lattice Dynamics and Semiconductor Physics*, World Scientific, Singapore, p. 507.
Lei X L (1992a), *Phys. Stat. Sol. (b)*, **170**, 519.
Lei X L (1992b), *J. Phys.: Condens. Matter*, **4**, 9367.
Lei X L (1992c), *Phys. Rev. B*, **45**, 14384.
Lei X L (1993), *J. Phys.: Condens. Matter*, **5**, 8579.
Lei X L (1994a), *J. Phys.: Condens. Matter*, **6**, L305.
Lei X L (1994b), *J. Phys.: Condens. Matter*, **6**, 3749.
Lei X L (1994c), *J. Phys.: Condens. Matter*, **6**, 9189.
Lei X L (1994d), *J. Phys.: Condens. Matter*, **6**, 10043.
Lei X L (1995a), *Phys. Rev. B*, **51**, 5184.
Lei X L (1995b), *Phys. Rev. B*, **51**, 5526.
Lei X L (1995c), *J. Phys.: Condens. Matter*, **7**, L429.
Lei X L (1995d), *Phys. Stat. Sol. (b)*, **192**, K1.
Lei X L (1996a), *J. Phys. Condens. Matter*, **8**, L117.
Lei X L (1996b), *J. Appl. Phys.* **79**, 3071.
Lei X L (1997), *J. Appl. Phys.* **82**, 718.
Lei X L (1998a), *J. Appl. Phys.* 1998, **86**, 1396.
Lei X L (1998b), *J. Phys. Condens. Matter*, **10**, 3201.
Lei X L (2004), *J. Phys.: Condens. Matter*, **16**, 4045.
Lei X L (2006), *Phys. Rev. B*, **73**, 235322.
Lei X L (2007a), *Appl. Phys. Lett.* **90**, 132119.
Lei X L (2007b), *Appl. Phys. Lett.* **91**, 112104.
Lei X L (2008), *Phys. Rev. B*, **77**, 033203.
Lei X L et al (1986), *Phys. Rev. B*, **33**, 4382.
Lei X L et al (1987), *Phys. Rev. B*, **36**, 9134.
Lei X L et al (1994), *Appl. Phys. Lett.* **65**, 2984.
Lei X L et al (1997), *Z. Phys. B*, **104**, 221.
Lei X L, Birman J L, Ting C S (1985), *J. Appl. Phys.* **58**, 2270.
Lei X L, Cai J (1990a), *Phys. Lett. A*, 1990, **144**, 316.
Lei X L, Cai J (1990b), *Phys. Rev. B*, **42**, 1574.
Lei X L, Cai W, Ting C S (1985), *J. Phys. C*, **18**, 4315.
Lei X L, Cai J, Xie L M (1988), *Phys. Rev. B*, **38**, 1529.
Lei X L, Cao J C, Dong B (1996), *J. Appl. Phys.* **80**, 1504.
Lei X L, Cui H L (1997), *J. Phys.: Condens. Matter*, **9**, 4853.
Lei X L, Cui H L (1998), *Eur. Phys. J. B*, **4**, 513.
Lei X L, Cui H L, Horing N J M (1987), *J. Phys. C*, **20**, L287.
Lei X L, Cui H L, Horing N J M (1988), *Phys. Rev. B*, **38**, 8230.
Lei X L, da Cunha Lima I C, Troper A (1997), *J. Appl. Phys.* **82**, 3906.
Lei X L, Dong B, Chen Y Q (1997), *Phys. Rev. B*, **56**, 12120.
Lei X L, Horing N J M (1987a), *Phys. Rev. B*, **35**, 6281.
Lei X L, Horing N J M (1987b), *Phys. Rev. B*, **36**, 4238.
Lei X L, Horing N J M (1988), *Solid State Electron.* **31**, 531.
Lei X L, Horing N J M (1989a), *Phys. Rev. B*, **40**, 3756.
Lei X L, Horing N J M (1989b), *Phys. Rev. B*, **40**, 5985.
Lei X L, Horing N J M, Cui H L (1991), *Phys. Rev. Lett.* **66**, 3277.

Lei X L, Horing N J M, Cui H L (1992), *J. Phys.: Condens Matter*, **4**, 9375.
Lei X L, Horing N J M, Cui H L (1993), *Superlatt. and Microstr.* **14**, 243.
Lei X L, Horing N J M, Cui H L (1995), *J. Phys.: Condens. Matter*, **7**, 9811.
Lei X L, Horing N J M, Cui H L (1998), *Commun. Theoret. Phys.* **29**, 515.
Lei X L, Horing N J M, Zhang J Q (1986a), *Phys. Rev. B*, **33**, 2912.
Lei X L, Horing N J M, Zhang J Q (1986b), *Phys. Rev. B*, **34**, 1139.
Lei X L, Horing N J M, Zhang J Q (1987), *Phys. Rev. B*, **35**, 2834.
Lei X L, Liu S Y (2000a), *J. Phys.: Condens. Matter*, **12**, 4655.
Lei X L, Liu S Y (2000b), *Eur. Phys. J. B*, **13**, 271.
Lei X L, Liu S Y (2003a), *Appl. Phys. Lett.* **82**, 3904.
Lei X L, Liu S Y (2003b), *Phys. Rev. B*, **67**, 233301.
Lei X L, Liu S Y (2003c), *Phys. Rev. Lett.* **91**, 226805.
Lei X L, Liu S Y (2005a), *Appl. Phys. Lett.* **86**, 262101.
Lei X L, Liu S Y (2005b), *Phys. Rev. B*, **72**, 075345.
Lei X L, Liu S Y (2006a), *Appl. Phys. Lett.* **88**, 212109.
Lei X L, Liu S Y (2006b), *Appl. Phys. Lett.* **89**, 182117.
Lei X L, Song W D (1992), *Commun. Theoret. Phys.* **18**, 129.
Lei X L, Ting C S (1984), *Phys. Rev. B*, **30**, 4809.
Lei X L, Ting C S (1985a), *Phys. Rev. B*, **32**, 1112.
Lei X L, Ting C S (1985b), *J. Phys. C*, **18**, 77.
Lei X L, Ting C S (1985c), *Solid State Commun.* **53**, 305.
Lei X L, Ting C S (1987), *Phys. Rev. B*, **36**, 8162.
Lei X L, Wang X F (1993), *J. Appl. Phys.* **73**, 3867.
Lei X L, Weng X M (1994), *J. Phys.: Condens. Matter*, **6**, L461.
Lei X L, Wu M W (1992), *Mod. Phys. Lett. B*, **6**, 1935.
Lei X L, Wu M W (1993), *Phys. Rev. B*, **47**, 13338.
Lei X L, Yang R Q, Tsai C H (1989), In: Tsai C H et al, ed. *Physics of Superlattices and Quantum Wells*, World Scientific, Singapore, p. 159.
Lei X L, Zhang J Q (1986), *J. Phys. C*, **19**, L73.
Levinson I B (1970), *Sov. Phys. JETP*, **30**, 362.
Lipavský P et al (1991), *Phys. Rev. B*, **43**, 4885.
Liu M et al (1988), *Phys. Rev. B*, **37**, 2997.
Liu M, Xing D Y, Ting C S (1989), *J. Phys.: Condens. Matter*, **1**, 407.
Liu S Y, Lei X L (1999), *Phys. Rev. B*, **60**, 10624.
Liu S Y, Lei X L (2003), *J. Phys.: Condens. Matter*, **15**, 4411.
Lu X J, Xie L M, Lei X L (1987), *Solid State Commun.* **64**, 593.
Lugli P, Goodnick S (1987), *Phys. Rev. Lett.* **59**, 716.
Maan J C et al (1982), *Appl. Phys. Lett.* **40**, 609.
Magarill L I, Sarvinykh S K (1970), *Sov. Phys. JETP*, **30**, 362.
Magnus W, Sala C, De Meyer K (1990), *Phys. Rev. B*, **41**, 5192.
Mahan G D (1972), In: Devreese J T, ed. *Polarons in Ionic Crystals and Polar Semiconductors*. North-Holland, Amsterdam, p. 533.
Mahan G D (1981), *Many Particle Physics*, Plenum, New York, p. 42.
Mani R G (2004), *Appl. Phys. Lett.* **85**, 4962.
Mani R G et al (2002), *Nature (London)* **420**, 646.
Mani R G et al (2004), *Phys. Rev. Lett.* **92**, 146801.

Marchetti M C, Cai W (1987), *Phys. Rev. B*, **36**, 8159.
Martin M J et al (1993), *Semicond. Sci. Technol.* **8**, 1291.
McGroddy J C, Nathan M I (1966), *J. Phys. Soc. Jpn.* **21**, 437.
McLean T P, Paige E G S (1960), *J. Phys. Chem. Solids*, **16**, 220.
McLennan Jr J A (1961), *Phys. Fluids*, **4**, 1319.
Mendez E E, Price P J, Heilblum M (1984), *Appl. Phys. Lett.* **45**, 294.
Mermin N D, Canel E (1964), *Ann. Phys. (NY)*, **26**, 247.
Mikhailov S A (2004), *Phys. Rev. B*, **70**, 165311.
Murayama Y, Ando T (1987), *Phys. Rev. B*, **35**, 2252.
Nash K J (1992), *Phys. Rev. B*, **46**, 7723.
Ng T K, Dai L X (2005), *Phys. Rev. B*, **72**, 235333.
Nimtz G et al (1974), *Phys. Rev. B*, **10**, 3302.
Niez J J, Ferry D K (1983), *Phys. Rev. B*, **28**, 889.
Noguchi H et al (1992), *Phys. Rev. B*, **45**, 12148.
Nougier J P (1980), In: Ferry D K, Barker J R, Jacoboni C, ed. *Physics of Nonlinear Transport in Semiconductors*, Plenum, New York, p. 415.
Nougier J P, Rolland M, (1973), *Phys. Rev. B*, **8**, 5728.
Onsager L (1931), *Phys. Rev.* **37**, 405.
Peeters F M, Devreese J T (1983), *Phys. Stat. Sol. (b)*, **115**, 539.
Peletminskii S, Yatsenko A (1968), *Sov. Phys. JETP*, **26**, 773.
Pogrebinskii M B (1977), *Sov. Phys. Semicond.* **11**, 372.
Price P J (1973), *IBM J. Res. Develop.* **17**, 39.
Price P J (1983), *Physica B*, **117**, 750.
Price P J (1985), *Physica B*, **134**, 164.
Rammer J, Smith H (1986), *Rev. Mod. Phys.* **58**, 323.
Reggiani L (1985), In: Reggiani L, ed. *Hot-Electron Transport in Semiconductors*, Springer-Verlag, Berlin, p.7.
Reigrotzki M (1998), Ph.D. thesis, University of Rostok.
Ridley B K (1988), *Quantum Processes in Semiconductor, 2nd ed.*, Oxford: Oxford University Press, Oxford.
Ridley B K (1989a), *Semicond. Sci, Tech.* **4**, 1142.
Ridley B K (1989b), *Phys. Rev. B*, **39**, 5282.
Ridley B K (1991), *Rep. Prog. Phys.* **54**, 169.
Rieger M et al (1989), *Phys. Rev. B*, **39**, 7866.
Rodriguez G A et al (1989), *Appl. Phys. Lett.* **49**, 458.
Rossi F, Jacoboni C (1991) In: *Proc. 7th Int. Conf. Hot Carriers in Semiconductors*, Nara, Japan, July 1-5, 1991.
Roth L M, Argyres P N (1966), In: Willardson R K, Beer A C, ed. *Semiconductors and Semimetals*, vol. 1, Academic, New York, p. 159.
Rousseau J S, Stoddart J C, March N H (1973), *J. Phys. C*, **5**, L175.
Rudan M, Odeh F, White J (1987), *COMPEL*, **6**, 151.
Ryzhii V I (1970), *Sov. Phys. Solid State*, **11**, 2078.
Ryzhii V I (2005), *Jpn. J. Appl. Phys.* **44**, 6600.
Ryzhii V I, Suris R (2003), *J. Phys.: Condens. Matter*, **15**, 6855.
Sadachi S (1985), *J. Appl. Phys.* **58**, R1.
Sano N et al (1990), *Phys. Rev. B*, **41**, 12122.

Sarker S K (1986), *Phys. Rev. B*, **33**, 7263.
Schomburg E *et al* (1996), *Appl. Phys. Lett.* **68**, 1096.
Seeger K (1982), *Semiconductor Physics*, Springer-Verlag, Berlin.
Sernelius B E (1989), *Phys. Rev. B*, **40**, 12438.
Sernelius B E, Söderström E (1991a), *J. Phys. C*, **3**, 1493.
Sernelius B E, Söderström E (1991b), *J. Phys. C*, **3**, 8425.
Shah J *et al* (1984), *Appl. Phys. Lett.* **44**, 322.
Shah J *et al* (1985), *Phys. Rev. Lett.* **47**, 264.
Shah J *et al* (1986), *IEEE: J. Quantum Electron.* **QE-22**, 1728.
Shu W M, Lei X L (1994), *Phys. Rev. B*, **50**, 17378.
Shubnikov L, de Haas W J (1930), *Leiden Comm.* No 207D.
Sibille A *et al* (1990a), *Phys. Rev. Lett.* **64**, 52.
Sibille A *et al* (1990b), *Europhys. Lett.* **13**, 279.
Smith P M, Inoue M, Frey J (1980), *Appl. Phys. Lett.* **37**, 797.
Solomon P M *et al* (1989), *Phys. Rev. Lett.* **63**, 2508.
Sood A K *et al* (1985), *Phys. Rev. Lett.* **54**, 2111.
Stein D, Ebert G, von Klitizing K (1984), *Surf. Sci.* **142**, 406.
Stern F, Howard W E (1964), *Phys. Rev.* **163**, 816.
Su Z B, Chen L Y, Birman J L (1987), *Phys. Rev. B*, **35**, 9744.
Su Z B, Sakita B (1989), *Phys. Rev. B*, **40**, 9959.
Swris R A, Shchamkhalova B S (1984), *Sov. Phys.: Semicond.* **18**, 738.
Tanatar B, Singh M, MacDonald A H (1991), *Phys. Rev. B*, **43**, 4308.
Tang T W, Ramaswamy S (1993), *IEEE: Tran Electron Devices*, **ED-40**, 1469.
Tao Z C, Ting C S, Singh, M (1993), *Phys. Rev. Lett.* **70**, 2467.
Thornber K K (1978), *Solid State Electron.* **21**, 259.
Thornber K K, Feynman R P (1970), *Phys. Rev. B*, **40**, 9959.
Thouless D J (1977), *Phys. Rev. Lett.* **39**, 1167.
Ting C S, Ganguly A K, Lai W Y (1981), *Phys. Rev. B*, **24**, 3371.
Ting C S, Nee T W (1986), *Phys. Rev. B*, **33**, 7056.
Ting C S, Ying S C, Quinn J J (1976a). *Phys. Rev. B*, **14**, 4439.
Ting C S, Ying S C, Quinn J J (1976b), *Phys. Rev. Lett.* **37**, 215.
Ting C S, Ying S C, Quinn J J (1977). *Phys. Rev. B*, **16**, 5394.
Tsuchiya T, Ando T (1992), *Semicond. Sci. Technol.* **7**, B73.
Tzoar N, Platzman P M, Simons A (1976), *Phys. Rev. Lett.* **36**, 1200.
Umansky V, de Picciotto R, Heiblum M (1997), *Appl. Phys. Lett.* **71**, 683.
Vasilopoulos *et al* (1989), *Phys. Rev. B*, **40**, 1810.
Vogl P (1981), In: Ferry D K, Barker J R, Jacoboni C, ed. *Physics of Nonlinear Transport in Semiconductors*, Plenum, New York, p. 75.
Wang X F *et al* (1998), *Phys. Rev. B*, **58**, 3529.
Wang X F, da Cunha Lima I C, Lei X L (1998), *J. Phys.: Condens. Matter*, **10**, 3743.
Wang X F, Lei X L (1993a), *Phys. Stat. Sol. (b)*, **175**, 433.
Wang X F, Lei X L (1993b), *Phys. Rev. B*, **47**, 16612.
Wang X F, Lei X L (1994a), *J. Phys.: Condens. Matter*, **6**, 5667.
Wang X F, Lei X L (1994b), *Phys. Rev. B*, **49**, 4780.
Wang X F, Lei X L (1995), *J. Phys.: Condens. Matter*, **7**, 7871.

Wang Z X, Guo D R (1989), *Special Functions*, World Scientific Publishing, Singapore.
Weng X M, Lei X L (1994), *J. Phys.: Condens. Matter*, **6**, 6287.
Weng X M and Lei X L (1995), *Phys. Stat. Sol* (*b*), **191**, 183.
Wingreen N S, Stanton C J, Wilkins J W (1986), *Phys. Rev. Lett.* **57**, 1084.
Winnerl S *et al* (1997), *Superlatt. and Microstr.* **21**, 91.
Wu C C, Spector H N (1971), *Phys. Rev. B*, **3**, 3979.
Wu H S, Huang X X, Weng M Q (1997), *Phys. Stat. Sol.* (*b*), **204**, 747.
Wu H S, Wu M W (1995), *Phys. Stat. Sol.* (*b*), **192**, 129
Wu M W, Cui H L, Horing N J M (1996), *Phys. Rev. B*, **54**, 2351.
Wu M W, Horing N J M, Cui H L (1996), *Phys. Rev. B*, **54**, 5438.
Wu M W, Yu Z G, Lei X L (1994), *Phys. Stat. Sol.* (*b*), **183**, 529.
Xia J B, Zhu B F (1995), *Semiconductor Superlattice Physics*, Shanghai Publishing of Science and Technology, Shanghai. In Chinese.
Xing D Y *et al* (1988), *J. Phys. C*, **21**, 2881.
Xing D Y *et al* (1995), *Phys. Rev. B*, **51**, 2193.
Xing D Y, Hu P, Ting C S (1987), *Phys. Rev. B*, **35**, 6379.
Xing D Y, Liu M (1992), *Int. J. Mod. Phys. B*, **6**, 1037; In: Ting C S ed. *Physics of Hot Electron Transport in Semiconductors*, World Scientific, Singapore, p.233.
Xing D Y, Liu M, Ting C S (1988), *Phys. Rev. B*, **37**, 10283.
Xing D Y, Ting C S (1987), *Phys. Rev. B*, **35**, 3971.
Xu W, Peeters F M, Devreese (1993), *J. Phys.: Condens. Matter*, **5**, 2307.
Yan F Q, Lei X L (2001), *J. Phys.: Condens. Matter*, **13**, 6625.
Yang C L *et al* (2002), *Phys. Rev. Lett.* **89**, 076801.
Yang C L *et al* (2003), *Phys. Rev. Lett.* **91**, 096803.
Yang R Q (1987), *J. Phys. C*, **21**, 987.
Yang R Q *et al* (1988), *Surface Science*, **196**, 487.
Ye P D *et al* (2001), *Appl. Phys. Lett.* **79**, 2193.
Zhang W *et al* (2007a), *Phys. Rev. B*, **75**, 041304(R).
Zhang W *et al* (2007b), *Phys. Rev. Lett.* **98**, 106804.
Ziman J M (1960), *Electrons and Phonons*, Oxford University Press, Oxford.
Ziman J M (1961), *Phil. Mag.* **6**, 1013.
Ziman J M (1972), *Principles of the Theory of Solids*, Cambridge University Press, London.
Zou Z Q, Lei X L (1995a), *Phys. Rev. B*, **50**, 9443.
Zou Z Q, Lei X L (1995b), *J. Phys. C*, **7**, 8587.
Zou Z Q, Lei X L (1996), *Commun. Theoret. Phys.* **25**, 391.
Zubarev D N (1961), *Dokl. Akad. Nauk. SSSR*, **144**, 92.
Zubarev D N (1974), *Nonequilibrium Statistical Thermodynamics*, Consultouts Bureau, New York.
Zudov M A (2004), *Phys. Rev. B*, **69**, 041304(R).
Zudov M A *et al* (2001), *Phys. Rev. B*, **64**, 201311(R).
Zudov M A *et al* (2003), *Phys. Rev. Lett.* **90**, 046807.
Zudov M A *et al* (2006), *Phys. Rev. B*, **73**, 041303(R).

Index

Γ–L model, 189
Γ–L–X model, 189
δ-doping, 206
γ coefficient, 451, 452, 455, 457, 460
$\mathbf{k} \cdot \mathbf{p}$-method, 462
3D phonon model, 135

absolute instability, 518
absorption coefficient, 550
absorption percentage, 555
absorption rate, 555
acoustic-phonon scattering, 41
adiabatic resistivity, 230, 231
advanced Green's function, 313
Auger recombination, 489, 558
autocorrelation function, 125
avalanche breakdown, 498

Büttiker–Thomas reduction tensor, 382, 385, 409
background impurities, 41, 47
band offset, 485
bichromatic radiation, 617
Bloch function, 364
Bloch–Grüneissen formula, 19
Bohr magneton, 252
Boltzmann equation, 349, 354, 397
Bragg scattering, 362, 382
Brillouin zone, 52, 362, 364, 367, 370, 382, 385

canonical conjugate variables, 2

carrier-transfer diffusion coefficient, 220
cell periodic function, 364
center of mass, 2
center-of-mass coordinate, 2, 530
center-of-mass momentum, 2, 365, 530
center-of-mass position, 365
center-of-mass variable, 2, 251, 529
center-of-mass velocity, 5, 530
close-time-path Green's function, 309
conductivity formula, 230, 231
confined phonons, 155
continuity equation, 337
convective instability, 515
correlation function, 125
Coulomb drag, 171
Coulomb interaction, 1
Coulomb lifetime, 242
cross diffusion coefficient, 220
cumulant expansion, 282
cyclotron resonance, 274, 584, 592

deformation potential coupling, 39
degenerate case, 19
density correlation function, 16, 18, 29, 35, 36, 48, 58, 62, 254, 582
density matrix, 6, 9, 310, 326
density–density correlation function, 16
diffusion coefficient, 112, 113, 218
displacement diffusion coefficient, 218

distribution function, 239, 332
Dopler shift, 128
drift-diffusion model, 354
drifted electron-temperature model, 330
Drude formula, 78
dynamic conductivity, 83, 113
dynamic mobility, 429
dynamic resistivity, 83
dynamic response theory, 230
dynamic scattering factor, 305, 306
dynamic screening, 29
Dyson equation, 58

effective g-factor, 252
effective-mass approximation, 31
effective-mass correction tensor, 72, 254
Einstein model, 28
Einstein relation, 114
elastic scattering time , 398
electro-optic coupling parameter, 532
electron cooling, 22, 541, 592
electron density correlation function, 72, 81, 84, 96
electron energy-loss rate, 380
electron heating, 592
electron life-time, 319
electron temperature, 7, 17, 122
electron–phonon correlation function, 72, 81, 84
energy balance equation, 6, 12, 15, 305, 338, 354, 379, 381, 387, 532, 581
energy relaxation time, 398
energy-flux balance equation, 351, 357
energy-loss rate, 15
energy-related memory function, 85, 96
energy-transfer rate, 12, 15
Esaki–Tsu model, 393
Euler-type equation, 337
evolution operator, 110

Fang–Howard–Stern variational functions, 38
Faraday configuration, 574
Faraday effect, 590
Faraday rotation, 590
Fermi liquid theory, 364
Feynman diagrams, 309
fluctuating force, 110, 112
fluctuating velocity, 112
fluctuating-force correlation function, 114, 115, 119
fluid elements, 334
Fokker-Planck equation, 250
force balance equation, 5, 10, 15, 305, 581
force correlation function, 232
force-balance theory, 230
form factor, 56
Fröhlich coupling, 39
free carrier absorption, 550
Fresnel integral, 309
frictional acceleration, 379
frictional force, 11, 15

Hall coefficient, 453, 455, 467
Hall factor, 467
Hall resistivity, 453
Heisenberg equation of motion, 328
Heisenberg operator, 110, 328
Heisenberg picture, 109
Herring-Vogt transformation, 479
high-field descreening effect, 30
high-frequency small-signal conductivity, 83, 113
high-frequency small-signal mobility, 83
high-frequency small-signal resistivity, 83
high-lying minibands, 484
hot phonons, 129
hot-phonon effect, 134
hydrodynamic balance equation, 333, 338, 511
hydrodynamic equation, 349
hydrodynamic model, 354

impact ionization, 489, 556, 558

inelastic scattering time, 398
initial density matrix, 7, 319
interaction picture, 110, 223
interaction representation, 303
interface phonons, 155
intracollisional field effect, 228, 301, 302, 307, 376
inverse effective mass tensor, 379, 385, 405, 450
iso-energy surface, 395
isothermal resistivity, 230, 231

Kane band, 462
Kogan formula, 318
Kramers–Kronig relation, 79, 85
Kronig-Penny model, 485
Kubo linear-response theory, 231

Landau level broadening, 281
Langevin-type equation, 109
laterally confined superlattice, 415
lattice momentum balance equation, 381, 385, 409
lattice momentum function, 369, 385
lattice temperature, 7
Liouville equation, 6, 303, 310, 326

magnetophonon resonance, 437
magnetoresistance oscillations, 597
maximally crossed diagram, 314
Maxwell–Boltzmann distribution, 26
memory effect, 76
memory function, 77, 83, 90
memory function approach, 83
miniband transport, 393, 403, 409
momentum balance equation, 337, 354, 379, 384, 403, 532, 581
momentum function, 367, 369, 384
momentum relaxation time, 225
momentum-related memory function, 95
monochromatic radiation, 598
Monte-Carlo method, 354
multiphoton process, 534, 535, 538, 553, 604
multiple photon process, 582, 609

multiple subband system, 195, 213
multivalley semicoductor, 189
multivalley semiconductor, 179, 183

noise conductivity, 114, 119
noise diffusion coefficient, 113, 219
noise temperature, 114, 119, 122
nondegenerate case, 25
nonequilibrium phonon effect, 321
nonequilibrium phonons, 127, 129, 136
nonequilibrium statistical operator method, 326
normal process, 54
Nyquist relation, 115

Onsager reciprocity principle, 349
Onsager relation, 349
optical deformation-potential scattering, 28
orthogonalized plane waves, 211

particle-number balance equation, 354
periodic Brillouin zone picture, 364
phonon drag, 134, 344
phonon life-time, 319
phonon–phonon interaction, 132
phonon–plasma coupled mode, 318
photoconductivity, 593
photoresistivity, 594
piezoelectric coupling, 39
Poisson equation, 340, 355
polar-optical-phonon scattering, 40

quantum Langevin equation, 328
quantum wire, 59
quantum-well superlattice, 45
quasi-2D phonon model, 142
quasi-one-dimensional system, 59
quasi-two-dimensional system, 31

radiative decay, 587
random force, 112
random phase approximation, 18, 38, 96, 254

random velocity, 112
real-space transfer, 206
relative electron system, 3
relative electrons, 3
relative-electron coordinate, 2
relative-electron momentum, 2
relative-electron variable, 2, 251, 529
relaxation time approximation, 78, 397, 443
remote impurities, 41, 47
renormalization, 323
resistivity formula, 230, 231
retarded Green's function, 313

Scharfetter-Gummel method, 359
Schrödinger picture, 6
self-consistent Born approximation, 281
single band subspace, 365
single photon process, 582
small-signal high-frequency resistivity, 90
superlattice, 391
superlattice miniband, 362, 391, 484
species, 179

thermal diffusion coefficient, 220
thermal equilibrium state, 7
thermal noise temperature, 112
thermalization, 8, 319

thermalization time, 223, 319
thermodynamic temperature, 7
thermoelectric power, 340
Thouless length, 316
tight-binding approximation, 391
transfer resistivity, 171
transmittance, 588
tunneling superlattices, 52
two-temperature model, 319
type-I superlattice, 162
type-II superlattice, 161, 163, 167, 175

umklapp process, 54

velocity function, 367, 384
virtual photon process, 535, 610
Voigt configuration, 573

warm-electron conduction, 21
weak localization, 314

Zeeman splitting, 252
zero resistance state, 597